Progress in Mathematics

Volume 195

Moduli of Abelian Varieties

Carel Faber
Gerard van der Geer
Frans Oort
Editors

Springer Basel AG

Editors:

Carel Faber
Institutionen för Matematik
Kungliga Tekniska Högskolan (KTH)
10044 Stockholm
Sweden

Gerard van der Geer
Korteweg de Vries Instituut
Universiteit van Amsterdam
Plantage Muidergracht 24
1018 TV Amsterdam
The Netherlands

Frans Oort
Mathematisch Instituut
Universiteit Utrecht
Budapestlaan 6
3508 TA Utrecht
The Netherlands

2000 Mathematics Subject Classification 14J60, 14D20, 11G15

A CIP catalogue record for this book is available from the Library of Congress, Washington D.C., USA

Deutsche Bibliothek Cataloging-in-Publication Data

Moduli of Abelian varieties / Carel Faber … ed. – Basel ; Boston ; Berlin : Birkhäuser, 2001
(Progress in mathematics ; Vol. 195)
ISBN 978-3-0348-9509-5 ISBN 978-3-0348-8303-0 (eBook)
DOI 10.1007/978-3-0348-8303-0

Originally published by Birkhäuser Verlag in 2001
Softcover reprint of the hardcover 1st edition 2001

Printed on acid-free paper produced of chlorine-free pulp. TCF ∞

ISBN 978-3-0348-9509-5

9 8 7 6 5 4 3 2 1

Contents

PARTICIPANTS

The following is the list of participants of the conference *Moduli of Abelian Varieties, Texel '99*, which was held on Texel Island during the last week of April 1999.

Dan Abramovich (Boston),
Fabrizio Andreatta (Utrecht),
Michael Bogouslavsky (Amsterdam),
Oliver Bültel (Köln),
Ciro Ciliberto (Roma),
Tim Dokshitzer (Utrecht),
Torsten Ekedahl (Stockholm),
Carel Faber (Stillwater, Stockholm),
Gavril Farkas (Amsterdam),
Eyal Goren (Montréal),
Joost van Hamel (Utrecht),
Everett Howe (San Diego),
Klaus Hulek (Hannover),
Johan de Jong (MIT, Cambridge),
David Lehavi (Jerusalem),
Andrea Miller (Bonn),
Ben Moonen (Utrecht),
Stefan Müller-Stach (Essen),
Wieslawa Niziol (Salt Lake City),
K. Oguiso (Tokyo),
Cathy O'Neil (MIT, Cambridge),
Sascha Orlik (Köln),
Gian-Pietro Pirola (Pavia),
Utpal Sarkar (Amsterdam),
Vasily Shabat (Amsterdam),
Tomasz Szemberg (MPI, Bonn),
Halszka Tutaj-Gasinska (Kraków),
Maja Volkov (Köln),
Mike Zieve (Los Angeles),
Valery Alexeev (Athens),
Christina Birkenhake (Erlangen),
Brad Brock (Malibu),
Ching-Li Chai (Philadelphia),
Elisabetta Colombo (Milano),
Bas Edixhoven (Rennes),
Hélène Esnault (Essen),
Gerd Faltings (MPI, Bonn),
Gerard van der Geer (Amsterdam),
Ulrich Görtz (Köln),
Günter Harder (Bonn),
Johan Huisman (Rennes),
Elham Izadi (Athens),
T. Katsura (Tokyo),
Ron Livné (Jerusalem),
A. Mokrane (Paris),
Laurent Moret-Bailly (Rennes),
I. Nakamura (Sapporo),
Rutger Noot (Rennes),
Arthur Ogus (Berkeley),
Frans Oort (Utrecht),
Pierre Parent (Rennes),
Gregory Sankaran (Bath),
René Schoof (Roma),
Bart de Smit (Leiden),
Valerio Talamanca (Amsterdam),
Eckart Viehweg (Essen),
Vitaly Vologodsky (Athens),
Thomas Zink (Bielefeld).

Contributors

V. Alexeev Department of Mathematics, University of Georgia, Athens, GA 30602, USA.
E-mail: valery@math.uga.edu

F. Andreatta Department of Mathematics and Statistics, McGill University, 805 Sherbrooke St. West, Montreal, Quebec, Canada H3A 2K6.
E-mail: andreatta@math.mcgill.ca

A. Beilinson Department of Mathematics, University of Chicago, 5734 S. University Avenue, Chicago, IL 60637, USA.
E-mail: sasha@math.uchicago.edu

B. Edixhoven IRMAR, Campus de Beaulieu, 35042 Rennes cedex, France.
E-mail: Sebastiaan.Edixhoven@univ-rennes1.fr

G. Faltings Max-Planck-Institut für Mathematik, Postfach 7280, D-53072 Bonn, Germany.
E-mail: gerd@mpim-bonn.mpg.de

G. van der Geer Korteweg-de Vries Instituut, Universiteit van Amsterdam, Plantage Muidergracht 24, 1018 TV Amsterdam, The Netherlands.
E-mail: geer@wins.uva.nl

E. Howe Center for Communications Research, 4320 Westerra Court, San Diego, CA 92121-1967, USA.
E-mail: however@alumni.caltech.edu

K. Hulek Institut für Mathematik (C), Universität Hannover, Postfach 6009, D-30060 Hannover, Germany.
E-mail: hulek@math.uni-hannover.de

T. Katsura Graduate School of Mathematical Sciences, University of Tokyo, 3-8-1 Komaba, Meguro, Tokyo 153-8914, Japan.
E-mail: tkatsura@ms.u-tokyo.ac.jp

Yu.I. Manin Max-Planck-Institut für Mathematik, Postfach 7280, D-53072 Bonn, Germany.
E-mail: manin@mpim-bonn.mpg.de

B. Moonen Mathematisch Instituut, Universiteit Utrecht, Postbus 80010, NL-3508 TA Utrecht, The Netherlands.
E-mail: moonen@math.uu.nl

I. Nakamura Department of Mathematics, Hokkaido University, Sapporo 060-810, Japan.
E-mail: nakamura@math.sci.hokudai.ac.jp

A. Ogus Department of Mathematics, University of California, Berkeley, CA 94720, USA.
E-mail: ogus@math.berkeley.edu

F. Oort Mathematisch Instituut, Universiteit Utrecht, Postbus 80010, NL-3508 TA Utrecht, The Netherlands.
E-mail: oort@math.uu.nl

A. Polishchuk Department of Mathematics and Statistics, Boston University, 111 Cummington Street, Boston, MA 02215, USA.
E-mail: apolish@math.bu.edu

T. Terasoma Graduate School of Mathematical Sciences, University of Tokyo, 3-8-1 Komaba, Meguro, Tokyo 153-8914, Japan.
E-mail: terasoma@ms.u-tokyo.ac.jp

T. Wedhorn Mathematisches Institut, Universität zu Köln, Weyertal 86–90, D-50931 Köln, Germany.
E-mail: wedhorn@mi.uni-koeln.de

Yu.G. Zarhin Department of Mathematics, Pennsylvania State University, University Park, PA 16802, USA.
Email: zarhin@math.psu.edu

Th. Zink Fakultät für Mathematik, Universität Bielefeld, Postfach 100131, D-33501 Bielefeld, Germany.
E-mail: zink@mathematik.uni-bielefeld.de

INTRODUCTION

That moduli spaces of abelian varieties are intriguing mathematical objects became clear in the 19th century through the work of Kronecker, Klein, Weber, and many others on the moduli of elliptic curves. Not only did the modular curves and their function theory illustrate the new concept of Riemann surfaces in a beautiful way, the theorem of Kronecker-Weber pointed to much deeper arithmetic properties of these modular curves. It took a long time before the tools were shaped that made it possible to study the arithmetic properties of modular curves and then these modular curves became the focus of intense study after the early 1970's, which resulted in Mazur's theorem on the rational points of $X_0(N)$. The insight that all motives coming from elliptic curves defined over $\mathbf{Q}$ are present in the motives defined by modular curves (Shimura-Taniyama-Weil) came slowly but developed into a main goal of research, culminating in the proof by Wiles of Fermat's Last Theorem.

In the meantime the moduli of higher-dimensional abelian varieties started to attract attention and the ground was prepared by work of Siegel, Satake, Igusa, Baily and others on the complex analytic case. A definitive advance was made by the work of Mumford who proved in the 1960's the existence of moduli spaces for polarized abelian varieties in the sense of Grothendieck. Moreover, Mumford and his collaborators designed the tools to construct compactifications of these moduli spaces over the complex numbers. By the work of Faltings and Chai we now have good compactifications of moduli spaces of abelian varieties over the integers.

This does not mean, however, that everything is now set for an exploration of the moduli of abelian varieties of higher dimension. We do not doubt that fascinating facts are waiting to be discovered by those who will venture into this terrain, but they will face great obstacles and will need endurance because progress will come slowly. The links with such fields as algebraic groups, representation theory, algebraic geometry, and number theory vote for the importance of the subject but the difficulties seem to repel many and the attention that these spaces got so far is limited. One of the insights obtained in recent years is that maybe the moduli spaces in positive characteristic are more accessible than their characteristic zero counterparts.

Since abelian varieties and their moduli constitute in our view one of the most important subjects in algebraic geometry and possess an enormous potential for number theory, we decided to devote the 3rd Texel conference to this topic. It was held during the last week of April 1999. This volume appears on the occasion of the conference, but the articles presented here are not necessarily connected to lectures given at the conference.

We would like to take the opportunity to thank the participants and the speakers who made the conference to a success. We also like to thank the institutions that financed the conference: NWO, het Korteweg-de Vries

Instituut van de Universiteit van Amsterdam, het Thomas Stieltjes Instituut, het MRI, en de Koninklijke Nederlandse Akademie van Wetenschappen.

July 2000

Carel Faber
Gerard van der Geer
Frans Oort

Progress in Mathematics, Vol. 195, © 2001 Birkhäuser Verlag Basel/Switzerland

ON EXTRA COMPONENTS IN THE FUNCTORIAL COMPACTIFICATION OF A_g

VALERY ALEXEEV

Recall the following from the theory of toroidal compactifications of moduli of polarized abelian varieties (Mumford et al. [AMRT75] over $\mathbb{C}$, Faltings and Chai [FC90] over $\mathbb{Z}$). Denote $X = \mathbb{Z}^g$ and let C be the convex hull in the space $\mathrm{Sym}^2(X^*_{\mathbb{R}})$ of semipositive symmetric matrices q with rational null-space. For any *admissible* $\mathrm{GL}(X)$-invariant decomposition τ of C (i.e. it is a face-fitting decomposition into finitely generated rational cones such that there are only finitely many cones modulo $\mathrm{GL}(X)$) there is a compactification $\overline{\mathrm{A}}_g^{\tau}$ of the moduli space A_g of principally polarized abelian varieties. $\overline{\mathrm{A}}_g^{\tau}$ comes with a natural stratification, and strata correspond in a 1-to-1 way to cones in τ modulo $\mathrm{GL}(X)$. There are infinitely many such decompositions τ and none of them seems to be better than another. True, some decompositions are smooth and projective but still there are infinitely many of these as well.

There is, however, a decomposition τ_{Vor} for the 2nd Voronoi decomposition which has a nicer geometric description. Strata of this compactification $\overline{\mathrm{A}}_g^{\mathrm{Vor}}$ still are in a bijection with cones of τ_{Vor}, however, these cones now correspond in a 1-to-1 way to *special* X-periodic face fitting decompositions of $X_{\mathbb{R}}$ with vertices in X, called *Delaunay.* A form q defines a distance function d_q on $X_{\mathbb{R}}$ and a cell of the Delaunay decomposition Del_q (Delaunay cell) is a convex hull of integral points circumscribed by an "empty sphere", a sphere that does not contain any integral points in its interior. As an example, for $g = 2$ there are 4 such decompositions: by 4-gons, by triangles, by infinite strips and finally the decomposition consisting of one big cell covering the whole plane. The decompositions appearing are either polytopal or the preimages of such from a lower dimension.

On the other hand, in [Ale99] I have constructed a functorial compactification $\overline{\mathrm{AP}}_g$ of A_g as the moduli of triples $G \curvearrowright P \supset \Theta$ whose geometric fibers have the following description: G is semiabelian, P is projective reduced connected and Θ is a Cartier divisor, all satisfying a few natural conditions:

1. P is seminormal,
2. there are only finitely many orbits,
3. Θ does not contain any orbit entirely,

4. for any $p \in P$, the stabilizer of p is connected and reduced and lies in the toric part of G.

This compactification comes with a stratification as well, with strata corresponding to *all* X-periodic face fitting decompositions of $X_{\mathbb{R}}$ with vertices in X. Just from this rough description we see that $\overline{\mathrm{A}}_g^{\mathrm{Vor}}$ and $\overline{\mathrm{AP}}_g$ must be very closely related and at the same time be different. The connection, according to [Ale99], is that $\overline{\mathrm{A}}_g^{\mathrm{Vor}}$ coincides with the main irreducible component of $\overline{\mathrm{AP}}_g$. The most obvious difference is that, unlike $\overline{\mathrm{A}}_g^{\mathrm{Vor}}$, $\overline{\mathrm{AP}}_g$ may have Extra Types of irreducible components, ETs for short, which will be the focus of this note. We would like to discuss:

1. Where and why ETs appear.
2. How to find ETs and how to study them.
3. Some concrete evidence of ETs.
4. Dimension 4 case in detail.
5. Relationship between ETs and the Jacobian locus.

1. Where and why ETs appear

There is actually a very simple reason for their appearance. Say, (P', Θ') is a pair with an abelian action and let (P, Θ) be its degeneration. Assume that (P, Θ) has several components P_i and set $\Theta_i = \Theta|_{P_i}$. Think of the deformations of (P, Θ) that keep the decomposition into the irreducible components. These deformations correspond to deformations of components (P_i, Θ_i) that are compatible on intersections. In the simplest case, when the intersections are elementary, there are, perhaps, no gluing conditions at all. Now, it is not hard to imagine the situation when the sum of the dimensions of the deformation spaces for pairs (P_i, Θ_i) is larger than the dimension of the deformation space of the constrained smooth pair (P', Θ'), i.e. $g(g+1)/2$. In this case there must be another irreducible component.

So, the answer to the "where" part is that ETs appear at the boundary of the world as we know it.

2. How to find ETs and how to study them

There are two basic methods:

1. Find a periodic non-Delaunay decomposition. The first such decomposition appears in dimension 4, and we will describe it in more detail below.
2. For a Delaunay decomposition $\widetilde{\Delta}$, compute the dimension of Stratum $\widetilde{\Delta}$ in $\overline{\mathrm{AP}}_g$. If it is higher than the dimension of the corresponding stratum in $\overline{\mathrm{A}}_g^{\mathrm{Vor}}$ then there must be an ET nearby. There is a remarkably simple formula for the first of these dimensions which we are now going to describe.

Let $\widetilde{\Delta}$ be an X-periodic face-fitting decomposition of $X_{\mathbb{R}}$ with vertices in X. We will identify $\widetilde{\Delta}$ with its quotient Δ which is a decomposition of the real torus $X_{\mathbb{R}}/X \simeq \mathbb{R}^g/\mathbb{Z}^g$. If we work over $\mathbb{C}$, this decomposition is directly related to the properties of each pair (P,Θ) in the Δ-stratum of $\overline{\mathrm{AP}}_g$, and this connection is described by using the *moment map.* (Even if the base field is not $\mathbb{C}$, this decomposition describes the main properties of (P,Θ) very faithfully.)

Say, $G \curvearrowright P \supset \Theta$ is a triple as before, and start with the case when $G = T = (\mathbb{C}^*)^g$ is a torus with the character group X. The moment map sends $P(\mathbb{C})$ to its quotient by the action of the compact torus $CT = (S^1)^g = U(1)^g \subset T$. If $L = \mathcal{O}(\Theta)$ is T-linearized then we can describe the moment map more directly. Let $\theta \in H^0(P,L)$ be an equation of Θ. The space $H^0(P,L)$ splits into the direct sum of T-eigenspaces and we can write θ as a finite sum $\sum_{x\in X}\xi_x$. Then

$$\mathrm{Mom}: P(\mathbb{C}) \ni p \mapsto \frac{\sum |\xi_x(p)|^2\cdot x}{\sum|\xi_x(p)|^2} \in X_{\mathbb{R}}\,.$$

It is well defined if at each $p\in P$ at least one of the ξ_x's is not zero, and that is one of the conditions on (G,P,Θ) that is satisfied by our definition of a triple.

If L is not linearized then (P,L) is in a canonical way the quotient by X of a pair $(\widetilde{P},\widetilde{L})$ with linearized $\widetilde{L}$ (the scheme $\widetilde{P}$ is only locally of finite type but the action of X is properly discontinuous in the Zariski topology). We take $\tilde{\theta}\in H^0(\widetilde{P},\widetilde{L})$ to be the pullback of θ. It is a fact that for any $\tilde{p}\in\widetilde{P}$ the sum $\sum_{x\in X}\xi_x$ is finite as almost all of ξ_x vanish at $\tilde{p}$. The moment map $\mathrm{Mom}_{\widetilde{P},\tilde{\theta}}$ commutes with the translation action of X, and we define $\mathrm{Mom}_{P,\theta}$ from the following diagram

$$\begin{array}{ccc}
\widetilde{P}(\mathbb{C}) & \xrightarrow{\mathrm{Mom}_{\widetilde{P},\tilde{\theta}}} & X_{\mathbb{R}} \\
{\scriptstyle /X}\big\downarrow & & \big\downarrow{\scriptstyle /X} \\
P(\mathbb{C}) & \xrightarrow{\mathrm{Mom}_{P,\theta}} & X_{\mathbb{R}}/X
\end{array}$$

Finally, if G is an arbitrary semiabelian variety then we can (there is a choice involved) write G as the quotient $(\mathbb{C}^*)^g/X'$, $X\supset X' = \mathbb{Z}^{g'}$ and $g' =$ the dimension of the abelian part of G, making P into a "toric" variety. The T-action here is not algebraic, of course, but we can still repeat the above definition. The sum $\sum|\xi_x(p)|^2\cdot x$ in this case is truly infinite, however, it is convergent because theta functions have exponential decline for large x.

With these definitions in mind, let (G,P,Θ) be a triple corresponding to a point in $\mathrm{Stratum}[\Delta]\subset\overline{\mathrm{AP}}_g$. Then the moment map sends P to $X_{\mathbb{R}}/X$ and for any $z\in\delta^0$, interior of δ, one has $\mathrm{Mom}^{-1}(z) = (S^1)^{\dim\delta}$. Moreover, P has as many irreducible components as there are cells in Δ, and these intersect exactly in the way the corresponding cells intersect. Moreover, if δ is a polytope then the irreducible component P_δ is the projective toric variety corresponding to δ, glued in a way that the decomposition suggests (in particular, all of its 0-dimensional orbits are glued together).

As triples degenerate, so do the moment maps. If the fiber over a particular point was 0-dimensional, it is not going to get bigger. On the other hand, a big fiber may get "squashed" to a smaller one in the limit, so the degeneration of triples of type Δ must correspond to a subdivision of Δ. A familiar example is that of an elliptic curve degenerating to a nodal curve. An elliptic curve has a moment map to a circle whose every fiber is S^1, and in the limit a fiber over one point becomes 0-dimensional. An elliptic curve corresponds to a decomposition $\widetilde{\Delta}$ with one big cell, and the nodal curve – to a decomposition of $\mathbb{R}$ into intervals. In higher dimensions exactly the same happens but the decompositions Δ that appear are more sophisticated.

Definition 2.1. For a "simple" cell δ we define $\widehat{\mathbb{L}}_{\delta,\mathbb{R}}$ to be the space of all $\mathbb{R}$-valued functions on $\delta \cap X$ modulo the subspace of linear nonhomogeneous functions. $C^0(\Delta, \widehat{\mathbb{L}})$ is defined as

$$\oplus_{\dim \delta = g} \widehat{\mathbb{L}}_{\delta,\mathbb{R}}$$

and $H^0(\Delta, \widehat{\mathbb{L}})$ as the subspace of C^0 consisting of functions that coincide on "intersections".

The reader will notice that we are computing here the space of global sections of a certain constructible sheaf $\widehat{\mathbb{L}}$ on $X_{\mathbb{R}}/X$ which is constant on locally closed strata defined by Δ.

Theorem 2.2 ([Ale99]). *One has* $\dim \mathrm{Stratum}[\Delta] = h^0(\Delta, \widehat{\mathbb{L}})$.

The meaning of "simple" and "intersections" should be clear from the following instructive examples.

Example 2.3. Say, $\widetilde{\Delta}$ consists of just one big cell, $\mathbb{R}^g$ – this is the stratum corresponding to triples with abelian G. Then $C^0(\widehat{\mathbb{L}})$ is the space of all real-valued functions on $X = \mathbb{Z}^g$ modulo the $(g+1)$ dimensional subspace of linear nonhomogeneous functions. The space $H^0(\widehat{\mathbb{L}})$ is the subspace of functions that are invariant under the translations by X. If $[f]$ is the equivalence class function of such a function then for any $y \in X$ the function $g(x) = f(y+x) - f(x)$ must be linear, so $f(x)$ must be quadratic. Therefore, $H^0(\widehat{\mathbb{L}})$ is the space of quadratic modulo linear functions and its dimension is $g(g+1)/2$, i.e. the dimension of A_g, as expected.

Example 2.4. Let $\widetilde{\Delta}$ be the decomposition of $\mathbb{R}^2$ into squares. There is only 1 maximal-dimensional cell modulo X, a square. It is easy to see that for any polytopal cell appearing in an X-periodic decomposition one has

$$l_\delta = \dim \widehat{\mathbb{L}}_{\delta,\mathbb{R}} = \#(\text{vertices of } \delta) - \dim \delta - 1$$

For the square we have $l = 1$ and for the intervals on the boundary: $l = 0$, so there are no gluing conditions. Hence, the dimension of the corresponding stratum is $h^0(\mathbb{L}) = 1$.

Example 2.5. Let us generalize the previous example slightly. Assume that all cells of $\widetilde{\Delta}$ are polytopal and that all maximal cells are simplicial, i.e. all of its proper faces are simplices. Then again $C^1(\widehat{\mathbb{L}}) = \oplus_{\dim\delta=g-1}\widehat{\mathbb{L}}_{\delta,\mathbb{R}} = 0$ and one obtains

$$h^0(\widehat{\mathbb{L}}) = \sum_{\dim\delta=g} l_\delta. \tag{1}$$

This very simple formula has many applications as we will see.

[Ale99] contains a more detailed information about Stratum[Δ]. In particular, if all the cells are polytopal then the normalization of the closure of this stratum is the quotient by the finite group $\operatorname{Sym}\Delta$ of the projective toric variety corresponding to the so called generalized secondary polytope $\Sigma(\Delta)$.

3. Some concrete evidence of ETs

Theorem 3.1. *There are no ETs in dimension $g \le 3$.*

Theorem 3.2. *For $g = 4$ there is exactly 1 ET and it is isomorphic to $\mathbb{P}^2$.*

Theorem 3.3. *For $g \ge 5$, $2^g - g - 3 \le \dim \overline{\mathrm{AP}}_g < g!$*

Hence, the maximal dimension of ETs grows at least exponentially. The dimension of the "main" component, of course, grows as a particular polynomial of degree 2, namely $g(g+1)/2$. One might say that in higher dimensions ETs are a dominating life form.

Theorems 3.1 and 3.2 follow from the concrete computations of [AE] where it is proved that all periodic decompositions in dimension ≤ 3 are Delaunay (easy) and that in dimension 4 there are exactly 2 non-Delaunay decompositions (hard). Both of them are subdecompositions of the unique maximal dicing with 9 hyperplanes, which I will denote by Δ_{RT}. The dimension of the stratum for Δ_{RT} in $\overline{\mathrm{A}}_g^{\mathrm{Vor}}$ is 1. On the other hand, $h^0(\Delta, \widehat{\mathbb{L}}) = 2$ and so there must be a second irreducible component. The secondary polytope $\Sigma(\Delta_{RT})$ is a square which corresponds to $\mathbb{P}^1 \times \mathbb{P}^1$, which gives $\mathbb{P}^2$ after dividing by $\operatorname{Sym}\Delta_{RT} = \mathbb{Z}_2 \times \mathbb{Z}_2$.

The lower bound in theorem 3.3 follows by computing a particular example, the Delaunay decomposition of the classical lattice D_n. One knows (see f.e. [CS93]) that there are 3 maximal cells in this decomposition, 2 copies of a hemicube $h\gamma_n$ (with 2^{g-1} vertices) and a crosspolytope β_n (with $2g$ vertices), and that these polytopes are simplicial. Applying formula 1 gives the bound. The upper bound follows from a simple observation that for a polytopal decomposition the dimension of a secondary polytope $\Sigma(\Delta)$ is always less than the volume of $|\Delta|$ (in the lattice units of volume). The volume of $X_{\mathbb{R}}/X$ is $g!$.

If the decomposition is not polytopal but is the pullback of a polytopal decomposition in $\mathbb{R}^{g-a}$, one also has to add the term $a(a+1)/2 + a(g-a)$ to $(g-a)!$ to account for the abelian part. However, it is easy to see that for $g \geq 3$ the bound $g!$ is greater.

I note another nice application of formula 1: the lattice E_8 whose cells are 135 copies of a crosspolytope β_8 and 1920 copies of a simplex α_8. The computation gives $\dim \text{Stratum}[E_8] = h^0(E_8, \widehat{\mathbb{L}}) = 945$. This gives the largest ET in dimension 8 that I am aware of. The normalization of the closure of this stratum is the quotient of $(\mathbb{P}^7)^{135}$ by a finite group.

4. Dimension 4 case in detail

The following is a classical description, due to Voronoi, of the 2nd Voronoi decomposition in dimension 4. The picture below gives a schematic view of a cross-section of this decomposition. First of all, in dimension 4 the 2nd Voronoi decomposition τ_{Vor} is a subdecomposition of the perfect decomposition. The perfect decomposition has cones of two types: the 1st domain cones, which are simplicial (with 10 sides) and cones with 64 sides. In τ_{Vor} the cones of the second type are subdivided into 64 simplicial cones, 48 of which belong to the so called 2nd domain, and 16 – to the 3rd domain. As we have indicated, the 2nd domain cones have cones of the 1st domains as "across-the-border" neighbors, and the 3rd domain cones – 3rd domain cones again.

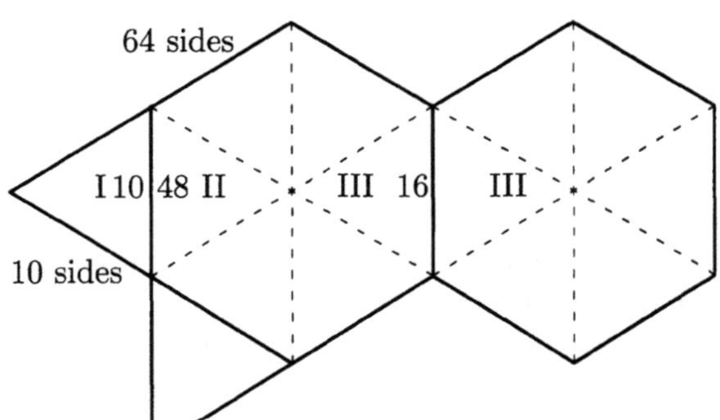

FIGURE 1. 2nd Voronoi decomposition in dimension 4

It is exactly the face between two 3rd type domains where an ET attaches itself. The way it happens is easier to see on a dual picture. A dual picture is the picture for a polarization function on τ_{Vor} ($\overline{\mathrm{A}}_g^{\text{Vor}}$ is projective, [Ale99]). Two maximal cones of 3rd type on the dual picture correspond to points, and the face between them – to an interval. In the moduli space this gives a $\mathbb{P}^1$ (which we have to divide by automorphisms which gives a $\mathbb{P}^1$ again). If we denote by Δ_{RT} the corresponding decomposition then $\dim \text{Stratum}[\Delta_{RT}]$ in $\overline{\mathrm{A}}_g^{\text{Vor}}$ is one.

An explicit description of this lattice can be found f.e. in [ER88], where this decomposition was called "red triangular". It contains:

1. a copy of a cyclic polytope C_6 with 6 vertices,
2. the inverse of it under the involution $x \mapsto -x$,
3. 18 simplexes.

Hence, according to the formula of example 2.5 the dimension of the corresponding stratum in $\overline{\mathrm{AP}}_g$ is $2 \cdot l_{C_6} = 2$. Hence, we must have an ET. This ET on the dual picture is represented by a square and is isomorphic to $\mathbb{P}^1 \times \mathbb{P}^1$. After dividing by the automorphism group $\mathbb{Z}^2 \times \mathbb{Z}^2$ this gives a $\mathbb{P}^2$ in the coarse moduli space. As the picture suggests, the intersection of the main component in $\overline{\mathrm{AP}}_g$ and the ET is a diagonal in $\mathbb{P}^1 \times \mathbb{P}^1$.

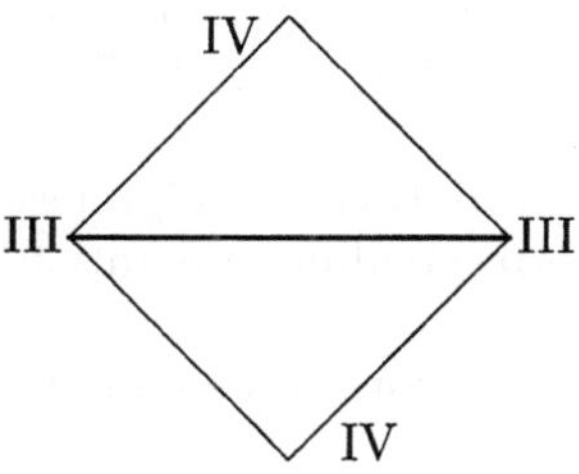

FIGURE 2. Part of the dual picture with an ET attached

A vertex marked *IV* corresponds to a non-Delaunay triangulation obtained by triangulating one copy of C_6 in one of the two possible ways, and the other – in the non-symmetric way. Clearly, this triangulation cannot be Delaunay, indeed it does not even have the basic symmetry $x \mapsto -x$. The same can be said about the corresponding variety P.

5. RELATIONSHIP BETWEEN ETS AND THE JACOBIAN LOCUS

The Torelli map $M_g \to A_g$ extends to a functorial morphism from the Mumford-Deligne compactification $\overline{M}_g$ to $\overline{\mathrm{AP}}_g$. On k-points, it sends a stable curve C to a triple $\mathrm{Pic}^0 C \curvearrowright \mathrm{Jac}^{g-1} C \supset \Theta_{g-1}$, where $\mathrm{Jac}^{g-1} C$ is the moduli space of semistable rank 1 sheaves of degree $g-1$ on C. An algorithm for computing this triple in terms of the dual graph of a stable curve C can be found in [Ale96].

By checking all genus 4 stable curves we find that precisely 1 of them maps to a point meeting an ET. It is a curve $C(K_{3,3})$ all of whose components are $\mathbb{P}^1$'s and the dual graph is the bipartite graph $K_{3,3}$.

Lemma 5.1. *The stable semiabelian variety* $\mathrm{Jac}_{g-1}(C(K_{3,3}))$ *has dimension 4 and corresponds to the decomposition* Δ_{RT}*, the "red triangular" decomposition introduced in the previous section.*

Proof. The dimension of this variety is, of course, $h_1(K_{3,3} + \sum g(C_i) = 4 + \sum 0 = 4$. We follow the algorithm of [Ale96] to compute the corresponding Delaunay decomposition. The maximal-dimensional cells correspond to the stable weightings on the graph $K_{3,3}$. A weighting (d_i) is semistable if it corresponds to a certain orientation of the graph, and the number d_i for a vertex v_i is the number of incoming arrows.

With this in mind, the stable weighings on the graph $K_{3,3}$ are (we are listing first the 3 vertices in the first group and then 3 vertices in the second group):

1. $(2,2,2,1,1,1)$. There are precisely 6 orientations of $K_{3,3}$ that give this weighting, which means that the corresponding cell is a polytope with 6 vertices. One easily computes its faces, and they are all simplicial. Therefore, this cell is isomorphic to a cyclic polytope C_6.
2. the symmetric weighting $(1,1,1,2,2,2)$. This gives a symmetric copy of C_6.
3. $(2,2,1,2,1,1)$ and its variations, 9 polytopes in total. For each of them there are exactly 5 corresponding orientations. Hence, these are all simplexes.
4. the symmetric 9 simplexes for the weighting $(2,1,1,2,2,1)$ and its variations.

Hence, we see that this is precisely the "red triangular" Delaunay decomposition Δ_{RT}. □

By the Kuratowski theorem this is the minimal (in terms of genus) graph that is not planar. The image of this curve in $\overline{\mathrm{AP}}_g$ is precisely the center $1 \in \mathbb{P}^1$ connecting two 3 type domains on the above figure. For any planar graph the corresponding Delaunay decomposition belongs to the 1st domain (or one of its faces) and so is away from an ET. Therefore, we obtain

Theorem 5.2. *For a stable curve of genus $g \le 4$ its Torelli image $[C] \in \overline{\mathrm{AP}}_g$ meets an ET if and only if the dual graph $C(\Gamma)$ is not planar.*

One can easily make this theorem a little stronger: it is true for a stable curve of arbitrary genus such that $h^1(C(\Gamma)) \le 4$. This certainly invites a

Guess 5.3. Theorem 5.2 holds in arbitrary genus.

It feels that the non-planarity of C somehow opens up new dimensions for deformations for its jacobian. Here is some additional evidence in favor of this guess:

1. If $\Gamma(C)$ is planar, the corresponding Delaunay decomposition belongs to the simplest part of the 2nd Voronoi decomposition, the so-called 1st domain. If there is any part of $\overline{\mathrm{A}}_g^{\mathrm{Vor}}$ where "nothing tricky happens", that should be it.
2. It is plausible that if $\Gamma(C)$ contains a subgraph $\Gamma(C')$ for a curve C' whose jacobian has extra deformations then the same must be true for the curve

C itself. By the Kuratowski theorem a graph Γ is non-planar if and only if it contains either a $K_{3,3}$ or a K_5. The case of K_5, therefore, becomes a crucial test for the validity of the conjecture. Therefore, let me give the description of the corresponding Delaunay decomposition.

The genus of $C(K_5)$ is $h_1(K_5) = 6$. The stable weightings in this case are:

1. $3+3+2+1+1$ and variations, 30 cells total. Each of these cells has 8 vertices and is easily seen to be $\mathrm{Pyr}^4(\alpha_1 \times \alpha_1)$, a fourfold pyramid over a 4-gon.
2. $3+2+2+2+1$ and variations. There are 20 such cells and each of them has 14 vertices.
3. $2+2+2+2+2$. This cell is unique and it has 24 vertices.

This is a fairly complicated decomposition and there are no *obvious* ETs in its neighborhood but some could still be hiding.

If the above guess were true, it would give one interesting explanation for the role of the extra components. On the other hand, let me note that the role of ETs is very similar to that of extra irreducible components in Hilbert schemes. These appear in the simplest cases, as soon as you look at a Hilbert scheme of curves of degree d in $\mathbb{P}^3$, for example. So, extra components in moduli spaces are, in my opinion, more of a rule than an exception, and the Deligne-Mumford compactification $\bar{M}_g$ in this sense is a very big exception indeed.

References

[AE] V. Alexeev and R. Erdahl, *Periodic tilings in dimension* ≤ 4, in preparation.

[Ale96] V. Alexeev, *Compactified jacobians*, preprint (1996), alg-geom/9608012.

[Ale99] V. Alexeev, *Complete moduli in the presence of semiabelian group action*, preprint (1999), math.AG/9905103.

[AMRT75] A. Ash, D. Mumford, M. Rapoport, and Y. Tai, *Smooth compactifications of locally symmetric varieties*, Lie groups: history, frontiers and applications, vol. IV, Math Sci Press, 1975.

[CS93] J.H. Conway and N.J.A. Sloane, *Sphere packings, lattices and groups*, 2nd ed., A Series of Comprehensive Studies in Mathematics, vol. 290, Springer-Verlag, 1993.

[ER88] R.M. Erdahl and S.S. Ryshkov, *The empty sphere, II*, Canad. J. Math. **40** (1988), 1058–1073.

[FC90] G. Faltings and C.-L. Chai, *Degenerations of abelian varieties*, vol. 22, Ergebnisse der Mathematik und ihrer Grenzgebiete, no. 3, Springer-Verlag, 1990.

Progress in Mathematics, Vol. 195, © 2001 Birkhäuser Verlag Basel/Switzerland

ON MUMFORD'S UNIFORMIZATION AND NERON MODELS OF JACOBIANS OF SEMISTABLE CURVES OVER COMPLETE RINGS

FABRIZIO ANDREATTA

Contents

1 Introduction

Fix a base scheme S, spectrum of a noetherian normal domain E which is complete and separated with respect to an ideal I. Let K be its function field and let S_0 be the closed subscheme defined by I.

1.1 *Assumptions and definitions.* Let $C \to S$ be a semistable curve, i.e., a proper and flat morphism with one dimensional, geometrically reduced and connected fibers having only ordinary double points as singularities.
We *assume* that the curve $C_0 \longrightarrow S_0$ obtained from $C \to S$ by base change to S_0 is the union of smooth connected curves $C_v \to S_0$ with $v \in \mathcal{V}_\mathcal{G}$ intersecting

transversally along sections $T_e\colon S_0 \to C_0$ with $e \in \mathcal{E}_{\mathcal{G}}$. In this case we can define the dual graph $\mathcal{G}$ of $C_0 \to S_0$; see 12.2. The set of vertices is $\mathcal{V}_{\mathcal{G}}$, the set of edges is $\mathcal{E}_{\mathcal{G}}$. We define in 12.3 the graph Γ by contracting the disconnecting edges of $\mathcal{G}$, i.e., the edges which do disconnect $\mathcal{G}$. The set of edges $\mathcal{E}$ of Γ is the set of non-disconnecting edges of $\mathcal{G}$.

We *assume* that for any $v \in \mathcal{V}_{\mathcal{G}}$ there exists a section of the smooth locus of C over S factoring over S_0 through C_v.

For any $e \in \mathcal{E}$ we *assume* that the completion of C at the singular section T_e of $C_0 \to S_0$ is of the form $E[\![X,Y]\!]/(XY - q_e)$ for some non-zero element q_e of E such that (q_e) is a prime ideal of E. In particular, C_K is smooth over K.

1.2 *The main example.* Consider S to be a versal deformation space of a semistable curve $C_0 \to S_0$ defined over the spectrum S_0 of an algebraically closed field and let $C \to S$ be the versal algebraic curve over S.

1.3 *Interpolating Néron models of Jacobians.* Let R be a discrete valuation ring. Let $V = \mathrm{Spec}(R)$. Let η be its generic point and V_0 be the closed point. Let $V \to S$ be a morphism such that C_V is a regular scheme, C_η is smooth over η and V_0 is a closed point of S_0. Denote by G_V the Néron model of $\mathrm{Pic}^0\big(C_\eta/k(\eta)\big)$ over V; it exists by [4]. The special fiber G_{V_0} is isomorphic as a scheme over V_0 to the disjoint union of finitely many copies of the semiabelian variety $\mathrm{Pic}^0(C_{V_0}/V_0)$. Under the regularity assumption on C_V it depends only on the closed point V_0 of S_0, rather than on V. See [4, Section 9.6, pages 282-283].

Question: Does there exist a scheme over S whose base change via any morphism $V \to S$ as above is the Néron model of $\mathrm{Pic}^0\big(C_\eta/k(\eta)\big)$ over V?

We give an affirmative answer to this question.

The idea is simple. For suitable elements φ of $\mathbf{R}^{\mathcal{V}}$, where $\mathcal{V}$ is the set of vertices of Γ, we construct compactifications $\bar{P}^{\varphi} \to S$ of $\mathrm{Pic}^0(C/S)$ such that their smooth locus $\bar{P}^{\varphi,\mathrm{sm}}$ has the wanted interpolating property. See 19.9 and 19.11. Moreover, $\bar{P}^{\varphi,\mathrm{sm}}$ contains $\mathrm{Pic}^0(C/S)$ as an open subscheme and there is an action of $\mathrm{Pic}^0(C/S)$ on $\bar{P}^{\varphi,\mathrm{sm}} \to S$ extending the action on $\mathrm{Pic}^0(C/S)$ induced by the group structure on $\mathrm{Pic}^0(C/S)$.

1.4 Remark. One can ask whether the scheme $\bar{P}^{\varphi,\mathrm{sm}} \to S$ is the Néron model of the abelian variety $\mathrm{Pic}^0(C_K/K)$ over S or, at least, if it has the structure of a group scheme over S extending the group scheme structure on $\mathrm{Pic}^0(C_K/K)$. In 19.12 we give an example which proves that both questions have a negative answer.

The dependence on φ is discussed in 19.13. In some cases, for example if $C_0 \to S_0$ is geometrically irreducible or Γ is simple in the sense of 5.3, the compactifications $\bar{P}^{\varphi}$, and hence the schemes $\bar{P}^{\varphi,\mathrm{sm}}$, do not depend on φ. In general, they depend on φ.

1.5 Remark. The theory of compactifications of relative Pic^0 of semistable curves has been extensively studied. For example, see [23] for curves over the spectrum of an algebraically closed field and [1] for base schemes over the complex numbers. See also [5]. In these papers the compactifications are constructed via *global* methods using moduli of coherent sheaves and methods of geometric invariant theory. The elements φ mentioned above appear as "generalized multidegrees" of line bundles. Our method is different and is based on deformation theory and uniformization techniques. It is more suited to get *local* informations allowing us to control the singularities and the smooth locus of our compactifications. In our case the elements φ play a purely combinatorial role. The geometric fibers of our compactifications over S_0 appear in [23]. We remark that our compactifications are not stable quasi-abelian varieties in the terminology of [22] and [1]. In fact, even if E is a discrete valuation ring with algebraically closed residue field, the special fibers of stable quasi-abelian varieties associated to $\mathrm{Pic}^0(C/S)$ have in general fewer irreducible components than the special fiber of the Néron model of $\mathrm{Pic}^0(C_K/K)$ over S. See [22, Section 19].

1.6 *Degenerating principally polarized abelian varieties.* Let $G \longrightarrow S$ be a semiabelian scheme over S. By this we mean a smooth and separated group scheme over S such that for any point s of S the group scheme G_s is the extension of an abelian variety over $k(s)$ by a torus over $k(s)$. Suppose that the generic fiber $G_K \longrightarrow \mathrm{Spec}(K)$ is an abelian variety over K. Let λ_K be a principal polarization of G_K.

The semiabelian schemes over S, whose base change to $\mathrm{Spec}(K)$ are principally polarized abelian varieties, naturally appear in the theory of degenerations of principally polarized abelian varieties. Indeed, a theorem of Grothendieck assures that the Néron model of any abelian variety over the fraction field of a discrete valuation ring over a finite extension of the base acquires semiabelian reduction, i.e., it is a semiabelian scheme.

In the last few decades a lot of work has been devoted to understanding and constructing compactifications of the moduli stack $\mathcal{A}_g$ of principally polarized abelian varieties of dimension g. The main technique has been the so called "Mumford uniformization".

1.7 *Mumford's uniformization.* With the notation of 1.6 suppose furthermore that we have the following exact sequence

$$0 \longrightarrow T_0 \longrightarrow G \times_S S_0 \longrightarrow B_0 \longrightarrow 0,$$

where T_0 is a torus over S_0 of rank r and B_0 is an abelian scheme. Assume that S satisfies the condition

(*) *for any étale E/I-algebra R_0, its unique lifting R to a formally étale E-algebra is normal.*

Under these assumptions, work of D. Mumford, G. Faltings and C.L. Chai shows that the pair (G, λ_K) defines the following data and can be recovered from the following data:

(1) a semiabelian scheme $\tilde{G}$ over S sitting in an exact sequence

$$0 \longrightarrow T \longrightarrow \tilde{G} \longrightarrow B \longrightarrow 0$$

with T a torus over S of rank r and B an abelian scheme over S with principal polarization α_B. The base change $\tilde{G} \times_S S_0$ of $\tilde{G}$ to S_0 is $G \times_S S_0$;

(2) a homomorphism $\iota: X(T) \to \tilde{G}_K$ lifting c_K^t. We denote by c^t the homomorphism from the character group $X(T) := \mathrm{Hom}_S(T, \mathbf{G}_{m,S})$ of T to B such that $\alpha_B \circ c^t: X(T) \to B^t := \mathrm{Ext}^1(B, \mathbf{G}_{m,S})$ defines $\tilde{G}$ as extension of B by T.

Datum (1) should be thought of as a "partial universal cover" of G. It is the Raynaud extension of G relative to $G \times_S S_0$; see [8, Section II.1]. Datum (2) should define the "lattice of periods". The object $\big((B, \alpha_B), c^t, \iota\big)$ is called Mumford's uniformization of (G, λ_K). See the Introduction to [8, Chapter II] for more details.

1.8 Remark. The advantages of Mumford's uniformization are many. From the point of view of moduli spaces it makes it possible to describe the boundaries (there are many compactifications!) of $\mathcal{A}_g$. Moreover it allows to interpret the q-expansion of modular forms as their evaluation on boundary families. For this last point the reader should keep in mind the role of the Tate curve (equivalent to Mumford's uniformization for g equal 1) in the theory of modular forms in one variable. See [6] and [8] for complete expositions.

1.9 *The* Pic^0 *of semistable curves.* Let $\mathrm{Pic}^0(C/S)$ be the subfunctor of the relative Picard functor of C over S consisting of line bundles L on C such that for any geometric point s of S the restriction of L to any irreducible component of C_s has degree 0. It is known that $\mathrm{Pic}^0(C/S)$ is represented by a smooth and separated scheme over S; see for example [4, Theorem 9.4/1]. Moreover the scheme $\mathrm{Pic}^0(C/S)$ is semiabelian. In fact, for any geometric point s of S the group scheme $\mathrm{Pic}^0(C/S)_s$ is an extension of the principally polarized abelian variety over $k(s)$ defined as the Pic^0 of the normalization of C_s, by a torus over $k(s)$.

1.10 *Mumford's uniformization of* Pic^0 *of semistable curves.* With the hypotheses on $C_0 \to S_0$ discussed in 1.1 we have an exact sequence

$$0 \longrightarrow T_0 \longrightarrow \mathrm{Pic}^0(C/S) \underset{S}{\times} S_0 \longrightarrow \mathcal{J}_0 \longrightarrow 0,$$

where T_0 is a torus over S_0 of rank r and $\mathcal{J}_0 = \prod_{v \in \mathcal{V}_{\mathcal{G}}} \mathrm{Pic}^0(C_v/S_0)$ is a principally polarized abelian scheme over S_0. Moreover $\mathrm{Pic}^0(C_K/K)$ is a principally polarized abelian variety. If S satisfies condition $(*)$, then $\mathrm{Pic}^0(C/S)$ defines its Mumford's uniformization $\big((\mathcal{J}, \alpha_{\mathcal{J}}), c^t, \iota\big)$ and $\mathrm{Pic}^0(C/S)$ can be recovered from it. See 1.7.

1.11 *Our approach to Mumford's uniformization of* Pic^0 *of semistable curves.* Let $C \to S$ be a semistable curve as in 1.1. The method to associate to the semiabelian scheme $\mathrm{Pic}^0(C/S)$ data (1) and (2) above, in the case that S satisfies $(*)$, is contained in [8, Chapter II]. The idea is that these data are encoded in the theta functions associated to $\mathrm{Pic}^0(C_K/K)$.

Question: Is it possible to get the uniformization data using the geometry of the curve $C \to S$?

The answer is affirmative. If E is a discrete valuation ring this is well documented; see [3] or [9]. If furthermore the curves C_v with $v \in \mathcal{V}_{\mathcal{G}}$ are isomorphic to $\mathbf{P}^1_{S_0}$, the construction is made even more explicit in [16] using the theory of Mumford's curves developed in [20]. We prove that the answer is still affirmative for general S, but we proceed along different lines. We also prove that the hypothesis $(*)$ on S can be dropped.

Fix a smooth section $\bar{0}$ of $C \to S$.

Assume first that $C_0 \to S_0$ is geometrically irreducible. In 14.7 we describe a compactification $\bar{P}_0$ of $\mathrm{Pic}^0(C/S) \times_S S_0$ and a Galois cover $\mathcal{P}_0$ of $\bar{P}_0$ with group $H_1(\Gamma, \mathbf{Z})$. We construct in 14.2 a Galois cover $\mathcal{C}_0$ of C_0 with group $H_1(\Gamma, \mathbf{Z})$ and an $H_1(\Gamma, \mathbf{Z})$-equivariant morphism $\mathcal{C}_0 \to \mathcal{P}_0$. Summarizing we get a commutative diagram

$$\begin{array}{ccc} \mathcal{C}_0 & \longrightarrow & \mathcal{P}_0 \\ \downarrow & & \downarrow \\ C_0 & \longrightarrow & \bar{P}_0. \end{array}$$

Then we study the deformation theory of the diagram. We prove in 15.9 that $C \to S$ induces a deformation (in the category of formal schemes) of the whole diagram to $\mathrm{Spf}(E)$. In other words it induces a deformation of $\bar{P}_0$, of the Galois covers $\mathcal{C}_0 \to C$ and $\mathcal{P}_0 \to \bar{P}_0$ and of the morphisms $\mathcal{C}_0 \to \mathcal{P}_0$ and $C_0 \to \bar{P}_0$ so that the associated diagram commutes.
We prove that the formal scheme deforming $\bar{P}_0$ can be algebraized to a scheme $\bar{P}$ over S. We obtain in 16.4 a morphism from C to $\bar{P}$ over S. We prove in 16.13 and in 17.7 that the smooth locus of $\bar{P} \to S$ has the structure of a group scheme isomorphic to $\mathrm{Pic}^0(C/S)$ having the image of $\bar{0}$ as identity element.
We prove in 10.2 and 10.11 that the deformation $\mathcal{P}$ of $\mathcal{P}_0$ over S is the formal scheme associated to a scheme $\widetilde{P} \to S$ which carries an action of $H_1(\Gamma, \mathbf{Z})$, contains $\tilde{G}$ as an open dense subscheme and has $\tilde{G}_K$ as fiber over $\mathrm{Spec}(K)$. In 16.18 we show how the induced action of $H_1(\Gamma, \mathbf{Z})$ on $\tilde{G}_K$ defines the morphism ι of (2) of 1.7. This defines Mumford's uniformization data, since it can be shown that $X(T)$ in 1.7 is canonically isomorphic to $H_1(\Gamma, \mathbf{Z})$.

Finally we use these results to get Mumford's uniformization of semistable curves $C \to S$ such that $C_0 \to S_0$ is not geometrically irreducible. We show in 18.12 how to reduce this case to the case of geometrically irreducible curves.

1.12 *Notation.* Fix an orientation of the edges $\mathcal{E}$ of Γ. There is a natural morphism $\mathbf{Z}^{\mathcal{E}} \to \mathbf{Z}^{\mathcal{V}}$ defined by sending an oriented edge e to the difference of its target and its source. The kernel of this homomorphism is $H_1(\Gamma, \mathbf{Z})$. It is isomorphic to the character group $X(T)$ of the torus T in (1) of 1.7 as already remarked; see 16.20 and 18.13. Due to the assumptions on C_0, there exists a $\mathbf{Z}$-basis of $X(T)$ consisting of S-valued points $l_1, \ldots, l_r$. Fix one. For any $a = 1, \ldots, r$ decompose

$$l_a = \bigoplus_{e \in \mathcal{E}} l_{a,e} \cdot e \qquad \text{via} \qquad H_1(\Gamma, \mathbf{Z}) \hookrightarrow \mathbf{Z}^{\mathcal{E}}.$$

The choice of the basis of $X(T)$ determines an isomorphism $T \cong \mathbf{G}^r_{m,S}$. In particular, T is split.

1.13 *The uniformization of 1.3.* We have announced in 1.3 the existence of schemes $\bar{P}^{\varphi,sm}$ over S interpolating the Néron models of $\mathrm{Pic}^0(C_\eta/k(\eta))$ for all spectra of discrete valuations V with generic fiber η and all maps $V \to S$ such that C_V is regular and C_η is smooth over $k(\eta)$. We have explained how the scheme $\mathrm{Pic}^0(C_K/K)$ is uniquely determined by the uniformization data (1) and (2) above, without the assumption $(*)$ on S. This leads to the following:

Question: Can we construct $\bar{P}^{\varphi,\mathrm{sm}}$ in terms of data (1) and (2)?

This is done as follows. For suitable elements φ of $\mathbf{R}^{\mathcal{V}}$ we construct a scheme $\widetilde{P}^{\varphi}$ over S, with an action of the group $H_1(\Gamma, \mathbf{Z})$. Moreover $\widetilde{P}^{\varphi}_K$ is isomorphic to $\tilde{G}_K$ and for any l in $H_1(\Gamma, \mathbf{Z}) \cong X(T)$ the induced action of l is translation by $\iota(l)$. See 19.7. The scheme $\widetilde{P}^{\varphi}$ is defined in terms of datum (1), of a refinement of datum (2) and of the combinatorics of the graph Γ. See 19.14. Let $\mathcal{P}^{\varphi} \to \mathrm{Spf}(E)$ be the I-adic completion of $\widetilde{P}^{\varphi} \to S$. The scheme $\bar{P}^{\varphi} \to S$ is constructed as the algebraization of the formal scheme obtained as the quotient of $\mathcal{P}^{\varphi}$ under the action of the group $H_1(\Gamma, \mathbf{Z})$. If $C \to S$ is geometrically irreducible, the schemes $\widetilde{P}^{\varphi}$ and $\mathcal{P}^{\varphi}$ and $\bar{P}^{\varphi}$ do not depend on φ and hence the subscript φ will be omitted. In this case these schemes have been introduced already in 1.11.

1.14 *The geometry of $C \to S$ and the period map ι.* Let a be an integer such that $1 \le a \le r$. The pullback T_a via $c^t(l_a)\colon S \to \mathcal{J}$ of $\tilde{G} \to \mathcal{J}$ is isomorphic to $T \to S$. The section $\iota(l_a)$ of $T_{a,K} \cong \mathbf{G}^r_{m,K}$ determines r fractional ideals $I_{a,1}, \ldots, I_{a,r}$. These define divisors on S; they should be thought of as the divisors along which the periods defined in (2) of 1.7 degenerate. In [8, Theorem II.8.3] it is shown that these divisors are uniquely determined by the discriminant locus of $C \to S$. More precisely, for any e in $\mathcal{E}$ the completion of C at the singularity T_e of C_0 is isomorphic by 1.1 to $E[\![X,Y]\!]/(XY - q_e)$ for some q_e in the ideal I of O_S. Let a and b be elements in $\{1,\dots,r\}$. Then

$$I_{a,b} = \Big(\prod_{e\in\mathcal{E}} q_e^{l_{a,e}\cdot l_{b,e}}\Big) E.$$

The fact that the two divisors have the same support is not a surprise; the lattice of periods $X(T)$ should degenerate when C acquires singularities. The difficult thing is to prove that the multiplicities are the same. The proof of this result in [8] relies on the study by Raynaud of the group of connected components of the Néron model of curves over discrete valuation rings.

Question: Is there a geometric proof of the fact that for any a and b the fractional ideal $I_{a,b}$ coincides with the ideal generated by $\prod_{e\in\mathcal{E}} q_e^{l_{a,e}\cdot l_{b,e}}$?

This is given in 19.16. The idea is the following. We can reduce to the case that $C_0 \to S_0$ is geometrically irreducible due to 18.12. In 16.4 we construct a compactification $\bar{P}$ of $\mathrm{Pic}^0(C/S)$ and a morphism $C \to \bar{P}$. The scheme $\bar{P}$ determines the uniformization data (1) and (2) defined above; see 16.7. The behavior of $\bar{P}$ at its singular points is determined by the the behavior of C at its singular points of C_0; see 15.9. The discriminant locus of $\bar{P} \to S$ is described by the ideals $I_{a,b}$ defined above.

1.15 *How to read this paper.* The first example the reader should have in mind is when $C_0 \to S_0$ is geometrically irreducible. In this case the schemes $\bar{P}_0$ and its infinite $H_1(\Gamma,\mathbf{Z})$-Galois cover $\mathcal{P}_0 \to \bar{P}_0$, the infinite $H_1(\Gamma,\mathbf{Z})$-Galois cover $\mathcal{C}_0 \to C_0$ and the morphisms $\mathcal{C}_0 \to \mathcal{P}_0$ and $C_0 \to \bar{P}_0$ take a simple form. See Section 14 and especially 14.7.

Let us illustrate the basic idea in the case that C_0 is obtained from a smooth curve $D_0 \to S_0$ identifying transversally two smooth sections P and Q. The relative Pic^0 of C_0 over S_0 has the structure of a $\mathbf{G}_m$-bundle over $\mathcal{J}_0 = \mathrm{Pic}^0(D_0/S_0)$. We construct $\bar{P}_0$ by first compactifying $\mathrm{Pic}^0(C_0/S_0)$ as a $\mathbf{P}^1$-bundle $\mathbf{P}_0$ over $\mathcal{J}_0$ and by then identifying transversally its zero section and its infinity section, which are isomorphic to $\mathcal{J}_0$, via translation by the point $P-Q$. Remark that even though $\mathbf{P}_0$ is a scheme over $\mathcal{J}_0$, the scheme $\bar{P}_0$ is a scheme over S_0 and does *not* admit a morphism to $\mathcal{J}_0$. We embed D_0 in $\mathbf{P}_0$ so that D_0 intersects transversally the zero (respectively infinity) section of $\mathbf{P}_0$ at P (respectively Q). The image of D_0 in $\bar{P}_0$ is by construction C_0. The scheme $\mathcal{P}_0$ is defined by taking countable many copies $(\mathbf{P}_0, z)_{z\in\mathbf{Z}}$ of $\mathbf{P}_0$ and identifying the zero section of the copy associated to $z+1$ with the infinity section of the copy associated to z via translation by $P-Q$. The pullback of C_0 defines $\mathcal{C}_0 \to \mathcal{P}_0$. Note that $\mathcal{P}_0$ is naturally a scheme over $\mathcal{J}_0$, it is endowed with an action of $\mathbf{Z}$ and we have that $\mathcal{P}_0/\mathbf{Z} \cong \bar{P}_0$.

The second step is the proof in 15.9 and 16.7 that a deformation $C \to S$ of $C_0 \to S_0$ together with a smooth section $S \to C$ defines *uniquely* a deformation $\bar{P} \to S$ of $\bar{P}_0 \to S_0$ as a scheme such that $C_0 \to S_0$ extends to a morphism $C \to \bar{P}$ and $\bar{P}$ contains $\mathrm{Pic}^0(C/S)$ as an open subscheme such that the image of the given section of C over S is the identity element. This is based on a detailed study of the local-global spectral sequence of the tangent sheaves relative to S_0 of C_0 and $\mathcal{P}_0$. This occupies Sections 7-9 and Section 15.

On the other hand the Galois cover $\mathcal{P}_0 \to \bar{P}_0$ extends to a Galois cover of formal schemes defining a deformation $\mathcal{P}$ of $\mathcal{P}_0$ over $\mathcal{S}$ (the formal scheme associated to S). The content of Section 10, especially 10.2, is to prove that $\mathcal{P} \to \mathcal{S}$ is the I-adic completion of a scheme $\widetilde{P} \to S$. In 11, especially 11.1, we prove that $\widetilde{P}_K$ is a semiabelian scheme $\tilde{G}_K$ and that the action of $H_1(\Gamma, \mathbf{Z})$ on $\mathcal{P}$ is algebraizable to an action on $\widetilde{P}$ and defines a 1-motive $H_1(\Gamma, \mathbf{Z}) \to \tilde{G}_K$ via the action by translations on $\tilde{G}_K = \widetilde{P}_K$. The scheme $\widetilde{P}$ over S can be recovered from the 1-motive. It is with the help of $\widetilde{P}$, or rather of the associated 1-motive, that we prove in 16.13 that the smooth locus $\bar{P}^{\mathrm{sm}}$ of $\bar{P}$ over S has the structure of a group scheme over S. We show in 17.7 that $\bar{P}^{\mathrm{sm}} \to S$ coincides with $\mathrm{Pic}^0(C/S)$. We prove in 17.2 that it has the wanted property of interpolating Néron models in the sense of 1.3. We should remark that the latter property, despite the fact that it concerns only $\mathrm{Pic}^0(C/S)$, is deduced from the knowledge of the singularities of $\bar{P}$ or equivalently of $\widetilde{P}$. Finally we prove in 16.21 that if S satisfies $(*)$ the 1-motive introduced above is Mumford's uniformization of $\mathrm{Pic}^0(C/S)$. This follows noting that $\widetilde{P} \to S$ is a relatively complete model in the sense of [8].

We explain now how to deduce the general case. We construct

1. the semiabelian scheme $\tilde{G}$ defined in 18.11;
2. the elements $x_e \in \tilde{G}(K)$ for all $e \in \mathcal{E}$ as defined in 18.17;
3. the Namikawa decomposition $\mathrm{Del}_\varphi\big(H^1(\Gamma, \mathbf{R})\big)$ defined in 3.6.

From (2) we obtain the 1-motive $H_1(\Gamma, \mathbf{Z}) \to \tilde{G}(K)$ defined in 18.12 as follows. Let l be an element of $H_1(\Gamma, \mathbf{Z})$ decomposing as $l = \oplus_{e\in\mathcal{E}}\, l_e \cdot e$ in $C_1(\Gamma, \mathbf{Z}) = \mathbf{Z}^{\mathcal{E}}$. The image of l in $\tilde{G}(K)$ is simply $\prod_{e\in\mathcal{E}} x_e^{l_e}$.

Data (1) and (2) are deduced reducing to the case of a geometrically irreducible curve. While the first two data depend on the family of curves $C \to S$, datum (3) is *discrete* and depends on the combinatorics of the graph Γ associated in 1.1 to $C_0 \to S_0$. It is a rational polyhedral decomposition of $H^1(\Gamma, \mathbf{R})$. Its definition occupies Section 3 and is originally due to [23]. In Section 6, especially 6.2, we explain how to use this datum to construct compactifications P_ξ of $\tilde{G}$ for $\xi \in H^1(\Gamma, \mathbf{Z})$. Using these data we define in 19.1

$$z_\xi \colon \tilde{G}_K \longrightarrow P_\xi.$$

From this we define $\widetilde{P}^\varphi$ in 19.2. The construction is rather delicate: one would like to define it as the Zariski closure of

$$\tilde{G}_K \xrightarrow{\prod z_\xi} \prod_{\xi\in H^1(\Gamma,\mathbf{Z})} P_\xi,$$

which is not well defined. Hence one has to define the Zariski closure taking only finitely many P_ξ's. One then constructs $\widetilde{P}^\varphi$ by patching suitable open subschemes of such Zariski closures. Fortunately, $\widetilde{P}^\varphi$ is the fppf quotient under the action of a torus of an open subscheme of a scheme of the type studied in Sections 10-11; see 10.2. This allows to study $\widetilde{P}^\varphi \to S$ in detail; see 19.7. In particular, its generic fiber is $\tilde{G}_K$, it is endowed with an action of $H_1(\Gamma, \mathbf{Z})$ extending the action on $\tilde{G}_K$. Finally we define $\bar{P}^\varphi \to S$ in 19.9 as the scheme algebraizing the quotient under the action of $H_1(\Gamma, \mathbf{Z})$ of the I-adic completion of $\widetilde{P}^\varphi$. From the study of $\widetilde{P}^\varphi$, and especially of the non-smooth locus of $\widetilde{P}^\varphi \to S$, we deduce in 19.11 that $\bar{P}^\varphi$ contains $\mathrm{Pic}^0(C/S)$ as an open subscheme, it is acted upon by $\mathrm{Pic}^0(C/S)$ and it has the interpolating property discussed in 1.3.

Acknowledgements. This paper is part of the author's PhD thesis completed under the supervision of Prof. Frans Oort. I would like to thank him for his valuable help. I thank B. Moonen for many stimulating discussions. I thank the many mathematicians with whom I could talk about the subject of the present paper. Especially M. Raynaud, A. Abbes, L. Barbieri-Viale, C.-L. Chai, B. Edixhoven, J. de Jong, E. Looijenga, M. van der Put. I thank Prof. W. Lütkebohmert for his comments. This project has been funded by the TMR-PreDoc fellowship of the European Community nr. ERBFMICT971961.

2 Notations

2.1 *Sheaves and bundles.* Let $f: X \to S$ be a morphism of schemes. If $\mathcal{L}$ is an invertible sheaf on X, denote by

i. $\mathbf{P}(\mathcal{L} \oplus O_X)$ the associated $\mathbf{P}^1$-bundle;
ii. $V(\mathcal{L})$ the associated line bundle coinciding with the complement of the infinity section of $\mathbf{P}(\mathcal{L} \oplus O_X)$;
iii. $G(\mathcal{L})$ the associated $\mathbf{G}_m$-bundle defined as the complement of the zero section of $V(\mathcal{L})$.

Recall that $\mathbf{P}(\mathcal{L} \oplus O_X)$ carries naturally two sections over X

$$0: X = \mathbf{P}(\mathcal{L}) \to \mathbf{P}(\mathcal{L} \oplus O_X) \qquad \text{and} \qquad \infty: X = \mathbf{P}(O_X) \to \mathbf{P}(\mathcal{L} \oplus O_X).$$

Let $\mathcal{L}_1, \dots, \mathcal{L}_r$ be invertible sheaves on X. Let $\mathbf{P}(\mathcal{L}_1 \oplus O_X), \dots, \mathbf{P}(\mathcal{L}_r \oplus O_X)$ be the associated $\mathbf{P}^1$-bundles. For any $s = 1, \dots, r$ the infinity and zero sections of $\mathbf{P}(\mathcal{L}_s) \to X$ define, by base change,

$$\infty_s: \prod_{s'=1,\dots,r | s' \neq s} \mathbf{P}(\mathcal{L}_{s'} \oplus O_X) \longrightarrow \prod_{t=1,\dots,r} \mathbf{P}(\mathcal{L}_t \oplus O_X)$$

and

$$0_s: \prod_{s'=1,\dots,r | s' \neq s} \mathbf{P}(\mathcal{L}_{s'} \oplus O_X) \longrightarrow \prod_{t=1,\dots,r} \mathbf{P}(\mathcal{L}_t \oplus O_X),$$

where the products are taken as schemes over X.

2.2 *Divisors and bundles.* Let $\mathcal{I}_1$ and $\mathcal{I}_2$ be invertible ideal sheaves of O_X defining effective Cartier divisors Y_1 and Y_2 relative to f. See [4, Section 8.2, page 213] for the notion of effective relative Cartier divisor. Define the $\mathbf{P}^1$-bundle

$$\mathbf{P}(Y_2 \oplus Y_1) := \mathbf{P}(\mathcal{I}_1 \oplus \mathcal{I}_2),$$

the invertible sheaf

$$\mathcal{L}(Y_2 - Y_1) := \mathcal{I}_1 \otimes \mathcal{I}_2^{-1}$$

and the $\mathbf{G}_m$-bundle

$$G(Y_2 \oplus Y_1) := G\big(\mathcal{L}(Y_2 - Y_1)\big).$$

If U is the open subscheme of X defined as the complement of $Y_1 \cap Y_2$, we have $(\mathcal{I}_1 + \mathcal{I}_2)_{|U} = O_U$. Hence we get a section σ over U of $\mathbf{P}\big(\mathcal{L}(Y_2 - Y_1) \oplus O_X\big)$ given by

$$\sigma: U = \mathbf{P}(O_U) = \mathbf{P}\big((\mathcal{I}_1 + \mathcal{I}_2)_{|U}\big) \longrightarrow \mathbf{P}(\mathcal{I}_{1|U} \oplus \mathcal{I}_{2|U}).$$

In particular, if Y_1 and Y_2 are disjoint, then $U = X$. In this case $\sigma(X)$ intersects the infinity section of $\mathbf{P}\big(\mathcal{L}(Y_2 - Y_1) \oplus O_X\big)$ along Y_2 and the zero section of $\mathbf{P}\big(\mathcal{L}(Y_2 - Y_1) \oplus O_X\big)$ along Y_1.

2.3 *Bundles on abelian varieties.* Suppose that X is an abelian scheme over S and $\mathcal{L}$ is an invertible sheaf algebraically equivalent to zero. The choice of an S-valued point of $G(\mathcal{L})$ over the identity element of X is equivalent to a trivialization as a $\mathbf{G}_m$-bundle of the fiber of $G(\mathcal{L})$ above the identity element of X. It is called a *rigidification* of the $\mathbf{G}_m$-bundle $G(\mathcal{L})$ over the identity element of X. As explained in [19, Theorem IV.23.1], it identifies $G(\mathcal{L})$ with the theta group of $V(\mathcal{L})$. In this way $G(\mathcal{L})$ acquires the structure of a group scheme over S. The group law is commutative since the commutator pairing $X \times_S X \to \mathbf{G}_{m,S}$ is trivial. The natural projection $G(\mathcal{L}) \to X$ is a homomorphism of group schemes and the kernel is isomorphic to $\mathbf{G}_{m,S}$.
Let $\mathcal{L}_1, \ldots, \mathcal{L}_r$ be invertible sheaves on X algebraically equivalent to zero. Let

$$G := G(\mathcal{L}_1) \underset{X}{\times} \cdots \underset{X}{\times} G(\mathcal{L}_r).$$

A *rigidification* of G is defined to be the choice of an S-valued point above the identity element of X and is equivalent to giving a rigidification of $G(\mathcal{L}_t)$ for any $t = 1, \ldots, r$. It determines uniquely a group scheme structure on G such that the morphism

$$G \to G(\mathcal{L}_1) \underset{S}{\times} \cdots \underset{S}{\times} G(\mathcal{L}_r)$$

is a group homomorphism considering on the right hand side the group structure induced by the rigidification of each of the $G(\mathcal{L}_t)$'s. In particular, the morphism $G \to X$ is a group homomorphism with kernel equal to $\mathbf{G}_{m,S}^r$.

3 Review of some results of T. Oda and C.S. Seshadri concerning graphs

This section is of technical nature. We collect a few definitions and results concerning graphs needed later. We assume our graphs to be *connected.* Our main reference is [23]. The main definition is contained in 3.5 and 3.6, where we define suitable polyhedral decompositions $\mathrm{Del}_\varphi\big(H^1(\Gamma,\mathbf{R})\big)$ associated to a graph Γ. In 3.9 we use these decompositions to define torus embeddings of the torus having character group $H_1(\Gamma,\mathbf{Z})$.

3.1 Definition. *Let Γ be a graph. Let $\mathcal{V}$ be the set of vertices and let $\mathcal{E}$ be the set of edges. Define*

$$C_1(\Gamma,\mathbf{Z}) := \mathbf{Z}^{\mathcal{E}} = \bigoplus_{e\in\mathcal{E}} \mathbf{Z}\cdot e \qquad \text{and} \qquad C_0(\Gamma,\mathbf{Z}) := \mathbf{Z}^{\mathcal{V}} = \bigoplus_{v\in\mathcal{V}} \mathbf{Z}\cdot v.$$

Fix an orientation of Γ. Let

$$\partial\colon C_1(\Gamma,\mathbf{Z}) \to C_0(\Gamma,\mathbf{Z})$$

be the morphism associating to an oriented edge the difference of its target and its source. Define

$$C^1(\Gamma,\mathbf{Z}) := \mathrm{Hom}_{\mathbf{Z}}\big(C_1(\Gamma,\mathbf{Z}),\mathbf{Z}\big) \qquad \text{and} \qquad C^0(\Gamma,\mathbf{Z}) := \mathrm{Hom}_{\mathbf{Z}}\big(C_0(\Gamma,\mathbf{Z}),\mathbf{Z}\big).$$

Denote by

$$\delta\colon C^0(\Gamma,\mathbf{Z}) \to C^1(\Gamma,\mathbf{Z})$$

the dual of ∂. The fact that the modules just constructed carry canonical $\mathbf{Z}$-bases induces canonical isomorphisms

$$a\colon C_0(\Gamma,\mathbf{Z}) \longrightarrow C^0(\Gamma,\mathbf{Z}) \qquad \text{and} \qquad b\colon C_1(\Gamma,\mathbf{Z}) \longrightarrow C^1(\Gamma,\mathbf{Z}).$$

Define

$$H_1(\Gamma,\mathbf{Z}) := \mathrm{Ker}\Big(C_1(\Gamma,\mathbf{Z})\xrightarrow{\partial}C_0(\Gamma,\mathbf{Z})\Big)$$

and

$$H^1(\Gamma,\mathbf{Z}) := \mathrm{Coker}\Big(C^0(\Gamma,\mathbf{Z})\xrightarrow{\delta}C^1(\Gamma,\mathbf{Z})\Big).$$

3.2 Remark. The maps ∂ and δ depend on the chosen orientation of Γ. The maps a and b and the groups $H_1(\Gamma,\mathbf{Z})$ and $H^1(\Gamma,\mathbf{Z})$ do not depend on the orientation.

3.3 Definition. *Let Γ be a graph. With the notation of 3.1 define a perfect pairing*

$$\langle\_,\_\rangle\colon H_1(\Gamma,\mathbf{Z})\times H^1(\Gamma,\mathbf{Z}) \longrightarrow \mathbf{Z}$$

as follows. Consider

$$\begin{array}{ccc} H_1(\Gamma,\mathbf{Z})\times H^1(\Gamma,\mathbf{Z}) & & \\ \downarrow \qquad\quad \uparrow & & \\ C_1(\Gamma,\mathbf{Z})\times C^1(\Gamma,\mathbf{Z}) & \xrightarrow{(-,-)} & \mathbf{Z} \\ \downarrow\partial \qquad\quad \uparrow\delta & & \\ C_0(\Gamma,\mathbf{Z})\times C^0(\Gamma,\mathbf{Z}) & \xrightarrow{[-,-]} & \mathbf{Z}. \end{array}$$

The pairings of the second and third rows are determined by the natural dualities defined in 3.1. If g is in $C_1(\Gamma,\mathbf{Z})$ and α is in $C^0(\Gamma,\mathbf{Z})$, then

$$\big[\partial(g),\alpha\big] = \big(g,\delta(\alpha)\big).$$

Let h be in $H_1(\Gamma,\mathbf{Z})$ and let β be in $H^1(\Gamma,\mathbf{Z})$. Let γ be an element of $C^1(\Gamma,\mathbf{Z})$ mapping to β. Define

$$\langle h,\beta\rangle := \big(h,\gamma\big).$$

3.4 *Polyhedral decompositions induced by* Γ. Let Γ be a graph. With the notations of 3.1 we have a diagram

$$\begin{array}{ccccc} H_1(\Gamma,\mathbf{Z}) & \longrightarrow & C_1(\Gamma,\mathbf{Z}) & \xrightarrow{\partial} & C_0(\Gamma,\mathbf{Z}) \\ \downarrow & & \downarrow b & & \downarrow a \\ H^1(\Gamma,\mathbf{Z}) & \longleftarrow & C^1(\Gamma,\mathbf{Z}) & \xleftarrow{\delta} & C^0(\Gamma,\mathbf{Z}). \end{array}$$

The square on the right does not commute in general. The left vertical map is defined as the unique one making the square on the left commute. There is a similar diagram when we replace $\mathbf{Z}$ by $\mathbf{R}$, i.e., when we tensor the present diagram with $\mathbf{R}$, in which all vertical arrows are isomorphisms. As in [23, Section I.1, page 9] the lattice $C^1(\Gamma,\mathbf{Z})$ of $C^1(\Gamma,\mathbf{R})$ induces two natural polyhedral decompositions of $C^1(\Gamma,\mathbf{R})$ called the Voronoi decomposition and the Delaunay decomposition.

Let $d(\_,\_)$ be the distance function between two points of $C^1(\Gamma,\mathbf{R})$ associated to the symmetric bilinear form $\big(b^{-1}(\_),\_\big)$ on $C^1(\Gamma,\mathbf{Z})$ defined in 3.3.

For any ξ in $C^1(\Gamma,\mathbf{Z})$ define

$$V(\xi) := \big\{x\in C^1(\Gamma,\mathbf{R}) \quad \big| \quad d(x,\xi)\le d(x,\eta) \qquad \forall\eta\in C^1(\Gamma,\mathbf{Z})\big\}.$$

The Voronoi polyhedra consist of the polyhedra $V(\xi)$ for ξ in $C^1(\Gamma,\mathbf{Z})$ and their faces.

For any x in $C^1(\Gamma,\mathbf{R})$ define

$$D(x) := \langle \xi_0,\ldots,\xi_r\rangle$$

as the convex hull of the subset $\{\xi_0,\ldots,\xi_r\}$ of $C^1(\Gamma,\mathbf{Z})$ defined by the elements ξ for which $d(x,\xi)$ attains its minimum. The Delaunay polyhedra consist of the polyhedra $D(x)$ with $x \in C^1(\Gamma, \mathbf{R})$. Denote by $\mathrm{Del}\big(C^1(\Gamma, \mathbf{Z})\big)$ the induced polyhedral decomposition of $C^1(\Gamma, \mathbf{R})$.

For any Delaunay polyhedron D there exists a unique Voronoi polyhedron V such that $D = D(x)$ if and only if x is in the relative interior of V. Following [23, Lemma I.1.2], denote

$$V = D^* \qquad \text{and} \qquad D = V^*$$

and call V the dual of D and D the dual of V.

3.5 Definition. *If φ is an element of $\partial\big(C_1(\Gamma, \mathbf{R})\big)$, define*

$$\mathrm{Del}_\varphi\big(H^1(\Gamma, \mathbf{R})\big)$$

to be the induced Namikawa decomposition of $H^1(\Gamma, \mathbf{R})$ as in [23, Section I.6, Definition on page 28]. *Call such decomposition non-degenerate if φ is non-degenerate according to* [23, Section I.7, Definition on page 34].

3.6 *Namikawa decompositions associated to* Γ. Fix φ in $\partial\big(C_1(\Gamma, \mathbf{R})\big)$. For the sake of completeness we recall how the polyhedra of $\mathrm{Del}_\varphi\big(H^1(\Gamma, \mathbf{R})\big)$ are defined. They are the images via $C^1(\Gamma, \mathbf{R}) \to H^1(\Gamma, \mathbf{R})$ of the Delaunay polyhedra D introduced in 3.4 such that φ belongs to the image via ∂ of the relative interior of the dual polyhedron D^*. The fact that these polyhedra define a polyhedral decomposition of $H^1(\Gamma, \mathbf{R})$ is proven in [23, Proposition I.6.1].

In the following proposition we summarize the most important properties of non-degenerate Namikawa decompositions.

3.7 Proposition. *Let φ be a non-degenerate element in $\partial\big(C_1(\Gamma, \mathbf{R})\big)$. Then*

i. *the 0-dimensional polyhedra of $\mathrm{Del}_\varphi\big(H^1(\Gamma, \mathbf{R})\big)$ coincide with $H^1(\Gamma, \mathbf{Z})$;*
ii. *the set $\mathrm{Del}_\varphi\big(H^1(\Gamma, \mathbf{R})\big)$ is invariant under the action of $H_1(\Gamma, \mathbf{Z})$ induced by the homomorphism $H_1(\Gamma, \mathbf{Z}) \to H^1(\Gamma, \mathbf{Z})$ defined in 3.4;*
iii. *for any polyhedron D in $\mathrm{Del}_\varphi\big(H^1(\Gamma, \mathbf{R})\big)$ there is a unique polyhedron D' in the Delaunay decomposition of $C^1(\Gamma, \mathbf{R})$ such that D is the image of D' via $C^1(\Gamma, \mathbf{R}) \to H^1(\Gamma, \mathbf{R})$ and φ belongs to the image via ∂ of the dual polyhedron D'^* of D'; see 3.4 for the notation. Moreover, the map $D' \to D$ is bijective;*
iv. *if D is a polyhedron in $\mathrm{Del}_\varphi\big(H^1(\Gamma, \mathbf{R})\big)$ and l is an element of $H_1(\Gamma, \mathbf{Z})$, let $D + l$ be the image of D under the action of l on $\mathrm{Del}_\varphi\big(H^1(\Gamma, \mathbf{R})\big)$ and let $D' + l$ be the image of D' under the action of l on the Delauny decomposition of $C^1(\Gamma, \mathbf{R})$ via $H_1(\Gamma, \mathbf{Z}) \subset C_1(\Gamma, \mathbf{Z}) = C^1(\Gamma, \mathbf{Z})$. Then $D' + l = (D + l)'$.*

Proof:

i. It follows from [23, Proposition I.7.6];
ii. it follows from [23, Proposition I.6.1];

iii. it follows from [23, Section I.7, Definition on page 34], from [23, Section I.6, Definition on page 28] and from [23, Proposition I.6.6];
iv. it follows from the uniqueness of the liftings D' of D and $(D+l)'$ of $D+l$ asserted in (iii).

3.8 Remark. Let φ be a non-degenerate element in $\partial\bigl(C_1(\Gamma,\mathbf{R})\bigr)$. Let D be a 1-dimensional cell of $\mathrm{Del}_\varphi\bigl(H^1(\Gamma,\mathbf{R})\bigr)$. Let D' be the lift to a 1-dimensional cell of $\mathrm{Del}\bigl(C^1(\Gamma,\mathbf{R})\bigr)$ defined by (iii) of 3.7. The 0-dimensional faces of D' are elements of $C^1(\Gamma,\mathbf{Z})$ and their difference defines an element $e\in\mathcal{E}$ since $C^1(\Gamma,\mathbf{Z}) = \mathbf{Z}^{\mathcal{E}}$. In this way we can label the 1-dimensional polyhedra of $\mathrm{Del}_\varphi\bigl(H^1(\Gamma,\mathbf{R})\bigr)$ with the elements of $\mathcal{E}$.

3.9 Definition. *Let φ be in $\partial\bigl(C_1(\Gamma,\mathbf{R})\bigr)$. For any ξ in $H^1(\Gamma,\mathbf{Z})$ let*

$$\Delta(\xi) = \Bigl\{\sigma(\xi,N)|\xi\in N\in \mathrm{Del}_\varphi\bigl(H^1(\Gamma,\mathbf{R})\bigr)\Bigr\}$$

be the finite rational polyhedral decomposition of $H^1(\Gamma,\mathbf{R})$ consisting of polyhedral cones $\sigma(\xi,N)$ generated by the set $\{x-\xi|x\in N\}$. See [23, Section II.13, page 77]. *Let*

$$\mathbf{G}_{m,S}\otimes H^1(\Gamma,\mathbf{Z}) \hookrightarrow \mathrm{Temb}(\xi)$$

be the torus embedding over a scheme S of the torus $\mathbf{G}_{m,S}\otimes H^1(\Gamma,\mathbf{Z})$ with cocharacter group $H^1(\Gamma,\mathbf{Z})$ associated to the polyhedral decomposition $\Delta(\xi)$.

3.10 Definition. *Suppose that $\varphi\in\partial\bigl(C_1(\Gamma,\mathbf{R})\bigr)$ is non-degenerate. Let $\xi\in H^1(\Gamma,\mathbf{Z})$. Let ξ' be the element of $C^1(\Gamma,\mathbf{Z})$ associated to ξ in* (iii) *of 3.7. It maps to ξ. As in 3.9, let*

$$\Delta(\xi') = \Bigl\{\sigma(\xi',N')|\xi\in N\in \mathrm{Del}_\varphi\bigl(H^1(\Gamma,\mathbf{R})\bigr)\Bigr\}.$$

For any $N\in\mathrm{Del}_\varphi\bigl(H^1(\Gamma,\mathbf{R})\bigr)$ the polyhedron $N'\in\mathrm{Del}\bigl(C^1(\Gamma,\mathbf{R})\bigr)$ is defined in (iii) *of 3.7. It contains ξ' as a 0-dimensional cell. Let*

$$\mathbf{G}_{m,S}\otimes C^1(\Gamma,\mathbf{Z}) \hookrightarrow \mathrm{Temb}(\xi')$$

be the torus embedding over a scheme S of the torus $\mathbf{G}_{m,S}\otimes C^1(\Gamma,\mathbf{Z})$ associated to the polyhedral decomposition $\Delta(\xi')$.
For any α in $C^1(\Gamma,\mathbf{Z})$ let

$$\Delta'(\alpha) = \Bigl\{\sigma(\alpha,M)|\alpha\in M\in \mathrm{Del}\bigl(C^1(\Gamma,\mathbf{R})\bigr)\Bigr\}.$$

Define

$$\mathbf{G}_{m,S}\otimes C^1(\Gamma,\mathbf{Z}) \hookrightarrow \mathrm{Temb}'(\alpha)$$

to be the torus embedding over the scheme S of the torus $\mathbf{G}_{m,S}\otimes C^1(\Gamma,\mathbf{Z})$

associated to the polyhedral decomposition $\Delta'(\alpha)$*. By construction it is the compactification*

$$\mathbf{G}_{m,S} \otimes C^1(\Gamma, \mathbf{Z}) \hookrightarrow \prod_{e \in \mathcal{E}} \mathbf{P}^1_S.$$

3.11 Remark. The torus $\mathbf{G}_{m,S} \otimes H^1(\Gamma, \mathbf{Z})$ is the quotient of the morphism of tori $\mathbf{G}_{m,S} \otimes C^0(\Gamma, \mathbf{Z}) \to \mathbf{G}_{m,S} \otimes C^1(\Gamma, \mathbf{Z})$. The map is induced by the map on cocharacter groups defined by

$$\delta: C^0(\Gamma, \mathbf{Z}) \longrightarrow C^1(\Gamma, \mathbf{Z}).$$

Since we work with connected graphs, the kernel of δ is isomorphic to $\mathbf{Z}$. Hence

$$\mathrm{Image}(\delta) = C^0(\Gamma, \mathbf{Z})/\mathbf{Z}.$$

Using 3.3 the character group of $\mathbf{G}_{m,S} \otimes C^0(\Gamma, \mathbf{Z})$ is $C_0(\Gamma, \mathbf{Z})$, the character group of $\mathbf{G}_{m,S} \otimes C^1(\Gamma, \mathbf{Z})$ is $C_1(\Gamma, \mathbf{Z})$ and the character group of $\mathbf{G}_{m,S} \otimes H^1(\Gamma, \mathbf{Z})$ can be identified with the kernel of the $\mathbf{Z}$-dual of δ, i.e., with the kernel of

$$\partial: C_1(\Gamma, \mathbf{Z}) \longrightarrow C^0(\Gamma, \mathbf{Z}).$$

Hence the character group of $\mathbf{G}_{m,S} \otimes H^1(\Gamma, \mathbf{Z})$ is $H_1(\Gamma, \mathbf{Z})$. The pairing $\langle \_, \_ \rangle$ between $H_1(\Gamma, \mathbf{Z})$ and $H^1(\Gamma, \mathbf{Z})$ is the one described in 3.3.

3.12 *The comparison among* Temb(ξ), Temb(ξ') *and* Temb$'$(ξ'). Let φ be a non-degenerate element of $\partial\big(C_1(\Gamma, \mathbf{R})\big)$; see 3.5. Let $\xi \in H^1(\Gamma, \mathbf{Z})$.

Let N be in $\mathrm{Del}_\varphi\big(H^1(\Gamma, \mathbf{R})\big)$ containing ξ. Let ξ' and N' be the associated polyhedra of $\mathrm{Del}\big(C^1(\Gamma, \mathbf{R})\big)$ defined using (iii) of 3.7. Then the image of ξ' in $H^1(\Gamma, \mathbf{R})$ is ξ and the image of N' is N. In particular, we obtain a morphism of the affine torus embedding of $\mathbf{G}_{m,s} \otimes C^1(\Gamma, \mathbf{Z})$ associated to $\sigma(\xi', N')$ to the affine torus embedding of $\mathbf{G}_{m,S} \otimes H^1(\Gamma, \mathbf{Z})$ associated to $\sigma(\xi, N)$. Using 3.11 it is clear that this morphism is the categorical quotient with respect to the action of $\mathbf{G}_{m,S} \otimes C^0(\Gamma, \mathbf{Z})$, i.e., on coordinate rings it is determined by taking $\mathbf{G}_{m,S} \otimes C^0(\Gamma, \mathbf{Z})$-invariants. The faces of N' and of N are in bijection via the projection $N' \to N$ by (iii) of 3.7. Hence over any geometric point x the fiber of this quotient map is isomorphic to $\mathbf{G}_{m,k(x)} \otimes \big(C^0(\Gamma, \mathbf{Z})/\mathbf{Z}\big)$. In particular, we obtain a natural map of torus embeddings

$$\begin{array}{ccc}
\mathbf{G}_{m,S} \otimes C^1(\Gamma, \mathbf{Z}) & \longrightarrow & \mathrm{Temb}(\xi') \\
\downarrow & & \downarrow \\
\mathbf{G}_{m,S} \otimes H^1(\Gamma, \mathbf{Z}) & \longrightarrow & \mathrm{Temb}(\xi),
\end{array}$$

such that $\mathrm{Temb}(\xi') \to \mathrm{Temb}(\xi)$ is affine and surjective, is invariant under the action of $\mathbf{G}_{m,S} \otimes C^0(\Gamma, \mathbf{Z})$ and for any geometric point x of Temb(ξ) its fiber is isomorphic to $\mathbf{G}_{m,k(x)} \otimes \big(C^0(\Gamma, \mathbf{Z})/\mathbf{Z}\big)$.

On the other hand we get a natural open immersion of torus embeddings of the torus $\mathbf{G}_{m,S} \otimes C^1(\Gamma, \mathbf{Z})$

$$\mathrm{Temb}(\xi') \hookrightarrow \mathrm{Temb}'(\xi') = \prod_{e \in \mathcal{E}} \mathbf{P}^1_S.$$

Consider the divisor defined as the sum over $e \in \mathcal{E}$ of the pullback to $\mathrm{Temb}'(\xi')$ of the zero section of the copy of $\mathbf{P}^1_S$ associated to e. It is ample relatively to S and it is invariant under the action of $\mathbf{G}_{m,S} \otimes C^1(\Gamma, \mathbf{Z})$. The same holds for its pullback to $\mathrm{Temb}(\xi')$. In particular, $\mathrm{Temb}(\xi')$ carries an invertible sheaf ample relatively to S and $\mathbf{G}_{m,S} \otimes C^0(\Gamma, \mathbf{Z})$-linearizable.

3.13 Proposition. *Let φ in $\partial\big(C_1(\Gamma, \mathbf{R})\big)$ be non-degenerate in the sense of 3.5. Let ξ be in $H^1(\Gamma, \mathbf{Z})$. Then, with the notation of 3.12,*

(i) *the schemes* $\mathrm{Temb}(\xi)$ *and* $\mathrm{Temb}(\xi')$ *are smooth over S;*

(ii) *the morphism* $\mathrm{Temb}(\xi') \to \mathrm{Temb}(\xi)$ *is a fppf* $\mathbf{G}_{m,S} \otimes \big(C^0(\Gamma, \mathbf{Z})/\mathbf{Z}\big)$*-torsor;*

(iii) *the scheme* $\mathrm{Temb}(\xi)$ *is projective over S;*

(iv) *for any l in $H_1(\Gamma, \mathbf{Z})$ the torus embeddings* $\mathrm{Temb}(\xi')$ *(resp.* $\mathrm{Temb}(\xi)$*) and* $\mathrm{Temb}\big((\xi + l)'\big)$ *(resp.* $\mathrm{Temb}(l + \xi)$*) are canonically isomorphic.*

Proof: The morphism $\mathrm{Temb}(\xi) \to S$ is smooth since the decomposition $\Delta(\xi)$ is saturated. This is shown in the proof of [23, Corollary II.13.3]. Clearly $\mathrm{Temb}(\xi')$ is smooth over S since it is an open of $\mathrm{Temb}'(\xi')$ and the latter is a product of copies of $\mathbf{P}^1_S$.

Next we prove (ii). Let s be a geometric point of S. The morphism of torus embeddings $\mathrm{Temb}(\xi') \times_S k(s) \to \mathrm{Temb}(\xi) \times_S k(s)$ is equidimensional and the two varieties are smooth. Hence it is flat. By 3.12 it is also surjective and fiberwise smooth. Hence the morphism $\mathrm{Temb}(\xi') \to \mathrm{Temb}(\xi)$ is faithfully flat and smooth by [12, Theorem 17.5.1]. Let x be a point of $\mathrm{Temb}(\xi)$. Let A_x be the completion of $\mathrm{Temb}(\xi)$ at x. Let B_x be the coordinate ring of the affine $\mathrm{Temb}(\xi') \times_{\mathrm{Temb}(\xi)} \mathrm{Spec}(A_x)$. After a possible faithfully flat base change and using the smoothness of $\mathrm{Temb}(\xi') \to \mathrm{Temb}(\xi)$, we can suppose that $\mathrm{Spec}(B_x) \to \mathrm{Spec}(A_x)$ admits a section γ. Let $\mathrm{Spec}(C_x) := \mathbf{G}_{m,A_x} \otimes \big(C^0(\Gamma, \mathbf{Z})/\mathbf{Z}\big)$. Consider the morphism $\mathrm{Spec}(C_x) \to \mathrm{Spec}(B_x)$ induced by the action of $\mathbf{G}_{m,S} \otimes C^0(\Gamma, \mathbf{Z})$ on γ. For any $\chi \in C^0(\Gamma, \mathbf{Z})$ we get a morphism of A_x-modules $B_x^\chi \to C_x^\chi$. These modules are flat and finite, since C_x and B_x are noetherian and A_x-flat. The morphism $B_x^\chi \to C_x^\chi$ is an isomorphism over $k(x)$. By Nakayama's lemma it is an isomorphism.

We prove (iii). The fact that $\mathrm{Temb}(\xi)$ is proper over S follows from the general theory of torus embeddings since the union of the polyhedra of $\Delta(\xi)$ is the whole $H^1(\Gamma, \mathbf{R})$. By 3.12 we know that $\mathrm{Temb}(\xi')$ carries a line bundle ample relatively to S and $\mathbf{G}_{m,S} \otimes C^0(\Gamma, \mathbf{Z})$-linearizable. This descends to a line bundle on $\mathrm{Temb}(\xi)$ which is S-ample.

The invariance under the action of $H_1(\Gamma, \mathbf{Z})$ stated in (iv) follows from (ii) and (iv) of 3.7.

4 Contraction of graphs and related constructions

In this section we continue our discussion concerning graphs. We define the *contraction* of a graph with respect to a subset of its edges. We study how the constructions introduced in the previous section change under contraction.

4.1 Definition. *Let Γ be a graph. The notation is as in 3.1. Let $H \subset \mathcal{E}$. Define*

$$\Gamma_H$$

to be the graph obtained from the graph Γ by contracting all the edges contained in H. More precisely, let

$$\mathcal{E}_H := \mathcal{E} - H \qquad \text{and} \qquad \varpi_H\colon \mathcal{E}_H \hookrightarrow \mathcal{E}.$$

Define

$$\mathcal{V}_H \qquad \text{and} \qquad \vartheta_H\colon \mathcal{V} \longrightarrow \mathcal{V}_H$$

as the quotient of $\mathcal{V}$ by the equivalence relation generated by the following relation. We say that two elements v_1 and v_2 of $\mathcal{V}$ are equivalent if and only if there is an edge in H whose end points are v_1 and v_2. Define Γ_H to be graph whose set of edges is $\mathcal{E}_H$ and whose set of vertices is $\mathcal{V}_H$ and such that for any edge e of $\mathcal{E}_H$ the end points of e are the images via ϑ_H of the end points of $\varpi_H(e)$.

4.2 Remark. Let H be a subset of $\mathcal{E}$. With the notations of 3.1 we get the following commutative diagram

$$\begin{array}{ccccccc}
0 \longrightarrow & H_1(\Gamma, \mathbf{Z}) & \longrightarrow & C_1(\Gamma, \mathbf{Z}) & \xrightarrow{\partial} & C_0(\Gamma, \mathbf{Z}) \\
& \downarrow & & \downarrow & & \downarrow \\
0 \longrightarrow & H_1(\Gamma_H, \mathbf{Z}) & \longrightarrow & C_1(\Gamma_H, \mathbf{Z}) & \xrightarrow{\partial_H} & C_0(\Gamma_H, \mathbf{Z}).
\end{array}$$

The right vertical map is induced from the map ϑ_H defined in 4.1. The central vertical map is defined to be the identity on $\mathcal{E}_H$, identified via ϖ_H to a subset of $\mathcal{E}$, and by sending to 0 any edge e in H . The morphism $H_1(\Gamma, \mathbf{Z}) \to H_1(\Gamma_H, \mathbf{Z})$ is induced from the commutativity of the square on the right hand side. The quotient of $C_0(\Gamma, \mathbf{Z}) = \mathbf{Z}^{\mathcal{V}}$ by the image of $\mathbf{Z}^H$ injects into $\mathbf{Z}^{\mathcal{V}_H}$. An application of the snake lemma implies that the morphism $H_1(\Gamma, \mathbf{Z}) \to H_1(\Gamma_H, \mathbf{Z})$ is surjective. Dualizing the diagram above we obtain the commutative diagram

$$\begin{array}{ccccccc}
0 \longleftarrow & H^1(\Gamma, \mathbf{Z}) & \longleftarrow & C^1(\Gamma, \mathbf{Z}) & \xleftarrow{\delta} & C^0(\Gamma, \mathbf{Z}) \\
& \uparrow & & \uparrow & & \uparrow \\
0 \longleftarrow & H^1(\Gamma_H, \mathbf{Z}) & \longleftarrow & C^1(\Gamma_H, \mathbf{Z}) & \xleftarrow{\delta_H} & C^0(\Gamma_H, \mathbf{Z}),
\end{array}$$

where the map $H^1(\Gamma_H, \mathbf{Z}) \to H^1(\Gamma, \mathbf{Z})$ is injective.
In particular, if the rank of $H_1(\Gamma_H, \mathbf{Z})$ equals the rank of $H_1(\Gamma, \mathbf{Z})$, the maps

$H_1(\Gamma,\mathbf{Z}) \to H_1(\Gamma_H,\mathbf{Z})$ and $H^1(\Gamma_H,\mathbf{Z}) \to H^1(\Gamma,\mathbf{Z})$ are isomorphisms.

4.3 *The behavior of Delaunay decompositions under contractions.* The notation is as in 4.1 and 4.2. Fix a subset H of $\mathcal{E}$. Consider the following commutative diagram

$$\begin{array}{ccc} C^1(\Gamma_H,\mathbf{Z}) & \xrightarrow{b_H^{-1}} & C_1(\Gamma_H,\mathbf{Z}) \\ \downarrow & & \uparrow \\ C^1(\Gamma,\mathbf{Z}) & \xrightarrow{b^{-1}} & C_1(\Gamma,\mathbf{Z}); \end{array}$$

where the map b (respectively b_H) is defined in 3.1. We get an isomorphism

$$\begin{array}{ccc} C^1(\Gamma,\mathbf{Z}) & \xrightarrow{f} & C^1(\Gamma_H,\mathbf{Z}) \times \mathbf{Z}^H \\ \xi & \mapsto & (\xi_H, \xi^H) \end{array}$$

as $\mathbf{Z}$-modules. Also the distance d on $C^1(\Gamma,\mathbf{R})$ defined in 3.4 is the sum of the distance d_H on $C^1(\Gamma_H,\mathbf{R})$ defined as in 3.4 and the Euclidean distance d' on $\mathbf{R}^H$. It is then clear that $f \otimes \mathbf{R}$ induces bijections between the Delaunay (respectively Voronoi) decomposition of $C^1(\Gamma,\mathbf{R})$ and the product of the Delaunay (respectively Voronoi) decompositions of $C^1(\Gamma_H,\mathbf{R})$ and $\mathbf{R}^H$. The latter is defined by the lattice $\mathbf{Z}^H$. See 3.4 for the definition of Delaunay and Voronoi decompositions. It is then clear that the following assertions hold

1. the Delaunay (Voronoi) decomposition of $C^1(\Gamma,\mathbf{R})$ maps surjectively to the Delaunay (Voronoi) decomposition of $C^1(\Gamma_H,\mathbf{R})$;
2. the duality between the Delaunay and Voronoi decompositions defined in 3.4 is preserved;
3. let φ be in $\partial\big(C_1(\Gamma,\mathbf{R})\big)$ and let φ_H be its image in $\partial_H\big(C_1(\Gamma,\mathbf{R})\big)$. Let D' be a Delaunay polyhedron of $C^1(\Gamma,\mathbf{R})$ and denote by D'_H its image in $C^1(\Gamma_H,\mathbf{R})$. If φ belongs to $\partial(D'^*)$, then φ_H belongs to $\partial_H(D'^*_H)$;
4. there exists φ in $\partial\big(C_1(\Gamma,\mathbf{R})\big)$ which is is non-degenerate in the sense of 3.5 and whose image φ_H in $\partial\big(C_1(\Gamma_H,\mathbf{R})\big)$ is non-degenerate for all subsets H of $\mathcal{E}$.

To prove (4) we use the criterion [23, Proposition I.7.6 (3)] for non-degeneracy. It is then clear that there exists an open non-empty subspace of $C_0(\Gamma,\mathbf{R})$ such that if φ belongs to it, condition (4) holds.

4.4 *The behavior of Namikawa decompositions under contractions.* We describe how the Namikawa decompositions of 3.5 change under the contraction defined in 4.1. For φ as in (4) of 4.3 there is a morphism

$$\begin{array}{ccc} \mathrm{Del}_\varphi\big(H^1(\Gamma,\mathbf{R})\big) & \longrightarrow & \mathrm{Del}_{\varphi_H}\big(H^1(\Gamma_H,\mathbf{R})\big) \\ D & \mapsto & D_H \end{array}$$

respecting the actions of $H_1(\Gamma,\mathbf{Z})$ and $H_1(\Gamma_H,\mathbf{Z})$. It is defined as follows. Let D be a polyhedron of $\mathrm{Del}_\varphi\big(H^1(\Gamma,\mathbf{R})\big)$. By 3.7 there exists a unique Delaunay polyhedron D' in $C^1(\Gamma,\mathbf{R})$ such that D is the image of D' and φ belongs to $\partial(D'^*)$. Let D'_H be the image of D' via

$$C^1(\Gamma,\mathbf{R})\xrightarrow{b}C_1(\Gamma,\mathbf{R})\longrightarrow C_1(\Gamma_H,\mathbf{R}).$$

Using (3) of 4.3 associate to D the polyhedron D_H in $\mathrm{Del}_{\varphi_H}\big(H^1(\Gamma_H,\mathbf{R})\big)$ defined as the image of D'_H via

$$C_1(\Gamma_H,\mathbf{R})\xrightarrow{b_H}C^1(\Gamma_H,\mathbf{Z})\longrightarrow H^1(\Gamma_H,\mathbf{R}).$$

By 4.3 there exist morphisms of polyhedra

$$\begin{array}{ccc} D' & \hookrightarrow & C_1(\Gamma,\mathbf{R}) \\ \downarrow & & \downarrow \\ D'_H & \hookrightarrow & C_1(\Gamma_H,\mathbf{Z}) \end{array} \quad\text{and}\quad \begin{array}{ccc} D' & \hookrightarrow & C^1(\Gamma,\mathbf{R}) \\ \uparrow & & \uparrow \\ D'_H & \hookrightarrow & C^1(\Gamma_H,\mathbf{R}). \end{array}$$

The notation is as in 4.2. Suppose that H is a subset of $\mathcal{E}$ such that the rank of $H_1(\Gamma_H,\mathbf{Z})$ is equal to the rank of $H_1(\Gamma,\mathbf{Z})$. By 4.2 the morphism $H^1(\Gamma_H,\mathbf{Z})\to H^1(\Gamma_H,\mathbf{Z})$ is an isomorphism.
Let ξ be a 0-dimensional cell in $\mathrm{Del}_\varphi\big(H^1(\Gamma,\mathbf{R})\big)$ or, equivalently by (ii) of 3.7, an element in $H^1(\Gamma,\mathbf{Z})$. Let

$$D\in\Delta(\xi)\qquad\text{and}\qquad D_H\in\Delta(\xi_H)$$

be a polyhedron of $H^1(\Gamma,\mathbf{R})$ defined as in 3.9 and the corresponding polyhedron in $\Delta(\xi_H)$. The morphism of polyhedra $D'_H\to D'$ induces an open immersion from the affine torus embedding of $\mathbf{G}_{m,S}\otimes H^1(\Gamma_H,\mathbf{Z})$ defined by D_H to the affine torus embedding of $\mathbf{G}_{m,S}\otimes H^1(\Gamma,\mathbf{Z})$ defined by D. Hence via the map from $\Delta(\xi)$ to $\Delta(\xi_H)$ we get a birational morphism

$$\mathrm{Temb}(\xi)\longrightarrow\mathrm{Temb}(\xi_H).$$

The definitions of Δ and of Temb are given in 3.9. By construction, the morphism is defined, and it is an isomorphism, on the open subscheme of $\mathrm{Temb}(\xi)$ associated to the polyhedra

$$\big\{D\in\Delta(\xi)\ \ |\ \ D'\subset C^1(\Gamma_H,\mathbf{R})\big\}.$$

5 Simple graphs

We introduce in this section the notion of a *simple* graph. We describe the constructions of the previous sections in the case of simple graphs.

5.1 Definition. *Let* Γ *be a graph. The notation is as in 3.1. Let* l *be an element of* $H_1(\Gamma, \mathbf{Z})$. *Let* $\oplus_{e\in\mathcal{E}} l_e\, e$ *be the image of* l *in* $C_1(\Gamma, \mathbf{Z})$. *Call the support of* l

$$\mathrm{Supp}(l) := \{e \in \mathcal{E} \mid l_e \neq 0\}.$$

We say that an edge e *of* Γ *belongs to* l *if* $e \in \mathrm{Supp}(l)$.
Call l *a cycle if for any* $e \in \mathrm{Supp}(l)$ *we have* $l_e = \pm 1$. *Call* l *an elementary cycle if* l *is a cycle and it can not be written as the sum of two non-zero cycles with disjoint support. If* l *is an elementary cycle, denote by* Γ_l *the unique subgraph of* Γ *whose edges consist of the edges of* Γ *contained in* $\mathrm{Supp}(l)$. *The graph* Γ_l *has a natural orientation defined by requiring that* $l_e = 1$ *for any* $e \in \mathrm{Supp}(l)$.

The next proposition, proven in [23, Section I.4, pag. 21-22], guarantees that a graph has enough elementary cycles.

5.2 Proposition. *Let* Γ *be a graph. Then* $H_1(\Gamma, \mathbf{Z})$ *admits a basis of elementary cycles.*

5.3 Definition. *A graph* Γ *is called simple if the supports of any two elementary cycles of* Γ *are either equal or disjoint.*

In the next proposition we collect the basic properties of simple graphs.

5.4 Proposition. *Let* Γ *be a simple graph. Let* $l_1, \ldots, l_r$ *be a basis of* $H_1(\Gamma, \mathbf{Z})$ *consisting of elementary cycles. For* $t = 1, \ldots, r$ *define* $\Gamma_t := \Gamma_{l_t}$. *Then*

1. *any non-disconnecting edge* e *of* Γ *belongs to a unique elementary cycle;*
2. *the elementary cycles* $l_1, \ldots, l_r$ *and their inverses are the only elementary cycles of* Γ*;*
3. *let* $i, t_1, \ldots, t_k$ *be* $k+1$ *distinct integers in* $\{1, \ldots, r\}$. *Suppose that the subgraph* Γ^k *of* Γ *spanned by* $\Gamma_{t_1}, \ldots, \Gamma_{t_k}$ *is connected. Then the subgraphs* Γ_i *and* Γ^k *of* Γ *have at most one vertex in common;*
4. *there is a unique orientation of* Γ *such that for any* $t = 1, \ldots, r$ *the induced orientation on the subgraph* Γ_t *coincides with the orientation defined by* l_t *in 5.1.*

Proof: The uniqueness in the first statement follows from the definition of Γ being simple. The existence is proven as follows. Let e be a non-disconnecting edge of Γ. Let Γ_e be the graph obtained from Γ by omitting e. It is connected since the edge e is non-disconnecting. Let $\mathcal{T}$ be a spanning tree of Γ_e in the sense of [23, Definition on pag. 21]. Then $\mathcal{T}$ is a spanning tree of Γ. By [23, pag. 21-22] there is an elementary cycle l of Γ such that e belongs to l.

Let l be an elementary cycle of Γ. Since $l_1, \ldots, l_r$ span $H_1(\Gamma, \mathbf{Z})$, there exist $t = 1, \ldots, r$ such that the support of l has non-empty intersection with the support of l_t. By the definition of simplicity, we conclude that $l = \pm l_t$. This proves the second statement.

To prove the third statement, we suppose that there are two vertices j_1 and j_2 of Γ which can be realized as vertices both of Γ^k and of Γ_i. Let Γ^{k+1} be the oriented connected subgraph of Γ spanned by $\Gamma_i, \Gamma_{t_1}, \ldots, \Gamma_{t_k}$. We construct a new graph. The vertices are j_1 and j_2. The edges are the subtrees $\mathcal{T}$ of Γ^{k+1} such that the sum over the edges e of $\mathcal{T}$, with induced orientation, of the difference between the target of e and the source of e is $\pm(j_1 - j_2)$. It is clear that there are two elementary cycles of this graph having one edge in common. This implies that there are two elementary cycles of Γ^{k+1} having different supports and at least one edge in common. This is a contradiction.

The fourth statement follows from the previous ones.

5.5 Proposition. *Suppose that Γ is a simple graph. Fix a $\mathbf{Z}$-basis $l_1, \ldots, l_r$ of elementary cycles of $H_1(\Gamma, \mathbf{Z})$. Let φ be a non-degenerate element contained in $\partial\big(C_1(\Gamma, \mathbf{R})\big)$. Then*

i. *there exist positive integers $n_1, \ldots, n_r$ such that $H_1(\Gamma, \mathbf{Z}) \subset H^1(\Gamma, \mathbf{Z})$ is defined by $\oplus_{t=1}^r \mathbf{Z} l_t \subset \oplus_{t=1}^r \mathbf{Z} f_t$ with $f_t := \frac{l_t}{n_t}$ for $t = 1, \ldots, r$;*

ii. *for any ξ in $H^1(\Gamma, \mathbf{Z})$ the torus embedding* $\mathrm{Temb}(\xi)$ *defined in 3.9 is the torus embedding $\mathbf{G}_{m,S} \otimes H^1(\Gamma, \mathbf{Z}) = \mathbf{G}_{m,S}^r \subset \big(\mathbf{P}_S^1\big)^r$.*

Proof: For any element l of $H_1(\Gamma, \mathbf{Z})$ let $l = \oplus_{e \in \mathcal{E}} l_e \cdot e$ as an element of $C_1(\Gamma, \mathbf{Z})$; see 3.1 for the notation. By the hypothesis on Γ being simple, if for the edge e we have $l_{t,e} \neq 0$ for some integer $t = 1, \ldots, r$, then $l_{t',e} = 0$ for $t' \neq t$. In particular, there exists a $\mathbf{Z}$-basis $f_1, \ldots, f_r$ of $H^1(\Gamma, \mathbf{Z})$ for which the morphism

$$\bigoplus_{e \in \mathcal{E}} \mathbf{Z} \cdot e = C^1(\Gamma, \mathbf{Z}) \longrightarrow H^1(\Gamma, \mathbf{Z}) = \bigoplus_{t=1,\ldots,r} \mathbf{Z} \cdot f_t$$

is defined by sending $e \in \mathcal{E}$ to $\oplus_{t=1}^r l_{t,e} f_t$. Statement (ii) is clear from the description of $C^1(\Gamma, \mathbf{Z}) \to H^1(\Gamma, \mathbf{Z})$ just given.

6 Compactifications of semiabelian schemes induced by graphs

In the previous sections we have defined torus embeddings associated to a graph Γ. In this section we fix a semiabelian scheme $\tilde{G}$, extension of an abelian scheme by a torus T. Supposing that the character group of T is $H_1(\Gamma, \mathbf{Z})$, we extend the constructions of the previous sections to get compactifications of $\tilde{G}$.

6.1 *Notation.* Let S be a scheme. Let Γ be a graph. Let $\tilde{G}$ be a semiabelian scheme over S, extension of an abelian scheme $\mathcal{J}$ over S by a torus T. Suppose that $T = \mathbf{G}_{m,S} \otimes H^1(\Gamma, \mathbf{Z})$. Fix an element φ of $\partial\big(C_1(\Gamma, \mathbf{R})\big)$ and suppose that φ is non-degenerate in the sense of 3.5. Finally let S_0 be a closed subscheme of S.

6.2 Definition. *Let ξ be in $H^1(\Gamma, \mathbf{Z})$. Define*

$$P_\xi := \tilde{G} \overset{T}{\times} \mathrm{Temb}(\xi).$$

The torus embedding $\mathrm{Temb}(\xi)$ *of T is defined in 3.9. By* $\tilde{G} \times^T \mathrm{Temb}(\xi)$ *we mean the fiber bundle over $\mathcal{J}$ defined as the pushout of*

$$0 \longrightarrow T \longrightarrow \tilde{G} \longrightarrow \mathcal{J} \longrightarrow 0$$

via $T \subset \mathrm{Temb}(\xi)$*. The construction also determines an inclusion of fiber bundles over $\mathcal{J}$*

$$r_\xi \colon \tilde{G} \hookrightarrow P_\xi.$$

By 3.9 for any element l of $H_1(\Gamma, \mathbf{Z})$ and any element ξ of $H^1(\Gamma, \mathbf{Z})$ the torus embeddings

$$\tilde{G} \hookrightarrow P_\xi \qquad \text{and} \qquad \tilde{G} \hookrightarrow P_{l+\xi}$$

are isomorphic.

6.3 Definition. *Let T' be the torus* $\mathbf{G}_m \otimes C^1(\Gamma, \mathbf{Z})$*. Let $\tilde{G}'$ be a semiabelian scheme sitting in an exact sequence*

$$0 \longrightarrow T' \longrightarrow \tilde{G}' \longrightarrow \mathcal{J} \longrightarrow 0$$

such that

$$\tilde{G}' / \big(\mathbf{G}_{m,S} \otimes C^0(\Gamma, \mathbf{Z})\big) \xrightarrow{\sim} \tilde{G}.$$

For any ξ in $H^1(\Gamma, \mathbf{Z})$ and any $\alpha \in C^1(\Gamma, \mathbf{Z})$ define

$$P_{\xi'} := \tilde{G}' \overset{T'}{\times} \mathrm{Temb}(\xi') \qquad \text{and} \qquad P'_\alpha := \tilde{G}' \overset{T'}{\times} \mathrm{Temb}'(\alpha).$$

The torus embeddings $\mathrm{Temb}(\xi')$ *and* $\mathrm{Temb}(\alpha)$ *are defined in 3.10. We also obtain natural inclusions of fiber bundles over $\mathcal{J}$*

$$r_{\xi'} \colon \tilde{G}' \hookrightarrow P_{\xi'} \qquad \text{and} \qquad r'_\alpha \colon \tilde{G}' \hookrightarrow P'_\alpha.$$

By 3.12 and 3.13 we get morphisms

$$\begin{array}{ccc} \tilde{G}' & \xrightarrow{r_{\xi'}} & P_{\xi'} \\ \downarrow & & \downarrow \\ \tilde{G} & \xrightarrow{r_\xi} & P_\xi \end{array}$$

and the second row is the fppf quotient of the first row by the action of the torus $\mathbf{G}_{m,S} \otimes \big(C^0(\Gamma, \mathbf{Z})/\mathbf{Z}\big)$*. We have an open immersion*

$$P_{\xi'} \hookrightarrow P'_{\xi'}.$$

6.4 Some notation. Let e be a 1-dimensional cell in $\mathrm{Del}_\varphi\big(H^1(\Gamma, \mathbf{R})\big)$ whose associated 0-dimensional faces are α and β. By 3.8 we consider e as an element of $\mathcal{E}$. By 3.7 we consider α and β as elements of $H^1(\Gamma, \mathbf{Z})$.

By the general theory of torus embeddings, e defines codimension one subschemes $e_{\alpha}^{\perp}$ and $e_{\beta}^{\perp}$ in $\mathrm{Temb}(\alpha)$ and $\mathrm{Temb}(\beta)$ respectively; see [8, Theorem IV.2.5]. They are isomorphic as schemes over S. By 6.2 they define codimension one closed subschemes

$$Z_{e,\alpha} \subset P_\alpha \qquad \text{and} \qquad Z_{e,\beta} \subset P_\beta,$$

which are isomorphic as schemes over S. Let

$$Z_{e,\alpha,0} := Z_{e,\alpha} \underset{S}{\times} S_0 \qquad \text{and} \qquad Z_{e,\beta,0} := Z_{e,\beta} \underset{S}{\times} S_0.$$

6.5 Definition. *Define the scheme*

$$\mathcal{P}_0^{\varphi} \longrightarrow \mathcal{J} \underset{S}{\times} S_0$$

by gluing, for any 1-dimensional polyhedron e of $\mathrm{Del}_{\varphi}(H^1(\Gamma, \mathbf{R}))$ with 0-dimensional faces α and β, the scheme $P_{\alpha,0} := P_\alpha \times_S S_0$ along $Z_{e,\alpha,0}$ transversally to $P_{\beta,0} := P_\beta \times_S S_0$ along $Z_{e,\beta,0}$. Define an action of $H_1(\Gamma, \mathbf{Z})$ on the scheme $\mathcal{P}_0^{\varphi}$ over S_0 using 3.9.

6.6 Remark. For any ξ in $H^1(\Gamma, \mathbf{Z})$ the morphism $P_\xi \to S$ is smooth by 3.13. Hence the complement of the smooth locus of $\mathcal{P}_0^{\varphi} \to S_0$ is the union over the 1-dimansional polehedra e of $\mathrm{Del}_{\varphi}(H^1(\Gamma, \mathbf{R}))$ of the images of the schemes $Z_{e,\alpha,0}$ (equivalently $Z_{e,\beta,0}$) defined in 6.4.

6.7 Definition. *Proceeding as in 6.4 and 6.5 and using the polyhedral decomposition $\mathrm{Del}(C^1(\Gamma, \mathbf{R}))$ of $C^1(\Gamma, \mathbf{R})$ introduced in 3.4 define the scheme*

$$\mathcal{P}'_0 \longrightarrow \mathcal{J}_0$$

by gluing the schemes $P'_\alpha \times_S S_0$ for $\alpha \in C^1(\Gamma, \mathbf{Z})$. Define

$$\mathcal{P}_0'^{o} \hookrightarrow \mathcal{P}'_0$$

to be the open subscheme defined as the image of the schemes $P_{\xi'} \times_S S_0$ for $\xi \in H^1(\Gamma, \mathbf{Z})$.

6.8 Remark. By 3.10, for any $\alpha \in C^1(\Gamma, \mathbf{Z})$ the scheme P'_α over $\mathcal{J}$ is isomorphic to a $\mathbf{P}^1_S \otimes C^1(\Gamma, \mathbf{Z})$-bundle over $\mathcal{J}$. By 6.3 there exists an action of $\mathbf{G}_{m,S_0} \otimes (C^0(\Gamma, \mathbf{Z})/\mathbf{Z})$ on $\mathcal{P}_0'^{o} \to S_0$ such that the fppf quotient exists and is defined by

$$\mathcal{P}_0'^{o} \longrightarrow \mathcal{P}_0^{\varphi}.$$

We conclude this section by describing the objects just defined in the case that the graph Γ is simple in the sense of 5.3.

6.9 *The case of simple graphs.* Let Γ be a simple graph in the sense of 5.3. Let φ be a non-degenerate element of $\partial\big(C_1(\Gamma,\mathbf{R})\big)$ as in 3.5. Fix a basis $l_1,\ldots,l_r$ of elementary cycles of $H_1(\Gamma,\mathbf{Z})$; see 5.1. Let

$$\tilde{G}\xrightarrow{\sim}\tilde{G}_1 \underset{\mathcal{J}}{\times}\cdots\underset{\mathcal{J}}{\times}\tilde{G}_r$$

be the decomposition of $\tilde{G}$ as product as schemes over $\mathcal{J}_0$ of $\mathbf{G}_m$-bundles over $\mathcal{J}$ associated to the isomorphism $H_1(\Gamma,\mathbf{Z})\xrightarrow{\sim}\mathbf{Z}l_1\oplus\ldots\oplus\mathbf{Z}l_r$. By 5.5 we get that

i. for any $\xi\in H^1(\Gamma,\mathbf{Z})$ the compactification P_ξ of $\tilde{G}$ defined in 6.2 is the product as schemes over $\mathcal{J}$ of the $\mathbf{P}^1$-bundles associated to $\tilde{G}_1,\ldots,\tilde{G}_r$. In particular, it is independent of ξ and of φ.
ii. the scheme $\mathcal{P}_0^\varphi\to\mathcal{J}\times_S S_0$ defined in 6.5 is independent of φ.

The scheme $\mathcal{P}_0^\varphi$, denoted simply $\mathcal{P}_0$ in view of (ii), can be described as follows. For any $t=1,\ldots,r$ let $\mathbb{P}_t$ be the $\mathbf{P}^1$-bundle over $\mathcal{J}_0=\mathcal{J}\times_S S_0$ associated to the $\mathbf{G}_m$-bundle $\tilde{G}_t\times_S S_0$. Let

$$\mathbf{P}_0:=\mathbb{P}_1\underset{\mathcal{J}_0}{\times}\cdots\underset{\mathcal{J}_0}{\times}\mathbb{P}_r.$$

For any $s=1,\ldots,r$ define 0_s and ∞_s as in 2.1. Then for any $\xi\in H^1(\Gamma,\mathbf{Z})$ we have $P_\xi\times_S S_0=\mathbf{P}_0$. Let

$$H^1(\Gamma,\mathbf{Z})=\mathbf{Z}f_1\oplus\ldots\oplus\mathbf{Z}f_r$$

be the splitting defined in 5.5. Then the scheme $\mathcal{P}_0$ is obtained by gluing for all $\xi\in H^1(\Gamma,\mathbf{Z})$ and all $s=1,\ldots,r$ the scheme $P_\xi\times_S S_0$ along the section ∞_s transversally to the scheme $P_{\xi+f_t}\times_S S_0$ along the section 0_s.

7 The scheme $\mathcal{P}_0$ and its local structure

Let S_0 be a noetherian affine scheme. In 7.1 we define a scheme $\mathcal{P}_0\to S_0$. It is of the type introduced in 6.9 and 6.7. There are two type of examples of such schemes the reader should keep in mind. The first is given by the "partial universal cover" of compactifications of Pic^0 of split semistable geometrically irreducible curves over S_0 to be studied in 14.7. The second is a more technical, but equally important, generalization of the first. It appears compactifying, as in 6.8, suitable torus torsors over the "partial universal cover" of compactifications of Pic^0 of more general split semistable curves $C_0\to S_0$; see 19.4-19.6. In 7.5 we study some of the cohomology groups involved in the deformation theory of $\mathcal{P}_0\to S_0$. Roughly speaking we study those responsible for the deformation of the "singularities" of $\mathcal{P}_0\to S_0$. The next sections deal with the global aspects of the deformation theory.

7.1 *Situation.* We begin by axiomatizing and slightly generalizing the construction given in 6.9 and 6.7. The reader should keep in mind Figure 1 while reading the definition. We start with the following data:

a. an abelian scheme $\mathcal{J}_0$ over S_0;

b. a semiabelian scheme $\mathbf{G}_0$ over S_0, extension of $\mathcal{J}_0$ by the split torus $\mathbf{G}^r_{m,S_0}$. Consider the induced product as schemes over $\mathcal{J}_0$ of $\mathbf{G}_m$-bundles over $\mathcal{J}_0$

$$\mathbf{G}_0 \xrightarrow{\sim} \mathbb{G}_1 \underset{\mathcal{J}_0}{\times} \cdots \underset{\mathcal{J}_0}{\times} \mathbb{G}_r;$$

c. the compactification $\mathbf{P}_0$ of $\mathbf{G}_0$ as $\{\mathbf{P}^1\}^r$-bundle over $\mathcal{J}_0$;

d. a lattice Λ of rank r and a $\mathbf{Z}$-basis $f_1, \dots, f_r$ of Λ. A sublattice

$$\Lambda' := \mathbf{Z}l_1 \oplus \ldots \oplus \mathbf{Z}l_r \subset \Lambda$$

such that for any $t = 1, \dots, r$ we have $l_t = n_t f_t$ for some positive integer n_t;

e. for any element $A \in \Lambda$ the scheme $\mathbf{P}_A = \mathbf{P}_0$. It is the product as schemes over $\mathcal{J}_0$ of $\mathbf{P}^1$-bundles over $\mathcal{J}_0$

$$\mathbf{P}_A \xrightarrow{\prod_{t=1}^r m_{A,t}} \mathbb{P}_{A,1} \underset{\mathcal{J}_0}{\times} \cdots \underset{\mathcal{J}_0}{\times} \mathbb{P}_{A,r}.$$

For any $t = 1, \dots, r$ define as in 2.1 the closed subschemes

$$0_{A,t} \qquad \text{and} \qquad \infty_{A,t}$$

induced by the zero and infinity sections of $\mathbb{P}_{A,t}$ over $\mathcal{J}_0$. Let

$$\mathbf{G}_A \hookrightarrow \mathbf{P}_A$$

be the $\{\mathbf{G}_m\}^r$-bundle associated to $\mathbf{P}_A$.

For any $A \in \Lambda$ and any $t = 1, \dots, r$ we have a canonical identification

$$\Phi_{A,t}\colon \infty_{A,t} = \infty_{0,t} = \prod_{s=1,\dots,r | s \neq t} \mathbb{P}_{0,s} = 0_{0,t} = 0_{A+f_t,t}.$$

Hence we can glue the scheme $\mathbf{P}_A$ transversally along the section $\infty_{A,t}$ to the scheme $\mathbf{P}_{A+f_t}$ along the section $0_{A+f_t,t}$. Repeating the gluing process for any A in Λ and any $t = 1, \dots, r$, obtain

1. a scheme locally of finite type over $\mathcal{J}_0$ containing $\mathbf{G}_0$ as an open subscheme:

$$\rho\colon \mathcal{P}_0 \longrightarrow \mathcal{J}_0.$$

Finally fix

f. a group homomorphism $\Lambda \longrightarrow \mathcal{J}_0$ and a lifting, called *rigidification*,

$$\amalg_{A \in \Lambda} S_0 \to \mathcal{P}_0$$

so that for any $A \in \Lambda$ the section corresponding to A factors through $\mathbf{G}_A$ and the image of the section associated to the element 0 of Λ is the identity element of $\mathbf{G}_0$.

For any A in Λ and any l in Λ' there is a *unique* isomorphism

$$\mu_{A,l}: \mathbf{P}_A \xrightarrow{\sim} \mathbf{P}_{A+l}$$

as $\{\mathbf{P}^1\}^r$-bundles over $\mathcal{J}_0$ lifting the translation by l on $\mathcal{J}_0$ and sending the S_0-section of $\mathbf{G}_A$ defined in (f) to the S_0-section of $\mathbf{G}_{A+l}$ defined in (f). Suppose that for any A in Λ, any l in Λ' and any $t = 1, \ldots, r$ we have

$$\Phi_{A+l,t}\ \mu_{A,l} = \mu_{A+f_t,l}\ \Phi_{A,t}.$$

Under this assumption we uniquely obtain

2. an action of Λ' on $\mathcal{P}_0 \to S_0$ respecting the rigidification and lifting the action by translation on $\mathcal{J}_0$. Denote the quotient scheme by

$$\mathcal{P}_0/\Lambda' =: \bar{P}_0 \longrightarrow S_0;$$

3.

   a group scheme structure on

$$\mathcal{P}_0^{\mathrm{sm},0} := \amalg_{A\in\Lambda'} \mathbf{G}_A \hookrightarrow \mathcal{P}_0$$

over S_0 such that the inclusion $\mathbf{G}_0 \to \mathcal{P}_0^{\mathrm{sm},0}$ is a group homomorphism and descending to a group scheme structure on the open subscheme of $\bar{P}_0$ defined as

$$\bar{P}_0^{\mathrm{sm},0} := \mathcal{P}_0^{\mathrm{sm},0}/\Lambda'.$$

In particular, $\mathcal{P}_0^{\mathrm{sm},0} \cong \tilde{G} \times \Lambda'$ as group schemes over $\mathcal{J}_0$;

4. an action of $\mathcal{P}_0^{\mathrm{sm},0}$ on $\mathcal{P}_0 \to S_0$ which descends to an action of $\bar{P}_0^{\mathrm{sm},0}$ on $\bar{P}_0 \to S_0$.

7.2 Remark. The definition above is very close to 6.9 and 6.7. The differences are that, first of all, we have omitted any reference to a graph and, second, we have introduced the concept of rigidification from which (2)-(4) follow.

To help the reader we draw an example in Figure 1. In this case the integer r in 7.1 is 2. Then $\mathbf{P}_0$ is a $\mathbf{P}^1 \times \mathbf{P}^1$-bundle over $\mathcal{J}_0$. The solid rectangles stand for the schemes $\mathbf{P}_A$. They are $\mathbf{P}^1 \times \mathbf{P}^1$-bundles over the abelian scheme $\mathcal{J}_0$. The picture should suggest how they are glued to obtain $\mathcal{P}_0$.

For later purposes we give the following

7.3 Definition. *The notation is as in 7.1. Let $A = a_1 + f_1 + \ldots + a_r f_r$ be an element of Λ. Define the contraction*

$$\tau_A: \mathcal{P}_0 \longrightarrow \mathbf{P}_A$$

to be the unique morphism such that

i. *the induced morphism $\mathbf{P}_A \hookrightarrow \mathcal{P}_0 \to \mathbf{P}_A$ is the identity;*

ii. *let* $B = b_1 f_1 + \ldots + b_r f_r$ *be an element of* Λ. *Let* $t = 1, \ldots, r$ *such that* $b_t \leq a_t - 1$ *(resp.* $b_t \geq a_t + 1$*). Then, the induced morphism* $\mathbf{P}_B \hookrightarrow \mathcal{P}_0 \to \mathbf{P}_A$ *factors via the projection of* $\mathbf{P}_B$ *to the section* $\infty_{B,t}$ *(resp.* $0_{B,t}$*).*

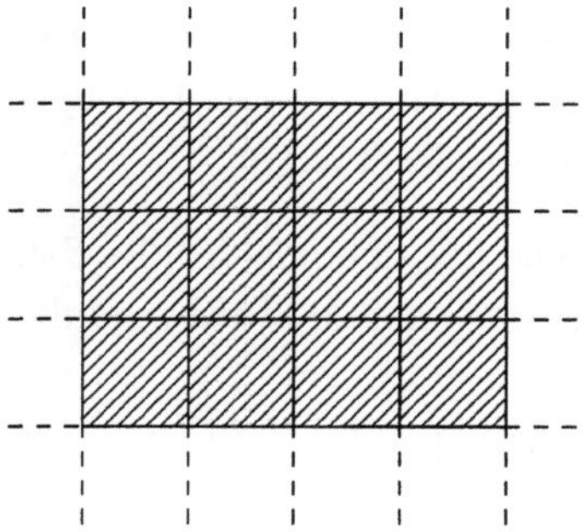

Figure 1.

7.4 *Some computations.* The notation is as in 7.1. Let x be a closed point of $\mathcal{P}_0$. Then there is an element A of Λ and an open affine neighborhood Spec($\mathbb{B}_0$) of x such that $\mathbb{B}_0$ is of the form

$$\mathbb{B}_0 = \mathbb{A}_0[z_1, \ldots, z_h, \gamma_1, \ldots, \gamma_v, \beta_1, \ldots, \beta_v]/(\gamma_1\beta_1, \ldots, \gamma_v\beta_v),$$

where the notation is as follows:

1. x belongs to $\mathbf{P}_A$. Let X be the image of x in $\mathcal{J}_0$ via the morphism
$$\mathbf{P}_A \hookrightarrow \mathcal{P}_0 \xrightarrow{\rho} \mathcal{J}_0;$$
2. Spec($\mathbb{A}_0$) is an open affine neighborhood of X in $\mathcal{J}_0$;
3. for any $i = 1, \ldots, h$ there exists an element s_i of $\{1, \ldots, r\}$ such that x is not contained either in ∞_{A,s_i} or in $0_{A,s_i}$ and z_i is a local parameter trivializing $\mathbb{P}_{A,s_i} \to \mathcal{J}_0$ at the image of x via the projection $m_{A,s_i} \colon \mathbf{P}_A \to \mathbb{P}_{A,s_i}$;
4. for any $w = 1, \ldots, v$ there exists an element t_w of $\{1, \ldots, r\}$ such that x is contained in ∞_{A,t_w} and γ_w is a local parameter trivializing $\mathbb{P}_{A,t_w} \to \mathcal{J}_0$ along the infinity section at X and β_w is a local parameter trivializing $\mathbb{P}_{A+f_{t_w},t_w} \to \mathcal{J}_0$ along the zero section at X;
5. the set $\{s_i\}$ with $i = 1, \ldots, h$ and the set $\{t_w\}$ with $w = 1, \ldots, v$ are disjoint subsets of $\{1, \ldots, r\}$ and their union coincides with the whole set $\{1, \ldots, r\}$.

It follows that $\Omega^1_{\mathbb{B}_0/\mathbb{A}_0}$ admits a resolution

$$\begin{array}{ccccccc} 0 \to & \mathbb{B}_0^v & \to & \mathbb{B}_0^h \times (\mathbb{B}_0 \oplus \mathbb{B}_0)^v & \to & \Omega^1_{\mathbb{B}_0/\mathbb{A}_0} & \to 0 \\ & (b_1, \ldots, b_v) & \mapsto & (\ldots, 0, \ldots, b_i(\gamma_i + \beta_i), \ldots) & . & & \end{array}$$

Since $\mathrm{Spec}(\mathbb{A}_0) \to S_0$ is smooth and the sequence

$$0 \longrightarrow \mathbb{B}_0 \otimes \Omega^1_{\mathrm{Spec}(\mathbb{A}_0)/S_0} \longrightarrow \Omega^1_{\mathrm{Spec}(\mathbb{B}_0)/S_0} \longrightarrow \Omega^1_{\mathrm{Spec}(\mathbb{B}_0)/\mathrm{Spec}(\mathbb{A}_0)} \longrightarrow 0$$

is locally split, we get

i. the restriction of $\mathcal{E}xt^i_{O_{\mathcal{P}_0}}(\Omega^1{}_{\mathcal{P}_0/S_0}, O_{\mathcal{P}_0})$ to $\mathrm{Spec}(\mathbb{B}_0)$ is equal to 0 if $i > 1$;
ii. the restriction of $\mathcal{E}xt^1_{O_{\mathcal{P}_0}}(\Omega^1{}_{\mathcal{P}_0/S_0}, O_{\mathcal{P}_0})$ to $\mathrm{Spec}(\mathbb{B}_0)$ is equal to the $\mathbb{B}_0$-module $\mathbb{B}_0/(\gamma_1\beta_1) \times \cdots \times \mathbb{B}_0/(\gamma_v\beta_v)$.

7.5 Proposition. 1. *The O_{S_0}-module $H^0\big(\mathcal{P}_0, \mathcal{E}xt^1_{O_{\mathcal{P}_0}}(\Omega^1_{\mathcal{P}_0/S_0}, O_{\mathcal{P}_0})\big)$ is isomorphic to the $\Gamma(S_0, O_{S_0})$-submodule of the module*

$$\prod_{A\in\Lambda}\Big(\prod_{t=1,\ldots,r}(\Gamma(0_{A,t}, O_{0_{A,t}}) \oplus \Gamma(\infty_{A,t}, O_{\infty_{A,t}}))\Big) \xrightarrow{\sim}$$

$$\xrightarrow{\sim} \prod_{A\in\Lambda}\Big(\prod_{t=1,\ldots,r}(\Gamma(S_0, O_{S_0}) \oplus \Gamma(S_0, O_{S_0}))\Big)$$

defined by those elements

$$\prod_{A\in\Lambda}\big((s^0_{A,1}, s^\infty_{A,1}), \ldots, (s^0_{A,r}, s^\infty_{A,r})\big)$$

such that for any element A of Λ and for any $t, v = 1, \ldots, r$ we have

$$\begin{cases} s^0_{A+f_t,v} = s^\infty_{A,v} & \text{if } t = v, \\ (s^0_{A+f_t,v}, s^\infty_{A+f_t,v}) = (s^0_{A,v}, s^\infty_{A,v}) & \text{if } t \neq v. \end{cases}$$

2. *There exists an isomorphism*

$$H^0\Big(\mathcal{E}xt^1_{O_{\bar{P}_0}}(\Omega^1_{\bar{P}_0/S_0}, O_{\bar{P}_0})\Big) \xrightarrow{\sim} H^0\Big(\mathcal{E}xt^1_{O_{\mathcal{P}_0}}(\Omega^1{}_{\mathcal{P}_0/S_0}, O_{\mathcal{P}_0})\Big)^{\Lambda'}.$$

Proof: This follows from globalizing the computations of 7.4 and from the construction of $\mathcal{P}_0$ in 7.1.

8 Cubical hyperresolutions and the cohomology of the relative tangent sheaf of $\mathcal{P}_0 \to S_0$

In this section we compute the cohomology of the relative tangent sheaf of the morphism $\mathcal{P}_0 \to S_0$, defined in 7.1. This will be essential for the study of the deformation theory of $\mathcal{P}_0 \to S_0$. The scheme $\mathcal{P}_0$ is constructed by gluing the schemes $\mathbf{P}_A$ with $A \in \Lambda$. It is natural to try to compute the cohomology groups of sheaves on $\mathcal{P}_0$ via the cohomology groups of their pullbacks to the constituents $\mathbf{P}_A$ of $\mathcal{P}_0$. The formal setting in which this can be phrased precisely is that of cubical hyperresolution. For this concept we refer to [14, Subsection I.2.11 and Definition I.2.12]. We do not make any attempt to summarize the general theory, hoping that the results we get justify its use. Figures 2 and 3 may give some idea of what this is about. The main results are the following. In 8.6 we prove that for any deformation $\mathcal{P}$ of $\mathcal{P}_0$ to a nilpotent thickening S of S_0 and for any deformation $\hat{0}$ of the section $S_0 \to \mathcal{P}_0$ associated to $0 \in \Lambda$, there is a unique deformation $\mathcal{J}$ of $\mathcal{J}_0$ as an abelian scheme over S such that the morphism $\mathcal{P}_0 \to \mathcal{J}_0$ extends to a morphism $\mathcal{P} \to \mathcal{J}$ and the image of $\hat{0}$ is the identity element of $\mathcal{J}$. In 8.10 and 8.11 we give an explicit description of the cohomology groups of the tangent sheaf of $\mathcal{P}_0$ relative to S_0 in terms of suitable invertible sheaves on the schemes $\mathbf{P}_A$ with $A \in \Lambda$ defined in 7.1.

8.1 Definition. *Define a r-cubical hyperresolution $Z_\bullet$ of $\mathcal{P}_0$ over $\mathcal{J}_0$ as follows:*

i. *consider the cartesian diagram*

$$\begin{array}{ccc} X^1_{1,1} & \longrightarrow & X^1_{0,1} \\ \downarrow & & \downarrow \\ X^1_{1,0} & \longrightarrow & X^1_{0,0} \end{array}$$

with

$$X^1_{0,0} := \mathcal{P}_0 \qquad \text{and} \qquad X^1_{0,1} := \amalg_{A\in\Lambda}\mathbf{P}_A;$$

the morphism $X^1_{0,1} \to X^1_{0,0}$ is the natural one and $X^1_{1,0} \to X^1_{0,0}$ is defined as the inclusion of the complement of the smooth locus of $\mathcal{P}_0 \to S_0$ in $\mathcal{P}_0$;

ii. *suppose that $X^k_\bullet$ is constructed with $k < r$. Suppose also that $X^k_{1,0,\dots,0}$ is the pure codimension k sublocus of $\mathcal{P}_0$ defined as the union over all vertices A of Λ of the intersection of k distinct closed subschemes of $\mathbf{P}_A$ among $\infty_{A,1},\dots,\infty_{A,r}$ (see 7.1). Define $X^{k+1}_{0,0,\bullet} := X^k_{1,\bullet}$ and construct the cartesian diagram*

$$\begin{array}{ccc} X^{k+1}_{1,1,0,\dots,0} & \longrightarrow & X^{k+1}_{0,1,0,\dots,0} \\ \downarrow & & \downarrow \\ X^{k+1}_{1,0,0,\dots,0} & \longrightarrow & X^{k+1}_{0,0,0,\dots,0}. \end{array} \tag{8.1.1}$$

The morphism

$$X^{k+1}_{0,1,0,\dots,0} \longrightarrow X^{k+1}_{0,0,0,\dots,0}$$

is given by the desingularization of $X^{k+1}_{0,0,0,\dots,0}$. More precisely, $X^{k+1}_{0,1,0,\dots,0}$ is the disjoint union over all vertices A of Λ of all intersections of k distinct subschemes among the subschemes $\infty_{A,1},\dots,\infty_{A,r}$ of $\mathbf{P}_A$. The morphism

$$X^{k+1}_{1,0,0,\dots,0} \longrightarrow X^{k+1}_{0,0,0,\dots,0}$$

is the inclusion of the complement of the smooth locus of $X^{k+1}_{0,0,0,\dots,0} \to S_0$ into $X^{k+1}_{0,0,0,\dots,0}$. Moreover, $X^{k+1}_{1,0,0,\dots,0}$ is the pure codimension $k+1$ sublocus of $\mathcal{P}_0$ defined as the union over all vertices A of Λ of the intersection of $k+1$ distinct closed subschemes of $\mathbf{P}_A$ among $\infty_{A,1},\dots,\infty_{A,r}$. Define $X^{k+1}_\bullet$ by pulling back *the diagram 8.1.1 to* $X^{k+1}_{0,0,\bullet}$.

Define

$$Z^k_\bullet := \mathrm{red}(X^1_\bullet, \dots, X^k_\bullet)$$

by induction on k *and call it the reduction of* $(X^1_\bullet, \dots, X^k_\bullet)$. *If* $k = 1$ *define*

$$Z^1_\bullet := X^1_\bullet.$$

Note that these schemes are indexed by the vertices of a two dimensional cube. By construction there exist natural morphisms between them indexed by the edges of the cube. If $k = 2$ *define*

$$Z^2_{\alpha,\beta} := \begin{cases} Z^1_{0,\beta} & \text{if } \alpha = (0,0) \text{ and } \beta = 0,1 \\ X^2_{\alpha,\beta} & \text{if } \alpha \neq (0,0) \text{ and } \beta = 0,1. \end{cases}$$

Note that these schemes are indexed by the vertices of a three dimensional cube. By construction there exist natural morphisms between them indexed by the edges of the cube. If $k > 2$ *with the same procedure of the case* $k = 2$ *define*

$$Z^k_\bullet := \mathrm{red}\big(Z^{k-1}_\bullet, X^k_\bullet\big);$$

these schemes are indexed by the vertices $\bullet$ *of the* $k+1$*-dimensional cube and there are natural morphisms between them indexed by the edges of the cube. Finally let*

$$Z_\bullet := Z^r_\bullet.$$

For any integer k *such that* $1 \leq k \leq r$ *define*

$$\pi^k_\bullet \colon Z^k_\bullet \longrightarrow \mathcal{P}_0 \qquad \text{and} \qquad \rho^k_\bullet := \pi^k_\bullet \circ \rho \colon Z^k_\bullet \longrightarrow \mathcal{J}_0.$$

Analogously define

$$\pi^r_\bullet =: \pi_\bullet \colon Z_\bullet \longrightarrow \mathcal{P}_0 \qquad \text{and} \qquad \rho^r_\bullet =: \rho_\bullet := \pi_\bullet \circ \rho \colon Z_\bullet \longrightarrow \mathcal{J}_0.$$

8.2 Remark. i. Each $Z^r_\bullet$ with $\bullet \neq (0, \dots, 0)$ is smooth over $\mathcal{J}_0$. Hence we get a cubical hyperresolution in the sense of [14, Definition I.2.12].
ii. For any integer k, if α and β are such that a morphism in $Z^k_\bullet$ from Z^k_α to Z^k_β exists then it is finite and consequently acyclic.
Consider the example in 7.2. In this case $\mathbf{P}_0$ is a $\mathbf{P}^1 \times \mathbf{P}^1$-bundle over $\mathcal{J}_0$. The integer r is 2 and $\Lambda \cong \mathbf{Z} \oplus \mathbf{Z}$. We draw a (symbolic) picture of the first step

$$\begin{array}{ccc} X^1_{1,1} & \longrightarrow & X^1_{0,1} \\ \downarrow & & \downarrow \\ X^1_{1,0} & \longrightarrow & X^1_{0,0} \end{array}$$

of the hyperresolution associated to $\mathcal{P}_0$ in Figure 2. The solid rectangles stand for the schemes $\mathbf{P}_A$ defined in 7.1. Recall that they are $\mathbf{P}^1 \times \mathbf{P}^1$-bundles over the abelian scheme $\mathcal{J}_0$. The empty rectangles stand for the union over the vertices A of Λ of the subschemes $0_{A,1}$, $0_{A,2}$, $\infty_{A,1}$ and $\infty_{A,2}$ of $\mathbf{P}_A$.

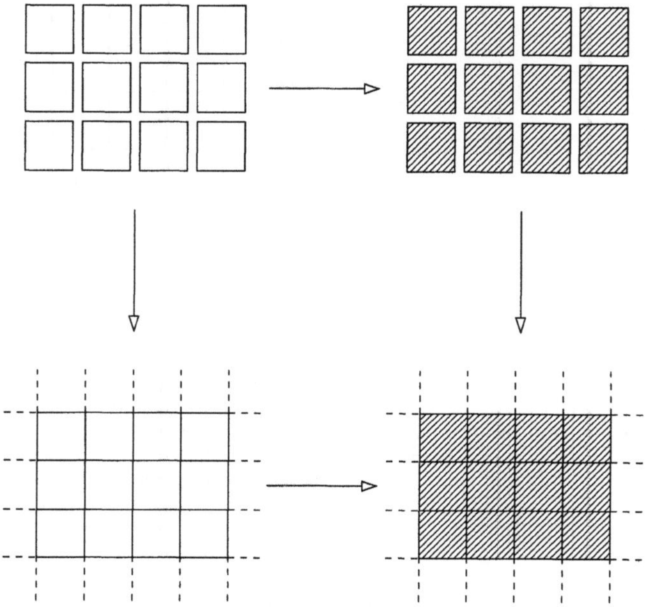

Figure 2.

In Figure 3 we draw

$$\begin{array}{ccc} & & X^2_{0,1,0} \\ & & \downarrow \\ X^2_{1,0,0} & \longrightarrow & X^2_{0,0,0}. \end{array}$$

The segments correspond to the $\mathbf{P}^1$-bundles over $\mathcal{J}_0$ defined by the subschemes $\infty_{A,1}$ and $\infty_{A,2}$ for all vertices A of Λ. The map $X^2_{0,1,0} \to X^2_{0,0,0}$ is the desingularization of $X^2_{0,0,0} := X^1_{1,0}$. The points in the picture correspond to $\mathcal{J}_0$ itself. The scheme $X^2_{1,0,0}$ is the complement of the smooth locus of $X^2_{0,0,0} \to S_0$ in $X^2_{0,0,0}$. It is the union over the vertices A of Λ of the intersection of the subschemes $\infty_{A,1}$ and $\infty_{A,2}$ of $\mathbf{P}_A$.

8.3 Proposition. *The cubical hyperresolution just constructed is of cohomological descent; see* [14, Definition I.5.16] *for the definition of cohomological descent. More precisely, for any sheaf of abelian groups F on $\mathcal{P}_0$ and any integer k such that $1 \leq k \leq r$ the sequence*

$$0 \longrightarrow F \longrightarrow \pi^k_{\bullet *}\left(\pi^{k*}_{\bullet}(F)\right)_{\bullet \neq (0,\dots,0)} \longrightarrow 0$$

is exact.

Proof: See the proof of [14, Theorem I.6.9].

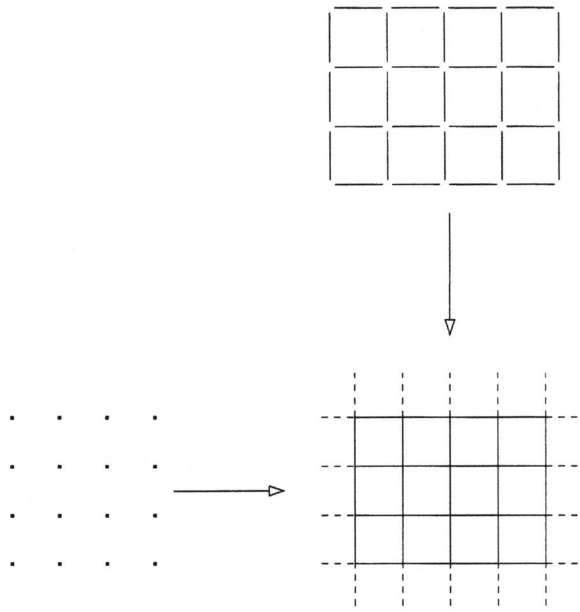

Figure 3.

We give a first application of the concept of cubical hyperresolution computing the obstructions to deform $\rho: \mathcal{P}_0 \to \mathcal{J}_0$; see 8.6.

8.4 Lemma. *For any integer k with $1 \leq k \leq r$, the long sequence $0 \to \rho^k_{\bullet *}(O_{Z^k_\bullet}) \to 0$ is exact. The notation is as in 8.1.*

Proof: By 8.3 we know that the long sequence $0 \to \pi^k_{\bullet *}(O_{Z^k_\bullet}) \to 0$ is exact. Proceeding by induction on k it suffices to prove that for any integer $h = 1, \ldots, r$ and for fixed α the 2-diagram

$$\begin{array}{ccc} X^h_{1,1,\alpha} & \longrightarrow & X^h_{0,1,\alpha} \\ \downarrow & & \downarrow \\ X^h_{1,0,\alpha} & \longrightarrow & X^h_{0,0,\alpha}. \end{array}$$

induces an exact sequence

$$0 \longrightarrow \rho^h_{0,0,\alpha *}(O_{X^h_{0,0,\alpha}}) \longrightarrow \bigoplus_{\delta=(0,1),(1,0)} \rho^h_{\delta,\alpha *}(O_{X^h_{\delta,\alpha}}) \longrightarrow \rho^h_{1,1,\alpha *}(O_{X^h_{1,1,\alpha}}) \longrightarrow 0.$$

This holds replacing ρ with π due to 8.3. Since $\rho^h_\bullet$ is defined by composing $\pi^h_\bullet$ with ρ, it suffices to prove that the short sequence above is exact on the right.

Fix a connected component Z of $X^h_{0,1,\alpha}$. By construction it is isomorphic to the product as schemes over $\mathcal{J}_0$ of $\mathbf{P}^1$-bundles $\mathbb{P}_{Z,1},\ldots,\mathbb{P}_{Z,r-h+1}$ over $\mathcal{J}_0$. The inclusion $X^h_{1,1,\alpha} \hookrightarrow X^h_{0,1,\alpha}$ is realized on Z by the inclusion of the union of the pullback to Z of the zero and infinity sections over $\mathcal{J}_0$ of each of the $\mathbf{P}^1$-bundles $\mathbb{P}_{Z,1},\ldots,\mathbb{P}_{Z,r-h+1}$. In particular, if $h \leq r-1$, any connected component of $X^h_{1,1,\alpha}$ injects in one and only one connected component of $X^h_{0,1,\alpha}$. Since these components are proper over $\mathcal{J}_0$, the exactness is trivial in this case. If $h = r$, the argument is slightly more involved. In this case the scheme $X^r_{0,0,\alpha}$ is a curve over $\mathcal{J}_0$ obtained by gluing $\mathbf{P}^1$-bundles over $\mathcal{J}_0$. For any geometric point of $\mathcal{J}_0$, consider the graph of the fiber of the curve defined by associating a vertex to each singularity and an edge to each irreducible component. The graph has trivial homology. In particular, $\mathrm{R}\rho^1_*(\pi^h_{0,0,\alpha*}(O_{X^h_{0,0,\alpha}}))$ is zero.

8.5 Proposition. *For any positive integer i the structure morphism $\rho\colon \mathcal{P}_0 \to \mathcal{J}_0$ defined in 7.1 induces an isomorphism*

$$H^i(\mathcal{J}_0, O_{\mathcal{J}_0}) \longrightarrow H^i(\mathcal{P}_0, O_{\mathcal{P}_0}).$$

Proof: Let Z_α be an element of the hyperresolution associated in 8.1 to $\mathcal{P}_0$ such that Z_α is different from $Z_{0,\ldots,0}$. Then Z_α is a product of $\mathbf{P}^1$-bundles over $\mathcal{J}_0$. In particular, O_{Z_α} is acyclic for the morphism $\rho_\alpha\colon Z_\alpha \to \mathcal{J}_0$. Hence 8.4 implies that also $O_{\mathcal{P}_0}$ is acyclic for $\rho\colon \mathcal{P}_0 \to \mathcal{J}_0$. Since the direct image of $O_{\mathcal{P}_0}$ via $\mathcal{P}_0 \to \mathcal{J}_0$ is $O_{\mathcal{J}_0}$ we are done.

8.6 Corollary. i. *Let $\mathcal{T}_{\mathcal{J}_0/S_0}$ denote the tangent bundle of $\mathcal{J}_0 \to S_0$. The structure morphism $\rho\colon \mathcal{P}_0 \to \mathcal{J}_0$ induces an isomorphism*

$$H^1(\mathcal{J}_0, \mathcal{T}_{\mathcal{J}_0/S_0}) \longrightarrow H^1(\mathcal{P}_0, \rho^*(\mathcal{T}_{\mathcal{J}_0/S_0})).$$

ii. *In particular, let S be a nilpotent thickening of S_0. Let $\mathcal{P} \to S$ be a deformation of $\rho\colon \mathcal{P}_0 \to S_0$ and let $\hat{0}\colon S \to \mathcal{P}$ be a deformation of the section $S_0 \to \mathcal{P}_0$ associated to $0 \in \Lambda$. Then there exists a unique deformation $\mathcal{J}$ of $\mathcal{J}_0 \to S_0$ as an abelian scheme over S such that ρ extends and the image of $\hat{0}$ in $\mathcal{J}$ is the identity element. Moreover, the extension of ρ is unique.*

Proof: Claim (i) is proven using 8.5 since $\mathcal{T}_{\mathcal{J}_0/S_0} \cong O_{\mathcal{J}_0} \otimes T_{\mathcal{J}_0,0}$, where $T_{\mathcal{J}_0,0}$ is the tangent space to $\mathcal{J}_0$ at the identity element. Claim (ii) follows from (i) by standard deformation theory.

We turn to the problem of computing the cohomology of the tangent sheaf on $\mathcal{P}_0$. In view of 8.9 we introduce the following

8.7 Definition. *For any element Z^k_α of the cubical hyperresolution defined in 8.1, let*

$$\mathcal{A}^k_\alpha := \mathrm{Im}\Big(\pi^{k*}_\alpha(\mathcal{H}om_{O_{\mathcal{P}_0}}(\Omega^1_{\mathcal{P}_0/\mathcal{J}_0}, O_{\mathcal{P}_0})) \longrightarrow \mathcal{H}om_{O_{Z^k_\alpha}}(\pi^{k*}_\alpha(\Omega^1_{\mathcal{P}_0/\mathcal{J}_0}), O_{Z^k_\alpha})\Big).$$

Define

$$\mathcal{B}^k_\alpha := \mathrm{Im}\Big(\pi^{k*}_\alpha\big(\mathcal{H}om_{\mathcal{O}_{\mathcal{P}_0}}(\Omega^1_{\mathcal{P}_0/S_0}, \mathcal{O}_{\mathcal{P}_0})\big) \longrightarrow \mathcal{H}om_{\mathcal{O}_{Z^k_\alpha}}\big(\pi^{k*}_\alpha(\Omega^1_{\mathcal{P}_0/S_0}), \mathcal{O}_{Z^k_\alpha}\big)\Big).$$

Let

$$\mathcal{A}_\alpha := \mathcal{A}^r_\alpha \qquad \text{and} \qquad \mathcal{B}_\alpha := \mathcal{A}^r_\alpha.$$

8.8 *The sheaf $\mathcal{A}_\alpha$ in local coordinates.* Let x be a closed point of Z_α for some α. Suppose that Z_α is of relative dimension $r-s$ over $\mathcal{J}_0$ for some positive integer s satisfying $0 \leq s \leq r$. Choose an open affine neighborhood $\mathrm{Spec}(\mathbb{A}_0)$ of $\rho_\alpha(x)$ and an open affine neighborhood of $\pi_\alpha(x)$

$$\mathrm{Spec}(\mathbb{B}_0) := \mathrm{Spec}\big(\mathbb{A}_0[z_1, \ldots, z_h, \gamma_1, \ldots, \gamma_v, \beta_1, \ldots, \beta_v]/(\gamma_1\beta_1, \ldots, \gamma_v\beta_v)\big)$$

as in 7.4. After possibly permuting the variables γ_u and β_u with $1 \leq u \leq v$, we can suppose by the very construction in 8.1 that Z_α is locally at x isomorphic to

$$\mathrm{Spec}(C_s) := \mathrm{Spec}\big(\mathbb{B}_0/(\beta_1, \ldots, \beta_v, \gamma_1, \ldots, \gamma_s)\big).$$

We have

$$\mathcal{H}om_{\mathcal{O}_{\mathcal{P}_0}}\big(\Omega^1_{\mathcal{P}_0/\mathcal{J}_0}, \mathcal{O}_{\mathcal{P}_0}\big)|_{\mathrm{Spec}(\mathbb{B}_0)} \cong \bigoplus_{i=1}^{h} \mathbb{B}_0 \frac{\partial}{\partial z_i} \bigoplus_{u=1}^{v} \Big(\mathbb{B}_0\, \gamma_u \frac{\partial}{\partial \gamma_u} \oplus \mathbb{B}_0\, \beta_u \frac{\partial}{\partial \beta_u}\Big).$$

Hence

$$\mathcal{A}_\alpha|_{\mathrm{Spec}(C_s)} \cong \bigoplus_{i=1}^{h} C_s \frac{\partial}{\partial z_i} \bigoplus_{u=s+1}^{v} C_s \gamma_u \frac{\partial}{\partial \gamma_u}.$$

8.9 Lemma. *For any non-negative integer k the long sequences*

$$0 \longrightarrow \mathcal{H}om_{\mathcal{O}_{\mathcal{P}_0}}\big(\Omega^1_{\mathcal{P}_0/\mathcal{J}_0}, \mathcal{O}_{\mathcal{P}_0}\big) \longrightarrow \pi^k_{\bullet *}(\mathcal{A}^k_\bullet)_{\bullet \neq (0,\ldots,0)} \longrightarrow 0$$

and

$$0 \longrightarrow \mathcal{H}om_{\mathcal{O}_{\mathcal{P}_0}}\big(\Omega^1_{\mathcal{P}_0/S_0}, \mathcal{O}_{\mathcal{P}_0}\big) \longrightarrow \pi^k_{\bullet *}(\mathcal{B}^k_\bullet)_{\bullet \neq (0,\ldots,0)} \longrightarrow 0$$

are exact.

Proof: First we note that by definition the sheaves $\mathcal{A}^k_\bullet$ and $\mathcal{B}^k_\bullet$ are sheaves on $Z^k_\bullet$ and that the two sequences are well defined. We prove the statement for the first sequence, the case of the second sequence being entirely parallel. We can reduce to prove that given the square

$$\begin{array}{ccc} X^k_{1,1,\alpha} & \longrightarrow & X^k_{0,1,\alpha} \\ \big\downarrow & & \big\downarrow \\ X^k_{1,0,\alpha} & \longrightarrow & X^k_{0,0,\alpha}, \end{array} \qquad 8.9.1$$

the following short sequence is exact

$$0 \longrightarrow \pi^k_{0,0,\alpha *}\big(\mathcal{A}^k_{0,0,\alpha}\big) \longrightarrow \bigoplus_{\delta=(1,0),(0,1)} \rho^k_{\delta,\alpha *}\big(\mathcal{A}^k_{\delta,\alpha}\big) \longrightarrow \rho^k_{1,1,\alpha *}\big(\mathcal{A}^k_{1,1,\alpha}\big) \longrightarrow 0.$$

This reduction is one of the main tools in the theory of cubical hyperresolutions. We have defined $\mathcal{A}^k_\bullet$ as the image of

$$D_\bullet := \pi^{k*}_\bullet\Big(\mathcal{H}om(\Omega^1_{\mathcal{P}_0/\mathcal{J}_0}, O_{\mathcal{P}_0})\Big) \longrightarrow \mathcal{H}om\Big(\pi^{k*}_\bullet(\Omega^1_{\mathcal{P}_0/\mathcal{J}_0}), O_{Z^k_\bullet}\Big) =: E_\alpha.$$

Evaluate these two terms on the 2-resolution 8.9.1. We get a commutative diagram

$$\begin{array}{ccccc}
\pi^k_{0,0,\alpha_*}(D_{0,0,\alpha}) & \rightarrow & \pi^k_{0,0,\alpha_*}(D_{0,1,\alpha}) \oplus \pi^k_{0,0,\alpha_*}(D_{1,0,\alpha}) & \rightarrow & \pi^k_{0,0,\alpha_*}(D_{1,1,\alpha}) \\
\downarrow & & \downarrow & & \downarrow \\
\pi^k_{0,0,\alpha_*}(\mathcal{A}^k_{0,0,\alpha}) & \rightarrow & \pi^k_{0,0,\alpha_*}(\mathcal{A}^k_{0,1,\alpha}) \oplus \pi^k_{0,0,\alpha_*}(\mathcal{A}^k_{1,0,\alpha}) & \rightarrow & \pi^k_{0,0,\alpha_*}(\mathcal{A}^k_{1,1,\alpha}) \\
\downarrow & & \downarrow & & \downarrow \\
\pi^k_{0,0,\alpha_*}(E_{0,0,\alpha}) & \rightarrow & \pi^k_{0,0,\alpha_*}(E_{0,1,\alpha}) \oplus \pi^k_{0,0,\alpha_*}(E_{1,0,\alpha}) & \rightarrow & \pi^k_{0,0,\alpha_*}(E_{1,1,\alpha}),
\end{array}$$

where , since any $\pi^k_\bullet$ is a finite morphism, the map on the left in the first row (1) is injective and the map on the right is surjective by 8.3, analogously the map on the left in the third row (3) is injective by 8.4 and the second row (2) is the image of the first one in the third one. Hence the map on the left in (2) is injective and the map on the right is surjective. If we prove that

$$K_\bullet := \mathrm{Ker}\big((1) \longrightarrow (3)\big)$$

is exact on the right, then (2) is exact and we are done.

The scheme $X^k_{0,0,\alpha}$ is necessarily not smooth over $\mathcal{J}_0$, but $X^k_{0,1,\alpha}$ is. Suppose that the two schemes have relative dimension $r-s$ over $\mathcal{J}_0$. In the local coordinates of 8.8 the scheme $X^k_{0,1,\alpha}$ can be covered by open schemes isomorphic to $\mathrm{Spec}(C_s)$. The closed immersion $X^k_{1,1,\alpha} \to X^k_{0,1,\alpha}$ is realized on such open subschemes by the morphism

$$C_s \longrightarrow C_s/\Big(\prod_{u=s+1,\dots,v} \gamma_u\Big) =: C'_s$$

and $K_{0,1,\alpha} \to K_{1,1,\alpha}$ restricted to it is described by

$$K_{0,1,\alpha} = \bigoplus_{u=1}^{v} C_s\beta_u\frac{\partial}{\partial\beta_u} \bigoplus_{u=1}^{s} C_s\gamma_u\frac{\partial}{\partial\gamma_u}$$

and

$$K_{1,1,\alpha} = \bigoplus_{u=1}^{v} C'_s\beta_u\frac{\partial}{\partial\beta_u} \bigoplus_{u=1}^{s} C'_s\gamma_u\frac{\partial}{\partial\gamma_u} \bigoplus_{u=s+1}^{v} \Big(\prod_{j\neq u}\gamma_j\Big)C'_s\gamma_u\frac{\partial}{\partial\gamma_u}.$$

Hence the cokernel is

$$\bigoplus_{u=s+1}^{v} \Big(\prod_{j\neq u}\gamma_j\Big)C'_s\gamma_u\frac{\partial}{\partial\gamma_u}.$$

Remark that $E_{1,0,\alpha}$ injects into $E_{1,1,\alpha}$ since $O_{X^1_{1,0,\alpha}}$ injects in $O_{X^1_{1,1,\alpha}}$. In particular, $K_{1,0,\alpha}$ is the inverse image of $K_{1,1,\alpha}$ in $D_{1,0,\alpha}$. For any integer u such that $s+1 \le u \le v$ the image of the element

$$(\prod_{j\neq u} \gamma_j) O_{X^k_{1,0,\alpha}} \gamma_u \partial/\partial\gamma_u$$

of $D_{1,0,\alpha}$ in $D_{1,1,\alpha,k}$ is $(\prod_{j\neq u}\gamma_j)C'_s\gamma_u\partial/\partial\gamma_u$. In particular, it is contained in $K_{1,0,\alpha}$ as wanted.

8.10 Proposition. *The notation is as in 8.1 and as in 8.7.*

1. *If $\alpha \neq (0,\ldots,0)$, the term Z_α of the cubical hyperresolution $Z_\bullet$ is smooth over $\mathcal{J}_0$. Its connected components are isomorphic to products as schemes over $\mathcal{J}_0$ of the $\mathbf{P}^1$-bundles $\mathbb{P}_t$ defined in 7.1.*

2. *Fix an $\alpha \neq (0,\ldots,0)$. Let a connected component Z of Z_α be*

$$Z \cong \mathbb{P}_{t_1} \underset{\mathcal{J}_0}{\times} \cdots \underset{\mathcal{J}_0}{\times} \mathbb{P}_{t_s}$$

for some integer s. Let m_{t_j} be the projection $Z \to \mathbb{P}_{t_j}$ and let $\mathcal{T}_{\mathbb{P}_{t_j}/\mathcal{J}_0}$ be the relative tangent bundle of $\mathbb{P}_{t_j}$ over $\mathcal{J}_0$. We have isomorphisms

$$\mathcal{A}_\alpha|_Z \cong \bigoplus_{j=1}^{s} m_{t_j}{}^*\big(\mathcal{T}_{\mathbb{P}_{t_j}/\mathcal{J}_0}[-0_{t_j} - \infty_{t_j}]\big) \cong \bigoplus_{j=1}^{s} m_{t_j}{}^*\big(O_{\mathbb{P}_{t_j}}\big) \cong \bigoplus_{j=1}^{s} O_Z.$$

3. *Let Z be as in (2). We have an exact sequence*

$$\begin{array}{ccccccc}
0 \longrightarrow & \rho_{\alpha *}(\mathcal{A}_\alpha)|_Z & \longrightarrow & \rho_{\alpha *}(\mathcal{B}_\alpha)|_Z & \longrightarrow & \rho_{\alpha *}\big(\rho_\alpha{}^*(\mathcal{T}_{\mathcal{J}_0/S_0})\big)|_Z & \longrightarrow 0 \\
& \uparrow \wr & & & & \uparrow \wr & \\
& \oplus_{j=1,\ldots,s} O_{\mathcal{J}_0} & & & & \mathcal{T}_{\mathcal{J}_0/S_0}. &
\end{array}$$

4. *We have an horizontal short exact sequence of vertical long exact sequences*

$$\begin{array}{ccccccc}
 & 0 & & 0 & & 0 & \\
 & \downarrow & & \downarrow & & \downarrow & \\
0 \to & \rho_*\big(\mathcal{T}_{\mathcal{P}_0/\mathcal{J}_0}\big) & \to & \rho_*\big(\mathcal{T}_{\mathcal{P}_0/S_0}\big) & \to & \mathcal{T}_{\mathcal{J}_0/S_0} & \to 0 \\
 & \downarrow & & \downarrow & & \downarrow & \\
0 \to & \rho_{\bullet *}(\mathcal{A}_\bullet) & \to & \rho_{\bullet *}(\mathcal{B}_\bullet) & \to & \rho_{\bullet *}\big(\rho_\bullet{}^*\mathcal{T}_{\mathcal{J}_0/S_0}\big) & \to 0 \\
 & \downarrow & & \downarrow & & \downarrow & \\
 & 0 & & 0 & & 0 &
\end{array}$$

where $\bullet \neq (0,\ldots,0)$ and

$$\mathcal{T}_{-/-} := \mathcal{H}om_{O_{\mathcal{P}_0}}\big(\Omega^1_{-/-}, O_{\mathcal{P}_0}\big).$$

Proof: Statement (1) is clear by the construction 8.1. Statement (2) follows from the computations in 8.8. Statement (3) follows from the fact that the higher direct images of $\mathcal{A}_\alpha$ with respect to ρ_α are trivial. Statement (4) follows from (3) and 8.9.

8.11 Proposition. *The notation is as in 8.1 and as in 8.7. For any non-negative integer i we have horizontal short exact sequences of vertical long exact sequences, with $\bullet \neq (0, \dots, 0)$,*

$$\begin{array}{ccccccccc} & & 0 & & 0 & & 0 & & \\ & & \downarrow & & \downarrow & & \downarrow & & \\ 0 \to & & H^i(\mathcal{P}_0, \mathcal{T}_{\mathcal{P}_0/\mathcal{J}_0}) & \to & H^i(\mathcal{P}_0, \mathcal{T}_{\mathcal{P}_0/S_0}) & \to & H^i(\mathcal{J}_0, \mathcal{T}_{\mathcal{J}_0/S_0}) & \to 0 \\ & & \downarrow & & \downarrow & & \downarrow & & \\ 0 \to & & H^i(Z_\bullet, \mathcal{A}_\bullet) & \to & H^i(Z_\bullet, \mathcal{B}_\bullet) & \to & \left(H^i(\mathcal{J}_0, \mathcal{T}_{\mathcal{J}_0/S_0})\right)_\bullet & \to 0 \\ & & \downarrow & & \downarrow & & \downarrow & & \\ & & 0 & & 0 & & 0 & & \end{array}$$

and

$$H^i(\mathcal{P}_0, \mathcal{T}_{\mathcal{P}_0/\mathcal{J}_0}) \longrightarrow H^i(Z_\bullet, \mathcal{A}_{0,\dots,0,1})$$

is injective.

Proof: First we prove the exactness of the short sequences appearing in the last row. It suffices to prove that the connecting homomorphism in the long exact sequence of H^i's associated to (3) of 8.10 is zero. We have an injective morphism of short exact sequences

$$\begin{array}{ccccccc} 0 \longrightarrow & \rho_{\alpha *}(\mathcal{A}_\alpha) & \longrightarrow & \rho_{\alpha *}(\mathcal{B}_\alpha) & \longrightarrow & \rho_{\alpha *}(\rho_\alpha{}^*(\mathcal{T}_{\mathcal{J}_0/S_0})) & \longrightarrow 0 \\ & \downarrow & & \downarrow & & \downarrow & \\ 0 \longrightarrow & \rho_{\alpha *}(\mathcal{T}_{Z_\alpha/\mathcal{J}_0}) & \longrightarrow & \rho_{\alpha *}(\mathcal{T}_{Z_\alpha/S_0}) & \longrightarrow & \rho_{\alpha *}(\rho_\alpha{}^*(\mathcal{T}_{\mathcal{J}_0/S_0})) & \longrightarrow 0. \end{array}$$

Since $Z_\alpha \to \mathcal{J}_0$ has a section in view of (1) of 8.10, the short exact sequence of the second row splits on the right and hence the connecting homomorphism of the associated long exact sequence of H^i's is zero. The same must hold true for the long exact sequence associated to the short exact sequence of the first row. To conclude it suffices to prove that the long sequences

$$H^i(Z_\bullet, \mathcal{A}_\bullet) \longrightarrow 0 \qquad \text{and} \qquad \left(H^i(\mathcal{J}_0, \mathcal{T}_{\mathcal{J}_0/S_0})\right)_\bullet \longrightarrow 0$$

in the statement of the proposition are exact Zariski locally on S_0. In particular we can suppose that each O_{S_0}-module $H^i(\mathcal{J}_0, O_{\mathcal{J}_0/S_0})$ is free. The cohomology groups appearing in the two long sequences above are direct sums of $H^i(\mathcal{J}_0, O_{\mathcal{J}_0})$. Choose a decomposition of the latter into rank 1 summands. Then the sequences are compatible with such decomposition. The statement follows as in (4) of 8.10 where we take $\mathcal{J}_0$ to be S_0.

9 From cohomology groups to deformation theory

In this section we translate the computations on cohomology groups made in the previous section, of which we use notations and assumptions, into the language of deformation theory of $\mathcal{P}_0 \to S_0$. The main result is contained in 9.9. It relates the deformation functors of various objects defined in 9.5. First of all we give a geometric meaning in terms of deformation theory to the second row of 8.11.

9.1 *The translation of 8.11 into deformation theory.* Let Z and Z_α be as in (1) and (2) of 8.10

$$Z \cong \mathbb{P}_{t_1} \underset{\mathcal{J}_0}{\times} \cdots \underset{\mathcal{J}_0}{\times} \mathbb{P}_{t_s}.$$

Below we interpret the exactness of the restriction to Z of

$$0 \longrightarrow H^i(Z_\alpha, \mathcal{A}_\alpha) \longrightarrow H^i(Z_\alpha, \mathcal{B}_\alpha) \longrightarrow H^i(\mathcal{J}_0, \mathcal{T}_{\mathcal{J}_0/S_0}) \longrightarrow 0$$

defined in 8.11 for $i = 0, 1, 2$. The expression "first order deformation of a scheme defined over S_0" stands for a deformation of the given scheme to the dual numbers over S_0 defined as $\mathcal{O}_{S_0}[\varepsilon/\varepsilon^2]$.

i. the first order deformations of $\mathcal{J}_0 \to S_0$ as an abelian scheme are a principal homogeneous space under $H^1(\mathcal{J}_0, \mathcal{T}_{\mathcal{J}_0/S_0})$;

ii. if we fix a first order deformation $\mathcal{J}$ of $\mathcal{J}_0$ as an abelian scheme, the first order deformations of $Z \to S_0$ as product as schemes over $\mathcal{J}$ of $\mathbf{P}^1$-bundles over $\mathcal{J}$ are a principal homogeneous space under

$$H^1(Z, \mathcal{A}_\alpha) = \bigoplus_{j=1}^{s} H^1(\mathcal{J}_0, \mathcal{O}_{\mathcal{J}_0}).$$

They coincide with the first order deformations over $\mathcal{J}$ of the $\mathbf{G}_m$-bundles $\mathbb{G}_{t_1}, \ldots, \mathbb{G}_{t_s}$ defined in 7.1 to which the $\mathbf{P}^1$-bundles $\mathbb{P}_{t_1}, \ldots, \mathbb{P}_{t_s}$ are associated to. In fact, for each i with $1 \leq i \leq s$ the first order deformations of $\mathbb{G}_{t_i}$ as a $\mathbf{G}_m$-bundle over $\mathcal{J}$ are a homogeneous space under $H^1(\mathcal{J}_0, \mathcal{O}_{\mathcal{J}_0})$ or, equivalently, under the tangent space of the dual abelian scheme $\mathcal{J}^\vee$ at the identity.

iii. The automorphisms of a first order deformation of $Z \to S_0$ as in (ii) are classified by $H^0(Z, \mathcal{B}_\alpha)$. This is a locally free $\mathcal{O}_{S_0}$-module of rank equal to r times the relative dimension of $\mathcal{J}_0 \to S_0$. Fix a morphism $S_0 \to Z_\alpha$ such that for each $i = 1, \ldots, s$ the map $S_0 \to \mathbb{P}_{t_i}$ factors via the $\mathbf{G}_m$-bundle $\mathbb{G}_{t_i}$. It is a rigidification in the sense of 2.3. Then the automorphism group of a first order deformation of $Z \to S_0$ as in (ii) and of the rigidification $S_0 \to Z$ is trivial. In fact, $H^0(Z, \mathcal{B}_\alpha)$ is the automorphism group of any first order deformation of the rigifdification.

9.2 Proposition. *Let $\mathcal{P} \to S$ be a deformation of the scheme $\mathcal{P}_0 \to S_0$ defined in 7.1 to a nilpotent thickening of S_0. Let $\hat{0}$ be a deformation of the* *section $S_0 \to \mathcal{P}_0$ fixed in 7.1. Due to 8.6 there is a unique deformation $\mathcal{J} \to S$ of $\mathcal{J}_0 \to S_0$ as an abelian scheme such that $\mathcal{P}_0 \to \mathcal{J}_0$ extends to a morphism $\mathcal{P} \to \mathcal{J}$ and $\hat{0}$ maps to the identity of $\mathcal{J}$. Let A be an element of Λ. Let $G_A \hookrightarrow \mathcal{P}$ be the open subscheme deforming the open immersion $\mathbf{G}_A \hookrightarrow \mathcal{P}_0$ defined in 7.1.*

i. *There exists a unique deformation $P_A \to \mathcal{J}$ of $\mathbf{P}_A \to \mathcal{J}_0$ as product as schemes over $\mathcal{J}_0$ of $\mathbf{P}^1$-bundles over $\mathcal{J}_0$ such that the contraction morphism $\tau_A\colon \mathcal{P}_0 \to \mathbf{P}_A$ defined in 7.3 extends to a morphism $\mathcal{P} \to P_A$.*

ii. *Moreover $G_A \to \mathcal{J}$ is a deformation of $\mathbf{G}_A \to \mathcal{J}_0$ as product as schemes over $\mathcal{J}_0$ of $\mathbf{G}_m$-bundles over $\mathcal{J}_0$. The product of associated $\mathbf{P}^1$-bundles is $P_A \to \mathcal{J}$.*

Proof: i. Let $S_0 \hookrightarrow \ldots \hookrightarrow S_i = S$ be a filtration of S by closed subschemes such that $S_k \hookrightarrow S_{k+1}$ is small in the sense that it is defined by an ideal I_{k+1} such that $I_{k+1} I$ is 0. Suppose that a deformation P_{n-1} of $\mathbf{P}_A$ as claimed in the theorem exists over S_n for some n such that $1 \leq n \leq i$. We prove that we can deform $\mathbf{P}_A$ to S_n in such a way that the properties of the theorem hold. First we deform P_{n-1} arbitrarily to P_n as a product of $\mathbf{P}^1$-bundles over $\mathcal{J} \times_S S_n$. The local existence of the extension of $\mathcal{P} \times_S S_{n-1} \to P_{n-1}$ to P_n as claimed in the theorem is clear since $P_n \to S_n$ is formally smooth. For $t = 1, \ldots, r$ let

$$\mathbf{P}_A \cong \mathbb{P}_{A,1} \underset{\mathcal{J}_0}{\times} \cdots \underset{\mathcal{J}_0}{\times} \mathbb{P}_{A,r} \qquad \text{and} \qquad \tau_{A,t}\colon \mathcal{P}_0 \xrightarrow{\tau_A} \mathbf{P}_A \xrightarrow{m_{A,t}} \mathbb{P}_{A,t},$$

where $\mathbb{P}_{A,1}, \ldots, \mathbb{P}_{A,r}$ are the $\mathbf{P}^1$-bundles over $\mathcal{J}_0$ defined in 7.1 and the morphism $m_{A,t}$ is the projection of $\mathbf{P}_A$ to $\mathbb{P}_{A,t}$. The obstruction to extend globally the morphism $\mathcal{P}_n \to P_n$ is in

$$\bigoplus_{t=1}^{r} H^1\Big(\mathcal{P}_0, \mathcal{H}om_{O_{\mathcal{P}_0}}\big(\tau_{A,t}^*(\Omega^1_{\mathbb{P}_{A,t}/\mathcal{J}_0}), O_{\mathcal{P}_0}\big)\Big) \underset{O_{S_0}}{\otimes} I_n.$$

Let $A = \oplus_i a_i f_i$. Let $t = 1, \ldots, r$. Consider the exact sequence

$$0 \to O_{\mathbf{P}_A} \to \mathcal{H}om_{O_{\mathbf{P}_A}}\big(m_{A,t}^*(\Omega^1_{\mathbb{P}_{A,t}/\mathcal{J}_0}), O_{\mathbf{P}_A}\big) \to O_{0_{A,t}} \oplus O_{\infty_{A,t}} \to 0.$$

The cokernel of

$$0 \longrightarrow \tau_A^*(O_{\mathbb{P}_A}) \longrightarrow \tau_A^*\Big(\mathcal{H}om_{O_{\mathbf{P}_A}}\big(m_{A,t}^*(\Omega^1_{\mathbb{P}_{A,t}/\mathcal{J}_0}), O_{\mathbf{P}_A}\big)\Big)$$

is the direct sum of the structure sheaves of the closed subschemes defined by union of $O_{B,t}$ and $\infty_{B,t}$ for $B = \oplus_i b_i f_i$ such that $a_t = b_t$. These subschemes are of the type defined in 7.1. Using 8.5 it follows that

$$H^1\Big(\mathbf{P}_A, \mathcal{H}om\big(m_{A,t}^*(\Omega^1_{\mathbb{P}_{A,t}/\mathcal{J}_0}), O_{\mathbf{P}_A}\big)\Big) \to H^1\Big(\mathcal{P}_0, \mathcal{H}om\big(\tau_{A,t}^*(\Omega^1_{\mathbb{P}_{A,t}/\mathcal{J}_0}), O_{\mathcal{P}_0}\big)\Big)$$

is an isomorphism. Proceeding by induction on n it is clear that there exist unique deformations $P_{A,1}, \ldots, P_{A,r}$ of $\mathbb{P}_{A,1}, \ldots, \mathbb{P}_{A,r}$ as $\mathbf{P}^1$-bundles over $\mathcal{J}_0$ such that if $P_A := P_{A,1} \times_{\mathcal{J}} \cdots \times_{\mathcal{J}} P_{A,r}$, then $\mathcal{P} \to P_A$ deforms $\mathcal{P}_0 \to \mathbf{P}_A$.

ii. Consider the morphism $\mathcal{P} \to P_A$. The induced morphism $G_A \to P_A$ is an open immersion since it is an open immersion after base change to S_0. For any $t = 1, \ldots, r$ the image of G_A in $P_{A,t}$ is open and defines a $\mathbf{G}_m$-bundle $G_{A,t}$ over $\mathcal{J}$. The zero and infinity sections of $P_{A,t}$ over $\mathcal{J}$ are defined as the complement of $G_{A,t}$. The morphism $G_A \to G_{A,1} \times_{\mathcal{J}} \cdots \times_{\mathcal{J}} G_{A,r}$ is an isomorphism over $\mathcal{J}_0$. Hence it is an isomorphism over $\mathcal{J}$.

9.3 Definition. *For any $A \in \Lambda$ define $\mathcal{A}_A$ to be the sheaf $\mathcal{A}_{0,\ldots,0,1}$ restricted to the connected component $\mathbf{P}_A$ of $Z_{0,\ldots,0,1}$:*

$$\mathcal{A}_A := \mathcal{A}_{0,\ldots,0,1}|_{\mathbf{P}_A} \cong \mathcal{T}_{\mathbf{P}_A/S_0}[-0_1 \ldots - 0_r - \infty_1 \ldots - \infty_r] \cong \bigoplus_{t=1}^{r} \mathcal{O}_{\mathbf{P}_A}.$$

See 8.10 for the notation and an explanation of the isomorphisms.

For S equal to the dual numbers over S_0, Proposition 9.2 has a cohomological counterpart.

9.4 Proposition. *Let A be in Λ. The contraction map $\tau_A \colon \mathcal{P}_0 \longrightarrow \mathbf{P}_A$ and the inclusion $\mathbf{P}_A \subset \mathcal{P}_0$ induce a map*

$$\operatorname{Ext}^1_{\mathcal{O}_{\mathcal{P}_0}}(\Omega^1_{\mathcal{P}_0/\mathcal{J}_0}, \mathcal{O}_{\mathcal{P}_0}) \longrightarrow H^1(\mathbf{P}_A, \mathcal{A}_A).$$

It has the property that the induced morphism

$$H^1(\mathcal{P}_0, \mathcal{H}om_{\mathcal{O}_{\mathcal{P}_0}}(\Omega^1_{\mathcal{P}_0/\mathcal{J}_0}, \mathcal{O}_{\mathcal{P}_0})) \longrightarrow \operatorname{Ext}^1_{\mathcal{O}_{\mathcal{P}_0}}(\Omega^1_{\mathcal{P}_0/\mathcal{J}_0}, \mathcal{O}_{\mathcal{P}_0}) \longrightarrow H^1(\mathbf{P}_A, \mathcal{A}_A)$$

is the map defined by the left vertical arrow in the diagram 8.11 identifying $Z_{0,\ldots,0,1}$ with $\amalg_{A\in\Lambda}\mathbf{P}_A$.

Proof: Clear.

9.5 Definition. *We define some functors on the category of nilpotent thickenings of S_0. The notation is as in 7.1. Let*

i. *$\mathcal{D}ef\mathcal{G}r(\mathcal{J}_0)$ be the functor of deformations of $\mathcal{J}_0 \to S_0$ as an abelian scheme;*

ii. *$\mathcal{D}ef\mathcal{G}r(\mathbf{G}_0)$ be the functor fibred over $\mathcal{D}ef(\mathcal{J}_0)$ of deformations of $\mathcal{J}_0$, of $\mathbf{G}_0 \to S_0$ as semiabelian scheme over S_0 and of the group homomorphism $\mathbf{G}_0 \to \mathcal{J}_0$. Due to 2.3 this is equivalent to the deformation functor of $\mathcal{J}_0$, of $\mathbf{G}_0 \to \mathcal{J}_0$ as product as schemes over $\mathcal{J}_0$ of $\mathbf{G}_m$-bundles over $\mathcal{J}_0$ and of the identity section $S_0 \to \mathbf{G}_0$.*

iii. *$\mathcal{D}ef\mathcal{G}r(\mathcal{P}_0)$ be the functor of deformations*

   a. *$\mathcal{P} \to S$ of $\mathcal{P}_0 \to S_0$ as a flat schemes over S,*

   b. *$\hat{0} \colon S \to \mathcal{P}$ of the section $S_0 \to \mathcal{P}_0$ associated to $0 \in \Lambda$ in 7.1, We remark that given $\mathcal{P}$ and $\hat{0} \colon S \to \mathcal{P}$ as in (a) and (b) there exists a unique abelian scheme $\mathcal{J}$ over S deforming $\mathcal{J}_0$ so that $\mathcal{P}_0 \to \mathcal{J}_0$ extends to $\mathcal{P} \to \mathcal{J}$ and the image of $\hat{0}$ in $\mathcal{J}$ is the identity; see 8.6. Let $\tilde{G}$ be the open subscheme of $\mathcal{P}$ deforming $\mathbf{G}_0$. Then 9.2 implies* *that it is a $\mathbf{G}_m^r$-torsor over $\mathcal{J}$. The section $\hat{0}$ factors through $\tilde{G}$ and defines canonically the structure of a semiabelian scheme over S on it. We ask for a deformation*

c. *of the action of the semiabelian scheme $\tilde{G}$ over S deforming $\mathbf{G}_0 \to S_0$ on the smooth locus of $\mathcal{P} \to S$.*

iv. *$\mathcal{D}ef\mathcal{G}r(\mathcal{P}_0, \Lambda')$ be the functor of deformations*
   a. *$\mathcal{P} \to S$ of $\mathcal{P}_0 \to S_0$ as a flat scheme over S,*
   b. *$\hat{0}: S \to \mathcal{P}$ of the section $S_0 \to \mathcal{P}_0$ associated to $0 \in \Lambda$ in 7.1,*
   c. *of the S-group structure on the deformation $\mathcal{P}^{\mathrm{sm},0}$ of $\mathcal{P}_0^{\mathrm{sm},0}$ having $\hat{0}$ as identity section and of the action of $\mathcal{P}^{\mathrm{sm},0}$ on the smooth locus of $\mathcal{P} \to S$ deforming those defined in 7.1,*
   d. *of the action of Λ' such that the group structure on $\mathcal{P}^{\mathrm{sm},0}$ and the action of $\mathcal{P}^{\mathrm{sm},0}$ on the smooth locus of $\mathcal{P} \to S$ descend under the action of Λ'.*

v. *Since $\mathcal{P}_0 \to \mathcal{P}_0/\Lambda'$ is a local isomorphism,* (iv) *is equivalent to the functor $\mathcal{D}ef\mathcal{G}r(\bar{P}_0)$ of deformations*
   a. *$\bar{P} \to S$ of $\bar{P}_0 \to S_0$ as a flat scheme over S,*
   b. *$\bar{0}: S \to \mathcal{P}$ of the induced section $S_0 \to \mathcal{P}_0 \to \bar{P}_0$ associated to $0 \in \Lambda$;*
   c. *of the group structure on the deformation $\bar{P}^{\mathrm{sm},0}$ of $\bar{P}_0^{\mathrm{sm},0}$ having $\bar{0}$ as identity section and of the action of $\bar{P}^{\mathrm{sm},0}$ on the smooth locus of $\bar{P} \to S$ deforming those defined in 7.1. Note that $\bar{P}^{\mathrm{sm},0} = \tilde{G}$ in the notation of* (iii).

9.6 Remark. The functor $\mathcal{D}ef\mathcal{G}r(\mathcal{P}_0, \Lambda')$ is fibred over $\mathcal{D}ef\mathcal{G}r(\mathcal{J}_0)$. Moreover, the uniqueness stated in 8.6 implies that for $\mathcal{P}$ as in (iv) the induced map $\Lambda' \to \mathcal{P} \to \mathcal{J}$ is a group homomorphism.

9.7 Lemma. *For any element A of Λ and with the notation of 9.3, consider*

$$\psi_A: \mathbf{P}_0 \longrightarrow \mathbf{P}_A \qquad \text{and} \qquad d\psi_A: H^1(\mathbf{P}_0, \mathcal{A}_0) \longrightarrow H^1(\mathbf{P}_A, \mathcal{A}_A),$$

where ψ_A is an isomorphism as product as schemes over $\mathcal{J}_0$ of $\mathbf{P}^1$-bundles over $\mathcal{J}_0$ acting as a translation on $\mathcal{J}_0$ and $d\psi_A$ is the induced map. Then

i. *$d\psi_A$ does not depend on the choice of the isomorphism ψ_A;*

ii. *the relative tangent space of the morphism of functors $\mathcal{D}ef\mathcal{G}r(\mathcal{P}_0) \to \mathcal{D}ef(\mathcal{J}_0)$ is isomorphic to the submodule of elements a of the O_{S_0}-module $\mathrm{Ext}^1_{O_{\mathcal{P}_0}}(\Omega^1_{\mathcal{P}_0/\mathcal{J}_0}, O_{\mathcal{P}_0})$ defined as follows. Let*

$$\mathrm{Ext}^1_{O_{\mathcal{P}_0}}(\Omega^1_{\mathcal{P}_0/\mathcal{J}_0}, O_{\mathcal{P}_0}) \longrightarrow H^1(Z_{0,\ldots,0,1}, \mathcal{A}_{0,\ldots,0,1}) \xrightarrow{\sim} \prod_{A\in\Lambda} H^1(\mathbf{P}_A, \mathcal{A}_A)$$

be defined as in 9.4. If $(a_A)_{A\in\Lambda}$ is the image of a we have $d\psi_A(a_0) = a_A$.

Proof: The first statement follows from the fact that the group of automorphisms of $\mathbf{P}_0$ as product of $\mathbf{P}^1$-bundles over $\mathcal{J}_0$ and acting as translations on $\mathcal{J}_0$ is isomorphic to $\mathbf{G}_0$. By 9.1 the module $H^1(\mathbf{P}_0, \mathcal{A}_0)$ is isomorphic to r times $H^1(\mathcal{J}_0, O_{\mathcal{J}_0})$. Hence $\mathbf{G}_0$ acts trivially on it. Using 9.2 and 9.4, the second statement is a reinterpretation in cohomological terms of the following. Let $\mathcal{P} \to S$ be a deformation of $\mathcal{P}_0 \to S_0$ to a nilpotent thickening of S_0. Let A be an element of Λ. Let G_A (resp. $\tilde{G}$) be the open subscheme of $\mathcal{P}$ deforming $\mathbf{G}_A$ (resp. $\mathbf{G}_0$) and let $\mathcal{J} \to S$ the induced deformation of $\mathcal{J}_0 \to S_0$ as an abelian scheme. Then the action of $\mathbf{G}_0$ on the smooth locus of $\mathcal{P}_0 \to S_0$ extends to an action of $\tilde{G}$ on the smooth locus of $\mathcal{P} \to S$ if and only if G_A and $\tilde{G}$ are isomorphic as product of $\mathbf{G}_m$-bundles over $\mathcal{J}$.

9.8 *A natural transformation between the functors defined in 9.5 and its splitting.* Define the natural transformation of functors

$$\mathcal{D}ef\mathcal{G}r(\mathcal{P}_0) \longrightarrow \mathcal{D}ef\mathcal{G}r(\mathbf{G}_0)$$

by associating to any nilpotent thickening S of S_0 and to any element $\mathcal{P} \to S$ of the set $\mathcal{D}ef\mathcal{G}r(\mathcal{P}_0)(S)$ the open subscheme $\tilde{G}$ of $\mathcal{P}$ with its natural structure of semiabelian scheme over S.

This morphism admits a splitting defined as follows. Let S be a nilpotent thickening of S_0. Let $\tilde{G} \to \mathcal{J}$ be a deformation of $\mathbf{G}_0 \to \mathcal{J}_0$ corresponding to an S-valued point of $\mathcal{D}ef\mathcal{G}r(\mathbf{G}_0)$. By 9.2 we know that $\tilde{G}$ is isomorphic to a product of $\mathbf{G}_m$-bundles $\tilde{G}_1, \ldots, \tilde{G}_r$ over $\mathcal{J}$. We let $P_1, \ldots, P_r$ be the corresponding $\mathbf{P}^1$-bundles over $\mathcal{J}$. For any element A of Λ and any integer t with $1 \leq t \leq r$ define

$$P_A := P_1 \times_{\mathcal{J}} \ldots \times_{\mathcal{J}} P_r \qquad \text{and} \qquad 0_{A,t}\colon \prod_{t' \neq t} P_{t'} \to P_A, \qquad \infty_{A,t}\colon \prod_{t' \neq t} P_{t'} \to P_A,$$

where $0_{A,t}$ and $\infty_{A,t}$ are defined as in 2.1. For any element A of Λ and any integer t with $1 \leq t \leq r$, glue P_A along the section $\infty_{A,t}$ of P_A transversally to P_{A+f_t} along the section $0_{A+f_t,t}$ of P_{A+f_t} as in 7.1. This defines a scheme $\mathcal{P} \to \mathcal{J}$ corresponding to an S-valued point of $\mathcal{D}ef\mathcal{G}r(\mathcal{P}_0)$.
For any $t = 1, \ldots, r$ and any integer c_t let $q_{c_t,t}$ be an element of the ideal I defining the closed immersion $S_0 \subset S$. Let

$$q := (q_{c_t,t})_{t=1,\ldots,r,c_t\in\mathbf{Z}}.$$

Let x be a closed point of $\mathcal{P}_0$. Let A be an element of Λ, let x be a point of $\mathbf{P}_A \subset \mathcal{P}_0$ and let $\mathrm{Spec}(\mathbb{B}^x)$ be an open neighborhood of x as in 7.4. Then $\mathbb{B}^x$ is of the form

$$\mathbb{B}^x = \mathbb{A}^x[\gamma_1, \ldots, \gamma_r, \beta_1, \ldots, \beta_r]/(\gamma_1\beta_1, \ldots, \gamma_r\beta_r),$$

where the coordinate β_t (respectively γ_t) trivializes $P_t \to \mathcal{J}$ along the zero (respectively infinity) section. The coordinates are *canonically* defined once we choose a trivialization of $P_1, \ldots, P_r$ above the open subscheme $\mathrm{Spec}(\mathbb{A}^x)$ of $\mathcal{J}$. Suppose that $A = a_1 f_1 + \ldots + a_r f_r$. Define

$$\mathbb{B}_q^x := \mathbb{A}^x[\gamma_1, \ldots, \gamma_r, \beta_1, \ldots, \beta_r]/(\gamma_1\beta_1 - q_{a_1,1}, \ldots, \gamma_r\beta_r - q_{a_r,r}).$$

By construction the affine schemes $\mathrm{Spec}(\mathbb{B}_q^x)$ glue to define a scheme

$$\mathcal{P}_q \to \mathcal{J}$$

defining an S-valued point of $\mathcal{D}ef\mathcal{G}r(\mathcal{P}_0)$ whose image in $\mathcal{D}ef\mathcal{G}r(\mathbf{G}_0)$ is $\tilde{G}$.

9.9 Theorem. *Let S_0 be the spectrum of a field. Let S be a nilpotent thickening of S_0. Let I be the ideal defining $S_0 \subset S$. For any $n \in \mathbf{N}$ let $S_n \subset S$ be the closed subscheme defined by I^{n+1}.*

1. *All the elements of $\mathcal{D}ef\mathcal{G}r(\mathcal{P}_0)(S)$ arise by the construction given in 9.8.*
2. *The natural map*

$$\mathcal{D}ef\mathcal{G}r(\mathcal{P}_0, \Lambda')(S) \to \mathcal{D}ef\mathcal{G}r(\mathcal{P}_0)(S) \times_{\mathcal{D}ef\mathcal{G}r(\mathcal{P}_0)(S_n)} \mathcal{D}ef\mathcal{G}r(\mathcal{P}_0, \Lambda')(S_n)$$

is surjective.

Proof: 1. We can assume that $S = S_n$ for some $n \in \mathbf{N}$. The proof follows by induction on n. For $n = 0$, there is nothing to prove. Now we prove the inductive step. By inductive hypothesis the elements of $\mathcal{D}ef\mathcal{G}r(\mathcal{P}_0)(S_{n+1})$ arising as in 9.8 map surjectively to $\mathcal{D}ef\mathcal{G}r(\mathcal{P}_0)(S_n)$. By Schlessinger's deformation theory they are a homogeneous space under $\mathrm{Ext}^1_{O_{\mathcal{P}_0}}(\Omega^1_{\mathcal{P}_0/S_0}, O_{\mathcal{P}_0})$. The conclusion follows from the local-global spectral sequence associated to the Ext groups, from 7.5, from 9.7 and from 8.11.

2. It suffices to prove the statement for $S = S_{n+1}$. Let $\Lambda' = \mathbf{Z}l_1 \oplus \ldots \oplus \mathbf{Z}l_r$. Let $\mathcal{P}_n$ be an element of $\mathcal{D}ef\mathcal{G}r(\mathcal{P}_0, \Lambda')(S_n)$. It is easy to prove by induction on n that, as element in $\mathcal{D}ef\mathcal{G}r(\mathcal{P}_0)(S_n)$, it arises from the construction in 9.8 for some q invariant under the natural actions of $\mathbf{Z}l_1, \ldots, \mathbf{Z}l_r$. We can lift $\mathcal{P}_n$ to an element $\mathcal{P}$ of $\mathcal{D}ef\mathcal{G}r(\mathcal{P}_0)(S)$ via (1) lifting q such that the invariance under the action of $\mathbf{Z}l_1, \ldots, \mathbf{Z}l_r$ is kept. It suffices to prove that for any $t = 1, \ldots, r$ the automorphisms of $\mathcal{P}_n$ over S_n associated to l_t can be lifted to an automorphism of $\mathcal{P}$ over S. By construction such lifting is unobstructed locally on $\mathcal{P}$. The global obstruction is in $H^1(\mathcal{P}_0, \mathrm{Hom}_{O_{\mathcal{P}_0}}(\Omega^1_{\mathcal{P}_0/S_0}, O_{\mathcal{P}_0}))$ since $l_t^*(\Omega^1_{\mathcal{P}_0/S_0}) = \Omega^1_{\mathcal{P}_0/S_0}$. For any $A \in \Lambda$ let P_A and G_A be as in 9.2. By construction $P_A \cong P_{A+l_t}$ since $G_A \cong G_{A+l_t} \cong \tilde{G}$. The conclusion follows from 8.11 since $H^1(\mathcal{P}_0, \mathrm{Hom}_{O_{\mathcal{P}_0}}(\Omega^1_{\mathcal{P}_0/S_0}, O_{\mathcal{P}_0}))$ injects in $\prod_{A \in \Lambda} H^1(\mathbf{P}_A, \mathcal{A}_A)$ with the terminology of 9.3.

9.10 Proposition. *Let S be a nilpotent thickening of S_0. Let $\mathcal{P}_0 \to S_0$ be as in 7.1. Let $\mathcal{P}$ be an S-valued point of the functor $\mathcal{D}ef\mathcal{G}r(\mathcal{P}_0)$ defined in 9.5. Let $\tilde{G} \to S$ be the associated S-valued point of $\mathcal{D}ef\mathcal{G}r(\mathbf{G}_0)$ via 9.8. For any A in Λ let G_A be the $\mathbf{G}_m^r$-bundle over $\mathcal{J}$ defined in 9.2. Then there is a unique* *action*

$$\mathcal{P} \underset{S}{\times} \tilde{G} \longrightarrow \mathcal{P}$$

extending the action of $\mathbf{G}_0$ *on* $\mathcal{P}_0 \to \mathcal{J}_0$ *defined in 7.1 and extending for any* $A \in \Lambda$ *the action of* $\tilde{G}$ *on* G_A *defined in 9.5. Moreover, if* $\mathcal{P} \in \mathcal{D}ef\mathcal{G}r(\mathcal{P}_0, \Lambda')$ *the action of* $\tilde{G}$ *on* $\mathcal{P}$ *descends to an action on* $\bar{P} = \mathcal{P}/\Lambda'$

$$\bar{P} \underset{S}{\times} \tilde{G} \longrightarrow \bar{P}.$$

Proof: Let I be the ideal defining S_0 in $S = \mathrm{Spec}(E)$. Let $x \in \mathcal{P}_0$. Let $\mathrm{Spec}(\mathbb{A})$ be an open neighborhood of the image of x in $\mathcal{J}$. Then there exists an open neighborhood $\mathrm{Spec}(\mathbb{B})$ of x in $\mathcal{P}$ of the form

$$\mathbb{B} := \mathbb{A}[\gamma_1, \ldots, \gamma_r, \beta_1, \ldots, \beta_r]/(\gamma_1\beta_1 - q_1, \ldots, \gamma_r\beta_r - q_r),$$

with $q_1, \ldots, q_r \in I$, deforming a neighborhood of x in $\mathcal{P}_0$ of the type described in 7.4. The intersection of $\mathrm{Spec}(\mathbb{B})$ with $\mathcal{P}^{\mathrm{sm}}$ is $\mathrm{Spec}(\mathbb{C})$ with

$$\mathbb{C} := \prod \mathbb{A}\left[z_1, \ldots, z_h, \beta_i, \frac{1}{\beta_i}, \gamma_k, \frac{1}{\gamma_k}\right]_{i \in J, k \notin J},$$

where the product is taken over all subsets J of $\{1, \ldots, r\}$. The basic remark is that $\mathbb{B}$ maps injectively in $\mathbb{C}$. In particular, the morphism $O_{\mathcal{P}} \to O_{\mathcal{P}^{\mathrm{sm}}}$ is injective. This implies that there exists at most one action of $\tilde{G}$ on $\mathcal{P}$ compatible with the action of $\tilde{G}$ on $\mathcal{P}^{\mathrm{sm}}$. It is then clear that if $\mathcal{P} \in \mathcal{D}ef\mathcal{G}r(\mathcal{P}_0, \Lambda')$ and an action of $\tilde{G}$ on $\mathcal{P}$ exists as claimed, it descends to an action of $\tilde{G}$ on $\bar{P}$. Since $\tilde{G}$ and $\mathcal{P}$ are schemes over $\mathcal{J}$ and $\tilde{G}$ acts on $\mathcal{J}$, it suffices to prove that the action of the torus T, defined as the kernel of $\tilde{G} \to \mathcal{J}$, extends from an action on $\mathcal{P}^{\mathrm{sm}}$ to an action on $\mathcal{P}$. By 9.2 the character group of T is isomorphic to $\mathbf{Z}^r = \mathbf{Z}e_1 \oplus \ldots \oplus \mathbf{Z}e_r$ and the action on $\mathbb{C}$ is given by the $\mathbf{Z}^r$-grading defined by requiring that for any $i = 1, \ldots, r$ the element γ_i (respectively β_i) is the $-e_i$ (respectively e_i) piece of the grading. This grading clearly extends to $\mathbb{B}$ defining an action of T on $\mathrm{Spec}(\mathbb{B})$ as wanted.

10 From the formal scheme $\mathcal{P}$ to the scheme $\widetilde{P}$

10.1 *Notation.* Let $S = \mathrm{Spec}(E)$ be the spectrum of a noetherian integral domain with fraction field K and let I be an ideal of E defining the closed subscheme S_0. Assume E to be complete and separated with respect to the I-adic topology and let $S_n := \mathrm{Spec}(E/I^{n+1})$. Let $\mathcal{S}$ be the affine formal scheme associated to E and the ideal I. If X is a scheme over S or a formal scheme over $\mathcal{S}$, denote by X_n the base change of X to S_n.

10.2 *The main result: the scheme* $\widetilde{P}$. Let $(\mathcal{P}_0, \Lambda')$ be as in 7.1. Let $(\mathcal{P}_n)_{n \in \mathbf{N}}$ be a compatible system of S_n-valued points of the functor $\mathcal{D}ef\mathcal{G}r(\mathcal{P}_0)$ defined in 9.5. In particular, $\mathcal{P}_n \to S_n$ is a compatible system of flat deformations of $\mathcal{P}_0 \to S_0$. Let $\mathcal{J}_n$ be the unique abelian scheme over S_n such that $\mathcal{P}_0 \to \mathcal{J}_0$ extends to a morphism $\mathcal{P}_n \to \mathcal{J}_n$ and the image in $\mathcal{J}_n$ of the section $\hat{0}\colon S_n \to \mathcal{P}_n$ is the identity of $\mathcal{J}_n$. Its existence is guaranteed by 8.6. Suppose that

i. the formal scheme $(\mathcal{J}_n)_{n\in\mathbf{N}}$ over $\mathcal{S}$ is the I-adic completion of an abelian scheme $\mathcal{J}$ over S. In particular, $\mathcal{J} \times_S S_0 \cong \mathcal{J}_0$;
ii. the formal scheme $(\mathcal{P}_n)_{n\in\mathbf{N}}$ over $\mathcal{S}$ does not admit a $\mathcal{S}$-valued section factoring through the complement of the smooth locus of $\mathcal{P}_0 \to S_0$.

The rest of this section is devoted to the construction of an integral scheme $\widetilde{P}$ flat and locally of finite type over $\mathcal{J}$ such that the associated I-adic completion is isomorphic to $(\mathcal{P}_n)_{n\in\mathbf{N}}$ as formal schemes over $(\mathcal{J}_n)_{n\in\mathbf{N}}$. See 10.10 for further important properties of $\widetilde{P}$.

10.3 *Step 1 of the construction of* $\widetilde{P}$. For any element A of Λ there exist schemes G_A and P_A over $\mathcal{J}$ such that

i. G_A is the product as schemes over $\mathcal{J}$ of $\mathbf{G}_m$-bundles $G_{A,1},\dots,G_{A,r}$;
ii. P_A is the product as schemes over $\mathcal{J}$ of $\mathbf{P}^1$-bundles $P_{A,1},\dots,P_{A,r}$ over $\mathcal{J}$ whose associated $\mathbf{G}_m$-bundles are $G_{A,1},\dots,G_{A,r}$;
iii. their base changes to S_0 are the schemes $\mathbf{G}_A$ and $\mathbf{P}_A$ respectively defined in 7.1;
iv. the natural contraction morphism $\mathcal{P}_0 \to \mathbf{P}_A$ defined in 7.3 extends to a morphism from $\mathcal{P}$ to the formal scheme associated to P_A.

Moreover if A and B are in Λ, then $G_{A,t} \cong G_{B,t}$ as $\mathbf{G}_m$-bundles over $\mathcal{J}$ for any $t = 1,\dots,r$. In particular, $G_A \cong G_B$ and $P_A \cong P_B$.

Proof: See 9.2.

Let A be in Λ. Define

$$N_A := \cup_{t=1,\dots,r}\big(0_{A,t} \cup \infty_{A,t}\big),$$

where the closed subschemes $0_{A,1},\dots,0_{A,r}$ and $\infty_{A,1},\dots,\infty_{A,r}$ of P_A are defined as in 2.1.

10.4 Definition. *Let* $\overline{a} := (a_1, c_1, \dots, a_r, c_r)$ *with* a_i *and* c_i *integers such that* $a_i \le c_i$ *for* $i = 1,\dots,r$. *Recall from 7.1 that* $\Lambda = \mathbf{Z}f_1 \oplus \cdots \oplus \mathbf{Z}f_r$. *Define* a *to be the finite set of elements of* Λ *of the form* $A = b_1 f_1 + \ldots + b_r f_r$ *where* b_i *is an integer such that* $a_i \le b_i \le c_i$ *for each* $i = 1,\dots,r$.

10.5 *Step 2.* The notation is as in 10.3. Let a be a finite set of elements of Λ. Let

$$P_a := \prod_{A\in a} P_A \qquad \text{and} \qquad G_a := \cup_{A\in a}\Big(G_A \times \big(\prod_{B\in a \mid B\neq A} P_B\big)\Big);$$

the products are taken as schemes over $\mathcal{J}$. Define the closed subscheme of P_a

$$N_a := \cap_{A\in a}\Big(N_A \times \big(\prod_{B\in a|B\neq A} P_B\big)\Big).$$

Remark that $G_a = P_a \backslash N_a$. For any $A \in a$ define the open subscheme P_A^c of P_A as the complement of all subschemes of the form $0_{A,i}$ or $\infty_{A,j}$ for $i,j = 1,\ldots,r$ such that their base changes to S_0 are in the intersection in $\mathcal{P}_0$ of $\mathbf{P}_A$ and $\cup_{B\notin a}\mathbf{P}_B$. Define

$$P_a^c := \prod_{A\in a} P_A^c.$$

The products are taken as schemes over $\mathcal{J}$. Let a be as in 10.4. There exists a closed subscheme

$$P_{\underline{a}} \hookrightarrow P_a$$

such that putting

$$U_{\underline{a}} := P_{\underline{a}} \cap P_a^c, \qquad N_{\underline{a}} := U_{\underline{a}} \cap N_a, \qquad G_{\underline{a}} := P_{\underline{a}} \cap G_a = U_{\underline{a}} \backslash N_{\underline{a}},$$

where the intersections are taken in P_a

i. for any non-negative integer n the morphism $\mathcal{P}_n \to P_{a,n}$ factors via $P_{\underline{a},n}$;

ii. the schemes $P_{\underline{a}}$ and $U_{\underline{a}}$ are flat over $\mathcal{J}$;

iii. for any non-negative integer n the map from the inverse image of $U_{\underline{a},n}$ in $\mathcal{P}_n$ to $U_{\underline{a},n}$ is an isomorphism;

iv. for any non-negative integer n the union over all subsets a of Λ as in 10.4 of the open subschemes $U_{\underline{a},n}$ is $\mathcal{P}_n$.

Proof of 10.5 Let x be a point of $\mathcal{P}_0$ such that x is in $\cup_{A\in a}\mathbf{P}_A \hookrightarrow \mathcal{P}_0$. Consider an affine formal neighborhood V^x of x in $\mathcal{P}$ of the form

$$V^x := \mathrm{Spf}(\mathbb{B}^x) \qquad \mathbb{B}^x := \mathbb{A}\{\gamma_1,\ldots,\gamma_r,\beta_1,\ldots,\beta_r\}/(\gamma_1\beta_1 - q_1,\ldots,\gamma_r\beta_r - q_r),$$

deforming over $\mathcal{S}$ an open subscheme of $\mathcal{P}_0$ of the type defined in 7.4. For any $B \in a$ the morphism from V^x to the formal scheme associated to P_B factors via a formal affine subscheme of the form V_B^x with

$$V_B^x := \mathrm{Spf}(\mathbb{B}_B^x) \qquad \mathbb{B}_B^x := \mathbb{A}\{\gamma_u^B, \beta_w^B\}_{u\in\Upsilon_B, w\in\Xi_B},$$

where Υ_B and Ξ_B are suitable disjoint subsets of $\{1,\ldots,r\}$ such that their union is the whole set $\{1,\ldots,r\}$. The convention is that for each $s = 1,\ldots,r$ the coordinate γ_s (resp. β_s) is a local parameter trivializing the $\mathbf{P}^1$-bundle $P_{B,s}$ along the infinity (resp. zero) section in a neighborhood of the image of x in $\mathcal{J}$. If $x \in \mathbf{P}_A$ with $A \in a$, the morphism from V^x to V_A^x is defined by $\gamma_u^A \mapsto \gamma_u$ and $\beta_w^A \mapsto \beta_w$. Let

$$W^x := \mathrm{Im}\Big(V^x \longrightarrow \prod_{A|x\in\mathbf{P}_A} V_A^x\Big).$$

Then

$$W^x := \mathrm{Spf}(\mathbb{B}^x_{\underline{a}}) \qquad \mathbb{B}^x_{\underline{a}} := \mathrm{Im}\Big(\hat{\otimes}_{A\in\underline{a}|x\in\mathbf{P}_A}\mathbb{B}^x_A \longrightarrow \mathbb{B}^x\Big);$$

the tensor product is taken over $\mathbb{A}$ and $\hat{\otimes}$ stands for the I-adically completed tensor product. In particular, W^x is flat over $\mathcal{S}$. Let

$$Z^x := \mathrm{Im}\Big(V^x \longrightarrow \prod_{B\in a} V^x_B\Big).$$

If x is in $U_{\underline{a},0}$, then $\mathbb{B}^x_{\underline{a}} = \mathbb{B}^x$ and the morphisms $V^x \to Z^x \to W^x$ are isomorphisms. If the set a is of the type described in 10.4, the lemma below shows that the morphism $\mathbb{B}^x_B \to \mathbb{B}^x$ factors via the subring $\mathbb{B}^x_{\underline{a}}$ of $\mathbb{B}^x$ for any B in a. In particular, the morphism

$$Z^x \longrightarrow W^x$$

is an isomorphism. Define

$$P_{\underline{a},n} = \cup_x\big(Z^x \times_{\mathcal{S}} S_n\big)$$

where the union is taken over all $x \in \mathcal{P}_0$ such that x is in the image of $\mathbf{P}_A \to \mathcal{P}_0$ for some $A \in a$. Then $P_{\underline{a},n}$ is a closed subscheme of $P_{a,n}$ since this holds for $n = 0$. Moreover, $P_{\underline{a},n}$ is flat over $\mathcal{J}_n$ and $P_{\underline{a},n+1} \times_{S_{n+1}} S_n = P_{\underline{a},n}$. The scheme $P_{\underline{a}}$ is obtained, using Grothendieck's existence theorem [11, Theorem 5.4.5], as the algebraization over S of the formal scheme associated to the system $\big(P_{\underline{a},n}\big)_{n\in\mathbf{N}}$. It is flat over $\mathcal{J}$ since this holds for the associated formal scheme.

10.6 Lemma. *Let a be a set of the type considered in 10.4. The notation is as in the proof of 10.5. Let $B \in a$. The morphism $\mathbb{B}^x_B \to \mathbb{B}^x$ is injective and factors via $\mathbb{B}^x_{\underline{a}}$.*

Proof: For any $D \in a$ there exists a unique $A \in a$ such that $x \in \mathbf{P}_A$ and $D = A + m_1 f_1 + \ldots + m_r f_r$ with $\sum_i |m_i|$ minimal. Define $d(D,x) := \sum_i |m_i|$. Proceed by induction on $d(B,x)$. If $d(B,x) = 0$, then $x \in \mathbf{P}_B$ and the claim is obvious. Suppose that $d(B,x) \geq 1$. There exist $t = 1,\ldots,r$ such that either $m_t \leq -1$ and if $C = B + f_t$, then $d(C,x) = d(B,x) - 1$ or $m_t \geq 1$ and if $C = B - f_t$, then $d(C,x) = d(B,x) - 1$. Fix such t. Suppose we are in the case $m_t \leq -1$. The proof in the other case is the same. Remark that $C \in a$. Let y be the image of x via $\tau_B\colon \mathcal{P}_0 \to \mathbf{P}_B$. Consider $y \in \mathbf{P}_B \cap \mathbf{P}_C \subset \mathcal{P}_0$. Then

$$\mathbb{B}^y_B = \mathbb{B}^x_B = \mathbb{A}\{\gamma^B_u, \beta^B_w\}_{u\in\Upsilon_B, w\in\Xi_B}$$

and

$$\mathbb{B}^y_C = \mathbb{A}\{\gamma^C_u, \beta^C_w\}_{u\in\Upsilon_C, w\in\Xi_C}$$

and

$$\mathbb{B}^x_C = \mathbb{A}\{\gamma^C_u, \beta^C_w, \gamma^C_t\}_{u\in\Upsilon_C, w\in\Xi_B} \qquad \text{with} \quad \gamma^C_t = \frac{1}{\beta^C_t}.$$

Let M be a field containing the fraction fields of $\mathbb{B}^y$, of $\mathbb{B}^x_C$ and of $\mathbb{B}^x_B$. The ring $\mathbb{B}^x_B$ coincides, via the inclusion in M, with the subring of $\mathbb{B}^x_C$, embedded in M, defined by $\mathbb{A}\{\gamma^C_u, \beta^C_w, q_t\gamma^C_t\}_{u\in\Upsilon_C, w\in\Xi_B}$ with the identifications $\Upsilon_B = \Upsilon_C \cup \{t\}$ and $\Xi_C = \Xi_B \cup \{t\}$ and $\beta^B_w = \beta^C_w$ if $w \neq t$ and $\gamma^B_u = \gamma^C_u$ if $u \neq t$, while $\gamma^B_t = \frac{q_u}{\beta^C_t}$. There exist elements $x = x_1, \ldots, x_n = y$ in $\cup_{A\in a}\mathbf{P}_A \subset \mathcal{P}_0$ and formal open neighborhoods $V^x = V^{x_1}, \ldots, V^{x_n} = V^y$ of $x_1, \ldots, x_n$ as in 10.5 such that for any $i = 1, \ldots, n-1$ the formal open subschemes V^{x_i} and $V^{x_{i+1}}$ of $\mathcal{P}$ have non-empty intersection and for any $i = 1, \ldots, n$ we have $d(C, x_i) \leq d(B, x) - 1$. Let L be the composite of M and of the function fields of the formal schemes V^{x_i}'s. Applying the induction hypothesis we conclude that $\mathbb{B}^x_C \to \mathbb{B}^{x_i} \to L$ is injective. By the choice of the V^{x_i}'s we conclude that $\mathbb{B}^x_B \to \mathbb{B}^{x_i} \to L$ factors via M and is the injection defined above composed with $M \subset L$. We conclude that $\mathbb{B}^x_B$ maps injectively to $\mathbb{B}^x$ and factors via the subring $\mathbb{B}^x_a$.

10.7 Remark. Let $\mathcal{A}$ be a subset of Λ. Let a be a finite subset of $\mathcal{A}$ and let $\langle a\rangle$ be the smallest subset of Λ of the type considered in 10.4 containing a. Suppose that $a = \mathcal{A} \cap \langle a\rangle$. For any A in a, define $P^{\mathcal{A}}_A$ as the complement in P_A of all subschemes of the form $0_{A,i}$ or $\infty_{A,j}$ for $i, j = 1, \ldots, r$ such that their base changes to S_0 are *not* in the intersection in $\mathcal{P}_0$ of $\mathbf{P}_A$ and $\mathbf{P}_B$ for some $B \in \mathcal{A}$, with B different from A. Let

$$P^{\mathcal{A}}_a := \prod_{A\in a} P^{\mathcal{A}}_A.$$

It contains P^c_a. For any $n \in \mathbf{N}$ let

$$P_{\underline{a},n} \qquad \text{and} \qquad U_{\underline{a},n}$$

be the image of $P_{\underline{\langle a\rangle},n}$ (resp. $U_{\underline{\langle a\rangle},n}$) in P_a. Let

$$P^{\mathcal{A}}_{\underline{a},n} := P_{\underline{a},n} \cap P^{\mathcal{A}}_a.$$

It follows as in the proof of 10.5 that $P^{\mathcal{A}}_{\underline{a},n}$ is flat over $\mathcal{J}_n$.

10.8 *Step 3.* Let a and a' be subsets of Λ as in 10.4. Suppose that a is contained in a'. The natural projection morphism $P_{a'} \to P_a$ induces a morphism

$$\alpha_{\underline{a'},\underline{a}}\colon P_{\underline{a'}} \to P_{\underline{a}}.$$

Moreover, the latter morphism is a blowing-up centered on a closed subscheme of the complement of P^c_a and it is an isomorphism over K.

Proof: Let $a' = (a'_1, c'_1, \ldots, a'_r, c'_r)$ and $a = (a_1, c_1, \ldots, a_r, c_r)$. After possibly renumbering and changing the signs of the f_i's in 7.1 and proceeding by induction, we can suppose that $c'_1 = c_1 + 1$, while $a'_i = a_i$ for any i and $c'_i = c_i$ for any $i \neq 1$. If A is in a, then $\mathbf{P}_A$, viewed as a subscheme of $\mathcal{P}_0$, is mapped isomorphically to its image both in $P_{\underline{a}',0}$ and in $P_{\underline{a},0}$. If A' is in a' but not in a, there exists a unique A in a such that $A' = A + f_1$. In this case $\mathbf{P}_{A'}$ maps isomorphically to its image in $P_{\underline{a}',0}$, while it is the contraction to the section $\infty_{A,1}$ of $\mathbf{P}_A$ defined in 10.5. Let x be a point on $\mathbf{P}_{A'} \to \mathcal{P}_0$. We use the notation of the proof of 10.5. On the thickening S_n of S_0, the image $V^x_{a,n}$ in $P_{\underline{a},n}$ of the open neighborhood V^x_n of x is the Spec of

$$\mathbb{B}^x_{a,n} := \mathbb{A}_n[\gamma_1, \beta_1, \gamma_u, \beta_u, \gamma_i, \beta_j]/(\beta_1, \gamma_u\beta_u - q_u)$$

for suitable disjoint subsets of u's, i's and j's of $\{2, \ldots, r\}$ whose union is the whole set $\{1, \ldots, r\}$. The image $V^x_{a',n}$ of V^x_n in $P_{a'}$ maps to $V_{a,n}$, it is the inverse image of $V_{a,n}$ in $P_{\underline{a}'}$ and the map $V_{a',n} \to V_{a,n}$ is described by

$$\mathbb{B}^x_{a,n} \longrightarrow \mathbb{A}_n[\gamma_1, \beta_1, \gamma_u, \beta_u, \gamma_i, \beta_j]/(\gamma_1\beta_1 - q_1, \gamma_u\beta_u - q_u)$$

$$\mathbb{B}^x_{a,n} \longrightarrow \mathbb{A}_n[\gamma_1, \frac{1}{\beta_1}, \gamma_u, \beta_u, \gamma_i, \beta_j]/(\gamma_1 - \frac{q_1}{\beta_1}, \gamma_u\beta_u - q_u).$$

The natural gluing of the Spec of the rings appearing on the right hand side of the morphisms above defines $V^x_{a',n}$. The element q_1 and the elements q_u's are in I. The crucial fact is that they are *not zero* by condition (ii) in 10.2. In fact, they determine the discriminant locus of $\mathcal{P} \to \mathcal{S}$. The morphism of formal schemes associated to $\alpha_{\underline{a}',\underline{a}}$ is a blowing-up centered above the closed subscheme of S defined by $q_1 = 0$. By [11, Theorem 5.4.1] the same must hold for $\alpha_{\underline{a}',\underline{a}}$.

10.9 *Step 4.* Let a and a' be finite subsets of Λ as in 10.4. Suppose that $a \subset a'$. Let $\alpha_{\underline{a}',\underline{a}}$ be as in 10.8. Let $U_{\underline{a}}$, of $N_{\underline{a}}$ and of $G_{\underline{a}}$ be as in 10.5. Then

i. $\alpha^{-1}_{\underline{a}',\underline{a}}(U_{\underline{a}})$ is contained in $U_{\underline{a}'}$ and $\alpha^{-1}_{\underline{a}',\underline{a}}(N_{\underline{a}})$ is contained in $N_{\underline{a}'}$;

ii. the inclusion $\alpha^{-1}_{\underline{a}',\underline{a}}(U_{\underline{a}}) \subset U_{\underline{a}'}$ in (i) is an isomorphism over K;

iii. the maps from $\alpha^{-1}_{\underline{a}',\underline{a}}(U_{\underline{a}})$ to $U_{\underline{a}}$ and from $\alpha^{-1}_{\underline{a}',\underline{a}}(N_{\underline{a}})$ to $N_{\underline{a}}$ are isomorphisms;

iv. the smooth locus of $U_{\underline{a}} \to S$ is the scheme $G_{\underline{a}}$ and the inclusion $G_{\underline{a}} \subset U_{\underline{a}}$ is an isomorphism over K.

In particular, for $A = a_1 f_1 + \ldots + a_r f_r \in \Lambda$, putting $\overline{a} := (a_1, a_1, \ldots, a_r, a_r)$, the morphism $U_{\underline{a}} \to G_A$ is an isomorphism and $N_{\underline{a}}$ is empty.

Proof: First we discuss the statements regarding $U_{\underline{a}}$ and $U_{\underline{a}'}$. The first statement follows from the definition of the open subschemes $U_{\underline{a}}$ and $U_{\underline{a}'}$. The second statement follows remarking that $\alpha_{\underline{a}',\underline{a}}$ is a blowing-up centered outside $U_{\underline{a}}$ due to 10.8. The third statement follows from the description of $\alpha_{\underline{a}',\underline{a}}$ in local coordinates given in 10.8. The statements about $N_{\underline{a}}$ and $N_{\underline{a}'}$ follow in the same way. Finally we prove that the smooth locus of $U_{\underline{a}} \to S$ is the complement of $N_{\underline{a}}$. Let $z \in U_{\underline{a}}$. Let y be a closed point of $P_{\underline{a}}$ lying in the closure of z.

Since $P_{\underline{a}}$ is proper over S, the point y is mapped to S_0. The completion of $P_{\underline{a}}$ at y is isomorphic to a ring of the form

$$\hat{R} := \hat{O}_{\mathcal{J},Y}[\![z_w, \gamma_u, \beta_u, \gamma_i, \beta_j]\!]/(\gamma_u\beta_u - q_u)$$

for suitable disjoint subsets of w's, of u's, of i's and of j's of $\{1, \ldots, r\}$ whose union is the whole $\{1, \ldots, r\}$ and Y is the image of y in $\mathcal{J}$. The completion at x of the ideal sheaf defining N_a is $\prod_u(\gamma_u\beta_u)$. The statement is now clear.

10.10 *Step 5.* Let a and a' be finite subsets of Λ as in 10.4. The open subscheme $U_{\underline{a \cap a'}}$ can be identified via 10.8 with an open subscheme of $U_{\underline{a}}$ and an open subscheme of $U_{\underline{a'}}$. Hence we can glue $U_{\underline{a}}$ and $U_{\underline{a'}}$ along such open subschemes. We construct the scheme $\widetilde{P}$ by gluing the schemes $U_{\underline{a}}$ for all finite subsets a of Λ via the procedure just explained. In the same way we can glue the closed subschemes $N_{\underline{a}}$ of $U_{\underline{a}}$ defined in 10.5 and we get a closed subscheme $\widetilde{N}$ of $\widetilde{P}$. From the previous Steps we obtain the following properties

i. $\widetilde{P}$ is an integral scheme flat and locally of finite type over $\mathcal{J}$ and the complement of $\widetilde{N}$ is the smooth locus of $\widetilde{P} \to S$;
ii. the base change of $\widetilde{P} \to S$ to S_n is isomorphic to $\mathcal{P}_n$;
iii. for any element A of Λ there is a morphism $\widetilde{P} \to P_A$ such that the base change of the morphism from $\widetilde{P} \to P_A$ to K is the inclusion $G_{A,K} \to P_{A,K}$;
iv. with the notation of 10.5, for any finite subset a of Λ of the type considered in 10.4, the image of $\widetilde{P}$ in P_a via the morphism induced by (iii) exists and coincides with the closed subscheme $P_{\underline{a}}$ of P_a. Moreover, $P_{\underline{a}}$ is integral and flat over S;
v. the inverse image of the open subscheme $U_{\underline{a}}$ of $P_{\underline{a}}$ in $\widetilde{P}$ maps isomorphically to $U_{\underline{a}}$. Moreover, $\widetilde{P}$ is covered by open subschemes of the form $U_{\underline{a}}$;
vi. with the notation of 10.3, if $a = \{A\}$ then $U_{\underline{a}} = G_A$ and $P_{\underline{a}} = P_A$. With these identifications the smooth locus of $\widetilde{P} \to S$ is $\cup_{A \in \Lambda} G_A$.

10.11 *Step 6: the action of Λ'.* Suppose that the schemes $(\mathcal{P}_n \to S_n)_{n \in \mathbf{N}}$ are a compatible system of S_n-valued points of the functor $\mathcal{D}ef\mathcal{G}r(\mathcal{P}_0, \Lambda')$ defined in 9.5 and satisfy the conditions of 10.2. Then there is an action of Λ' on $\widetilde{P} \to S$ and on $\widetilde{N} \to S$ which extends the natural action on their I-adic completions.

Proof: It is enough to show how any element $l = c_1 l_1 + \ldots + c_r l_r$ of Λ' acts on the system $\left(P_{\underline{a}}, N_{\underline{a}}\right)_a$. It suffices to remark that for any element A of Λ and any non-negative integer n, there is a unique isomorphism $P_{A,n} \to P_{A+l,n}$ such that the following diagram commutes:

$$\begin{array}{ccc} \mathcal{P}_n & \xrightarrow{+l} & \mathcal{P}_n \\ \downarrow & & \downarrow \\ P_{A,n} & \xrightarrow{\sim} & P_{A+l,n}, \end{array}$$

where the vertical arrows are defined as in (iv) of 10.3.

11 The generic fiber of $\widetilde{P} \to S$ as a 1-motive

The notation is as in 10.1. Consider as in 10.2 a formal scheme $(\mathcal{P}_n \to S_n)_{n\in\mathbf{N}}$, deforming the scheme $\mathcal{P}_0 \to S_0$ defined in 7.1, associated to compatible S_n-valued points of the functor $\mathcal{D}ef\mathcal{G}r(\mathcal{P}_0)$. Under the assumptions of 10.2 we have shown in the previous section that any such formal scheme is the I-adic completion of a scheme $\widetilde{P}$ over S. See 10.2 and 10.10 for its properties. In this section we are concerned with the properties of the fiber of $\widetilde{P}$ over the function field K of S. This is a semiabelian scheme $\tilde{G}_K$ over Spec(K). We show in 11.1 how $\widetilde{P}$, and hence $(\mathcal{P}_n \to S_n)_n$, can be recovered from $\tilde{G}_K$. In 11.6 we refine the equivalence in 11.1 taking into account actions of the sublattice Λ' of Λ, defined in 7.1, on $(\mathcal{P}_n \to S_n)_n$ and on $\tilde{G}_K$. It turns out that the actions of Λ' on $\tilde{G}_K$ which arise in this way come from homomorphisms, also called 1-motives, $\Lambda' \to \tilde{G}_K$.

11.1 Theorem. *The notation is as in 7.1. The following two types of data determine uniquely one another.*

I *The data consisting of*

Ia. *a compatible system $(\mathcal{P}_n)_{n\in\mathbf{N}}$ of S_n-points of the functor $\mathcal{D}ef\mathcal{G}r(\mathcal{P}_0)$ defined in 9.5 satisfying conditions* (i) *and* (ii) *of 10.2.*

Ib. *a compatible system of rigidifications $\big((\amalg_\Lambda S_n) \to \mathcal{P}_n\big)_{n\in\mathbf{N}}$ deforming the rigidification defined in 7.1 for $n=0$. We require that the section corresponding to $0 \in \Lambda$ is the section $\hat{0}\colon S_n \to \mathcal{P}_n$.*

II *The data consisting of*

IIa. *an abelian scheme $\mathcal{J}$ over S and a semiabelian scheme $\tilde{G}$ over S deforming the semiabelian scheme $\mathbf{G}_0$ and sitting in an exact sequence*

$$0 \longrightarrow \mathbf{G}^r_{m,S} \longrightarrow \tilde{G} \longrightarrow \mathcal{J} \longrightarrow 0.$$

Let $\tilde{G}$ split as product of $\mathbf{G}_m$-bundles $\tilde{G}_1 \times \ldots \times \tilde{G}_r$ over $\mathcal{J}$. For $1 \le t \le r$, let $p_t\colon \tilde{G} \to \tilde{G}_t$ be the projection.

IIb. *points $x_A \in \tilde{G}(K)$ for all $A \in \Lambda$ satisfying the following conditions;*

i. *for any $A \in \Lambda$ there is a section $y_A\colon S \to \mathcal{J}$ such that $y_{A,K}$ is the image of x_A in $\mathcal{J}_K$;*

ii. *for integers s,t with $1 \le s,t \le r$ and for any $A \in \Lambda$ denote by $I_{A,t,s}$ the fractional ideal of S determined by the section $p_{s,K}(x_{A+f_t}x_A^{-1})$ over K of the $\mathbf{G}_m$-bundle $(y_{A+f_t} - y_A)^*(\tilde{G}_s)$ over S. Then $I_{A,t,s}$ is a non-trivial ideal of O_S contained in I if $s=t$, it coincides with O_S if $s \ne t$ and it is equal to $I_{A+f_{t'},t,s}$ for any $t' = 1,\ldots,r$ different from t.*

If we change datum (Ib) *datum* (IIa) *does not change.*

Proof: First we describe how to obtain data (II) from data (I). Let $\widetilde{P}$ be the scheme over S algebraizing $(\mathcal{P}_n \to S_n)_n$ as in 10.2. Let $\amalg_\Lambda S \to \widetilde{P}$ be the sections algebraizing those defined in (Ib). Let A be in Λ. Composing the section of $\widetilde{P} \to S$ corresponding to A with the morphism $\widetilde{P} \to P_A$ defined in (iii) of 10.10, we get a section $0_A: S \to P_A$ factoring through G_A. Define $\tilde{G}$ as G_0, where the subscript 0 stands for the element $0 \in \Lambda$. In particular, $\tilde{G}$ is a $\mathbf{G}^r_{m,S}$-bundle over $\mathcal{J}$ by 10.3. By 2.3, there is a unique group structure on $\tilde{G}$ having the section $0_0: S \to \tilde{G}$ as identity element and deforming the group structure on $\tilde{G} \times_S S_0 := \mathbf{G}_0$. This determines datum (IIa). Let A be a element of Λ. By (iii) of 10.10 we get a morphism s_A

$$\tilde{G}_K = G_{0,K} \overset{\sim}{\longleftarrow} \widetilde{P}_K \overset{\sim}{\longrightarrow} G_{A,K} \longrightarrow G_A.$$

Define $x_A := s_A^{-1}(0_{A,K})$ and y_A as the image in $\mathcal{J}$ of the section of $\widetilde{P} \to S$ corresponding to $A \in \Lambda$. By 9.9 and 9.8 condition (ii) in (IIb) is satisfied. Using the construction in 9.8 it follows that all data (II) appear in this way.

We describe a map in the opposite direction. Define P_0, with $0 \in \Lambda$, as the product as schemes over $\mathcal{J}$ of the rigidified $\mathbf{P}^1$-bundles over $\mathcal{J}$ associated to $\tilde{G}_1, \ldots, \tilde{G}_r$. For any $A \in \Lambda$ define $P_A = P_0$. Define

$$t_A: P_{0,K} \overset{\sim}{\longrightarrow} P_A$$

to be the composition of the unique automorphism of $P_{0,K}$ extending multiplication by x_A on $\tilde{G}_K$ and the identification $P_0 = P_A$. Let a be as in 10.4 and $P_{\underline{a}}$ as in 10.5. Then $P_{\underline{a}}$ is flat over S and by 10.8 its generic fiber is the generic fiber of $P_0 \to S$, where 0 is the zero element of Λ. Hence the scheme $P_{\underline{a}}$ is the join of the isomorphisms $t_{A,K}$ with $A \in \underline{a}$. By join we mean the scheme theoretic closure of

$$P_{0,K} \xrightarrow{\prod t_A} P_a := \prod_{A \in a} P_A.$$

This allows us to reconstruct the scheme $\widetilde{P}$ as in 10.10. If A is in Λ, the element $t_A(1)$ extends to an S-point of G_A. This defines a section of $\widetilde{P} \to S$ since the inverse image of G_A in $\widetilde{P}$ maps isomorphically to G_A. Consequently it defines a S_n-section of $\widetilde{P} \times_S S_n =: \mathcal{P}_n \to S_n$. In this way we get back the sections $\amalg_\Lambda S_n \to \mathcal{P}_n$.

11.2 Definition. *Define*

$$\widetilde{P} \longrightarrow \mathcal{J} \qquad \text{and} \qquad \amalg_\Lambda S \longrightarrow \widetilde{P}$$

to be the scheme associated in 10.2 to data (Ia) *in 11.1 and the sections algebraizing the sections fixed in* (Ib) *in 11.1. For its basic properties see 10.10.*

11.3 *Some notation needed in 11.4.* Let $\mathcal{P} = (\mathcal{P}_n)_n$ be as in (I) of 11.1. Let $\hat{0}: \mathcal{S} \to \mathcal{P}$ be the formal section associated to the element 0 of Λ. By construction it also defines the identity element of the formal group scheme structure on the smooth locus of $\mathcal{P} \to \mathcal{S}$. Denote by $\tilde{0}: S \to \widetilde{P}$ the section algebraizing $\hat{0}$. Due to 11.1 the scheme $\tilde{G}$ can be realized canonically as an open subscheme of $\widetilde{P}$ via

$$\tilde{G} := G_0 \xleftarrow{\sim} U_{\underline{\{0\}}} \hookrightarrow \widetilde{P};$$

for the isomorphism see 10.9 and for the inclusion see 10.10. By construction the identity element of $\tilde{G}$ coincides with the section $\tilde{0}$.

11.4 Proposition. *There is a unique action of the semiabelian scheme $\tilde{G}$ on the scheme $\widetilde{P}$ over S extending the action on $\tilde{G}$ induced by the group scheme structure on $\tilde{G}$.*

Proof: The notation is as in 10.5. By 10.9, for any finite subset a of Λ the smooth locus of the open subscheme $U_{\underline{a}} \to S$ of $\widetilde{P}$ is $G_{\underline{a}}$. Let $A \in a$. As explained in 10.10 the scheme G_A can be naturally identified with an open subscheme of $\widetilde{P}$. With these identifications and since $\widetilde{P} = \cup_a U_{\underline{a}}$ the smooth locus of $\widetilde{P} \to S$ is $\cup_{A \in \Lambda} G_A$. In particular, we get a well defined action of $\tilde{G}$ on the smooth locus of $\widetilde{P} \to S$. We conclude the proof remarking that by 9.10 the action of the I-adic completion of $\tilde{G}$ on the I-adic completion of the smooth locus of $\widetilde{P} \to S$ extends to the I-adic completion of $\widetilde{P}$.

11.5 Proposition. *Let v be a valuation of the function field of $\tilde{G}$ such that $v(E) \geq 0$. Let c_v be the center of v on $\mathcal{J}$. Then v has center on the closed subscheme $\mathbf{P}_A$ of $\mathcal{P}_0 = \widetilde{P} \times_S S_0$ for $A = a_1 f_1 + \ldots + a_r f_r$ in Λ if and only for any $t = 1, \ldots, r$ we have*

$$v\Big(c_v^*(\gamma_{A,t})\Big) \geq 0 \geq v\Big(c_v^*(\gamma_{A+f_t,t})\Big).$$

Here $\gamma_{A,t}$ (respectively $\gamma_{A+f_t,t}$) is a local coordinate of the $\mathbf{P}^1$-bundle $P_{A,t}$ over $\mathcal{J}$ defined in 10.3 (respectively $P_{A+f_t,t}$) along the infinity section over $\mathcal{J}$ at c_v.

Proof: Let x_A be the element of $\tilde{G}(K)$ defined in 11.1. Considering $x_{A*}(v)$ instead of v, we reduce to the case $A = 0$. In this case the statement follows from 10.8 with $a = \{0\}$.

11.6 Corollary. *The notation is as in 11.1. The following two types of data determine uniquely one another.*

I *The data consisting of*

- Ia. *a compatible system $(\mathcal{P}_n, \Lambda')$ of S_n-points of the functor $\mathcal{D}ef\mathcal{G}r(\mathcal{P}_0, \Lambda')$ defined in 9.5 satisfying conditions* (i) *and* (ii) *of 10.2.*
- Ib. *a compatible system of rigidifications $\big((\amalg_\Lambda S_n) \to \mathcal{P}_n\big)_{n \in \mathbf{N}}$ deforming the rigidification defined in 7.1 for $n = 0$. We require them to be invariant under the action of Λ' and that the section corresponding to $0 \in \Lambda$ is the section $\hat{0}: S_n \to \mathcal{P}_n$.*

II *The data consisting of*

IIa. *an abelian scheme $\mathcal{J}$ over S and a semiabelian scheme $\tilde{G}$ over S deforming the semiabelian scheme $\mathbf{G}_0$ and sitting in an exact sequence*

$$0 \longrightarrow \mathbf{G}_{m,S}^r \longrightarrow \tilde{G} \longrightarrow \mathcal{J} \longrightarrow 0.$$

Let $\tilde{G}$ split as product of $\mathbf{G}_m$-bundles $\tilde{G}_1 \times \ldots \times \tilde{G}_r$ over $\mathcal{J}$. For $1 \leq t \leq r$, let $p_t: \tilde{G} \to \tilde{G}_t$ be the projection.

IIb. *points $x_A \in \tilde{G}(K)$ for all $A \in \Lambda$ satisfying the following conditions. For $1 \leq t \leq r$ and $0 \leq k \leq n_t - 1$ let $x_{t,k} := x_{(k+1)f_t} x_{kf_t}^{-1}$;*

i. *for any $A \in \Lambda$ there is a section $y_A: S \to \mathcal{J}$ such that $y_{A,K}$ is the image of x_A in $\mathcal{J}_K$. Denote $y_{t,k} := y_{(k+1)f_t} - y_{kf_t}$;*

ii. *let $A \in \Lambda$ with $A = \oplus_{t=1,\ldots,r} a_t f_t$. For any $t = 1, \ldots, r$ let $b_t \in \mathbf{N}$ with $b_t = a_t \bmod n_t$ and $0 \leq b_t \leq n_t$. Then $x_{A+f_t} x_A^{-1} x_{b_t f_t}^{-1} \in \tilde{G}(S)$;*

iii. *for any $A \in \Lambda$ we have $x_{A+l_t} = x_A \cdot x_{l_t}$ and $y_{A+l_t} = y_A + y_{l_t}$;*

iv. *for any integer s with $1 \leq s \leq r$ denote by $I_{t,k,s}$ the fractional ideal of S determined by the section $p_{s,K}(x_{t,k})$ over K of the $\mathbf{G}_m$-bundle $y_{t,k}^*(\tilde{G}_s)$ over S. Then $I_{t,k,s}$ is a non-trivial ideal of O_S contained in I if $s = t$ and coincides with O_S otherwise.*

If we change datum (Ib), *neither datum* (IIa) *nor for any $t = 1, \ldots, r$ the product $x_{t,0} \cdots x_{t,r-1} = x_{l_t}$ do change.*

Proof: To go from data (I) to data (II) one proceeds as in 11.1 taking in account the action of Λ' and using 10.11. To go from data (II) to data (I) one proceeds as in 11.1 using 11.8.

11.7 Definition. *Let $\widetilde{P}$ be the scheme over S defined in 11.2 and arising from data of type* (II) *in 11.6. Define*

$$\cup_{A \in \Lambda'} G_A =: \widetilde{P}^{\mathrm{sm},0} \hookrightarrow \widetilde{P}.$$

11.8 Lemma. *Let $\widetilde{P}$ be the scheme over S defined in 11.2 and arising from data of type* (II) *in 11.6.*

1. *The scheme $\widetilde{P} \to S$ is endowed with a unique action of the lattice $\Lambda' = \mathbf{Z}l_1 \oplus \ldots \oplus \mathbf{Z}l_r$ characterized by the fact that for $t = 1, \ldots, r$ the element l_t acts on $\mathcal{J}$ as translation by y_{l_t} and acts on $\widetilde{P}_K = \tilde{G}_K$ as translation by x_{l_t}. Moreover,*

$$y_{l_t} = \sum_{k=0,\ldots,n_t-1} y_{t,k} \quad \text{and} \quad x_{l_t} = \prod_{k=0,\ldots,n_t-1} x_{t,k}.$$

2. *If the data of type* (II) *we started with come from data of type* (I) *of 11.6, the induced action of Λ' on the I-adic completion $\mathcal{P} \to \mathcal{S}$ of $\widetilde{P} \to S$ coincides with the action defined in* (I) *of 11.6.*

3. *There exists a unique S-group structure on $\widetilde{P}^{\mathrm{sm},0}$ and a unique action of $\widetilde{P}^{\mathrm{sm},0}$ on $\widetilde{P} \to S$ extending the group scheme structure on $\widetilde{P}_K = \widetilde{P}_K^{\mathrm{sm},0} = \tilde{G}_K$. The inclusion $\tilde{G} \hookrightarrow \widetilde{P}^{\mathrm{sm},0}$ defined in 11.3 is a group homomorphism.*
4. *If the data of type (II) we started with come from data of type (I) of 11.6, the formal $\mathcal{S}$-group scheme $\mathcal{P}^{\mathrm{sm},0}$ defined as the I-adic completion of $\widetilde{P}^{\mathrm{sm},0}$ and the action on the I-adic completion $\mathcal{P}^{\mathrm{sm}}$ of $\widetilde{P}^{\mathrm{sm}}$ coincides with the formal group structure on $\mathcal{P}^{\mathrm{sm},0}$ and the formal action on $\mathcal{P}^{\mathrm{sm}}$ defined in (I) of 11.6 and in 9.5.*

Proof: (1), (2) As in 10.11.

(3), (4) Use 11.4 and the action of Λ' to get a group law on $\widetilde{P}^{\mathrm{sm},0}$ and an action of $\widetilde{P}^{\mathrm{sm},0}$ on $\widetilde{P}^{\mathrm{sm}}$ satisfying the requirements of the lemma.

12 The setup

In this section we fix a split semistable curve $C_0 \to S_0$ as in 12.1 and we collect a few definitions and notations needed later. In 12.2 we associate to $C_0 \to S_0$ the dual graph $\mathcal{G}$. In 12.3 we construct the graph Γ by contracting the disconnecting edges of $\mathcal{G}$. Finally, we describe in 12.6 the principally polarized abelian scheme $\mathrm{Pic}^0(C_0/S_0)$ in a way which is convenient for our purposes.

12.1 *Some notation.* Let S_0 be a connected, noetherian, affine scheme. Let $f_0: C_0 \to S_0$ be a *semistable* curve over S_0, i.e., a proper and flat morphism such that any geometric fiber is a connected, reduced, one dimensional scheme having only ordinary double points as singularities. Suppose that $C_0 \to S_0$ is *split.* By this we mean, first of all that the complement of the smooth locus of f_0 consists of a disjoint union of sections of f_0. Second, we require that for any such section $T: S_0 \to C_0$ the branches of C_0 at T are rational over S_0. In other words we require that the completion of C_0 along T is Zariski locally on S_0 isomorphic to $O_{S_0}[\![X,Y]\!]/(XY)$. In particular, the curve $C_0 \to S_0$ has constant topological type.

12.2 Definition. *Let $\mathcal{S}$ing be the complement of the smooth locus of f_0. Let*

$$N: \tilde{C}_0 \longrightarrow C_0$$

be the morphism to C_0 from the blowing-up of C_0 along $\mathcal{S}$ing. Call the dual graph of C_0 the connected graph $\mathcal{G}$ defined as follows. Associate a vertex to each connected component of $\tilde{C}_0$. Associate an edge to each connected component of $\mathcal{S}$ing according to the following rule. Let T be a connected component of $\mathcal{S}$ing. There are two possibilities

i. *if $N^{-1}(T)$ lies on two different connected components Δ_1 and Δ_2 of $\tilde{C}_0$ we associate to T an edge whose end points coincide with the vertices corresponding to Δ_1 and Δ_2;*

ii. *if $N^{-1}(T)$ lies on a connected component Δ of $\tilde{C}_0$ we associate to T an edge whose end points coincide with the vertex corresponding to Δ.*

12.3 Definition. *Define Γ as the graph obtained from the dual graph of C_0 by collapsing the disconnecting edges, i.e., those which, if removed, do disconnect the dual graph itself. More precisely, the edges $\mathcal{E}$ of Γ consist of the non-disconnecting edges of the graph $\mathcal{G}$. The set $\mathcal{V}$ of vertices of Γ is defined as the set of equivalence classes of vertices of $\mathcal{G}$ under the equivalence relation generated by the following relation. We say that two vertices v_1 and v_2 are equivalent if and only if there is a disconnecting edge e of $\mathcal{G}$ whose end points are v_1 and v_2. The end points of an edge e in $\mathcal{E}$ are the elements of $\mathcal{V}$ corresponding to the equivalence classes of the end points of e considered as an edge of $\mathcal{G}$.*

Denote by

$$T_e \qquad \text{with } e \in \mathcal{E}$$

the connected components of $\mathcal{S}ing$ whose associated edges in $\mathcal{G}$ are not disconnecting. We define

$$p_0 \colon D_0 \longrightarrow C_0$$

to be given by separating the branches at the sections T_e with $e \in \mathcal{E}$, i.e., to be the morphism from the blowing-up of C_0 along the sections T_e with $e \in \mathcal{E}$ to C_0. Define

$$n \colon \tilde{C}_0 \to D_0$$

to be the unique morphism such that $N = p_0 \circ n$. Define the curves

$$\mathbb{D}_j \to S_0 \qquad \text{with } j \in \mathcal{V}$$

to be the connected components of the curve D_0 containing the image of a (equivalently any) connected component of $\tilde{C}_0$ corresponding to a vertex of $\mathcal{G}$ in the equivalence class defined by j.

With the notation of 3.1 and 5.1, choose elementary cycles

$$l_1, \ldots, l_r$$

of Γ such that $H_1(\Gamma, \mathbf{Z})$ is equal to $\mathbf{Z}l_1 \oplus \ldots \oplus \mathbf{Z}l_r$. For any $t = 1, \ldots, r$ let Γ_t be the oriented subgraph Γ_{l_t} spanned by l_t with the orientation induced by l_t as in 5.1.

12.4 Remark. Each curve $\mathbb{D}_j$ is a nice curve, i.e., its associated dual graph as defined in 12.2 has trivial homotopy. We can construct the graph Γ also in terms of $p_0 \colon D_0 \to C_0$. To this end associate to each connected component $\mathbb{D}_j$ of D_0 a vertex of Γ. To each section T_e of C_0 as in 12.3 associate an edge such that, if $p_0^{-1}(T_e)$ lie on the connected components $\mathbb{D}_a$ and $\mathbb{D}_b$ of D_0, its end points are the vertex associated to $\mathbb{D}_a$ and the vertex associated to $\mathbb{D}_b$ respectively.

Figure 4 shows how to get the graph Γ from the dual graph. The graph on the left hand side is the dual graph. The graph on the right hand side is Γ. The curve associated to the vertex f of Γ is a nice curve over S_0. It can be constructed by gluing transversally the smooth curves over S_0 associated to the vertices b, c, d and e as prescribed by the edges of the dual graph which are contracted in Γ.

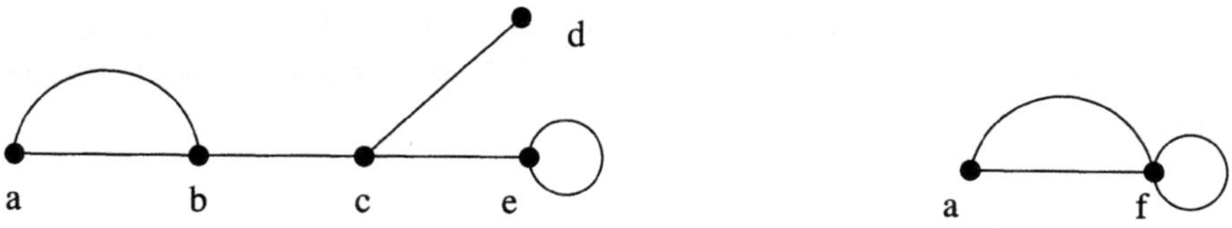

Figure 4.

12.5 *Assumption.* We assume that for each with j in $\mathcal{V}$ the curve $\mathbb{D}_j$ admits a section

$$0_j: S_0 \longrightarrow D_j,$$

whose image via $p_0: D_0 \to C_0$ is contained in the smooth locus of $f_0: C_0 \to S_0$. Fix such sections.

12.6 Definition. *Using 12.5 define for any j in $\mathcal{V}$ the morphism*

$$i_j: \mathbb{D}_j \longrightarrow \mathrm{Pic}^0(\mathbb{D}_j/S_0)$$

as follows. Each connected component Δ of $n^{-1}(\mathbb{D}_j)$ is smooth over S_0 and has at least one section 0_Δ over S_0. Define the morphism $i_\Delta: \Delta \to \mathrm{Pic}^0(\Delta/S)$ by sending, for each scheme T over S, a T-valued point P of $\Delta \times_S T$ to the invertible sheaf associated to the divisor $P - 0_\Delta \times_S T$. The morphism i_j is the unique morphism sending the section 0_j of $\mathbb{D}_j$ to the identity element of $\mathrm{Pic}^0(\mathbb{D}_j/S_0)$ and satisfying the following. For each connected component Δ of $n^{-1}(\mathbb{D}_j)$ the restriction to Δ of the composition of i_j with the projection $\mathrm{Pic}^0(\mathbb{D}_j/S_0) \to \mathrm{Pic}^0(\Delta/S_0)$ is the morphism i_Δ up to translation.

Define the effective Cartier divisor Θ_j of $\mathrm{Pic}^0(\mathbb{D}_j/S_0)$ relative to S_0 as the image of the symmetric power of $\mathbb{D}_j$ of order equal to $g(\mathbb{D}_j) - 1$. It defines a principal polarization on the abelian scheme $\mathrm{Pic}^0(\mathbb{D}_j/S_0)$.

Finally define

$$\mathcal{J}_0 := \mathrm{Pic}^0(D_0/S_0) \cong \prod_{j \in \mathcal{V}} \mathrm{Pic}^0(\mathbb{D}_j/S_0).$$

The product of the principal polarizations on each factor $\mathrm{Pic}^0(\mathbb{D}_j/S_0)$ defines a principal polarization

$$\lambda_0: \mathcal{J}_0 \xrightarrow{\sim} \mathcal{J}_0^\vee.$$

12.7 Remark. If $\mathbb{D}_j \to S_0$ is smooth, the fact that Θ_j is an effective Cartier divisor relative to S_0 is proven in [4, Proposition 9.4/4]. In general, the fact that Θ_j is a Cartier divisor relative to S_0 follows from the smooth case. In fact, it is easy to prove that the divisor Θ_j is the sum over the components Δ of $n^{-1}(\mathbb{D}_j)$ of the pullbacks via $\mathrm{Pic}^0(\mathbb{D}_j/S_0) \to \mathrm{Pic}^0(\Delta/S_0)$ of the divisors Θ_Δ. The latter is defined as the image of the symmetric power of Δ of order equal to $g(\Delta) - 1$.

The divisor Θ_j depends on the choice of the section 0_j used to define i_j. But the principal polarization induced by Θ_j on $\mathrm{Pic}^0(\mathbb{D}_j/S_0)$ does not. See [4, Proposition 9.4/4].

13 The Jacobian of $C_0 \to S_0$

In this section we describe Pic^0 of C_0 over S_0. In 12.3 we have defined a partial normalization $\amalg_{j\in\mathcal{V}}\mathbb{D}_j$ of C_0. We describe in 13.14 an explicit $\mathbf{G}_m^r$-torsor over the abelian scheme $\mathcal{J}_0 = \prod_{j\in\mathcal{V}} \mathrm{Pic}^0(\mathbb{D}_j/S_0)$. We prove in 13.16 that it represents $\mathrm{Pic}^0(C_0/S_0)$.

13.1 Definition. *The notation is as in 12.3. Denote by W the set of pairs (s,k) with $k \in \mathcal{V}$ and $s \in \{1,\ldots,r\}$ such that the vertex of Γ associated to $\mathbb{D}_k$ belongs to the cycle l_s of Γ. If j is in $\mathcal{V}$ denote by*

$$W_j = \{(s,k) \in W \mid k = j\}.$$

Analogously, if t is an element of $\{1,\ldots,r\}$ define

$$W^t = \{(s,k) \in W \mid s = t\}.$$

Let (t,j) be in W. With respect to the orientation on l_t defined in 12.3, there is a unique edge belonging to l_t leaving the vertex j. Analogously, there is a unique edge belonging to l_t getting to the vertex j:

i. *if these edges are distinct, they correspond to two sections of C_0 whose inverse image via $\mathbb{D}_j \to C_0$ determine uniquely S_0-valued points $P_{(t,j)}$ and $Q_{(t,j)}$ respectively of $\mathbb{D}_j$;*
ii. *if they are the same, they define a section of C_0 whose inverse image via $\mathbb{D}_j \to C_0$ determine S_0-valued points $P_{(t,j)}$ and $Q_{(t,j)}$ of $\mathbb{D}_j$.*

Finally for any j in $\mathcal{V}$ define

$$\mathbb{D}_j \longrightarrow \mathbb{D}'_j$$

to be the morphism which identifies transversally $P_{(t,j)}$ and $Q_{(t,j)}$ for any (t,j) in W_j as schemes over S_0.

13.2 Remark. The curve D_0 carries the S_0-sections defined as the inverse images via $p_0\colon D_0 \to C_0$ of the non-disconnecting singularities T_e of C_0. In 13.1 we have labeled such sections in a convenient way. We remark that the labeling in (i) is equivalent to the choice of the orientations on $l_1, \ldots, l_r$ defined in 12.3. Instead the labeling in (ii) is not unique and we have chosen one.

The existence of the curve $D'_j \to S_0$ in the category of schemes over S_0 is proven in the following lemma. As remarked by B. Edixhoven it is essential that the sections $P_{(t,j)}$ and $Q_{(t,j)}$ are identified as schemes over S_0.

13.3 Lemma. *The curve $\mathbb{D}'_j \to S_0$ as claimed in 13.1 exists.*

Proof: We proceed by induction on the cardinality of W_j. If the cardinality is 0, there is nothing to prove. Suppose that $\mathbb{D}''_j \to S_0$ has been constructed identifying the sections P_w and Q_w with w in a subset W'' of W_j and let $v \in W_j \backslash W''$. We show that the curve obtained from $\mathbb{D}''_j$ identifying transversally P_v and Q_v as schemes over S_0 exists. The curve $\mathbb{D}^v_j \to S_0$, constructed by gluing transversally two copies of $\mathbb{D}''_j$ so that the section P_v (respectively Q_v) on the first copy is identified to the section Q_v (respectively P_v) on the second copy as schemes over S_0, exists since it exists on open subschemes S_0 where the relative tangent spaces of $\mathbb{D}_j \to S_0$ at P_v and Q_v are free as O_{S_0}-modules. There is a free action of $\mathbf{Z}/2\mathbf{Z}$ on $\mathbb{D}^v_j \to S_0$ defined by switching the two copies of $\mathbb{D}''_j$ used to construct $\mathbb{D}^v_j$. Since $\mathbb{D}^v_j \to S_0$ is projective locally on S_0 the quotient $\mathbb{D}^v_j \to \mathbb{D}'_j$ exists and is faithfully flat, finite and étale.

13.4 Definition. *Let j be in $\mathcal{V}$. Define ${\Theta_j}^-$ as the image via the inverse map of the theta divisor Θ_j on $\mathrm{Pic}^0(\mathbb{D}_j/S_0)$ defined in 12.6. For each w in W_j define $t_{Q_w}(\Theta_j^-)$ (resp. $t_{P_w}(\Theta_j^-)$) as the image of the divisor Θ_j^- via translation by Q_w (resp. P_w). Define as in 2.2*

$$\mathbb{P}_w := \mathbf{P}\big(t_{Q_w}(\Theta_j^-) \oplus t_{P_w}(\Theta_j^-)\big).$$

Define

$$\mathcal{L}_w := \mathcal{L}\big(t_{Q_w}(\Theta_j^-) - t_{P_w}(\Theta_j^-)\big) \qquad \text{and} \qquad \mathbb{G}_w := G(\mathcal{L}_w).$$

13.5 Remark. By the theorem of the square the invertible sheaf $\mathcal{L}_w$ in the definition is isomorphic to $\mathcal{L}\big(t_{P_w - Q_w}(\Theta_j) - \Theta_j\big)$. Hence, it is algebraically equivalent to 0. Furthermore, if P_w and Q_w lie on a rational component of $\mathbb{D}_j$, then $\mathcal{L}_w$ is trivial.

The choice of ${\Theta_j}^-$ is of a technical nature; see 13.8.

13.6 Definition. *Fix j in $\mathcal{V}$. Let*

$$\mathbf{G}_j := \prod_{w \in W_j} \mathbb{G}_w \qquad \text{and} \qquad \mathbf{P}_j := \prod_{w \in W_j} \mathbb{P}_w;$$

the products are taken in the category of schemes over $\mathrm{Pic}^0(\mathbb{D}_j/S_0)$. In particular, we have morphisms

$$\mathbf{G}_j \hookrightarrow \mathbf{P}_j \longrightarrow \mathrm{Pic}^0(\mathbb{D}_j/S_0).$$

The set W_j is defined in 13.1.

13.7 Definition. *Fix j in $\mathcal{V}$ and w in W_j. As in 2.1 define*

$$\infty_w\colon \prod_{w'\neq w} \mathbb{P}_{w'} \longrightarrow \mathbf{P}_j \qquad \text{and} \qquad 0_w\colon \prod_{w'\neq w} \mathbb{P}_{w'} \longrightarrow \mathbf{P}_j.$$

Here the products are taken as schemes over $\mathrm{Pic}^0(\mathbb{D}_j/S_0)$.

13.8 Lemma. *Let j be in $\mathcal{V}$. For any w in W_j the pullback via $i_j\colon \mathbb{D}_j \to$* $\mathrm{Pic}^0(\mathbb{D}_j/S_0)$ *of $\mathbb{P}_w$ is isomorphic to the $\mathbf{P}^1$-bundle $\mathbf{P}(Q_w \oplus P_w)$ over $\mathbb{D}_j$.*

Proof: If P_w and Q_w lie in the same rational component of the inverse image of $\mathbb{D}_j$ via the map n defined in 12.3, the assertion is obvious because both the $\mathbf{P}^1$-bundles mentioned in the statement are trivial. If not, the ideal sheaf defining P_w (resp. Q_w) contains the pullback via i_j of the ideal sheaf defining $t_{P_w}(\Theta_j^-)$ (resp. $t_{Q_w}(\Theta_j^-)$). Moreover, P_w and Q_w lie in the same non-rational component Δ of $n^{-1}(\mathbb{D}_j)$. Then [17, Lemma 6.8] applied to Δ implies that two ideal sheaves are equal fiberwise and hence are equal by Nakayama's lemma.

13.9 Lemma. *Let j be in $\mathcal{V}$. The morphism $\prod_{w\in W_j} \mathbf{P}(Q_w \oplus P_w) \to \mathbb{D}_j$ admits a section. Here the product is taken as schemes over $\mathbb{D}_j$.*

Proof: See 2.2.

13.10 Proposition. *Let j be in $\mathcal{V}$. Let $i_j\colon \mathbb{D}_j \to$* $\mathrm{Pic}^0(\mathbb{D}_j/S_0)$ *be the morphism defined in 12.6. Then there is a lifting*

$$\mathbb{D}_j \xrightarrow{u_j} \mathbf{P}_j$$

of the morphism i_j with the property that for any w in the set W_j defined in 13.1, the image of $\mathbb{D}_j$ via u_j intersects the section ∞_w (resp. the section 0_w) of $\mathbb{P}_w$ defined in 13.7 at P_w (resp. Q_w) with multiplicity 1.

Proof: Use 13.8 and 13.9.

13.11 Remark. The map u_j contracts the rational components of $\mathbb{D}_j$ which do not contain any of the sections P_w, Q_w with $w \in W$.

13.12 *Some notation needed in 13.13*. Fix j in $\mathcal{V}$. Let $n_j\colon \tilde{\mathbb{D}}_j \to \mathbb{D}_j$ be the pullback of n via the morphism $\mathbb{D}_j \to D_0$; see 12.3 for the notation. Let $\Delta_1, \ldots, \Delta_s$ denote the connected components of $\tilde{\mathbb{D}}_j$. They are proper and smooth curves over S_0 of genus $d_1, \ldots, d_s$. For any integer α such that $1 \leq \alpha \leq s$, let Δ'_α be the image of Δ_α in the curve $\mathbb{D}'_j$ defined in 13.1. Define $d'_\alpha = g(\Delta'_\alpha)$. By construction Δ_α carries at least one S_0-section 0_α whose image factors through the smooth locus of $\Delta'_\alpha \to S_0$. We require that if the inverse image via n_j of the section 0_j of $\mathbb{D}_j \to S_0$ defined in 12.5 belongs to Δ'_α, then 0_α is $n_j^{-1}(0_j)$. Then $\mathrm{Pic}^0(\mathbb{D}_j/S_0)$ is isomorphic to the product of $\mathrm{Pic}^0(\Delta_\alpha/S_0)$ with $\alpha = 1, \ldots, s$. Also, Θ_j is the sum over $\alpha = 1, \ldots, s$ of translates of the pullbacks via $\mathrm{Pic}^0(\mathbb{D}_j/S_0) \to \mathrm{Pic}^0(\Delta_\alpha/S_0)$ of the theta divisors Θ_α defined on $\mathrm{Pic}^0(\Delta_\alpha/S_0)$ using the section 0_α.

For any $\alpha = 1, \dots, s$ define W_α as the set of elements w of W_j such that the sections P_w and Q_w defined in 13.1 factor via Δ_α. For any $w \in W_\alpha$ define the $\mathbf{P}^1$-bundle $\mathbb{P}_{\alpha,w}$ over $\mathrm{Pic}^0(\Delta_\alpha/S_0)$ to be $\mathbf{P}\big(t_{Q_w}(\Theta_\alpha^-) \oplus t_{P_w}(\Theta_\alpha^-)\big)$ as in 13.4. Let $\mathbb{G}_{\alpha,w}$ be the associated $\mathbf{G}_m$-bundle as in 2.1. Define $\mathbf{P}_\alpha$ (resp. $\mathbf{G}_\alpha$) as the product over $\mathrm{Pic}^0(D_\alpha/S_0)$ of $\mathbb{P}_{\alpha,w}$ (resp. $\mathbb{G}_{\alpha,w}$) for all $w \in W_\alpha$. As in 13.10, there is a morphism $u_\alpha \colon D_\alpha \to \mathbf{P}_\alpha$.

Let j be in $\mathcal{V}$. The image via u_j of the section 0_j of $\mathbb{D}_j \to S_0$ fixed in 12.5 defines an S_0-valued point of $\mathbf{G}_j$ above the identity element of $\mathrm{Pic}^0(\mathbb{D}_j/S_0)$. As explained in 2.3, this S_0-valued point of $\mathbf{G}_j$ defines a rigidification of $\mathbf{G}_j$ and defines on $\mathbf{G}_j$ the structure of group scheme over S_0.

13.13 Proposition. *Let $j \in \mathcal{V}$. Let $\mathbb{D}'_j$ be the curve defined in 13.1. Define $\mathrm{Pic}^0(\mathbb{D}'_j/S_0)$ to be the functor of invertible sheaves on $\mathbb{D}'_j$ whose restriction to the fiber $\mathbb{D}'_{j,c}$ for any geometric point c of S_0 belongs to $\mathrm{Pic}^0\big(\mathbb{D}'_{j,c}/k(c)\big)$. Then the group scheme $\mathbf{G}_j$ over S_0 represents $\mathrm{Pic}^0(\mathbb{D}'_j/S_0)$.*

Proof: By 13.12 and 13.6 we have that $\mathbf{G}_j \cong \prod_{\alpha=1,\dots,s} \mathbf{G}_\alpha$; the product is taken as schemes over S_0. For each α with $1 \leq \alpha \leq s$, the image of 0_α via u_α rigidifies $\mathbf{G}_\alpha$ as product of $\mathbf{G}_m$-bundles over $\mathrm{Pic}^0(\Delta_\alpha/S_0)$ defining, via 2.3, a unique group structure on $\mathbf{G}_\alpha$. The isomorphism $\mathbf{G}_j \cong \prod_{\alpha=1,\dots,s} \mathbf{G}_\alpha$ is then an isomorphism of group schemes. The morphism $\amalg_{\alpha=1,\dots,s}\Delta'_\alpha \to \mathbb{D}'_j$ induces a morphism between the associated Picard functors which is an isomorphism. In particular, it suffices to consider the case that $\mathbb{D}_j \to S_0$ is smooth. Let $\mathcal{H}^h$ be the symmetric power of order h of the smooth locus of $\mathbb{D}'_j \to S_0$. The map u_j defines a morphism from $\mathcal{H}^h$ to $\mathbf{G}_j$. Hence we get an equivalence relation $\mathcal{R}_1$ on $\mathcal{H}^h$ by requiring that for any scheme T over S_0 two T-valued points of $\mathcal{H}^h$ are equivalent if and only if they define the same T-valued point of $\mathbf{G}_j$. The section 0_j defined in 12.5 induces a morphism from $\mathcal{H}^h$, viewed as in [4, Section 9.3, pag. 254-255] as the Hilbert scheme of h points of the smooth locus of $\mathbb{D}'_j \to S_0$, to the functor $\mathrm{Pic}^0(\mathbb{D}'_j/S_0)$. Since $\mathbb{D}'_j \to S$ is projective, flat and geometrically integral we know, for example by [4, Theorem 8.2/5] and [4, Proposition 8.4/2], that $\mathrm{Pic}^0(\mathbb{D}'_j/S_0)$ is represented by a smooth scheme over S_0. As before we get an equivalence relation $\mathcal{R}_2$.

By [4, Section 8.2, pag. 222], we have that $\mathrm{Pic}^0(\mathbb{D}'_j/S_0)$ represents effectively the quotient of $\mathcal{H}^h$ modulo $\mathcal{R}_2$ for h large enough. If $\mathcal{R}_1$ and $\mathcal{R}_2$ coincide, we get a morphism $\mathrm{Pic}^0(\mathbb{D}'_j/S_0)$ to $\mathbf{G}_j$ which is a homomorphism of group schemes and is injective on T-valued points for any scheme T over S_0. Since $\mathrm{Pic}^0(\mathbb{D}'_j/S_0)$ and $\mathbf{G}_j$ are smooth over S_0 and have the same relative dimension d'_j over S_0, the morphism must be an isomorphism.

Hence, it suffices to prove that for any noetherian scheme T over S_0 the T valued points of $\mathcal{R}_1$ and $\mathcal{R}_2$ coincide. Let U_1 and U_2 be two T-valued points of $\mathcal{H}^h$. We need to prove that they define the same T-valued point of $\mathrm{Pic}^0(\mathbb{D}'_j/S_0)$ if and only if they define the same T-valued point in $\mathbf{G}_j$. In particular, it suffices to check it at the completion of T of any closed point. Hence, we can assume that S_0 is a versal deformation space of $\mathbb{D}'_j$ at a geometric point of S_0. In particular, S_0 is complete, local and regular. Let $\mathcal{W}$ be the maximal open subscheme of $\mathcal{H}^{d'_j}$ with the property that the map to $\mathrm{Pic}^0(\mathbb{D}'_j/S_0)$ is an open immersion; see [4, Section 9.3, page 255]. For any geometric point c of S_0 the complement of the fiber of the image of $\mathcal{W}_c$ in $\mathrm{Pic}^0\big(\mathbb{D}'_{j,c}/k(c)\big)$ consists of invertible sheaves associated either to effective Cartier divisors of degree d'_j whose support is contained in the singular locus of $\mathbb{D}'_{j,c} \to k(c)$ or to effective Cartier divisors of degree d'_j whose $k(c)$-vector space of global sections has dimension at least 2. In particular, $\mathcal{W}_c$ has codimension at least 2 in $\mathrm{Pic}^0\big(\mathbb{D}'_{j,c}/k(c)\big)$. Since S_0 is normal, by [4, Theorem 4.4/1], this induces a unique morphism from $\mathrm{Pic}^0(\mathbb{D}'_j/S_0)$ to $\mathbf{G}_j$ compatible with the morphism from $\mathcal{W}$ to $\mathrm{Pic}^0(\mathbb{D}'_j/S_0)$ and to $\mathbf{G}_j$. Fiberwise it is a morphism of group schemes by [25, Theorem V.2.9.2]. Hence the morphism is a morphism of group schemes. In particular, we get a commutative diagram

$$\begin{array}{ccc} \mathrm{Pic}^0(\mathbb{D}'_j/S_0) & \longrightarrow & \mathrm{Pic}^0(\mathbb{D}_j/S_0) \\ \downarrow & & \downarrow\wr \\ \mathbf{G}_j & \longrightarrow & \mathrm{Pic}^0(\mathbb{D}_j/S_0). \end{array}$$

The isomorphism on the right hand side follows since the claim holds if $\mathbb{D}'_j = \mathbb{D}_j$. Moreover, the morphism $\{\mathbb{D}'_j\}^{\mathrm{sm}} \to \mathrm{Pic}^0(\mathbb{D}'_j/S_0) \to \mathbf{G}_j$ is simply the morphism u_j of 13.10. In fact, on a nonempty open subscheme of $\{\mathbb{D}'_j\}^{\mathrm{sm}}$ it coincides with u_j. In particular, the homomorphism from $\mathrm{Pic}^0(\mathbb{D}'_j/S_0)$ to $\mathbf{G}_j$ is compatible with the morphisms from $\mathcal{H}^h$ to $\mathbf{G}_j$ and to $\mathrm{Pic}^0(\mathbb{D}'_j/S_0)$. To conclude, it suffices to prove that $\mathrm{Pic}^0(\mathbb{D}'_j/S_0) \to \mathbf{G}_j$ is an isomorphism. Since the two group schemes are smooth of the same relative dimension over S_0, it suffices to prove that it is a monomorphism. We can reduce to the case that S_0 is an algebraically closed field, where the result is well known.

So far we have constructed and described some objects associated to each component $\mathbb{D}_j$ of D_0. We want to relate them to C_0. For any j in $\mathcal{V}$ denote by

$$p_j\colon \mathcal{J}_0 \to \mathrm{Pic}^0(\mathbb{D}_j/S_0)$$

the natural projection.

13.14 Definition. *For any integer t such that $1 \le t \le r$ and any cycle l_t as fixed in 12.3, define the following two effective relative Cartier divisors:*

$$Y_2 := \sum_{w=(t,j)\in W^t} {p_j}^*\big(t_{Q_w}(\Theta_j^-)\big) \qquad \textit{and} \qquad Y_1 := \sum_{w=(t,j)\in W^t} {p_j}^*\big(t_{P_w}(\Theta_j^-)\big).$$

Define as in 2.2 the $\mathbf{G}_m$*-bundle over* $\mathcal{J}_0$

$$\mathbb{G}_t := G(Y_2 \oplus Y_1).$$

Define

$$\mathbf{G}_0 := \prod_{t=1,\dots,r} \mathbb{G}_t,$$

the product being taken as schemes over $\mathcal{J}_0$.

The scheme $\mathbf{G}_0$ *is an extension of the abelian scheme* $\mathcal{J}_0$ *by a torsor under a constant split torus* T_0 *over* S_0 *with character group* $H_1(\Gamma, \mathbf{Z})$. *The extension class is defined by the morphism*

$$\begin{array}{rcl} \mathbf{Z}\cdot l_1 + \ldots + \mathbf{Z}\cdot l_r = H_1(\Gamma,\mathbf{Z}) & \longrightarrow & \mathcal{J}_0{}^\vee(S_0) \cong \mathcal{J}_0(S_0) \\ l_t & \mapsto & \sum_{w\in W^t}(P_w - Q_w); \end{array}$$

where the isomorphism $\mathcal{J}_0{}^\vee \cong \mathcal{J}_0$ *is the inverse of the principal polarization defined in 12.6.*

13.15 Remark. For any j in $\mathcal{V}$ the restriction of $\mathbf{G}_0$ to $\mathrm{Pic}^0(\mathbb{D}_j/S_0)$ is the scheme $\mathbf{G}_j$ defined in 13.6. In fact, $\mathbf{G}_0$ is the product as $\mathbf{G}_m$-bundles of all $p_j^*(\mathbf{G}_j)$ with j in $\mathcal{V}$. In particular, $\mathbf{G}_0$ is rigidified above the zero section of $\mathcal{J}_0$ and acquires the structure of a semiabelian scheme as explained in 2.3.

Due to 13.10 for any j in $\mathcal{V}$ there is a morphism

$$u_j: \mathbb{D}_j \backslash \{P_w, Q_w | w \in W_j\} \longrightarrow \mathbf{G}_0$$

factoring through $\mathbf{G}_j$.

13.16 Proposition. *Let* $\mathrm{Pic}^0(C_0/S_0)$ *be the functor of invertible sheaves on* C_0 *such that for any geometric point* c *of* S_0 *their restriction to* $C_{0,c}$ *belongs to* $\mathrm{Pic}^0(C_{0,c}/k(c))$. *The group scheme* $\mathbf{G}_0$ *represents the functor* $\mathrm{Pic}^0(C_0/S_0)$.

Proof: The proof is very close to the one of 13.13. Let h be a positive integer. Let $\mathcal{H}^h$ be a connected component of the hth symmetric product of the smooth locus of $C_0 \to S_0$. The morphisms $u_j: \mathbb{D}_j \backslash \{P_w, Q_w | w \in W_j\} \to \mathbf{G}_0$ for j in $\mathcal{V}$ induce a morphism from $\mathcal{H}^h$ to $\mathbf{G}_0$. For any j in $\mathcal{V}$ we can define a contraction morphism $\gamma_j: C_0 \to \mathbb{D}'_j$. Consider an effective Cartier divisor on C_0 relative to S_0 whose support is on $\mathbb{D}_j \backslash \{P_w, Q_w | w \in W_j\}$ and which is very ample on the image of $\mathbb{D}_j$ in C_0. The contraction is simply the morphism from C_0 to its image via the linear system associated to the divisor we have constructed. In particular, we get a diagram

$$\begin{array}{ccccc} \mathcal{H}^h & \longrightarrow & \prod_{j\in\mathcal{V}} \mathbf{G}_j & \xrightarrow{\sim} & \prod_{j\in\mathcal{V}} \mathrm{Pic}^0(\mathbb{D}'_j/S_0) \\ & & \downarrow & & \downarrow \prod \gamma_j^* \\ & & \mathbf{G}_0 & & \mathrm{Pic}^0(C_0/S_0); \end{array} \qquad 13.16.1$$

the products are taken over S_0. The left vertical map is deduced from 13.15.

The existence of the isomorphism, as group schemes, is a consequence of 13.13. To interpret the morphism from $\mathcal{H}^h$ to $\mathrm{Pic}^0(C_0/S_0)$, we identify $\mathcal{H}^h$ as in [4, Section 9.3, pages 254-255] with a connected component of the Hilbert scheme of h points of the smooth locus of $C_0 \to S_0$. Let T be a scheme over S_0. With the notations of 12.3 and 13.1, we have an equivalence between the categories consisting of

i. invertible sheaves $\mathcal{L}$ on $C_{0,T}$ defining T-valued points of $\mathrm{Pic}^0(C_0/S_0)$;
ii. for any j in $\mathcal{V}$ invertible sheaves $\mathcal{L}_j$ on $\mathbb{D}_{j,T}$ defining T-valued points of $\mathrm{Pic}^0(\mathbb{D}_j/S_0)$ plus isomorphisms Φ_t for any $t = 1, \ldots, r$. For each t the isomorphism Φ_t is defined by composing over all edges e belonging to l_t the following isomorphisms Φ_e. For any oriented edge e of l_t such that the source of e is the vertex j_1 and the target of e is the vertex j_2, define

$$\Phi_e \colon P^*_{(j_1,t),T}(\mathcal{L}_{j_1}) \xrightarrow{\sim} Q^*_{(j_2,t),T}(\mathcal{L}_{j_2}).$$

Analogously, for any j in $\mathcal{V}$ we have an equivalence between the categories consisting of

i. invertible sheaves $\mathcal{L}'_j$ on $\mathbb{D}'_{j,T}$ defining T-valued points L'_j of $\mathrm{Pic}^0(\mathbb{D}'_j/S_0)$;
ii. invertible sheaves $\mathcal{L}_j$ on $\mathbb{D}_{j,T}$ defining T-valued points L_j of $\mathrm{Pic}^0(\mathbb{D}_j/S_0)$ plus isomorphisms for all w in W_j

$$\Psi_w \colon P^*_{w,T}(\mathcal{L}_j) \xrightarrow{\sim} Q^*_{w,T}(\mathcal{L}_j).$$

Via the isomorphism of 13.13 between $\mathrm{Pic}^0(\mathbb{D}'_j/S_0)$ and $\mathbf{G}_j$, the point L_j of $\mathrm{Pic}^0(\mathbb{D}_j/S_0)$ is the image of L'_j via the morphism $\mathbf{G}_j \to \mathrm{Pic}^0(\mathbb{D}_j/S_0)$. Let w be in W_j. The fiber of the scheme $\mathbb{G}_w$ defined in 13.4 above the point L_j is a $\mathbf{G}_m$-torsor. If we fix the point L'_j above L_j the action of $a \in \mathbf{G}_{m,T}(T)$ on the fiber of $\mathbf{G}_{w,T}$ above L_j is expressed by multiplication of Ψ_w by the element a. We conclude that the morphism as sheaves for the fppf topology on S_0

$$\prod_{j\in\mathcal{V}} \mathrm{Pic}^0(\mathbb{D}'_j/S_0) \longrightarrow \mathrm{Pic}^0(C_0/S_0)$$

is surjective on T-valued points of $\mathrm{Pic}^0(C_0/S_0)$ corresponding to invertible sheaves on $C_{0,T}$. Hence it is surjective. For $j \in \mathcal{V}$ and $w \in W_j$ denote by K_w the kernel of $\mathbb{G}_w \to \mathrm{Pic}^0(\mathbb{D}_j/S_0)$. For $t = 1, \ldots, r$ denote by K_t with $t = 1, \ldots, r$ the kernel of the morphism $\mathbb{G}_t \to \mathrm{Pic}^0(\mathbb{D}_j/S_0)$. We also conclude that the kernel of

$$\prod_{j\in\mathcal{V}} \mathbf{G}_j \longrightarrow \mathrm{Pic}^0(C_0/S_0)$$

is the product over $t = 1, \ldots, r$ of the kernels of $\prod_{w\in W^t} K_w \longrightarrow K_t$. In particular, we get the isomorphism

$$\mathbf{G}_0 \xrightarrow{\sim} \mathrm{Pic}^0(C_0/S_0)$$

completing the diagram 13.16.1.

14 The universal cover of $\mathrm{Pic}^0(C_0/S_0)$ in the geometrically irreducible case

The main result of this section is 14.7. Let $C_0 \to S_0$ be as in 12.1. We assume that $C_0 \to S_0$ is geometrically irreducible. We construct in 14.4 a compactification $\mathbf{P}_0$ of the semiabelian scheme $\mathbf{G}_0$ defined in 13.14. In 14.7, in 14.8 and in 14.9 we use it to construct a $H_1(\Gamma, \mathbf{Z})$-Galois cover $\mathcal{P}_0$ of a compactification $\bar{P}_0$ of $\mathrm{Pic}^0(C_0/S_0)$. We construct a $H_1(\Gamma, \mathbf{Z})$-equivariant morphism u_V from a $H_1(\Gamma, \mathbf{Z})$-Galois cover $\mathcal{C}_0$ of C_0 to $\mathcal{P}_0$. In the next sections we will study the infinitesimal deformations of $\mathcal{C}_0$, of $\mathcal{P}_0$ and of u_V.

14.1 Definition. *Let $C_0 \to S_0$ be as in 12.1. Let Γ be the graph associated to C_0 in 12.3. Define the graph*

$$\epsilon\colon \Gamma^{\mathrm{ab}} \longrightarrow \Gamma$$

as the maximal abelian cover of Γ. Identify the Galois group of ϵ with $H_1(\Gamma, \mathbf{Z})$. Let $\mathcal{V}^{\mathrm{ab}}$ be the set of vertices of Γ^{ab} and let $\mathcal{E}^{\mathrm{ab}}$ be the set of edges of Γ^{ab}.

14.2 Definition. *The notation is as in 14.1. Define the $H_1(\Gamma, \mathbf{Z})$-Galois cover*

$$\mathcal{C}_0 \longrightarrow C_0$$

as follows. To any element $j \in \mathcal{V}^{\mathrm{ab}}$ we let $D_j \to S_0$ be the curve $D_{\epsilon(j)} \to S_0$ associated in 12.3 to the element $\epsilon(j) \in \mathcal{V}$. For any oriented edge $e \in \mathcal{E}^{\mathrm{ab}}$ with source $s(e)$ and target $t(e)$ we glue transversally $D_{t(e)} = D_{t(\epsilon(e))}$ and $D_{s(e)} = D_{s(\epsilon(e))}$ according to the singularity of C_0 associated to $\epsilon(e)$ in 12.3.

Assumption: In the rest of this section we will assume $C_0 \to S_0$ to be geometrically irreducible.

14.3 *Consequences of the assumption.* The assumption implies that the graph Γ associated to C_0 in 12.3 has a unique vertex j. In particular, Γ is simple in the sense of 5.3. Let $l_1, \dots, l_r$ be the basis of elementary cycles of $H_1(\Gamma, \mathbf{Z})$ fixed in 12.3. For any $t = 1, \dots, r$ the support of the elementary cycle l_t consists of a unique edge e_t of Γ. In particular, the set W^t defined in 13.1 consists of the element (t, j). With the notations of 13.1 define

$$P_t := P_{(t,j)} \qquad \text{and} \qquad Q_t := Q_{(t,j)}.$$

14.4 Definition. *For any integer s define the following effective Cartier divisors on $\mathcal{J}_0$ relative to S_0*

$$Y_{2,s} := p_j^*(t_{Q_s}(\Theta_j^-)) \qquad \text{and} \qquad Y_{1,s} := p_j^*(t_{P_s}(\Theta_j^-)),$$

where $p_j: \mathcal{J}_0 \to \mathrm{Pic}^0(\mathbb{D}_j/S_0)$ *is the natural projection. See 14.3 and 13.4 for the notation. As in 2.2 define the* $\mathbf{P}^1$*-bundle over* $\mathcal{J}_0$

$$\mathbb{P}_s := \mathbf{P}\big(Y_{2,s} \oplus Y_{1,s}\big).$$

Define

$$\mathbf{P}_0 := \prod_{t=1,\dots,r} \mathbb{P}_t,$$

where the products are taken as schemes over $\mathcal{J}_0$*. Let*

$$m_t: \mathbf{P}_0 \longrightarrow \mathbb{P}_t$$

be the natural projection. Let $\mathbf{G}_0$ *be the scheme over* $\mathcal{J}_0$ *defined in 13.14. By definition there is a natural inclusion*

$$\mathbf{G}_0 \hookrightarrow \mathbf{P}_0.$$

14.5 Definition. *For any* $t = 1, \dots, r$ *define as in 2.1*

$$0_t: \prod_{t' \neq t} \mathbb{P}_{t'} \longrightarrow \mathbf{P}_0 \qquad \text{and} \qquad \infty_t: \prod_{t' \neq t} \mathbb{P}_{t'} \longrightarrow \mathbf{P}_0.$$

14.6 Remark. The scheme $\mathbf{P}_j$ defined in 13.6 is canonically isomorphic to the restriction of $\mathbf{P}_0$ to $\mathrm{Pic}^0(\mathbb{D}_j/S_0)$. Due to 13.10 there is a morphism

$$u_j: \mathbb{D}_j \longrightarrow \mathbf{P}_0$$

factoring through $\mathbf{P}_j$. The image of $\mathbb{D}_j$ via u_j meets the section ∞_t (resp. 0_t) of $\mathbf{P}_0$ only at P_w (resp. Q_w). The multiplicities of these intersections are 1. The morphism $\mathbb{D}_j \xrightarrow{u_j} \mathbf{P}_0 \xrightarrow{m_t} \mathbb{P}_t$ for $t = 1, \dots, r$ can be described as follows. There exists an open covering $(U_i)_{i \in I}$ of $\mathcal{J}_0$ trivializing the $\mathbf{P}^1$-bundle $\mathbb{P}_t$ as $\big(\mathbf{P}^1_{S_0} \times_{S_0} U_i\big)_{i \in I}$. For each U_i intersecting the image of $\mathbb{D}_j$ the rational map from $\mathbb{D}_j$ to $\mathbf{P}^1_{S_0} \times_{S_0} U_i$ is defined by a morphism $\mathbb{D}_j \to \mathbf{P}^1_{S_0}$ which is unramified at P_t and such that P_t lies above 0 and which is unramified at Q_t and such that Q_t lies above ∞.

14.7 *The universal cover of the compactification of* $\mathrm{Pic}^0(C_0/S_0)$. Fix a vertex V of Γ^{ab}. We shall construct

1. a scheme $\mathcal{P}_0 \to \mathcal{J}_0$ of the type introduced in 7.1 with $\Lambda = \Lambda' := H_1(\Gamma, \mathbf{Z})$ and $\mathbf{G}_0 := \mathrm{Pic}^0(C_0/S_0)$. In particular, we have
   i. a morphism $\mathcal{P}_0 \to \mathcal{J}_0$ flat and locally of finite type,
   ii. an open immersion $\alpha_V: \mathbf{G}_0 \to \mathcal{P}_0$ such that the action of $\mathbf{G}_0$ on itself extends to an action on $\mathcal{P}_0 \to S_0$,
   iii an action of $H_1(\Gamma, \mathbf{Z})$ on $\mathcal{P}_0 \to S_0$ such that the quotient $\bar{P}_0 \to S_0$ exists and the action of $\mathbf{G}_0$ descends to an action on $\bar{P}_0 \to S_0$.

2. a morphism $u_V : \mathcal{C}_0 \to \mathcal{P}_0$ which is equivariant with respect to the action of $H_1(\Gamma, \mathbf{Z})$. See 14.2 for the definition of $\mathcal{C}_0$;
3. an ample effective Cartier divisor $\tilde{\Theta}_0$ on $\mathcal{P}_0$ relative to S_0 and an an action of $H_1(\Gamma, \mathbf{Z})$ on it compatible with the action on $\mathcal{P}_0$.

Summarizing we get

$$\begin{array}{ccccc} \mathcal{C}_0 & \xrightarrow{u_V} & \mathcal{P}_0 & \longleftarrow & \tilde{\Theta}_0 \\ \downarrow & & \downarrow & & \downarrow \\ C_0 & \longrightarrow & \bar{P}_0 & \longleftarrow & \bar{\Theta}_0. \end{array}$$

We try to visualize the objects just mentioned in Figure 5. We represent $\mathcal{C}_0 \to \mathcal{P}_0$ where $\mathcal{E}$ has cardinality two, i.e., $C_0 \to S_0$ has two singular sections. The rectangles stand for the compactification $\mathbf{P}_0$ of $\mathbf{G}_0 \cong \mathrm{Pic}^0(C_0/S_0)$; see 14.4. It is a $\mathbf{P}^1 \times \mathbf{P}^1$-bundle over $\mathcal{J}_0$, which is the Pic^0 of the "normalization" of C_0. The union of the rectangles (infinitely many of them!) stands for $\mathcal{P}_0$. If two rectangles have one edge in common then the corresponding copies of $\mathbf{P}_0$ will be glued transversally along a $\mathbf{P}^1$-bundle over $\mathcal{J}_0$. Inside we have also drawn the curve $\mathcal{C}_0$. Remark that $\Lambda = H_1(\Gamma) \cong \mathbf{Z}l_1 \oplus \mathbf{Z}l_2$ where $\mathcal{E} = \{l_1, l_2\}$; see 12.3. The action of $H_1(\Gamma, \mathbf{Z})$ on $\mathcal{C}_0 \to \mathcal{P}_0$ is defined on the picture as follows. The element l_1 acts by shifting the picture from left to right by one block. The element l_2 acts by shifting the picture vertically by one block.

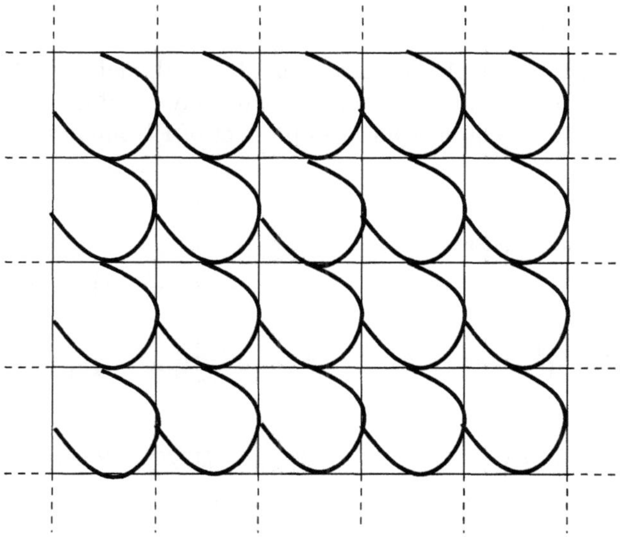

Figure 5.

14.8 Definition. *Let $A = A_1 l_1 + \ldots + A_r l_r \in H_1(\Gamma, \mathbf{Z})$. Let $\mathbf{P}_0 \to \mathcal{J}_0$ be the* *scheme defined in 14.4. Define*

$$\mathbf{P}_A \longrightarrow \mathcal{J}_0 \qquad \text{and} \qquad \nu_A : \mathbf{P}_A \xrightarrow{\sim} \mathbf{P}_0$$

by induction on the integers $A_1, \ldots, A_r$. If $A = 0$, put $\nu_0 = \mathrm{id}$. Suppose that $\mathbf{P}_A \to S_0$ and ν_A have been constructed. Let t be an integer such that $1 \le t \le r$ and let $h = \pm 1$. Let A' be in $H_1(\Gamma, \mathbf{Z})$ defined as $A + hf_t$. If $h = 1$ we define $\mathbf{P}_{A'} \to \mathcal{J}_0$ to be the pullback of $\mathbf{P}_A \to \mathcal{J}_0$ via the isomorphism

$$\begin{array}{ccc} \mathcal{J}_0 & \longrightarrow & \mathcal{J}_0 \\ x & \mapsto & x - \Big(\big[i_j(Q_t) - i_j(P_t)\big]\Big), \end{array}$$

where the sections Q_t and P_t are defined in 14.3. If $h = -1$, the definition is analogous switching the roles of P_t and Q_t. Define $\nu_{A'}$ by composing ν_A with the natural isomorphism $\mathbf{P}_{A'} \to \mathbf{P}_A$.
The pullback via the isomorphism $\nu_A : \mathbf{P}_A \to \mathbf{P}_0$ of the splitting $\mathbf{P}_0$ as product $\mathbb{P}_1 \times_{\mathcal{J}_0} \cdots \times_{\mathcal{J}_0} \mathbb{P}_r$ of 14.4 defines a splitting

$$\mathbf{P}_A \xrightarrow{\sim} \mathbb{P}_{A,1} \underset{\mathcal{J}_0}{\times} \cdots \underset{\mathcal{J}_0}{\times} \mathbb{P}_{A,r}$$

into a product of $\mathbf{P}^1$-bundles over $\mathcal{J}_0$. Define $\mathbf{G}_A$ to be the open subscheme of $\mathbf{P}_A$ corresponding via the isomorphism between $\mathbf{P}_A$ and $\mathbf{P}_0$ to $\mathbf{G}_0 \hookrightarrow \mathbf{P}_0$ as defined in 14.4. For any t in $\{1, \ldots, r\}$ we get subschemes $0_{A,t}$ and $\infty_{A,t}$ of $\mathbf{P}_A$ corresponding to the subschemes 0_t and ∞_t on $\mathbf{P}_0$ defined in 14.5.

14.9 *The construction.* 1. For any t in $\{1, \ldots, r\}$ and any A in $H_1(\Gamma, \mathbf{Z})$ glue $\mathbf{P}_A$ along the subscheme $\infty_{A,t}$ transversally to $\mathbf{P}_{A+f_t}$ along the subscheme $0_{A+f_t,t}$ via the isomorphism as product of $\mathbf{P}^1$-bundles over $\mathcal{J}_0$

$$\Phi_{A,t} : \infty_{A,t} \longrightarrow 0_{A+f_t,t}$$

defined as follows. Consider

$$\begin{array}{ccccc} \infty_{A,t} & \xrightarrow{\sim} & \infty_t & \xrightarrow{\sim} & \prod_{t' \neq t} \mathbb{P}_t \\ \downarrow & & \downarrow & & \\ \mathbf{P}_A & \xrightarrow{\nu_A} & \mathbf{P}_0 & \xrightarrow{\nu_{A+f_t}^{-1}} & \mathbf{P}_{A+f_t} \\ & & \uparrow & & \uparrow \\ \prod_{t' \neq t} \mathbb{P}_t & \xrightarrow{\sim} & 0_t & \xrightarrow{\sim} & 0_{A+f_t,t}. \end{array}$$

Then $\Phi_{A,t}$ is induced by the unique isomorphism between ∞_t and 0_t, as product of $\mathbf{P}^1$-bundles over $\mathcal{J}_0$, sending

$$\prod_{t' \neq t} \Big[m_{t'}\big(u_j(P_t)\big) \Big] \qquad \text{to} \qquad \prod_{t' \neq t} \Big[m_{t'}\big(u_j(Q_t)\big) \Big];$$

where the points Q_t and P_t are defined in 14.3 and u_j is defined in 14.6 and

$m_{t'}$ is defined in 14.4.

Proceeding with this gluing for all $A \in H_1(\Gamma, \mathbf{Z})$ we get $\mathcal{P}_0$ as a scheme over S_0. Define

$$\rho: \mathcal{P}_0 \longrightarrow \mathcal{J}_0$$

to be the unique morphism of schemes over S_0 such that for any $A \in \Lambda$ its restriction to $\mathbf{P}_A \subset \mathcal{P}_0$ is the natural morphism $\mathbf{P}_A \xrightarrow{\nu_A} \mathbf{P}_0 \to \mathcal{J}_0$. The action of $H_1(\Gamma, \mathbf{Z})$ is clearly defined. Define α_V to be the natural inclusion

$$\alpha_V: \mathbf{G}_0 \hookrightarrow \mathbf{P}_0 \longrightarrow \mathcal{P}_0$$

corresponding to the element 0 of Λ. The scheme $\mathcal{P}_0 \to S_0$ is constructed by gluing the schemes $\mathbf{P}_A$ for A in $H_1(\Gamma, \mathbf{Z})$. The scheme $\mathrm{Pic}^0(C_0/S_0) = \mathbf{G}_0$ is the group of automorphisms of $\mathbf{P}_A$ as $\{\mathbf{P}^1\}^r$-bundle over $\mathcal{J}_0$ lifting translations on $\mathcal{J}_0$. The action respects the gluing and determines an action on $\mathcal{P}_0 \to S_0$. It is compatible with the group scheme structure on $\mathrm{Pic}^0(C_0/S_0)$ and descends to an action on $\bar{P}_0 \to S_0$.

2. The curve $\mathcal{C}_0$ is constructed in 14.2. Remark that

$$\mathcal{V}^{\mathrm{ab}} = H_1(\Gamma, \mathbf{Z}) \cdot V.$$

Identify $\mathcal{V}^{\mathrm{ab}}$ with $H_1(\Gamma, \mathbf{Z})$. For any A in $\mathcal{V}^{\mathrm{ab}}$ define the morphism

$$u_A: \mathbb{D}_A \longrightarrow \mathbf{P}_A$$

by pulling back the morphism $\mathbb{D}_A = \mathbb{D}_j \xrightarrow{u_j} \mathbf{P}_0$ defined in 14.6 via the isomorphism ν_A between $\mathbf{P}_A$ and $\mathbf{P}_0$ defined in 14.8. If e is an edge of Γ^{ab} connecting the vertices A_1 and A_2 of Γ^{ab} corresponding to curves $\mathbb{D}_{A_1}$ and $\mathbb{D}_{A_2}$, the way $\mathbb{D}_{A_1}$ and $\mathbb{D}_{A_2}$ are glued according to e is compatible by construction via u_{A_1} and u_{A_2} with the way $\mathbf{P}_{A_1}$ and $\mathbf{P}_{A_2}$ are glued in $\mathcal{P}_0$. Hence we get a global morphism

$$u_V: \mathcal{C}_0 \longrightarrow \mathcal{P}_0.$$

The compatibility with the action of $H_1(\Gamma)$ is clear.

1. For any vertex A of Γ^{ab} the curve $\mathbb{D}_A \to S_0$ has a marked section

$$0_A: S_0 \to \mathbb{D}_A;$$

defined as by the section 0_j of $\mathbb{D}_j$ defined in 12.5 via $\mathbb{D}_j = \mathbb{D}_A$. Its image via u_A defines a rigidification of $\mathbf{P}_A \to \mathcal{J}_0$ as product of $\mathbf{P}^1$-bundles over $\mathcal{J}_0$. Hence we get an $H_1(\Gamma, \mathbf{Z})$-equivariant morphism

$$\amalg_{A \in \mathcal{V}^{\mathrm{ab}}} S_0 \to \mathcal{P}_0$$

factoring through the smooth locus of $\mathcal{P}_0$. The smooth locus $\mathcal{P}_0^{\mathrm{sm}}$ of $\mathcal{P}_0$ over S_0 is the disjoint union of the schemes $\mathbf{G}_A$ for $A \in H_1(\Gamma, \mathbf{Z})$. Since Γ has only one vertex, $\mathcal{P}_0^{\mathrm{sm}}$ is endowed with a unique group scheme structure descending to a group scheme structure on the smooth locus of $\bar{P}_0 \to S_0$. In fact, $\mathcal{P}_0^{\mathrm{sm}} = \mathbf{G}_0 \times H_1(\Gamma, \mathbf{Z})$

3. Let g be the relative dimension of $\mathcal{P}_0$ over $\mathcal{J}_0$. Consider the morphism

$$u_V^{(g)}: \left(\mathcal{C}_0^{\mathrm{sm}}\right)^{(g)} \to \mathcal{P}_0,$$

where (g) denotes the gth symmetric product and $\mathcal{C}_0^{\mathrm{sm}}$ stands for the smooth locus of $\mathcal{C}_0 \to S_0$. Define $\overline{\Theta}_0$ as the scheme theoretic closure via $u_V^{(g)}$ of the union over B in $H_1(\Gamma, \mathbf{Z})$ of the closed subschemes

$$\{0_B\} \underset{S_0}{\times} \left(\mathcal{C}_0^{\mathrm{sm}}\right)^{(g-1)} \subset \left(\mathcal{C}_0^{\mathrm{sm}}\right)^{(g)}.$$

The invariance under the action of $H_1(\Gamma, \mathbf{Z})$ is clear from the construction. For any element A of Λ define Θ_A to be the intersection in $\mathcal{P}_0$ of $\overline{\Theta}_0$ and $\mathbf{P}_A$. We have to verify that for each element A of Λ the closed subscheme Θ_A of $\mathbf{P}_A$ is an effective Cartier divisor relative to S_0. This is done in 14.10 and 14.11.

14.10 *The scheme Θ_A is an effective Cartier divisor in $\mathbf{P}_A$.* The statement holds if we can prove it for the completion of any local ring of S_0. It even suffices to prove it when S_0 is the spectrum of a versal deformation space of one of the fibers of $\tilde{C}_0 \to C_0 \to S_0$. In particular, we can assume S_0 to be the scheme associated to a local regular ring with algebraically closed residue field. Let $g := g(C_0)$ and let $\mathcal{H}^g$ be a connected component of $\left(\mathcal{C}_0^{\mathrm{sm}}\right)^{(g)}$. It can be identified via $\mathcal{C}_0 \to C_0$ with a connected component of the Hilbert scheme of g smooth sections of $C_0 \to S_0$.

Let A be in Λ such that $\mathcal{H}^g$ maps to $\mathbf{G}_A$. Via the isomorphism between $\mathbf{G}_A$ and $\mathbf{G}_0$ of 14.8 and via the isomorphism between $\mathbf{G}_0$ and $\mathrm{Pic}^0(C_0/S_0)$, the morphism $\mathcal{H}^g \to \mathbf{G}_0$ is induced by the map from the Hilbert scheme of smooth sections of $C_0 \to S_0$ to $\mathrm{Pic}^0(C_0/S_0)$. Let j be the only element in $\mathcal{V}$. Denote by d_j the genus of $\mathbb{D}_j$. In this case C_0 is the curve $\mathbb{D}'_j$ defined in 13.1 and $g = d_j + r$. Let A be in Λ and let ν_A be the isomorphism between $\mathbf{P}_A$ and $\mathbf{P}_0$ defined in 14.8. The intersection in $\mathcal{P}_0$ of the scheme theoretic image of $(\mathcal{C}_0^{\mathrm{sm}})^{(g)}$ and of the closed subscheme $\mathbf{P}_A$ is the inverse image via ν_A of the scheme theoretic image of

$$U := \left\{\mathbb{D}_j^{\prime\,\mathrm{sm}}\right\}^{(g)} \longrightarrow \mathbf{P}_0.$$

The map is the one induced from the morphisms $u_j: \mathbb{D}_j \to \mathbf{P}_0$ of 14.6. From now on we identify $\mathbf{G}_0$ with the subfunctor of the Picard functor of C_0 relative to S_0 consisting of elements whose restriction to any geometric point c of S_0 lie in the component of $\mathrm{Pic}\big(C_{0,c}/k(c)\big)$ of degree g. In this way the morphism $U \to \mathbf{G}_0$ is the natural one. Then Θ_A corresponds to the scheme theoretic image of the scheme Z defined as

$$Z := \{0_j\} \times \left\{\mathbb{D}_j^{\prime\,\mathrm{sm}}\right\}^{(g-1)},$$

where the section 0_j of $\mathbb{D}_j \to S_0$ is defined in 12.5. We claim that the scheme theoretic image of Z defines an effective Cartier divisor in $\mathbf{P}_0$ relative to S_0.

As in [4, Lemma 9.3/4], there exists a non-empty open subscheme W of the scheme U mapping finite étale to an open subscheme of $\mathbf{G}_0$ of which it is the inverse image in U. The open subscheme W is defined by taking all points u of U such that if c is the image of u in S_0, the first cohomology group of the universal Cartier divisor M_u on $C_{0,c}$ associated to u is zero. Let c be a geometric point of S_0. The complement of the set theoretic image of W_c in $\mathbf{G}_{0,c}$ consists either of invertible sheaves associated to Cartier divisors supported on the complement of the smooth locus of $C_{0,c} \to k(c)$ or of invertible sheaves associated to Cartier divisors M_c supported on the smooth locus of $C_{0,c} \to k(c)$ such that the dimension of the $k(c)$-vector space $H^0(C_{0,c}, O_{C_{0,c}}(M_c))$ is ≥ 2. Hence the complement of the set theoretic image of W_c in $\mathbf{G}_{0,c}$ is closed and has codimension at least 2. The intersection of Z with W is an effective Cartier divisor relative to S_0.

The group scheme $\mathbf{G}_0$ is the automorphism group scheme of $\mathbf{P}_0$ as product of rigidified $\mathbf{P}^1$-bundles. Let us fix an integer $t \in \{1, \ldots, r\}$. The morphism u_j of 13.10 induces a commutative diagram

$$\begin{array}{ccc} \mathbf{P}_0 \times \mathbf{G}_0 & \longrightarrow & \mathbf{P}_j \\ \uparrow & & \uparrow \\ \mathbb{D}_j \times \{\mathbb{D}_j'^{\,\mathrm{sm}}\}^{(g-1)} & \longrightarrow & \mathbf{P}_j . \end{array}$$

Let R be P_t or Q_t; see 14.3 for the notation. We can find as above a maximal open subscheme V_R of $\{\mathbb{D}_j'^{\,\mathrm{sm}}\}^{(g-1)}$ such that the morphism from V_R to $\prod_{s\neq t} \mathbb{P}_s$ is finite and étale; see 14.4 for the notation. By 14.6 it is clear that there exists an open subscheme U_R, product of an open subscheme of $\mathbb{D}_j$ containing R and the scheme V_R, such that the map from U_R to $\mathbf{P}_0$ is étale. We conclude that the pullback of Z to U_R is an effective Cartier divisor relative to S_0.

Hence Θ_A is an effective Cartier divisor relative to S_0 fiberwise in codimension 2 in $\mathbf{P}_0$. Since $\mathbf{P}_0$ is a regular scheme, it is locally factorial. Hence Θ_A is an effective Cartier divisor. Moreover, the formation of Θ_A commutes with taking the fibers above geometric points c of S_0 since this holds in codimension 2 in $\mathbf{P}_{0,c}$ and $\Theta_{A,c}$ is locally defined by one equation. But then Θ_A is also an effective Cartier divisor relative to S_0 for example applying [4, Lemma 8.2/6(iii)].

14.11 Lemma. *Let A be an element of $\Lambda = H_1(\Gamma, \mathbf{Z})$. Let Θ_A be the restriction to $\mathbf{P}_A$ of the divisor on $\mathcal{P}_0$ defined in (3) of 14.7. The divisor Θ_A defines the same invertible sheaf as the sum of the following two effective Cartier divisors. One is the pullback of the theta divisor Θ_j on $\mathrm{Pic}^0(\mathbb{D}_j/S_0)$ defined in 12.6. The other is the sum over $t = 1, \ldots, r$ of the pullbacks via m_t of $O_{\mathbb{P}_t}(1)$; see 14.4 for the notation.*

Proof: We proceed by induction on the relative dimension r of $\mathbf{P}_j$ over the abelian scheme $\mathrm{Pic}^0(\mathbb{D}_j/S_0)$. The case $r = 0$ is trivial. Now we proceed with the inductive step. We choose t between 1 and r. By the study of Θ_A in a neighborhood of the subschemes 0_t and ∞_t of $\mathbf{P}_A$ in 14.10, we know that the intersection of Θ_A with the section ∞_t (resp. 0_t) is the translation by P_t (resp. Q_t) of a Cartier divisor $\Theta_{A,t}$ on $\prod_{s\neq t}\mathbb{P}_s$. We have proven in 14.10 that the projection of Θ_A to $\prod_{s\neq t}\mathbb{P}_s$ is birational. Hence Θ_A is ample relatively to $\prod_{s\neq t}\mathbb{P}_s$ and locally isomorphic to the pullback of $O_{\mathbb{P}_t}(1)$ via m_t. Let $C_{0,t}$ be obtained from C_0 by separating the branches at the singularities corresponding to edge e_t defined in 14.3. The scheme $\prod_{s\neq t}\mathbb{P}_s$ is the compactification of the Pic^0 of the curve $C_{0,t}$ constructed as in in 14.4. By the discussion in 14.10 the divisor $\Theta_{A,t}$ is the divisor defined by $C_{0,t}$ analogously to the definition of the divisor Θ_A in 14.10. The inductive hypothesis applied to $\prod_{s\neq t}\mathbb{P}_s$ and $\Theta_{A,t}$ allows us to finish the proof.

15 The deformation theory of $C_0 \to \bar{P}_0$ in the geometrically irreducible case

As in the previous section we assume that $C_0 \to S_0$ is *geometrically irreducible.* We study the deformation theories of $C_0 \to S_0$, of $\bar{P}_0 \to S_0$ and of $C_0 \to \bar{P}_0$. See 14.7 for the notation. The main results of this section are 15.4 and 15.9. They state the following. Let $C \to S$ be a deformation of the curve $C_0 \to S_0$ over a nilpotent thickening S of S_0. Let $\mathcal{C} \to C$ be the $H_1(\Gamma,\mathbf{Z})$-cover deforming $\mathcal{C}_0 \to C_0$ defined in 14.2. Let $\hat{0}_V$ be a section of $\mathcal{C} \to S$ deforming the section 0_V of $\mathcal{C}_0 \to S_0$. Then there exists a unique deformation $\mathcal{P} \to S$ of $\mathcal{P}_0 \to S_0$ in $\mathcal{D}ef\mathcal{G}r(\mathcal{P}_0, H_1(\Gamma,\mathbf{Z}))(S)$, see 9.5, such that the morphism $\mathcal{C}_0 \to \mathcal{P}_0$ extends to a $H_1(\Gamma,\mathbf{Z})$-equivariant morphism $\mathcal{C} \to \mathcal{P}$ and $\hat{0}_V$ is sent to the distinguished section of $\mathcal{P} \to S$. Moreover, in this case $\mathcal{P}$ carries an ample effective Cartier divisor relative to S deforming the effective Cartier divisor on $\mathcal{P}_0$ relative to S_0 defined in 14.7. Finally, the deformation of the "singularities" of $\mathcal{P}_0 \to S_0$ is completely determined by the deformation of the "singularities" of $\mathcal{C}_0 \to S_0$.

15.1 *A recall of 14.7*. In what follows we assume we are in the specific situation of 14.7. In this case $\Lambda = \Lambda' = H_1(\Gamma,\mathbf{Z})$. There exists a $H_1(\Gamma,\mathbf{Z})$-Galois cover $\mathcal{C}_0 \to C_0$ and a map $u_V\colon \mathcal{C}_0 \to \mathcal{P}_0$ which is $H_1(\Gamma,\mathbf{Z})$-equivariant.

15.2 Remark. Let $T\colon S_0 \to \mathcal{C}_0$ be a section of $\mathcal{C}_0 \to S_0$ associated to an edge of Γ^{ab}; see 14.7. In 14.9 we have described a morphism

$$u_V\colon \mathcal{C}_0 \longrightarrow \mathcal{P}_0.$$

From its construction it follows that the local chart of $\mathcal{P}_0$ at $u_V \circ T(S_0)$ as in 7.4 can be chosen of the form $\mathbb{A}[z_1,\ldots,z_{r-1},\gamma_1,\beta_1]/(\gamma_1\beta_1)$.Indeed, if $u_V \circ T$ factors via $\mathbf{P}_A \subset \mathcal{P}_0$ then it factors via one and only one of the subschemes $\infty_{A,1},\ldots,\infty_{A,r}$ or $0_{A,1},\ldots,0_{A,r}$ of $\mathbf{P}_A$. The completion of $\mathcal{C}_0$ at T is locally on S_0 of the form $O_{S_0}[\![\gamma_1,\beta_1]\!]/(\gamma_1\beta_1)$.

15.3 Proposition. *By 15.2 we get an isomorphism*

$$\prod_{e\in\mathcal{E}^{ab}} H^0\Big(\mathcal{C}_0, \mathcal{E}xt^1_{O_{\mathcal{C}_0}}(\Omega^1_{\mathcal{C}_0/S_0}, O_{\mathcal{C}_0})|_{T_e}\Big) \xrightarrow{\sim} H^0\Big(\mathcal{P}_0, \mathcal{E}xt^1_{O_{\mathcal{P}_0}}(\Omega^1_{\mathcal{P}_0/S_0}, O_{\mathcal{P}_0})\Big),$$

where $\mathcal{E}^{ab}$ is the set of edges of the maximal abelian cover Γ^{ab} of Γ defined in 14.1. We recall that $\mathcal{E}^{ab}$ is also the set of non-disconnecting edges of the graph Γ^{ab} associated to $\mathcal{C}_0$ in 14.2. The notation $|_{T_e}$ stands for the restriction to the section of $\mathcal{C}_0 \to S_0$ associated to the edge e.

15.4 Corollary. *We have an isomorphism*

$$\prod_{e\in\mathcal{E}} H^0\Big(\mathcal{E}xt^1_{O_{C_0}}(\Omega^1_{C_0/S_0}, O_{C_0})|_{T_e}\Big) \longrightarrow H^0\Big(\mathcal{E}xt^1_{O_{\bar{P}_0}}(\Omega^1_{(\bar{P}_0)/S_0}, O_{\bar{P}_0})\Big),$$

where $\mathcal{E}$ are the non-disconnecting edges of the graph Γ associated in 12.3 to $C_0 \to S_0$.

Proof: The isomorphism follows from the isomorphism:

$$H^0(\mathcal{E}xt^1_{O_{C_0}}(\Omega^1_{C_0/S_0}, O_{C_0})) \xrightarrow{\sim} H^0(\mathcal{E}xt^1_{O_{\mathcal{C}_0}}(\Omega^1_{\mathcal{C}_0/S_0}, O_{\mathcal{C}_0}))^{H_1(\Gamma)}.$$

15.5 Definition. *Let Γ^{ab} be the graph defined in 14.1. The choice of the elementary cycles $l_1, \ldots, l_r$ in 12.3 determines by (4) of 5.4 a unique orientation of Γ and consequently of Γ^{ab}. For any $e \in \mathcal{E}^{ab}$ let*

$$T_e\colon S_0 \longrightarrow \mathcal{C}_0$$

be the singular section associated to e. Consider on e the orientation induced from the one on Γ^{ab}. Denote by $s(e)$ the source of e and by $t(e)$ the target of e. Let P_e and Q_e be the inverse images of T_e via $\amalg_A \mathbb{D}_A \to \mathcal{C}_0$ so that P_e belongs to $\mathbb{D}_{s(e)}$ and Q_e belongs to $\mathbb{D}_{t(e)}$.

15.6 *The hyperresolution of a curve.* In 8.1 we have constructed a cubical hyperresolution of $\mathcal{P}_0 \to S_0$. It served the purpose of computing the cohomology the tangent sheaf of $\mathcal{P}_0 \to S_0$. The first step of the hyperresolution was defined by the natural morphism from $X^1_{0,1} := \amalg_{A\in\Lambda}\mathbf{P}_A$ to $\mathcal{P}_0 =: X^1_{0,0}$. If we pullback it back via u_V, we get the standard cubical hyperresolution of $\mathcal{C}_0$ over S_0 defined by the cartesian square

$$\begin{array}{ccc} \amalg_{A\in\mathcal{V}^{ab}}\Big(\amalg_{e|A=s(e)}P_e\ \amalg_{e|A=t(e)}\ Q_e\Big) & \longrightarrow & \amalg_{A\in\mathcal{V}^{ab}}\mathbb{D}_A \\ \big\downarrow & & \big\downarrow \\ \amalg_{e\in\mathcal{E}^{ab}}T_e & \longrightarrow & \mathcal{C}_0. \end{array}$$

15.7 Lemma. *The notation is as in 15.6.*

1. *We have the following short exact sequence:*

$$0 \to \mathcal{H}om_{O_{\mathcal{P}_0}}\left(\Omega^1_{\mathcal{P}_0/\mathcal{J}_0}, u_{V*}(O_{\mathcal{C}_0})\right) \to \underset{\beta=\alpha,\delta}{\oplus} \mathcal{A}_\beta \underset{O_{\mathcal{P}_0}}{\otimes} O_{\mathcal{C}_0} \to \mathcal{A}_\mu \underset{O_{\mathcal{P}_0}}{\otimes} O_{\mathcal{C}_0} \to 0$$

where $\mathcal{A}_\alpha = \mathcal{A}_{0,\ldots,0,0,1}$ *and* $\mathcal{A}_\delta = \mathcal{A}_{0,\ldots,0,1,0}$ *and* $\mathcal{A}_\mu = \mathcal{A}_{0,\ldots,0,1,1}$ *are the sheaves defined in 8.7.*

2. *We have isomorphisms:*

$$H^1\Big(\mathcal{C}_0, \mathcal{H}om_{O_{\mathcal{C}_0}}\big(u_V{}^*(\Omega^1_{\mathcal{P}_0/\mathcal{J}_0}), O_{\mathcal{C}_0}\big)\Big) \cong$$
$$\cong H^1\Big(\mathcal{P}_0, \mathcal{H}om_{O_{\mathcal{P}_0}}\big(\Omega^1_{\mathcal{P}_0/\mathcal{J}_0}, u_{V*}(O_{\mathcal{C}_0})\big)\Big) \cong \prod_{A\in\mathcal{V}^{\mathrm{ab}}} H^1(\mathbb{D}_A, O_{\mathbb{D}_A})^r.$$

Proof: 1. The proof is local on $\mathcal{P}_0$. It suffices to prove the exactness on the stalks at every closed point of $u_V(\mathcal{C}_0)$. Let x be a closed point of $\mathcal{C}_0$. If $u_V(x)$ is in the smooth locus of $\mathcal{P}_0 \to S_0$, this is trivial. If $u_V(x)$ is not contained in the smooth locus of $\mathcal{P}_0 \to S_0$, we choose an affine neighborhood $\mathrm{Spec}(\mathbb{B}_0)$ of $u_V(x)$ as in 7.4 of which we use the notations. We let $\mathbb{B}_0 \to \mathbb{F}_0$ represent $O_{\mathcal{P}_0} \to u_{V*}(O_{\mathcal{C}_0})$ on $\mathrm{Spec}(\mathbb{B}_0)$. Since $u_V(x)$ lies on the image of $\mathcal{C}_0$, the remark 15.2 implies that the integer v in 7.4 is equal to 1. Moreover the completion of $\mathbb{F}_0$ at the complement of the smooth locus of $\mathrm{Spec}(\mathbb{F}_0) \to S_0$ is isomorphic to $O_{S_0}[\![\gamma_1, \beta_1]\!]/(\gamma_1\beta_1)$. The lemma follows since on the open subscheme $\mathrm{Spec}(\mathbb{B}_0)$ of $\mathcal{P}_0$ we have

$$\mathcal{H}om_{O_{\mathcal{P}_0}}\big(\Omega^1_{\mathcal{P}_0/\mathcal{J}_0}, O_{\mathcal{C}_0}\big) \cong \overset{h}{\underset{i=1}{\oplus}} \mathbb{F}_0 \frac{\partial}{\partial z_i} \oplus \mathbb{F}_0\gamma_1 \frac{\partial}{\partial \gamma_1} \oplus \mathbb{F}_0\beta_1 \frac{\partial}{\partial \beta_1}$$

$$\mathcal{A}_\alpha \underset{O_{\mathcal{P}_0}}{\otimes} O_{\mathcal{C}_0} \cong \overset{h}{\underset{i=1}{\oplus}} \mathbb{F}_0/(\gamma_1) \frac{\partial}{\partial z_i} \oplus \mathbb{F}_0/(\gamma_1)\beta_1 \frac{\partial}{\partial \beta_1} \overset{h}{\underset{i=1}{\oplus}} \mathbb{F}_0/(\beta_1) \frac{\partial}{\partial z_i} \oplus \mathbb{F}_0/(\beta_1)\gamma_1 \frac{\partial}{\partial \gamma_1}$$

$$\mathcal{A}_\delta \underset{O_{\mathcal{P}_0}}{\otimes} O_{\mathcal{C}_0} \cong \overset{h}{\underset{i=1}{\oplus}} O_{S_0} \frac{\partial}{\partial z_i} \quad \text{and} \quad \mathcal{A}_\mu \underset{O_{\mathcal{P}_0}}{\otimes} O_{\mathcal{C}_0} \cong \overset{h}{\underset{i=1}{\oplus}} O_{S_0} \frac{\partial}{\partial z_i} \overset{h}{\underset{i=1}{\oplus}} O_{S_0} \frac{\partial}{\partial z_i}.$$

2. The first isomorphism follows from 13.11, i.e., from the fact that u_V contracts only some of the rational components of $\mathcal{C}_0$. The second isomorphism follows from (1) by taking the long exact sequence of cohomology groups and computing that the sequence of H^0's is exact.

15.8 *A comparison of tangent spaces of deformation functors.* The notation is as in 15.6.

With the notation $\mathcal{D}_{\mathcal{P}_0/\_} := \mathcal{H}om_{O_{\mathcal{P}_0}}(\Omega^1_{\mathcal{P}_0/\_}, O_{\mathcal{C}_0})$ and denoting by $\mathcal{T}$ the tangent sheaf we have the following diagram with exact rows

$$\begin{array}{ccccc} H^1(\mathcal{P}_0, \mathcal{T}_{\mathcal{P}_0/\mathcal{J}_0}) & \longrightarrow & H^1(\mathcal{P}_0, \mathcal{T}_{\mathcal{P}_0/S_0}) & \longrightarrow & H^1(\mathcal{J}_0, \mathcal{T}_{\mathcal{J}_0/S_0}) \\ \downarrow & & \downarrow & & \downarrow \\ H^1(\mathcal{C}_0, \mathcal{D}_{\mathcal{P}_0/\mathcal{J}_0}) & \longrightarrow & H^1(\mathcal{C}_0, \mathcal{D}_{\mathcal{P}_0/S_0}) & \longrightarrow & H^1(\mathcal{C}_0, O_{\mathcal{C}_0} \otimes \mathcal{T}_{\mathcal{J}_0/S_0}). \end{array}$$

In the first row we have made use of 8.5. The group

$$H^1\big(\mathcal{C}_0, \mathcal{H}om(\Omega^1_{\mathcal{P}_0/S_0}, O_{\mathcal{C}_0})\big)$$

is the space of "global obstructions" to deform $u_{\mathcal{V}}\colon \mathcal{C}_0 \to \mathcal{P}_0$. Remark that the induced morphism

$$H^1(\mathcal{J}_0, \mathcal{T}_{\mathcal{J}_0/S_0}) \longrightarrow H^1(\mathcal{C}_0, O_{\mathcal{C}_0} \otimes \mathcal{T}_{\mathcal{J}_0/S_0})^{H_1(\Gamma,\mathbf{Z})}$$

is an isomorphism since

$$H^1(\mathcal{C}_0, O_{\mathcal{C}_0} \otimes \mathcal{T}_{\mathcal{J}_0/S_0}) \cong \bigoplus_{H_1(\Gamma,\mathbf{Z})} H^1(D_0, \mathcal{T}_{\mathcal{J}_0/S_0}).$$

Define $\overline{\mathrm{Ext}}^1(\Omega^1_{\mathcal{C}_0/S_0}, O_{\mathcal{C}_0})$ as

$$\mathrm{Ker}\Big(\mathrm{Ext}^1_{O_{\mathcal{C}_0}}(\Omega^1_{\mathcal{C}_0/S_0}, O_{\mathcal{C}_0}) \longrightarrow \prod_{e\in\Gamma^{\mathrm{ab}}} H^0\big(\mathcal{C}_0, \mathcal{E}xt^1_{O_{\mathcal{C}_0}}(\Omega^1_{\mathcal{C}_0/S_0}, O_{\mathcal{C}_0})|_e\big)\Big).$$

The local-global spectral sequence implies that the group just defined is the tangent space relative to S_0 of the functor of deformations of $C_0 \to S_0$ to nilpotent thickenings of S_0 preserving the singularities T_e with $e \in \mathcal{E}$ defined in 12.3.

With the notation $\mathcal{D}_{\mathcal{P}_0/\_} := \mathcal{H}om_{O_{\mathcal{P}_0}}(\Omega^1_{\mathcal{P}_0/\_}, O_{\mathcal{C}_0})$ we have the following diagram with exact rows

$$\begin{array}{ccccc}
\bigoplus_{e\in\mathcal{V}^{\mathrm{ab}}}(O_{P_e}\oplus O_{Q_e}) & \longrightarrow & \overline{\mathrm{Ext}}^1(\Omega^1_{\mathcal{C}_0/S_0}, O_{\mathcal{C}_0}) & \longrightarrow & \prod_{A\in\mathcal{V}^{\mathrm{ab}}} \mathrm{Ext}^1(\Omega^1_{\mathbb{D}_A/S_0}, O_{\mathbb{D}_A}) \\
\downarrow & & \downarrow & & \downarrow \\
H^1(\mathcal{C}_0, \mathcal{D}_{\mathcal{P}_0/\mathcal{J}_0}) & \longrightarrow & H^1(\mathcal{C}_0, \mathcal{D}_{\mathcal{P}_0/S_0}) & \longrightarrow & H^1(\mathcal{C}_0, O_{\mathcal{C}_0}\otimes\mathcal{T}_{\mathcal{J}_0/S_0}).
\end{array}$$

The first row is induced by the map $\amalg_{A\in\mathcal{V}^{\mathrm{ab}}}\mathbb{D}_A \to \mathcal{C}_0$ identifying

$$\mathcal{H}om(\Omega^1_{\mathcal{C}_0/S_0}, \amalg_{A\in\mathcal{V}^{\mathrm{ab}}} O_{\mathbb{D}_A}) \xrightarrow{\sim} \prod_{A\in\mathcal{V}^{\mathrm{ab}}} \mathcal{H}om(\Omega^1_{\mathbb{D}_A/S_0}, O_{\mathbb{D}_A}).$$

The central vertical map is induced comparing the local global spectral sequences of the tangent sheaves on $\mathcal{C}_0$ and $\mathcal{P}_0$ relative to S_0 via the map $u_{\mathcal{V}}\colon \mathcal{C}_0 \to \mathcal{P}_0$ and using 15.3. Analogously the vertical map on the right is defined by the morphism $\amalg_{A\in\mathcal{V}^{\mathrm{ab}}}\mathbb{D}_A \to \mathcal{J}_0$. The diagram can also be induced from maps on tangent spaces among suitable functors. Indeed, to a deformation $\mathcal{C}$ of $\mathcal{C}_0$ to the scheme of dual numbers $S_0[\epsilon/\epsilon^2]$ over S_0 preserving the singular sections T_e with $e \in \mathcal{E}$ we can associate deformations D_A of the components $\mathbb{D}_A$ with A in $\mathcal{V}^{\mathrm{ab}}$. To $\mathcal{C}$ we can associate the obstruction to deform $\mathcal{C}_0 \to \mathcal{P}_0$ (resp. $\mathcal{C}_0 \to \mathcal{J}_0$) to a map from $\mathcal{C}$ to the canonical deformation of $\mathcal{P}_0$ (resp. $\mathcal{J}_0$) to $S_0[\epsilon/\epsilon^2]$. On the other hand for each $A \in \mathcal{V}^{\mathrm{ab}}$ we can associate to D_A the obstruction to deform $\mathbb{D}_A \to \mathcal{J}_0$ to a map from D_A to the canonical deformation of $\mathcal{J}_0$ to S. The obstructions to deform $\mathcal{C}_0 \to \mathcal{J}_0$ and $\amalg_{A\in\mathcal{V}^{ab}}\mathbb{D}_A \to \mathcal{J}_0$ are the same. This also proves that the diagram above commutes.

15.9 Theorem. *The notation is as in 14.7. Let S be a nilpotent thickening of S_0. Let $C \to S$ be a deformation of $C_0 \to S_0$. Let $\mathcal{C} \to C$ be the $H_1(\Gamma, \mathbf{Z})$-Galois cover deforming the cover $\mathcal{C}_0 \to C_0$ defined in 14.2. There exists a unique deformation $\bar{P} \to S$ of $\bar{P}_0 \to S_0$ with the following properties.*

i. *$\bar{P} \to S$ defines an S-valued point of the functor $\mathcal{D}ef\mathcal{G}r(\bar{P}_0)$ defined in 9.5.*

ii. *The morphism $C_0 \to \bar{P}_0$ defined in (2) of 14.7 extends to a morphism $C \to \bar{P}$.*

iii. *The scheme $\bar{P}$ contains a canonical effective Cartier divisor $\overline{\Theta}$ relative to S deforming the divisor $\overline{\Theta}_0$ on $\bar{P}_0$ defined in (3) of 14.7.*

Equivalently there exists a unique deformation $\mathcal{P} \to S$ of $\mathcal{P}_0 \to S_0$ with the following properties.

i'. *$\mathcal{P} \to S$ defines an S-valued point of the functor $\mathcal{D}ef\mathcal{G}r(\mathcal{P}_0, H_1(\Gamma, \mathbf{Z}))$.*

ii'. *The morphism $\mathcal{C}_0 \to \mathcal{P}_0$ extends to a $H_1(\Gamma, \mathbf{Z})$-equivariant morphism $\mathcal{C} \to \mathcal{P}$.*

iii'. *The scheme $\mathcal{P}$ contains a $H_1(\Gamma, \mathbf{Z})$-invariant canonical effective Cartier divisor $\tilde{\Theta}$ relative to S deforming the divisor defined in 14.7.*

Moreover, if $\bar{0}_j: S \to C$ is a deformation of the section $0_j: S_0 \to C_0$ defined in 12.5, there is a morphism $C \to \bar{P}$ extending to S the morphism $C_0 \to \bar{P}_0$ so that the induced section $\bar{0}: S \to C \to \bar{P}$ is the distinguished section defined in 9.5. Finally if $e \in \mathcal{E}$ and T_e is the section of $C_0 \to S_0$ associated to e in 12.3 the morphism from the completion of $\bar{P}$ at T_e to the completion of C at T_e is of the form

$$\mathbb{A}[\![z_1, \ldots, z_{r-1}, \beta, \gamma]\!]/(\beta\gamma - q_e) \longrightarrow R[\![\beta, \gamma]\!]/(\beta\gamma - q_e).$$

Proof: By (iii) and descent theory, we can reduce to the case that S_0 is the spectrum of a field. Let $\hat{0}_V$ be the unique section of of $\mathcal{C} \to S$ above $\bar{0}_j$ deforming the section 0_V of $\mathcal{C}_0 \to S_0$ defined in 14.7.

(i), (ii) It suffices to prove the theorem assuming that $S_n \hookrightarrow S_{n+1} = S$ is defined by an ideal I_{n+1} such that $I_{n+1}I = 0$ and that the theorem holds for S_n. By assumption there exists a S_n-valued point $\mathcal{P}_n$ of $\mathcal{D}ef\mathcal{G}r(\mathcal{P}_0, H_1(\Gamma, \mathbf{Z}))$ and a $H_1(\Gamma, \mathbf{Z})$-equivariant morphism $u_n: \mathcal{C}_n \to \mathcal{P}_n$ over S_n. Let $\mathcal{P} \to S$ be an S-valued point of $\mathcal{D}ef\mathcal{G}r(\mathcal{P}_0, H_1(\Gamma, \mathbf{Z}))$ deforming $\mathcal{P}_n$. It exists by 9.9. Suppose that the morphism $\mathcal{C}_n \to \mathcal{P}_n$ extends $H^1(\Gamma, \mathbf{Z})$-equivariantly locally on a neighborhood of the sections T_e of $\mathcal{C}_n$ defined in 15.5. This can be achieved because of 15.3 and 14.6. This fixes the structure of $\mathcal{P} \to S$ at the complement of its smooth locus. Hence by formal smoothness, u_n can be extended $H_1(\Gamma, \mathbf{Z})$-equivariantly to morphisms from open subschemes of $\mathcal{C}$, covering $\mathcal{C}$, to $\mathcal{P}$. Due to 15.8 the obstruction to extend globally u_n lives in

$$H^1\big(\mathcal{C}_0, \mathcal{H}om(\Omega^1_{\mathcal{P}_0/S_0}, O_{\mathcal{C}_0})\big) \otimes I_{n+1}.$$

The image of this obstruction in $H_1(\mathcal{C}_0, \mathcal{O}_{\mathcal{C}_0} \otimes \mathcal{T}_{\mathcal{J}_0/S_0}) \otimes I_{n+1}$ is invariant under $H_1(\Gamma, \mathbf{Z})$. This determines a *unique* deformation $\mathcal{J}$ of $\mathcal{J}_n$ as an abelian scheme over S such that the composition of u_n with $\mathcal{P}_n \to \mathcal{J}_n$ extends to a morphism $\mathcal{C} \to \mathcal{J}$ and the image of $\hat{0}_V$ is the identity element of $\mathcal{J}$. After a modification of $\mathcal{P} \to S$ we can suppose that $\mathcal{P}_n \to \mathcal{J}_n$ extends to a morphism $\mathcal{P} \to \mathcal{J}$. The fact that such a modification exists is deduced from the surjectivity of the morphism

$$H^1(\mathcal{P}_0, \mathcal{T}_{\mathcal{P}_0/S_0}) \to H^1(\mathcal{J}_0, \mathcal{T}_{\mathcal{J}_0/S_0})$$

in 15.8. The obstruction b to deform u_n to S is in

$$H^1(\mathcal{C}_0, \mathcal{H}om(\Omega^1_{\mathcal{P}_0/\mathcal{J}_0}, \mathcal{O}_{\mathcal{C}_0})) \otimes I_{n+1}$$

and is $H_1(\Gamma, \mathbf{Z})$-invariant. Hence by (2) of 15.7 it lies in the module of $H_1(\Gamma, \mathbf{Z})$-invariants of

$$\prod_{A \in \mathcal{V}^{\mathrm{ab}}} H^1(\mathbb{D}_A, \mathcal{O}_{\mathbb{D}_A})^r \otimes I_{n+1}$$

and can be written as $(b_A)_{A \in \mathcal{V}^{\mathrm{ab}}}$. The composite map from $H^1(\mathbf{P}_0, \mathcal{A}_0)$ to the $H_1(\Gamma, \mathbf{Z})$-invariants of $\oplus_{A \in \mathcal{V}^{\mathrm{ab}}} H^1(\mathbb{D}_A, \mathcal{O}_{\mathbb{D}_A})^r$ is an isomorphism. Indeed, by (ii) of 9.1 this is the map from r times the tangent space of the dual of $\mathcal{J}_0$ to r times the product of the tangent spaces of the duals of $\mathrm{Pic}^0(\mathbb{D}_j/S_0)$ with $j \in \mathcal{V}$. Hence there exists a unique element $a_0 \in H^1(\mathbf{P}_0, \mathcal{A}_0) \otimes I_{n+1}$ such that for any vertex A of Γ^{ab} considering

$$\begin{array}{ccccc} H^1(\mathbf{P}_0, \mathcal{A}_0) \otimes I_{n+1} & \xrightarrow{d\psi_A} & H^1(\mathbf{P}_A, \mathcal{A}_A) \otimes I_{n+1} & \longrightarrow & H^1(\mathbb{D}_A, \mathcal{O}_{\mathbb{D}_A})^r \otimes I_{n+1} \\ a_0 & \mapsto & a_A. & & \end{array}$$

we have $a_A \mapsto b_A$. Using (ii) of 9.7 we get a unique S-valued point $\mathcal{P} \to S$ of the functor $\mathcal{D}ef\mathcal{G}r(\mathcal{P}_0)$ defined in 9.5 such that $u_n \colon \mathcal{C}_n \to \mathcal{P}_n$ extends to a morphism $\mathcal{C} \to \mathcal{P}$. A slight modification of the argument just given shows that we can find $\mathcal{P}$ and a morphism $\mathcal{C} \to \mathcal{P}$ such that the identity element for the group structure on the smooth locus of $\mathcal{P} \to S$ is the image of $\hat{0}_V$. Let $l \in H_1(\Gamma, \mathbf{Z})$. The obstruction to deform the action of l on $\mathcal{P}_n$ to an action on $\mathcal{P} \to S$ is trivial by (2) of 9.9. The requirement that $\mathcal{C} \to \mathcal{P}$ is $H_1(\Gamma, \mathbf{Z})$-equivariant forces the lifting to be unique.

15.10 *The proof of* (iii). Let g be the genus of $C_0 \to S_0$. For any element A of Λ let G_A be the open subscheme of $\mathcal{P}$ deforming the open subscheme $\mathbf{G}_A$ as in 9.2. Denote by $\tilde{G}$ the open subscheme associated to $0 \in \Lambda$. Let $\bar{0}_j$ be a section of C over S. Let $\amalg_{A \in H_1(\Gamma, \mathbf{Z})} \hat{0}_A$ be the inverse image of $\bar{0}_j$ via $\mathcal{C} \to C$. Since $\mathcal{P}^{\mathrm{sm}} = \mathcal{P}^{\mathrm{sm},0}$ is a group scheme over S by 9.5, there is a morphism

$$(\mathcal{C}^{\mathrm{sm}})^{(g)} \to \mathcal{P}^{\mathrm{sm}}.$$

Let W^0_{g-1} be the set theoretic image of the union over $A \in H_1(\Gamma, \mathbf{Z})$ of

$$\{\hat{0}_A\} \underset{S}{\times} (\mathcal{C}^{\mathrm{sm}})^{(g-1)}$$

in $\mathcal{P}^{\mathrm{sm}}$. It is an open subscheme of its closure. For every B in Λ consider the open subscheme defined by the points

$$(\hat{0}_B, R_1, \ldots, R_{g-1}) \in \{\hat{0}_B\} \underset{S}{\times} (\mathcal{C}^{\mathrm{sm}})^{(g-1)}$$

such that for all geometric points $c \in S$

$$\dim_{k(c)} H^0(C_c, O_{C_c}(\bar{0}_j + R_1 + \ldots R_{g-1})) = 1.$$

Define W^1_{g-1} to be the closed subset of W^0_{g-1} whose complement is the union over $B \in H_1(\Gamma, \mathbf{Z})$ of the open subschemes just constructed.

We claim two things. The first is that the map from the inverse image of the scheme $W^0_{g-1} \backslash W^1_{g-1}$ in the union over $A \in \Lambda$ of $\{\hat{0}_A\} \times_S (\mathcal{C}^{\mathrm{sm}})^{(g-1)}$ to $W^0_{g-1} \backslash W^1_{g-1}$ is an isomorphism. The second is that W^1_{g-1} has fiberwise over S dimension $\leq g-3$. To prove these claims we can reduce to S_0 since S is a deformation of S_0. We can also restrict to $\mathbf{G}_A$ for any element A of Λ. The closure of W^0_{g-1} in $\mathbf{G}_A$ is the restriction to $\mathbf{G}_A$ of the divisor Θ_A of $\mathbf{P}_A$ defined in 14.9. By 14.10, of which we use the notation, we can identify $\mathbf{G}_A$ with the subfunctor of the Picard functor of C_0 relative to S_0 consisting of elements whose restriction to any geometric point c of S_0 lie in the component of $\mathrm{Pic}(C_{0,c}/k(c))$ of degree g. The first claim is standard; for example one can deduce it from [4, Proposition 8.2/7]. The second claim would be classical if C_0 were smooth over $\mathbf{C}$; see [2, Theorem IV.5.1]. In our case we can reduce to the case that S_0 is the spectrum of an algebraically closed field. We can compute the tangent space at points of W^1_{g-1} in $\mathbf{G}_A$ as in [2, Proposition IV.4.2(i)] since the argument given there is based on Serre's duality and the theorem of Riemann-Roch which hold in our case due to [4, Theorem 9.1/1] and the discussion after [7, Definition 1.1]. A simple computation as in [15, Lemma IV.5.5] gives the wanted estimate on the dimension of the tangent space.

Let $T := T_e$ be one of the sections of $\mathcal{C}_0 \to S_0$ defined in 15.5. Let B be in $H_1(\Gamma, \mathbf{Z})$. We claim that there exists an open subscheme

$$M_T \subset \{\hat{0}_B\} \underset{S}{\times} \mathcal{C} \underset{S}{\times} (\mathcal{C}^{\mathrm{sm}})^{(g-2)}$$

containing a closed subscheme of the form

$$\{\hat{0}_B\} \underset{S}{\times} \{T\} \underset{S}{\times} N_T$$

for some open subscheme $N_T \subset (\mathcal{C}_0^{\mathrm{sm}})^{(g-2)}$, an open subscheme I_T of the scheme theoretic image of $(\mathcal{C}^{\mathrm{sm}})^{(g-1)}$ in $\mathcal{P}$ and an étale morphism $M_T \to I_T$

extending

$$\{\hat{0}_B\} \underset{S}{\times} (\mathcal{C}^{\mathrm{sm}})^{(g-1)} \to \mathcal{P}.$$

Using 9.10 there is a morphism $\mathcal{P} \times_S \mathcal{P}^{\mathrm{sm}} \to \mathcal{P}$ extending the group scheme structure on $\mathcal{P}^{\mathrm{sm}}$. In particular, we get a morphism

$$\mathcal{C} \underset{S}{\times} (\mathcal{C}^{\mathrm{sm}})^{(g-1)} \to \mathcal{P}$$

extending $\mathcal{C}^{\mathrm{sm}} \times_S (\mathcal{C}^{\mathrm{sm}})^{(g-1)} \to \mathcal{P}$. This allows us to assume S equal to S_0. We can also restrict to $\mathbf{P}_A \subset \mathcal{P}_0$ for each A in $H_1(\Gamma, \mathbf{Z})$. From the discussion of 14.10 there exist an open dense subscheme

$$V_T \subset \{\mathbb{D}_j'^{\,\mathrm{sm}}\}^{(g-1)},$$

whose complement has fiberwise codimension 2, and an open subscheme U_T of the scheme $\mathbb{D}_j \times (\mathbb{D}_j'^{\,\mathrm{sm}})^{(g-1)}$ product of an open neighborhood of T in $\mathbb{D}_j$ and V_T mapping étale to its image. The open subschemes M_T and N_T are the restrictions of U_T and V_T respectively and I_T is the image of M_T.

Finally we get that the scheme theoretic image in $\mathcal{P}$ of the union over $B \in \Lambda$ of

$$\{\hat{0}_B\} \underset{S}{\times} (\mathcal{C}^{\mathrm{sm}})^{(g-1)}$$

is flat over S fiberwise in codimension at least 3. Hence it is a Cartier divisor relative to S fiberwise in codimension at least 3 by [4, Lemma 8.2/6(ii)]. But then [13, Corollary XII.4.8] guarantees that it is a Cartier divisor since $\mathcal{P}_0 \to S_0$ is fiberwise parafactorial in codimension 3. The fact that it is an effective Cartier divisor relative to S can be checked fiberwise using [4, Lemma 8.2/6(iii)].

15.11 Remark. By 9.5 there is a map $\mathcal{D}ef\mathcal{G}r(\mathcal{P}_0, H_1(\Gamma, \mathbf{Z})) \to \mathcal{D}ef\mathcal{G}r(\mathcal{J}_0)$. Let $\mathcal{D}ef(C_0)$ be the deformation functor of $C_0 \to S_0$ to nilpotent thickenings of S_0. By 15.9 there is a map $\mathcal{D}ef(C_0) \to \mathcal{D}ef\mathcal{G}r(\mathcal{P}_0)$. Hence there is composite map $\mathcal{D}ef(C_0) \to \mathcal{D}ef\mathcal{G}r(\mathcal{J}_0)$. In [10] the question whether the image of this map consists of Jacobians is investigated. The answer is negative already for deformations to the dual numbers over S_0. As an exercise the reader can interpret the morphism "cotan" in the commutative diagram on [10, pag. 373] in terms of the map on tangent spaces induced by $\mathcal{D}ef(C_0) \to \mathcal{D}ef\mathcal{G}r(\mathcal{J}_0)$.

15.12 Theorem. *The notation is as in 14.7. Let S be a nilpotent thickening of S_0. Let $\mathcal{C}$ be a deformation to S of the scheme $\mathcal{C}_0 \to S_0$ together with the action of $H_1(\Gamma, \mathbf{Z})$. Let $(\mathcal{P}, H_1(\Gamma, \mathbf{Z}))$ be an S-valued point of the functor $\mathcal{D}ef\mathcal{G}r(\mathcal{P}_0, H_1(\Gamma, \mathbf{Z}))$ defined in 9.5. Let $\tilde{G}$ be the S-valued point of $\mathcal{D}ef\mathcal{G}r(\mathbf{G}_0)$ associated to $\mathcal{P}$ as in 9.8. Suppose that the morphism $\mathcal{C}_0 \to \mathcal{P}_0$ extends to an $H_1(\Gamma, \mathbf{Z})$-equivariant morphism $\mathcal{C} \to \mathcal{P}$. Then the semiabelian scheme $\tilde{G}$ represents the subfunctor* $\mathrm{Pic}^0(C/S)$ *of* $\mathrm{Pic}(C/S)$ *consisting of line bundles whose restrictions to all fibers C_s with $s \in S$ belong to* $\mathrm{Pic}^0(C_s/k(s))$.

Proof: Let $\tilde{G} := G_0$, with $0 \in \Lambda$. Let

$$u: C \longrightarrow \bar{P} \qquad \text{and} \qquad \alpha: \tilde{G} \longrightarrow \bar{P}$$

be the deformations of u_V and α_V of 14.7. Call $\overline{\Theta}$ the divisor on $\bar{P}$ defined in (iii) of 15.9. By 9.10 there is an action $c: \bar{P} \times_S \tilde{G} \to \bar{P}$ of $\tilde{G}$ on $\bar{P}$. Let

$$m: C \underset{S}{\times} \tilde{G} \to \bar{P}$$

be the morphism obtained composing the map $u \times 1$ and c. Let M be the line bundle on $C \times_S \tilde{G}$ associated to

$$m^*(\overline{\Theta}) \otimes \left(u^*(\overline{\Theta})\right)^{-1} \otimes \left(\alpha^*(\overline{\Theta})\right)^{-1}.$$

Let a be a geometric point of $\tilde{G}$. Let s be the image of a in S_0. Then

$$M_a := M|_{C_s \times \{a\}} = u^*\Big(t_{-\alpha(a)}(\overline{\Theta}_s) - \overline{\Theta}_s\Big),$$

where $t_{-\alpha(a)}$ is translation by $-\alpha(a)$. Let j be the unique element of $\mathcal{V}$. Let $u_j: D_j \to \mathbf{P}_0$. By 14.11 the pullback of $t_{-\alpha(a)}(\overline{\Theta}_0) - \overline{\Theta}_0$ to $\mathbf{P}_0 \subset \mathcal{P}_0$ is the pullback of a line bundle on $\mathcal{J}_0$ algebraically equivalent to zero. Hence the restriction of M_a to $D_{j,s}$ has degree 0. By [4, Theorem 9.4/1] the sheaf $\mathrm{Pic}^0(C/S)$ is represented by a semiabelian scheme over S. The line bundle M on $C \times_S \tilde{G}$ defines a morphism

$$\beta: \tilde{G} \to \mathrm{Pic}^0(C/S).$$

Recall that $\tilde{G}_0 \cong \mathbf{G}_0 \cong \mathrm{Pic}^0(C_0/S_0)$. Arguing as in [17, Lemma 6.9], it is easy to see that β_0 induces an isomorphism between $\tilde{G}_0$ and $\mathrm{Pic}^0(C_0/S_0)$. Consequently β is an isomorphism.

16 Uniformization and compactification of $\mathrm{Pic}^0(C/S)$ in the geometrically irreducible case

16.1 *Notation.* Let $S = \mathrm{Spec}(E)$ be the spectrum of a noetherian integral normal domain with fraction field K and let I be an ideal defining the closed subscheme S_0. Assume E to be complete and separated with respect to the I-adic topology and let $S_n := \mathrm{Spec}(E/I^{n+1})$. Let $\mathcal{S}$ be the affine formal scheme associated to E and the ideal I. If X is a scheme over S or a formal scheme over $\mathcal{S}$, denote by X_n the scheme over S_n obtained as the base change of X to S_n. Let $f_0: C_0 \to S_0$ be a split semistable curve as in 12.1. Assume $C_0 \to S_0$ to be *geometrically irreducible.* Let

$$f: C \longrightarrow S$$

be a proper and flat morphism such that its base change to S_0 is $f_0: C_0 \to S_0$ and its base change to K is smooth. With the notation of 14.1 fix a vertex V of Γ^{ab}. Fix

i. a section
$$\bar{0}_j: S \longrightarrow C$$
deforming the section $0_j: S_0 \to C_0$ fixed in 12.5;

ii. the $H_1(\Gamma, \mathbf{Z})$-formal Galois cover $\mathcal{C} \to \mathcal{S}$ of the I-adic completion of C deforming the morphism $\mathcal{C}_0 \to C_0$ defined in 14.2;

iii. the section $\hat{0}_V: \mathcal{S} \to \mathcal{C}$ over $\bar{0}_j$ whose base change to S_0 is the section 0_V defined in 14.7;

iv. the deformation $(\mathcal{P}, H_1(\Gamma, \mathbf{Z}))$ to $\mathcal{S}$ of $(\mathcal{P}_0, H_1(\Gamma, \mathbf{Z}))$, as defined in 14.7, determined by $\mathcal{C} \to \mathcal{S}$ via 15.9;

v. the $H_1(\Gamma, \mathbf{Z})$-equivariant morphism $\mathcal{C} \to \mathcal{P}$ deforming the map $u_V: \mathcal{C}_0 \to \mathcal{P}_0$ defined in 15.9 so that the image of $\hat{0}_V$ is the identity element for the formal group law on the smooth locus of $\mathcal{P} \to \mathcal{S}$.

16.2 *An assumption.* Let t be an integer. Since C_0 is assumed to be geometrically irreducible, the map $H_1(\Gamma, \mathbf{Z}) \to C_1(\Gamma, \mathbf{Z})$ is an isomorphism. In particular, to the elementary cycle l_t fixed in 12.3 corresponds a unique edge e_t of Γ. Assume that the completion of C at the singularity associated to e_t in 12.3 is isomorphic to $E[\![X, Y]\!]/(XY - q_{e_t})$ for some $q_{e_t} \in I$.

16.3 *The main results.* In this section we extend the results of the previous sections from the nilpotent thickenings S_n of S_0 to the scheme S. In 16.4 we prove that the scheme $\mathcal{P}/H_1(\Gamma, \mathbf{Z})$ can be algebraized to a scheme $\bar{P} \to S$ and hence we obtain a morphism $C \to \bar{P}$. By 10.2 the formal scheme $\mathcal{P}$ over $\mathcal{S}$ is the I-adic completion of a scheme $\widetilde{P}$ over S. The scheme $\widetilde{P} \to S$ is the link between $\bar{P} \to S$ (and $\mathrm{Pic}^0(C/S)$ due to 17.7) and the uniformization of $\bar{P} \to S$ defined in 16.7. It plays an essential role in the proof of the fact that the smooth locus $\bar{P}^{\mathrm{sm}}$ of $\bar{P} \to S$ has a group scheme structure having $\bar{0}_j$ as the identity element; see 16.13. Finally we show in 16.22 that the uniformization of the semiabelian scheme $\bar{P}^{\mathrm{sm}} \to S$ we obtain is the one constructed in [8].

16.4 Proposition. *There exist*

i. *a flat morphism $\bar{P} \to S$ whose base change to S_0 is the morphism $\bar{P}_0 \to S_0$ defined in 14.7;*

ii. *an ample effective Cartier divisor $\overline{\Theta}$ on $\bar{P}$ relative to S deforming the Cartier divisor $\overline{\Theta}_0$ on $\bar{P}_0$ defined in 14.7;*

iii. *a morphism $C \to \bar{P}$ whose base change to S_0 is the morphism $C_0 = \mathcal{C}_0/H_1(\Gamma, \mathbf{Z}) \to \mathcal{P}_0/H_1(\Gamma, \mathbf{Z}) = \bar{P}_0$ induced from the morphism u_V defined in 14.7.*

Moreover the morphism $\bar{P} \to S$ has connected fibers.

Proof: Use 15.9, 15.12 and finally Grothendieck's existence theorem [11, Theorem 5.4.5]. For the last statement use Zariski's main theorem.

16.5 *The scheme $\widetilde{P} \to S$.* Let $\mathcal{P}$ be the formal scheme over $\mathcal{S}$ defined in 16.1. By 9.6 for any $n \in \mathbf{N}$ there exists a unique abelian scheme $\mathcal{J}_n$ over S_n whose base change to S_0 is $\mathcal{J}_0$, such that $\mathcal{P}_0 \to \mathcal{J}_0$ extends to a morphism $\mathcal{P}_n \to \mathcal{J}_n$ and such that the image of $\hat{0}_V \times_{\mathcal{S}} S_n$ is the identity element of $\mathcal{J}_n$. By 16.6 the formal abelian scheme associated to $(\mathcal{J}_n)_{n\in\mathbf{N}}$ is the I-adic completion of an abelian scheme $\mathcal{J}$ over S. By 15.9 the discriminant locus of $\mathcal{P} \to \mathcal{S}$ is determined by the discriminant locus of $C \to S$. By assumption C_K is smooth over Spec(K). In particular, $(\mathcal{P}, H_1(\Gamma, \mathbf{Z}))$ satisfies the conditions in 10.2. Define

$$\widetilde{P} \longrightarrow \mathcal{J} \qquad \text{and} \qquad \tilde{0}\colon S \to \widetilde{P}$$

to be the scheme constructed in 10.2 and the section algebraizing the image of $\hat{0}_V$.

16.6 Proposition. *The notation is as in 16.5. The formal abelian scheme associated to $(\mathcal{J}_n)_{n\in\mathbf{N}}$ is the I-adic completion of an abelian scheme $\mathcal{J}$ over S.*

Proof: For any $n \in \mathbf{N}$ let $\tilde{G}_n$ be as in 15.12. Then $\tilde{G}_n$ is the extension of $\mathcal{J}_n$ by a split torus and $\tilde{G}_n \cong \mathrm{Pic}^0(C_n/S_n)$. By [4, Theorem 9.4/1] the functor $\mathrm{Pic}^0(C/S)$ is representable and carries an S-ample line bundle $\mathcal{L}$. Since S is normal, by [18, Proposition I.2.6] the choice of a rigidification of the associated $\mathbf{G}_m$-bundle $G(\mathcal{L})$ is equivalent to define a cubical structure on $G(\mathcal{L})$. Fix a rigidification. By [18, Theorem IV.2.1] the cubical $\mathbf{G}_m$-torsor $\left(G(\mathcal{L}\otimes[-1]^*(\mathcal{L})) \times_S S_n\right)_n$ with $n \in \mathbf{N}$ descends to a formal cubical $\mathbf{G}_m$-torsor on the formal scheme associated to $(\mathcal{J}_n)_{n\in\mathbf{N}}$. Hence the latter carries an ample line bundle relative to $\mathcal{S}$. In particular, it can be algebraized to an abelian scheme $\mathcal{J}$.

16.7 Corollary. *The notation is as in 16.4. Let $\bar{0}$ be the section of $\bar{P} \to S$ induced by the section $\bar{0}_j$ of $C \to \bar{P}$ via $C \to \bar{P}$. To the pair consisting of the* *scheme $\bar{P}$ over S and the section $\bar{0}$ we can canonically associate the following data and the pair $\bar{P} \to S$ and the section $\bar{0}$ can be reconstructed from the following data*

a. *an abelian scheme $\mathcal{J}$ over S and a semiabelian scheme $\tilde{G} \to S$ deforming the semiabelian scheme $\mathbf{G}_0 = \mathrm{Pic}^0(C_0/S_0)$ and sitting in an exact sequence*
$$0 \to \mathbf{G}^r_{m,S} \longrightarrow \tilde{G} \longrightarrow \mathcal{J} \longrightarrow 0;$$
b. *a point*
$$x_t \in \tilde{G}(K) \qquad \text{for all } \; 1 \le t \le r.$$

Moreover, the image of x_t in $\mathcal{J}(K)$ comes from an S-valued point y_t of $\mathcal{J}$. Let $\tilde{G}$ split as product of $\mathbf{G}_m$-bundles $\tilde{G}_1 \times \ldots \times \tilde{G}_r$ over $\mathcal{J}$ as implied by (a). *For any integer t satisfying $1 \le t \le r$, let $p_t \colon \tilde{G} \to \tilde{G}_t$ be the projection. For any integer s with $1 \le s \le r$ denote by $I_{t,s}$ the fractional ideal of S determined by the section $p_{s,K}(x_t)$ over K of the $\mathbf{G}_m$-bundle $y_t^*(\tilde{G}_s)$ over S. Then, with the notation above, we have the equality of fractional ideals*
$$I_{t,s} = \begin{cases} (q_{e_t}) & \text{if } s = t \\ O_S & \text{if } s \ne t. \end{cases}$$

Proof: The proof follows immediately from 11.6 using 16.5 and the following remarks. The S-scheme $\bar{P}$ is uniquely determined by its I-adic completion and the latter is uniquely determined by its $H_1(\Gamma, \mathbf{Z})$-Galois cover $\mathcal{P}$. Via this correspondence, S-points of $\bar{P}$ uniquely determine $\mathcal{S}$-valued points of the I-adic completion of $\bar{P}$ and the latter uniquely determine $H_1(\Gamma, \mathbf{Z})$-orbits of $\mathcal{S}$-points of $\mathcal{P}$. The last statement follows from 15.9.

16.8 *The divisor $\tilde{\Theta}$.* Suppose we are in the case 16.5. By 15.9 for any $n \in \mathbf{N}$ the scheme $\mathcal{P}_n$ carries an ample effective formal Cartier divisor $\tilde{\Theta}_n$ relative to S_n invariant under the action of $H_1(\Gamma, \mathbf{Z})$. Then

16.9 Proposition. *The formal divisor $(\tilde{\Theta}_n)_{n \in \mathbf{N}}$ is the I-adic completion of an effective Cartier divisor $\tilde{\Theta}$ on $\widetilde{P}$ relative to S invariant under the action of $H_1(\Gamma, \mathbf{Z})$.*

Proof: Let $\overline{a}$ be as in 10.4. Let P_a and $P_{\underline{a}}$ and $U_{\underline{a}}$ be as in 10.5. Fix $n \in \mathbf{N}$. Since the inverse image of the open subscheme $U_{\underline{a},n}$ of $P_{\underline{a},n}$ maps isomorphically to $U_{\underline{a},n}$, the divisor $\tilde{\Theta}_n$ defines a divisor $\tilde{\Theta}_{\underline{a},n}$ in $U_{\underline{a},n}$. Let x be a point of $\mathcal{P}_0$, lying in the intersection of $\tilde{\Theta}_n$ and $\mathbf{P}_A \subset \mathcal{P}_0$ with $A \in a$, but not contained in $U_{\underline{a},n}$. With the notation of 10.5 and 7.4 and using the description of $\tilde{\Theta}_n$ given in 15.12, we know that in codimension ≥ 3 in $\mathcal{P}_0$ the closed immersion $\tilde{\Theta}_n \subset \mathcal{P}_n$ at x is defined by
$$\mathbb{A}^x_n[z_1, \ldots, z_{r-1}][\![\beta_t, \gamma_t]\!]/(\gamma_t\beta_t - q_t) \longrightarrow \mathbb{F}^x_n[\![\gamma_t, \beta_t]\!](\gamma_t\beta_t - q_t),$$

where $\mathbb{F}_n^x$ defines an effective Cartier divisor on $\mathrm{Spec}(\mathbb{A}_n^x[z_1,\ldots,z_{r-1}])$ relative to S_n and t is an integer with $1 \le t \le r$. By 10.5 we also know that, up to interchanging γ_t and β_t, the morphism $\mathcal{P}_n \to P_{\underline{a},n}$ at x is described by

$$\mathbb{A}_n^x[z_1,\ldots,z_{r-1}][\![\beta_t]\!] \hookrightarrow \mathbb{A}_n^x[z_1,\ldots,z_{r-1}][\![\beta_t,\gamma_t]\!]/(\gamma_t\beta_t - q_t).$$

This local description guarantees that the closure of $\tilde{\Theta}_{\underline{a},n}$ in $P_{\underline{a},n}$ defines an effective Cartier divisor relative to S_n in codimension ≥ 3 on $P_{\underline{a},n}$. Remark that $P_{\underline{a},0} = \cup_{A\in a}\mathbf{P}_A$ is naturally a subscheme of $\mathcal{P}_0$. Since $\mathcal{P}_0 \to S_0$ is fiberwise parafactorial, [13, Corollary XII.4.8] guarantees that $\tilde{\Theta}_{\underline{a},n}$ extends to a Cartier divisor $\underline{\tilde{\Theta}}_{\underline{a},n}$ on $P_{\underline{a},n}$. By [4, Lemma 8.2/6(iii)] such extension is an effective Cartier divisor relative to S_n since $\underline{\tilde{\Theta}}_{\underline{a},0}$ is the divisor $\cup_{A\in a}\Theta_A$ in the notation of 14.11. By [11, Theorem 5.4.5] the formal Cartier divisor $\underline{\tilde{\Theta}}_{\underline{a},n}$ is the I-adic completion of a unique effective Cartier divisor $\underline{\tilde{\Theta}}_{\underline{a}}$ on $P_{\underline{a}}$ relative to S. The intersection of $\underline{\tilde{\Theta}}_{\underline{a}}$ with $U_{\underline{a}}$ defines an effective Cartier divisor $\tilde{\Theta}_{\underline{a}}$ relative to S. Proceeding as in 10.10 the divisors $\tilde{\Theta}_{\underline{a}}$ glue to form the wanted effective Cartier divisor $\tilde{\Theta}$ on $\widetilde{P}$ relative to S. In view of 10.11 the invariance under the action of $H_1(\Gamma,\mathbf{Z})$ is clear.
The next step is the important 16.13. To prove it we need some preliminary lemmas and a better knowledge of the smooth locus of $\bar{P} \to S$ defined in 16.4. The main tool is the scheme $\widetilde{P} \to S$ defined in 16.5.

16.10 Lemma. *Let W be the spectrum of a noetherian ring complete and separated with respect to an ideal J. Let $f_1: X_1 \to W$ and $f_2: X_2 \to W$ be morphisms of schemes locally of finite type over S with X_2 proper over W. Suppose that there exists a morphism of associated J-adic completions*

$$\mathcal{X}_1 \xrightarrow{\pi} \mathcal{X}_2$$

over W with π étale and surjective. Let $B_1 \subset X_1$ and $B_2 \subset X_2$ be closed subschemes such that we have an inclusion of associated J-adic completions $\mathcal{B}_1 \subset \pi^{-1}\mathcal{B}_2$. If the restriction of f_1 to $X_1\backslash B_1$ is flat (resp. smooth of relative dimension d), then also the restriction of f_2 to $X_2\backslash B_2$ is flat (resp. smooth of relative dimension d).

Proof: This is proven in [21, Proposition 4.1].

16.11 Definition. *The notation is as in 16.5. Let $\tilde{N}$ be the closed subscheme of $\widetilde{P}$ defined in 10.10. Its complement is the smooth locus of $\widetilde{P} \to S$. Let*

$$\tilde{\mathcal{N}} \hookrightarrow \mathcal{P}$$

be the completion of $\tilde{N}$ along $\tilde{N}_0$. The group $H_1(\Gamma,\mathbf{Z})$ acts on it since it acts on $\tilde{N}$. Let

$$\mathcal{N} := \tilde{\mathcal{N}}/H_1(\Gamma,\mathbf{Z}) \hookrightarrow \mathcal{P}/H_1(\Gamma,\mathbf{Z}).$$

Then there is a unique closed subscheme of $\bar{P}$ algebraizing $\mathcal{N}$

$$\bar{N} \hookrightarrow \bar{P}.$$

16.12 Proposition. *The complement of the subscheme $\bar{N}$ of $\bar{P}$ is the smooth locus of $\bar{P} \to S$. The scheme $\bar{N}_K$ is empty and consequently $\bar{P}_K$ is smooth over* Spec(K). *Moreover, for any $z \in \bar{N}$ the completion of $\bar{P}$ at z is of the form*

$$\hat{O}_{\bar{P},z} := F[\![\gamma_u, \beta_u]\!]_{u \in \Xi_z} / (\gamma_u \beta_u - q_u)$$

where Ξ_z is a set of indices, while F is a formally smooth E-algebra and $\{q_u\}_{u \in \Xi_z}$ is a subset of the set $\{q_{e_t}\}_{t=1,\dots,r}$ defined in 16.2.

Proof: The complement of $\widetilde{N}$ in $\widetilde{P}$ is the smooth locus of $\widetilde{P} \to S$. From 16.10 applied with $X_1 = \widetilde{P}$ and $B_1 = \widetilde{N}$, while $X_2 = \bar{P}$ and $B_2 = \bar{N}$ we deduce that $\bar{P} \backslash \bar{N}$ is smooth over S. By 10.9 the ideal sheaf defining $\widetilde{N}$ in $\widetilde{P}$ is killed by some non-zero element of E. But then this holds also for the ideal sheaf defining $\mathcal{N}$ and consequently for the one defining $\bar{N}$. In particular, the scheme $\bar{N}_K$ is empty and consequently $\bar{P}_K$ is smooth over K. Let $z \in \bar{N}$ and let x be a closed point of $\bar{P}$ in the closure of z. Since $\bar{P} \to S$ is proper, x lies in $\bar{P}_0$. Let y be a closed point of $\mathcal{P}_0$ mapping to x. The completion $\hat{R}$ of $\bar{P}$ at x is isomorphic to the completion of $\widetilde{P}$ at y since the I-adic completion of $\widetilde{P}$ is an étale cover of the I-adic completion of $\bar{P}$. In the notations of 10.5 there exists a set a such that $y \in U_{\underline{a}}$ and $\hat{R}$ is the completion of $U_{\underline{a}}$ at y. The ideal $\hat{I}_N$ defining $\bar{N}$ in $\hat{R}$ is the ideal sheaf defining $N_{\underline{a}}$ in $\hat{R}$. The conclusion follows from the explicit description of $U_{\underline{a}}$ and $N_{\underline{a}}$ in the proof of 10.9.

16.13 Corollary. *Let $\bar{P}$ and $\bar{0}\colon S \to \bar{P}$ be as in 16.7. The smooth locus*

$$\bar{P}^{\mathrm{sm}} \longrightarrow S$$

of $\bar{P} \to S$ has a unique structure of smooth and separated group scheme over S which has $\bar{0}$ has identity element, which extends the group structure on the smooth locus of $\bar{P}_0 \to S_0$ defined in 14.7 and which is geometrically connected. Moreover $\bar{P}_K$ is smooth over K and hence it is an abelian variety.

Proof: The smoothness of $\bar{P}_K$ over K follows from 16.12. The idea of the proof of the first statement is borrowed from [21]. Define the following objects

i $\tilde{Z}^o$ as the closure of $\widetilde{P}^{\mathrm{sm}} \times_S \widetilde{P}^{\mathrm{sm}}$ in the scheme $\widetilde{P} \times_S \widetilde{P} \times_S \bar{P}$ via the morphism which is the natural inclusion in $\widetilde{P} \times_S \widetilde{P}$ on the first two components and is the composition of the group law defined in 11.8 on $\widetilde{P}^{\mathrm{sm}}$ and the inclusion of $\widetilde{P}^{\mathrm{sm}}$ into $\widetilde{P}$ on the last component. Let $\tilde{Z}^{o\prime}$ be the intersection of $\tilde{Z}^o$ with $\widetilde{P}^{\mathrm{sm}} \times_S \widetilde{P}^{\mathrm{sm}} \times_S \bar{P}$. It is isomorphic to $\widetilde{P}^{\mathrm{sm}} \times_S \widetilde{P}^{\mathrm{sm}}$;

ii. $\tilde{Z}$ as the union of the translates by $H_1(\Gamma, \mathbf{Z}) \times H_1(\Gamma, \mathbf{Z}) \times H_1(\Gamma, \mathbf{Z})$ of $\tilde{Z}^o$. Let $\tilde{Z}'$ be the intersection of $\tilde{Z}$ with $\widetilde{P}^{\mathrm{sm}} \times_S \widetilde{P}^{\mathrm{sm}} \times_S \widetilde{P}$. It is isomorphic to a disjoint union of copies of $\widetilde{P}^{\mathrm{sm}} \times_S \widetilde{P}^{\mathrm{sm}}$ indexed by $H_1(\Gamma, \mathbf{Z})$;

iii. $\tilde{\mathcal{Z}}$ as the completion of $\tilde{Z}$ along its base change to S_0;

iv. $\mathfrak{Z}$ as $\tilde{\mathfrak{Z}}/(H_1(\Gamma,\mathbf{Z})\times H_1(\Gamma,\mathbf{Z})\times H_1(\Gamma,\mathbf{Z}))$;
v. the subscheme $\bar{Z}$ of $\bar{P}\times_S\bar{P}\times_S\bar{P}$ algebraizing $\mathfrak{Z}$;
vi. the open subscheme $\bar{Z}'$ of $\bar{Z}$ defined as the intersection of $\bar{Z}$ with the scheme $\bar{P}^{\rm sm}\times_S\bar{P}^{\rm sm}\times_S\bar{P}$.

16.14 *$\bar{Z}$ is well defined.* First of all we prove that $\tilde{\mathfrak{Z}}$ is a closed formal subscheme of the I-adic completion of $\widetilde{P}\times_S\widetilde{P}\times_S\widetilde{P}$ so that $\mathfrak{Z}$ in (iv) is well defined. Let a and b be subsets of $H_1(\Gamma,\mathbf{Z})$ of the type considered in 10.4 and let $U_{\underline{a}}$ and $U_{\underline{b}}$ be as in 10.5. Define c to be the smallest subset of $H_1(\Gamma,\mathbf{Z})$ as in 10.4 containing $a+b$. Assume that a and b are chosen so that $\widetilde{P}^3$ is covered by translates by $H_1(\Gamma,\mathbf{Z})^3$ of $U_{\underline{a}}\times_S U_{\underline{b}}\times_S U_{\underline{c}}$. Let $(l,h,g)\in H_1(\Gamma,\mathbf{Z})^3$ such that $g+c$ and $(h+l)+c$ are disjoint. For $t=1,\ldots,r$ let

$$p_t\colon \mathbf{Z}f_1\oplus\ldots\oplus\mathbf{Z}f_r\to\mathbf{Z}f_t\cong\mathbf{Z}$$

be the natural projection. Then there exist $A\in H_1(\Gamma,\mathbf{Z})$ and $t=1,\ldots,r$ such that $p_t(g+c)$ and $p_t(l+h+c)$ are disjoint and $p_t(g+c)\le p_t(A)\le p_t(l+h+c)$ (or $p_t(l+h+c)\le p_t(A)\le p_t(g+c)$). Consider the morphism $\widetilde{P}\to P_A$ defined in 10.10. Then $U_{\underline{g+c}}\times_S S_0$ is mapped to $0_{A,t}$ (respectively $\infty_{A,t}$) and $U_{\underline{l+h+c}}\times_S S_0$ is mapped to $\infty_{A,t}$ (respectively $0_{A,t}$). In particular, $\tilde{Z}^o\times_S S_0$ is disjoint from $U_{\underline{l+a}}\times_S U_{\underline{h+b}}\times_S U_{\underline{g+c}}$. In particular, $\tilde{Z}_0$ is locally a finite union of translates of $\tilde{Z}_0^o$ and hence it is a closed subscheme of $\widetilde{P}_0$.

16.15 *The natural projection $\delta\colon\bar{Z}'\to\bar{P}^{\rm sm}\times_S\bar{P}^{\rm sm}$ is an isomorphism.*
Proof: The claim holds for $\tilde{Z}^{o\prime}$ instead of $\bar{Z}'$ and $\widetilde{P}$ instead of $\bar{P}$. The projection map δ is a projective morphism and it induces an isomorphism on I-adic completions. As in [21, Theorem 4.10] we conclude that δ is unramified over the whole $\bar{P}^{\rm sm}\times_S\bar{P}^{\rm sm}$ and is étale of degree 1 over an open subscheme mapping surjectively to S_0. Since $\bar{P}^{\rm sm}$ is normal and by construction $\bar{Z}$ is reduced the claim follows from Zariski's main theorem; see [4, Theorem 2.3.2'].

16.16 *The natural projection $\bar{Z}'\to\bar{P}$ to the third component in the product of the schemes $\bar{P}^{\rm sm}\times_S\bar{P}^{\rm sm}\times_S\bar{P}$ factors via $\bar{P}^{\rm sm}$.*
Proof: Let $\tilde{N}$ and $\bar{N}$ be defined as in 16.11 and

$$B_1:=\tilde{Z}\cap\Big((\tilde{N}\underset{S}{\times}\widetilde{P}\underset{S}{\times}\widetilde{P})\cup(\widetilde{P}\underset{S}{\times}\tilde{N}\underset{S}{\times}\widetilde{P})\Big)$$

and

$$B_2:=\bar{Z}'\cap\Big((\bar{N}\underset{S}{\times}\bar{P}\underset{S}{\times}\bar{P})\cup(\bar{P}\underset{S}{\times}\bar{N}\underset{S}{\times}\bar{P})\Big).$$

It suffices to verify that a power of the ideal defining the inverse image of $\bar{N}$ in $\bar{Z}$ is contained in the ideal defining B_1. This can be checked after taking completions with respect to I since $\bar{P}\to S$ is projective. This can be verified locally after pulling back to the étale cover $\mathfrak{P}$ of the I-adic completion of $\bar{P}$. But the pullback of the I-adic completion $\mathfrak{N}$ of $\bar{N}$ is $\tilde{\mathfrak{N}}$ by 16.11 and the pullback of the I-adic completion $\mathcal{B}_2$ of B_2 is the I-adic completion $\mathcal{B}_1$ of B_1. The claim holds for $\tilde{Z}'$ instead of $\bar{Z}'$ and $\widetilde{P}$ instead of $\bar{P}$. This implies that locally a power of the ideal defining the inverse image of $\tilde{N}$ in $\tilde{Z}$ is contained in the ideal defining B_2. Hence the same must hold for their I-adic completions as wanted.

16.17 *The end of the proof of 16.13.* We have got the wanted morphism

$$\bar{P}^{\mathrm{sm}} \underset{S}{\times} \bar{P}^{\mathrm{sm}} \to \bar{P}^{\mathrm{sm}}.$$

It is now straightforward to check that this group law is associative, commutative and that $\bar{0}$ is the identity element. In fact, this holds for on I-adic completions since it holds for $\widetilde{P}^{\mathrm{sm}}$ instead of $\bar{P}^{\mathrm{sm}}$ and $\tilde{0}$ instead of $\bar{0}$.

16.18 Definition. *Call the* 1*-motive*

$$\begin{array}{ccc} H_1(\Gamma, \mathbf{Z}) = \mathbf{Z}l_1 \oplus \ldots \oplus \mathbf{Z}l_r & \xrightarrow{\ \iota\ } & \tilde{G}(K) \\ l_t & \mapsto & x_{t,1} \cdots x_{t,r} \end{array}$$

associated in 16.7 to $\bar{P}$ and the section $\bar{0}$ the degenerating 1*-motive uniformizing the abelian variety $\bar{P}_K$. See 16.22 for justification of the terminology.*

16.19 Lemma. *Let $H_1(\Gamma, \mathbf{Z}) \to \tilde{G}(K)$ be the* 1*-motive defined in 16.18. There exists a line bundle $\mathcal{L}$ on $\tilde{G}$ such that the action by translation of $H_1(\Gamma, \mathbf{Z})$ on $\tilde{G}_K$ lifts to an action on $\mathcal{L}_K$ and $\mathcal{L}$ descends to a line bundle $\mathcal{M}$ on $\mathcal{J}$ defining a principal polarization*

$$\lambda : \mathcal{J} \longrightarrow \mathcal{J}^{\vee}$$

deforming the principal polarization on $\mathcal{J}_0$ defined in 12.6.

Proof: Let $\widetilde{P}$ be as in 16.5. Let $\tilde{\Theta}$ be the effective Cartier divisor on $\widetilde{P}$ relative to S defined in 16.8. Let $\tilde{\Theta}_{\tilde{G}}$ be the intersection of $\tilde{\Theta}$ with $\tilde{G}$. Let $\mathcal{L}$ be the associated line bundle. Choose a rigidification of the associated $\mathbf{G}_m$-bundle $G(\mathcal{L})$. By [18, I.2.6] the latter has a unique structure of cubical $\mathbf{G}_m$-torsor. By [18, I.7.2.4] it is clear that $G(\mathcal{L})$ descends to a cubical $\mathbf{G}_m$-torsor $G(\mathcal{M})$ on $\mathcal{J}$. The base change of $\tilde{\Theta}_{\tilde{G}}$ to S_0 is the intersection of the divisor Θ_0 of $\mathbf{P}_0$ defined in 14.10 with $\mathbf{G}_0 = \tilde{G} \times_S S_0$. In particular, $\mathcal{M} \times_S S_0$ defines the given principal polarization on $\mathcal{J}_0$ by 14.11. Since $\tilde{G}_K \subset \widetilde{P}_K$ is an isomorphism, we have $\tilde{\Theta}_{\tilde{G},K} = \tilde{\Theta}_K$. By 16.8 the group $H_1(\Gamma, \mathbf{Z})$ acts on $\tilde{\Theta}$. In particular, it induces an action on $\mathcal{L}_K$ lifting the action on $\tilde{G}_K$ as wanted.

16.20 *The connection with the work of* [8]. Let $\tilde{G}$ be the semiabelian scheme defined in 16.7. It is an extension of an abelian scheme $\mathcal{J}$ by a split torus T. Hence it is uniquely defined by a morphism

$$c : X(T) \to \mathcal{J}^{\vee}$$

from the character group $X(T)$ of T to the dual abelian scheme of $\mathcal{J}$. In 16.19 we have constructed a line bundle $\mathcal{L}$ on $\tilde{G}$ which descends to a line bundle $\mathcal{M}$

on $\mathcal{J}$ and defines a principal polarization on $\mathcal{J}$. Let

$$c^t: H_1\big(\Gamma, \mathbf{Z}\big) \longrightarrow \mathcal{J}$$

be defined as the composite of $H_1(\Gamma)\cdot\tilde{0} \to \widetilde{P}$ with the projection $\widetilde{P} \to \mathcal{J}$ where $\tilde{0}$ corresponds to the identity element of $\tilde{G}$ via $\tilde{G} \hookrightarrow \widetilde{P}$. In particular, c^t extends to the normal scheme S the composite of ι defined in 16.18 and the projection $\tilde{G} \to \mathcal{J}$. As explained in [6, Discussion II.2.3.3], to define ι is equivalent to define a trivialization τ of the pullback via $c^t \times c$ of the biextension on $\mathcal{J}_K \times_K \mathcal{J}_K^\vee$ defined by the Poincaré bundle. In 12.3 we have fixed a decomposition $H_1(\Gamma, \mathbf{Z})$ as $\mathbf{Z}l_1 \oplus \cdots \oplus \mathbf{Z}l_r$. On the other hand, $\tilde{G}$ splits as product of the $\mathbf{G}_m$-bundles over $\mathcal{J}$ given by $\tilde{G}_1, \ldots, \tilde{G}_r$. This induces a decomposition $X(T) \cong \mathbf{Z}^r$. Over S_0 we have $\tilde{G} \times_S S_0 = \mathbf{G}_0$ and $\tilde{G}_t \times_S S_0 = \mathbb{G}_t$ for $t = 1, \ldots, r$; see 13.14. By construction, for any $t = 1, \ldots, r$ the $\mathbf{G}_m$-bundle $\mathbb{G}_t$ is obtained as the pullback of the theta divisor by the image of l_t in $\mathcal{J}_0$ minus the theta divisor itself. Consider the isomorphism

$$\begin{array}{ccc} H_1(\Gamma, \mathbf{Z}) & \stackrel{\Phi}{\longrightarrow} & X(T) \\ \downarrow & & \downarrow \\ \mathbf{Z}l_1 \oplus \ldots \oplus \mathbf{Z}l_r & & \mathbf{Z}^r, \end{array}$$

respecting the two decompositions. Consider the diagram

$$\begin{array}{ccc} H_1(\Gamma, \mathbf{Z}) & \stackrel{c^t}{\longrightarrow} & \mathcal{J} \\ \downarrow{\scriptstyle \Phi} & & \downarrow{\scriptstyle \lambda} \\ X(T) & \stackrel{c}{\longrightarrow} & \mathcal{J}^\vee. \end{array}$$

It commutes over S_0. The existence of an action of $H_1(\Gamma, \mathbf{Z})$ on $\mathcal{L}_K$ lifting the action by translation on $\tilde{G}_K$ guarantees that the image of $H_1(\Gamma, \mathbf{Z})$ via $\lambda\, c^t$ is contained in the image of $X(T)$ via c. This implies that the diagram commutes if $c^t \times_S S_0$ is injective. This implies that the diagram *always* commutes. In fact, it suffices to check it after base change to the completion of S at a closed point s. The latter embeds in the universal deformation space S' of C_s and of the section $0_{j,s}$. It suffices to consider $S = S'$ and S_0 equal to the discriminant locus of the universal curve over S. In this case $c^t \times_S S_0$ is injective.
Reversing the construction in [6, II.2.7.3-II.2.7.5], the action of $H_1(\Gamma, \mathbf{Z})$ on $\mathcal{L}_K$, lifting the action on $\tilde{G}_K$, and the isomorphism Φ just constructed define an ample sheaf data $\psi: H_1(\Gamma, \mathbf{Z}) \to {c^t}^*(\mathcal{L}^{-1})$ with respect to $(\iota, \Phi, \mathcal{L})$ in the sense of [6, Definition II.3.3].

16.21 Proposition. 1. *The data* $\Big(\tilde{G}, X(T), H_1\big(\Gamma, \mathbf{Z}\big), c, c^t, \iota, \lambda, \Phi, \tau\Big)$ *are in the category* $\mathrm{DD}_{\mathrm{pol}}$ *defined in* [8, Section III.2, pag. 57-58].

2. *The data* $\Big(\tilde{G}, X(T), H_1\big(\Gamma, \mathbf{Z}\big), c, c^t, \iota, \lambda, \Phi, \tau, \mathcal{L}, \psi, \mathcal{M}\Big)$ *defined in 16.20 are in the category* $\mathrm{DD}_{\mathrm{ample}}$ *defined in* [8, Section III.2, pag. 57].

Proof: 1. To prove this we need to verify that the symmetry and positivity conditions hold. The positivity condition is a consequence of the last statement of 16.7. There is an action of $H_1(\Gamma, \mathbf{Z})$ on $\mathcal{L}_K$ lifting the action by translation on $\tilde{G}_K$ induced by ι. By 16.19 the line bundle $\mathcal{L}$ descends to a line bundle $\mathcal{M}$ on $\mathcal{J}$ inducing the principal polarization λ. The symmetry condition follows from [6, Proposition II.2.6.b]. The second part of the proposition is clear.

16.22 Proposition. *Let* $\mathrm{DEG}_{\mathrm{pol}}$ *(respectively* $\mathrm{DEG}_{\mathrm{ample}}$*) be the category of degenerating principally polarized abelian varieties over* S *as defined in* [8, Section II.0, pag. 32-33]. *Assume that* S *satisfies the condition that for any étale affine* S_0*-scheme, its unique formally étale lifting over* S *is normal. Then the image in* $\mathrm{DEG}_{\mathrm{pol}}$ *(respectively in* $\mathrm{DEG}_{\mathrm{ample}}$*) of the data* (1) *(respectively* (2)*) in 16.21 via the functor* M *defined in* [8, Section III.6, page 71] *is* $\bar{P}^{\mathrm{sm}}$ *(respectively* $\bar{P}^{\mathrm{sm}}$ *and the line bundle associated to* $\overline{\Theta}$ *in 16.4).*

Proof: It follows from [8, III.5.5] and 16.23.

16.23 *Relatively complete models.* Consider the integral scheme $\widetilde{P}$ over $\mathcal{J}$, the line bundle $\tilde{\mathcal{L}}$ associated to the divisor $\tilde{\Theta}$ defined in 16.8 and the action of $H_1(\Gamma, \mathbf{Z})$ both on $\widetilde{P}$ and $\tilde{\Theta}$. By 11.4 there is an action of $\tilde{G}$ on $\widetilde{P}$. By construction, if we choose the set $\overline{a}$ as in 10.4 large enough, then $\widetilde{P}$ is covered by translates by $H_1(\Gamma, \mathbf{Z})$ of the $\tilde{G}$-invariant open subscheme $U_{\underline{a}}$ defined in 10.5. Moreover, the divisor $\tilde{\Theta}$ is ample relatively to S by 14.11. Finally, we remark that there exists an action of $\tilde{G}$ on $\tilde{\mathcal{L}}$ tensored with the pullback of $\mathcal{M}^{-1}$ to $\widetilde{P}$. This action lifts the action on $\widetilde{P}$. In fact, for any $A \in H_1(\Gamma, \mathbf{Z})$ the restriction of this line bundle to the open subscheme $G_A = U_{\underline{A}}$ of $\widetilde{P}$ is isomorphic to the pull back to G_A of $c^t(A)^*(\mathcal{M}) \otimes \mathcal{M}^{-1}$ which is trivial on G_A. Hence, if we rigidify $\mathcal{M}$, the claimed action exists and is unique over $\widetilde{P}^{\mathrm{sm}}$. In particular, if an extension to an action over $\widetilde{P}$ exists, it is unique. Hence, it suffices to construct such an extension in the universal case, i.e., supposing that S is a versal deformation space of $C_s \to \mathrm{Spec}(k(s))$ for some $s \in S_0$. In this case S is regular, $\widetilde{P}$ is normal and the action extends being defined in codimension 2 . By 11.5, the data above define a relatively complete model in the sense of [8, III.3.1] for the data defined in (2) of 16.21.

17 $\mathrm{Pic}^0(C/S)$ as family of Néron models in the geometrically irreducible case

The notation is as in 16.1. This section contains two major results which conclude the discussion of geometrically irreducible curves. In 16.4 we have associated to $C \to S$ a proper scheme $\bar{P} \to S$. In 16.13 we have proven that its smooth locus $\bar{P}^{\mathrm{sm}} \to S$ is endowed with the structure of commutative group scheme over S which is geometrically connected. Moreover, by 16.12, the scheme $\bar{P}_K$ is smooth over K. Under some assumptions on S, we prove in 17.7 that $\bar{P}^{\mathrm{sm}}$ represents the functor $\mathrm{Pic}^0(C/S)$. Let V be the Spec of a discrete valuation ring with generic fiber η. Let $V \to S$ be a morphism such that $C \times_S V$ is regular and has smooth fiber over η. In 17.1 we prove that $\bar{P}^{\mathrm{sm}} \times_S V$ is the Néron model of the abelian variety $\bar{P}_\eta = \mathrm{Pic}^0(C_\eta/k(\eta))$ over V.

17.1 Theorem. *Let $S = \mathrm{Spec}(E)$. Let $\mathcal{E}$ be the set of edges of graph Γ defined in 12.3. Let the completion of $C \to S$ at the singular section T_e be isomorphic to $E[\![X,Y]\!]/(XY - q_e)$. Assume that for each $e \in \mathcal{E}$ the ring $E/(q_e)$ is a domain. Let $\bar{P}^{\mathrm{sm}}$ be the smooth locus of the scheme $\bar{P} \to S$ defined in 16.4 with the group structure defined in 16.13. Then $\bar{P}^{\mathrm{sm}}$ satisfies the Néron mapping property for smooth E-algebras which are discrete valuation rings.*

Proof: Let R be a ring endowed with a discrete valuation v normalized so that the value of a uniformizer is 1. Suppose that R is smooth as an E-algebra and that its fraction field contains the fraction field K of E. Let q be an element of I such that $E/(q)$ is a domain. If $v(q) \neq 0$, then $v(q) = 1$ since $R/(q)$ is smooth over the domain $E/(q)$. Let

$$y\colon \mathrm{Spec}(R) \longrightarrow \bar{P}$$

be a R-point of $\bar{P}$ and let z be the point of $\bar{P}$ determined by the closed point of $\mathrm{Spec}(R)$. Suppose that y does not factor through the smooth locus of $\bar{P} \to S$. By 16.12 the completion of the local ring of $\bar{P}$ at z is isomorphic to

$$\hat{O}_{\bar{P},z} := F[\![\gamma_u, \beta_u]\!]_{u \in \Xi_z}/(\gamma_u\beta_u - q_u),$$

where Ξ_z is a set of indices and F is a formally smooth E-algebra. For any q_u with $u \in \Xi_z$ the ring $E/(q_u)$ is a domain since by 15.9 the element q_u corresponds to one of the elements q_e with $e \in \mathcal{E}$ of the hypothesis. In particular, if $\hat{v}$ is the valuation induced from v on the completion of R with respect to v, we get $1 \geq \hat{v}(q_u) = v(\gamma_u) + v(\beta_u) \geq 2$. The evident contradiction forces y to factor through the smooth locus of $\bar{P} \to S$.

17.2 Corollary. *The notation is as in 17.1. Let $V = \mathrm{Spec}(R)$ be the spectrum of a discrete valuation. Let η be its generic point. Suppose that the valuation is normalized so that the value of a uniformizer is 1. Let $V \to S$ be a morphism such that for any $e \in \mathcal{E}$ the element q_e in R has valuation ≤ 1. Then $\bar{P}^{\mathrm{sm}} \times_S V$ is the Néron model of the abelian variety $\bar{P}_\eta$ over V.*

Proof: The same proof of 17.1 shows that $\bar{P}^{\mathrm{sm}} \times_S V$ satisfies the Néron mapping property for smooth R-algebras which are discrete valuation rings. Let Z be a smooth scheme over V and let $\phi_\eta\colon Z_\eta \to \bar{P}_\eta$ be a morphism. The localization $O_{Z,\zeta}$ of Z at an irreducible component of the special fiber of $Z \to V$ is a discrete valuation ring and it is smooth as R-algebra. Hence, ϕ_η extends to a morphism from $\mathrm{Spec}(O_{Z,\zeta})$ to $\bar{P}^{\mathrm{sm}} \times_S V$. Hence ϕ_η can be extended to a V-rational morphism from Z to $\bar{P}^{\mathrm{sm}} \times_S V$. We conclude using [4, Theorem 4.4/1].

17.3 Corollary. *The notation is as in 17.1. Let Z be a smooth scheme over S. Then any S-rational map from Z to $\bar{P}^{\mathrm{sm}}$ extends uniquely to a morphism* *from Z to $\bar{P}^{\mathrm{sm}}$.*

Proof: Use 17.1 and [4, Theorem 4.4/1].

17.4 Remark. The assumption of 17.1 on E holds for example if S_0 is the Spec of a field, S is a versal deformation space of $C_0 \to S_0$ and $C \to S$ is the versal algebraic curve deforming C_0 over S.

17.5 *Covers of degenerating abelian schemes and covers of* 1*-motives.* Let α be a finite flat commutative group scheme over S. Suppose there are exact sequences of finite flat group schemes over S

$$0 \longrightarrow \alpha_1 \longrightarrow \alpha \longrightarrow \alpha_2 \longrightarrow 0,$$

$$0 \longrightarrow \beta_1 \longrightarrow \alpha_1 \longrightarrow \beta_2 \longrightarrow 0.$$

Let $\tilde{G}$ and $\iota\colon H_1(\Gamma, \mathbf{Z}) \to \tilde{G}(K)$ be the semiabelian scheme and the 1-motive defined in 16.18. Define an α-cover of $(\mathcal{J}, \tilde{G}, H_1(\Gamma, \mathbf{Z}), H_1(\Gamma, \mathbf{Z}) \to \tilde{G}(K))$

$$(\mathcal{J}', G', H, H_K \to G'(K)) \longrightarrow (\mathcal{J}, \tilde{G}, H_1(\Gamma, \mathbf{Z}), H_1(\Gamma, \mathbf{Z}) \to \tilde{G}(K))$$

to be an extension

$$0 \longrightarrow T' \longrightarrow G' \longrightarrow \mathcal{J}' \longrightarrow 0,$$

with $\mathcal{J}'$ an abelian scheme and T' a torus over S together with a homomorphism from the generic fiber of an étale sheaf H of free abelian groups over S of rank r to G'_K, an isogeny $G' \to \tilde{G}$ with kernel α_1 such that the induced isogeny $\mathcal{J}' \to \mathcal{J}$ has kernel β_2 and a monomorphism $H \to H_1(\Gamma, \mathbf{Z})$ with cokernel α_2 such that the following diagram commutes

$$\begin{array}{ccc} H_K & \longrightarrow & G'(K) \\ \big\downarrow & & \big\downarrow \\ H_1(\Gamma)_K & \stackrel{\iota}{\longrightarrow} & \tilde{G}(K). \end{array}$$

Define a connected α-cover of $\bar{P}^{\mathrm{sm}}$

$$\gamma\colon \bar{P}^0_* \longrightarrow \bar{P}^{\mathrm{sm}}$$

to be a semiabelian scheme $\bar{P}^0_*$ over S with an isogeny γ from $\bar{P}^0_*$ to $\bar{P}^{\mathrm{sm}}$ such that the kernel of the formal isogeny induced by γ on I-adic completions is α_1 and the kenel of γ_K is α_K. Finally define an α-cover of $\bar{P}^{\mathrm{sm}}$ to be

$$\tau\colon \bar{P}_* \longrightarrow \bar{P}^{\mathrm{sm}},$$

with $\bar{P}_*$ a smooth group scheme over S and τ an isogeny with kernel α such that the induced isogeny $\bar{P}^0_* \to \bar{P}^{\mathrm{sm}}$ is a connected α-cover.

Assume that $S = \mathrm{Spec}(E)$ satisfies the condition (*) that for any étale E/I-algebra R_0 the unique formally étale E-algebra R lifting R_0 is normal. Under this assumption 16.22 and [8, Theorem III.5.5], [8, Corollary III.5.6] and [8, Corollary III.5.11] guarantee that there exists a fully faithful functor from the category of α-covers of $\big(\mathcal{J}, \tilde{G}, H_1(\Gamma), H_1(\Gamma) \to \tilde{G}(K)\big)$ to the category of connected α-covers of $\bar{P}^{\mathrm{sm}}$.

We give an explicit description of this functor in a special case.

17.6 Proposition. *Let the notations and assumptions be those of 17.5. Assume that α_1 is trivial. There is a fully faithful functor from the category of α-covers of $\big(\mathcal{J}, \tilde{G}, H_1(\Gamma, \mathbf{Z}), H_1(\Gamma, \mathbf{Z}) \to \tilde{G}(K)\big)$ to the category of α-covers of $\bar{P}^{\mathrm{sm}}$. If (*) holds, the composition of this functor with the functor associating to an α-cover $\bar{P}_*$ of $\bar{P}^{\mathrm{sm}}$ the connected α-cover $\bar{P}_*^0$ of $\bar{P}^{\mathrm{sm}}$ is the functor defined in 17.5.*

Proof: Let $\big(\tilde{G}, H_K \to \tilde{G}(K)\big)$ be an α-cover of $\big(\tilde{G}, H_1(\Gamma, \mathbf{Z}) \to \tilde{G}(K)\big)$. In particular, H is naturally identified with a subgroup of $H_1(\Gamma, \mathbf{Z})$. Let $\mathcal{P}$ be the I-adic completion of $\widetilde{P} \to S$; see 16.5. Then $\bar{P} \to S$ is the scheme whose I-adic completion coincides with $\mathcal{P}/H_1(\Gamma, \mathbf{Z})$. Define $\bar{P}_{**} \to S$ as the scheme algebraizing $\mathcal{P}/H$. By construction $\bar{P}_{**}$ is an α-torsor over $\bar{P}$. Define $\bar{P}_*$ as the smooth locus of $\bar{P}_{**} \to S$. Using 16.13 the scheme $\bar{P}_*$ is endowed with a group scheme structure. Moreover, the morphism from $\bar{P}_*$ to $\bar{P}^{\mathrm{sm}}$ is a group scheme homomorphism with kernel α. Suppose that (*) holds. Applying 16.22 to the object of $\mathrm{DD}_{\mathrm{pol}}$ defined by $\big(\tilde{G}, H_K \to \tilde{G}(K)\big)$ we conclude that $\bar{P}_*^0$ is the image of $\big(\tilde{G}, H_K \to \tilde{G}(K)\big)$ defined in 17.5.

17.7 Corollary. *The assumptions are as in 17.1. Then $(\bar{P}^{\mathrm{sm}}, \bar{0})$, with the group scheme structure defined in 16.13, represents the subfunctor* $\mathrm{Pic}^0(C/S)$ *of the relative Picard functor* $\mathrm{Pic}(C/S)$ *consisting of elements whose restrictions to all fibers of C_s with $s \in S$ belong to* $\mathrm{Pic}^0\big(C_s/k(s)\big)$*; see* [4, Section 8.4, page 233]. *In particular, the abelian variety $\bar{P}_K$ with the morphism $C_K \to \bar{P}_K$ defined in 16.4 is isomorphic to the Albanese variety of C_K over K.*

Proof: Let g be the genus of $C \to S$. By [4, Theorem 9.4/1] the sheaf $\mathrm{Pic}^0(C/S)$ is represented by a semiabelian scheme over S. Let W' be the connected open subscheme of $(C^{\mathrm{sm}})^{(g)}$ and R the S-section of W'as in [4, Lemma 9.3/6]. We recall that R exists after an étale cover of S. The basic property of W' is that it can be realized as an S-rational open immersion $W' \hookrightarrow \mathrm{Pic}^0(C/S)$ mapping R to the identity element. The morphism $C \to \bar{P}$ induces a unique morphism from W' to $\bar{P}^{\mathrm{sm}}$ sending R to 0. In particular, we deduce the existence of an S-rational map from $\mathrm{Pic}^0(C/S)$ to $\bar{P}^{\mathrm{sm}}$. By 17.3 it extends uniquely to a morphism from $\mathrm{Pic}^0(C/S)$ to $\bar{P}^{\mathrm{sm}}$ sending the identity section to the identity section. Uniqueness implies that this morphism is a group homomorphism. To prove that this morphism is an isomorphism we can reduce to the universal case, i.e., we may assume that S_0 is the spectrum of a field and S is a versal deformation space of $C_0 \to S_0$. In particular S satisfies condition (*) in 17.5. We know that $\bar{P}^{\mathrm{sm}} \times_S S_0$ is the scheme $\mathbf{G}_0$ defined in 13.16. Hence the morphism $\mathrm{Pic}^0(C/S) \to \bar{P}^{\mathrm{sm}}$ is an isomorphism over S_0. The morphism $\mathrm{Pic}^0(C_K/K) \to$

$\bar{P}_K$ is the generic fiber over $\mathrm{Spec}(K)$ of a morphism $\bar{P}_*^0 \to \bar{P}^{\mathrm{sm}}$ defined in 17.5. By [24, Corollary IX.1.4] the isomorphism over K between $\mathrm{Pic}^0(C/S)$ and $\bar{P}_*^0$ extends to an isomorphism $\bar{P}_*^0 = \mathrm{Pic}^0(C/S)$. In particular, the morphism $\bar{P}_*^0 \to \bar{P}^{\mathrm{sm}}$ can be realized as in 17.6 by a finite and étale isogeny $\bar{P}_* \to \bar{P}^{\mathrm{sm}}$. Let V be the spectrum of a discrete valuation ring with generic point η. Let $V \to S$ be a morphism such that the closed point of V is sent to S_0 and $C \times_S V$ is regular and C_η is smooth over $k(\eta)$. Let s be the image of η in S. To conclude it suffices to prove that $\bar{P}_* \to \bar{P}^{\mathrm{sm}}$ is an isomorphism for the point s of S. We are reduced to consider the case $S = V$. The morphism $\left(C^{\mathrm{sm}}\right)^{(g)} \to \bar{P}^{\mathrm{sm}}$ factors via $\bar{P}_{*,K} = \mathrm{Pic}^0(C_K/K)$. By 17.2 and the fact that $\bar{P}_* \to \bar{P}^{\mathrm{sm}}$ is finite and étale, we conclude that $\bar{P}^{\mathrm{sm}}$ is the Néron model over V of $\bar{P}_K^{\mathrm{sm}}$. Since $C \to S$ is geometrically irreducible we conclude.

18 1-motives uniformizing Jacobians of semistable curves

In this section we drop the condition of $C_0 \to S_0$ being geometrically irreducible. Assume that $C \to S$ satisfies the conditions of 18.1 and 18.2. The main results of this section are contained in 18.11. We define a semiabelian scheme $\tilde{G}$ and a 1-motive $H_1(\Gamma, \mathbf{Z}) \to \tilde{G}_K$. We will prove in 19.17 that they define the uniformization of $\mathrm{Pic}^0(C/S)$. To construct them we heavily rely on the study of the case of geometrically irreducible curves.

18.1 *Notation.* Let $S = \mathrm{Spec}(E)$ be the spectrum of a noetherian integral normal domain with fraction field K, complete and separated with respect to an ideal I defining the closed subscheme S_0 of S. For any non-negative integer n let $S_n := \mathrm{Spec}(E/I^{n+1})$. Let $\mathcal{S}$ be the affine formal scheme associated to E and the ideal I. If T is a scheme over S or a formal scheme over $\mathcal{S}$, denote by T_n the base change of T to S_n. Let

$$f: C \longrightarrow S$$

be a semistable curve over S, i.e., a proper and flat morphism whose geometric fibers are connected, reduced, one dimensional schemes having only ordinary double points as singularities. Suppose that the base change of f to S_0 satisfies the conditions of 12.1, of which we use notations and results. Suppose also that f_K is a smooth morphism. Recall that we have associated to $C_0 \to S_0$ the graph Γ; see 12.3. We have defined $\mathcal{V}$ to be the set of vertices of Γ and $\mathcal{E}$ to be the set of edges of Γ. We have fixed a decomposition

$$H_1(\Gamma, \mathbf{Z}) \cong \mathbf{Z}l_1 \oplus \ldots \oplus \mathbf{Z}l_r$$

in elementary cycles. For each j in $\mathcal{V}$ fix a S-section

$$\bar{0}_j: S \longrightarrow C$$

of f deforming the section 0_j of f_0 chosen in 12.5. *Choose* one of them and label it by $\bar{0}$.

18.2 *Assumption.* Assume that for each $e \in \mathcal{E}$ there exists a non-zero ideal (q_e) such that the completion of C at the singular S_0-valued point T_e defined in 12.3 is of the form

$$O_C^{\wedge T_e} \xrightarrow{\sim} E[\![x_e, y_e]\!]/(x_e y_e - q_e).$$

Assume furthermore that the algebra $E/(q_e)$ is a domain.

18.3 *A few reasons why the case of irreducible curves is easier, but useful in the general case.* In this and the following sections we consider the case when $C_0 \to S_0$ is not geometrically irreducible. In this case the constructions of the previous sections are considerably more involved. The first problem is that 15.9 and 16.4 are not available.

In (1) of 14.7 we have defined a Galois cover $\mathcal{P}_0$ of a compactification of the scheme $\mathrm{Pic}^0(C_0/S_0)$ assuming $C_0 \to S_0$ to be geometrically irreducible. Its simple structure has allowed us to study the deformation theory of $\mathcal{P}_0 \to S_0$. In 14.7 we have constructed a morphism from the Galois cover $\mathcal{C}_0 \to C_0$, defined in 14.2, to $\mathcal{P}_0$. This has allowed us to compare the deformation theories of C_0 and $\mathcal{P}_0$. In the general case $\mathcal{P}_0 \to S_0$ exists, but has a much more complicated structure. For example, it is not canonical. Moreover, the computation of its infinitesimal deformation theory becomes extremely involved. The author failed to construct objects $\mathcal{P}_0$ receiving the Galois cover $\mathcal{C}_0$ of C_0 and satisfying the fundamental 15.4. In particular, the comparison of the first order deformation theories of C_0 and $\mathcal{P}_0$ over S_0 is not available.

Still something can be done. In fact, in the universal case, i.e., if S_0 is the spectrum of a field and S is a versal deformation space of $C_0 \to S_0$, the complement of the open subscheme of points of S where $C \to S$ is geometrically irreducible has high codimension, so that some of the results obtained in the irreducible case will hold also in this case. It is then easy to descend those results from the universal case to general schemes S as in 18.1. In particular, we can recover the 1-motive uniformizing $\mathrm{Pic}^0(C/S)$. See 18.12. We can also conclude the existence and describe the uniformization of a scheme $\bar{P}^{\varphi,\mathrm{sm}}$, smooth over S, containing $\mathrm{Pic}^0(C/S)$ as an open subscheme, whose fiber over K is $\mathrm{Pic}^0(C_K/K)$, acted upon by $\mathrm{Pic}^0(C/S)$ and satisfying the following. Let V be the spectrum of a discrete valuation ring with generic point η. Let $V \to S$ be a morphism such that $C \times_S V$ is regular and C_η is smooth over $k(\eta)$. Then $\bar{P}^{\varphi,\mathrm{sm}} \times_S V$ is the Néron model of $\mathrm{Pic}^0\big(C_\eta/k(\eta)\big)$. See 19.11.

18.4 Definition. *Let H be a subset of $\mathcal{E}$. Let J_H be the ideal of E generated by the elements q_h with h not in H. Let*

$$S_H^{\mathrm{o}} := \mathrm{Spec}(E_H^{\mathrm{o}}) \qquad \text{and} \qquad S_{H,0} := \mathrm{Spec}(E_{H,0})$$

be respectively the open subscheme of S defined by inverting all q_e with e in H

and the closed subscheme of S_H^o defined by the ideal $I_H^o = J_H\, E_H^o$. Let E_H be the completion of E_H^o with respect to I_H^o. Define

$$S_H := \mathrm{Spec}(E_H) \qquad \text{and} \qquad I_H := I_H^o\, E_H.$$

If E_H is a domain, let K_H be its fraction field.

18.5 Definition. *The notation is as in 18.4. Define*

$$C_{H,0} := C \times_S S_{H,0} \qquad \text{and} \qquad C_H := C \times_S S_H.$$

The curve $C_{H,0} \to S_{H,0}$ is split semistable in the sense of 12.1. Let Γ_H be the graph associated to $C_{H,0}$ as in 12.3. Denote by $\mathcal{V}_H$ (respectively $\mathcal{E}_H$) the set of vertices (respectively edges) of Γ_H.

18.6 Remark. The graph Γ_H is obtained from the graph Γ by contracting all the edges $e \in \mathcal{E}$ contained in H via the procedure described in 4.1.

18.7 Definition. *A subset H of $\mathcal{E}$ is said to be good if the ring E_H defined in 18.4 is a normal integral domain.*

A subset H of $\mathcal{E}$ as in 18.4 is said to be irreducible if the curve $C_{H,0} \to S_{H,0}$ is geometrically irreducible. We say that H is irreducible of maximal rank if the rank of $H_1(\Gamma_H, \mathbf{Z})$ is equal to the rank of $H_1(\Gamma, \mathbf{Z})$ or, equivalently by 4.2, if the morphism $H_1(\Gamma, \mathbf{Z}) \to H_1(\Gamma_H, \mathbf{Z})$ is an isomorphism.

18.8 Remark. If S_0 in 18.1 is a field and S is a versal deformation space of $C_0 \to S_0$, then any subset H of $\mathcal{E}$ is good. In fact, since E is the spectrum of a complete local ring, $S_H \to S_H^o$ has regular fibers by [12, Proposition 7.4.6]. Hence S_H is normal by [12, Corollary 6.5.4]. It is easy to verify that S_H is also integral.

18.9 Definition. *Suppose that H is good and irreducible of maximal rank. Let*

$$\tilde{G}_H \qquad \text{and} \qquad H_1(\Gamma_H, \mathbf{Z}) \longrightarrow \tilde{G}_H(K_H)$$

be the semiabelian scheme over S_H and the 1-motive uniformizing the semiabelian scheme $\mathrm{Pic}^0(C_H/S_H)$ defined in 16.7 and in 16.18. Let $\mathcal{J}_H$ be the abelian scheme over S_H obtained as the abelian part of $\tilde{G}_H$ as in 16.7. It is naturally endowed with a principal polarization by 16.19.

Define

$$\bar{P}_H \longrightarrow S_H \qquad \text{and} \qquad \overline{\Theta}_H$$

to be the scheme and the divisor constructed in 16.4. By 16.13 and 17.7 the smooth locus $\bar{P}_H^{\mathrm{sm}}$ of $\bar{P}_H \to S_H$ coincides with $\mathrm{Pic}^0(C_H/S_H)$. There is a morphism $C_H \to \bar{P}_H$ sending the section $\bar{0} \times_S S_H$ of C_H fixed in 18.1 to the identity of $\mathrm{Pic}^0(C_H/S_H)$. The divisor $\overline{\Theta}_H$ is an an ample effective Cartier divisor relative to S_H.

Let

$$\widetilde{P}_H \longrightarrow \mathcal{J}_H \qquad \text{and} \qquad \tilde{\Theta}_H$$

be the scheme over $\mathcal{J}_H$ and the divisor associated to C_H in 16.5 and 16.8. The fiber of $\widetilde{P}_H$ over K_H is $\widetilde{G}_{H,K_H}$ and the I_H-adic completion of $\widetilde{P}_H$ is a formal Galois cover with group $H_1(\Gamma_H, \mathbf{Z})$ of the I_H-adic completion of $\bar{P}_H$. By 11.8 there is a group scheme structure on the smooth locus of $\widetilde{P}_H \to S_H$. By 11.8 and 16.8 the group $H_1(\Gamma, \mathbf{Z})$ acts on $\widetilde{P}_H$ and on $\tilde{\Theta}_H$ over S_H.

18.10 Theorem. *Let* Pic(C/S) *be the relative Picard functor. Let* Pic$^0(C/S)$ *be the subfunctor consisting of elements whose restrictions to all fibers of C_s with $s \in S$ belong to* Pic$^0(C_s/k(s))$*; see* [4, Section 8.4, page 233]. *Then the functor* Pic$^0(C/S)$ *is represented by a semiabelian scheme over S.*

Proof: This follows from [4, Theorem 9.4/1].

18.11 Proposition. *Let $\tilde{G}$ be the Raynaud extension of* Pic$^0(C/S)$ *associated as in* [8, Section II.1] *to* Pic$^0(C_0/S_0)$. *Let $\mathcal{J}$ be the abelian scheme over S and T the torus over S, such that we have a short exact sequence*

$$0 \longrightarrow T \longrightarrow \tilde{G} \longrightarrow \mathcal{J} \longrightarrow 0.$$

Then

1. *$\mathcal{J}$ carries a principal polarization*

$$\lambda \colon \mathcal{J} \longrightarrow \mathcal{J}^\vee$$

   extending the principal polarization on the abelian scheme $\mathcal{J} \times_S S_0 = \mathcal{J}_0$ defined in 12.6.
2. *let H be a good irreducible subset of $\mathcal{E}$ of maximal rank in the sense of 18.7.*
   i. *The pullback of $\tilde{G}$ to S_H is isomorphic to the semiabelian scheme $\tilde{G}_H$ over S_H defined in 18.9.*
   ii. *The pullback to S_H of the principally polarized abelian scheme $\mathcal{J}$ coincides with the principally polarized abelian scheme $\mathcal{J}_H$ defined in 18.9.*

Proof: Let H be as in (2). Let J_H be the ideal of E defined in 18.4. We introduce some notation. The base change of Pic$^0(C/S)$ to Spec(E/J_H) is isomorphic to Pic$^0\big(C \times_S \text{Spec}(E/J_H)/\text{Spec}(E/J_H)\big)$. Hence it is a semiabelian scheme over Spec(E/J_H), extension

$$0 \longrightarrow T_0^H \longrightarrow G_0^H \longrightarrow B_0^H \longrightarrow 0$$

of an abelian scheme B_0^H and a torus T_0^H of rank r. For any $h \notin H$ the singular section of the curve $C_0 \to S_0$ associated to h defined in 18.1 survives on the curve $C \times_S \text{Spec}(E/J_H)$. In particular, B_0^H is the Pic^0 of the curve obtained from $C \times_S \text{Spec}(E/J_H)$ by blowing-up the singularities associated to the elements h not in H. Let $\mathcal{G}^H$ be the completion of $\mathrm{Pic}^0(C/S)$ along G_0^H. By the discussion in [8, Section II.1] it sits in an exact sequence

$$0 \longrightarrow \mathcal{T}^H \longrightarrow \mathcal{G}^H \longrightarrow \mathcal{B}^H \longrightarrow 0,$$

where $\mathcal{T}^H$ is a formal torus and $\mathcal{B}^H$ is a formal abelian scheme. The base change of this exact sequence to $\mathrm{Spec}(E/J_H)$ coincides with the exact sequence above. By descent theory the statements of the proposition hold if they hold for the completion of S at all closed points of S_0. In particular, it suffices to consider the universal case, i.e., we can assume that S is a versal deformation space of a semistable curve $C_0 \to S_0$ over the spectrum of a field. By 18.8 there exists a subset H of $\mathcal{E}$ which is good and irreducible of maximal rank. Suppose that the principal polarization on B_0^H extends to a principal polarization on $\mathcal{B}^H \times_{\mathcal{S}} \mathrm{Spec}(E/J_H^n)$. The obstruction b to lift the principal polarization from $\mathrm{Spec}(E/J_H^n)$ to $\mathrm{Spec}(E/J_H^{n+1})$ lives in

$$H^1\left(B_0^H, \mathcal{H}om_{O_{B_0^H}}(\Omega^1_{B_0^H/\mathrm{Spec}(E/J_H)}, O_{B_0^H})\right) \underset{E/J_H}{\otimes} (J_H^n/J_H^{n+1}).$$

The curve obtained by base change of $C \to S$ to $S_{H,0}$ is the geometrically irreducible curve $C_{H,0} \to S_{H,0}$ so that the principal polarization extends on S_H by 16.19. Hence $b \otimes_E E_H$ is trivial. Hence $b \otimes_E E_H^o$ is trivial. Since S is the versal deformation of a curve over a field, multiplication by the elements q_h with $h \in H$ is injective on J_H^n/J_H^{n+1}. Hence b is trivial. In particular, the formal abelian scheme $\mathcal{B}^H$ can be algebraized to an abelian scheme B^H over S by Grothendieck's existence theorem. Moreover, the principal polarization on B_0^H extends to a principal polarization on B^H. Due to [8, Section II.1] the sequence of formal group schemes above can be algebraized to a sequence

$$0 \longrightarrow T^H \longrightarrow G^H \longrightarrow B^H \longrightarrow 0,$$

where T^H is a torus over S. Since the completion of G^H along its base change to S_0 is the further completion of $\mathcal{G}^H$ along its base change to S_0, it is then clear that G^H is the Raynaud extension of $\mathrm{Pic}^0(C/S)$ associated to $\mathrm{Pic}^0(C_0/S_0)$. To conclude we remark that the base change of $\tilde{G}$ to S_H is the base change of G^H to S_H. By construction this is $\tilde{G}_H$.

18.12 *The construction of the uniformizing* 1*-motive.* We aim to construct a homomorphism

$$\iota\colon H_1(\Gamma, \mathbf{Z}) \longrightarrow \tilde{G}\left(S\left[\frac{1}{q_e}\right]_{e\in\mathcal{E}}\right) \hookrightarrow \tilde{G}(K).$$

The notation is as in 18.2. The homomorphism will have the property that for any good irreducible subset H of $\mathcal{E}$ of maximal rank in the sense of 18.7 the following diagram commutes

$$\begin{array}{ccc} H_1(\Gamma,\mathbf{Z}) & \xrightarrow{\iota} & \tilde{G}(K) \\ \downarrow & & \downarrow \\ H_1(\Gamma_H,\mathbf{Z}) & \longrightarrow & \tilde{G}_H(K_H). \end{array}$$

The left vertical map is the one defined in 4.2, the right vertical map is induced from 18.11 and the lower horizontal map is defined in 18.9.

First of all we construct the homomorphism in the case that S_0 is the spectrum of a field and S is a versal deformation space of $C_0 \to S_0$. By 18.8 any subset H of $\mathcal{E}$ is good. We will show that the property of compatibility with the 1-motives defined in 18.9 for H irreducible of maximal rank characterizes the homomorphism uniquely. It suffices to prove that for any H irreducible of maximal rank the map induced composing the isomorphism $H_1(\Gamma,\mathbf{Z}) \cong H_1(\Gamma_H,\mathbf{Z})$ with $H_1(\Gamma_H,\mathbf{Z}) \to \tilde{G}(K_H)$ factors via $\tilde{G}(K)$ and does not depend on H. Since S is excellent, the theory of [6] is available. Consider two irreducible subsets H_1 and H_2 of $\mathcal{E}$ of maximal rank. By 16.22 the 1-motives $H_1(\Gamma_{H_1},\mathbf{Z}) \to \tilde{G}_{H_1}(K_{H_1})$ and $H_1(\Gamma_{H_2},\mathbf{Z}) \to \tilde{G}_{H_2}(K_{H_2})$ define uniformizations of the same principally polarized abelian variety $\mathrm{Pic}^0(C_K/K)$ over S_{H_1} and S_{H_2} respectively. After extending the principal polarization to a theta level structure of type $(2^d,\ldots,2^d,2^{2d},\ldots,2^{2d})$ with $d \geq 3$, the two 1-motives uniformize the same abelian variety $\mathrm{Pic}^0(C_K/K)$ over K with theta level structure. The values of the theta nulls of $\mathrm{Pic}^0(C_K/K)$ are expressed via their Fourier-Jacobi expansion

a. in terms of the values of the sections of the canonical rigidified line bundle L_{H_1} associated to the given theta structure on $\mathcal{J}_{H_1,K_{H_1}}$, which is the abelian part of $\tilde{G}_{H_1,K_{H_1}}$, at the r sections
$$\mathbf{Z}l_1 \oplus \ldots \oplus \mathbf{Z}l_r \xrightarrow{\Phi_{H_1}^{-1}} H_1(\Gamma_{H_1},\mathbf{Z}) \xrightarrow{c_{H_1}^t} \mathcal{J}_{H_1,K_{H_1}}$$
defined in 16.20;
b. in terms of the liftings of the r sections of $\mathcal{J}_{H_1,K_{H_1}}$ over K_{H_1} mentioned in (a) to
$$H^1(\Gamma_{H_1},\mathbf{Z}) \longrightarrow L_{H_1}(K_{H_1}).$$

The same holds for the set H_2 instead of the set H_1. See [6, Discussion in III.2.0-2.3, especially Lemma III.2.3.1] or [8, Theorem III.6.2]. Data (a) must coincide for the sets H_1 and H_2 by 18.11. Also data (b) must coincide since they define the Fourier-Jacobi expansion of the theta nulls of the same abelian variety $\mathrm{Pic}^0(C_K/K)$ over K with theta level structure. Since the data (a) and (b) determine uniquely the 1-motives associated to H_1 and H_2, they determine the same 1-motive $H_1(\Gamma,\mathbf{Z}) \to \tilde{G}(K_{H_1} \cap K_{H_2})$. Since $K = \cap_H K_H$, where the intersection is taken over all irreducible subsets H of $\mathcal{E}$ of maximal rank, we are done. The fact that the 1-motive is defined by a homomorphism $H_1(\Gamma,\mathbf{Z}) \to$

$\tilde{G}(K)$ factoring via $H_1(\Gamma, \mathbf{Z}) \to \tilde{G}\left(S\left[\frac{1}{q_e}\right]_{e\in\mathcal{E}}\right)$ follows from 16.7.

In the general case, the completion $S^{\backslash s}$ of S at a point s of S_0 maps to a versal defomation space of $C_s \to \text{Spec}(k(s))$. The homomorphism ι is the unique one compatible with those induced on $S^{\backslash s}\left[\frac{1}{q_e}\right]_{e\in\mathcal{E}}$ for all $s \in S_0$ via the construction in the versal case.

18.13 *Degeneration data associated to $\tilde{G}$ and ι.* Let T be the split torus over S defined as the kernel of $\tilde{G} \to \mathcal{J}$ and let $X(T)$ be its character group. As in 16.20 define the isomorphism

$$\Phi\colon H_1(\Gamma, \mathbf{Z}) \longrightarrow X(T).$$

Consider the following diagram

$$\begin{array}{ccc} H_1(\Gamma, \mathbf{Z}) & \xrightarrow{c^t} & \mathcal{J} \\ \downarrow{\scriptstyle \Phi} & & \downarrow{\scriptstyle \lambda} \\ X(T) & \xrightarrow{c} & \mathcal{J}^{\vee}, \end{array}$$

where c^t is the extension to S of the composite of the 1-motive $H_1(\Gamma, \mathbf{Z}) \to \tilde{G}(K)$ with the projection $\tilde{G}_K \to \mathcal{J}_K$, while c is the homomorphism defining $\tilde{G}$ as the extension of $\mathcal{J}$ by the torus T and finally λ is the principal polarization defined in 18.11. By 16.20 and 18.8 the diagram is commutative if S_0 is a field and S is the versal deformation of $C_0 \to S_0$. Hence the diagram is commutative. Factor $\tilde{G}$ as product as schemes over $\mathcal{J}$ of $\mathbf{G}_m$-bundles

$$\tilde{G} = \tilde{G}_1 \underset{\mathcal{J}}{\times} \ldots \underset{\mathcal{J}}{\times} \tilde{G}_r$$

induced by the isomorphism

$$X(T) \xrightarrow{\Phi^{-1}} H_1(\Gamma, \mathbf{Z}) \xrightarrow{\sim} \mathbf{Z}l_1 \oplus \ldots \oplus \mathbf{Z}l_r;$$

the last isomorphism is the one fixed in 12.3. By [6, Section II.2.3.3], to define ι is equivalent to define a trivialization τ of the pullback via $c^t \times c$ of the biextension on $\mathcal{J}_K \times_K \mathcal{J}_K^{\vee}$ defined by the Poincaré bundle.

18.14 Corollary. *The data $\left(\tilde{G}, X(T), H_1(\Gamma, \mathbf{Z}), c, c^t, \iota, \lambda, \Phi, \tau\right)$ are in the category* DD_{pol} *defined in* [8, Section III.2, pag. 57-58].

Proof: It suffices to consider the case where S_0 is the spectrum of a field and S is a versal deformation space of $C_0 \to S_0$. In this case the corollary holds after base change to S_H for irreducible subsets H of $\mathcal{E}$ of maximal rank in the sense of 18.7 by 16.20 and 16.21. By 18.12 the data we get agree for various sets H's. Since S is normal, since the complements of the union of the images of the schemes S_H's in S have codimension at least 2 and since $K = \cap_H K_H$, the corollary is proven.

18.15 *The definition of $x_e \in \tilde{G}(K)$ for e in $\mathcal{E}$.* The notation is as in 18.1. Let H be a good irreducible subset of $\mathcal{E}$ of maximal rank as defined in 18.7. As explained in 16.4 the choice of the section $\bar{0}$ defines a morphism $C_H \to \bar{P}_H$. Let $\mathcal{C}_H \to \mathcal{P}_H$ be the morphism of formal Galois covers with group $H_1(\Gamma, \mathbf{Z})$ of the I_H-adic completion of $C_H \to \bar{P}_H$ deforming the diagram

$$\begin{array}{ccc} \mathcal{C}_{H,0} & \longrightarrow & \mathcal{P}_{H,0} \\ \big\downarrow & & \big\downarrow \\ C_{H,0} & \longrightarrow & \bar{P}_{H,0} = \mathcal{P}_{H,0}/H_1(\Gamma) \end{array}$$

defined in 14.7. We view the maximal abelian cover Γ_H^{ab} of Γ_H as a contraction of the dual graph of $\mathcal{C}_{H,0}$ as in 12.3. Let e be an oriented edge of Γ whose source corresponds to the curve $\mathbb{D}_{j_{s(e)}}$ and whose target corresponds to the curve $\mathbb{D}_{j_{t(e)}}$ in 12.3. Let d be an oriented edge in the maximal abelian cover Γ^{ab} of Γ mapping to e compatibly with the orientations. The sections

$$\bar{0}_{j_{s(e)}},\quad \bar{0}_{j_{t(e)}}\colon S \longrightarrow C$$

define sections

$$\bar{0}_{j_{s(e)},H},\quad \bar{0}_{j_{t(e)},H}\colon S_H \longrightarrow C_H.$$

The contraction map from Γ to Γ_H defines a contraction map between their maximal abelian covers, i.e., from Γ^{ab} to Γ_H^{ab}. Denote by $s(d)$ (resp. $t(d)$) the image of the source (resp. the target) of d in Γ^{ab}. Let

$$\hat{0}_{s(d),H},\quad \hat{0}_{t(d),H}\colon \mathcal{S}_H \longrightarrow \mathcal{C}_H$$

be the elements in the inverse images of $\bar{0}_{j_{s(e)},H}$ and $\bar{0}_{j_{t(e)},H}$ in $\mathcal{C}_H$ whose base change to $S_{H,0}$ lie in the components of $\mathcal{C}_{H,0}$ associated to $s(d)$ and $t(d)$. Their images in $\mathcal{P}_H$ uniquely algebraize to sections

$$\widetilde{0}_{s(d),H},\quad \widetilde{0}_{t(d),H}\colon S_H \longrightarrow \widetilde{P}_H$$

of the scheme $\widetilde{P}_H$ algebraizing $\mathcal{P}_H$ defined in 18.9. Using 16.7 define

$$x_e := \widetilde{0}_{t(d),H} \cdot \left(\widetilde{0}_{s(d),H}\right)^{-1} \in \tilde{G}\left(S_H\left[\frac{1}{q_e}\right]\right) \hookrightarrow \tilde{G}(K_H).$$

The element $q_e \in E$ is defined in 18.2. As the notation suggests, the element x_e does not depend on d, but only on e. The definition might seem involved. In the case that $C \to S$ is geometrically irreducible, the definition of the elements x_e coincides with the definition of the elements x_t for $t = 1, \ldots, r$ given in 16.7.

18.16 Lemma. *Suppose that S_0 is the spectrum of a field and S is a versal* *deformation space of $C_0 \to S_0$. Then there exists a unique element*

$$x_e \in \tilde{G}\Big(S\Big[\frac{1}{q_e}\Big]\Big)$$

such that for any (good) irreducible subset H of $\mathcal{E}$ of maximal rank in the sense of 18.7 the image $x_e \in \tilde{G}\Big(S_H\Big[\frac{1}{q_e}\Big]\Big)$ is the element constructed in 18.15.

Proof: Recall that by 18.8 any subset H of $\mathcal{E}$ is good. The notation is as in 18.15. Fix a theta level structure on $\mathrm{Pic}^0(C_K/K)$ of type $(2^d, \ldots, 2^d, 2^{2d}, \ldots, 2^{2d})$ for some $d \geq 3$. The image of $\bar{0}_{j_{s(e)}}$ (respectively $\bar{0}_{j_{t(e)}}$) in $\mathrm{Pic}^0(C_K/K)$ is determined via the projective embedding given by the sections of the very ample line bundle associated to the given theta structure. For H irreducible of maximal rank, this can be expressed, by evaluating the Fourier-Jacobi expansion of these sections as in [6, III.2.2-2.3], in terms of the sections of the pullback to $\tilde{G}_K$ of a suitable rigidified line bundle L on $\mathcal{J}$ at the image of $\widetilde{0}_{s(d),H}$ (resp. $\widetilde{0}_{t(d),H}$) in $\tilde{G}(K_H)$. The just mentioned evaluations do not depend on H. This forces the image of $\tilde{0}_{s(d),H}$ (resp. $\tilde{0}_{t(d),H}$) to belong to $\tilde{G}(K)$ and to be independent of H as claimed.

18.17 Definition. *Using 18.16 and proceeding as in 18.12 define*

$$x_e \in \tilde{G}\Big(S\Big[\frac{1}{q_e}\Big]\Big) \hookrightarrow \tilde{G}(K).$$

It has the property that for any good irreducible subset H of $\mathcal{E}$ the induced element in $\tilde{G}(K_H)$ is the element constructed in 18.15.

18.18 Proposition. *Let $\tilde{G}$ split as products of $\mathbf{G}_m$-bundles $\tilde{G}_1, \ldots, \tilde{G}_r$ over the abelian scheme $\mathcal{J}$ as in 18.13.*

1. *The image of x_e in $\mathcal{J}(K)$ extends uniquely to an element y_e of $\mathcal{J}(S)$.*
2. *For any e in $\mathcal{E}$ and for any l_t with $1 \leq t \leq r$ as in 18.1, if e does not belong to l_t, the projection of x_e to $\tilde{G}_t(K)$ lies in $\tilde{G}_t(S)$.*
3. *Let e be in $\mathcal{E}$ such that e belongs to l_t for some $1 \leq t \leq r$. The fractional ideal in K defined by the section x_e over K of the $\mathbf{G}_m$-bundle $y_e^*(\tilde{G}_t)$ over S coincides with the ideal (q_e) defined in 18.1.*
4. *For $1 \leq t \leq r$ let Γ_t be the subgraph of Γ defined in 12.3. The image of l_t in $\tilde{G}(K)$ (resp. in $\mathcal{J}(K)$) via ι is the product over the positively oriented edges e of Γ_t of the elements x_e (resp. y_e).*

Proof: As in 18.12 it suffices to consider the universal case. We first prove (1). Let H be an irreducible subset of $\mathcal{E}$ of maximal rank. Let J_H be the ideal of E defined in 18.4. Let

$$C_0^H := C \times_S \mathrm{Spec}(E/J_H).$$

In 14.2 we have defined $H_1(\Gamma_H, \mathbf{Z})$-Galois covers

$$\mathcal{C}_0^H \longrightarrow C_0^H \qquad \text{and} \qquad \mathcal{C}_{H,0} \longrightarrow C_{H,0}.$$

Moreover, $C_{H,0}$ is the base change of C_0^H to $S_{H,0}$. Let $\mathcal{C}$ be the $H_1(\Gamma, \mathbf{Z})$ cover of the J_H-adic completion of $C \to S$ extending $\mathcal{C}_0^H \to C_0^H$. In the notation of 18.15 the formal curve $\mathcal{C}^H$ carries unique sections $\hat{0}_{s(d)}$ and $\hat{0}_{t(d)}$ over the J_H-adic completion of S lifting the sections $\bar{0}_{j_{s(e)}}$ and $\bar{0}_{j_{t(e)}}$ of $C \to S$ and extending the sections $\hat{0}_{s(d),H}$ and $\hat{0}_{t(d),H}$ of $\mathcal{C}_{H,0}$. The morphism

$$\mathcal{C}_{H,0} \longrightarrow \mathcal{J}_H \underset{S_H}{\times} S_{H,0}$$

defined in 14.7 extends uniquely to a morphism

$$\mathcal{C}_0^H \to \mathcal{J} \underset{S}{\times} \operatorname{Spec}(E/J_H).$$

Consider the problem of extending the latter morphism to a morphism from $\mathcal{C}^H$ to the J_H-adic completion of $\mathcal{J}$. Suppose this can be done over $\operatorname{Spec}(E/J_H^n)$. The obstruction b to lift it to $\operatorname{Spec}(E/J_H^{n+1})$ lives in

$$H^1(\mathcal{C}_0^H, O_{\mathcal{C}_0^H}) \underset{E/J_H}{\otimes} (J_H^n/J_H^{n+1})$$

tensored over E/J_H with the global sections of the tangent bundle of the abelian scheme $\mathcal{J} \times_S \operatorname{Spec}(E/J_H)$ relative to $\operatorname{Spec}(E/J_H)$. The existence of the morphism $\mathcal{C}_H \to \mathcal{P}_H$ as in 18.17 and of the morphism from $\mathcal{P}_H$ to the I_H-completion of $\mathcal{J}_H$ defined in 8.6 imply that b is trivial after tensoring with E_H. In particular, b is zero after tensoring with E_H^o. Hence b is trivial since $H^1(\mathcal{C}_0^H, O_{\mathcal{C}_0^H})$ is a locally free E/J_H-module and S is supposed to be a versal deformation space of a curve over a field so that multiplication by the elements q_h with $h \in H$ on J_H^n/J_H^{n+1} is injective. Define y_e as the S-section of $\mathcal{J}$ algebraizing the difference of the images of $\hat{0}_{t(d)}$ and $\hat{0}_{s(d)}$. By 16.7 it is clear that y_{e,K_H} is the image of $x_{e,H}$ in $\mathcal{J}$. This finishes the proof of (1). The other properties follow from similar properties in the irreducible case described in 16.7 and in 16.18 using 18.12 and the last sentence of 18.15.

18.19 Remark. Let H be a good irreducible subset of $\mathcal{E}$ of maximal rank in the sense of 18.7. Let j be an element of $\mathcal{V}$ and let $\bar{0}_j$ be the associated section of $C \to S$. It defines a section of the curve over S_H associated to $\vartheta_H(j)$; see 4.1 for the notation. Hence for any element v of $\mathcal{V}_H$ the sections $\bar{0}_j \times_S S_H$ with j in $\vartheta_H^{-1}(v)$ meet the curve associated to v. For any $v \in \mathcal{V}_H$ *choose* one of them. Suppose that the section $\bar{0}$ is among them. Then for any oriented edge f of Γ_H we can define the elements $x_f \in \tilde{G}(K)$ and $y_f \in \mathcal{J}_H(S_H)$ as in 18.17 and 18.18.

19 Families of Néron models of Jacobians of semistable curves

The notation is as in 18.1 and in 18.2. So far we have constructed a semiabelian scheme $\tilde{G}$ in 18.11 and a 1-motive $H_1(\Gamma, \mathbf{Z}) \to \tilde{G}(K)$ in 18.12. In 18.17 we have defined a homomorphism $C_1(\Gamma, \mathbf{Z}) \to \tilde{G}$ refining the 1-motive. Using the graph Γ we obtain as in 6.2 compactifications P_ξ of $\tilde{G}$ for all $\xi \in H^1(\Gamma, \mathbf{Z})$. In this section we combine these ingredients to construct suitable compactifications $\bar{P}^\varphi \to S$ of $\mathrm{Pic}^0(C/S)$ interpolating the Néron models of $\mathrm{Pic}^0(C_\eta/k(\eta))$ for all spectra of discrete valuations V with generic point η such that C_V is regular and C_η is smooth over $k(\eta)$. See 19.9 and 19.11. The compactifications, and also their smooth loci over S, are not canonical in general, depending on suitable elements φ defined in terms of the graph Γ; see 19.13. We conclude in 19.17 that the 1-motive defined in 18.12 is the uniformization of $\mathrm{Pic}^0(C/S)$. The main ingredient of the proof is the construction in 19.2 of a scheme $\widetilde{P}^\varphi \to S$ with an action of $H_1(\Gamma, \mathbf{Z})$. Its generic fiber is $\tilde{G}_K$. The action of $H_1(\Gamma, \mathbf{Z})$ allows one to recover the 1-motive $H_1(\Gamma, \mathbf{Z}) \to \tilde{G}(K)$. See 19.7. On the other hand, the I-adic completion $\mathcal{P}^\varphi$ of $\widetilde{P}^\varphi$ is an $H_1(\Gamma, \mathbf{Z})$-Galois cover of the I-adic completion of $\bar{P}^\varphi$. See 19.8 and 19.9.

19.1 Definition. *Let φ be a non-degenerate element in $\partial\big(C_1(\Gamma, \mathbf{R})\big)$ in the sense of 3.5. Let $\tilde{G}$ be the semiabelian scheme over S defined in 18.11. Let ξ be an element of $H^1(\Gamma, \mathbf{Z})$. Let*

$$r_\xi \colon \tilde{G} \hookrightarrow P_\xi$$

be the inclusion of fiber bundles over $\mathcal{J}$ defined in 6.2. Define

$$s_\xi \colon \tilde{G}_K \longrightarrow \tilde{G} \qquad \text{and} \qquad t_\xi \colon \mathcal{J} \longrightarrow \mathcal{J}$$

as follows. If $\xi = 0$, then s_ξ is the natural morphism $\tilde{G}_K \longrightarrow \tilde{G}$ and t_ξ is the identity. Suppose that e is a 1-dimensional polyhedron of $\mathrm{Del}_\varphi\big(H^1(\Gamma, \mathbf{R})\big)$ with associated 0-dimensional faces μ and ξ. By 3.8 we consider e as an element of $\mathcal{E}$. Suppose that s_μ is defined and orient e so that its source is μ and its target is ξ. Define

$$s_\xi \colon \tilde{G}_K \xrightarrow{\cdot x_e} \tilde{G}_K \xrightarrow{s_\mu} \tilde{G}.$$

Define

$$t_\xi \colon \mathcal{J} \xrightarrow{+y_e} \mathcal{J} \xrightarrow{t_\mu} \mathcal{J}.$$

The notation $\cdot x_e$ (resp. $+y_e$) stands for multiplication by the element x_e of $\tilde{G}(K)$ (resp. y_e of $\mathcal{J}(S)$) defined in 18.18. It is easy to see that for any element ξ of $H^1(\Gamma, \mathbf{Z})$ the morphism s_ξ (respectively t_ξ) does not depend on the choice of μ. Finally for any element ξ of $H^1(\Gamma, \mathbf{Z})$ define

$$z_\xi \colon \tilde{G}_K \xrightarrow{s_\xi} \tilde{G} \xrightarrow{r_\xi} P_\xi.$$

Let H be a good irreducible subset of $\mathcal{E}$ of maximal rank in the sense of 18.7.

An analogous definition is given replacing Γ with Γ_H, replacing S with S_H and replacing $\tilde{G}$ with $\tilde{G}_H$ using 18.9 and 18.19.

19.2 *The scheme $\widetilde{P}^\varphi \to S$.* Let φ be a non-degenerate element in $\partial\big(C_1(\Gamma,\mathbf{R})\big)$ in the sense of 3.5. From the data

$$\tilde{G}_K \xrightarrow{z_\xi} P_\xi,$$

with $\xi \in H^1(\Gamma,\mathbf{Z})$, we shall define a scheme $\widetilde{P}^\varphi$ and a map

$$\rho^\varphi \colon \widetilde{P}^\varphi \longrightarrow \mathcal{J}.$$

We proceed as follows

1. we construct a scheme $\widetilde{P}'$ of the type considered in 11.2;
2. we construct an open subscheme $\widetilde{P}'^o$ of $\widetilde{P}'$;
3. we construct $\widetilde{P}^\varphi$ as a suitable fppf quotient of $\widetilde{P}'^o$ by a torus T''.

19.3 *Some notation.* With the notation of 3.1 define $T'' := \mathbf{G}_{m,S}\otimes C^0(\Gamma,\mathbf{Z})$ and $T' := \mathbf{G}_{m,S}\otimes C^1(\Gamma,\mathbf{Z})$ and $T = \mathbf{G}_{m,S}\otimes H^1(\Gamma,\mathbf{Z})$. We have an exact sequence of tori

$$1 \longrightarrow \mathbf{G}_{m,S} \longrightarrow T'' \longrightarrow T' \longrightarrow T \longrightarrow 1.$$

Recall that $\tilde{G}$ is a semiabelian scheme over S extension of $\mathcal{J}$ by T. Let

$$0 \longrightarrow T' \longrightarrow \tilde{G}' \longrightarrow \mathcal{J} \longrightarrow 0$$

be the semiabelian scheme over S defined by

$$\begin{array}{ccc} C_1(\Gamma,\mathbf{Z}) = \mathbf{Z}^{\mathcal{E}} & \longrightarrow & \mathcal{J} \xrightarrow{\lambda} \mathcal{J}^\vee = \mathrm{Ext}^1(\mathcal{J},\mathbf{G}_{m,S}) \\ \sum_{e\in\mathcal{E}} \alpha_e\, e & \mapsto & \sum_{e\in\mathcal{E}} \alpha_e\, y_e, \end{array}$$

where the elements y_e are defined in 18.18 and λ is the principal polarization defined in 18.11. By (4) of 18.18 and 6.3 for any $\xi \in H^1(\Gamma,\mathbf{Z})$ we have a smooth, affine and surjective morphism

$$\begin{array}{ccc} \tilde{G}' & \xrightarrow{r_{\xi'}} & P_{\xi'} \\ \downarrow & & \downarrow \\ \tilde{G} & \xrightarrow{r_\xi} & P_\xi. \end{array}$$

Moreover, the second row is the fppf quotient of the first by the torus $T''/\mathbf{G}_{m,S}$. Let $l_1,\dots,l_r$ be the basis of elementary cycles of $H_1(\Gamma,\mathbf{Z})$ fixed in 12.3. By 18.13 each element l_t defines a character of the torus T. Analogously each element e of $\mathcal{E}$ defines a character of T'. We get decompositions

$$T \xrightarrow{\sim} \prod_{t=1,\dots,r} (\mathbf{G}_{m,S},t) \qquad \text{and} \qquad T' \xrightarrow{\sim} \prod_{e\in\mathcal{E}} (\mathbf{G}_{m,S},e),$$

where $(\mathbf{G}_{m,S}, t)$ (resp. $(\mathbf{G}_{m,S}, e)$) is the torus $\mathbf{G}_{m,S}$ and the morphism $T \to (\mathbf{G}_{m,S}, t)$ (resp. $T' \to (\mathbf{G}_{m,S}, e)$) is defined by the character l_t (resp. e). The morphism $T' \to T$ respects the two decompositions. Let $l_t = \oplus_e l_{t,e}$ be the decomposition of l_t via the natural inclusion $H_1(\Gamma, \mathbf{Z}) \hookrightarrow C_1(\Gamma, \mathbf{Z})$ defined in 3.1. Since l_t is an elementary cycle by definition, $l_{t,e}$ is 0 or ± 1. If $e \in \mathcal{E}$, then $(\mathbf{G}_{m,S}, l_t) \to (\mathbf{G}_{m,S}, e)$ is trivial if $l_{t,e} = 0$ and is the identity (resp. the inverse map) if $l_{t,e} = 1$ (resp. $l_{t,e} = -1$). In particular, for any oriented edge e of Γ there exists a lifting

$$x'_e \in \tilde{G}'(K)$$

of the element x_e defined in 18.17. Choose the lifting so that for any oriented edge f of Γ different from e the projection of x'_e in $\big(\mathbf{G}_{m,S}(K), f\big)$ lies in $\big(\mathbf{G}_{m,S}(S), f\big)$. For any α in $C^1(\Gamma, \mathbf{Z})$ define morphisms

$$s_\alpha: \tilde{G}'_K \longrightarrow \tilde{G}' \qquad \text{and} \qquad z_\alpha: \tilde{G}'_K \longrightarrow P'_\alpha$$

analogously to the definitions given in 19.1 with $\tilde{G}'$ instead of $\tilde{G}$ and x'_e instead of x_e for $e \in \mathcal{E}$. By 6.3 we get a commutative diagram

$$\begin{array}{ccc} \tilde{G}'_K & \xrightarrow{z_{\xi'}} & P_{\xi'} \\ \downarrow & & \downarrow \\ \tilde{G}_K & \xrightarrow{z_\xi} & P_\xi; \end{array}$$

for all $\xi \in H^1(\Gamma, \mathbf{Z})$ and the second row is the fppf quotient of the first row by the action of $T''/\mathbf{G}_{m,S}$. By 6.3 we also have

$$\tilde{G}'_K \xrightarrow{z_{\xi'}} P_{\xi'} \hookrightarrow P'_{\xi'}$$

for all $\xi \in H^1(\Gamma, \mathbf{Z})$.

19.4 *The definition of $\widetilde{P}'$* . Consider the data

$$\tilde{G}'_K \xrightarrow{z_\alpha} P'_\alpha$$

for all $\alpha \in C^1(\Gamma, \mathbf{Z})$. By construction

IIa. the semiabelian scheme $\tilde{G}' \to S$;

IIb. the homomorphism

$$\begin{array}{ccc} C^1(\Gamma, \mathbf{Z}) = \mathbf{Z}^{\mathcal{E}} & \longrightarrow & \tilde{G}'(K) \\ \alpha = \sum_{e \in \mathcal{E}} \alpha_e\, e & \mapsto & z_\alpha(1) = \prod_{e \in \mathcal{E}} \big(x'_e\big)^{\alpha_e}; \end{array}$$

define data of type (II) in 11.6 with $\Lambda = \Lambda' = C^1(\Gamma, \mathbf{Z})$. Let

$$\widetilde{P}' \longrightarrow \mathcal{J} \qquad \text{and} \qquad 0_\alpha: S \to \widetilde{P}' \quad \alpha \in C^1(\Gamma, \mathbf{Z})$$

be the scheme and the sections defined as in 11.2. Let b be a finite subset of $C^1(\Gamma, \mathbf{Z})$ of the type considered in 10.4 with $\Lambda = C^1(\Gamma, \mathbf{Z})$, i.e., suppose that there exist integers $m_e \leq n_e$ for $e \in \mathcal{E}$ such that

$$b = \left\{\alpha = \sum \alpha_e\, e \in C^1(\Gamma, \mathbf{Z}) = \mathbf{Z}^{\mathcal{E}} \mid m_e \leq \alpha_e \leq n_e\right\}.$$

Let

$$U_{\underline{b}}, \qquad P_b, \qquad P_{\underline{b}} \qquad \text{and} \qquad P_b^c$$

be defined as in 10.5. Their basic properties are given in 10.10. Note that

a. we have $P_b = \prod_{\alpha \in b} P'_\alpha$;

b. $\tilde{G}'$ can be identified with the open subscheme $U_{\underline{\{0\}}}$ of $\widetilde{P}'$. The identity section of $\tilde{G}'$ coincides with the section of $\widetilde{P}' \to S$ associated to $0 \in C^1(\Gamma, \mathbf{Z})$;

c. by 11.4 there is a unique action of $\tilde{G}'$ on $\widetilde{P}'$ extending the action of $\tilde{G}'$ on itself and such that $\widetilde{P}' \to P'_\alpha$ is $\tilde{G}'$-equivariant for any $\alpha \in C^1(\Gamma, \mathbf{Z})$;

d. from (c) we deduce an action of T'' on $P_{\underline{b}}$ and $U_{\underline{b}}$ extending the action of T'' on $\tilde{G}'$ and such that for any $\alpha \in b$ the morphisms from $U_{\underline{b}}$ and $P_{\underline{b}}$ to P'_α are T''-equivariant;

e. the scheme $\widetilde{P}'$ is endowed with a unique action of $C^1(\Gamma, \mathbf{Z})$ over S compatible with the action on $\tilde{G}'_K$ induced by the homomorphism in (IIb). This follows from 11.8;

f. let $\mathcal{P}'_0 := \mathcal{P}' \times_S S_0$. It is isomorphic to the scheme defined in 6.7. By construction it is of the type considered in 7.1.

19.5 *The definition of* $\widetilde{P}'^o$. Define

$$\widetilde{P}'^o \hookrightarrow \widetilde{P}'$$

as the open subscheme defined as the intersection of the inverse images of the open subschemes $P_{\xi'}$ of $P'_{\xi'}$ via the morphisms $\widetilde{P}' \to P'_{\xi'}$ defined in 10.10 over all $\xi \in H^1(\Gamma, \mathbf{Z})$. On the open subschemes $U_{\underline{b}}$ of $\widetilde{P}'$ defined in 19.4 the intersection is finite and consequently the definition is well posed. Let

$$a \subset H^1(\Gamma, \mathbf{Z}) \qquad \text{and} \qquad \langle a \rangle \subset C^1(\Gamma, \mathbf{Z})$$

be respectively a finite subset of $H^1(\Gamma, \mathbf{Z})$ and the smallest subset of $C^1(\Gamma, \mathbf{Z})$ of the type considered in 19.4 and containing ξ' for all ξ in a. Suppose that if ξ' is in $\langle a \rangle$ for some ξ in $H^1(\Gamma, \mathbf{Z})$, then ξ is in a. Let

$$P^o_{\underline{\langle a \rangle}} := P_{\underline{\langle a \rangle}} \cap \Big(\prod_{\xi \in a} P_{\xi'} \times \prod_{\alpha \in \langle a \rangle \setminus a} P'_\alpha\Big), \qquad U^o_{\underline{\langle a \rangle}} := U_{\underline{\langle a \rangle}} \cap \Big(\prod_{\xi \in a} P_{\xi'} \times \prod_{\alpha \in \langle a \rangle \setminus a} P'_\alpha\Big),$$

where $P_{\underline{\langle a \rangle}}$ and $U_{\underline{\langle a \rangle}}$ are defined in 19.4. For any positive integer n, define

$$U^o_{\underline{a},n} := U_{\underline{a},n} \cap \prod_{\xi \in a} P_{\xi'} \qquad P^o_{\underline{a},n} := P^{H_1(\Gamma)}_{\underline{a},n} \cap \prod_{\xi \in a} P_{\xi'} \qquad P^{c,o}_a := P^c_a \cap \prod_{\xi \in a} P_{\xi'},$$

where the intersections are taken in $\prod_{\xi \in a} P'_{\xi'}$, the schemes $U_{\underline{a},n}$ and $P^{H_1(\Gamma)}_{\underline{a},n}$ are defined in 10.7 putting $\mathcal{A} = H_1(\Gamma, \mathbf{Z})$. Then

i. by construction the scheme $P^o_{\underline{a},n}$ is closed in $\prod_{\xi \in a} P_{\xi'}$ and by 10.7 it is flat over $\mathcal{J}_n$;

ii. the natural projection $P^o_{\underline{\langle a \rangle},n} \xrightarrow{\sim} P^o_{\underline{a},n}$ is an isomorphism. In fact, this holds for $n = 0$ as can be seen using the definition of $P_{\xi'}$ in 6.3 for $\xi \in a$ and the definition of the contraction $P_{\underline{\langle a \rangle},0} \to P'_\alpha \times_S S_0$ in 7.3 for $\alpha \in C^1(\Gamma, \mathbf{Z})$;

iii. by (d) of 19.4 the schemes $P^o_{\underline{a},n}$ and $U^o_{\underline{a},n}$ are endowed with a unique action of T'' compatible with the action of T'' on $P_{\underline{\langle a \rangle}}$ and $U_{\underline{\langle a \rangle}}$ and the diagonal action of T'' on $\prod_{\xi \in a} P_{\xi'}$. In particular, for any $\xi \in a$ the morphisms from $P^o_{\underline{a},n}$ and $U^o_{\underline{a},n}$ to $P_{\xi'}$ are T''-equivariant;

iv the scheme $\widetilde{P}'^o$ is covered by open subschemes of the type $U^o_{\underline{a}}$ as defined above and in particular it inherits the action of $T''/\mathbf{G}_{m,S}$;

v. the open subscheme $\widetilde{P}'^o$ is invariant under the action of $H_1(\Gamma, \mathbf{Z})$ induced by the action of $C^1(\Gamma, \mathbf{Z})$ on $\widetilde{P}'$ defined in (e) of 19.4 via the inclusion $H_1(\Gamma, \mathbf{Z}) \subset C_1(\Gamma, \mathbf{Z}) \cong C^1(\Gamma, \mathbf{Z})$ defined in 3.1;

vi. the scheme $\widetilde{P}'^o \times_S S_0$ is isomorphic to the scheme $\mathcal{P}'^o_0$ defined in 6.7.

19.6 *The definition of* $\widetilde{P}^\varphi$. Define

$$\widetilde{P}'^o \longrightarrow \widetilde{P}^\varphi$$

as the fppf quotient under the action of the torus $T''/\mathbf{G}_{m,S}$. We need to prove that such quotient exists. We use the notation of 19.4 and of 19.5. Let a be a finite subset of $H^1(\Gamma, \mathbf{Z})$ of the type considered in 19.5. Define

$$P_{\underline{a}^o,n} := P^o_{\underline{a},n}/T'' \qquad \text{and} \qquad U_{\underline{a}^o,n} := U^o_{\underline{a},n}/T''$$

as the categorical quotients by T''. These exist since the two schemes $P^o_{\underline{a},n}$ and $U^o_{\underline{a},n}$ can be covered by T''-invariant open affine subschemes. They are defined on T''-invariant affine open subschemes by taking the subalgebras where T'' acts through the trivial character. In particular, these quotients commute with base change and the quotient maps admit a splitting as $O_{\mathcal{J}_n}$-modules. This implies that $P_{\underline{a}^o,n}$ is flat over $\mathcal{J}_n$. Moreover, the base change of $P_{\underline{a}^o,n}$ to S_0 is the quotient of $P^o_{\underline{a},0}$ by $T'' \times_S S_0$. By 6.8 this is the closed subscheme of the scheme $\mathcal{P}^\varphi_0$ defined in 6.5 corresponding to the union of the images of $P_\xi \times_S S_0$ for all ξ in a. By Grothendieck's existence theorem, the formal subscheme of the I-adic completion of $\prod_{\xi \in a} P_\xi$ associated to $(P_{\underline{a}^o,n})_{n \in \mathbf{N}}$

algebraizes to a closed subscheme

$$P_{\underline{a}^o} \hookrightarrow \prod_{\xi \in a} P_\xi$$

and $P_{\underline{a}^o}$ is flat over $\mathcal{J}$. Let $P^c_{\underline{a}^o}$ be the quotient under $\prod_{\xi\in a} T''$ of $P^{c,o}_{\underline{a}}$. It is an open subscheme of $\prod_{\xi\in a} P_\xi$. Define

$$U_{\underline{a}^o} := P_{\underline{a}^o} \cap P^c_{\underline{a}^o}.$$

Then $U_{\underline{a}^o}$ algebraizes the formal scheme $(U_{\underline{a}^o,n})_n$ which is flat over $\mathcal{J}$. By construction the projection

$$P^o_{\underline{\langle a\rangle}} \longrightarrow \prod_{\xi\in a} P_\xi$$

factors over S_n via $P_{\underline{a}^o,n}$ for any n. Hence it factors via $P_{\underline{a}^o}$. The morphism

$$P^o_{\underline{\langle a\rangle}} \longrightarrow P_{\underline{a}^o}$$

is T''-equivariant. Over S_0 it is the faithfully flat quotient with respect to the action of $T'' \times_S S_0/\mathbf{G}_{m,S_0}$. Since $P_{\underline{a}^o}$ is proper over S, the local criterion of smoothness holds at closed points of $P_{\underline{a}^o}$. Hence the morphism is smooth and faithfully flat. The conclusion is that it is the $T''/\mathbf{G}_{m,S}$-fppf quotient. In particular,

$$U^o_{\underline{\langle a\rangle}} \longrightarrow U_{\underline{a}^o}$$

is the fppf quotient under $T''/\mathbf{G}_{m,S}$. We conclude that

a. $P_{\underline{a}^o}$ is an integral closed subscheme of $P_{a^o} := \prod_{\xi\in a} P_\xi$. In particular, it is the Zariski closure of $\prod_{\xi\in a} z_\xi \colon \tilde{G}_K \longrightarrow \prod_{\xi\in a} P_\xi$;

b. $\widetilde{P}^\varphi$ admits a morphism to P_a whose image is $P_{\underline{a}^o}$;

c. the inverse image of the open subscheme $U_{\underline{a}^o}$ of $P_{\underline{a}^o}$ in $\widetilde{P}^\varphi$ maps isomorphically to $U_{\underline{a}^o}$;

d. the scheme $\widetilde{P}^\varphi$ is covered by open subschemes of the form $U_{\underline{a}^o}$.

19.7 Proposition. *Let φ be a non-degenerate element in $\partial\big(C_1(\Gamma,\mathbf{R})\big)$ in the sense of 3.5. The scheme $\widetilde{P}^\varphi$ defined in 19.2 has the following properties:*

i. *it is integral and flat and locally of finite type over $\mathcal{J}$;*

ii. *it contains $\tilde{G}$ as an open subscheme and the inclusion $\tilde{G} \subset \widetilde{P}$ is an isomorphism over K;*

iii. *the action of the semiabelian scheme $\tilde{G}$ over S on itself extends to an action of $\tilde{G}$ on $\widetilde{P}^\varphi$;*

iv. *it is endowed with an action of $H_1(\Gamma,\mathbf{Z})$ over S compatible with the action on $\tilde{G}_K$ defined by the homomorphism $\iota\colon H_1(\Gamma,\mathbf{Z}) \to \tilde{G}(K)$ in 18.12;*

v. *its base change* $\widetilde{P}_0^\varphi$ *to* S_0*, with the induced action of* $H_1(\Gamma, \mathbf{Z})$*, is isomorphic over* S_0 *to* $\mathcal{P}_0^\varphi$*, with the action of* $H_1(\Gamma, \mathbf{Z})$*, as defined in 6.5;*

vi. *let* x *be a point of* $\widetilde{P}_0^\varphi$*. Assume that* x *belongs to the intersection of the closed subschemes* $Z_{e_i,\alpha_i,0}$ *with* $i = 1, \ldots, v$ *for suitable elements* $e_i \in \mathcal{E}$ *and* $\alpha_i \in H^1(\Gamma, \mathbf{Z})$*; see 6.4. The completion of the local ring of* $\widetilde{P}^\varphi$ *at* x *is isomorphic to a power series ring of the form*

$$E[\![z_1, \ldots, z_h, \gamma_1, \ldots, \gamma_v, \beta_1, \ldots, \beta_v]\!]/\big(\gamma_1\beta_1 - q_{e_1}, \ldots, \gamma_v\beta_v - q_{e_v}\big),$$

where the elements q_e*'s are defined as in 18.1;*

vii. *let* φ *satisfy the condition in (4) of 4.3. Let* H *be a good irreducible subset of* $\mathcal{E}$ *as defined in 18.7. Let* S_H *and* Γ_H *be as in 18.4. The base change of* $\widetilde{P}^\varphi \to S$ *to* S_H *together with the action of* $H_1(\Gamma, \mathbf{Z})$ *is isomorphic to* $\widetilde{P}_H$ *together with the action of* $H_1(\Gamma_H, \mathbf{Z})$ ($\cong H_1(\Gamma, \mathbf{Z})$ *by 4.2) as defined in 18.9.*

Proof: By 19.6 the scheme $\widetilde{P}^\varphi$ is the fppf quotient of an open subscheme $\widetilde{P}'^o$ of $\widetilde{P}'$ under $T''/\mathbf{G}_{m,S}$. Properties (i)-(vi) follow from analogous properties holding for $\widetilde{P}'$ replacing $\tilde{G}$ by $\tilde{G}'$ and $H_1(\Gamma, \mathbf{Z})$ by $C^1(\Gamma, \mathbf{Z})$. See 19.4. To deduce (v) use 6.8. To deduce (iv) use (v) of 19.5. Finally we prove (vii). Let φ and H be as in (vii) of 19.7. Let $\widetilde{P}_H \to S_H$ be the scheme defined in 18.9. Let ξ be an element of $H^1(\Gamma, \mathbf{Z})$. Viewing it as a 0-dimensional cell of $\mathrm{Del}_\varphi\big(H^1(\Gamma, \mathbf{R})\big)$ via 3.7 and using 4.4 we can associate to it an element ξ_H of $H^1(\Gamma_H, \mathbf{Z})$ viewed as the set of 0-dimensional cells of $\mathrm{Del}_\varphi\big(H^1(\Gamma, \mathbf{R})\big)$. By 4.4 for all $\xi \in H^1(\Gamma, \mathbf{Z})$ there exists a unique *birational* map $P_\xi \times_S S_H \to P_{\xi_H}$ such that

1. it is defined on the complement of the closed subschemes $Z_{e,\xi}$ defined in 6.4 for all 1-dimensional cells e of $\mathrm{Del}_\varphi\big(H^1(\Gamma, \mathbf{R})\big)$ such that ξ is a face of e and e, considered as element of $\mathcal{E}$ as in 6.4, is contracted in Γ_H;
2. the following diagram commutes

$$\begin{array}{ccc} \tilde{G}_{K_H} & \xrightarrow{z_\xi} & P_\xi \times_S S_H \\ \| & & \downarrow \\ \tilde{G}_{H,K_H} & \xrightarrow{z_{\xi_H}} & P_{\xi_H}. \end{array}$$

By construction the scheme $\widetilde{P}_H$ defined in 18.9 and the scheme obtained from the data in the second row of this diagram, proceeding as in 19.2, are the same. Hence the commutative diagram above induces a birational map

$$\widetilde{P}^\varphi \times_S S_H \longrightarrow \widetilde{P}_H$$

respecting the action of $H_1(\Gamma, \mathbf{Z}) \cong H_1(\Gamma_H, \mathbf{Z})$. Note that if $e \in \mathcal{E}$ and e is contracted in Γ_H, then q_e is invertible in E_H and $x_e \in \tilde{G}(S)$. One concludes from (vi) that the birational map is a morphism and that this morphism is an isomorphism.

19.8 Definition. *Let $\mathcal{P}^\varphi$ be the formal scheme over $\mathcal{S}$ defined as the I-adic completion of the scheme $\widetilde{P}^\varphi$ over S constructed in 19.2. It is endowed with an action of $H_1(\Gamma, \mathbf{Z})$ induced from the action of $H_1(\Gamma, \mathbf{Z})$ on $\widetilde{P}^\varphi \to S$ defined in 19.7. Let $\hat{0}$ be the section of $\mathcal{P}^\varphi \to \mathcal{S}$ determined by the section $\tilde{0}$ of $\widetilde{P}^\varphi \to S$ defining the identity of $\tilde{G}$.*

19.9 Theorem. *The formal scheme $\mathcal{P}^\varphi/H_1(\Gamma, \mathbf{Z})$ over $\mathcal{S}$ is uniquely algebraizable to a scheme*

$$\bar{P}^\varphi \to S.$$

The section $\hat{0}$ of $\mathcal{P}^\varphi \to \mathcal{S}$ is uniquely algebraizable to a section $\bar{0}$ of $\bar{P}^\varphi \to S$. The scheme $\bar{P}^\varphi$ has the following properties

i. *it is flat over S with connected fibers and is endowed with a line bundle ample relatively to S;*
ii. *let x be a closed point of $\bar{P}^\varphi$. The completion of the local ring of $\bar{P}^\varphi$ at x is isomorphic to a power series ring of the form*

$$E[\![z_1, \ldots, z_h, \gamma_1, \ldots, \gamma_v, \beta_1, \ldots, \beta_v]\!]/\big(\gamma_1\beta_1 - q_{e_1}, \ldots, \gamma_v\beta_v - q_{e_v}\big)$$

the elements q_e's being defined as in 18.1;
iii. *let φ satisfy the condition in (4) of 4.3. Let H be a good irreducible subset of $\mathcal{E}$ of maximal rank as defined in 18.7. Let S_H and Γ_H be as in 18.4. The base change of $\bar{P}^\varphi$ to S_H is isomorphic to the scheme$\bar{P}_H \to S_H$ defined in 18.9.*

In particular, $\bar{P}^\varphi_K$ is smooth and connected over K.

Proof: Suppose that the algebraizability holds. Then statements (ii) and (iii) are obvious. Statement (iii) follows from the definition of $\bar{P}_H$ in 18.9 and from (vii) of 19.7. We are left with the proof of the fact that the proper formal scheme $\mathcal{P}^\varphi/H_1(\Gamma, \mathbf{Z})$ is algebraizable. This follows from 19.10 and Grothendieck's existence theorem.

19.10 *The construction of the invertible sheaf $\bar{\mathcal{L}}^\varphi$.* We construct an invertible sheaf $\bar{\mathcal{L}}^\varphi$ on $\mathcal{P}^\varphi/H_1(\Gamma, \mathbf{Z})$ ample relatively to $\mathcal{S}$. By descent theory it suffices to construct it canonically in the universal case, i.e., we can assume that S is a versal deformation space of the pullback of $C_0 \to S_0$ to a geometric point of S_0. Hence, we can assume that S_0 is the spectrum of a field. In particular, S is regular. Hence $\widetilde{P}^\varphi$ is regular by (vii) of 19.7. By [12, Corollary 21.6.14] it is parafactorial in codimension 2, i.e., an invertible sheaf (a divisor) defined in codimension 2 on $\widetilde{P}^\varphi$ can be uniquely extended to an invertible sheaf (a divisor) on $\widetilde{P}^\varphi$.

Let H be a good irreducible subset of $\mathcal{E}$ of maximal rank. Let $\tilde{\Theta}_H$ be the ample Cartier divisor relative to S_H on $\widetilde{P}_H$ and let $\overline{\Theta}_H$ be the ample Cartier divisor on $\bar{P}_H$ relative to S_H defined in 18.9. Recall that the smooth locus of $\bar{P}_H$ over S_H is $\mathrm{Pic}^0(C_H/S_H)$. Moreover, the choice of the section $\bar{0}$ in 18.1

of C over S defines a morphism $C_H \to \bar{P}_H$ sending $\bar{0} \times_S S_H$ to the identity element of $\mathrm{Pic}^0(C_H/S_H)$. Consider the theta divisor on $\mathrm{Pic}^0(C_H/S_H)$ defined as the scheme theoretic image of the $(g-1)$th symmetric power of the smooth locus of C_H over S_H. This is well defined since $\mathrm{Pic}^0(C_H/S_H)$ is parafactorial in codimension 2. Since $\bar{P}_H$ is parafactorial in codimension 2, it extends uniquely to a divisor on $\bar{P}_H$. By construction it contains the divisor $\overline{\Theta}_H$. Since the theta divisor over K_H is irreducible, we conclude that $\overline{\Theta}_H$ is the extension of the theta divisor on $\mathrm{Pic}^0(C_{K_H}/K_H)$.

Given the section $\bar{0}$ of C over S, the theta divisor on $\mathrm{Pic}^0(C_S/S)$ is canonically defined in codimension 2 over S. Hence it uniquely defines an ample divisor on $\mathrm{Pic}^0(C/S)$ relative to S. By [18, II.3.2.2] it defines a cubical invertible sheaf on $\mathrm{Pic}^0(C/S)$ which is unique up to rigidification. This defines a unique ample cubical invertible sheaf $\mathcal{L}$ on the Raynaud extension $\tilde{G}$ of $\mathrm{Pic}^0(C/S)$ relative to S. It is clear that $\mathcal{L} \times_S S_H$ is isomorphic to the invertible sheaf defined by the intersection of $\tilde{\Theta}_H$ with $\tilde{G}_H$. By 16.23 and [8, III.6.2] for sets H and H' satisfying the property above, the actions of $H_1(\Gamma, \mathbf{Z})$ on the fiber of $\mathcal{L}$ over the generic point of $S_H \times_S S_{H'}$ induced by the actions of $H_1(\Gamma_H, \mathbf{Z})$ (resp. $H_1(\Gamma_{H'}, \mathbf{Z})$) on $\tilde{\Theta}_H$ (resp. $\tilde{\Theta}_{H'}$) are compatible since they uniformize the invertible sheaf defined by the theta divisor on $\mathrm{Pic}^0(C/S)$. Since $\cap_H K_H = K$, this action is induced by an action on $\mathcal{L}_K$.

Let H be a good subset of $\mathcal{E}$, irreducible of maximal rank. Consider the invertible sheaf $\mathcal{L}_H$ associated to the divisor $\tilde{\Theta}_H$ restricted to the smooth locus of $\widetilde{P}_H$ over S_H. Then $\mathcal{L}_H$ extends $\mathcal{L} \otimes K_H$ over $\widetilde{P}^\varphi \times_S S_H$ which is $\widetilde{P}_H$ by (vii) of 19.7. The morphism $S_H \to S_H^o$ defined in 18.4 is faithfully flat on a neighborhood Y of $S_{H,0}$. Then $\mathcal{L}_H$ descends to $\widetilde{P}^\varphi \times_S Y$ since $\mathcal{L}_H \otimes K_H$ is defined over K. In particular, $\mathcal{L}$ with the action of $H_1(\Gamma, \mathbf{Z})$ on $\mathcal{L}_K$ extends in codimension ≥ 2 on $\widetilde{P}^\varphi$. This implies that there is a unique invertible sheaf $\tilde{\mathcal{L}}$ on $\widetilde{P}$ with an induced action of $H_1(\Gamma, \mathbf{Z})$ extending $\mathcal{L}$ and the action of $H_1(\Gamma, \mathbf{Z})$ on $\mathcal{L}_K$. Remark that $\mathcal{L}$ on $\tilde{G}$ is ample relatively to S.

Analogously, for any ξ in $H^1(\Gamma, \mathbf{Z})$ one defines an invertible sheaf $\tilde{\mathcal{L}}_\xi$ on $\widetilde{P}$ with an action of $H_1(\Gamma, \mathbf{Z})$ extending the push forward of $\mathcal{L}_K$ by the element $z_\xi(1) \in \tilde{G}(K)$ defined in 19.1. In fact, for any H as above we have that $z_\xi(1)$ extends to an S_H-section of the smooth locus of $\widetilde{P}_H$ over S_H and it acts by translation on $\widetilde{P}_H$ over S_H by 11.8. For any $\xi \in H^1(\Gamma, \mathbf{Z})$ the I-adic completion of $\tilde{\mathcal{L}}_\xi$ descends to $\mathcal{P}^\varphi/H_1(\Gamma, \mathbf{Z})$. Consider the pullback of $\tilde{\mathcal{L}}_\xi$ to the scheme $\widetilde{P}'^o \times_S S_0$ defined in 19.5. Its restriction to the closed subscheme $P'_{\xi'} \times_S S_0$ is associated to a non trivial effective Cartier divisor relative to S_0 by construction. By 14.11, applied to all irreducible subsets H of $\mathcal{E}$ such that the rank of $H_1(\Gamma_H, \mathbf{Z})$ is r and using the rational map $P_\xi \times_S S_{H,0} \to P_{\xi_H} \times_{S_H} S_{H,0}$ defined in 4.4, such restriction contains the pullback of $\prod_{e \in \mathcal{E}} O(1)$ on the $\mathbf{P}^1 \otimes C^1(\Gamma, \mathbf{Z})$-bundle $P_{\xi'} \times_S S_0$ over $\mathcal{J}_0$ which is ample relatively to $\mathcal{J}_0$. Hence the sum $\tilde{\mathcal{L}}^\varphi$ over representatives $\{\xi\} \subset H^1(\Gamma, \mathbf{Z})$ of $H^1(\Gamma, \mathbf{Z})/H_1(\Gamma, \mathbf{Z})$ of the I-adic completions of $\tilde{\mathcal{L}}_\xi$

descended to $\mathcal{P}^{\varphi}/H_1(\Gamma,\mathbf{Z})$ defines an ample invertible sheaf on $\mathcal{P}^{\varphi}/H_1(\Gamma,\mathbf{Z})$ relatively to $\mathcal{S}$ as wanted.

19.11 Theorem. *The smooth locus $\bar{P}^{\varphi,\mathrm{sm}}$ of $\bar{P}^{\varphi} \to S$ has the following properties*

i. *it contains* $\mathrm{Pic}^0(C/S)$ *as an open subscheme. Moreover, there is an action of* $\mathrm{Pic}^0(C/S)$ *on it extending the group scheme structure on* $\mathrm{Pic}^0(C/S)$*;*
ii. *let V be the spectrum of a discrete valuation ring with generic fiber η. Let $V \to S$ be a morphism such that $C \times_S V$ is regular and C_η is smooth over $k(\eta)$. Then $\bar{P}^{\varphi,\mathrm{sm}} \times_S V$ is the Néron model of* $\mathrm{Pic}^0\big(C_\eta/k(\eta)\big)$.

Proof: Let a be the subset of $H^1(\Gamma,\mathbf{Z})$ defined by the zero element. Let $U_{\underline{a}^o}$ be the open subscheme of $\widetilde{P}^{\varphi,\mathrm{sm}}$ defined in 19.6. Let U be the union of the translates of $U_{\underline{a}^o}$ by elements of $H_1(\Gamma,\mathbf{Z})$. Let B be its complement in $\widetilde{P}^{\varphi}$. As in 11.8 the scheme U inherits the structure of smooth and separated group scheme over S and it acts on $\widetilde{P}^{\varphi,\mathrm{sm}}$. The I-adic completion of B descends to a closed formal subscheme of the I-adic completion of $\bar{P}^{\varphi}$ and hence can be algebraized to a subscheme $\bar{B}$ of $\bar{P}^{\varphi}$. Let G be the complement of $\bar{B}$ in $\bar{P}^{\varphi}$. By 16.10 it is a smooth scheme over S since U is. Moreover $G_K = \bar{P}^{\varphi}_K$. Proceeding as in 16.13 it is endowed with the structure of group scheme over S since U is and G acts on $\bar{P}^{\varphi,\mathrm{sm}}$. Since $G \times_S S_0$ is $\tilde{G} \times_S S_0$, the group scheme G is semiabelian by Zariski's main theorem. Let H be a good irreducible subset of $\mathcal{E}$ of maximal rank in the sense of 18.7. The choice of the section $\bar{0}$ of C over S in 18.1 determines a *unique* morphism $C_H \to \bar{P}^{\mathrm{sm}}_H$ sending $\bar{0} \times_S S_H$ to the identity element. By 17.7 this induces a unique isomorphism between the identity component of $\bar{P}^{\mathrm{sm}}_H$ and $\mathrm{Pic}^0(C_H/S_H)$. This isomorphism descends to an isomorphism from $\mathrm{Pic}^0(C_K/K)$ to G_K. Using [8, Proposition I.2.7] we conclude the proof of (i) at least if a good irreducible subset H of $\mathcal{E}$ exists. By 18.8 this holds if S_0 is the spectrum of a field and S is a versal deformation space of $C_0 \to S_0$. Reducing to this case we conclude.

We prove (ii). Let M_V be the Néron model of $\mathrm{Pic}^0\big(C_\eta/k(\eta)\big)$. There is a unique map α from $\bar{P}^{\varphi,\mathrm{sm}} \times_S V$ to M_V extending the isomorphism $\bar{P}^{\varphi,\mathrm{sm}}_\eta = \mathrm{Pic}^0\big(C_\eta/k(\eta)\big)$. The identity component of M_V is $\mathrm{Pic}^0(C_V/V)$. By uniqueness α is $\mathrm{Pic}^0(C_V/V)$-equivariant. By (i) and [8, Proposition I.2.7] it induces an isomorphism as semiabelian schemes over V between $\mathrm{Pic}^0(C_V/V)$, viewed as an open subscheme of $\bar{P}^{\varphi,\mathrm{sm}}$, and the identity component M_V^0 of M_V. Proceeding as in 17.2 it is clear that $\bar{P}^{\varphi,\mathrm{sm}} \times_S V$ satisfies the Néron mapping property for discrete valuations R which are smooth over V. In particular, we can take R to be the localization of M_V at the generic point of an irreducible component of its special fiber. By construction the number of connected components of the special fiber of $\bar{P}^{\varphi,\mathrm{sm}} \times_S V$ is $H^1(\Gamma_0,\mathbf{Z})/H_1(\Gamma_0,\mathbf{Z})$ where Γ_0 is the dual graph of the special fiber of C_V. By [4, Discussion on page 283] and by the Kirchoff-Trent's theorem stated in [23, pag. 21] this is also the number of connected components of the special fiber of M_V. Hence α induces a bijection on the set of connected components of the special fibers. Each connected component is a torsor under the special fiber of $\mathrm{Pic}^0(C_V/V)$. We conclude applying Zariski's main theorem.

19.12 *An example.* One can ask whether the scheme $\bar{P}^{\varphi,\mathrm{sm}}$ over S constructed in 19.11 is the Néron model of $\mathrm{Pic}^0(C_K/K)$ or, at least, if it has the structure of a group scheme extending the group structure on $\mathrm{Pic}^0(C_K/K)$. We give an example to prove that this can not be expected.

Let k be an algebraically closed field of characteristic different from 2. Let $S_0 :=$ Spec(k). Let $D_1 := \mathbf{P}^1_k$ and $D_2 := \mathbf{P}^1_k$. Let C_0 be the semistable curve over S_0 defined by gluing D_1 and D_2 transversally along the zero sections and the infinity sections respectively. Let $C \to S$ be the universal deformation in equal characteristic of $C_0 \to S_0$ with two sections $P_1, P_2 : S \to C$ such that P_1 meets D_1 over S_0, while P_2 meets D_2 over S_0 and P_1, P_2 factors through the smooth locus of $C \to S$. In particular, S is isomorphic to a power series ring in two variables $E := k[\![x,y]\!]$ over k. Choose $\bar{0}: S \to C$ to be the section P_1. In this case, $\bar{P}$ is isomorphic to C itself and $\bar{P}^{\mathrm{sm},0} = \mathrm{Pic}^0(C/S)$ is the degenerating elliptic curve defined by the genus one curve C and the marking $\bar{0}$. Suppose that there would exist a group structure on C^{sm} extending the group scheme structure on $\mathrm{Pic}^0(C/S)$. First of all, the induced group structure on C_0^{sm} is split. In fact, consider an étale $2:1$ cover of the Tate curve $T := \mathbf{G}_m(k[\![q]\!])/q^{\mathbf{Z}}$ over Spec$(k[\![q]\!])$ with the two markings defined as the pullback of the identity of T^{sm}. The choice of one of the markings determine a unique map Spec$(k[\![q]\!]) \to S$ by the definition of S such that Spec(k) is sent to S_0 and the induced group scheme structure on the special fiber, coming from the group scheme structure on E^{sm}, is a split extension of $\mathbf{Z}/2\mathbf{Z}$ by $\mathbf{G}_{m,k}$.In particular, multiplication by 2 on C^{sm} is étale and finite. Since k is algebraically closed and E is henselian, all 2-torsion points of C^{sm} are rational over S. Choose a two torsion point $\eta \in C^{\mathrm{sm}}[2]$ not lying in $C^{\mathrm{sm},0}$. Translation by η on C^{sm} determines an involution $\tau: C^{\mathrm{sm}} \longrightarrow C^{\mathrm{sm}}$ which over S_0 interchanges the two connected components. The formal completion of C^{sm} along $D_1\backslash\{0,\infty\}$ (resp. $D_2\backslash\{0,\infty\}$) is isomorphic to $E[\![X_1, 1\backslash X_1]\!]$ (resp. $E[\![X_2, 1\backslash X_2]\!]$) and τ induces an isomorphism between the two. We can suppose that $\tau(X_1) = 1\backslash X_2$. Let Q_1 and Q_2 be the singular points of C_0. Then we can choose the topological generators x and y of E so that the completion of C at Q_1 is $E[\![X_1, Z_2]\!](X_1 Z_2 - x)$ and the completion of C at Q_2 is $E[\![X_2, Z_1]\!](X_2 Z_1 - y)$ with $Z_2 = 1\backslash X_2$ and $Z_1 := 1\backslash X_1$. Let u be a non constant map from Spec$(k[\![q]\!])$ to S such that $x(q), y(q)$ are in (q) but not in (q^2) and $x(q) \neq y(q)$. Then the pull back of τ extends uniquely to an automorphism of $C \times_S \mathrm{Spec}(k[\![q]\!])$ sending Q_1 to Q_2. This implies that $x(q) = y(q)$. Our assumption that C^{sm} should have a group scheme structure extending the one on $\mathrm{Pic}^0(C/S)$ lead us to a contradiction.

19.13 Remark. Example (ii) of [23, pag. 42] shows that $\bar{P}^{\varphi,\mathrm{sm}} \to S$ *depends* on the choice of the non-degenerate element $\varphi \in \partial(C_1(\Gamma, \mathbf{R}))$. By [23, 7.2-

7.3] there are only finitely many non-degenerate elements φ in $\partial(C_1(\Gamma,\mathbf{R}))$ defining non-isomorphic schemes $\bar{P}^{\varphi}$ over S. Their number admits an upper bound depending uniquely on the graph Γ. By 5.5, if Γ is simple in the sense of 5.3 then $\bar{P}^{\varphi}$ is independent of φ. In this case it is possible to show that the morphism $C_K \to \bar{P}^{\varphi}_K = \mathrm{Pic}^0(C_K/K)$ defined by the section $\bar{0}_K$ of C_K over Spec(K) fixed in 18.1 extends to a morphism $C \to \bar{P}^{\varphi}$.

19.14 Remark. Summarizing the long construction of the scheme $\widetilde{P}^{\varphi}$ and of $\bar{P}^{\varphi}$ and hence of the interpolating family of Néron models of $\mathrm{Pic}^0(C/S)$, we want to stress that three ingredients were needed:

1. the semiabelian scheme $\tilde{G}$ defined in 18.11;
2. the elements $x_e \in \tilde{G}(K)$ for all $e \in \mathcal{E}$ as defined in 18.17;
3. the Namikawa decomposition $\mathrm{Del}_{\varphi}(H^1(\Gamma,\mathbf{R}))$ defined in 3.5.

We want to connect data (2) with the 1-motive $H_1(\Gamma,\mathbf{Z}) \to \tilde{G}(K)$ defined in 18.12. Let l be an element of $H_1(\Gamma,\mathbf{Z})$ decomposing as $l = \oplus_{e\in\mathcal{E}} l_e \cdot e$ in $C_1(\Gamma,\mathbf{Z}) = \mathbf{Z}^{\mathcal{E}}$. The image of l in $\tilde{G}(K)$ is simply $\prod_{e\in\mathcal{E}} x_e^{l_e}$; see (4) of 18.18.

19.15 Definition. *Let $l_1,\ldots,l_r$ be a $\mathbf{Z}$-basis of $H_1(\Gamma,\mathbf{Z})$. Let T be the kernel of the morphism $\tilde{G} \to \mathcal{J}$. By 18.13 the choice of the basis of $H_1(\Gamma,\mathbf{Z})$ determines a decomposition $T \cong \mathbf{G}^r_{m,S}$. Let a be an integer $1 \le a \le r$. The pullback T_a of $\tilde{G}$ via the morphism $c^t(l_a)\colon S \to \mathcal{J}$ defined in 18.13 is isomorphic to T. The morphism $H_1(\Gamma,\mathbf{Z}) \to \tilde{G}_K$ defines a section of $T_a \cong \mathbf{G}^r_{m,s}$ over K and consequently fractional ideals $I_{a,1},\ldots,I_{a,r}$ of S. We define the bilinear form with target the set of fractional ideals of S:*

$$\begin{array}{ccc} H_1(\Gamma,\mathbf{Z}) \times H_1(\Gamma,\mathbf{Z}) & \xrightarrow{B} & \mathrm{Div}(S) \\ (l_a,l_b) & \mapsto & I_{a,b}. \end{array}$$

19.16 Corollary. *(See* [8, Theorem III.8.3]*) The notations are as in 19.15. For any integers a and b such that $1 \le a, b \le r$ the fractional ideal $I_{a,b}$ is generated by the product of the elements $q_e^{l_{a,e}\cdot l_{b,e}}$ for all e in $\mathcal{E}$.*

Proof: This is a consequence of (3) and (4) of 18.18.

19.17 Corollary. *Let* DEG *be the category of degenerating principally polarized abelian varieties over S as defined in* [8, Section II.0, pag. 32-33]. *We assume that S satisfies the condition that for any étale affine S_0-scheme, its unique formally étale lifting to S is normal. Then the image in* DEG *of the data in 18.14 via the functor M defined in* [8, Section III.6, page 71] *is the connected component of* $\mathrm{Pic}^0(C/S)$. *In other words the semiabelian scheme $\tilde{G}$ and the 1-motive*

$$H_1(\Gamma,\mathbf{Z}) \to \tilde{G}(K)$$

define the uniformization of $\mathrm{Pic}^0(C/S)$.

Proof: We reduce to the universal case. The corollary follows from 16.22 and [8, I.2.7].

References

[1] V. Alexeev: Compactified Jacobians. Preprint alg-geom/9608012.

[2] E. Arbarello, M. Cornalba, P.A. Griffiths, J. Harris: Geometry of algebraic curves I. Grundlehren der mathematischen Wissenschaften 267, Springer-Verlag (1985).

[3] S. Bosch, W. Lütkebohmert: Stable reduction and uniformization of abelian varieties II. Invent. math. 78 (1984), 257-297.

[4] S. Bosch, W. Lütkebohmert, M. Raynaud: Néron Models. Ergebnisse der Mathematik und ihrer Grenzgebiete, 3. Folge, Band 21, Springer-Verlag (1990).

[5] L. Caporaso: A compactification of the universal Picard variety over the moduli space of stable curves. J. Amer. Math. Soc. 7 (1994), 589-660.

[6] C.-L. Chai: Compactification of Siegel moduli schemes. London Mathematical Society LNS 107, Cambridge University Press (1985).

[7] P. Deligne, D. Mumford: The irreducibility of the space of curves of given genus. Publ. Math. IHES 36 (1969), 75-110.

[8] G. Faltings, C.-L. Chai: Degeneration of abelian varieties. Ergebnisse der Mathematik und ihrer Grenzgebiete, 3. Folge, Band 22, Springer-Verlag (1990).

[9] J. Fresnel, M. van der Put: Uniformisation de variétés abéliennes. Ann. Fac. Scie. Toulouse, Vol. 97 (1989), 9-42.

[10] J. Fresnel, M. van der Put: Uniformisation de variétés de Jacobi et déformations de courbes. Ann. Fac. Scie. Toulouse, Vol. 3 (1994), 363-385.

[11] A. Grothendieck, J. Dieudonné: Étude cohomologique des faisceaux cohérents. Publ Math. IHES 11 (1961) and 17 (1963).

[12] A. Grothendieck, J. Dieudonné: Étude locale des schémas et des morphismes de schémas. Publ Math. IHES 20 (1964), 24 (1965), 28 (1966) and 32 (1967).

[13] A. Grothendieck et al.: Cohomologie locale des faisceaux cohérents et théorèmes de Lefschetz locaux et globaux. North-Holland (1968).

[14] F. Guillén, V. Navarro Aznar, P. Pascual-Gainza, F. Puerta: Hyperrésolutions cubiques et descente cohomologique. LNM 1335, Springer-Verlag (1983).

[15] R. Hartshorne: Algebraic geometry. Graduate Texts in Mathematics 52, Springer-Verlag (1977).

[16] Y. Manin, V.G. Drinfel'd: Periods of p-adic Schottky groups. J. Reine. Angew. Math. 262/263 (1973), 239-247.

[17] J.S. Milne: Jacobian varieties, in Arithmetic Geometry, G. Cornell, J.H. Silverman editors. Springer-Verlag (1985).

[18] L. Moret-Bailly: Pinceaux de variétés abéliennes. Astérisque 129 (1985).
[19] D. Mumford: Abelian varieties. Tata Inst. Fund. Research Studies in Math. vol. 5, Oxford University Press (1970).
[20] D. Mumford: An analytic construction of degenerating curves over complete local rings. Comp. Math. 24 (1972), 129-174.
[21] D. Mumford: An analytic construction of degenerating abelian varieties over complete rings. Comp. Math. 24 (1972), 239-272.
[22] Y. Namikawa: A new compactification of the Siegel space and degeneration of abelian varieties I, II. Math Ann. 221 (1976), 97-141, 201-241.
[23] T. Oda, C.S. Seshadri: Compactifications of the generalized Jacobian variety. Trans. Amer. Math. Soc. 253 (1979), 1-90.
[24] M. Raynaud: Faisceaux amples sur les schémas en groupes et les espaces homogènes. Lecture Notes in Math. 119, Springer-Verlag (1970).
[25] J.-P. Serre: Groupes algébriques et corps de classes. Hermann (1959).

Progress in Mathematics, Vol. 195, © 2001 Birkhäuser Verlag Basel/Switzerland

TORELLI THEOREM VIA FOURIER-MUKAI TRANSFORM

A. BEILINSON AND A. POLISHCHUK

ABSTRACT. We show that the Fourier transform on the Jacobian of a curve interchanges "δ functions" on the curve and the theta divisor. The Torelli theorem is an immediate consequence.

1. STATEMENT OF THE THEOREM.

1.1. We live over an algebraically closed base field k. Let J be an abelian variety equipped with a principal polarization $\theta : J \xrightarrow{\sim} J^\circ = \mathrm{Pic}^0(J)$, so we have the corresponding Fourier transform $\mathcal{F}$ on the derived category of quasi-coherent sheaves $D(J, \mathcal{O})$ (see [6]).

Let Θ be the theta divisor in J. Notice that Θ is defined up to translation, and any non-trivial translation does not preserve Θ. So we may consider Θ as a canonically defined algebraic variety equipped with a J-torsor of embeddings $j : \Theta \hookrightarrow J$; we call these j's *standard embeddings*. Denote by Θ^{ns} the open subset of smooth points of Θ. For a standard embedding j let $j^{ns} : \Theta^{ns} \hookrightarrow J$ be its restriction to Θ^{ns}.

Our Θ carries a canonical involution $x \mapsto x^\nu$; this is the unique involution such that for any standard embedding j the embedding $j^\nu : x \mapsto -j(x^\nu)$ is also standard. For a line bundle L on Θ or Θ^{ns} set $L^\nu := \nu^* L$.

The pull-back j^*F of an $\mathcal{O}_J$-module F does not change if we translate both j and F by the same element of J. Thus the image of $j^{ns*} : \mathrm{Pic}(J) \to \mathrm{Pic}(\Theta^{ns})$ is a canonically defined subgroup of $\mathrm{Pic}(\Theta^{ns})$ (it does not depend on j). Denote by $A(J)$ the corresponding quotient group.

Let $\mathcal{T} \subset \mathrm{Pic}(\Theta^{ns})$ be the subset of line bundles L such that

(i) $L \cdot L^\nu = \omega_{\Theta^{ns}}$

(ii) $A(J)$ is generated by the image of L.

Remark. Since the tangent bundle to J is trivial, one has $\omega_{\Theta^{ns}} = j^{ns*}\mathcal{O}_J(j(\Theta))$. Thus, if $\mathcal{T}$ is non-empty then ν acts on $A(J)$ as -1.

1.2. From now on we assume that J is the Jacobian of a smooth projective curve C of genus $g \geq 2$ equipped with the canonical polarization. There is a standard embedding $i : C \hookrightarrow J$ defined up to translation; the standard embeddings i form a J-torsor isomorphic to $\mathrm{Pic}^{-1}(C)$.

1.3. **Theorem.** The set $\mathcal{T}$ is non-empty. For any $L \in \mathcal{T}$ and a standard embedding $j : \Theta \hookrightarrow J$ the Fourier transform $\mathcal{F}(j^{ns}_* L)$ equals $(\pm i)_*(M)[1-g]$

where $i : C \hookrightarrow J$ is a standard embedding, M is a line bundle of degree $g-1$ on C.

2. Proof of the theorem.

2.1. Let us start with some preliminaries.

Consider J as the moduli space[1] of line bundles of degree 0 on C. A line bundle E of degree $1-g$ yields a standard embedding $j = j_E : \Theta \hookrightarrow J$. Namely, there is a canonical morphism $\sigma : \mathrm{Sym}^{g-1} C \to \Theta$ such that $j_E\sigma$ sends a divisor D to $E(D)$. Thus the J-torsor of j's equals $\mathrm{Pic}^{1-g} C$. The above σ is an isomorphism over Θ^{ns}; we denote by α the inverse open embedding $\Theta^{ns} \hookrightarrow \mathrm{Sym}^{g-1} C$.

For a group T equipped with an involution ν we denote by $\widetilde{T}$ the corresponding semi-direct product of T and $\mathbb{Z}/2\mathbb{Z}$. Our $\mathcal{T}$ carries a canonical action of $\widetilde{J}$ (here ν acts on J as -1). Namely, an element $l \in J$ acts as tensor product by the line bundle $j^{ns*}\theta(l)$ (notice that, since $\theta(l)$ is translation invariant, this line bundle does not depend on j and is ν-anti-invariant), and $\mathbb{Z}/2\mathbb{Z}$ acts by ν.

2.2. **Proposition.** The $\widetilde{J}$-action on $\mathcal{T}$ is transitive.

We prove 2.2 in 2.7. Notice that for $g = 2$ we have $\Theta = \Theta^{ns} = C$, so here the proposition is clear.

The map $(L, j) \mapsto j^{ns}_* L$ commutes with the action of $\widetilde{J}$; here $\widetilde{J}$ acts on $\mathcal{O}$-modules on J by twists by degree 0 line bundles and -1 symmetry. The Fourier transform interchanges twists by degree 0 line bundles and translations and commutes with the -1 symmetry. Since the set of embeddings $\pm i : C \hookrightarrow J$ is a $\widetilde{J}$-torsor, 2.2 implies that it suffices to prove our theorem for a single pair (L, j).

Take any pair (i, M) where $i : C \hookrightarrow J$ is a standard embedding, $M \in \mathrm{Pic}^{g-1} C$. The theorem follows immediately from the involutivity property of $\mathcal{F}$ and the following fact:

2.3. **Proposition.** One has $\mathcal{F}(i_* M)[1] = j^{ns}_* L$ where $j := j_{M^{-1}}$ and $L \in \mathcal{T}$.

We prove 2.3 in 2.8.

For every point $x \in C$ and every $d \geq 2$ consider the embedding

$$a^d_x : \mathrm{Sym}^{d-1} C \to \mathrm{Sym}^d C : D \mapsto D + x.$$

Let us denote by R^d_x the image of a^d_x.

2.4. **Lemma.** For $d \geq 2$ there is an exact sequence of abelian groups

$$0 \longrightarrow \mathrm{Pic}(J) \xrightarrow{(j\sigma)^*} \mathrm{Pic}(\mathrm{Sym}^d C) \xrightarrow{\deg} \mathbb{Z} \longrightarrow 0$$

where the homomorphism deg is normalized by the condition $\deg(\mathcal{O}(R^d_x)) = 1$ for any $x \in C$. If $\mathbb{P} \subset \mathrm{Sym}^d C$ is a complete linear system of positive dimension then $\deg(L) = \deg(L|_{\mathbb{P}})$.

---

[1] We consider J as a plain variety ignoring the stack structure.

Proof of the lemma[2]. For d sufficiently large the statement is true since $\operatorname{Sym}^d C$ is a projective bundle over J. Let $\pi_d : C^d \to \operatorname{Sym}^d C$ be the canonical projection. Then we have $\pi_d^* \mathcal{O}(R_x^d) \simeq \mathcal{O}_C(x) \boxtimes \ldots \boxtimes \mathcal{O}_C(x)$. Therefore, $\mathcal{O}(R_x^d)$ is ample and $H^i(\operatorname{Sym}^d C, \mathcal{O}(-nR_x^d)) = 0$ for $n > 0$, provided that $i \leq 1$, $d \geq 2$ or $i \leq 2$, $d \geq 3$ (this follows from the symmetric Künneth formula proved in [7], exp. XVII, 5.5.34). It follows from the Lefschetz theorem for Picard groups (see [5], exp.11 (3.12), exp.12 (3.4)) that the map $a_x^{d*} : \operatorname{Pic}(\operatorname{Sym}^d C) \to \operatorname{Pic}(\operatorname{Sym}^{d-1} C)$ is an isomorphism for $d > 3$ and is an embedding for $d = 3$. It remains to check that a_x^{3*} is surjective. Let us denote by $K \subset \operatorname{Pic}(C \times C)$ the subgroup of line bundles L such that the restrictions $L|_{x \times C}$ and $L|_{C \times x}$ are trivial. Then there is an isomorphism $u : \operatorname{End}(J) \xrightarrow{\sim} K : \phi \mapsto (i_x \times \phi i_x)^* \mathcal{P}$ where $i_x : C \to J$ is the embedding corresponding to x, $\mathcal{P}$ is the Poincaré line bundle on $J \times J$. Let $\operatorname{Pic}^+(C \times C)$ be the subgroup of line bundles stable under the involution $(x_1, x_2) \mapsto (x_2, x_1)$, let $K^+ = K \cap \operatorname{Pic}^+(C \times C)$. Then u induces an isomorphism of $\operatorname{End}^+(J)$ with K^+ where $\phi \in \operatorname{End}^+(J)$ if and only if ϕ is self-dual. Let $r : \operatorname{Pic}(C \times C) \to K$ be the homomorphism given by $r(F) = F \otimes [(F^{-1}|_{C \times x}) \boxtimes (F^{-1}|_{x \times C})]$. Then $r(\operatorname{Pic}^+(C \times C)) = K^+$ and we have the following commutative diagram

$$\begin{array}{ccccc} \operatorname{Pic}(J) & \xrightarrow{\sigma_2^*} & \operatorname{Pic}(\operatorname{Sym}^2 C) & \xrightarrow{\pi_2^*} & \operatorname{Pic}^+(C \times C) \\ \downarrow{\scriptstyle s} & & & & \downarrow{\scriptstyle r} \\ \operatorname{End}^+(J) & & \xrightarrow{u} & & K^+ \end{array} \tag{1}$$

where $\sigma : \operatorname{Sym}^2 C \to J$ maps D to $\mathcal{O}(D - 2x)$, $s : \operatorname{Pic}(J) \to \operatorname{End}^+(J)$ is the standard homomorphism $L \mapsto \phi_L$ where $\phi_L(a) = t_a^* L \otimes L^{-1}$. Since s is surjective it follows that the composition $r \circ \pi_2^* \circ \sigma_2^*$ is surjective. Thus, in proving that some line bundle $L \in \operatorname{Pic}(\operatorname{Sym}^2 C)$ comes from $\operatorname{Pic}(\operatorname{Sym}^3 C)$ we may assume that $\pi_2^* L$ belongs to the kernel of r. In other words, $\pi_2^* L \simeq L_1 \boxtimes L_1$ for some line bundle L_1 on C. The S_2-action on $L_1 \boxtimes L_1$ either coincides with the standard one or differs from it by -1. In accordance with this dichotomy we equip $\widetilde{L} = L_1 \boxtimes L_1 \boxtimes L_1$ either with the standard S_3-action or with the standard action twisted by the sign character. Then if we consider $\widetilde{L}$ as a line bundle on $\operatorname{Sym}^3 C$ we have $a_x^{3*} \widetilde{L} = L$.

2.5. **Lemma.** Assume that C is hyperelliptic. Let $Q \subset \operatorname{Sym}^{g-1} C$ be the complement to the image of α. Then Q is an irreducible divisor and $\deg(Q) = -2$.

Proof of the lemma.

Let $\tau : C \to C$ be the hyperelliptic involution. For every $d > 1$ let us denote by $Q_d \subset \operatorname{Sym}^d C$ the reduced effective divisor consisting of D such that D contains a divisor of the form $x + \tau x$. Note that $Q_2 \simeq \mathbb{P}^1$ while Q_d is just the

[2] An alternative proof can be found in [4].

image of $Q_2 \times \mathrm{Sym}^{d-2} C$ under the natural map $\mathrm{Sym}^2 C \times \mathrm{Sym}^{d-2} C \to \mathrm{Sym}^d C$, so it is irreducible. We have $Q = Q_{g-1}$. It is easy to check that

$$a_x^{d*}\mathcal{O}(Q_d) \simeq \mathcal{O}(Q_{d-1} + R^{d-1}_{\tau(x)}). \tag{2}$$

On the other hand, for any points $x, y \in C$ one has

$$a_x^{d*}\mathcal{O}(R^d_y) \simeq \mathcal{O}(R^{d-1}_y). \tag{3}$$

Consider the embedding $a : \mathbb{P}^1 \simeq Q_2 \hookrightarrow \mathrm{Sym}^d C$ given by $D \mapsto D + D_0$ where $D_0 = x_1 + \ldots + x_{d-2}$ is a fixed effective divisor of degree $d-2$. Then by induction we derive from (2) and (3) that $a^*\mathcal{O}(R^d_y) \simeq \mathcal{O}_{\mathbb{P}^1}(1)$ while $a^*\mathcal{O}(Q_d) \simeq \mathcal{O}_{\mathbb{P}^1}(Q_2 \cdot Q_2 + d - 2)$, where $Q_2 \cdot Q_2$ is the self-intersection index of Q_2 in $\mathrm{Sym}^2 C$. Since $Q_2 \cdot Q_2 = 1 - g$ we obtain $a^*\mathcal{O}(Q_d) \simeq \mathcal{O}_{\mathbb{P}^1}(d-g-1)$. In particular, $a^*\mathcal{O}(Q_{g-1}) \simeq \mathcal{O}_{\mathbb{P}^1}(-2)$ as required.

2.6. **Corollary.** Assume that $g \geq 3$. Then the map $j^{ns*} : \mathrm{Pic}(J) \to \mathrm{Pic}(\Theta^{ns})$ is injective. The group $A(J) = \mathrm{Pic}(\Theta^{ns})/j^{ns*}(\mathrm{Pic}(J))$ is isomorphic to $\mathbb{Z}$ if C is non-hyperelliptic, and to $\mathbb{Z}/2\mathbb{Z}$ otherwise.

Proof of the corollary. (i) Assume that C is non-hyperelliptic. Then by Martens's theorem the complement to the image of α has codimension > 1 (see [2], IV 5.1)[3], so $\alpha^* : \mathrm{Pic}(\mathrm{Sym}^{g-1} C) \xrightarrow{\sim} \mathrm{Pic}(\Theta^{ns})$. We are done by Lemma 2.4.

(ii) Assume that C is hyperelliptic. Then Lemma 2.5 implies that $\mathrm{Pic}(\Theta^{ns})$ is isomorphic to $\mathrm{Pic}(\mathrm{Sym}^{g-1} C)/\mathbb{Z}[Q]$ where $\deg([Q]) = 2$, so our statement follows easily from Lemma 2.4.

2.7. *Proof of Proposition 2.2.*

Choose j to be symmetric, $j^\nu = j$; then $j^{ns*} : \mathrm{Pic} J \to \mathrm{Pic}(\Theta^{ns})$ commutes with the involution. Take $L, L' \in \mathcal{T}$. Since L generates $A(J)$ and $L^\nu \equiv L^{-1} \mod j^{ns*}\mathrm{Pic}(J)$ we have either $L \equiv L' \mod j^{ns*}\mathrm{Pic}(J)$ or $L^\nu \equiv L' \mod j^{ns*}\mathrm{Pic}(J)$. Replacing L by L^ν if necessary we obtain that $L^{-1} \cdot L' \simeq j^{ns*}\xi$ for some $\xi \in \mathrm{Pic}(J)$. Since $\xi^\nu = \xi^{-1}$ we deduce that $\xi \in \mathrm{Pic}^0(J) = J$ which implies the proposition.

Remark. We actually proved that if C is non-hyperelliptic then $\mathcal{T}$ is a $\widetilde{J}$-torsor, while for hyperelliptic C, it is a J-torsor (we did not prove that it is non-empty as yet).

2.8. *Proof of Proposition 2.3.* Our $F := \mathcal{F}(i_*M)[1]$ vanishes outside of $j(\Theta)$ where $j = j_{M^{-1}}$. Since F is the push-forward of a (shifted) line bundle on $C \times J$ one may represent it as the cone of a morphism $f : V_1 \to V_0$ of vector bundles on J. Therefore f is injective (so $F = \mathrm{Coker}\, f$), and for any closed subset $Y \subset J$ of codimension > 2 one has $\mathcal{H}^1_Y F = 0$. Note also that $j(\Theta)$ is precisely the zero locus of $\det(f)$ (this follows from the well-known determinantal description of Θ, see e.g. [2]). On the other hand, it is easy to see that $\det(f)$ annihilates $\mathrm{Coker}\, f$. Therefore, $F = j_*j^*F$. Since the codimension in J of the singular locus of Θ is > 2 this shows that $F = j^{ns}_* L$ where $L := j^{ns*}F$. By Riemann's

[3]The proof given in [2] works in arbitrary characteristic.

theorem on singularities of Θ the fibers of L at all closed points of Θ^{ns} are one-dimensional. Since Θ is reduced (see [2] IV.4.5) we see that L is a line bundle on Θ^{ns}.

It remains to prove that $L \in \mathcal{T}$. It is clear that the derived pull-back $\Phi := Lj^{ns*}F$ is a complex with $H^0\Phi = L$, $H^{-1}\Phi = L \otimes \mathcal{O}_J(-j(\Theta))$, other cohomology 0. On the other hand, computing Φ using base change and the definition of F we see that for $x \in \Theta^{ns}$ the corresponding fibers are $H^0\Phi_{j(x)} = H^1(C, E_x \otimes M)$, $H^{-1}\Phi_{j(x)} = H^0(C, E_x \otimes M)$ where $E_x := i^*\theta(j(x))$ is the line bundle on C of degree 0 that corresponds to $j(x)$. Since $E_{x^\nu} \simeq E_x^{-1} \otimes \omega_C \otimes M^{-2}$, Serre duality yields $(H^0\Phi_{j(x^\nu)})^* = H^{-1}\Phi_{j(x)}$. Therefore, $L^{\nu*} = L \otimes \mathcal{O}_J(-j(\Theta))$. We see that condition (i) from 1.1 is satisfied.

Let us check condition (ii). We may assume that i corresponds to a line bundle $\mathcal{O}_C(-x)$, $x \in C$. Consider the universal divisor $\mathcal{D} \subset C \times \mathrm{Sym}^{g-1} C$. Then the pull-back of the Poincaré line bundle on $J \times J$ by the morphism $(i \times (j\sigma)) : C \times \mathrm{Sym}^{g-1} C \to J \times J$ is isomorphic to $p_1^* M^{-1}(\mathcal{D} - C \times R_x)$. It follows that the line bundle L^{-1} on Θ^{ns} is $\alpha^* p_{2*}(\mathcal{O}(\mathcal{D}))(-R_x)$ where $p_2 : C \times \mathrm{Sym}^{g-1} C \to \mathrm{Sym}^{g-1} C$ is the projection. The canonical morphism $\mathcal{O}_{\mathrm{Sym}^{g-1} C} \to p_{2*}(\mathcal{O}(\mathcal{D}))$ is an isomorphism over $\alpha(\Theta^{ns})$. Therefore, $L^{-1} = \alpha^*\mathcal{O}(-R_x)$ which generates $A(J)$, so we are done.

3. Concluding remarks.

3.1. From the Lefschetz theorem for Picard groups (see [5], exp.11 (3.12), exp.12 (3.4)) one can easily derive that the restriction map $j^* : \mathrm{Pic}(J) \to \mathrm{Pic}(\Theta)$ is an isomorphism for an arbitrary principally polarized abelian variety J of dimension $g \geq 4$. Furthermore, if the dimension of the singular locus $\mathrm{Sing}\,\Theta$ is $< g - 4$ then Θ is locally factorial as follows from [5], exp.11 (3.14). Hence, in this case $\mathrm{Pic}(\Theta) = \mathrm{Pic}(\Theta^{ns})$ (since the notions of Cartier divisors and Weil divisors on Θ coincide) and $A(J) = 0$. Notice that if J is a Jacobian then the dimension of $\mathrm{Sing}\,\Theta$ is $\geq g-4$. Moreover, Andreotti and Mayer proved in [1] that the closure of the locus of Jacobians constitute an irreducible component of the locus N_{g-4} of principally polarized abelian varieties with $\dim \mathrm{Sing}\,\Theta \geq g-4$. In [3] Beauville established that in the case $g = 4$ the locus N_0 has two irreducible components. He also proved (assuming that the characteristic is zero) that a generic point of N_0 which is not contained in the closure of the locus of Jacobians corresponds to an abelian variety J with $\mathrm{Sing}\,\Theta$ consisting of one ordinary double point (see [3], 7.5). It follows that the corresponding group $A(J)$ is either zero or isomorphic to $\mathbb{Z}$ and the involution acts on $A(J)$ as identity. Therefore, in this case either $A(J) = 0$ or the set $\mathcal{T}$ is empty. The natural question is whether in higher dimensions one still has either $A(J) = 0$ or $\mathcal{T} = \emptyset$ for principally polarized abelian varieties which are not in the closure of the locus of Jacobians.

Acknowledgments. We are grateful to Hélène Esnault for pointing out some errors in the first draft of the paper. The second author would like to thank J. Harris and O. Debarre for helpful conversations and A. Collino for pointing out the reference [4]. A. P. is partially supported by an NSF grant.

References

[1] A. Andreotti, A. Mayer, *On period relations for abelian integrals on algebraic curves*, Ann. Scuola Norm. Sup. Pisa 21 (1967), 189–238.

[2] E. Arbarello, M. Cornalba, P. A. Griffiths, J. Harris, *Geometry of algebraic curves*, vol. 1. Springer-Verlag, 1984.

[3] A. Beauville, *Prym varieties and the Schottky problem*, Inventiones Math. 41 (1977), 149–196.

[4] A. Collino, *The rational equivalence ring of symmetric products of curves*, Illinois J. Math. 19 (1975), 567–583.

[5] A. Grothendieck, *Cohomologie locale des faisceaux cohérents et théorèmes de Lefschetz locaux et globaux* (SGA 2). Amsterdam, North-Holland Publishing Co., 1969.

[6] S. Mukai, *Duality between $D(X)$ and $D(\hat{X})$ with its application to Picard sheaves*, Nagoya Math. J. 81 (1981), 153–175.

[7] *Théorie des topos et cohomologie étale des schémas* (SGA 4), t. 3, Lecture Notes in Math., vol. 305, Springer-Verlag, 1973.

Progress in Mathematics, Vol. 195, © 2001 Birkhäuser Verlag Basel/Switzerland

ON THE ANDRÉ-OORT CONJECTURE FOR HILBERT MODULAR SURFACES

BAS EDIXHOVEN

1. INTRODUCTION.

In order to state the conjecture mentioned in the title, we need to recall some terminology and results on Shimura varieties; as a general reference for these, we use [20, Sections 1–2]. So let $\mathbb{S} := \mathrm{Res}_{\mathbb{C}/\mathbb{R}}\mathbb{G}_{\mathrm{m},\mathbb{C}}$ be the algebraic group over $\mathbb{R}$ obtained by restriction of scalars from $\mathbb{C}$ to $\mathbb{R}$ of the multiplicative group. For V an $\mathbb{R}$-vector space, it is then equivalent to give an $\mathbb{R}$-Hodge structure or an action by $\mathbb{S}$ on it. A Shimura datum is a pair (G, X), with G a connected reductive affine algebraic group over $\mathbb{Q}$, and X a $G(\mathbb{R})$-conjugacy class in the set of morphisms of algebraic groups $\mathrm{Hom}(\mathbb{S}, G_{\mathbb{R}})$, satisfying the three conditions of [20, Def. 1.4] (i.e., the usual conditions (2.1.1–3) of [13]). These conditions imply that X has a natural complex structure (in fact, the connected components are hermitian symmetric domains), such that every representation of G on a $\mathbb{Q}$-vector space defines a polarizable variation of Hodge structure on X. For (G, X) a Shimura datum, and K a compact open subgroup of $G(\mathbb{A}_{\mathrm{f}})$, we let $\mathrm{Sh}_K(G,X)(\mathbb{C})$ denote the complex analytic variety $G(\mathbb{Q})\backslash(X \times G(\mathbb{A}_{\mathrm{f}})/K)$, which has a natural structure of quasi-projective complex algebraic variety, denoted $\mathrm{Sh}_K(G,X)_{\mathbb{C}}$; the projective limit $\mathrm{Sh}(G,X)_{\mathbb{C}}$ over all K of the $\mathrm{Sh}_K(G,X)_{\mathbb{C}}$ is a scheme on which $G(\mathbb{A}_{\mathrm{f}})$ acts continuously. (The action being continuous means that the scheme has a cover by open affines $U_i = \mathrm{Spec}(A_i)$ such that each U_i is stabilized by some open subgroup K_i of $G(\mathbb{A}_{\mathrm{f}})$ and each f in A_i has open stabilizer in K_i.) A morphism of Shimura data from (G_1, X_1) to (G_2, X_2) is a morphism $f \colon G_1 \to G_2$ that maps X_1 to X_2; for K_1 and K_2 compact open subgroups of $G_1(\mathbb{A}_{\mathrm{f}})$ and $G_2(\mathbb{A}_{\mathrm{f}})$ with $f(K_1)$ contained in K_2, such an f induces a morphism $\mathrm{Sh}(f)$ from $\mathrm{Sh}_{K_1}(G_1,X_1)_{\mathbb{C}}$ to $\mathrm{Sh}_{K_2}(G_2,X_2)_{\mathbb{C}}$.

Definition 1.1. *Let (G,X) be a Shimura datum, K an open compact subgroup of $G(\mathbb{A}_{\mathrm{f}})$, and Z an irreducible closed subvariety of $\mathrm{Sh}_K(G,X)_{\mathbb{C}}$. Then Z is a*

Partially supported by the Institut Universitaire de France, and by the European TMR Network Contract ERB FMRX 960006 "arithmetic algebraic geometry". A preprint version of this article together with an appendix that could be considered as the author's scratch paper while working on this subject can be downloaded from the author's home page. This appendix contains details that the author does not find interesting enough to publish, but that may be helpful for readers who got lost.

subvariety of Hodge type if there is a Shimura datum (G', X'), *a morphism of Shimura data* $f\colon (G', X') \to (G, X)$, *and an element* g *of* $G(\mathbb{A}_{\mathrm{f}})$ *such that* Z *is an irreducible component of the image of the map:*

$$\mathrm{Sh}(G', X')_{\mathbb{C}} \xrightarrow{\mathrm{Sh}(f)} \mathrm{Sh}(G, X)_{\mathbb{C}} \xrightarrow{\cdot g} \mathrm{Sh}(G, X)_{\mathbb{C}} \longrightarrow \mathrm{Sh}_K(G, X)_{\mathbb{C}}.$$

This definition is equivalent to [20, 6.2], which uses only closed immersions $f\colon G' \to G$. In [21, Prop. 2.8] it is explained that the subvarieties of Hodge type are precisely the loci where certain given classes in certain variations of Hodge structures (obtained from representations of G) are Hodge classes; hence the terminology.

Definition 1.2. *Let* (G, X) *be a Shimura datum. For* h *in* X *we let* $\mathrm{MT}(h)$ *be the Mumford-Tate group of* h, *i.e., the smallest algebraic subgroup* H *of* G *such that* h *factors through* $H_{\mathbb{R}}$. *A point* h *in* X *is called special if* $\mathrm{MT}(h)$ *is commutative (in which case it is a torus). For* K *a compact open subgroup of* $G(\mathbb{A}_{\mathrm{f}})$, *a point in* $\mathrm{Sh}_K(G, X)_{\mathbb{C}}$ *is special if its preimages in* $\mathrm{Sh}(G, X)_{\mathbb{C}}$ *are of the form* (h, g) *with* h *in* X *special. Equivalently, the special points in* $\mathrm{Sh}_K(G, X)_{\mathbb{C}}$ *are the zero dimensional subvarieties of Hodge type.*

Conjecture 1.3 (André-Oort). *Let* (G, X) *be a Shimura datum. Let* K *be a compact open subgroup of* $G(\mathbb{A}_{\mathrm{f}})$ *and let* S *be a set of special points in* $\mathrm{Sh}_K(G, X)(\mathbb{C})$. *Then every irreducible component of the Zariski closure of* S *in* $\mathrm{Sh}_K(G, X)_{\mathbb{C}}$ *is a subvariety of Hodge type.*

Some remarks are in order at this point. André stated this conjecture as a problem for curves containing infinitely many special points in general Shimura varieties in [2, X.4]. Independently, Oort raised the question for general subvarieties of the moduli spaces of principally polarized abelian varieties in [26]. In [2, X.4] André mentions the similarity with the Manin-Mumford conjecture (proved by Raynaud, see [1]), and [4] contains a version generalizing both the conjecture above and the Manin-Mumford conjecture; see also [20, 6.7.2].

Let us now discuss the results on the conjecture above that have been obtained until now. All of them deal with moduli spaces of abelian varieties. Moonen proved in his thesis (see [22, §5], in particular the equivalence between Conjectures 5.1 and 5.3, and [19, IV]) that the conjecture is true for sets S for which there exists a prime number p at which all s in S have an ordinary reduction of which they are the canonical lift. Needless to say, his methods use reduction modulo a prime number p. This gives a quite general result, but it has the disadvantage that one neglects most of the Galois action on the special points, and that one has to work with one Frobenius element simultaneously for all s in S.

In [14], the conjecture was proved for the moduli space of pairs of elliptic curves, assuming the generalized Riemann hypothesis (GRH) for imaginary quadratic fields. In a few words, the proof exploits the Galois action on CM-points and considers intersections of the subvarieties in question with images of them under suitable Hecke operators. In this approach, we work with a different Frobenius element for each s in S; GRH comes in via the existence of small primes with suitable properties. The same case of Conjecture 1.3 was proved unconditionally by André in [5]. He uses the Galois action on the CM-points, and a Diophantine approximation result of Masser on the j-function.

More recently, Yafaev has generalized the result in [14] to the case of products of two Shimura curves that are associated to quaternion algebras over $\mathbb{Q}$, see [31], and B. Belhaj Dahman, a student of André, is working on the families of jacobians of the curves

$$y^n = x(x-1)(x-\lambda).$$

The question about these families of jacobians is whether or not the various isogeny factors coming from the decomposition for the action of $\mu_n(\mathbb{C})$ are simultaneously of CM type for infinitely many complex numbers λ.

Recently, Clozel and Ullmo have proved ([10]), for G among GSp_{2n} and GL_n, that sets of the form $T_p x$, with x in $G(\mathbb{Q})\backslash G(\mathbb{A})/K$ and T_p certain Hecke operators with p tending to infinity, are equidistributed. The idea behind this is that one would like to imitate and apply the equidistribution results for Galois orbits of strict sequences of points of small height in abelian varieties as in [1]. A sequence of closed points of an algebraic variety is called strict if every proper closed subset contains only finitely many elements of the sequence. Phrased in this terminology, the André-Oort conjecture says that a sequence of special points is strict if no proper subvariety of Hodge type contains an infinite subsequence. Of course, to prove the André-Oort conjecture in this way, one has to replace $T_p x$ by the Galois orbit of x, which seems to be a hard problem, and moreover, one has to deal with the fact that the heights of CM points tend to infinity and not to zero.

In this article, we prove Conjecture 1.3, assuming GRH, for Hilbert modular surfaces. The method of proof is basically the same as in [14], but now we do use more advanced techniques. The two main results of the article are described in Section 2. The reason for which we state and prove Theorem 2.2 is that it has an interesting application to transcendence of special values of certain hypergeometric functions via work of Wolfart, Cohen and Wüstholz, see [11], *without* having to assume GRH.

Let us briefly describe the contents of this article. Section 2 introduces the Hilbert modular surfaces that we work with in terms of a Shimura datum, gives their interpretation as moduli spaces of abelian surfaces with multiplications by the ring of integers of a real quadratic field K, and states the main results.

Section 3, which is not so essential, discusses the difference between working with abelian surfaces with or without a given polarization. In group theoretical terms, the choice is between working with $\mathrm{GL}_2(K)$ or its subgroup $\mathrm{GL}_2(K)'$ consisting of the elements of $\mathrm{GL}_2(K)$ whose determinant is in $\mathbb{Q}^*$. The reason for considering both cases is that with a polarization (and a suitable level structure), the variation of Hodge structure provided by the lattices of the abelian surfaces comes from a representation of the group in the Shimura datum, which is not true without given polarizations. We need variations of Hodge structure in Section 4. On the other hand, the size of Galois orbits of special points, studied in Section 6, is simpler to understand in terms of class groups when working without polarizations. We could have chosen to work throughout the article with $\mathrm{GL}_2(K)'$, but we think that it is instructive to see the consequences of such a choice in the relatively easy case of Hilbert modular surfaces, before trying to treat general Shimura varieties completely in group theoretical terms.

In Section 4 we recall an important result of André, relating the generic Mumford-Tate group of a variation of Hodge structure to its algebraic monodromy group (i.e., the Zariski closure of the image of monodromy). We use it to prove that for a curve in a Hilbert modular surface that is not of Hodge type and that does contain a special point, the connected algebraic monodromy group is maximal, i.e., $\mathrm{SL}_{2,K}$.

Section 5 introduces the Hecke correspondence T_p associated to a prime number p. We use a very powerful result of Nori in order to prove that for C a curve with maximal algebraic monodromy group, T_pC is irreducible if C is large enough.

The main result of Section 6 says that the size of the Galois orbit of a special point x grows at least as a positive power of the discriminant $\mathrm{discr}(R_x)$ of the ring of endomorphisms (commuting with the real multiplications) of the corresponding abelian variety. This section is quite long, and contains some messy computations, depending on the structure of the Galois group of the normal closure of the CM field in question. The problem is that one has to give a lower bound for the image under the reflex type norm of one class group in another.

Section 7 gives an upper bound for the number of points in intersections of the form $Z_1 \cap T_g Z_2$, with Z_1 and Z_2 fixed subvarieties of a general Shimura variety, and with T_g a varying Hecke correspondence.

Finally, Section 8 combines all these preliminary results as follows. One supposes that C is a curve in a Hilbert modular surface S, containing infinitely many special points, and not of Hodge type. If p is large enough (depending only on C), then T_pC is irreducible by Section 5. Since the T_p-orbits in S are dense, one cannot have $C = T_pC$. Hence the intersections $C \cap T_pC$ are finite, and hence bounded above (Section 7) by a constant times p^2. Let now x be a special point on C. If p is a prime that is split in R_x, then $C \cap T_pC$ contains the Galois orbit of x, hence $|C \cap T_pC|$ grows at least as a positive power of $|\mathrm{discr}(R_x)|$. But this lower bound for primes that are split in R_x contradicts the conditional effective Chebotarev theorem (this is where GRH comes in). Hence, assuming GRH, one has proved that if C does contain infinitely many special points, then C is of Hodge type. The reason that one can prove Thm. 2.2 unconditionally is that in that case the CM field $\mathbb{Q} \otimes R_x$ is independent of x, and hence Chebotarev's theorem itself is sufficient.

In April 1999, we have proved Conjecture 1.3, assuming GRH, for arbitrary products of modular curves, extending the methods of [14]. A detailed proof, which is quite elementary, will be written up in the near future. One can hope that combining the techniques used for these last two results will make it possible to treat more general higher dimensional cases of Conjecture 1.3. Of course, eventually everything should be expressed in terms of "(G,X)-language". In fact, in this article we could already have worked without mentioning abelian varieties.

Before we really start, let us first mention two obvious general principles. The first is that level structures don't matter in Conjecture 1.3: for (G,X) a Shimura datum, K and K' open compact in $G(\mathbb{A}_f)$ with $K \subset K'$, an irreducible subvariety Z of $\mathrm{Sh}_K(G,X)_{\mathbb{C}}$ is of Hodge type if and only if its image in $\mathrm{Sh}_{K'}(G,X)_{\mathbb{C}}$ is. The second principle says that the irreducible components of intersections of subvarieties of Hodge type are again of Hodge type (this is clear from the interpretation of subvarieties of Hodge type given right after Definition 1.1).

2. The main results.

Let K be a real quadratic extension of $\mathbb{Q}$, let O_K be its ring of integers, and let G be the $\mathbb{Z}$-group scheme $\mathrm{Res}_{O_K/\mathbb{Z}}(\mathrm{GL}_{2,O_K})$. After numbering the two embeddings of K in $\mathbb{R}$, we have $\mathbb{R}\otimes K = \mathbb{R}^2$, and hence $G(\mathbb{R}) = \mathrm{GL}_2(\mathbb{R})^2$. We will study the Shimura variety:

$$S(\mathbb{C}) := G(\mathbb{Q})\backslash(X \times G(\mathbb{A}_f)/G(\hat{\mathbb{Z}})),$$

where $X = (\mathbb{H}^{\pm})^2$, and where $\mathbb{H}^{\pm}$ is the usual $\mathrm{GL}_2(\mathbb{R})$-conjugacy class of morphisms from $\mathbb{S}$ to $\mathrm{GL}_{2,\mathbb{R}}$, i.e., the class of $a+bi \mapsto \left(\begin{smallmatrix} a & -b \\ b & a \end{smallmatrix}\right)$. The surface $S_{\mathbb{C}}$, called a Hilbert modular surface, is the coarse moduli space for pairs (A,α) with A an abelian surface and α a morphism from O_K to $\mathrm{End}(A)$ (see [15, Ch. X], and the end of Section 3 for the moduli interpretation for a closely related Shimura datum). This implies that the reflex field of (G,X) is $\mathbb{Q}$ and that the canonical model $S_{\mathbb{Q}}$ (see [20, Section 2] for this notion) is simply the coarse moduli space for pairs $(A/S/\mathbb{Q},\alpha)$ with S a $\mathbb{Q}$-scheme, A/S an abelian scheme of relative dimension two, and α a morphism from O_K to $\mathrm{End}_S(A)$. The set of geometrically connected components of $S_{\mathbb{Q}}$ is $K^*\backslash\mathbb{A}_K^*/(\mathbb{R}\otimes K)^{*,+}O_K^{\wedge *} = \mathrm{Pic}(O_K)^+$, the group of isomorphism classes of invertible O_K-modules with orientations at the two infinite places, and has trivial action by $\mathrm{Gal}(\overline{\mathbb{Q}}/\mathbb{Q})$ ([15, Ch. I, Cor. 7.3]). The main objective of this article is to prove the following two theorems.

Theorem 2.1. *Assume GRH. Let $C \subset S_{\mathbb{C}}$ be an irreducible closed curve containing infinitely many CM points. Then C is of Hodge type.*

Theorem 2.2. *Let $C \subset S_{\mathbb{C}}$ be an irreducible closed curve containing infinitely many CM points corresponding to abelian varieties that lie in one isogeny class (the isogenies are not required to be compatible with the multiplications by O_K). Then C is of Hodge type.*

Let us note immediately that these theorems apply in fact to all Hilbert modular surfaces, because the André-Oort conjecture is insensitive to level structure. Before proving the theorems we need to discuss some of the tools we will use.

3. Choosing a suitable Shimura variety.

For a variation of Hodge structure on a complex variety, one has the notions of generic Mumford-Tate group and that of monodromy. A relation between these two notions will be very useful for us. In order to get a suitable variation of Hodge structure on $S(\mathbb{C})$ as above, there is a little complication, and at least two options to get around it. The problem is that the tautological representation of $G_{\mathbb{Q}}$ on the $\mathbb{Q}$-vector space K^2 does not induce a variation of Hodge structure on $\mathrm{Sh}_H(G_{\mathbb{Q}}, X)(\mathbb{C})$, even if H is an arbitrary small open subgroup of $G(\mathbb{A}_f)$; just consider the action of O_K^* in $G(\mathbb{Q})$ on $X \times G(\mathbb{A}_f)/H$ (see [21, Section 2.3] for a general statement).

The first possible way out is to use an other representation, and have the monodromy take place in the image of G under this representation. For example, one can take the representation $\mathrm{Sym}^2(\rho_0) \otimes \det(\rho_0)^{-1}$, with ρ_0 the tautological representation on O_K^2. This representation induces a faithful representation ρ of the quotient $G^{\mathrm{ad}} = \mathrm{Res}_{O_K/\mathbb{Z}}\mathrm{PGL}_{2,O_K}$. The morphism $G \to G^{\mathrm{ad}}$ induces an isomorphism from X to a conjugacy class X^{ad} in $\mathrm{Hom}_{\mathbb{R}}(\mathbb{S}, G^{\mathrm{ad}}_{\mathbb{R}})$, and gives a morphism of Shimura data from (G, X) to $(G^{\mathrm{ad}}, X^{\mathrm{ad}})$. Let $S^{\mathrm{ad}}(\mathbb{C})$ be the Shimura variety $\mathrm{Sh}_{G^{\mathrm{ad}}(\hat{\mathbb{Z}})}(G^{\mathrm{ad}}, X^{\mathrm{ad}})(\mathbb{C})$. The natural morphism from $S_{\mathbb{C}}$ to $S^{\mathrm{ad}}_{\mathbb{C}}$ is finite and surjective (this follows directly from the definition); one can show that it is the quotient for a faithful action of $\mathrm{Pic}(O_K)$, but that will not be used. Conjecture 1.3 is then true for $S_{\mathbb{C}}$ if and only if it is true for $S^{\mathrm{ad}}_{\mathbb{C}}$, and ρ induces a variation of Hodge structure on $\mathrm{Sh}_H(G^{\mathrm{ad}}, X^{\mathrm{ad}})(\mathbb{C})$ for suitable H. The disadvantage of working with $S^{\mathrm{ad}}_{\mathbb{C}}$ is that it does not seem to have an interpretation as a moduli space of abelian varieties; this is not a real problem, but we prefer to work with Shimura varieties that are as simple as possible.

Another way out is to replace the group G by its subgroup G' given by the following Cartesian diagram:

$$\begin{array}{ccc} G' & \hookrightarrow & G \\ \downarrow & \Box & \downarrow\mathrm{det} \\ \mathbb{G}_{\mathrm{m},\mathbb{Z}} & \hookrightarrow & \mathrm{Res}_{O_K/\mathbb{Z}}\mathbb{G}_{\mathrm{m},O_K} \end{array}$$

Loosely speaking, G' is the subgroup of G consisting of those elements whose determinant is in $\mathbb{Q}$. As the morphism det in the diagram above is smooth, G' is smooth over $\mathbb{G}_{\mathrm{m},\mathbb{Z}}$, hence over $\mathbb{Z}$. It follows that G' is the scheme-theoretic closure in G of its generic fibre. We note that $G'(\mathbb{R})$ is the subgroup of (x, y) in $\mathrm{GL}_2(\mathbb{R})^2$ with $\det(x) = \det(y)$. All h in X factor through $G'_{\mathbb{R}}$, but X consists of two $G'(\mathbb{R})$-conjugacy classes. The conjugacy class X' we work with is the disjoint union of $(\mathbb{H}^+)^2$ and $(\mathbb{H}^-)^2$. This gives a morphism of Shimura data from (G', X') to (G, X), and a morphism of Shimura varieties $S'_{\mathbb{C}} \to S_{\mathbb{C}}$ with $S'_{\mathbb{C}} = \mathrm{Sh}_{G'(\hat{\mathbb{Z}})}(G', X')_{\mathbb{C}}$. One can prove that the Shimura variety $S'_{\mathbb{C}}$ is connected, and that the morphism to its image in $S_{\mathbb{C}}$ is the quotient by a faithful action of the finite group $O_K^{*,+}/O_K^{*,2}$, i.e., by the group of totally positive global units modulo squares of global units. We will only use that the morphism $S'_{\mathbb{C}} \to S_{\mathbb{C}}$ is finite and that its image is open and closed; these two facts follow directly from the definitions. It follows that Conjecture 1.3 is true for $S_{\mathbb{C}}$ if and only if it is for $S'_{\mathbb{C}}$, and similarly for the two theorems above that we want to prove. Moreover, the tautological representation of G' does induce a variation of Hodge structure on $\mathrm{Sh}_H(G', X')(\mathbb{C})$ for H sufficiently small.

The option we choose is the last. The variety $S'_{\mathbb{C}}$ is the (coarse) moduli space for triplets (A, α, λ) where:

$$(3.1)\qquad \begin{cases} A \text{ is a complex abelian surface,} \\ \alpha\colon O_K \to \mathrm{End}(A) \text{ a morphism of rings,} \\ \text{and } \lambda\colon A \to A^* \text{ a principal } O_K\text{-polarization,} \end{cases}$$

a notion that we will now explain. First of all, A^* is the dual of A in the category of abelian varieties with O_K-action: $A^* := \mathrm{Ext}^1(A, O_K \otimes \mathbb{G}_{\mathrm{m}})$. One verifies that $A^* = \delta \otimes_{O_K} A^t$, where δ is the different of the extension $\mathbb{Z} \to O_K$, and where $A^t = \mathrm{Ext}^1(A, \mathbb{G}_{\mathrm{m}})$, the dual of A in the usual sense. The inclusion $\delta \subset O_K$ induces a morphism $A^* \to A^t$, which is an isogeny. A principal O_K-polarization is then an isomorphism $\lambda\colon A \to A^*$ such that the induced morphism from A to A^t is a polarization. Interpreted in Hodge-theoretical terms, a triplet (A, α, λ) as in (3.1) corresponds to a triplet (V, h, ψ) with V a locally free O_K-module of rank two, $h\colon \mathbb{S} \to (\mathrm{GL}_{\mathbb{Z}}(V))_{\mathbb{R}}$ a Hodge structure of type $(-1,0), (0,-1)$, and $\psi\colon V \times V \to O_K$ a perfect antisymmetric O_K-bilinear form such that $\mathrm{tr} \circ \psi\colon V \times V \to \mathbb{Z}$ is a polarization. Note that for such a triplet (V, h, ψ), the pair (V, ψ) is isomorphic to the standard pair $(O_K \oplus O_K, \left(\begin{smallmatrix} 0 & 1 \\ -1 & 0 \end{smallmatrix}\right))$. In order to prove that the set of isomorphism classes of (A, α, λ) as in (3.1) is $S'(\mathbb{C})$ one uses the following two facts: 1: $G'(\mathbb{A}_{\mathrm{f}})/G'(\hat{\mathbb{Z}})$ is the set of O_K-lattices in K^2 on which $\psi = \left(\begin{smallmatrix} 0 & 1 \\ -1 & 0 \end{smallmatrix}\right)$ induces a perfect pairing of O_K-modules, up to a factor in $\mathbb{Q}^*$; and 2: X' is the set of Hodge structures of type $(-1,0), (0,-1)$ on the K-vector space K^2 such that, up to sign, $\mathrm{tr} \circ \psi$ is a polarization. The moduli space over $\mathbb{Q}$ of triplets $(A/S, \alpha, \lambda)$ with S a $\mathbb{Q}$-scheme, $\alpha\colon O_K \to \mathrm{End}_S(A)$ a morphism of rings, and λ a principal O_K-polarization, is then the canonical model $S'_{\mathbb{Q}}$ of $S'_{\mathbb{C}}$ (see also [28, 1.27] and [12, 4.11]).

For $n \geq 1$, let H_n be the kernel of the morphism $G'(\hat{\mathbb{Z}}) \to G'(\mathbb{Z}/n\mathbb{Z})$, and let $S'_{\mathbb{Q},n}$ denote the Shimura variety $\mathrm{Sh}_{H_n}(G', X')_{\mathbb{Q}}$. Then $S'_{\mathbb{Q},n}$ is the moduli space for 4tuples $(A/S, \alpha, \lambda, \phi)$, with S a $\mathbb{Q}$-scheme, $(A/S, \alpha, \lambda)$ an abelian scheme over S with multiplications by O_K and a principal O_K-polarization, and with ϕ an isomorphism of S-group schemes with O_K-action:

$$\phi\colon (O_K/nO_K)^2_S \longrightarrow A[n],$$

such that there exists a (necessarily unique) isomorphism $\overline{\phi}\colon (\mathbb{Z}/n\mathbb{Z})_S \to \mu_{n,S}$ making the diagram:

$$\begin{array}{ccc} ((O_K/nO_K)_S^2)^2 & \xrightarrow{\phi} & A[n]^2 \\ \downarrow \psi_n & & \downarrow e_{\lambda,n} \\ O_K \otimes (\mathbb{Z}/n\mathbb{Z})_S & \xrightarrow{\mathrm{id}\otimes\overline{\phi}} & O_K \otimes \mu_{n,S} \end{array}$$

commutative. In this diagram, ψ_n is the pairing given by $\left(\begin{smallmatrix} 0 & 1 \\ -1 & 0 \end{smallmatrix}\right)$, and $e_{\lambda,n}$ is the perfect pairing on $A[n]$ induced by λ. For $n \geq 3$ the objects $((A/S, \alpha, \lambda))$ have no non-trivial automorphisms (see [23, IV, Thm. 5]), and $S'_{\mathbb{Q},n}$ is a fine moduli space. (Representability by an algebraic space can be found in [28, §1.23]. Quasi-projectiveness follows from [7].) In particular, for $n \geq 3$ we do have a polarized variation of $\mathbb{Z}$-Hodge structure on $S'_n(\mathbb{C})$, given by the first homology groups of the fibers of the universal family.

4. Monodromy and generic Mumford-Tate groups.

We recall some results that can be found in [21, Sections 1.1–1.3], with references to [9] and [3].

The Mumford-Tate group $\mathrm{MT}(V)$ of a $\mathbb{Q}$-Hodge structure V, given by $h\colon \mathbb{S} \to \mathrm{GL}(V)_{\mathbb{R}}$, is defined to be the smallest algebraic subgroup H of $\mathrm{GL}(V)_{\mathbb{Q}}$ such that h factors through $H_{\mathbb{R}}$. Equivalently, $\mathrm{MT}(V)$ is the intersection in $\mathrm{GL}(V)$ of all stabilizers of all lines generated by Hodge classes (i.e., of some type (p,p)) in all $\mathbb{Q}$-Hodge structures of the form $\oplus_i V^{\otimes n_i} \otimes (V^*)^{\otimes m_i}$.

For S a smooth complex algebraic variety with a polarizable variation of $\mathbb{Q}$-Hodge structure V on the associated analytic variety $S(\mathbb{C})$, there is a countable union Σ of proper algebraic subvarieties such that $s \mapsto \mathrm{MT}(V_s)$ is locally constant outside Σ (this makes sense because V is a locally constant sheaf on $S(\mathbb{C})$). The smallest such Σ is called the Hodge exceptional locus, and its complement the Hodge generic locus. For s in $S(\mathbb{C})$ and not in Σ, $\mathrm{MT}(V_s) \subset \mathrm{GL}(V_s)$ is called the generic Mumford-Tate group (at s).

Assume now that S is connected, and that we have an element s of $S(\mathbb{C})$. Then the locally constant sheaf V corresponds to a representation

$$\rho\colon \pi_1(S(\mathbb{C}), s) \to \mathrm{GL}(V_s),$$

called the monodromy representation. The algebraic monodromy group is defined to be the smallest algebraic subgroup H of $\mathrm{GL}(V_s)$ such that ρ factors through H, i.e., it is the Zariski closure of the image of ρ; its connected component of identity is called the connected algebraic monodromy group, and denoted $\mathrm{AM}(V_s)$. With these hypotheses, we have the following theorem.

Theorem 4.1 (André). *Assume moreover that V admits a $\mathbb{Z}$-structure, that s in $S(\mathbb{C})$ is Hodge generic, and that there is a point t in $S(\mathbb{C})$ such that $\mathrm{MT}(V_t)$ is abelian (i.e., t is special). Then $\mathrm{AM}(V_s)$ is the derived subgroup $\mathrm{MT}(V_s)^{\mathrm{der}}$ of $\mathrm{MT}(V_s)$, i.e., the algebraic subgroup generated by commutators.*

Let us now consider what this theorem implies for the variation of Hodge structure that we have on $S'_n(\mathbb{C})$ ($n \geq 3$), and, more importantly, for its restrictions to subvarieties of $S'_n(\mathbb{C})$. The Hodge exceptional locus of $S'_n(\mathbb{C})$ is by construction the union of all lower dimensional subvarieties of Hodge type. The generic Mumford-Tate group on $S'_n(\mathbb{C})$ is G' (use that it contains a subgroup of finite index of $G'(\mathbb{Z})$, and that for all $h = (h_1, h_2)\colon \mathbb{C}^* \to \mathrm{GL}_2(\mathbb{R})^2$ in X' one has $\det(h_1(z)) = z\bar{z} = \det(h_2(z))$ for all z).

Proposition 4.2. *Let $n \geq 3$. Let C be an irreducible curve in $S'_{\mathbb{C},n}$ (i.e., an irreducible closed subvariety of dimension one); let C^{nor} denote its normalization and C^{sm} its smooth locus. Then C is of Hodge type if and only if the generic Mumford-Tate group on C^{sm} is strictly smaller than $G'_{\mathbb{Q}}$. If C is not of Hodge type and contains a special point, then the connected algebraic monodromy group on C^{nor} equals $G'^{\mathrm{der}}_{\mathbb{Q}} = \mathrm{Res}_{K/\mathbb{Q}}\mathrm{SL}_{2,K}$.*

Proof. Suppose that C is of Hodge type. Then some element in some tensor construction of the variation of Hodge structure on $S'_{\mathbb{C},n}$ is a Hodge class on C, but not on $S'_{\mathbb{C},n}$. The interpretation of the Mumford-Tate group as stabilizer of lines generated by Hodge classes shows that the generic Mumford-Tate group on C^{sm} is strictly smaller than $G'_{\mathbb{Q}}$. Now suppose that the generic Mumford-Tate group on C^{sm} is strictly smaller than $G'_{\mathbb{Q}}$. Then C does carry an extra Hodge class. The locus where this class is a Hodge class is necessarily of dimension one, hence, C, being an irreducible component of it, is of Hodge type. The second statement follows now from André's theorem above. □

5. Irreducibility of images under Hecke correspondences.

For (G, X) a Shimura datum, K_1 and K_2 open subgroups of $G(\mathbb{A}_f)$, and g in $G(\mathbb{A}_f)$, one has the so-called Hecke correspondence T_g that is defined as follows. Consider the diagram:

$$\mathrm{Sh}_{K_1}(G,X)_{\mathbb{C}} \xleftarrow{\pi_1} \mathrm{Sh}(G,X)_{\mathbb{C}} \xrightarrow{\cdot g} \mathrm{Sh}(G,X)_{\mathbb{C}} \xrightarrow{\pi_2} \mathrm{Sh}_{K_2}(G,X)_{\mathbb{C}},$$

where π_1 and π_2 are the quotient maps for the actions by K_1 and K_2, respectively. The morphism $\pi_2 \circ \cdot g$ is the quotient for the action of gK_2g^{-1}, hence π_1 and $\pi_2 \circ \cdot g$ both factor through the quotient by $K := K_1 \cap gK_2g^{-1}$, and T_g is the correspondence:

$$\mathrm{Sh}_{K_1}(G,X)_{\mathbb{C}} \xleftarrow{\overline{\pi_1}} \mathrm{Sh}_K(G,X)_{\mathbb{C}} \xrightarrow{\overline{\pi_2 \circ \cdot g}} \mathrm{Sh}_{K_2}(G,X)_{\mathbb{C}}.$$

Of course, T_g exists already over the reflex field E of (G,X). In particular, for Z a closed subvariety of $\mathrm{Sh}_{K_1}(G,X)_E$, its image T_gZ is a closed subvariety of $\mathrm{Sh}_{K_2}(G,X)_E$.

We now specialize to our situation, i.e., to the Shimura datum (G', X') as above. For p a prime number, we let T_p be the Hecke correspondence on $S'_{\mathbb{Q}}$ given by the element $g(p)$ in $G'(\mathbb{A}_f)$ with $g(p)_p = \left(\begin{smallmatrix} p & 0 \\ 0 & 1 \end{smallmatrix}\right)$ and $g(p)_l = 1$ for l different from p. Note that $g(p)^{-1}$ gives the same correspondence as $g(p)$

does, because $g(p) = \left(\begin{smallmatrix} p & 0 \\ 0 & p \end{smallmatrix}\right)\left(\begin{smallmatrix} 0 & 1 \\ 1 & 0 \end{smallmatrix}\right)g(p)^{-1}\left(\begin{smallmatrix} 0 & 1 \\ 1 & 0 \end{smallmatrix}\right)$. The modular interpretation of T_p is the following. Let $[(A,\lambda)]$ in $S'(\mathbb{C})$ denote the isomorphism class of a complex abelian surface A with multiplications by O_K and with a principal O_K-polarization λ. Then, as a cycle, the image of $[(A,\lambda)]$ is given by:

$$T_p[(A,\lambda)] = \sum_H [(A/H, \overline{p\lambda})],$$

where H ranges through the O_K/pO_K-submodules of $A[p](\mathbb{C})$ that are free of rank one, and where $\overline{p\lambda}$ is the principal O_K-polarization induced by $p\lambda$ on A/H. In order to see this, one uses, as in Section 3, that $G'(\mathbb{A}_f)/G'(\hat{\mathbb{Z}})$ is the set of O_K-lattices in K^2 on which $\psi = \left(\begin{smallmatrix} 0 & 1 \\ -1 & 0 \end{smallmatrix}\right)$ induces a perfect pairing of O_K-modules, up to a factor in $\mathbb{Q}^*$, and that the correspondence on it induced by $g(p)^{-1}$ sends such a lattice to the set of lattices containing it with quotient free of rank one over O_K/pO_K.

Proposition 5.1. *Let C be an irreducible curve in $S'_{\mathbb{C}}$. Suppose that C is not of Hodge type and that it contains a special point. Then, for all primes p large enough, T_pC is irreducible.*

Proof. Let $n \geq 3$ be some integer, and let C_n be an irreducible component of the inverse image of C in $S'_{\mathbb{C},n}$. Irreducibility of T_pC_n implies that of T_pC. Let V denote the polarized variation of $\mathbb{Z}$-Hodge structure on $S'_n(\mathbb{C})$ that we considered before, let s be in $C_n(\mathbb{C})$. We choose an isomorphism of O_K-modules from O_K^2 to V_s. Let $\rho\colon \pi_1(C_n(\mathbb{C}), s) \to \mathrm{SL}_2(O_K)$ be the monodromy representation. Proposition 4.2 implies that the Zariski closure in G' of $\rho(\pi_1(C_n^{\mathrm{nor}}(\mathbb{C}), s))$ is the subgroup $\mathrm{Res}_{O_K/\mathbb{Z}}\mathrm{SL}_{2,O_K}$. For p prime, the correspondence T_p on $S'_{\mathbb{C},n}$ is given by a diagram:

$$S'_{\mathbb{C},n} \xleftarrow{\pi_1} S'_{\mathbb{C},n,p} \xrightarrow{\pi_2} S'_{\mathbb{C},n}.$$

For $T_pC_n = \pi_2\pi_1^{-1}C_n$ to be irreducible, it suffices that $C_{n,p}$ be irreducible, with $C_{n,p}$ the covering of C_n^{nor} obtained from π_1. But this covering corresponds to the $\pi_1(C_n^{\mathrm{nor}}(\mathbb{C}), s)$-set $\mathbb{P}^1(O_K/pO_K)$ of O_K/pO_K-submodules of $(O_K/pO_K)^2$ that are free of rank one. Nori's Theorem [25, Thm. 5.1] (Theorem 5.2 below) implies that for p large enough, the reduction map from $\pi_1(C_n^{\mathrm{nor}}(\mathbb{C}), s)$ to $\mathrm{SL}_2(O_K/pO_K)$ is surjective. Since $\mathrm{SL}_2(O_K/pO_K)$ acts transitively on $\mathbb{P}^1(O_K/pO_K)$, irreducibility follows. □

Theorem 5.2 (Nori). *Let π be a finitely generated subgroup of $\mathrm{GL}_n(\mathbb{Z})$, let H be the Zariski closure of π, and for p prime, let $\pi(p)$ be the image of π in $\mathrm{GL}_n(\mathbb{F}_p)$. Then, for almost all p, $\pi(p)$ contains the subgroup of $H(\mathbb{F}_p)$ that is generated by the elements of order p.*

6. Galois action.

The aim of this section is to show that the Galois orbits of special points in $S'(\overline{\mathbb{Q}})$ are big, in a suitable sense. For A and B abelian surfaces (over some field) with O_K-action, we let $\mathrm{Hom}_{O_K}(A,B)$ be the O_K-module of morphisms from A to B that are compatible with the O_K-actions.

Lemma 6.1. *Let x in $S'(\overline{\mathbb{Q}})$ be a special point, corresponding to a triplet (A,α,λ) with A an abelian surface over $\overline{\mathbb{Q}}$, $\alpha\colon O_K\to\mathrm{End}(A)$ and λ a principal O_K-polarization. Then $\mathrm{End}_{O_K}(A)$ is an order, containing O_K, of a totally imaginary quadratic extension of K.*

Proof. Let R be the endomorphism algebra $\mathbb{Q}\otimes\mathrm{End}(A)$ of A. Then R is a semi-simple $\mathbb{Q}$-algebra containing a commutative semi-simple subalgebra of dimension 4. Suppose that A is simple. Then R is a division algebra. Since R acts faithfully on $\mathrm{H}_1(A(\mathbb{C}),\mathbb{Q})$, it has dimension dividing 4, hence R is a quadratic extension of K. Since $\mathbb{R}\otimes R$ has a complex structure commuting with the R-action, R is a totally imaginary. Suppose now that A is not simple. Then A is isogenous to the product of two elliptic curves, B_1 and B_2, say. These elliptic curves are in fact isogenous to each other, because otherwise K does not admit a morphism to the endomorphism algebra of $B_1\times B_2$. So A is isogenous to B^2, with B some elliptic curve. Since A is of CM-type, $\mathbb{Q}\otimes\mathrm{End}(B)$ is an imaginary quadratic field E, and $R=\mathrm{M}_2(E)$. In this case $\mathrm{End}_{O_K}(A)$ is an order in the totally imaginary extension $K\otimes E$ of K. □

Theorem 6.2. *There exist real numbers $\varepsilon>0$ and $c>0$ such that for (A,α,λ) corresponding to a special point x in $S'(\overline{\mathbb{Q}})$ one has:*

$$|\mathrm{Gal}(\overline{\mathbb{Q}}/\mathbb{Q})\cdot x| > c\,|\mathrm{discr}(R_x)|^{\varepsilon},$$

where $R_x=\mathrm{End}_{O_K}(A)$.

Remark 6.3. *The proof will show that one can take ε to be any number less than 1/4. (To get this, one also has to optimize Theorem 6.4, noting that we only apply Stark's result to fields L of degree at least 4.) Assuming the generalized Riemann hypothesis at this point does not improve this exponent (this is caused by the case where $\mathbb{Q}\otimes R_x$ is Galois over $\mathbb{Q}$ with group $(\mathbb{Z}/2\mathbb{Z})^2$).*

Proof. Let $f\colon S'_{\mathbb{Q}}\to S_{\mathbb{Q}}$ be the morphism induced by the closed immersion of the Shimura data $(G'_{\mathbb{Q}},X')\to(G_{\mathbb{Q}},X)$. Since f is finite, and since the Hecke correspondences on $S_{\mathbb{Q}}$ permute the irreducible components transitively, the statement we want to prove is equivalent to its analog for $S_{\mathbb{Q}}$. So we will show in fact that there are positive ε and c such that for x special in $S(\overline{\mathbb{Q}})$ corresponding to (A,α), we have:

$$|\mathrm{Gal}(\overline{\mathbb{Q}}/\mathbb{Q})\cdot x| > c\,|\mathrm{discr}(R_x)|^{\varepsilon}.$$

For x special in $S(\overline{\mathbb{Q}})$, let L_x be $\mathbb{Q}\otimes R_x$, and let M_x be the Galois closure in $\overline{\mathbb{Q}}$ of L_x. Since M_x is of degree at most 8 over $\mathbb{Q}$, the statement we want to prove is equivalent to the existence of positive ε and c such that for all special x in $S(\overline{\mathbb{Q}})$:

$$|\mathrm{Gal}(\overline{\mathbb{Q}}/M_x)\cdot x| > c\,|\mathrm{discr}(R_x)|^{\varepsilon}.$$

So let now x be special in $S(\overline{\mathbb{Q}})$, corresponding to some (A_x, α_x). To study the $\mathrm{Gal}(\overline{\mathbb{Q}}/M_x)$-orbit of x, we construct a zero-dimensional subvariety of Hodge type over M_x, containing x, and we use the theory of Shimura-Taniyama on complex multiplication, rephrased in the language of Shimura varieties (see [20, Section 2.2]). Let H_x be $\mathrm{H}_1(A(\mathbb{C}), \mathbb{Z})$; it is an R_x-module that is locally free of rank two as O_K-module. Let $f\colon K^2 \to \mathbb{Q}\otimes H_x$ be an isomorphism of K-vector spaces. The Hodge structure on H_x given by A_x gives an element h_x of X. The lattice $f^{-1}H_x$ in K^2 corresponds to an element $\overline{g_x}$ of $G(\mathbb{A}_{\mathrm{f}})/G(\hat{\mathbb{Z}})$. By construction, x is the image of (h_x, g_x). Let $T_x := \mathrm{Res}_{L_x/\mathbb{Q}}\mathbb{G}_{\mathrm{m},L_x}$. Then f gives a closed immersion $T_x \to G_{\mathbb{Q}}$. Since $\mathbb{Q}\otimes H_x$ is a one-dimensional L_x-vector space, T_x is its own centralizer in $G_{\mathbb{Q}}$. It follows that h_x factors through $T_{x,\mathbb{R}}$. Hence we have a closed immersion of Shimura data: $(T_x, \{h_x\}) \to (G_{\mathbb{Q}}, X)$. The reflex field of $(T_x, \{h_x\})$ is contained in M_x, hence we have a canonical model $\mathrm{Sh}(T_x, \{h_x\})_{M_x}$ over M_x. We put $U_x := T_x(\mathbb{A}_{\mathrm{f}}) \cap g_x G(\hat{\mathbb{Z}}) g_x^{-1}$. Then one easily verifies that we have an injective morphism of Shimura varieties $\mathrm{Sh}_{U_x}(T_x, \{h_x\})_{M_x} \to S_{M_x}$, which, on $\mathbb{C}$-valued points, is given by $\overline{t} \mapsto \overline{(h_x, tg_x)}$. By construction, U_x is the stabilizer in $T_x(\mathbb{A}_{\mathrm{f}})$ of the lattice $f^{-1}H_x$; it follows that $U_x = R_x^{\wedge,*}$, hence:

$$\mathrm{Sh}_{U_x}(T_x, \{h_x\})_{M_x}(\overline{\mathbb{Q}}) = L_x^*\backslash(\mathbb{A}_{\mathrm{f}} \otimes L_x)^*/R_x^{\wedge,*} = \mathrm{Pic}(R_x).$$

Our next objective is to describe in sufficient detail the action of $\mathrm{Gal}(\overline{\mathbb{Q}}/M_x)$ on $\mathrm{Pic}(R_x)$ induced by the above bijections. Class field theory gives a continuous surjection from $M_x^*\backslash\mathbb{A}_{M_x,\mathrm{f}}^*$ to $\mathrm{Gal}(\overline{\mathbb{Q}}/M_x)^{\mathrm{ab}}$, characterized by the following property. In a representation of $\mathrm{Gal}(\overline{\mathbb{Q}}/M_x)^{\mathrm{ab}}$ that is unramified at a finite place v of M_x, the arithmetic Frobenius element is the image of the class of an idèle that is trivial at all places other than v, and the inverse of a uniformizer at v. Let $\mu\colon \mathbb{G}_{\mathrm{m},\mathbb{C}} \to \mathbb{S}_{\mathbb{C}}$ be the cocharacter obtained by composing $\mathbb{C}^* \to \mathbb{C}^* \times \mathbb{C}^*$, $z \mapsto (z, 1)$ with the inverse of the isomorphism $\mathbb{S}(\mathbb{C}) = (\mathbb{C}\otimes_{\mathbb{R}}\mathbb{C})^* \to \mathbb{C}^*\times\mathbb{C}^*$, $x\otimes y \mapsto (xy, x\overline{y})$. Then $h_x \circ \mu$ is defined over M_x, and one defines:

$$r_x\colon T'_x := \mathrm{Res}_{M_x/\mathbb{Q}}\mathbb{G}_{\mathrm{m},M_x} \to T_x$$

to be the morphism $\mathrm{Res}_{M_x/\mathbb{Q}}(h_x \circ \mu)$ composed with the norm map from $\mathrm{Res}_{M_x/\mathbb{Q}}T_{x,M_x}$ to T_x. With these definitions, the quotient $\mathrm{Gal}(\overline{\mathbb{Q}}/M_x)^{\mathrm{ab}}$ of $T'_x(\mathbb{A}_{\mathrm{f}})$ acts on $\mathrm{Pic}(R_x)$ via the morphism r_x, where we view $\mathrm{Pic}(R_x)$ as $T_x(\mathbb{Q})\backslash T_x(\mathbb{A}_{\mathrm{f}})/R_x^{\wedge,*}$. It follows that:

$$|\mathrm{Gal}(\overline{\mathbb{Q}}/M_x)\cdot x| = |\text{image of } r_x(T'_x(\mathbb{A}_{\mathrm{f}})) \text{ in } \mathrm{Pic}(R_x)|.$$

We will need a more explicit description of r_x, in terms of the CM type associated to h_x. The morphism $h_x\colon \mathbb{C}^* \to (\mathbb{R}\otimes L_x)^*$ extends to a morphism of $\mathbb{R}$-algebras $h\colon \mathbb{C} \to \mathbb{R}\otimes L_x$. Extending scalars from $\mathbb{R}$ to $\mathbb{C}$ gives a morphism of $\mathbb{C}$-algebras $\mathrm{id}\otimes h\colon \mathbb{C}\otimes_{\mathbb{R}}\mathbb{C} \to \mathbb{C}\otimes L_x$. Via the isomorphisms:

$$\mathbb{C}\otimes_{\mathbb{R}}\mathbb{C} \to \mathbb{C}\times\mathbb{C}, \quad x\otimes y \mapsto (xy, x\overline{y}),$$

and

$$\mathbb{C} \otimes L_x \to \mathbb{C}^{\mathrm{Hom}(L_x,\mathbb{C})}, \quad x \otimes y \mapsto (\phi \mapsto x\phi(y)),$$

the idempotent $(1,0)$ of $\mathbb{C} \times \mathbb{C}$ gives an idempotent in $\mathbb{C}^{\mathrm{Hom}(L_x,\mathbb{C})}$, i.e., a partition of $\mathrm{Hom}(L_x, \mathbb{C})$ into two sets Φ_x and $\iota\Phi_x$, where ι is the complex conjugation on $\mathbb{C}$. The set Φ_x is the CM type corresponding to h_x. Since M_x is the Galois closure of L_x in $\mathbb{C}$, $\mathrm{Hom}(L_x, M_x) = \mathrm{Hom}(L_x, \mathbb{C})$. With these notations, we have, for any $\mathbb{Q}$-algebra R:

$$r_x \colon (R \otimes M_x)^* \longrightarrow \prod_{\phi\in\Phi_x} (R \otimes L_x)^*, \quad u \longmapsto \prod_{\phi\in\Phi_x} \mathrm{Norm}_\phi(u),$$

where Norm_ϕ is the norm map of the extension $\phi\colon R \otimes L_x \to R \otimes M_x$. Finally, let ϕ_0 be in $\mathrm{Hom}(L_x, M_x)$, and define $\Sigma_{x,\phi_0} := \{g \in \mathrm{Gal}(M_x/\mathbb{Q}) \mid g\phi_0 \in \Phi_x\}$. Then we have:

$$\phi_0 \circ r_x \colon T'_x \longrightarrow T_x \hookrightarrow T'_x, \quad u \mapsto \prod_{g\in\Sigma_{x,\phi_0}} g^{-1}u,$$

for all $\mathbb{Q}$-algebras R and all u in $(R \otimes M_x)^*$. This is the description of r_x that we work with.

Since M_x is generated over K by the extension L_x and its conjugate, M_x has degree 4 or 8 over $\mathbb{Q}$, and its Galois group $\mathrm{Gal}(M_x/\mathbb{Q})$ is isomorphic to $\mathbb{Z}/4\mathbb{Z}$, $\mathbb{Z}/2\mathbb{Z} \times \mathbb{Z}/2\mathbb{Z}$, or D_4, the dihedral group of order 8. We define T to be $\mathrm{Res}_{K/\mathbb{Q}}\mathbb{G}_{\mathrm{m},K}$; note that T is a subtorus of T_x, equal to the center of $G_\mathbb{Q}$. We will see below that $r_x \circ \phi_0 \colon T_x \to T_x$ induces an endomorphism of T_x/T whose image, after passing to $\mathbb{A}_\mathrm{f}$-valued points, in $\mathrm{Pic}(R_x)/\mathrm{Pic}(O_K)$ is big enough for our purposes.

Suppose first that $\mathrm{Gal}(M_x/\mathbb{Q})$ is isomorphic to $\mathbb{Z}/4\mathbb{Z}$, say with generator σ. Then $M_x = L_x$, σ^2 is the complex conjugation and $K = L_x^{\langle\sigma^2\rangle}$. After changing ϕ_0, if necessary, one has that $\Sigma_{x,\phi} = \{1, \sigma\}$. The formula above for $\phi_0 \circ r_x$ shows that r_x is simply given by the element $1+\sigma^{-1}$ of $\mathbb{Z}[\mathrm{Gal}(M_x/\mathbb{Q})]$. Since σ^2 acts as -1 on T_x/T, we have $r_x \circ (1 - \sigma^{-1}) = 2$ on T_x/T. It follows that:

$$|\mathrm{Gal}(\overline{\mathbb{Q}}/M_x){\cdot}x| \geq |\text{image of } {\cdot}2\colon \mathrm{Pic}(R_x)/\mathrm{Pic}(O_K) \to \mathrm{Pic}(R_x)/\mathrm{Pic}(O_K)|.$$

Theorem 6.4 below finishes the proof in this case.

Suppose now that $\mathrm{Gal}(M_x/\mathbb{Q})$ is isomorphic to $\mathbb{Z}/2\mathbb{Z} \times \mathbb{Z}/2\mathbb{Z}$. After changing ϕ_0, if necessary, one has $\Sigma_{x,\phi_0} = \{1, \sigma\}$, with σ of order two and $K_x := L_x^{\langle\sigma\rangle} \neq K$. Let R'_x be the order $O_{K_x} \cap R_x$ of K_x. Since r_x is given by $1 + \sigma$, the induced map $T'_x(\mathbb{A}_\mathrm{f}) \to \mathrm{Pic}(R_x)$ factors through $\mathrm{Pic}(R'_x) \to \mathrm{Pic}(R_x)$ induced by the inclusion $R'_x \to R_x$. The fact that σ acts as 1 on $T''_x := \mathrm{Res}_{K_x/\mathbb{Q}}\mathbb{G}_{\mathrm{m},K_x}$ and as -1 on T_x/T''_x implies that the kernel of the map $\mathrm{Pic}(R'_x) \to \mathrm{Pic}(R_x)$ is killed by multiplication by 2. Since $1 + \sigma$ acts as multiplication by 2 on T''_x, we get:

$$|\mathrm{Gal}(\overline{\mathbb{Q}}/M_x){\cdot}x| \geq |\text{image of } {\cdot}4\colon \mathrm{Pic}(R'_x) \to \mathrm{Pic}(R'_x)|.$$

Since the order $O_K \otimes R'_x$ of L_x is contained in R_x, and has discriminant $\mathrm{discr}(R'_x)^2\mathrm{discr}(O_K)^2$, we have:

$$|\mathrm{discr}(R'_x)| \geq |\mathrm{discr}(O_K)|^{-1}|\mathrm{discr}(R_x)|^{1/2}.$$

The proof in this case is finished by Theorem 6.4.

Suppose that $\mathrm{Gal}(M_x/\mathbb{Q})$ is isomorphic to D_4. Let τ and σ be generators of $\mathrm{Gal}(M_x/\mathbb{Q})$, with $M_x^{\langle\tau\rangle} = L_x$, and with σ of order 4. Then σ^2 is the complex conjugation, and $\tau\sigma^{-1} = \sigma\tau$. After changing ϕ_0, if necessary, we have $\Sigma_{x,\phi_0} = \{1, \tau, \sigma, \sigma\tau\}$. It follows that $\phi_0 \circ r_x$ is given by the element $t := 1 + \tau + \sigma^3 + \sigma\tau$ of $\mathbb{Z}[\mathrm{Gal}(M_x/\mathbb{Q})]$. Using that $\mathrm{Norm}_{\phi_0} = 1 + \tau$, a simple computation gives:

$$\phi_0 \circ r_x \circ \phi_0 \circ \mathrm{Norm}_{\phi_0} = 2(1+\tau) + \sigma \mathrm{Norm}_{M_x/K}.$$

It follows that $r_x \circ \phi_0$ acts as 2 on T_x/T. We conclude that:

$$|\mathrm{Gal}(\overline{\mathbb{Q}}/M_x){\cdot}x| \geq |\text{image of } {\cdot}2\colon \mathrm{Pic}(R_x)/\mathrm{Pic}(O_K) \to \mathrm{Pic}(R_x)/\mathrm{Pic}(O_K)|.$$

Theorem 6.4 finishes the proof in this last case. □

Theorem 6.4. *Let K be a totally real number field. There exists $c > 0$ such that for all orders R, containing O_K, in totally complex quadratic extensions L of K, one has:*

$$|\textit{image of } {\cdot}4 \textit{ on } \mathrm{Pic}(R)/\mathrm{Pic}(O_K)| \geq c\,|\mathrm{discr}(R)|^{1/8}.$$

If one assumes the generalized Riemann hypothesis, then one can replace the exponent $1/8$ by any number less than $1/2$.

Proof. We will use the following lower bound for class numbers:

let K be a totally real number field; there exists $c > 0$ such that for all totally complex quadratic extensions L of K, one has:

$$|\mathrm{Pic}(O_L)| \geq c\,|\mathrm{discr}(O_L)|^{1/6}.$$

In order to prove this, one distinguishes two cases: $K = \mathbb{Q}$ and $K \neq \mathbb{Q}$, and one notes that the regulator $\mathrm{Reg}(O_L)$ is at most $\mathrm{Reg}(O_K)$. In the case $K \neq \mathbb{Q}$ one uses the following consequence of Stark's Theorem 2 in [30]:

for K a totally real number field, there exists $c > 0$ such that for all totally complex quadratic extensions L of K, one has:

$$|\mathrm{Pic}(O_L)| \geq c\,|\mathrm{discr}(O_L)|^{1/2-1/[L:\mathbb{Q}]}.$$

In the case $K = \mathbb{Q}$ one applies the Brauer-Siegel theorem (see for example [18, Ch. XVI]):

for $N > 0$ and $\varepsilon > 0$, there exists $c > 0$ such that for all Galois extensions L of $\mathbb{Q}$ of degree at most N one has:

$$|\mathrm{Pic}(O_L)|{\cdot}\mathrm{Reg}(O_L) \geq c\,|\mathrm{discr}(O_L)|^{1/2-\varepsilon}.$$

Combining these two results, and using that $[L:\mathbb{Q}] \geq 3$ if $K \neq \mathbb{Q}$ gives the inequality we want. We could replace the exponent $1/6$ by $1/4$ if we would just use that $[L:\mathbb{Q}] \geq 4$ if $K \neq \mathbb{Q}$.

Let now K, R and L be as in the theorem. Then R is the inverse image of a subring $\overline{R}$ of some finite quotient $\overline{O_L}$ of O_L. We have an exact sequence:

$$0 \longrightarrow R^* \longrightarrow O_L^* \longrightarrow \overline{O_L}^*/\overline{R}^* \longrightarrow \mathrm{Pic}(R) \longrightarrow \mathrm{Pic}(O_L) \longrightarrow 0.$$

The torsion of O_L^* is bounded in terms of the degree of K, and by Dirichlet's theorem on units the quotient O_L^*/O_K^* is finite. The long exact cohomology sequence obtained by taking $\mathrm{Gal}(L/K)$-invariants of the short exact sequence:

$$0 \longrightarrow \mathrm{tors}(O_L^*) \longrightarrow O_L^* \longrightarrow O_L^*/\mathrm{tors}(O_L^*) \longrightarrow 0$$

gives an injection from $(O_L^*/\mathrm{tors}(O_L^*))/(O_K^*/\mathrm{tors}(O_K^*))$ into

$$\mathrm{H}^1(\mathrm{Gal}(L/K), \mathrm{tors}(O_L^*)),$$

showing that $(O_L^*/\mathrm{tors}(O_L^*))/(O_K^*/\mathrm{tors}(O_K^*))$ is of order at most two. We conclude that there exists $c > 0$, depending only on the degree of K, such that:

$$|\mathrm{Pic}(R)| \geq c \frac{|\overline{O_L}^*|}{|\overline{R}^*|} |\mathrm{Pic}(O_L)|.$$

On the other hand, we have:

$$\mathrm{discr}(R) = \left(\frac{|\overline{O_L}|}{|\overline{R}|}\right)^2 \mathrm{discr}(O_L).$$

We claim that for every $\varepsilon > 0$ there exists $c > 0$, depending only on the degree of K, such that:

$$\frac{|\overline{O_L}^*|}{|\overline{R}^*|} \geq c \left(\frac{|\overline{O_L}|}{|\overline{R}|}\right)^{1-\varepsilon}.$$

To prove this claim, one notes that:

$$\frac{|\overline{R}|}{|\overline{R}^*|} = \prod_{k \text{ res field of } \overline{R}} \frac{|k|}{|k^*|}, \quad \text{and} \quad \frac{|\overline{O_L}|}{|\overline{O_L}^*|} = \prod_{k \text{ res field of } \overline{O_L}} \frac{|k|}{|k^*|}.$$

A simple computation then shows:

$$\frac{|\overline{O_L}^*|}{|\overline{R}^*|} \geq n \prod_{p|n} \left(1 - \frac{1}{p}\right)^{[L:\mathbb{Q}]} \geq n \left(\frac{1}{5\log(n)}\right)^{[L:\mathbb{Q}]},$$

where $n = |\overline{O_L}|/|\overline{R}|$ is assumed to be at least 2. We conclude that there exists $c > 0$, depending only on K, such that:

$$|\mathrm{Pic}(R)| \geq c\,|\mathrm{discr}(R)|^{1/7}.$$

In order to finish the proof of the theorem, it suffices to prove that for every $\varepsilon > 0$ there exists $c > 0$, depending only on K, such that

$$|\mathrm{Pic}(R)[2]| \leq c\,|\mathrm{discr}(R)|^{\varepsilon}.$$

To do this, we proceed in the same way as we did in [14, Lemma 3.4]. As $\mathrm{Pic}(R)$ is a finite commutative group, the two $\mathbb{F}_2$-vector spaces $\mathrm{Pic}(R)[2]$ and $\mathbb{F}_2 \otimes \mathrm{Pic}(R)$ have the same dimension. The cover of $\mathrm{Spec}(R)$ by the disjoint union of $\mathrm{Spec}(\mathbb{Z}_2 \otimes R)$ and $\mathrm{Spec}(R[1/2])$ gives an exact sequence:

$$(\mathbb{Q}_2 \otimes L)^* \longrightarrow \mathrm{Pic}(R) \longrightarrow \mathrm{Pic}(R[1/2]) \longrightarrow 0.$$

It follows that $\dim_{\mathbb{F}_2} \mathbb{F}_2 \otimes \mathrm{Pic}(R)$ is bounded by $\dim_{\mathbb{F}_2} \mathbb{F}_2 \otimes \mathrm{Pic}(R[1/2])$ plus a number depending only on the degree of K. We put $S := \mathrm{Spec}(R[1/2])$ and $T := \mathrm{Spec}(O_K[1/2])$. The long exact sequence associated to the multiplication by two on the sheaf $\mathbb{G}_{\mathrm{m}}$ on the etale site S_{et} of S shows that $\dim_{\mathbb{F}_2} \mathbb{F}_2 \otimes \mathrm{Pic}(S)$ is at most $\dim_{\mathbb{F}_2} \mathrm{H}^1(S_{\mathrm{et}}, \mathbb{F}_2)$. Let $\pi\colon S \to T$ be the morphism induced by the inclusion of O_K in R. Then $\mathrm{H}^1(S_{\mathrm{et}}, \mathbb{F}_2)$ is the same as $\mathrm{H}^1(T_{\mathrm{et}}, \pi_*\mathbb{F}_2)$, and we have a short exact sequence:

$$0 \longrightarrow \mathbb{F}_{2,S_{\mathrm{et}}} \longrightarrow \pi_*\pi^*\mathbb{F}_{2,T} \longrightarrow j_!\mathbb{F}_{2,U} \longrightarrow 0,$$

where $j\colon U \to T$ is the maximal open immersion over which π is etale. Let $i\colon Z \to T$ be closed immersion giving the complement of U, with Z reduced. Then the long exact sequences of cohomology groups associated to the exact sequence above and to the exact sequence:

$$0 \longrightarrow j_!\mathbb{F}_{2,U_{\mathrm{et}}} \longrightarrow \mathbb{F}_{2,S_{\mathrm{et}}} \longrightarrow i_*\mathbb{F}_{2,Z_{\mathrm{et}}} \longrightarrow 0$$

show that there exists an integer c, depending only on K, such that:

$$\dim_{\mathbb{F}_2} \mathbb{F}_2 \otimes \mathrm{Pic}(R) \leq c + [K : \mathbb{Q}]\, |\{p \text{ prime dividing } \mathrm{discr}(R)\}| .$$

As $2^{|\{p|n\}|} = n^{o(1)}$, for $n \to \infty$, we have proved the first statement of the theorem. If one assumes GRH, then the Brauer-Siegel theorem as stated above is true without the condition that the extension $\mathbb{Q} \to L$ be Galois, see [18, XIII, §4]. □

7. Intersection numbers.

The aim of this section is to give a bound on intersections of subvarieties of Shimura varieties, provided that they are finite. In particular, we need to study the intersection of a subvariety with its images under Hecke correspondences. As our arguments work for general Shimura varieties, we give such a result in the general case. The main tool used in proving the result is the Baily-Borel compactification, together with its given ample line bundles. We start by recalling some properties of these line bundles, that follow immediately from the results in [7] (see also [6]).

Theorem 7.1. *Let (G, X) be a Shimura datum. For $K \subset G(\mathbb{A}_{\mathrm{f}})$ a compact open subgroup, let $S_K := \mathrm{Sh}_K(G,X)_{\mathbb{C}}$ the corresponding complex Shimura variety, and $\overline{S}_K$ its Baily-Borel compactification. For every such K, and for every sufficiently divisible positive integer n, the nth power of the line bundle of holomorphic forms of maximal degree of X descends to S_K, and extends uniquely to a very ample line bundle $\mathcal{L}_{K,n}$ on $\overline{S}_K$, such that, at the generic points of the* *boundary components of codimension one, it is given by nth powers of forms with logarithmic poles. Let* K_1 *and* K_2 *be compact open subgroups of* $G(\mathbb{A}_f)$, *and* g *in* $G(\mathbb{A}_f)$ *such that* $K_2 \subset gK_1g^{-1}$. *Then the morphism from* S_{K_2} *to* S_{K_1} *induced by* g *extends to a morphism* $f\colon \overline{S}_{K_2} \to \overline{S}_{K_1}$. *If* n *is positive and sufficiently divisible so that* $\mathcal{L}_{K_1,n}$ *exists, then* $\mathcal{L}_{K_2,n}$ *exists, and is canonically isomorphic to* $f^*\mathcal{L}_{K_1,n}$.

Proof. Let us briefly recall how the compactification $\overline{S}_K$ is defined. Let X^+ be a connected component of X. Then each connected component of $S_K(\mathbb{C})$ is of the form $\Gamma_i\backslash X^+$, with Γ_i an arithmetic subgroup of $G^{\mathrm{ad}}(\mathbb{Q})$ (G^{ad} being the quotient of G by its center). The compactification $\overline{S}_K(\mathbb{C})$ is then defined as the disjoint union of the $\Gamma_i\backslash\overline{X^+}$, where $\overline{X^+}$ is the union of X^+ with its so-called rational boundary components, endowed with the Satake topology. It follows that we can write $\overline{S}_K(\mathbb{C})$ as $G(\mathbb{Q})\backslash(\overline{X} \times G(\mathbb{A}_f)/K)$, with $\overline{X}$ the disjoint union of the $\overline{X^+}$.

Let X^{ad} be the $G^{\mathrm{ad}}(\mathbb{R})$-conjugacy class of morphisms from $\mathbb{S}$ to $G^{\mathrm{ad}}_{\mathbb{R}}$ containing the image of X. Each connected component of X maps isomorphically to one of X^{ad} (see [20, 1.6.7]). We first prove the Theorem above for the Shimura datum $(G^{\mathrm{ad}}, X^{\mathrm{ad}})$. The group G^{ad} is a product of simple algebraic groups G_j over $\mathbb{Q}$, and X^{ad} decomposes as a product of X_j's. For compact open subgroups K, K_1 and K_2 that are products of compact open subgroups of the $G_j(\mathbb{A}_f)$, the corresponding Shimura varieties decompose as a product, so that it suffices to treat the G_j separately. If (G_j, X_j) gives compact Shimura varieties, Kodaira's theorem ([16, Section 1.4]) implies what we want, for compact open subgroups K_j that are sufficiently small; for arbitrary K_j one takes quotients by finite groups. Suppose now that (G_j, X_j) does give Shimura varieties that are not compact. If G_j is of dimension 3, then it is isomorphic to $\mathrm{PGL}_{2,\mathbb{Q}}$, and we are in the case of modular curves, where the Theorem we are proving is well known (the canonical line bundle with log poles at the cusps on the modular curve $X(n)$, $n \geq 3$, has degree > 0). Suppose now that G_j has dimension > 3. Then the boundary components are of codimension > 1, and the results we want are given in [7, Thm. 10.11].

The case of arbitrary open compact subgroups of $G^{\mathrm{ad}}(\mathbb{A}_f)$ follows by considering quotients by finite groups. The theorem for (G, X) itself follows from the fact that the connected components of the $S_K(\mathbb{C})$ are of the $\Gamma\backslash X^+$, with Γ an arithmetic subgroup of $G^{\mathrm{ad}}(\mathbb{Q})$. □

Theorem 7.2. *Let* (G, X) *be a Shimura datum, let* K_1 *and* K_2 *be compact open subgroups of* $G(\mathbb{A}_f)$, *and let* Z_1 *and* Z_2 *be closed subvarieties of the Shimura varieties* $S_1 := \mathrm{Sh}_{K_1}(G, X)_{\mathbb{C}}$ *and* $S_2 := \mathrm{Sh}_{K_2}(G, X)_{\mathbb{C}}$, *respectively. Suppose that* Z_1 *or* Z_2 *is of dimension at most one. Then there exists an integer* c *such that for all* g *in* $G(\mathbb{A}_f)$ *for which* $T_gZ_1 \cap Z_2$ *is finite, one has:*

$$|T_gZ_1 \cap Z_2| \leq c \deg(\overline{\pi_1}\colon S_g \to S_1),$$

where $S_g = \mathrm{Sh}_{K_g}(G,X)_{\mathbb{C}}$ with $K_g = K_1 \cap gK_2g^{-1}$, and with T_g and $\overline{\pi_1}$ as in the beginning of Section 5.

Proof. We start with two reductions. First of all, writing Z_1 and Z_2 as the unions of their irreducible components, one sees that we may suppose that Z_1 and Z_2 are irreducible. Secondly, for g in $G(\mathbb{A}_f)$, let $p_{1,g}$ and $p_{2,g}$ be the morphisms from S_g to S_1 and S_2, respectively. Then one has:

$$T_g Z_1 \cap Z_2 = p_{2,g}\left(p_{1,g}^{-1} Z_1 \cap p_{2,g}^{-1} Z_2\right),$$

which shows that $T_g Z_1 \cap Z_2$ is finite if and only if $p_{1,g}^{-1} Z_1 \cap p_{2,g}^{-1} Z_2$ is, and that $|T_g Z_1 \cap Z_2|$ is at most $|p_{1,g}^{-1} Z_1 \cap p_{2,g}^{-1} Z_2|$. This also shows that we may replace K_1 and K_2 by smaller compact open subgroups. Hence we may suppose, by the previous theorem, that we have very ample line bundles $\mathcal{L}_1$ and $\mathcal{L}_2$ on the Baily-Borel compactifications $\overline{S_1}$ and $\overline{S_2}$ such that, for each g, $\overline{p_{1,g}}^* \mathcal{L}_1$ and $\overline{p_{2,g}}^* \mathcal{L}_2$ are isomorphic to the same line bundle $\mathcal{L}_g$ on $\overline{S_g}$.

We let $\overline{Z_1}$ and $\overline{Z_2}$ be the closures of Z_1 and Z_2 in $\overline{S_1}$ and $\overline{S_2}$, respectively. Let m denote the degree of $\overline{Z_2}$ with respect to $\mathcal{L}_2$. Let g be in $G(\mathbb{A}_f)$, such that $T_g Z_1 \cap Z_2$ is finite. If the intersection is empty, there is nothing to prove, so we suppose that the intersection is not empty. Then the codimension of Z_2 is at least the dimension d of Z_1, and we can choose $f_1, \ldots, f_d$ in $\mathrm{H}^0(\overline{S_2}, \mathcal{L}_2^{\otimes m})$ such that $\overline{Z_2}$ is contained in $V_{S_2}(f_1, \ldots, f_d)$, and $\overline{T_g Z_1} \cap V_{S_2}(f_1, \ldots, f_d)$ is finite (because of our assumption on the dimensions of Z_1 and Z_2, $\overline{Z_1} \cap \overline{Z_2}$ is finite). It then follows that $|\overline{p_{1,g}^{-1} Z_1} \cap \overline{p_{2,g}^{-1} Z_2}|$ is at most m^d times the degree of $\overline{p_{1,g}^{-1} Z_1}$ with respect to $\mathcal{L}_g$. But this degree is $\deg(p_{1,g})$ times the degree of $\overline{Z_1}$ with respect to $\mathcal{L}_1$, hence we have:

$$|T_g Z_1 \cap Z_2| \leq \deg(p_{1,g}) m^d \deg_{\mathcal{L}_1}(\overline{Z_1}).$$

This finishes the proof. □

8. Proof of the main results.

We will now prove Theorems 2.1 and 2.2. We first deal with Thm. 2.1. As we have already noticed, we may as well replace $S_{\mathbb{C}}$ by $S'_{\mathbb{C}}$, so let C be an irreducible closed curve in $S'_{\mathbb{C}}$ that contains infinitely many CM points. We have to show that C is of Hodge type.

Since C has infinitely many points in $S'(\overline{\mathbb{Q}})$, it is, as a reduced closed subscheme of $S'_{\mathbb{C}}$, defined over $\overline{\mathbb{Q}}$. To be precise, there is a unique closed subscheme $C_{\overline{\mathbb{Q}}}$ of $S'_{\overline{\mathbb{Q}}}$ that gives C after base change from $\overline{\mathbb{Q}}$ to $\mathbb{C}$. But then $C_{\overline{\mathbb{Q}}}$ has only finitely many conjugates under $\mathrm{Gal}(\overline{\mathbb{Q}}/\mathbb{Q})$; we let $C_{\mathbb{Q}}$ be the reduced closed subscheme of $S'_{\mathbb{Q}}$ that, after base change to $\overline{\mathbb{Q}}$, gives the union of these conjugates. In other words, we simply let $C_{\mathbb{Q}}$ be the image of C under the morphism of schemes $S'_{\mathbb{C}} \to S'_{\mathbb{Q}}$.

Let x in $C(\overline{\mathbb{Q}})$ be a CM point, corresponding to a pair (A,λ) with A an abelian variety over $\overline{\mathbb{Q}}$ with multiplications by O_K and with λ a principal O_K-polarization. As before, we let R_x denote $\mathrm{End}_{O_K}(A)$, $L_x := \mathbb{Q}\otimes R_x$ and M_x the Galois closure of L_x in $\mathbb{C}$. Let $\overline{x}$ be the image of x in $S(\overline{\mathbb{Q}})$. In the proof of Theorem 6.2 we have seen that the quotient $\mathrm{Gal}(\overline{\mathbb{Q}}/M_x)^{\mathrm{ab}}$ of $(\mathbb{A}_{\mathrm{f}}\otimes M_x)^*$ acts on the subset $\mathrm{Gal}(\overline{\mathbb{Q}}/M_x)\cdot\overline{x}$ of $L_x^*\backslash(\mathbb{A}_{\mathrm{f}}\otimes L_x)^*/R_x^{\wedge,*}$ via the morphism

$$r_x\colon (\mathbb{A}_{\mathrm{f}}\otimes M_x)^* \longrightarrow (\mathbb{A}_{\mathrm{f}}\otimes L_x)^*, \quad u\mapsto \prod_{\phi\in\Phi_x}\mathrm{Norm}_\phi(u),$$

where the subset Φ_x of $\mathrm{Hom}(L_x,M_x)$ is the CM type of x. Since Φ_x is a set of representatives for the action of $\mathrm{Gal}(L_x/K)$ on $\mathrm{Hom}(L_x,M_x)$, it follows that this map r_x factors through the subgroup of elements of $(\mathbb{A}_{\mathrm{f}}\otimes L_x)^*$ whose norm to $(\mathbb{A}_{\mathrm{f}}\otimes K)^*$ is in $\mathbb{A}_{\mathrm{f}}^*$. Hence, in the notation of the proof of Theorem 6.2, the morphism r_x factors through the intersection of the subtorus T_x of $G_{\mathbb{Q}}$ and $G'_{\mathbb{Q}}$. It follows that the action of $\mathrm{Gal}(\overline{\mathbb{Q}}/M_x)$ on $\mathrm{Gal}(\overline{\mathbb{Q}}/M_x)\cdot x$ is given by the same morphism r_x, taking values in $G'(\mathbb{A}_{\mathrm{f}})$.

Lemma 8.1. *Suppose that p is a prime that is split in R_x, i.e., such that $\mathbb{F}_p\otimes R_x$ is isomorphic to a product of copies of $\mathbb{F}_p$. Then* $\mathrm{Gal}(\overline{\mathbb{Q}}/\mathbb{Q})x$ *is contained in* $C_{\mathbb{Q}}(\overline{\mathbb{Q}})\cap(T_pC_{\mathbb{Q}})(\overline{\mathbb{Q}})$.

Proof. The localization $\mathbb{Z}_{(p)}\otimes R_x$ of R_x is the same as that of O_K. Hence, if we let H_x denote $\mathrm{H}_1(A(\mathbb{C}),\mathbb{Z})$, then $\mathbb{Z}_{(p)}\otimes H_x$ is free of rank one over $\mathbb{Z}_{(p)}\otimes R_x$. It follows that we can choose the isomorphism $f\colon K^2\to\mathbb{Q}\otimes H_x$ to preserve the integral structures on both sides at p, i.e., such that it induces an isomorphism from $(\mathbb{Z}_{(p)}\otimes R_x)^2$ to $\mathbb{Z}_{(p)}\otimes H_x$. We note that p is split in M_x (i.e., $\mathbb{F}_p\otimes O_{M_x}$ is a product of copies of $\mathbb{F}_p$), because M_x is the Galois closure of L_x. Consider now an element u of $(\mathbb{A}_{\mathrm{f}}\otimes M_x)^*$ that is equal to p at one place above p and equal to 1 at all other finite places of M_x. Then $r_x(u)$, viewed as an element of $(\mathbb{A}_{\mathrm{f}}\otimes L_x)^*$, is equal to p at exactly two places of L_x above p that are not in the same $\mathrm{Gal}(L_x/K)$-orbit, and equal to 1 at all other finite places of L_x. It follows that $r_x(u)$ is conjugated in $G'(\mathbb{A}_{\mathrm{f}})$, by some element in $G'(\hat{\mathbb{Z}})$, to the element $g(p)$ that induces T_p (use that $G'(\mathbb{Z}_p)$ acts transitively on the set of free rank one $\mathbb{Z}_p\otimes O_K$-submodules of $\mathbb{Z}_p\otimes O_K^2$). We conclude that $r_x(u)x$ is in T_px. But since x is in $C_{\mathbb{Q}}(\overline{\mathbb{Q}})$, $r_x(u)x$ is also in $C_{\mathbb{Q}}(\overline{\mathbb{Q}})$. It follows that $\mathrm{Gal}(\overline{\mathbb{Q}}/\mathbb{Q})x$ is contained in $C_{\mathbb{Q}}(\overline{\mathbb{Q}})\cap(T_pC_{\mathbb{Q}})(\overline{\mathbb{Q}})$. □

Theorem 6.2 gives a lower bound for $|\mathrm{Gal}(\overline{\mathbb{Q}}/\mathbb{Q})x|$, whereas Theorem 7.2 gives an upper bound for $|C_{\mathbb{Q}}(\overline{\mathbb{Q}})\cap(T_pC_{\mathbb{Q}})(\overline{\mathbb{Q}})|$, assuming that the intersection is finite. What we want, of course, is to show that we can choose x and then p such that the lower bound exceeds the upper bound, and conclude that $C_{\mathbb{Q}}$ and $T_pC_{\mathbb{Q}}$ do not intersect properly. We note that if x varies over the infinite set of CM points of $C(\overline{\mathbb{Q}})$, then $|\mathrm{discr}(R_x)|$ tends to infinity because there are only finitely many orders of degree 2 over O_K with a given discriminant, and for each such order there are only finitely many x in $S(\overline{\mathbb{Q}})$ with

R_x isomorphic to that order. Since our lower bound for $|\mathrm{Gal}(\overline{\mathbb{Q}}/\mathbb{Q})x|$ is a positive constant times a positive power of $|\mathrm{discr}(R_x)|$, and our upper bound for $|C_{\mathbb{Q}}(\overline{\mathbb{Q}}) \cap (T_pC_{\mathbb{Q}})(\overline{\mathbb{Q}})|$ is some fixed power of p, we get what we want if we can take, for $|\mathrm{discr}(R_x)|$ big, p of size something polynomial in $\log|\mathrm{discr}(R_x)|$. We note that $|\mathrm{discr}(O_{M_x})| \leq |\mathrm{discr}(R_x)|^4$ because M_x is the composite of the extension L_x of K and its conjugate. We also note that the number of primes dividing $\mathrm{discr}(R_x)$ is at most $\log_2(|\mathrm{discr}(R_x)|)$.

At this point we invoke the effective Chebotarev theorem of Lagarias, Montgomery and Odlyzko, assuming GRH, as stated in [29, Thm. 4] and the second remark following that theorem. A simple computation shows that this theorem implies the following result.

Proposition 8.2. *For M a finite Galois extension of $\mathbb{Q}$, let n_M denote its degree, d_M its absolute discriminant $|\mathrm{discr}(O_M)|$, and for x in $\mathbb{R}$, let $\pi_{M,1}(x)$ be the number of primes $p \leq x$ that are unramified in M and such that the Frobenius conjugacy class* Frob_p *contains just the identity element of* $\mathrm{Gal}(M/\mathbb{Q})$*. Then for M a finite Galois extension of $\mathbb{Q}$ and x sufficiently big (i.e., bigger than some absolute constant), and bigger than* $2(\log(d_M)^2(\log(\log(d_M)))^2$, *one has:*

$$\pi_{M,1}(x) \geq \frac{x}{3n_M \log(x)}.$$

This result shows that there exist infinitely many primes p such that $C_{\mathbb{Q}}$ and $T_pC_{\mathbb{Q}}$ do not intersect properly. Since $C_{\mathbb{Q}}$ is irreducible, it follows that, for such primes p, $C_{\mathbb{Q}}$ is contained in $T_pC_{\mathbb{Q}}$.

Assume now that C is not of Hodge type. Then Proposition 5.1 tells us that for all primes p large enough, T_pC is irreducible. Since the correspondence T_p is defined over $\mathbb{Q}$, i.e., is given by a correspondence on $S'_{\mathbb{Q}}$, it follows that $T_pC_{\mathbb{Q}}$ is irreducible for p large enough. But then we see that there exist infinitely many prime numbers p such that $C_{\mathbb{Q}}$ is equal to $T_pC_{\mathbb{Q}}$. But this is absurd, since by Lemma 8.3 below, for each x in $S'(\mathbb{C})$, the Hecke orbit $\cup_{n\geq 0}T_p^n x$ is dense in $S'(\mathbb{C})$ if p is unramified in K. This finishes the proof of Theorem 2.1.

Lemma 8.3. *Let x be in $S'(\mathbb{C})$ and let p be a prime number that is not ramified in K. Then the Hecke orbit $\cup_{n\geq 0}T_p^n x$ is dense in $S'(\mathbb{C})$ for the archimedean topology.*

Proof. By Lemma 8.4 below, $g_0 := \left(\begin{smallmatrix} p & 0 \\ 0 & 1 \end{smallmatrix}\right)$ and $G'(\mathbb{Z}_p)$ generate $G'(\mathbb{Q}_p)$ (here we use that p is not ramified in K). Let now x be in $S'(\mathbb{C})$, and let (y, g) be a preimage of it in $X \times G'(\mathbb{A}_{\mathrm{f}})$ under the quotient map for the action by $G(\mathbb{Q}) \times G(\hat{\mathbb{Z}})$. The fact that T_p is then given by right multiplication on $G'(\mathbb{A}_{\mathrm{f}})$ by the element g_0 at the place p shows that the T_p-orbit of x is the image in $S'(\mathbb{C})$ of the $G'(\mathbb{Q}_p)$-orbit of (y, g). Let now Γ be the subgroup of $G'(\mathbb{Q})$ consisting of γ such that γg is in $gG'(\hat{\mathbb{Z}})G'(\mathbb{Q}_p)$. Then the T_p-orbit of x is the image in $S'(\mathbb{C})$ of the subset Γy of $X \times \{g\}$. Now one notes that Γ contains a congruence subgroup of $G'(\mathbb{Z}[1/p])$. It follows that the intersection of

Γ with $\mathrm{SL}_2(O_K[1/p]) = G^{\mathrm{der}}(\mathbb{Z}[1/p])$ is dense in $G^{\mathrm{der}}(\mathbb{R})$ (for the archimedean topology) because G^{der} is generated by additive subgroups. Since $G^{\mathrm{der}}(\mathbb{R})$ acts transitively on $X^+ = (\mathbb{H}^+)^2$, the lemma is proved. $\square$

Lemma 8.4. *Let p be a prime that is not ramified in K. Then $g_0 := \left(\begin{smallmatrix} p & 0 \\ 0 & 1 \end{smallmatrix}\right)$ and $G'(\mathbb{Z}_p)$ generate $G'(\mathbb{Q}_p)$.*

Proof. In order to minimize notation, let $O_{K,p}$ denote $\mathbb{Z}_p \otimes O_K$, let K_p denote $\mathbb{Q}_p \otimes O_K$, and let H denote the subgroup of $G'(\mathbb{Q}_p)$ generated by $g_0 := \left(\begin{smallmatrix} p & 0 \\ 0 & 1 \end{smallmatrix}\right)$ and $G'(\mathbb{Z}_p)$. Let Y be the set of $O_{K,p}$-lattices in K_p^2 on which the K_p-bilinear form ψ given by $\left(\begin{smallmatrix} 0 & 1 \\ -1 & 0 \end{smallmatrix}\right)$ is a perfect pairing of $O_{K,p}$-modules, up to a factor in $\mathbb{Q}_p^*$. The map $G'(\mathbb{Q}_p) \to Y$, $g \mapsto gO_{K,p}^2$ induces a bijection from $G'(\mathbb{Q}_p)/G'(\mathbb{Z}_p)$ to Y. Hence, in order to prove our claim, it suffices to show that H acts transitively on Y. So let L be in Y. We note that $O_{K,p}$ is either a product of two copies of $\mathbb{Z}_p$, or the ring of integers $\mathbb{Z}_{p^2}$ in the unramified quadratic extension $\mathbb{Q}_{p^2}$ of $\mathbb{Q}_p$; in both cases, $O_{K,p}$ is a product of discrete valuation rings with uniformizer p. The theory of finitely generated modules over a discrete valuation ring says that there exists r in $\mathbb{Z}$ and d_1 and d_2 in $O_{K,p}$ such that $p^r L$ is contained in $O_{K,p}^2$ and has an $O_{K,p}$-basis of the form (d_1e_1, d_2e_2), with (e_1, e_2) the standard basis of $O_{K,p}^2$. We note that conjugating g_0 by suitable elements of $G'(\mathbb{Z}_p)$ shows that $\left(\begin{smallmatrix} 1 & 0 \\ 0 & p \end{smallmatrix}\right)$ and $\left(\begin{smallmatrix} p & 0 \\ 0 & p \end{smallmatrix}\right)$ are in H, and that, in the split case, $(\left(\begin{smallmatrix} 1 & 0 \\ 0 & p \end{smallmatrix}\right), \left(\begin{smallmatrix} p & 0 \\ 0 & 1 \end{smallmatrix}\right))$ is in H. Since the element d_1d_2 of $O_{K,p}$ is the factor by which ψ differs from a perfect pairing on p^rL, it is actually in $\mathbb{Z}_p$. It follows that $\left(\begin{smallmatrix} d_1 & 0 \\ 0 & d_2 \end{smallmatrix}\right)$ is in H. This finishes the proof that H is $G'(\mathbb{Q}_p)$. $\square$

Let us now prove Theorem 2.2. We keep the notations of the proof Theorem 2.1, and we assume again that C is not of Hodge type. So now we may suppose moreover that C contains infinitely many CM points that have the same CM type. In particular, we have infinitely many CM points x such that L_x and Φ_x are constant, say L and Φ. Of course, the orders R_x are such that $|\mathrm{discr}(R_x)|$ tends to infinity. The classical Chebotarev theorem (see for example [18, Ch. VIII, §4]) asserts that the set of primes p that are split in M has natural density $1/[M : \mathbb{Q}]$ (actually, Dirichlet density is good enough here). Also, recall that the number of primes dividing some $\mathrm{discr}(R_x)$ is at most $\log_2(|\mathrm{discr}(R_x)|)$. Hence there do exist x and p such that p is split in M, split in R_x, sufficiently large so that $T_pC_{\mathbb{Q}}$ is irreducible, and such that the lower bound for $|\mathrm{Gal}(\overline{\mathbb{Q}}/\mathbb{Q})x|$ of Theorem 6.2 exceeds the upper bound for the intersection $C_{\mathbb{Q}}(\overline{\mathbb{Q}}) \cap (T_pC_{\mathbb{Q}})(\overline{\mathbb{Q}})$, if it is finite. Then we have $C_{\mathbb{Q}} = T_pC_{\mathbb{Q}}$, hence a contradiction because of Lemma 8.3.

Acknowledgements. It is a pleasure to thank Rutger Noot, for teaching a very useful course on Shimura varieties, for answering my questions concerning them, and for his useful comments on this manuscript. Without the influence of Johan de Jong, Ben Moonen and Frans Oort I would not even have started working on this subject. I thank Paula Cohen for sending me a preliminary version of [11], and for pointing out to me which version of the André-Oort conjecture was exactly needed in it. Peter Stevenhagen gave me the reference to Stark's results in [30]. I thank Andrei Yafaev for useful remarks on this manuscript; his numerous questions on the subject have caused me to learn a good deal about Shimura varieties. I thank the organizers of the Texel conference for their excellent work. The referee deserves much credit for pointing out a serious mistake in Section 7 of the preprint version, and for a long list of detailed comments. Last but not least I thank my wife Reinie Erné for her influence on this article via conversations both at work and at home, and for letting me sleep when I am too tired to fetch a bottle for Tom at five o'clock in the morning.

References

[1] A. Abbes. *Hauteurs et discrétude (d'après L. Szpiro, E. Ullmo et S. Zhang).* Séminaire Bourbaki, Vol. 1996/97. Astérisque No. **245**, (1997), Exp. No. 825, 4, 141–166.

[2] Y. André. *G-functions and geometry.* Aspects of mathematics, Vol. **E13**, Vieweg, Braunschweig, 1989.

[3] Y. André. *Mumford-Tate groups of mixed Hodge structures and the theorem of the fixed part.* Compositio Mathematica **82** (1992), pp. 1–24.

[4] Y. André. *Distribution des points CM sur les sous-variétés des variétés de modules de variétés abéliennes.* Manuscript, April 1997.

[5] Y. André. *Finitude de couples d'invariants modulaires singuliers sur une courbe algébrique plane non modulaire.* J. Reine Angew. Math. **505** (1998), pp. 203–208.

[6] W.L. Baily and A. Borel. *On the compactification of arithmetically defined quotients of bounded symmetric domains.* Bull. Amer. Soc. **70** (1964), 588–593.

[7] W.L. Baily and A. Borel. *Compactification of arithmetic quotients of bounded symmetric domains.* Ann. Math. (2) **84** (1966), 442–528.

[8] J-L. Brylinski and J-P. Labesse. *Cohomologie d'intersection et fonctions L de certaines variétés de Shimura.* Ann. Sci. E.N.S. 4ème série, **17** (1984), 361–412.

[9] E. Cattani, P. Deligne and A. Kaplan. *On the locus of Hodge classes.* J. Amer. Math. Soc. **8** (1995), pp. 483–506.

[10] L. Clozel and E. Ullmo. Talk at the Journées Arithmétiques, July 1999, Roma.

[11] P.B. Cohen and G. Wüstholz. *Application of the André-Oort conjecture to some questions in transcendence.* Preprint.

[12] P. Deligne. *Travaux de Shimura.* Séminaire Bourbaki, 23ème année (1970/71), Exp. No. **389**, pp. 123–165. Lecture Notes in Math., Vol. 244, Springer, Berlin, 1971.

[13] P. Deligne. *Variétés de Shimura: interprétation modulaire, et techniques de construction de modèles canoniques*, in Automorphic forms, representations, and L-functions, Part 2, A. Borel and W. Casselman, eds., Proc. of Symp. in Pure Math., Vol. 33, American Mathematical Society, 1979, pp. 247–290.

[14] S.J. Edixhoven. *Special points on the product of two modular curves.* Compositio Mathematica **114**, 315–328, 1998.

[15] G. van der Geer. *Hilbert modular surfaces.* Ergebnisse der Mathematik und ihrer Grenzgebiete, 3. Folge, Band 16. Springer-Verlag, 1988.

[16] P. Griffiths and J. Harris. *Principles of algebraic geometry.* Pure and applied mathematics. A Wiley-Interscience publication, John Wiley and sons, New York, Chichester, Brisbane, Toronto, 1978.

[17] A. Grothendieck et. al. *Cohomologie locale des faisceaux cohérents et théorèmes de Lefschetz locaux et globaux.* North-Holland, Amsterdam (1968).

[18] S. Lang. *Algebraic number theory.* GTM 110, Springer-Verlag, second edition, 1994.

[19] B.J.J. Moonen. *Special points and linearity properties of Shimura varieties.* Thesis, Utrecht, September 1995.

[20] B.J.J. Moonen. *Models of Shimura varieties in mixed characteristic.* Pages 267–350 in "Galois representations in arithmetic algebraic geometry", edited by A.J. Scholl and R.L. Taylor. Cambridge University Press 1998.

[21] B.J.J. Moonen. *Linearity properties of Shimura varieties, I.* Journal of Algebraic Geometry **7** (1998), 539–567.

[22] B.J.J. Moonen. *Linearity properties of Shimura varieties, II.* Compositio Math. **114** (1998), no. 1, 3–35.

[23] D. Mumford. *Abelian varieties.* Oxford University Press, 1970.

[24] D. Mumford. *The topology of normal singularities of an algebraic surface and a criterion for simplicity.* Publications Mathématiques de l'IHES **9** (1961), 5–22.

[25] M.V. Nori. *On subgroups of* $\mathrm{GL}_n(\mathbb{F}_p)$. Invent. math. **88** (1987), pp. 257–275.

[26] F. Oort. *Canonical lifts and dense sets of CM-points.* Arithmetic Geometry, Proc. Cortona symposium 1994, F. Catanese, ed., Symposia Math., Vol. XXXVII, Cambridge University Press, 1997, pp. 228–234.

[27] V.P. Platonov and S. Rapinchuk. *Algebraic groups and number theory.* Moscow, 1991 (in Russian). English translation: Academic Press, 1993.

[28] M. Rapoport. *Compactifications de l'espace de modules de Hilbert-Blumenthal.* Compositio Mathematica **36** (1978), 255–335.

[29] J-P. Serre. *Quelques applications du théorème de densité de Chebotarev.* Publications Mathématiques de l'IHES **54** (1981), 123–202.

[30] H.M. Stark. *Some effective cases of the Brauer-Siegel theorem.* Inventiones math. **23**, 135–152 (1974).

[31] A. Yafaev. *Special points on products of two Shimura curves.* To appear in Manuscripta Mathematica.

Progress in Mathematics, Vol. 195, © 2001 Birkhäuser Verlag Basel/Switzerland

TOROIDAL RESOLUTIONS FOR SOME MATRIX SINGULARITIES

GERD FALTINGS

1. INTRODUCTION

In the classical theory of modular curves parametrising abelian varieties with level structures an important role is played by Drinfeld level structures. These allow us to find reasonably good models over primes of bad characteristics, see for example the book [KM] by Katz and Mazur. Unfortunately we have no such device for abelian varieties of higher dimensions. As a modest replacement one can use explicit equations for the relevant singularities. These are given by identities between matrices, but they cover only the case where the level is divisible by the characteristic p but not by its square. Here the pioneers are Chai/Norman and Deligne/Pappas ([CN], [DP]). In a previous paper [F] I had constructed explicit resolutions for such singularities which are associated to symmetric spaces. Here we extend this (in some cases) to the level structures associated to an Iwahori subgroup, which seems to be the maximum possible accessible by these methods. However we can only show that our models have toroidal singularities.

The previous methods do not work because of difficulties arising from linear algebra over discrete valuation rings. Namely in the previous paper the key idea was to use normal forms for certain such linear algebra data. For example, a quadratic form over a discrete valuation ring can be diagonalised (assuming for simplicity $p > 2$, that is, 2 is invertible in the base), which gives a rather explicit parametrisation of such objects via a regular model. By the valuative criterion we obtain a proper resolution of the space of symmetric matrices.

However here there is no such well-known (that is, to the author) normal form. In fact, our method gives us such an object which depends on certain combinatorial data (a piecewise linear convex function on a simplex). Unfortunately we could not decide which of these data may actually occur. Also, we obtain a compactification of $PGL(d)^r/PGL(d)$ which we hope to be toroidal. This can be shown for lattices of rank 2, and in general we obtain at least a log structure induced by local maps to a toroidal scheme. We hope that these maps are smooth. The preprint version of this article inspired A. Genestier to construct a toroidal resolution in the general case (see [G]) based on work of G. Lafforgue which however (as it later turned out) needs some overhaul. Aside from this probably temporary reason I hope that my paper still retains some interest because it gives a functorial interpretation of the resolution for lattices of rank 2, and because of some general insights into lattice arrangements. After completing it I found out that my toroidal structure coincides with G. Lafforgue's.

The plan of the paper is as follows: The first chapter introduces notations, explains the matrix singularities under consideration and why they are related to Shimura varieties. Next we consider as an example the special case of rank 2, which already exhibits the general features, but where the toroidal singularities can still be made explicit. After that we construct the minimal model also for higher ranks, and construct a parameter space for it which will hopefully define good resolutions.

The first half of the paper represents the talk given at Texel. The theory of the minimal model has been developed later, and replaces the original second half. I thank the organisers for allowing me to rewrite the article at a rather late stage in the publishing process. I also thank the referee for many useful comments. The research on this paper has been supported by the Deutsche Forschungsgemeinschaft through its Leibniz-Preis.

2. Deformation problems and crystalline cohomology

In [CN] one considers the moduli space of isogenies

$$\phi : A \to B,$$

with A and B g-dimensional principally polarised abelian varieties and ϕ of degree p^g compatible with polarisations (i.e. $\phi^*(\mathrm{pol}_B) = p \cdot \mathrm{pol}_A$). As in [dJ] we study the related space where in addition one factors ϕ into a sequence of isogenies of degree p:

$$A = A_0 \to A_1 \to \ldots \to A_g = B.$$

This moduli problem is defined and representable over the p-adic numbers $\mathbb{Z}_p$, and its local singularities can be described as follows: Consider the contravariant Dieudonné-module $M(A) = H^1(A)$, the first crystalline cohomology. For A defined over any base S it is a crystal of vector bundles of rank $2g$, and admits a Hodge-filtration $F(M) \subset M$ by a subbundle of rank g. Furthermore, a principal polarisation induces a non-degenerate symplectic form on $M(A)$ which is parallel, and for which $F(M)$ is maximal isotropic. Thus a family

$$A = A_0 \to A_1 \to \ldots \to A_g = B$$

over S induces a family

$$M(B) = M_g \to M_{g-1} \to \ldots \to M_0 = M(A).$$

One can check that the cokernel of $M_i \to M_{i-1}$ can be locally generated by one element (reduce to $S = \mathrm{Spec}(k), k$ a perfect field, and use the crystalline cohomology over the Witt-vectors $W(k)$). It then follows that locally the sequence above is isomorphic to the following:

Choose a free $\mathbb{Z}_p$-module $N \simeq \mathbb{Z}_p^g$, with a complete flag

$$N_g = p \cdot N \subset N_{g-1} \subset \ldots \subset N_0 = N.$$

Then $M_0 = N_0 \oplus N^*$ as well as $M_g = N_g \oplus N^*$ admit non degenerate symplectic forms (both summands are isotropic), and if $\mathbb{M}_i = N_i \oplus N^*$ there is a sequence $\mathbb{M}_g \to \mathbb{M}_{g-1} \to \ldots \to \mathbb{M}_0$ as above first over $\mathbb{Z}_p$ and then by base extension over any $\mathbb{Z}_p$-scheme S. Also, any other such sequence is isomorphic to it (respecting symplectic forms), and the isomorphism is unique up to transformation by the smooth group scheme G consisting of symplectic automorphisms of $\mathbb{M}_0$ respecting all $\mathbb{M}_i$.

However in $\mathbb{M}_i = M(A_i)$ we also have Hodge-filtrations $F(M_i) \subset M_i$. These describe the deformations, namely we have by [DP]: Consider a closed point x in the moduli space of sequences

$$A = A_0 \to A_1 \to \ldots \to A_g = B.$$

Choose a symplectic isomorphism of the universal sequence of crystals

$$M(B) = M_g \to M_{g-1} \to \ldots \to M_0 = M(A)$$

with the constant sequence induced from $\mathbb{M}$, in a neighbourhood U of x and such that the derivative of the isomorphism vanishes in x. Then the Hodge filtrations on the M_i define a map from U into the variety $X/\mathbb{Z}_p$ classifying such sequences of isotropic subbundles. This map is étale in x.

For the proof one considers as usual a thickening R of the local ring R_0 in x. Now R has an ideal $I \subset R, \mathfrak{M} \cdot I = I^2 = (0), R_0 = R/I$. For a family of A's over R_0 there is a canonical lift of $M(A)$'s over R, because I has a divided power structure. This system is isomorphic to the constant system of $\mathbb{M} \otimes R$'s, and in fact the given isomorphism over R_0 can be lifted to an isomorphism over R (by smoothness of G). Now by crystalline deformation theory lifting the A's to R amounts to lifting the Hodge-filtrations. Using this, one first derives that the map is unramified in x (this uses the assumption on derivatives), and then that it is also smooth.

Note that we have used that without the Hodge-filtrations the crystals have a canonical form, and that the automorphism group G of the situation is smooth. Both these assumptions fail for more complicated level structures, like level p^2.

Thus finally we have to study the singularities of the variety X. Now X is a closed subscheme of a product of the Grassmannians $\mathrm{Grass}_g(\mathbb{M}_i)$, parametrising compatible families of $F_i \subset \mathbb{M}_i$ with F_0 and F_g maximal isotropic. The group scheme G acts on X. Note that in the special fibre $G \otimes \mathbb{F}_p$ is solvable and has a unique fixed point. Namely extend the sequence of $\mathbb{M}_i$ by

$$\mathbb{M}_{-i} = \mathbb{M}_i^* \supset \mathbb{M}_0^* = \mathbb{M}_0,$$

and then the fixed point is given by

$$F_i = \mathrm{Kernel}(\mathbb{M}_i \otimes \mathbb{F}_p \to \mathbb{M}_{i-g} \otimes \mathbb{F}_p) \subset \mathbb{M}_i \otimes \mathbb{F}_p.$$

As $X \otimes \mathbb{F}_p$ is proper by Borel's theorem this fixed point lies in any G-invariant closed subset. Thus if a certain property is G-invariant and holds in a neighbourhood of this fixed point, it holds on all of X. For example, if we define a G-equivariant modification $Y \to X$ and show that it has only toroidal singularities over the preimage of the fixed point, then it has such singularities everywhere. A simpler variant of X is obtained by starting with our complete flag

$$N_g = p \cdot N \subset N_{g-1} \subset \ldots \subset N_0 = N,$$

and considering just compatible systems of direct summands $F_i \subset N_i \otimes R$ of a fixed rank a. Here compatible means that there are induced maps, and that F_g and F_0 are identified under the isomorphism "multiplication by p" $N_g \cong N_0$. Here the natural symmetry group is the Iwahori in $G \subset GL(N)$ stabilising all N_i. Of course this model also describes local singularities in a suitable Shimura variety. For example, one can construct such Shimura varieties using unitary groups.

These methods also apply to $\Gamma_1(p)$-level structures. For example, one adds to the data of a chain of isogenies

$$A = A_0 \to A_1 \to \ldots \to A_g = B$$

those of Drinfeld-generators of the finite flat group schemes of order p which are the kernels of $A_i \to A_{i+1}$. These correspond to the level structure given by a p-Sylow subgroup of the p-adic group, and the representing scheme is obtained from the previous one by adjoining certain $(p-1)$st roots. Adjoining such roots in our previous toroidal resolution Y will give such a resolution in this more general case.

3. The case of rank two

In this chapter we try to explain our strategy in the first non trivial case, where we can be more explicit than in general. Consider a discrete valuation ring $V, K \supset V$ its ring of fractions, $\pi \in V$ a uniformiser, $k = V/\pi \cdot V$ the residue field. Fix a complete flag

$$N_g = \pi \cdot N \subset N_{g-1} \subset \ldots \subset N_0 = N = V^g, g = a + b,$$

and study the projective scheme X over V parametrising compatible families of direct summands $F_i \subset N_i \otimes R$, of rank a. Equivalently we could consider compatible systems of quotients of rank b, and also duality exchanges the role of a and b. If one of them is equal to 1 the resulting x will have semistable singularities over V, and there is no need for further resolution (this is quite the same as in [D]). Thus the next case is (say) $a = 2$, and we assume this from now on. We shall construct a projective birational map $f : Y \to X$, isomorphic on the generic fibre, such that Y has rational singularities. In fact, Y will represent a certain moduli problem. To explain and motivate which one, we first study points with values in a valuation ring $W \supset V$ dominating V. Note that properness of f is essentially equivalent to the fact that any such W-valued point of X lifts uniquely to Y. Now a point in $X(W)$ is a sequence of compatible subbundles (which are locally direct summands of rank two) $F_i \subset N_i \otimes W$. If L denotes the fraction field of W these all become isomorphic over L, thus defining a finite collection of lattices in a two dimensional L-vector space isomorphic to L^2. For each lattice $F \subset L^2$ we have the projective line $\mathbb{P}(F)$ over W, which depends only on F up to homothety. That is, replacing F by $a \cdot F$ with $a \in L^*$ does not change $\mathbb{P}(F)$. Furthermore, for any finite collection $\{F_\nu\}$ of such lattices we define a flat curve C over W as the closure of $\mathbb{P}(L^2)$ embedded diagonally into the product of the $\mathbb{P}(F_\nu)$. Then C is a form of $\mathbb{P}^1$, and the irreducible components of its special fibre correspond to the homothety classes of F_ν's. More precisely for each such component the projections onto the factors $\mathbb{P}(F_\nu)$ are either constant or isomorphisms. Enlarging the collection $\{F_\nu\}$ replaces C by a modification $C' \to C$. We can add a minimal number of new F's to achieve the following:

a) C' has only simple double points as singularities.
b) Any irreducible component of C' corresponding to a new F meets at least three other components.

In fact, the newly added F's are all lattices of the form

$$D = a \cdot A + b \cdot B + c \cdot C$$

with A, B, C in the original collection, the coefficients a, b and c are elements of L^*, the three inclusions of the summands (like $a \cdot A \subset D$) have the property that after tensoring with the residue field of W they have rank one, and the images are three different lines in the (two-dimensional) reduction of D. Later we shall call these essential linear combinations.

To see all this one can for example use the theory of stable pointed curves: For each of the original F_ν, choose W-points of C specialising to different points on the irreducible F_ν-component of the special fibre of C, and not on other components. Choose so many of those such that each F_ν-component has at least three points which are either intersections with other components, or on the new list. Then C' is the stable model for the corresponding punctured curve (of genus zero). The components of the special fibre of C' and their intersections define as usual the vertices, respectively edges, of a graph which is a connected tree. All vertices at the ends of the tree must be in the original list $\{F_\nu\}$, since vertices corresponding to new F's have at least three edges. They are then also the midpoint of three ends which implies their description as above.

Furthermore, the edges e correspond to double points in the special fibre. At each such double point we associate an ideal $(d_e) \subset W$ if the local singularity looks like $W[u,v]/(u \cdot v - d_e)$.

Theorem 1. *C' admits a versal deformation, with base a power series ring over V. As indeterminates in the power series one can use a set of parameters including the d_e's.*

The proof is standard.

In terms of lattices the d_e's describe the distances in the building, and they admit the following description: Suppose F and G are two such lattices in our collection. Replacing one of them by a homothetic one, we may find elements $\{\phi, \psi\} \in L^2$ and $d \in W$ such that $F = W \cdot (d \cdot \phi) + W \cdot \psi, G = W \cdot \phi + W \cdot \psi$. Then d is up to units the product of the d_e with e on the minimal path (in our tree) from the vertex F to the vertex G.

Apply this to the original lattices $F_i \subset N_i \otimes W$: The inclusions $F_i \subset F_{i-1}$ are already of the form above (no need for a homothety), and give an invariant d which is the product of the d_e's on the path from F_i to F_{i-1}. Composing all these paths $(F_g \to F_{g-1} \to \ldots \to F_0 = \pi^{-1} \cdot F_g)$ gives a path (with repetitions) from F_0 to F_0 which passes each edge e an even number $2 \cdot n_e$ of times, namely n_e times in one direction and n_e times in the other. Then checking determinants the product must be equal to π, up to a unit:

$$\prod d_e^{n_e} = \pi \cdot \text{unit}.$$

Now our modification $Y \to X$ will parametrise such curves C', together with maps $C' \to \mathbb{P}(F_i)$ of degree one. That is, C' is a proper flat curve over the base. All geometric fibres are trees of $\mathbb{P}^1$'s, and on each component the maps to $\mathbb{P}(F_i)$ are either constant or isomorphisms. Furthermore, if they are all constant the component contains at least three double points. Finally for each i precisely one of the components maps onto $\mathbb{P}(F_i)$. It is easily seen that this is representable by a projective scheme over X, for example using the previous construction of C' via stable pointed curves. To see that Y has toroidal singularities one uses deformation theory: The special fibre of a curve like C' has a versal deformation over a power series ring, for example using as indeterminates the d_e's. The equation

$$\prod d_e^{n_e} = \pi \cdot \text{unit}$$

then defines a toroidal modification of this ring, which is formally smooth over

$$V[\{d_e\}, u, u^{-1}]/(\prod d_e^{n_e} - \pi \cdot u).$$

As the curve C' may have automorphisms this does not represent a good moduli problem. Note however that an automorphism is trivial if this holds on the components corresponding to the original F_i. After choice of a model (in its homothety class) for F_0 we obtain over our base a sequence

$$\pi \cdot F_0 = F_g \to F_{g-1} \to \ldots \to F_0$$

of bundles. Injecting these into the flag induced from

$$N_g = \pi \cdot N \subset N_{g-1} \subset \ldots \subset N_0 = N = \mathbb{Z}_p^g$$

defines a scheme which is smooth over our versal deformation. For example, to get local coordinates, choose the standard basis $\{l_0, \ldots, l_{g-1}\}$ of linear forms on N. Then choose freely the restriction of l_i/π to F_{i+1}, to define a map from the F-flag to the N-flag. That this gives embeddings is an open condition. Now there are no more automorphisms, and the new bases of versal deformations are the formal completion of local rings of Y. Thus:

Theorem 2. *The above modification Y has toroidal singularities, an explicit local equation being given above.*

For completeness we also explain what happens for the original problem of isotropic subspaces. Now $g = 2$, we have symplectic forms on $M_0 = N_0 \oplus N^*$ and $M_2 = N_2 \oplus N^*$, and X parametrises isotropic subspaces. As before the versal deformations of the stable $\mathbb{P}^1$-model for the lattice arrangement of the F_i's are parametrised by a scheme with toroidal singularities. Over that scheme, embeddings into the flag of M_i's define a smooth extension, and isotropic embeddings form a closed subscheme. We claim that this subscheme is also formally smooth. For that it suffices if this holds in a neighbourhood of the preimage of the unique G-fixed point in X, with G the Iwahori which acts on X and Y. In such a neighbourhood the projections of F_0 onto the second direct summand

N^* as well as the projection of F_2 onto $N_2 = N$ are isomorphisms. Using these identifications we obtain a bilinear pairing between F_0 and F_2, and isotropy is equivalent to the fact that the inclusion $F_0 \subset F_2$ is symmetric.

Thus isotropic embedding amounts to an isomorphism between F_2 and N as well as a non-degenerate pairing between F_0 and F_2 which makes the inclusion symmetric. Giving the isomorphism defines obviously a smooth extension. The same holds for the pairing: Namely in suitable bases the inclusion $F_0 \subseteq F_2$ can be represented by a diagonal matrix $\mathrm{diag}(\alpha, \beta)$ with regular functions α and β, with β dividing π and α dividing β. In this basis the pairing is given by a matrix with diagonal terms u and v and off-diagonal w and $w \cdot \beta/\alpha$, with u, v, w indeterminates. That its determinant is invertible defines an open subscheme.

4. Lattice arrangements over discrete valuation rings

We now come to the core of our strategy. As illustrated by the examples in the last chapter we shall study certain types of arrangements of lattices and try to show that their deformation theory is either unobstructed, respectively leads to toroidal singularities. Here we shall consider lattice arrangements corresponding to "convex" subsets of the Bruhat-Tits building. Our notion of "convex" does not coincide with the usual one. This is so because the emphasis is on sets of lattices (that is, vertices of the building), not on general subsets of the building. Having made this remark we drop from now on the quotation marks in "convex".

Thus let V denote a discrete valuation ring, with uniformiser π, residue field $k = V/\pi \cdot V$, and fraction field $K \supset V$. Fix a finite dimensional K-vector space $E = K^d$. Recall that the Bruhat-Tits building parametrises norms on the dual space E^*, up to scalar factors. For example, for any V-lattice $M \subset E$ such a norm is defined by $\| \ l \ \| = \sup\{\| \ l(M) \ \|\}$. For simplicity we prefer to use additive notation and use negative logarithms of norms, or valuations $v : E - \{0\} \to \mathbb{R}$. We also normalise such that π gets valuation one. In general for a complete flag

$$M_0 = \pi \cdot M_d \subset M_1 \subset \ldots \subset M_d$$

and real numbers

$$1 = x_d \geq x_{d-1} \geq \ldots \geq x_0 = 0$$

we obtain such a valuation

$$v(l) = \inf\{v(l(M_i)) + x_i \mid 0 \leq i \leq d\}.$$

Furthermore, any valuation is of this form, and the resulting simplices give the Bruhat-Tits building some type of linear structure. We also remark that we consider two v's as equal if they differ by a constant.

For two valuations v_1, v_2 the "geodesic" (in the future without quotation marks) between v_1 and v_2 consists of the valuations

$$v(l) = \inf\{v_1(l), x + v_2(l)\},$$

$x \in \mathbb{R}$ a real number. It is a shortest path if we define the distance between v_1 and v_2 as the variation (= maximum−minimum) of the difference $v_1 - v_2$ (on $E^* - \{0\}$). The distance function is convex on geodesics ("negative curvature"), and the distance between two lattices is always an integer. Also, a collection of lattices forms the vertices of a simplex if and only if the distance between any two of them is one. A subset of the Bruhat-Tits building is called convex if it contains the geodesic between any two of its elements. For a finite set of valuations v_i their convex hull consists of all v's of the form $v(l) = \inf\{v_i(l) + x_i\}, x_i \in \mathbb{R}$ real numbers. For example, for a flag

$$M_0 = \pi \cdot M_d \subset M_1 \subset \ldots \subset M_d$$

as above the simplex of valuations associated to it is equal to the convex hull of the valuations belonging to the vertices M_i. We have:

Lemma 3. *Suppose $\{M_i\}$ is a finite collection of V-lattices. Then their convex hull is the union of all simplices associated to flags*

$$N_0 = \pi \cdot N_a \subset N_1 \subset \ldots \subset N_a$$

for which each vertex N_j is a linear combination

$$N_j = \sum x_i \cdot M_i,$$

with $x_i \in K^*$. Furthermore, this set is finite.

Proof. It is clear that all such N_j, and thus also any simplex formed by them, lie in the convex hull. Conversely we can use induction over the number of M_i's and reduce to examine what happens if we add one further lattice N to a simplex

$$N_0 = \pi \cdot N_a \subset N_1 \subset \ldots \subset N_a.$$

Then any valuation in the convex hull is of the form

$$v(l) = \inf\{v(l(N_j)) + x_j, v(l(N)) + y\},$$

with $1 \geq x_a \geq x_{a-1} \geq \ldots \geq x_0$ and (after replacing N by a multiple) $1 \geq y \geq 0$. Next we replace each M_i by $M_i + \pi \cdot N$, and add to N the sum of all $\pi \cdot M_i$. Finally we add N to M_i if $y \geq x_i$, and M_i to N if $x_i \geq y$. The result is a new simplex (now including N among its vertices) which contains v. For the finiteness choose an integer m such that $\pi^m \cdot M_i \subset M_{i'}$ for any two elements $M_i, M_{i'}$ of our list of representatives. Then we need only consider linear combinations

$$\sum \pi^{n_i} \cdot M_i \quad \text{with} \quad 0 \le n_i \le m.$$

Call a linear combination $\sum x_i \cdot M_i$ essential if any endomorphism of $\bar{M} = M \otimes_V k$ is a scalar if it is a scalar on the image of each summand $x_i \cdot M_i$. Another way to formulate this: Define a split torus T as a product of multiplicative groups G_m, one for each basic lattice M_i. Then for each M we get a subtorus $T(M) \subset T \otimes k$ consisting of endomorphisms of $\bar{M}$ scalar on the image of $x_i \cdot M_i$ (choose x_i such that this image is non-zero). For each inclusion $M \subset N$ the induced map $M \to N$ commutes with the action of $T(M) \cap T(N)$. Also, M is essential if and only if $T(M)$ is reduced to scalars.

5. Deligne-schemes

Deligne-schemes are a useful tool to study lattice arrangements. For example, lemma 9 is basically a result of linear algebra, but its proof uses geometry (of course one can now reformulate to get an elementary proof, see lemma 10). Some of the following material can be found in [M].

To a convex collection of lattices $\{M_\nu\}$ (in a fixed K-vector space) we associate a scheme X. The simplest definition of X is that it is the join of the projective spaces $\mathbb{P}(M_\nu)$. A better definition is as follows:

Definition 4. *The Deligne-scheme X represents the functor on V-algebras R such that an R-valued point of X is a collection of quotient line bundles $L(M_\nu)$ of $M_\nu \otimes R$, one for each homothety class of M_ν's, such that each inclusion $M \subset N$ maps $L(M)$ to $L(N)$.*

It is clear that such a scheme X exists, and that it is projective over V. Also, over the fraction field K, we have that $X \otimes K$ is isomorphic to projective space $\mathbb{P}(E)$. That it is in fact the join of the $\mathbb{P}(M_\nu)$ can be easily seen from the following. However, as we make no use of this fact we leave the details to the reader.

For an extension field k' of k the k'-valued points of X have the following description:

Suppose given such a point, that is, compatible quotients $L(M_\nu)$ of $M_\nu \otimes k'$. Call a lattice M maximal if for any strict inclusion $M \subset N$ (with M and N both homothetic to members of the collection $\{M_\nu\}$) the induced map from $L(M)$ to $L(N)$ vanishes. If M and N are both maximal, any linear combination $x \cdot M + y \cdot N$ (with $x, y \in K^*$) must be homothetic to either M or N, since $L(x \cdot M + y \cdot N)$ is generated by either $L(M)$ or $L(N)$. It follows easily that the maximal M's form a simplex σ, that is, can be put into the form

$$M_0 = \pi \cdot M_n \subset M_1 \subset \ldots \subset M_{n-1} \subset M_n.$$

Furthermore, each M can be modified by a homothety such that for some i with $0 < i \leq n$ it is contained in M_i but not in M_{i-1}, and then $L(M)$ is induced from $L(M_i)$ via the inclusion. It follows that the projection from M_i to $L(M_i)$ vanishes on M_{i-1} but not on any strictly bigger sublattice $M \subset M_i$. Conversely, for any simplex σ as above and any family $L(M_i)$ satisfying these conditions, we get a point in $X(k')$ having σ as simplex of maximal lattices. These points are parametrised by an open subscheme of the product of projective spaces $\mathbb{P}(M_i/M_{i-1})$, which has dimension $d-n$. In more generality we obtain a covering of X by open sets $U(\sigma)$, such that a point of $X(k')$ lies in $U(\sigma)$ if its simplex of maximal lattices is contained in σ. We also require that the generic fibre of $U(\sigma)$ is all of $X \otimes K$. Also, $U(\sigma) \subset U(\sigma')$ if $\sigma \supset \sigma'$.

By rather explicit computation $U(\sigma)$ is shown to have semistable singularities. Namely choose a point in $U(\sigma)(k')$, that is, quotients $M_i \otimes k'' \to L(M_i)$. An explicit open covering of $U(\sigma)$ can be defined as follows: Choose a k-basis in each M_i/M_{i+1} and lift its elements to elements of M_i. This defines as usual an adapted basis for each M_i. Choose one among these basis elements, and consider systems of quotients where the chosen basis elements generate $L(M_i)$. For these, $L(M_i)$ is identified with R, the maps $L(M_{i-1}) \to L(M_i)$ are multiplication by some $t_i \in R$ with $t_1 \cdot \ldots \cdot t_n = \pi$, and the projections of other basis elements (projections to $L(M_i)$ for the lifts of a basis of M_i/M_{i-1}) are just elements of R. Thus our subfunctor is represented by an open subscheme of the product of an affine space with $\mathrm{Spec}(V[t_1, \ldots, t_n]//\prod t_i - \pi)$.

Also, the union of these subfunctors contains all points in the closed fibre for which σ is the simplex of maximal lattices. Thus varying σ we get a covering of all of X, which has semistable singularities and is flat over V. Especially X is regular, and the special fibre is a divisor with simple normal crossings. Its irreducible components are parametrised by the lattices M_ν. The irreducible component $X(M)$ corresponding to M contains all points whose simplex of maximal lattices contains M. It is isomorphic to the blow-up of the projective space $\mathbb{P}(\bar{M})$ in all subspaces $\mathbb{P}(M/N)$, for proper sublattices N with $\pi \cdot M \subset N \subset M$. It represents the functor (over k) of compatible systems of quotient lines of $\bar{M}$ as well as all M/N above. More generally, the intersections of the $X(M)$'s define a stratification of $X \otimes k$. The (closed) strata $X(\sigma)$ are indexed by simplices σ and represent points for which σ is contained in the simplex of maximal lattices. If σ has the form

$$M_0 = \pi \cdot M_n \subset M_1 \subset \ldots \subset M_{n-1} \subset M_n$$

then $X(\sigma)$ is the product of $\mathbb{P}(M_i/M_{i-1})$, blown up in all subspaces given by lattices N with $M_{i-1} \subset N \subset M_i$. Furthermore, $X(\sigma)$ is contained in $X(\sigma')$

if and only if $\sigma \supset \sigma'$. The scheme X comes equipped with the line bundles $L(M)$, for each lattice M in the arrangement. There is the small subtlety that homothetic M's give isomorphic but not canonically isomorphic $L(M)$'s. One can compute their cohomology:

Lemma 5. *Suppose X is defined by a convex lattice arrangement. Then*

a) $H^0(X, L(M)^{\otimes n}) = S^n(M)$, *for* $n \geq 0$.

b) $H^i(X, L(M)^{\otimes n})$ *vanishes for* $i > 0, n \geq 0$.

c) *All* $H^i(X, L(M)^{\otimes -n})$ *vanish for* $0 < n < d$.

Proof. This holds if $X = \mathbb{P}(M)$, that is, M is the only lattice in the arrangement. In general we show that X can be obtained from $\mathbb{P}(M)$ by successive blow-ups of regular closed subschemes. To see this, find a lattice N in the arrangement which is not a linear combination of multiples of other lattices, that is, it is not in the convex hull of the remaining lattices. To get such an N, choose a linear form l on E, and N minimal among (multiples of) the lattices of the arrangement such that $l(N) = V$. Then for any proper sublattice $N' \subset N$ the image $l(N')$ is contained in the maximal ideal $\pi \cdot V$, and their sum cannot be N. Furthermore, for a suitable choice of l we can achieve that M is not minimal, thus $N \neq M$ (unless M is the only lattice).

But adding N to the remaining lattices amounts to a blow-up as claimed. More precisely, if $N' \subset N$ denotes the sum of all proper sublattices, the center of the blow-up is the proper transform in $X(N')$ of $\mathbb{P}(N'/\pi \cdot N) \subset \mathbb{P}(\bar{N}')$. This is so because its ideal is locally isomorphic to the image of $\pi \cdot N$ in $L(N')$.

Next denote by ω_X the relative dualising sheaf on X. Its restriction to $X(M)$ is equal to the line bundle of $(d-1)$ differentials with simple poles along the boundary, and its restriction to the generic fibre $\mathbb{P}(E)$ is $\det(E)(-d)$. For lattices $M_1, ..., M_d$ and elements $m_i \in M_i$ we define a global section $\gamma(m_1, ..., m_d)$ of $\omega_X \otimes \bigotimes_{i=1}^n L(M_i)$ as follows: Over the generic fibre identify each $L(M_i)$ with $\mathcal{O}(1)$, and our line bundle with the twist of $\mathcal{O}_X$ by $\det(E)$. Thus the wedge product of the m_i defines a global section. To see that it extends to all of X it suffices to check in generic points of $X(M)$'s. Scale all M_i such that they are contained in M but not in $\pi \cdot M$. Then near the generic point of $X(M)$ all $L(M_i)$ are isomorphic to $L(M)$, and $\gamma(m_1, ..., m_d)$ is the pullback of a regular section on $\mathbb{P}(M)$.

We remark that for any $(d+1)$ elements $m_i \in M_i$ we have the relation

$$(*) \sum_{i=0}^{d} (-1)^i \gamma(m_0, ..., m_{i-1}, m_{i+1}, ..., m_d) \cdot m_i = 0.$$

(Check on the generic fibre.)

Also, these regular sections are used in the following result:

Lemma 6. *Suppose X is defined by a convex lattice arrangement. Choose lattices $M_1, ..., M_r$, with $r > 0$. Then*

a) $H^0(X, \omega_X \otimes \bigotimes_{i=1}^r L(M_i))$ *vanishes, if* $r < d$.

b) $H^0(X, \omega_X \otimes \bigotimes_{i=1}^r L(M_i))$ *is generated by sections of the form* $\gamma(m_1, ..., m_d) \cdot m_{d+1} \cdots m_r$ $(m_i \in M_i)$*, and their analogues obtained by permuting the* M_i*'s.*

c) $H^j(X, \omega_X \otimes \bigotimes_{i=1}^r L(M_i))$ *vanishes for* $j > 0$.

Proof. We use an inductive argument. Obviously we may assume $d > 1$. If $r = 1$ the cohomology in question is dual to that of $L(M_1)^{\otimes -1}$ and vanishes. Now assume the assertion shown for smaller values of either d or r.

Call an element $m_1 \in M_1$ "generic" if the following holds: Suppose M is a lattice in our arrangement with $M_1 \subseteq M$, $M_1 \not\subseteq \pi \cdot M$. Then the image of m_1 in $\bar{M} = M/\pi \cdot M$ does not vanish. "Generic" elements exist after (if necessary) passage to a finite unramified extension of V.

Thus choose a "generic" m_1. It defines a global section of $L(M_1)$ which does not vanish identically on any irreducible component $X(M)$ of the special fibre. Its zero-set $X' \subset X$ is obviously equal to the Deligne-scheme associated to the convex lattice arrangement in $E' = E/K \cdot m_1$ defined by the images of all lattices M (just check the functor). Furthermore, the restriction of $\omega_X \otimes \bigotimes_{i=1}^d L(M_i)$ to X' is isomorphic (by adjunction) to $\omega_{X'} \otimes \bigotimes_{i=2}^d L(M_i)$, with $\gamma(m_1, m_2, ..., m_d)$ restricting to $\gamma(m_2, ..., m_d)$. Thus if we apply induction to the restriction of $\omega_X \otimes \bigotimes_{i=1}^r L(M_i)$ we get that it has no higher cohomology, and that all its global sections lift. Finally the kernel of the restriction map is the line bundle $\omega_X \otimes \bigotimes_{i=2}^r L(M_i)$, to which we can apply induction again.

Do these global sections generate the line bundles? Consider a closed stratum $X(\sigma)$, where σ is a simplex of the form

$$N_0 = \pi \cdot N_n \subset N_1 \subset \ldots \subset N_{n-1} \subset N_n,$$

and $X(\sigma)$ is the product of all $\mathbb{P}(N_i/N_{i-1})$, blown up in all subspaces given by lattices N with $N_{i-1} \subset N \subset N_i$. Again by adjunction the restriction of ω_X to $X(\sigma)$ is the product of the top differentials on the factors, with simple poles along the exceptional divisors. Then one checks that the restriction of $\gamma(m_1, ..., m_d)$ either vanishes on all of $X(\sigma)$, or it generates the restriction of $\omega_X \otimes \bigotimes_{i=1}^d L(M_i)$ on the open stratum $U(\sigma)$, that is, on the complement of the exceptional divisors. The latter happens if and only if the following condition holds:

For each $1 \leq i \leq d$, denote by $j = j(i) \geq 1$ the index such that, after replacing M_i (and m_i) by a multiple, $N_{j-1} \subset M_i \subseteq N_j$. Thus m_i defines an element of $N_{j(i)}/N_{j(i)-1}$. Now $\gamma(m_1, ..., m_d)$ does not vanish on the open stratum precisely if one obtains this way a basis for all quotients $N_{j(i)}/N_{j(i)-1}$.

(Especially the number of i's with $j(i) = j$ must be equal to the dimension of N_j/N_{j-1}).

For a proof we may replace M_i by $N_{j(i)}$ and may assume that our arrangement consists only of the simplex σ. Then use an easy explicit calculation.

As a corollary we get that $\omega_X \otimes \bigotimes_{i=1}^r L(M_i)^{\otimes d}$ is globally generated if the M_i generate the convex lattice arrangement, that is, each M is a sum of multiples of the M_i.

6. The minimal model

The minimal model X_{min} will be a blow-down of the Deligne-scheme X. The name "minimal" is purely heuristic, justified because the construction generalises what is obviously minimal for $d = 2$. It will have "toroidal singularities", by which we mean that our V-schemes are locally isomorphic to the following:

For a split torus $T = \mathbb{G}_m^n$, consider finitely generated normal T-stable V-subalgebras (the sum is over characters χ of T)

$$\mathcal{A} = \bigoplus I_\chi \cdot \chi \subset \mathcal{O}_T \otimes_V K,$$

where each $I_\chi \subset K$ is a fractional ideal. Then Spec($\mathcal{A}$) is a normal T-equivariant scheme, with generic fibre T_K. It can be constructed using usual torus embeddings (as in [KKMS]):

If each I_χ is generated by π^{n_χ} (allowing $n_\chi = \infty$), and $N(T) = \mathbb{Z}^n$ denotes the one-parameter groups in T, consider the cone in $N(T)_{\mathbb{R}} \times \mathbb{R}$ generated by $(0,1)$ as well as all (χ, n_χ). This defines an affine torus embedding of $T \times \mathbb{G}_m$, mapping equivariantly to $\mathbb{G}_m \subset \mathbb{A}^1$. It induces Spec($\mathcal{A}$) by pullback via $\pi \in \mathbb{A}^1(V)$.

We assume given lattices $M_1, ..., M_r$ and consider the convex lattice arrangement generated by them, consisting of all linear combinations $\sum x_i \cdot M_i$. Let X denote the associated Deligne-scheme. We show:

Theorem 7. *There exists a normal V-scheme X_{min} and a birational map $X \to X_{min}$ such that*

a) *There exists a very ample line bundle L on X_{min} with pullback $\omega_X \otimes \bigotimes_{i=1}^r L(M_i)^{\otimes(d+1)}$ on X.*
b) *X_{min} has toroidal singularities.*
c) *The map $X \to X_{min}$ is locally isomorphic to a toroidal resolution defined by subdivisions.*

Proof. The line bundles $L(M_i)$ as well as $\omega_X \otimes \bigotimes_{i=1}^r L(M_i)^d$ are globally generated and (assuming $r > 1$) very ample on the generic fibre. Thus their global sections define a map from X to some product of projective spaces over V. We define X_{min} as the normalisation of its image. Thus property a) holds obviously. In fact, we shall see that the image itself is already normal.

Next we choose a trivialisation of det(E). We use

Definition 8. *A basic vector v is an r-tuple $v = (d_1, \ldots, d_r)$ of non negative integers, with sum d (these form the integral points in a certain convex subset of $\mathbb{R}^r$, of dimension $r-1$). For each basic vector v the fractional ideal $I_v \subset K$ is generated by the coefficients of the map*

$$\Lambda^{d_1} M_1 \otimes \Lambda^{d_2} M_2 \otimes \ldots \otimes \Lambda^{d_r} M_r \to \det(E) = K.$$

The ideals satisfy some identities:

Lemma 9. *Suppose we have for basic vectors $v, v_1, ..., v_s$ an identity*

$$v_1 + v_2 + ... + v_s = s \cdot v.$$

Then $I_{v_1} \cdot I_{v_2} \cdot ... \cdot I_{v_s} \subseteq I_v^s$.

Proof of lemma. The product on the left is contained in the ideal I_{sv} defined by tensoring everything with V^s. That is, we replace E by E^s, and also each M_i by M_i^s. The convex span then consists of all lattices M^s. In fact, I_{sv} is the sum of the ideals arising from all possible decompositions of $s \cdot v$.

Now the proof of lemma 6 gives us a way to find generators of these ideals: Namely denote by $M_1, ..., M_d$ the lattices M_i repeated with multiplicity d_i. Then choose a "generic" element $m_1 \in M_1$, divide by it. In the quotient choose a generic element in the projection of M_2, and lift it to $m_2 \in M_2$. Proceeding this way we obtain $m_i \in M_i$ such that $\gamma(m_1, ..., m_d)$ generates the global sections of the appropriate line bundle on X. Especially it divides all other possible γ's, which means that $m_1 \wedge m_2 \wedge ... \wedge m_d$ generates I_v.

For I_{sv} we have to do the same with E replaced by E^s. However one checks that the elements $(m_1, 0, ..., 0)$, $(0, m_1, 0, ..., 0)$,..., $(0, 0, ..., 0, m_1)$, $(m_2, 0, ..., 0)$, $(0, m_2, 0, ..., 0)$,..., $(0, 0, ..., 0, m_2)$, etc., are "generic" in this situation: It suffices to check that for the first s (involving m_1). If for some $t \leq s$ one has already divided by the vectors which have m_1 in position $< t$, the projection of a multiple $x \cdot M_1^s$ $(x \in K^*)$ is contained in the projection of M^s, but not in that of $\pi \cdot M^s$, precisely if this holds for $x \cdot M_1$ and M. Then m_1 is also not contained in $\pi \cdot M$, and the projection of $(0, ..., m_1, ..., 0)$ not in that of M^s (check t-th component). Now finally I_{nv} is generated by the wedge product of all $(0, ..., m_i, ..., 0)$, thus equal to I_v^s.

Now we proceed with the proof of the theorem. Consider the torus $T = \mathbb{G}_m^r$. Its group of characters is $\mathbb{Z}^r$, and each basic vector v defines such a character χ_v. Next consider the subring $\mathcal{A}$ of its affine coordinate ring $K[T]$ over K generated by all $\{I_v \cdot \chi_v\}$. Its normalisation is the direct sum of its T-eigenspaces, and a multiple $x \cdot \chi$ lies in such an eigenspace if and only if some power of it lies in $\mathcal{A}$, and its restriction to the diagonal is divisible by d. By lemma 9 (applied to E^s, M_i^s etc., if χ is the sum of s basic v's) $x \cdot \chi$ itself already lies in $\mathcal{A}$, that is, $\mathcal{A}$ is normal. The corresponding affine scheme Z_{min} over V has generic fibre isomorphic to T_K/μ_d (embedded diagonally), and admits a T-action extending translations on the generic fibre.

Define a $T \times \mathbb{G}_m$-torsor over X by adding trivialisations of all $L(M_i)$ as well as of ω_X. It is canonically the pullback of such a torsor over X_{min}. We shall construct a map from this torsor to Z_{min} and shall show that it is smooth of relative dimension d. It will be equivariant for the T-action, and also for $\mathbb{G}_m$ if we let it act on Z_{min} via the d-th root of the inverse of the diagonal embedding into T. Note that the torsor is a quasi-affine scheme.

Namely under our map the elements of $I_v \cdot \chi_v$ pullback to global sections $\gamma(m_1, ..., m_d)$ for elements $m_j \in M_{j(i)}$, where a given M_i occurs with total multiplicity d_i: On the one hand they define functions on X_{min} (because of the trivialisations), and on the other hand they define elements of I_v. These two should correspond. It is easily seen that this factors over a map to the normalisation $\bar{T}$, and has the claimed equivariance. Namely it suffices to check that over K.

For smoothness, consider the open subset of (our torsor over) X_{min} where a certain $\gamma(m_1, ..., m_d)$ does not vanish. If $m_i \in M_{j(i)}$ the images of M_i in $L(M_{j(i)})$ define functions (using trivialisations), and thus a map to $\bar{T} \times \mathbb{A}^d$. We claim that this is an open immersion:

All other $\gamma(...)$'s are products of $\gamma(m_1, ..., m_n)$ with pullbacks of functions on $\bar{T}$. Furthermore, for each $m_i \in M_i$ the image of m_i in $L(M_i)$ can be computed from the relation $(*)$ applied to m_i and $m_1, ..., m_d$. As our open subset of X_{min} is defined by taking the subring generated by these elements it follows that it immerses into $\bar{T} \times \mathbb{A}^d$. Also, the map is an open immersion on the generic fibre, thus the image is dense, and as $\bar{T} \times \mathbb{A}^d$ is normal we obtain indeed an open immersion. Thus b) has been shown. Also, as γ's and m_i's suffice to define coordinates on the normalisation of the image, the image itself is already normal.

For c) we need to consider the fibres of our map. These are the maximal geometrically connected closed subschemes of X on which the line bundles $L(M_i)$ as well as ω_X are trivial. An example of such a subscheme can be obtained as follows: Consider a stratum $X(\sigma)$ corresponding to a simplex

$$N_0 = \pi \cdot N_n \subset N_1 \subset \ldots \subset N_{n-1} \subset N_n.$$

For each index j write $N_j = \sum_{i=1}^r x_i \cdot M_i$ $(x_i \in K^*)$ such that all the maps $M_i \to \bar{N}_j = N_j/\pi \cdot N_j$ do not vanish. Thus N_j/N_{j+1} becomes a quotient of $\bigoplus_{i=1}^r M_i$. Consider the subtorus $T(\sigma) \subseteq T$, defined over k, of endomorphisms scalar on each M_i which respect all these quotients. It can be described as follows:

Consider a basis of N_j/N_{j+1} where each basis element is in the image of some M_i. Call two indices i and i' equivalent if one can find two such bases (with some j) which differ by changing one element of M_i into an element of $M_{i'}$. This generates an equivalence relation, and $T(\sigma)$ consists of families of scalars constant on each equivalence class. Equivalently it is the intersection of the kernels of $\chi_v/\chi_{v'}$, with v, v' running through all basic vectors for which the direct sum $\bigoplus N_j/N_{j+1}$ has bases of types v and of type v'.

Now $T(\sigma)$ operates on $X(\sigma)$, and the restrictions to $X(\sigma)$ of all global sections of $L(M_i)$ as well as of the $\gamma(m_1, ..., m_d)$ are eigenvectors (with the obvious character). It follows that the closures of the $T(\sigma)$-orbits are contained in the fibres of our map. By induction over the dimension of σ it follows that using these degenerations we always can reach a point in an open σ-stratum such that the whole stratum is fixed by $T(\sigma)$, that is, $T(\sigma)$ acts by scalars on each N_j/N_{j+1}. Also, the torus $T(\sigma)$ is constant on each fibre: It is determined by the set of basic v such that $\bigoplus N_j/N_{j+1}$ has a basis of type v, which means that some $\gamma(m_1, .., m_d)$ does not vanish for suitable choices of $m_j \in M_{i(j)}$.

Now we study the fibre through such a degenerate point in the σ-stratum, σ the simplex

$$N_0 = \pi \cdot N_n \subset N_1 \subset \ldots \subset N_{n-1} \subset N_n.$$

We normalise the M_i so that they are contained in N_n but not in N_0. Denote $T' = T(\sigma) \cong \mathbb{G}_m^n$ the corresponding torus. Now assume that some point in the σ'-stratum has the same image in X_{min}, and is also fixed by $T' = T(\sigma')$. Then σ' has the form

$$N'_0 = \pi \cdot N'_n \subset N'_1 \subset \ldots \subset N'_{n-1} \subset N'_n.$$

(The numbers $n = \dim(T')$ are equal). There exists an element $\underline{x} = (x_1, ..., x_r)$ in $T(K)$, all of whose entries are powers of π, such that $x_i \cdot M_i$ is contained in N'_n but not in N'_0. We claim that this element lies in $T'(K)$: Suppose v and v' are two basic vectors such that $\bigoplus N_j/N_{j+1}$ has a basis of type v as well as of type v'. Such bases are given by elements $m_l \in M_{i(l)}$ such that $\gamma(m_1, ..., m_d)$ does not vanish on our fibre. They also form a basis of N_n, and their multiples $x_{i(l)} \cdot m_l$ form a basis of N'_n. Now the transition from a basis of type v for N_n to one of type v' is given by a matrix $a_{l,l'} \in GL_d(V)$. For N'_n the corresponding matrix is $x_{i(l)} a_{l,l'} x_{i(l)}^{-1}$. Thus if $a_{l,l'}$ is invertible it follows that $x_{i(l)}$ and $x_{i(l')}$ have the same valuation, and are equal because they are powers of π. As any two bases can be connected by a chain where one exchanges one basis element at a time, the assertion follows easily.

We now choose one such basis $\{m_1, ..., m_d\}$ and let T' act on E such that each m_l is an eigenvector of character $x_i(l)$. Then all N_j and N'_j are T'-stable, and have a decomposition into T'-eigenspaces $N_n = \bigoplus_{j=1}^n N_n^{(j)}$, etc.

Next for any i we define $j(i)$ as the index j such that $M_i \subseteq N_j$, $M_i \nsubseteq N_{j-1}$, and similarly for $j'(i)$ ($x_i \cdot M_i \subseteq N_{j(i)}$). Then $T(\sigma) = T(\sigma') \subseteq T$ consists of families constant on the fibres of either of these maps, and thus there exists a permutation δ of $\{1,...,n\}$ with $j'(i) = \delta(j(i))$. Also, we denote elements of $T' \cong \mathbb{G}_m^n$ by coordinates $(y_1,...,y_n)$, with $y_j = x_{j(i)}$. Then N'_j is obtained from N_n by applying an element of $T'(K)$ which is the product of $\underline{x}$ and an element with only entries 1 at position j if $\delta(j) \leq j'$, and π else.

Furthermore, for each i we have that $x_{j(i)} \cdot M_i$ is contained in $N'_{\delta(j(i))}$. Thus the projection of $x_i \cdot M_i$ onto the l-th coordinate in $E = K^l$ is a fractional ideal contained in the ideal generated by the projection of $N'_{\delta(j)}$, $j = j(i)$. If the element m_l is of type i, and $\tilde{j} = j(i)$, this means that the projection of M_i is contained in the ideal generated by $y_{\tilde{j}}/y_j$, and even π times that ideal if $\delta(j) < \delta(\tilde{j})$. This amounts to a certain upper bound on the valuation of $y_{\tilde{j}}/y_j$.

Finally multiplication by a suitable element of $T'(K)$ defines an isomorphism between N_j/N_{j-1} and $N_{\delta(j)}/N_{\delta(j)-1}$ (consider the grading for the T'-action). Our two points define elements in the associated projective spaces, that is, one-dimensional quotients. Furthermore, for each i with $j(i) = j$ the sublattice $M_i \subseteq N_j$ surjects onto the quotient (because our point does not lie in a smaller stratum), and similarly for $x_i \cdot M_i \subseteq N'_{\delta(j)}$. Furthermore, as our points have the same image in $\mathbb{P}(M_i)$ the two quotients of M_i coincide. It follows that under the isomorphism above between N_j/N_{j-1} and $N'_{\delta(j)}/N'_{\delta(j)-1}$ the two quotients differ by an automorphism which is scalar on the image of each M_i, that is, they are conjugate under $T(\sigma)$. As this operates trivially on the stratum the two points coincide. Using the eigenspace decomposition we can also consider them as points p_j of $\mathbb{P}(N_n^j)$.

We are now ready to give a complete description of the fibre. Namely consider the quotient by the diagonal $T^\dagger = T'/\mathbb{G}_m$. Its space of one-parameter groups is $N(T^\dagger) = \mathbb{Z}^n/\mathbb{Z}$. In $N(T^\dagger)_\mathbb{R}$ define a convex region $\mathcal{C}$ by the inequalities $z_{\tilde{j}} - z_j \leq a_{\tilde{j},j}$ where $a_{\tilde{j},j}$ is the minimum of the valuations of $pr_l(M_i)$, $(j = j(i), \tilde{j} = j(\tilde{i})$ with $m_l \in M_{\tilde{i}})$. It has non empty interior (it contains the "basic standard simplex", see below) and defines an affine torus embedding $\bar{T}^\dagger_{min}$ over V, with affine algebra the normalisation of the subring of $K[T^\dagger]$ generated by the elements $\pi^{a_{\tilde{j},j}} \cdot y_j/y_{\tilde{j}}$ (the z's correspond to valuations of the y's). Furthermore, we obtain a proper modification of torus embeddings $\bar{T}^\dagger \to \bar{T}^\dagger_{min}$ by subdividing $\mathcal{C}$. Namely define the basic standard simplex as the region defined by $z_1 \geq z_2 \geq ... \geq z_n \geq z_1 - 1$, and define a standard simplex as a transform of this by integral translation and permutation of coordinates. Our convex region then becomes subdivided by the standard simplices contained in it. These also parametrise an affine covering of the modification $\bar{T}^\dagger$. For example, to the standard simplex corresponds the affine torus embedding given by $V[y_1/y_2,..., y_{n-1}/y_n, \pi \cdot y_n/y_1]$, and its faces are obtained by inverting some of the variables. Also, the basic standard simplex is in no way distinguished: By a simple change of coordinates we can make any other standard simplex basic.

Now we claim that our map is locally (in the base) isomorphic to

$$\bar{T}^\dagger \times \prod_{j=1}^{n} \mathbb{P}(N_n^{(j)}) \to \bar{T}^\dagger_{min} \times \prod_{j=1}^{n} \mathbb{P}(N_n^{(j)}),$$

which obviously suffices to show assertion c). Our fibre will correspond to the fibre over the point in the target whose $\bar{T}^\dagger_{min}$-projection is the origin, and whose other coordinates are the p_j.

We first map a neighbourhood of this fibre to X. In each $\mathbb{P}(N_n^{(j)})$ choose an open neighbourhood U_j of p_j and in it a generator for the line bundle $\mathcal{O}(1)$. Our map will be defined on an open neighbourhood U of $\bar{T}^\dagger$ in the product of $\bar{T}^\dagger$ and these neighbourhoods U_j.

We have to construct it over each affine open subset of $\bar{T}^\dagger$ given by a standard simplex. For the model standard simplex we map to the system of rank one quotients $N_j \twoheadrightarrow L(N_j)$ which on $N_n^{(j)}$ are the quotients defined by the projections on $\mathbb{P}(N_n^{(j)})$, and for which the transition maps $L(N_{j-1}) \to L(N_j)$ are in the chosen trivialisations given by y_{j-1}/y_j. One checks that this is well defined near $\bar{T}^\dagger$ (the image has to avoid boundary strata) and an open immersion, with image contained in $U(\sigma)$, the union of strata defined by faces τ of σ.

Any other standard simplex is the transform of the model standard simplex by a permutation δ and a translation. The translation is given by an element $\bar{y} \in T'(K)$ all whose entries are powers of π. Then these define a new simplex σ' as above, with $N_n' = \bar{y} \cdot N_n$ and N_l' the sum of multiples $\bar{y} \cdot N_n^{(j)}$ (if $j \geq \delta(l)$), respectively $\pi \cdot \bar{y} \cdot M_n^{(jl)}$ (if $j < \delta(l)$): One checks that if that standard simplex is contained in our convex region, than these are sums of multiples of M_i's. Namely the inequalities defining our region become the previous inequalities on $\bar{y}$ and δ. Also, one builds an open immersion X using quotients on N_n^l's, and explicit maps on local generators. We also may assume that U is the preimage of an open $U_{min} \subseteq \bar{T}^\dagger_{min} \times \prod_{j=1}^n U_j$, as the subdivision map is proper.

Next we check that the fibres of $\bar{T}^\dagger \to \bar{T}^\dagger_{min}$ are contracted in X_{min}. It suffices to check this on each stratum ($T^\dagger$-orbit), since the fibres are connected and union of the closures of their intersection with strata. Now such a stratum is given by a face F of a standard simplex, and we may assume that it is the basic standard simplex σ. Then F corresponds to a face τ of σ defined by deleting some of the lattices N_j defining σ, and we may assume that these are different from N_n. Otherwise use cyclic permutations. In the affine space $N(T^\dagger)_\mathbb{R}$ the basic standard simplex is the set of z's with $z_1 \geq z_2 \geq ... \geq z_n \geq z_1 - 1$, and F is defined by setting $z_j = z_{j+1} = ... = z_{\tilde{j}}$ for certain pairs $j < \tilde{j}$. Each point of it is fixed by the subtorus of $T^\dagger$ defined by equality of the corresponding coordinates y_j.

Furthermore, F is contained in the face of $\mathcal{C}$ which is the intersection of all hyperplanes $z_{\tilde{j}} - z_j = a_{\tilde{j},j}$ containing it, that is all hyperplanes $z_{\tilde{j}} = z_j$ which contain it and for which $a_{\tilde{j},j} = 0$. Again we obtain a subtorus of $T^\dagger$ by requiring equality of the corresponding y_j's, and our fibres are the orbits of the subtorus on the F-stratum in $\bar{T}^\dagger$. We claim that this subtorus is contained in $T(\tau)$, so that its orbits are indeed contracted in X_{min}: The assertion means that for any subquotient $N_{\tilde{j}}/N_j$ occurring in τ, the action of our subtorus acts by scalars on the image of each M_i for which $j < j(i) \leq \tilde{j}$. However a basis of $N_{\tilde{j}}/N_j$ is given by the $m_l \in M_{\tilde{i}}$ for which $j < j(\tilde{i} \leq \tilde{j}$, and if the l-projection of M_i does not vanish modulo π then $a_{j(\tilde{i}),j(i)} = 0$. Thus our subtorus of $T^\dagger$ acts by the same character $y_{j(i)}$ on all non-vanishing l-projections, and we are done.

It follows that our map $U \to X_{min}$ factors over U_{min}: We have such a map on the generic fibre. The closure of its graph is proper over the product, and dominated by U. Thus the first projection is a universal homeomorphism, hence finite and birational, hence an isomorphism as U_{min} is normal. Also, U contains all $T^\dagger$-fixed points of the fibre of $X \to X_{min}$, hence the whole fibre. Thus U is an open neighbourhood of this complete fibre, and its image in X_{min} contains an open neighbourhood of the image. Also, the map $U_{min} \to X_{min}$ has a one-point fibre over this image, thus is finite over a (perhaps) smaller neighbourhood. As it is birational it is a local isomorphism.

Remark. If I_v is generated by π^{n_v}, consider in $(r+1)$-dimensional space the cone C generated by the elements (v, n_v) and by $(0,1)$. Its dual cone is spanned by its extremal rays. These are generated by either the r coordinate functions of the v's, or elements $(\rho, 1)$, where ρ is a linear combination of the coordinate functions of v such that $\rho(v) + n_v$ takes its minimum on a set of v's which contains a basis of $\mathbb{R}^d$. These ρ's have rational coefficients.

Over a suitably ramified extension V' of V we can form the lattice $M_\rho = \sum_{i=1}^{r} \pi^{\rho_i} \cdot M_i$. Then M_ρ has a basis of type v for each v on which $\rho(v) + n_v$ attains its minimum. By the condition on ρ, M_ρ must be an essential linear combination of the M_i's, hence the differences of the "minimum"-v's span $\mathbb{Z}^{r-1}$, ρ is integral modulo the diagonal, and M_ρ is up to homothety defined over V. Conversely any essential linear combination $\sum_{i=1}^{r} x_i \cdot M_i$ defines such a ρ, namely its vector of valuations translated by a suitable diagonal vector. Thus ρ's, essential lattices, and irreducible components of the special fibre of X_{min} are all in bijection.

Also, lemma 9 means that the semigroup of integral elements in C is generated by (v, n_v)'s and $(0,1)$. Thus each rational face is spanned by a subset of these. Especially this applies to the zero-set of $(\rho, 1)$. It is isomorphic to its projection Δ_ρ to $\mathbb{R}^r$, which is the convex hull of a certain set of v's. These span the sublattice of $\mathbb{Z}^r$ where the sum of coordinates is divisible by d, that is the character group of T/μ_d. Also, $v = (d_1, ..., d_r)$ lies in Δ_ρ if and only if the lattice M_ρ has a basis of type v, that is, d_i of its elements lie in a multiple $x_I \cdot M_i \subseteq M_\rho$. This means that the vector space $\bar{M}_\rho$ has a basis with d_i elements in the image of $x_i \cdot M_i$. Thus the following variant of lemma 9 gives important information about Δ_ρ:

Lemma 10. *Suppose $\bar{M}$ is a k-vector space of dimension d, $\bar{M}_i \subseteq \bar{M}$ subspaces, $v = (d_1, ..., d_r)$ a basic vector.*

a) $\bar{M}$ has a basis of type v if and only if for each subset of the i's, the sum of the corresponding d_i is not smaller than the dimension of the sum of the corresponding M_i's.

b) The semigroup generated by the v's satisfying the inequalities in a) is saturated, that is, equal to the intersection of a polyhedral cone and of the group generated by them.

Proof. The condition in a) is obviously necessary. If it holds, assume (say) that $d_1 > 0$. We may assume that k is infinite, and then find an element $m_1 \in \bar{M}_1$ which does not lie in any sum of M_i's with $i > 1$, unless this sum already contains $\bar{M}_1$. Then apply induction to $\bar{M}/k \cdot m_1$ and the images of the $\bar{M}_i$ in it.

For b) note that $\bar{M}^s$ has a basis of type w, thus also such a basis with each element contained in one of the factors.

From the lemma we derive that all boundary strata of Δ_ρ are defined by equations $d_i = 0$ or $\sum_{i\in A} d_i = \dim(\sum_{i\in A} \bar{M}_i)$, for certain subsets A of the indices. For a given v in such a stratum the A's for which this holds form a lattice, that is, are closed under unions and intersections. As the stratum is spanned by v's we thus may assume that the A's appearing in the equations defining it form a lattice. It then follows that the set of indices i can be decomposed into disjoint non empty subsets A_ν such that each A is the union of some of them, and $\sum_{i\in A_\nu} d_i = a_\nu$ is constant on the stratum. As $\bar{M}_i$ does not vanish, we always have $a_\nu > 0$. Thus there are at most d such A_ν. We derive that each boundary stratum meeting the interior of Δ has codimension $< d$.

6. Universal families

We want to study families of lattice arrangements, say over a scheme S. This means the following: We have an open scheme-theoretically dense subscheme $S^\circ \subseteq S$, and a vector bundle $\mathcal{E}^\circ$ on S°. Furthermore, there is a finite family of vector bundles $\mathcal{E}_i$ $(1 \leq i \leq r)$ on S, and isomorphisms of their restrictions $\mathcal{E}_i|S^\circ \cong \mathcal{E}^\circ$, or better of the corresponding $PGL(d)$-torsors. Furthermore, we require given a projective scheme $X_{\min} \to S \times \prod_{i=1}^r \mathbb{P}(\mathcal{E}_i)$, flat over S, such that after base change to a discrete valuation ring V via a V-point of S, generically in S°, $X_{\min}$ becomes isomorphic to the minimal model for the Deligne-scheme of the induced lattice arrangement over V. We shall construct such a family for which the base S is a proper scheme over $\mathrm{Spec}(\mathbb{Z})$, and for which each $\mathcal{E}_i$ is trivialised. Furthermore, any arrangement over a discrete valuation ring V, with trivialised M_i, will be the pullback by a unique V-point of S (generically in S°).

S will be stratified, with the strata parametrised by the following data: In affine r-space $\mathbb{R}^r$ consider the simplex Δ spanned by all basic v's ("$\sum z_i = d$"). Consider decompositions of Δ into finitely many convex subsets Δ_ρ each of which is spanned by a subset of v's which spans the sublattice of $\mathbb{Z}^r$ where the sum of coordinates is divisible by d. To each such decomposition corresponds one of the strata of S. The order of strata corresponds to refinement of decompositions.

Next fix r free $\mathbb{Z}$-modules $M_i = \mathbb{Z}^d$. We first construct a scheme S' which is a $T/\mathbb{G}_m$-torsor over S. It is a blow-up of ideals in the Grassmannian of rank-d-quotients M of $\bigoplus_{i=1}^r M_i$. For each basic $v = (d_1, ..., d_r)$ denote by $L_v \subseteq \bigwedge^d(\bigoplus_{i=1}^r M_i)$ the submodule $\bigwedge^{d_1} M_1 \wedge ... \wedge \bigwedge^{d_r} M_r$ of $\bigwedge^d(\bigoplus_{i=1}^r M_i)$. Denote by I_v the ideal generated by the coefficients of $L_v \to \det(M)$. Furthermore, for each w, I_w is the sum of all $I_{v_1} \cdot ... \cdot I_{v_s}$, for all representations $w = v_1 + ... + v_s$. This vanishes unless w has all components ≥ 0, and their sum is divisible by d. The number s is called the height of w. Now finally S' is the blow-up of all such I_w.

We show next that finitely many blow-ups suffice: Namely consider all possible decompositions Δ into Δ_ρ which can arise from lattice arrangements. There are only finitely many of them. Choose a number N such that for all possible Δ_ρ the following holds:

a) The module of relations (in $\mathbb{Z}^r$) $\sum_{v \in \Delta_\rho} a_v \cdot v = 0$ is generated by relations with $\sum_{v \in \Delta_\rho} |a_v| \leq 2N$.

b) Each basic v' is a linear combination $v' = \sum_{v \in \Delta_\rho} a_v \cdot v$ satisfying $\sum_{v \in \Delta_\rho} |a_v| \leq 2N - 1$.

Then we claim that it suffices to blow up all I_w with w of height $\leq N$.

Namely define S' as this blow-up. We have to show that all I_w become invertible on S'. Choose a point in S', and a map $\mathrm{Spec}(V) \to S'$, with V a discrete valuation ring, such that the closed point maps to our chosen point, and the open point $\mathrm{Spec}(K)$ maps into the open S'° which is the complement of all exceptional divisors. After pullback to V we get a lattice arrangement over V, thus a function ϕ and a decomposition Δ_ρ.

If R denotes the local ring of S', the I_w (w of height $\leq N$) are generated by regular elements f_w of R. Let f denote the product of all f_v, so $\mathrm{Spec}(R_f)$ is the intersection with the open stratum. For each relation $\sum_{v \in \Delta_\rho} a_v \cdot v = 0$

between $v \in \Delta_\rho$ the product $\prod_{v\in\Delta_\rho} f_v^{a_v}$ is a unit:

It suffices to consider relations with $\sum_{v\in\Delta_\rho} |a_v| \leq 2N$. These can be written as

$$\sum_{v\in\Delta_\rho} a_v \cdot v = w = \sum_{v\in\Delta_\rho} b_v \cdot v,$$

with $a_v, b_v \geq 0$, and the sums $\leq N$. But then the products $\prod_{v\in\Delta_\rho} f_v^{a_v}$ and $\prod_{v\in\Delta_\rho} f_v^{b_v}$ both generate the invertible ideal I_w, as can be checked by pullback to V. It follows that there exists an element $t_\rho \in T/\mu_d(R_f)$ such that $\chi_v(t_\rho)\cdot f_v$ is a unit for all $v \in \Delta_\rho$.

Then for all v', $\chi_{v'}(t_\rho)\cdot f_{v'} \in R$: Namely consider a relation

$$v' + \sum_{v\in\Delta_\rho} a_v \cdot v = w = \sum_{v\in\Delta_\rho} b_v \cdot v,$$

with $a_v, b_v \geq 0$, and the sums $\leq N-1, N$. Then $\prod_{v\in\Delta_\rho} f_v^{b_v}$ generates I_w (pullback to V), so divides the product of $f_{v'}$ and $\prod_{v\in\Delta_\rho} f_v^{a_v}$. Applying t_ρ gives the claim.

It implies that all $\chi_w(t_\rho)\cdot I_w$ are integral, and equal to R if w lies in the span of Δ_ρ. As Δ_ρ is arbitrary the claim follows. It also implies that $M_\rho = t_\rho(M)$ extends over R to a quotient of $\sum_{i=1}^r M_i$, as its Plücker coordinates are integral. Conversely, if we have such quotients, with Plücker coordinates indexed by $v \in \Delta_\rho$ units, it follows easily that all I_w are invertible. Namely if w lies in $s\cdot\Delta_\rho$, $M_\rho^{\oplus s}$ has a basis of type w. Thus our blow-ups are essentially equivalent to requiring the existence of these M_ρ.

Next we describe the logarithmic structure. Define H' as the torus with characters the free abelian group generated by all basic v. It contains as subtorus (corresponding to linear functions) T/μ_d. In the real vector space spanned by cocharacters of H' consider the set C' of convex functions $\phi \geq 0$ which are linear on the Δ_ρ's of some decomposition arising from a lattice arrangement (and thus determined by their values on v's). It is a union of rational polyhedral cones and defines a torus embedding $H' \subset \bar{H}'$ over $\mathbb{Z}$. The functions $\chi_v(t_\rho)\cdot f_v$ above define locally a map from S' to $\bar{H}'$, unique modulo H'-action.

Similarly denote by H the quotient H'/T. In the real vector space spanned by the cocharacters of H consider the image C of C'. It defines a torus embedding $\bar{H}$, with an equivariant map $\bar{H}' \to \bar{H}$, and local maps $S' \to \bar{H}$, unique modulo H-action. Thus globally a map into the quotient stack $\bar{H}/H$, that is a logarithmic structure defining the stratification on S'. Note that the map into $\bar{H}$ avoids the stratum defined by strictly positive ϕ's. Namely this translates into the fact that the f_v generate the unit ideal.

The projection $\bar{H}' \to \bar{H}$ is flat: Choose a subset $\{v_1, ..., v_r\}$ which is a basis of the characters of T/μ_d. Then any ϕ can be uniquely decomposed as

$\phi = \phi_0 + \lambda$, with λ linear and $\phi_0(v_j) = 0$. For each w the function $\phi_0(w)$ is a character of H and thus a meromorphic function on $\bar{H}$, which generates a fractional ideal J_w. We claim that $\bar{H}'$ is defined by the affine algebra (over $\mathcal{O}_{\bar{H}}$) $\bigoplus_w J_w \cdot \chi_w$:

Namely given a decomposition Δ_ρ the corresponding affine in $\bar{H}'$ has a ring with basis all positive integral linear combinations $\sum a_v \phi(v)$, considered as linear forms in ϕ (which is linear on each Δ_ρ). The w-eigenspace is spanned by such linear combinations with $\sum a_v \cdot v = w$. By convexity the corresponding linear form is $\geq \phi(w)$, which means that our element is a multiple of $g_w \cdot \chi_w$.

Now define for each w of height $\leq N$ L_w as the direct sum of all tensor products $L_{v_1} \otimes ... \otimes L_{v_s}$, for all representations $w = v_1 + ... + v_s$. The line bundles $I_s \cdot \det(M)^{\otimes s}$ define a T-invariant map p from S' into the product of all $\mathbb{P}(L_w)$. Define S as the scheme defined by $p_*(\mathcal{O}_{S'})$, finite over the product of projective spaces. If $\mathcal{L}_w$ are the universal quotients of L_w, we have over S maps $\bigotimes \mathcal{L}_{v_j} \to \mathcal{L}_w$ if $w = \sum v_j$. At least one of them is surjective, and the decomposition into Δ_ρ's is determined by which are surjective and which are not. It is thus constant on the fibres of $S' \to S$. The image of the open stratum in S' consists of the open subset $S^\circ \subset S$ where all $\bigotimes \mathcal{L}_{v_j} \to \mathcal{L}_w$ are isomorphic. If one chooses generators for r $\mathcal{L}_v$'s corresponding to a basis of the characters of T/μ_d, these isomorphisms define a compatible system of generators for all $\mathcal{L}_v$. Identifying them via this basis defines a quotient of $\bigwedge^d(\sum_{i=1}^r M_i)$ satisfying the Plücker equations (as S' dominates S). It is then given by a quotient M of $\sum_{i=1}^r M_i$ isomorphic to each M_i. It follows that S° is isomorphic to $PGL(d)^r/PGL(d)$.

Define a H'-torsor over S by choosing local generators l_v for each v. Then the ratios of different products $\prod l_{v_j}$ with the same $\sum v_j = w$ define a map from S° into H which pulls back to the previously defined map on S'°. It thus defines locally a map $S \to \bar{H}$, unique up to H-action. We claim that (locally) the $\mathbb{G}_m$-torsor over S defining trivialisations of $\det(M)$ becomes equal to the fibered product $S' \times_{\bar{H}} U'$, $U' \subset \bar{H}$ the complement of the stratum defined by strictly positive ϕ's:

Namely if in the torsor over S' we have chosen generators of $\det(M)$ as well as all $p^*\mathcal{L}_v = I_v \cdot \det(M)$, their quotients define generators f_v of I_v, which are the coordinates of a map into U'. Thus a map from the torsor over S' into the fibered product.

Conversely, locally in S the line bundles $J_v^{-1} \cdot \mathcal{L}_v$ (J_v is an invertible fractional ideal on $\bar{H}$, trivial on the open stratum, which pulls back to S) have generators $g_v^{-1} \cdot l_v$ which respect all relations $\sum v_j = w$. Over the fibered product the elements $f_v = g_v \cdot \chi_v$ become regular functions generating an invertible ideal I_v, and the line bundles $I_v^{-1} \cdot \mathcal{L}_v$ become trivialised by $f_v^{-1} \cdot l_v$, compatible with relations. We thus get a quotient line of $\bigwedge^d(\sum_{i=1}^r M_i)$ satisfying the Plücker relations over the open stratum, which is scheme-theoretically dense by flatness. Also, the sum of the I_v is the unit ideal. Hence a quotient M of $\sum_{i=1}^r M_i$ with trivial $\det(M)$, invertible ideals I_v etc. That is, we have defined a map from the fibered product into the $\mathbb{G}_m$-torsor over S'.

The compositions either way of these maps are the identities: It suffices to check that over the open strata. There $\bar{H}'$ becomes an affine space over $\bar{H}$, U' is obtained by removing the origin, and the assertion becomes pretty obvious.

Next we construct a scheme X_{min}, extending $\mathbb{P}(\mathcal{E}^\circ)$ on S'°, as follows:

The spectrum of the affine algebra $\mathcal{A} = \bigoplus_w \mathcal{L}_w$ defines a torus embedding Z_{min} of T/μ_r, flat over S'. It is locally isomorphic to the pullback of $\bar{H}'$. Next we have maps γ_v from $\bigwedge^{d_1} M_1 \wedge ... \wedge \bigwedge^{d_r} M_r$ to $\mathcal{L}_v$. Finally, in the affine space defined by the tensor product of $\mathcal{A}$ and $S(\bigoplus_{i=1}^r M_i)$, consider the closed subscheme X_{min} defined by relations (*). Over the complement of the origin in Spec($\mathcal{A}$) it embeds locally into the product of Z_{min} and an affine space $\mathbb{A}^d$ (as in the proof of theorem 7). This embedding is open over S'° and thus in general. The resulting scheme will be a $T \times \mathbb{G}_m$-torsor over X_{min}.

Conversely, start with a lattice arrangement over a discrete valuation ring V. We obtain a decomposition by Δ_ρ's, quotients M_ρ (the essential linear combinations), and ideals I_v as above, and thus a map from Spec(V) to the S' corresponding to the decomposition, sending the special point into the closed stratum.

Theorem 11. *The construction above defines an equivariant embedding $S^\circ \subset S$, with $S^\circ = PGL(d)^r/PGL(d)$, and S projective. S is smooth if $d = 2$.*

Proof. Only the assertion for $d = 2$ remains. Then S maps locally to the versal deformation of X_{min}. On the other hand over this versal deformation all I_w are already invertible because the lattices M_ρ exist locally, and we obtain an inverse.

Of course we hope that S is smooth over $\bar{H}/H$. The reason for this is the case $d = 2$, and the fact that the toroidal structure given by $\bar{H}/H$ seems to be the only reasonable candidate.

7. Applications

The moduli problems in the introduction lead to certain open subschemes of S. For example, we want to consider lattice arrangements generated by $M_1, ..., M_r$ such that for some i there exists an element $x_i \in K^*$ with $x_i \cdot M_{i-1} \subseteq M_i$, and the quotient is generated by one element. This is equivalent to the fact that for each $s > 0$, $\bigwedge^{d-s}(x_i \cdot M_{i-1}) \wedge \bigwedge^s M_i$ coincides with $\bigwedge_d M_i$. Such an x_i exists if the valuation of I_v (the image of γ_v) is linear on the line between $v = (d-1, 1)$ and $v' = (0, d)$ (all other components vanish), that is, if this line is contained in a Δ_ρ.

Hence if we require this condition for all i (setting $M_0 = M_r$), we obtain an open subscheme of S. On it the product of the x_i generates a Cartier-divisor. Locally adjoining a unit u and requiring that this product is $\pi \cdot u$ then defines a toroidal scheme over V. If we require in addition that the M_i become quotients of the standard sequence of lattices in $\mathbb{Z}_p^r$ we get a smooth scheme over it (at the most degenerate point choose elements $m_i \in M_i$ which generate $M_i/x_i \cdot M_{i-1}$). On it the group $PGL(d)^r$ operates freely, and we may form the quotient Y (the construction provides us with local bases for each M_i). Hence we obtain (using the valuative criterion):

Theorem 12. *Consider the standard chain of lattices*

$$N_g = \pi \cdot N \subset N_{g-1} \subset \ldots \subset N_0 = N = V^g, \quad g = a + b,$$

and the scheme X/V which classifies compatible systems of quotients of rank b (equivalently we can consider subbundles of rank a).

The above defines a proper birational map $Y \to X$. If $a = 2$ then Y has toroidal singularities.

For the symplectic variant consider lattices $T \cdot M_n \subset M_0 \subset \ldots \subset M_g$, together with linear forms l_i on M_i $(0 \leq i \leq g)$ and a bilinear pairing $M_0 \times M_g \to$ scalars. These should have the property that the l_i generate the quotient $M_i^{\text{dual}}/M_{i+1}^{\text{dual}}$, that all of them form a basis of M_0^{dual}, and that the bilinear form on $M_0 \times M_g$ is non degenerate and symmetric on $M_0 \times M_0$. This defines a smooth covering: use that the inclusion $M_0 \to M_n$ can be written in standard diagonal form, since we have added all essential linear combinations of M_0 and M_n. After base change $T \to \pi = p$ we obtain an open subset of the symplectic moduli problem. Namely we have trivialisations of M_0 and M_g. Embed M_g into the standard symplectic space R^{2g} with the first g components the $p \cdot l_i$, and the rest given by the trivialisation. The image is isotropic, and the M_i become intersections of M_g with the standard sublattices of R^{2g} where the components in degrees between i and $g-1$ are divisible by p.

Theorem 13. *If $g = 2$ this defines a model with toroidal singularities.*

In addition the cover defined by a $\Gamma_1(p)$-level structure is etale locally given by adjoining $(p-1)$st roots of generators of the annihilators of M_i/M_{i-1}, and thus hopefully also toroidal. However unfortunately I cannot give an explicit description of the model.

Examples.

a) As a first example consider bundles of rank two. We have seen that they give rise to trees, which can be classified and easily described. For example, if we start with three lattices we have to add at most one additional essential linear combination. The resulting graph looks like a star with three rays, that is, it has three edges connecting this new vertex to the three original lattices. If the edges define divisors generated by elements x, y, z then the map $M_0 \to M_2$ can be locally represented by the diagonal matrix $Diag(x, xyz)$. The entries must divide π, and the local equation for the resolution is $x \cdot y \cdot z \cdot t = \pi$. That is, our model is regular and semistable over V. For example, this applies to the Siegel case for two-dimensional abelian varieties.

b) If we start with two lattices $M \subset N$, adding essential linear combinations amounts to the following: Blow up the ideals generated by $i \times i$-minors of the inclusion, for $0 < i < d$ (these will be the ideals I_w). Then locally we can choose a basis in which the inclusion is given by a diagonal matrix with entries $(t_1, t_1 \cdot t_2, \dots, t_1 \cdot t_2 \cdot \dots \cdot t_d)$. The essential linear combinations are $M + x \cdot N$ with $x = t_1 \cdot \dots \cdot t_i$ a coefficient in this matrix. After adding these the result is closed under essential linear combinations.

c) (This example led to the definition of minimal models.) In general one could repeat the process of adding essential linear combinations, and ask whether this stops at some stage. This holds if $d = 2$, or if we start with only two lattices, or if the base is a discrete valuation ring. However we cannot expect this in general. For example, consider three lattices L, M, N over V, where L is generated by a basis $\{e, f, g\}$, M by $\{\pi^{-1} \cdot e, f, g\}$ and N by $\{e, \pi^{-n} \cdot f, g\}$. Then we obtain as essential linear combination $M + N$ and after that $N' = L + \pi \cdot (M + N) = L + \pi \cdot N$. In the triple (L, M, N') we have replaced n by $n - 1$, and by induction we obtain all lattices $L + \pi^i \cdot N, M + \pi^i \cdot N$. Their number is approximately $2n$ thus unbounded. It follows that the process of adding essential linear combinations cannot stop over any base from which we can obtain all these examples (with n varying) by base change. This applies to the base $R = V[x, y]$ (polynomial ring) with the lattices L, M, N generated by $\{e, f, g\}, \{x^{-1} \cdot e, f, g\}$, and $\{e, y^{-1} \cdot f, g\}$.

Bibliography

[CN] C.L. Chai, P. Norman, *Singularities of $\Gamma_0(p)$-level structures*, J. Alg. Geom. 1 (1992), 251-277.

[D] V.G. Drinfeld, *Coverings of p-adic symmetric regions*, Funct. Anal. and its Appl. 10 (1976), 107-115.

[DP] P. Deligne, G. Pappas, *Singularités des espaces de modules de Hilbert, en les caractéristiques divisant le discriminant*, Comp. Math. 90 (1994), 59-79.

[F] G. Faltings, *Explicit resolution of local singularities of moduli-spaces*, Journ. reine angew. Math. 483 (1997), 183-196.

[G] A. Genestier, *A toroidal resolution for the bad reduction of some Shimura varieties*, preprint (1999).

[dJ] A.J. de Jong, *Moduli of abelian varieties and Dieudonné-modules of finite group schemes*, Thesis Nijmegen (1992).

[K] K. Kato, *Logarithmic structures of Fontaine-Illusie*, Global analysis, geometry and number theory (J.I. Igusa, editor), 191-224.

[KKMS] G. Kempf, F. Knudsen, D. Mumford, B. Saint-Donat, *Toroidal embeddings I*, Springer Lecture Notes 339, Springer-Verlag (1991).

[KM] N. Katz, B. Mazur, *Arithmetic moduli of elliptic curves*, Princeton Univ. Press, Princeton (1985).

[M] G.A.Mustafin, *Nonarchimedean uniformization*, Math. USSR Sbornik 34 (1978), 187-214.

Progress in Mathematics, Vol. 195, © 2001 Birkhäuser Verlag Basel/Switzerland

FORMAL BRAUER GROUPS AND MODULI OF ABELIAN SURFACES

G. VAN DER GEER AND T. KATSURA

1. Introduction

Let X be an algebraic surface over an algebraically closed field k of characteristic $p > 0$. We denote by Φ_X the formal Brauer group of X and by $h = h(\Phi_X)$ the height of Φ_X. In a previous paper, [6], we examined the structure of the stratification given by the height h in the moduli space of K3 surfaces, and we determined the cycle class of each stratum. We also showed that the final stratum is non-reduced. In this paper, we use the methods of [6] to treat the case of abelian surfaces. In this case, the situation is more concrete, and so we can more easily determine the structure of the stratification given by the height $h(\Phi_A)$ in the moduli of abelian surfaces. For the local structure we refer to [20].

On the moduli of principally polarized abelian varieties in positive characteristic there is another natural stratification, called the Ekedahl-Oort stratification, cf. [17], and one can calculate the corresponding cycle classes [5]. Although our three strata coincide set-theoretically with strata of the Ekedahl-Oort stratification, there is a subtle difference: one of the strata comes with multiplicity 2.

We will here summarize our results. We consider the moduli stack $M = \mathcal{A}_2$ of principally polarized abelian surfaces over k; alternatively, we can consider the moduli spaces $M = \mathcal{A}_{2,n}$ ($n \geq 3$, $p \not| n$) of principally polarized abelian surfaces with level n-structure. We know that M is a 3-dimensional algebraic stack (variety). We let $\pi : \mathcal{X} \to M$ be the universal family over M. We set

$$M^{(h)} := \{s \in M : h(\Phi_{\mathcal{X}_s}) \geq h\}.$$

Note that $M^{(3)} = M^{(\infty)}$. The moduli stack M possesses a natural compactification $\tilde{M}$ which is an example of a smooth toroidal compactification, cf. [3] or [1]. It carries a universal family $\tilde{\pi} : \tilde{\mathcal{X}} \to \tilde{M}$. We can extend in a natural way the loci $M^{(h)}$ to loci in $\tilde{M}$ which are again denoted by $M^{(h)}$.

1991 *Mathematics Subject Classification.* Primary 14K02; Secondary 14K15.
Key words and phrases. Abelian surface, formal Brauer group.

For $h = 2, 3$, we denote by $M_F^{(h)}$ the scheme-theoretic zero locus in $\tilde{M}$ of the Frobenius action

$$F : \mathrm{H}^2(A, W_{h-1}(\mathcal{O}_A)) \longrightarrow \mathrm{H}^2(A, W_{h-1}(\mathcal{O}_A)).$$

(See Section 7 for details.) We set $M_F^{(\infty)} = M_F^{(3)}$. We will show $M^{(2)} = M_F^{(2)}$ and $M^{(\infty)} = (M_F^{(\infty)})_{\mathrm{red}}$ in $\tilde{M}$, and the following theorem describes the Chow classes of these loci in $\tilde{M}$. We denote by v the first Chern class of the coherent sheaf $R^0\tilde{\pi}_*\Omega^2_{\tilde{\mathcal{X}}/\tilde{M}}$ in the Chow group $CH^1_{\mathbf{Q}}(\tilde{M})$.

Theorem. *The classes of the loci $M_F^{(h)}$ in the Chow group $CH^*_{\mathbf{Q}}(\tilde{M})$ are given by*

$$M_F^{(2)} = (p-1)v, \; M_F^{(\infty)} = (p-1)(p^2-1)v^2.$$

To prove these results, we use the characterization of the height of Φ_A by the action of the Frobenius morphism on $\mathrm{H}^2(W_i(\mathcal{O}_A))$ which was obtained in [6] (cf. Section 2). We also investigate the natural images of $\mathrm{H}^1(A, B_i)$ and $\mathrm{H}^1(A, Z_i)$ in $\mathrm{H}^1(A, \Omega^1_A)$, from which the usual invariants of abelian surfaces can be read off and which are related to the tangent space to $M^{(h)}$. (For the case of K3 surfaces, see [6] and [15].)

Comparing our results with those in [5], we have the following relation between $M^{(h)}$ and $M_F^{(h)}$.

Theorem. *In the Chow ring $CH^*_{\mathbf{Q}}(\tilde{M})$ we have*

$$M_F^{(2)} = M^{(2)}, \; M_F^{(\infty)} = 2M^{(\infty)}.$$

To examine the structure of abelian surfaces A, one usually employs the first cohomology group $\mathrm{H}^1(A, \mathcal{O}_A)$. Instead, in this paper, we mainly use $\mathrm{H}^2(A, W_i(\mathcal{O}_A))$ and we will show that our techniques developed in [6] also work for the study of abelian surfaces. We hope and expect that these techniques can be used more widely for other varieties.

Most part of this work was done at the Max-Planck-Institut für Mathematik in Bonn. The authors would like to thank the Max-Planck-Institut for excellent working conditions during their stay in 1998/99.

2. Preliminaries

Let X be a non-singular complete algebraic variety defined over an algebraically closed field k of characteristic $p > 0$. We denote by $W_n(\mathcal{O}_X)$ the sheaf of Witt vectors of length n (cf. J.-P. Serre, [19]). The sheaf $W_n(\mathcal{O}_X)$ is a coherent sheaf of rings. Sometimes, we write W_n instead of $W_n(\mathcal{O}_X)$ for the sake of simplicity. We denote by F (resp. V, resp. R) the Frobenius map (resp. the Verschiebung, resp. the restriction map). They satisfy relations $RVF = FRV = RFV = p$.

The cohomology groups $\mathrm{H}^i(X, W_n(\mathcal{O}_X))$ are finitely generated $W_n(k)$-modules. The projective system $\{W_n(\mathcal{O}_X), R\}_{n=1,2,\ldots}$ defines the cohomology group

$$\mathrm{H}^i(X, W(\mathcal{O}_X)) = \text{proj. lim } \mathrm{H}^i(X, W_n(\mathcal{O}_X)).$$

This is a $W(k)$-module, but not necessarily a finitely generated $W(k)$-module. The semi-linear operators F and V act on it with relations $FV = VF = p$.

For any artinian local k-scheme S with residue field k, we consider the kernel

$$\Phi_X^i(S) = \text{Ker } \mathrm{H}^i(X \times S, \mathbf{G}_m) \longrightarrow \mathrm{H}^i(X, \mathbf{G}_m).$$

Here $\mathbf{G}_m$ is the multiplicative group of dimension 1 and cohomology is étale cohomology. This gives a contravariant functor $\Phi_X^i : \text{Art} \to \text{Ab}$ from the category of artinian local k-schemes with residue field k to the category of abelian groups. This functor was introduced and investigated by Artin and Mazur [2]. They showed that the tangent space of Φ_X^i is given by $\mathrm{H}^i(X, \mathcal{O}_X)$ and proved the following crucial theorem.

Theorem 2.1. *The Dieudonné module of the Artin-Mazur formal group* Φ_X^i *is given by*

$$D(\Phi_X^i) \cong \mathrm{H}^i(X, W(\mathcal{O}_X)).$$

If A is an abelian surface over k then Φ_A^1 is isomorphic to the formal completion of the Picard scheme $\mathrm{Pic}^0(A)$ of A, which is a formal Lie group. Moreover, we have $\mathrm{H}^3(A, \mathcal{O}_A) = 0$ and $\mathrm{H}^2(A, \mathcal{O}_A) \cong k$. Therefore, by a criterion of pro-representability in [2], Cor. 4.1, p. 109, we know that $\Phi_A = \Phi_A^2$ is pro-representable by a formal Lie group of dimension 1, which is called a formal Brauer group of A. One-dimensional formal groups are classified by their height h, which is a positive integer or $h = \infty$ for the case of the additive group.

3. Abelian Surfaces

Recall that we are working over an algebraically closed field of characteristic p. For an abelian surface A, we denote by $[p]_A$ the homomorphism given by multiplication by p on A, and by Ker $[p]_{A,\mathrm{red}}$ the reduced group scheme associated with the kernel Ker $[p]_A$. The dimension of Ker $[p]_{A,\mathrm{red}}$ over $\mathbf{F}_p$ is called the p-rank of A, and we denote it by $r(A)$. As is well-known, we have $0 \leq r(A) \leq 2$. Since any regular 1-form on an abelian surface is d-closed, the Cartier operator C acts on $\mathrm{H}^0(A, \Omega_A^1)$. We have also a natural action of the Frobenius map on $\mathrm{H}^1(A, \mathcal{O}_A)$, which is dual to the Cartier operator C on $\mathrm{H}^0(A, \Omega_A^1)$. The abelian surface A is said to be *ordinary* if $r(A) = 2$, which is equivalent to F (resp. C) being bijective on $\mathrm{H}^1(A, \mathcal{O}_A)$ (resp. $\mathrm{H}^0(A, \Omega_A^1)$). Furthermore, A is said to be *supersingular* if A is isogenous to a product of two supersingular elliptic curves, which is equivalent to the effect that F (resp. C) is nilpotent on $\mathrm{H}^1(A, \mathcal{O}_A)$ (resp. $\mathrm{H}^0(A, \Omega_A^1)$). Finally, A is said to be *superspecial* if A is isomorphic to a product of two supersingular elliptic curves, which is equivalent to F (resp. C) being the zero map on $\mathrm{H}^1(A, \mathcal{O}_A)$ (resp. $\mathrm{H}^0(A, \Omega_A^1)$). The superspecial case can be characterized numerically by the a-number of A. Here $a(A) = \dim_k \mathrm{Hom}(\alpha_p, A)$ and we know

- $a = 0 \iff r = 2$,
- $a \geq 1 \iff r \leq 1$,
- $a = 2 \iff A$ is superspecial.

For details on abelian surfaces, see [10], [16] and [13]. The following lemma is well-known.

Lemma 3.1. *The height h of the formal Brauer group Φ_A of an abelian surface A is as follows:*

(1) $h = 1$ *if* $r(A) = 2$, *i.e.* A *is ordinary,*
(2) $h = 2$ *if* $r(A) = 1$,
(3) $h = \infty$ *if* $r = 0$, *i.e.* A *is supersingular.*

Proof. We denote by $\mathrm{H}^i_{cris}(A)$ the i-th cristalline cohomology of A and as usual by $\mathrm{H}^i_{cris}(A)_{[\ell,\ell+1[}$ the additive group of elements in $\mathrm{H}^i_{cris}(A)$ whose slopes are in the interval $[\ell, \ell+1[$. By the general theory in Illusie [8], we have

$$\mathrm{H}^2(A, W(\mathcal{O}_A)) \otimes_W K \cong (\mathrm{H}^2_{cris}(A) \otimes_W K)_{[0,1[}$$

with K the quotient field of W. The theory of Dieudonné modules implies

$$h = \dim_K D(\Phi_A) = \dim_K \mathrm{H}^2(A, W(\mathcal{O}_A)) \otimes_W K \quad \text{if } h < \infty,$$

and $\dim_K D(\Phi_A) = 0$ if $h = \infty$. We know the slopes of $\mathrm{H}^1_{cris}(A)$ for each case. Since we have

$$\mathrm{H}^2_{cris}(A) \cong \wedge^2 \mathrm{H}^1_{cris}(A),$$

counting the number of slopes in $[0, 1[$ of $\mathrm{H}^2_{cris}(A)$ gives the result. Cf. also formula (i) in [2], p. 129. □

We shall need the following lemma.

Lemma 3.2. *For an abelian surface A the following sequence is exact.*

$$0 \to \mathrm{H}^2(A, W_{n-1}(\mathcal{O}_A)) \xrightarrow{V} \mathrm{H}^2(A, W_n(\mathcal{O}_A)) \xrightarrow{R^{n-1}} \mathrm{H}^2(A, \mathcal{O}_A) \to 0.$$

Proof. By the exact sequence

$$0 \to W_{n-1}(\mathcal{O}_A) \xrightarrow{V} W_n(\mathcal{O}_A) \xrightarrow{R^{n-1}} \mathcal{O}_A \to 0,$$

we have the long exact sequence

$$\to \mathrm{H}^2(A, W_{n-1}(\mathcal{O}_A)) \xrightarrow{V} \mathrm{H}^2(A, W_n(\mathcal{O}_A)) \xrightarrow{R^{n-1}} \mathrm{H}^2(A, \mathcal{O}_A) \to 0.$$

Since the Picard scheme of A is reduced, all Bockstein operations vanish by Mumford [11], p. 196. The result follows from this fact. □

Using this lemma and $FV = VF$, we have the following lemma.

Lemma 3.3. *If F acts as zero on $\mathrm{H}^2(A, W_n(\mathcal{O}_A))$ then F acts as zero on $\mathrm{H}^2(A, W_i(\mathcal{O}_A))$ for any $i < n$.*

The following theorem is an analogue of a result in [6].

Theorem 3.4. *The height satisfies $h(\Phi_A) \geq i+1$ if and only if the Frobenius map $F : \mathrm{H}^2(A, W_i(\mathcal{O}_A)) \to \mathrm{H}^2(A, W_i(\mathcal{O}_A))$ is the zero map. In particular, we have the following characterization of the height:*

$$h(\Phi_A) = \min\{i \geq 1 \mid [F : \mathrm{H}^2(W_i(\mathcal{O}_A)) \to \mathrm{H}^2(W_i(\mathcal{O}_A))] \neq 0\}.$$

Corollary 3.5. *Let A be an abelian surface over k. Then A is supersingular if and only if the Frobenius endomorphism on $\mathrm{H}^2(A, W_2(\mathcal{O}_A))$ is zero.*

Proof. If the height is finite, then we know $h \leq 2$ by Lemma 3.1. So this corollary follows from Theorem 3.4 and Lemma 3.1 . □

Corollary 3.6. *Let A be an abelian surface over k. Set $\mathrm{H} = \mathrm{H}^2(A, W_2(\mathcal{O}_A))$. If the height h of Φ_A is finite, then $F(\mathrm{H}) = R^{h-1}V^{h-1}(\mathrm{H})$.*

Proof. The $(h-1)$-th step $V^{h-1}\mathrm{H}^2(W(\mathcal{O}_A))$ in the filtration

$$V^2\mathrm{H}^2(W(\mathcal{O}_A)) \subset V\mathrm{H}^2(W(\mathcal{O}_A)) \subset \mathrm{H}^2(W(\mathcal{O}_A))$$

maps surjectively to the corresponding step $R^{h-1}V^{h-1}\mathrm{H}$ of the filtration on H. Therefore, this corollary follows from

$$V^{h-1}\mathrm{H}^2(W(\mathcal{O}_A)) = F\mathrm{H}^2(W(\mathcal{O}_A)).$$

□

For an element $\omega \in \mathrm{H}^2(W(\mathcal{O}_A))$, we denote by $\bar{\omega}$ the natural restriction of ω in $\mathrm{H}^2(A, W_i(\mathcal{O}_A))$. The following corollary follows immediately from $FV = VF$.

Corollary 3.7. *Let A be an abelian surface over k. If $h(\Phi_A) = h < \infty$ and if $\{\omega, V^{h-1}\omega\}$ is a W-basis of $\mathrm{H}^2(X, W(\mathcal{O}_X))$, then F acts as zero on $\mathrm{H}^2(A, W_i(\mathcal{O}_A))$ if and only if $F(\bar{\omega}) = 0$, where $\bar{\omega}$ is the image of ω in $\mathrm{H}^2(A, W_i(\mathcal{O}_A))$.*

Corollary 3.8. *Let A be an abelian surface over k. Putting $h(\Phi_A) = h$, we have*

$$\dim_k \mathrm{Ker}\ [F : \mathrm{H}^2(W_i) \to \mathrm{H}^2(W_i)] = \begin{cases} 0 & \text{if } h = 1 \\ 1 & \text{if } h = 2 \\ i & \text{if } h = \infty \end{cases}$$

Proof. If $h = 1$ then $F : \mathrm{H}^i(A, \mathcal{O}_A) \to \mathrm{H}^i(A, \mathcal{O}_A)$ is a p-linear isomorphism for any i. Therefore, using exact sequences in Lemma 3.2 we can inductively show that $F : \mathrm{H}^2(W_i) \to \mathrm{H}^2(W_i)$ is an isomorphism for any i.

If $h = 2$, then by Corollary 3.7 we have $\dim_k \mathrm{Ker}\ [F : \mathrm{H}^2(\mathcal{O}_A) \to \mathrm{H}^2(\mathcal{O}_A)] = 1$. Assume $i \geq 2$. Using the notation in Corollary 3.7, we know that $\langle V^{i-1}R^{i-1}\bar{\omega}\rangle$ is a basis of Ker F. Therefore, it is one-dimensional.

If $h = \infty$, then by Theorem 3.4, we see that $\mathrm{Ker}\ [F : \mathrm{H}^2(W_i) \to \mathrm{H}^2(W_i)] = \mathrm{H}^2(W_i)$. Therefore, using Lemma 3.2 we get the result by induction. □

Let A be an abelian surface and assume that F is zero on $\mathrm{H}^2(\mathcal{O}_A)$. Then we have $F\mathrm{H}^2(W_2) \subset V\mathrm{H}^2(\mathcal{O}_A)$ and F vanishes on $V\mathrm{H}^2(\mathcal{O}_A)$. Since we have the natural (σ^{-1})-isomorphism $\mathrm{H}^2(\mathcal{O}_A) \cong V\mathrm{H}^2(\mathcal{O}_A)$, we have the induced homomorphism

$$\phi_2 : \mathrm{H}^2(\mathcal{O}_A) \cong \mathrm{H}^2(W_2)/V\mathrm{H}^2(\mathcal{O}_A) \to V\mathrm{H}^2(\mathcal{O}_A) \cong \mathrm{H}^2(\mathcal{O}_A).$$

This map is σ^2-linear. The following theorem is clear by the construction.

Theorem 3.9. *Suppose F is zero on $\mathrm{H}^2(\mathcal{O}_A)$. Then F is zero on $\mathrm{H}^2(W_2)$ if and only if $\phi_2 = 0$.*

4. Cohomology Groups of Abelian Surfaces

In this section we introduce the 'Illusie' sheaves $B_i\Omega_A^1$ and $Z_i\Omega_A^1$ and we show that the invariants of abelian surfaces considered in the preceding section can be read off from the cohomology of these sheaves. Another motivation for the study of these cohomology groups is the relation with the tangent spaces of our strata on the moduli space.

Let A be an abelian surface defined over k. We define sheaves $B_i\Omega_A^1$ inductively by $B_0\Omega_A^1 = 0$, $B_1\Omega_A = d\mathcal{O}_A$ and $C^{-1}(B_i\Omega_A^1) = B_{i+1}\Omega_A^1$. Similarly, we define sheaves $Z_i\Omega_A^1$ inductively by $Z_0\Omega_A^1 = \Omega_A^1$, $Z_1\Omega_A^1 = \Omega_{A,\mathrm{closed}}^1$, the sheaf of d-closed forms and by setting

$$Z_{i+1}\Omega_A^1 := C^{-1}(Z_i\Omega_A^1).$$

Sometimes we simply write B_i (resp. Z_i) instead of $B_i\Omega_A^1$ (resp. $Z_i\Omega_A^1$). Note that we have the inclusions

$$0 = B_0 \subset B_1 \subset \ldots \subset B_i \subset \ldots \subset Z_i \subset \ldots \subset Z_1 \subset Z_0 = \Omega_X^1.$$

The sheaves B_i and Z_i can be viewed as locally free subsheaves of $(F^i)_*\Omega_A^1$ on $A^{(p^i)}$. The inverse Cartier operator gives rise to an isomorphism

$$C^{-i} : \Omega_A^1 \xrightarrow{\sim} Z_i\Omega_A^1/B_i\Omega_A^1.$$

We also have an exact sequence

$$0 \to Z_{i+1}\Omega_A^1 \longrightarrow Z_i\Omega_A^1 \xrightarrow{dC^i} d\Omega_A^1 \to 0.$$

The proofs of the results in this section are similar to those in [6].

Lemma 4.1. *Let A be an abelian surface over k. Then the dimension $h^j(A, B_i)$ for $i \geq 1$ is as in the following table.*

type	$h^0(B_i)$	$h^1(B_i)$	$h^2(B_i)$
$h=1$	0	0	0
$h=2$	1	2	1
$h=\infty,\ a=1$	$\begin{cases}1 & i=1\\ 2 & i\geq 2\end{cases}$	$\begin{cases}2 & i=1\\ 2+i & i\geq 2\end{cases}$	i
$h=\infty,\ a=2$	2	$2+i$	i

The exact sequence

$$0 \to d\Omega_A^1 \longrightarrow \Omega_A^2 \xrightarrow{C} \Omega_A^2 \to 0.$$

implies:

Lemma 4.2. *Let A be an abelian surface over k. Then the dimension $h^i(d\Omega_A^1)$ of $\mathrm{H}^i(A, d\Omega_A^1)$ is as in the following table.*

type	$h^0(d\Omega_A^1)$	$h^1(d\Omega_A^1)$	$h^2(d\Omega_A^1)$
$h=1$	0	0	0
$h=2$	1	2	1
$h=\infty,\ a=1$	1	2	1
$h=\infty,\ a=2$	1	3	2

Similarly, the exact sequence $0 \to B_i \longrightarrow Z_i \xrightarrow{C^i} \Omega_A^1 \to 0$ implies

Lemma 4.3. *If $h=1$ we have $h^0(Z_i)=2$, $h^1(Z_i)=4$ and $h^2(Z_i)=2$.*

We also need the following lemma.

Lemma 4.4. *Consider the natural inclusions $Z_i \hookrightarrow Z_{i-1}$ ($i \geq 1$) and the induced homomorphisms $\mathrm{H}^1(A, Z_i) \to \mathrm{H}^1(A, Z_{i-1})$. The surjectivity of the map $\mathrm{H}^1(A, Z_i) \to \mathrm{H}^1(A, Z_{i-1})$ implies that the map $\mathrm{H}^1(A, Z_j) \to \mathrm{H}^1(A, Z_{j-1})$ is surjective for any $j \geq i$.*

Lemma 4.5. *Assume that $h \geq 2$, and that A is not superspecial. Then, either $\dim \mathrm{H}^1(A, Z_i) = \dim \mathrm{H}^1(A, Z_{i-1})$ or $\dim \mathrm{H}^1(A, Z_i) = \dim \mathrm{H}^1(A, Z_{i-1}) + 1$. Moreover, the latter case occurs if and only if the natural homomorphism $\mathrm{H}^1(A, Z_i) \to \mathrm{H}^1(A, Z_{i-1})$ is surjective.*

Proof. By the exact sequence $0 \to Z_i \longrightarrow Z_{i-1} \xrightarrow{dC^{i-1}} d\Omega_A^1 \to 0$ we have an exact sequence

$$0 \to k \longrightarrow \mathrm{H}^1(A, Z_i) \longrightarrow \mathrm{H}^1(A, Z_{i-1}) \longrightarrow k.$$

The result follows from this exact sequence. □

Proposition 4.6. *Assume that* $h = 2$. *Then* $\dim \mathrm{H}^1(A, Z_i) = 4$ *for* $i \geq 0$.

Proof. The proof is similar to proofs in [6]. □

5. The de Rham Cohomology

The de Rham cohomology of an abelian surface A is the hypercohomology of the complex $(\Omega_A^\bullet, d)$. On H^2_{dR} we have a perfect pairing $\langle\, , \rangle$ given by Poincaré duality. The Hodge spectral sequence with $E_1^{ij} = \mathrm{H}^j(A, \Omega^i)$ converges to $\mathrm{H}^*_{dR}(A)$. The second spectral sequence of hypercohomology has $E_2^{ij} = \mathrm{H}^i(\mathcal{H}^j(\Omega^\bullet))$ abutting to $\mathrm{H}^{i+j}_{dR}(A/k)$. But the Cartier operator yields an isomorphism of sheaves

$$C^{-1} : \Omega^i_{A^{(p)}} \xrightarrow{\sim} \mathcal{H}^i(F_*(\Omega^\bullet_{A/k})),$$

so that we can rewrite this as

$$E_2^{ij} = \mathrm{H}^i(A', \mathcal{H}_j(\Omega^\bullet)) = \mathrm{H}^i(A', \Omega^j_{X'}) \Rightarrow \mathrm{H}^*_{dR}(A),$$

where A' is the base change of A under Frobenius. We thus get two filtrations on the de Rham cohomology: the Hodge filtration

$$(0) \subset F^2 \subset F^1 \subset \mathrm{H}^2_{dR},$$

and the conjugate filtration

$$(0) \subset G_1 \subset G_2 \subset \mathrm{H}^2_{dR}.$$

We have rank $(F^1) =$ rank $(G_2) = 5$, rank $(F^2) =$ rank $(G_1) = 1$ and

$$(F^1)^\perp = F^2 \quad \text{and} \qquad G_1^\perp = G_2.$$

We have also

$$F^1/F^2 \cong \mathrm{H}^1(A, \Omega^1_A).$$

Note that the image of $\mathrm{H}^2(A, O_A)$ under Frobenius is G_1.

Now we consider the Frobenius map on the de Rham cohomology $\mathrm{H}^2_{dR}(A)$. On F^1 the Frobenius map F is zero and so it induces a homomorphism

$$F : \mathrm{H}^2(A, \mathcal{O}_A) \cong \mathrm{H}^2_{dR}(A)/F^1 \longrightarrow \mathrm{H}^2_{dR}(A).$$

The image of F coincides with G_1.

The following result can be found in [14].

Lemma 5.1. *An abelian surface* A *over* k *is superspecial if and only if the Hodge filtration satisfies* $F^2 = G_1$.

6. Subspaces of $\mathrm{H}^1(A, \Omega_A^1)$

For an abelian surface A, we consider the natural inclusions

$$B_i \hookrightarrow \Omega_A^1, \ Z_i \hookrightarrow \Omega_A^1.$$

These inclusions induce homomorphisms

$$\mathrm{H}^1(B_i) \to \mathrm{H}^1(\Omega_A^1), \ \mathrm{H}^1(Z_i) \to \mathrm{H}^1(\Omega_A^1).$$

We denote by $\mathrm{Im}\ \mathrm{H}^1(B_i)$ and $\mathrm{Im}\ \mathrm{H}^1(Z_i)$, respectively, the images of these homomorphisms. In this section, we determine the image $\mathrm{Im}\ \mathrm{H}^1(A, B_i)$ (resp. $\mathrm{Im}\ \mathrm{H}^1(A, Z_i)$) of $\mathrm{H}^1(A, B_i)$ (resp. $\mathrm{H}^1(A, Z_i)$) in $\mathrm{H}^1(A, \Omega_A^1)$. For the proof of the following lemma, see [6], Lemma 9.3.

Lemma 6.1. $\mathrm{Im}\ \mathrm{H}^1(B_i)$ *and* $\mathrm{Im}\ \mathrm{H}^1(Z_i)$ *are orthogonal subspaces.*

Lemma 6.2. *Assume* $h = 1$. *Then we have* $\mathrm{Im}\ \mathrm{H}^1(A, B_i) = 0$ *and moreover* $\mathrm{Im}\ \mathrm{H}^1(A, Z_i) \cong \mathrm{H}^1(A, \Omega_A^1)$ *for* $i \geq 1$.

Proof. This follows from $\mathrm{H}^\ell(A, d\Omega_A^1) = 0$ for any ℓ, and the exact sequence

$$0 \to Z_{i+1} \longrightarrow Z_i \xrightarrow{dC^i} d\Omega_A^1 \to 0.$$

□

Lemma 6.3. *Assume* $h = 2$. *Then the subspace* $\mathrm{Im}\ \mathrm{H}^1(A, B_i) \subset \mathrm{H}^1(A, \Omega_A^1)$ *has dimension* 1 *for any* $i \geq 1$.

We now consider the homomorphisms

$$\begin{array}{rcccc} c_1^{(i)}: & \mathrm{Pic}(A) & = & \mathrm{H}^1(A, \mathcal{O}_A^*) & \longrightarrow & \mathrm{H}^1(A, Z_i) \\ & D & = & \{h_{ij}\} & \mapsto & \{d \log h_{ij}\}. \end{array}$$

If a divisor D is algebraically equivalent to a divisor E, then there exists an element a of A such that $T_a^*E \sim D$ (linearly equivalent) with T_a the translation by a. Since A is a complete variety, the linear representation of A is trivial. Therefore, A acts on $\mathrm{H}^1(A, Z_i)$ trivially. Therefore, we have $c_1^{(i)}(D) = c_1^{(i)}(T_a^*E) = c_1^{(i)}(E)$. Hence, we have the induced homomorphisms

$$\begin{array}{rccc} c_1^{(i)}: & NS(A) & \longrightarrow & \mathrm{H}^1(A, Z_i) \\ & D = \{h_{ij}\} & \mapsto & \{d \log h_{ij}\}. \end{array}$$

Note that $c_1 = c_1^{(0)} : NS(A) \longrightarrow \mathrm{H}^1(A, \Omega_A^1)$ gives the usual Chern class. We denote by c_1^{dR} the Chern mapping $NS(A) \longrightarrow \mathrm{H}^2_{dR}(A)$.

The following lemma is due to Ogus [14].

Lemma 6.4. *The abelian surface* A *is superspecial if and only if we have* $\mathrm{Im}\ c_1^{dR} \cap F^2 \neq \{0\}$.

Lemma 6.5. *Assume that A is not a superspecial (resp. is a superspecial) abelian surface. Then the homomorphism $c_1^{(n)} : NS(A)/pNS(A) \longrightarrow \mathrm{H}^1(A, Z_n)$ is injective for any $n \geq 0$ (resp. $n \geq 1$).*

Proof. Assume that A is not superspecial. It suffices to prove

$$c_1 : NS(A)/pNS(A) \longrightarrow \mathrm{H}^1(A, \Omega_A^1)$$

is injective. Suppose there exists an element $D = \{h_{ij}\} \in NS(A)/pNS(A)$ such that $\{d \log h_{ij}\}$ is zero in $\mathrm{H}^1(A, \Omega_A^1)$. Then, there exists an element $\omega_i \in \Gamma(U_i, \Omega_A^1)$ such that $d \log h_{ij} = \omega_j - \omega_i$. If $d\omega_i \neq 0$, then we have $\mathrm{Im}\ c_1^{dR} \cap F^2 \neq \{0\}$, which contradicts our assumption. Therefore, we have $d\omega_i = 0$. Applying the Cartier operator to both sides, we have $d \log h_{ij} = C(\omega_j) - C(\omega_i)$. Therefore, we have $C(\omega_j) - \omega_j = C(\omega_i) - \omega_i$. Since $C - \mathrm{id} : \mathrm{H}^0(A, \Omega_A^1) \to \mathrm{H}^0(A, \Omega_A^1)$ is surjective, there exists an element $\omega \in \mathrm{H}^0(A, \Omega_A^1)$ such that $C(\omega_i) - \omega_i = C\omega - \omega$. Replacing ω_i by $\omega_i - \omega$, we have $d \log h_{ij} = \omega_j - \omega_i$ with $C(\omega_i) = \omega_i$. Therefore, there exists $h_i \in \Gamma(U_i, \mathcal{O}_A^*)$ such that $\omega_i = d \log h_i$ and we see $D \in pNS(A)$.

If A is superspecial it suffices to prove that $c_1^{(1)} : NS(A)/pNS(A) \longrightarrow \mathrm{H}^1(A, Z_1)$ is injective. The proof is completely similar to the latter part of the above proof. □

Lemma 6.6. *The natural homomorphism $\mathrm{H}^1(A, Z_1) \longrightarrow \mathrm{H}^2_{dR}(A)$ is injective.*

Proof. Suppose there exists a Čech cocycle $\alpha = \{\omega_{ij}\} \in \mathrm{H}^1(A, Z_1)$ w.r.t. an affine open covering $\{U_i\}$ of A such that α is zero in $\mathrm{H}^2_{dR}(A)$. Then by the definition of de Rham cohomology, there exists $\omega_i \in \Gamma(U_i, \Omega_A^1)$ such that $\omega_{ij} = \omega_j - \omega_i$ on $U_i \cap U_j$ and that $d\omega_i = 0$. This means that $\alpha = 0$ in $\mathrm{H}^1(A, Z_1)$. □

The following proposition follows from Lemmas 6.5 and 6.6.

Proposition 6.7. *Let A be an abelian surface over k. Then the Chern mapping*

$$c_1^{dR} : NS(A)/pNS(A) \longrightarrow \mathrm{H}^2_{dR}(A)$$

is injective.

7. The Supersingular Case

If A is a supersingular abelian surface we denote by σ_0 the Artin-invariant of A. This is defined as $\sigma_0 = \mathrm{ord}_p(\Delta)$, where Δ is the discriminant of the intersection form $\langle\, ,\, \rangle$ on $NS(A)$. Then, σ_0 is equal to either 1 or 2. Moreover, A is superspecial if and only if $\sigma_0 = 1$. We denote by $\langle \mathrm{Im}\ c_1 \rangle$ the subspace of $\mathrm{H}^1(A, \Omega_A^1)$ generated by $\mathrm{Im}\ c_1$ over k.

Lemma 7.1. *Assume that A is a superspecial abelian surface. Then*

$$\langle \mathrm{Im}\ c_1 \rangle = \mathrm{H}^1(A, \Omega_A^1).$$

Proof. For a superspecial abelian surface the endomorphism ring is $M_2(R)$ with R an order in a quaternion division algebra over $\mathbf{Q}$. It follows that the rank of $NS(A)$ equals 6. From $\sigma_0 = 1$ it follows that the image of c_1^{dR} has rank at least 5 in $\mathrm{H}^2_{dR}(A)$. Since $\langle \mathrm{Im}\ c_1^{dR}\rangle \subset F^1$ and $\dim F^1 = 5$, we have $\langle \mathrm{Im}\ c_1^{dR}\rangle = F^1$. Hence, the result follows from $F^1/F^2 \cong \mathrm{H}^1(A, \Omega^1_A)$. □

Corollary 7.2. *Assume A is superspecial. Then the natural mapping*

$$\mathrm{H}^1(A, Z_i) \to \mathrm{H}^1(A, Z_{i-1})$$

is surjective for $i \geq 1$. In particular, the equality $\mathrm{Im}\ \mathrm{H}^1(A, Z_i) = \mathrm{H}^1(A, \Omega^1_A)$ *holds for $i \geq 1$.*

Proof. By Lemma 4.4, it suffices to show this corollary for $\mathrm{H}^1(A, Z_1) \to \mathrm{H}^1(A, \Omega^1_A)$. We have $\langle \mathrm{Im}\ c_1\rangle = \mathrm{H}^1(A, \Omega^1_A)$ by the assumption, and $\mathrm{Im}\ c_1 \subset \mathrm{Im}\ \mathrm{H}^1(A, Z_1)$. The result follows from these facts. □

Corollary 7.3. *Assume A is superspecial. Then* $\dim \mathrm{Im}\ \mathrm{H}^1(A, B_i) = 0$ *for any $i \geq 1$. Moreover,* $\dim \mathrm{H}^1(A, Z_i) = 4 + i$.

Lemma 7.4. *Assume that A is a supersingular abelian surface which is not superspecial. Then* $\dim \mathrm{Im}\ \mathrm{H}^1(A, Z_1) \geq 3$.

Proof. Use the exact sequence $0 \to d\mathcal{O}_A \longrightarrow Z_1 \xrightarrow{C} \Omega^1_A \to 0$. □

Lemma 7.5. *If A is a supersingular abelian surface which is not superspecial then* $\dim \mathrm{Im}\ \mathrm{H}^1(A, B_1) = 1$.

Proof. By Lemmas 7.4 and 6.1, we have $\dim \mathrm{Im}\ \mathrm{H}^1(A, B_1) \leq 1$. With respect to a suitable affine open covering $\mathcal{U} = \{U_i\}_{i \in I}$ of A, the following 2-cocycles give a basis of $\mathrm{H}^1(A, \mathcal{O}_A)$: $\{f_{ij}\}$, $\{f_{ij}^p\}$ such that $\{f_{ij}^{p^2}\} \sim 0$ and that $\{f_{ij}^p\}$ is not cohomologous to zero. Therefore, there exists $f_i \in \Gamma(U_i, \mathcal{O}_A)$ such that $f_{ij}^{p^2} = f_j - f_i$. Now $\omega_1 = df_i$ gives a non-zero regular 1-form on A. We thus have a non-zero element $f_{ij}^p \omega_1 = d(f_{ij}^p f_i)$ in $\mathrm{H}^1(A, B_1)$ which is not zero in $\mathrm{H}^1(A, \Omega^1_A) = \mathrm{H}^1(A, \mathcal{O}_A) \otimes_k \mathrm{H}^0(A, \Omega^1_A)$. Therefore, we have $\dim \mathrm{Im}\ \mathrm{H}^1(A, B_1) \geq 1$. Hence, we have $\dim \mathrm{Im}\ \mathrm{H}^1(A, B_1) = 1$. □

Corollary 7.6. *If A is a supersingular abelian surface which is not superspecial then* $\dim \mathrm{H}^1(A, Z_1) = 4$ *and* $\dim \mathrm{Im}\ \mathrm{H}^1(A, Z_1) = 3$.

Proof. The latter part follows from Lemma 7.4, $\dim \mathrm{Im}\ \mathrm{H}^1(A, B_1) = 1$ and $\mathrm{Im}\ \mathrm{H}^1(A, Z_1) \subset \mathrm{Im}\ \mathrm{H}^1(A, B_1)^{\perp}$. The former part follows from the latter part and the exact sequence $0 \to Z_1 \longrightarrow \Omega^1_X \xrightarrow{d} d\Omega^1_X \to 0$. □

Let A be a supersingular abelian surface, and let ℓ_n be the smallest non-negative integer ℓ such that there exist elements $D_1, \ldots, D_\ell$ in $NS(A)/pNS(A)$ with $c_1^{(n)}(D_1), \ldots, c_1^{(n)}(D_\ell)$ linearly independent over $\mathbf{F}_p$ and linearly dependent over k in $\mathrm{H}^1(A, Z_n)$. Since we have, by Lemma 6.5,

$$\dim_{\mathbf{F}_p} c_1^{(1)}(NS(A)/pNS(A)) = 6 \text{ and } \dim_k \mathrm{H}^1(A, Z_1) \leq 5,$$

we see that $\ell_1 \geq 2$.

Lemma 7.7. *Let A be a supersingular abelian surface. For $n \geq 1$ we have $\ell_{n+1} > \ell_n$ if $\ell_{n+1} \neq 0$.*

Proof. Assume $\ell_{n+1} \neq 0$. Then in $NS(A)/pNS(A)$ there exist elements $D_1, \ldots, D_{\ell_{n+1}}$ such that $c_1^{(n+1)}(D_1), \ldots, c_1^{(n+1)}(D_{\ell_{n+1}})$ are linearly independent over $\mathbf{F}_p$, and such that

$$(1) \qquad a_1 c_1^{(n+1)}(D_1) + \ldots + a_{\ell_{n+1}} c_1^{(n+1)}(D_{\ell_{n+1}}) = 0 \text{ in } \mathrm{H}^1(A, Z_{n+1})$$

is a non-trivial linear relation over k. We may assume $a_1 = 1$. By the minimality of ℓ we see $a_i/a_j \notin \mathbf{F}_p$ for any i, j, $i \neq j$. Applying the Cartier operator C to both sides in (1), we get

$$(2) \quad a_1 c_1^{(n)}(D_1) + a_2^{p^{-1}} c_1^{(n)}(D_2) + \ldots + a_{\ell_{n+1}}^{p^{-1}} c_1^{(n)}(D_{\ell_{n+1}}) = 0 \text{ in } \mathrm{H}^1(A, Z_n)$$

by using $C(c_1^{(n+1)}(D_j)) = c_1^{(n)}(D_j)$. Subtracting (2) from (1) in $\mathrm{H}^1(A, Z_n)$, we get a non-trivial linear relation over k whose length is smaller than or equal to $\ell_{n+1} - 1$. Hence we have $\ell_{n+1} > \ell_n$. □

Corollary 7.8. *For a supersingular abelian surface A, there exists a integer $n \geq 2$ such that the natural homomorphism*

$$(3) \qquad NS(A)/pNS(A) \otimes_{\mathbf{F}_p} k \to \mathrm{H}^1(A, Z_n)$$

is injective.

Proof. Since $\dim_{\mathbf{F}_p} NS(A)/pNS(A) = 6$, we have $\ell_n \leq 6$. Hence, by Lemma 7.7, there exists a positive integer n such that $\ell_n = 0$. □

Lemma 7.9. *Assume that A is a supersingular abelian surface which is not superspecial. Let s be the smallest integer such that (3) is injective and let $\varphi_t^s : \mathrm{H}^1(A, B_t) \to \mathrm{H}^1(A, Z_s)$ be the homomorphism induced by the natural inclusion $B_t \to Z_s$. Then $\mathrm{Im}\, \varphi_t^s \subset \langle c_1^{(s)}(NS(A)/pNS(A))\rangle$ for any $t \geq 1$.*

Proof. Since $\dim_{\mathbf{F}_p} NS(A)/pNS(A) = 6$ and $\dim \mathrm{H}^1(A, Z_1) = 4$ by Corollary 7.6, we see $s \geq 3$. We consider an exact sequence

$$0 \to B_1 \longrightarrow Z_s \xrightarrow{C} Z_{s-1} \to 0.$$

By the long exact sequence, we have $\dim \mathrm{Im}\, \varphi_1^s = 1$.

By the assumption on s there exist elements $D_1, \dots, D_{\ell_{s-1}}$ in the $\mathbf{F}_p$-vector space $NS(A)/pNS(A)$ such that $c_1^{(s-1)}(D_1), \dots, c_1^{(s-1)}(D_{\ell_{s-1}})$ are linearly independent over $\mathbf{F}_p$, and such that

$$(4) \qquad a_1 c_1^{(s-1)}(D_1) + \dots + a_{\ell_{s-1}} c_1^{(s-1)}(D_{\ell_{s-1}}) = 0 \text{ in } \mathrm{H}^1(A, Z_{s-1})$$

is a non-trivial linear relation over k. We may assume $a_1 = 1$. Then, by the minimality of ℓ_{s-1}, we see $a_i/a_j \notin \mathbf{F}_p$ for any i, j, $i \neq j$. We apply the Cartier inverse to (4) and find

$$a_1 c_1^{(s)}(D_1) + a_2^p c_1^{(s)}(D_2) + \dots + a_{\ell_{s-1}}^p c_1^{(s)}(D_{\ell_{s-1}}) = \{dh_{ij}\} \text{ in } \mathrm{H}^1(A, Z_s)$$

with $\{dh_{ij}\} \in \mathrm{H}^1(A, B_1)$. If $\{dh_{ij}\} = 0$ in $\mathrm{H}^1(A, B_1)$ we see that

$$NS(A)/pNS(A) \otimes_{\mathbf{F}_p} k \to \mathrm{H}^1(A, Z_s)$$

is not injective, a contradiction. Therefore, $\{dh_{ij}\} \neq 0$ in $\mathrm{H}^1(A, B_1)$, and $\{dh_{ij}\}$ gives a basis of $\mathrm{Im}\ \varphi_1^s$. So we have $\mathrm{Im}\ \varphi_1^s \subset \langle c_1^{(s)}(NS(A)/pNS(A))\rangle$.

Suppose that there exists a j such that $\mathrm{Im}\ \varphi_j^s$ is not contained in the k-vector space $\langle c_1^{(s)}(NS(A)/pNS(A))\rangle$. We take the smallest such j. Then, we have $j \geq 2$ as we showed above. We take an element $\alpha \in \mathrm{Im}\ \varphi_j^s$ such that $\alpha \notin \langle c_1^{(s)}(NS(A)/pNS(A))\rangle$. Then, there exists an element $\tilde{\alpha} \in \mathrm{H}^1(A, B_j)$ such that $\varphi_j^s(\tilde{\alpha}) = \alpha$. We have $C(\tilde{\alpha}) \in \mathrm{H}^1(A, B_{j-1})$ and $\varphi_{j-1}^s(C(\tilde{\alpha})) \in \langle c_1^{(s)}(NS(A)/pNS(A))\rangle$ by the assumption on j. Applying the inverse Cartier, we see that there exists an element $\{dk_{ij}\} \in \mathrm{H}^1(A, B_1)$ such that $\alpha + \{dk_{ij}\}$ is contained in $\langle c_1^{(s)}(NS(A)/pNS(A))\rangle$. Since

$$\{dk_{ij}\} \in \mathrm{H}^1(A, B_1) \subset \langle c_1^{(s)}(NS(A)/pNS(A))\rangle,$$

we have $\alpha \in \langle c_1^{(s)}(NS(A)/pNS(A))\rangle$, which is a contradiction. This completes the proof. □

Proposition 7.10. *For a supersingular abelian surface A which is not superspecial* $\mathrm{H}^1(A, Z_{i+1}) \to \mathrm{H}^1(A, Z_i)$ *is surjective for* $i \geq 1$, *and* $\mathrm{H}^1(A, Z_1) \to \mathrm{H}^1(A, \Omega_A^1)$ *is not surjective. Moreover, we have* $\dim \mathrm{H}^1(A, Z_i) = i + 3$ *and* $\dim \mathrm{Im}\ \mathrm{H}^1(A, Z_i) = 3$ *for* $i \geq 1$.

Proof. We take the smallest integer s such that (3) is injective. Since the images of $NS(A)/pNS(A) \otimes_{\mathbf{F}_q} k$ by C^s and by the natural homomorphism from $\mathrm{H}^1(A, Z_s)$ to $\mathrm{H}^1(A, \Omega_A^1)$ coincide with each other, we have an exact sequence

$$\begin{aligned} 0 \to k^{\oplus 2} &\longrightarrow \mathrm{H}^1(A, B_s) \\ \longrightarrow \langle c_1^{(s)}(NS(A)/pNS(A))\rangle &\xrightarrow{C^s} \langle c_1(NS(A)/pNS(A))\rangle \to 0. \end{aligned}$$

Here we use the result in Lemma 7.9. Since we have

$$\dim \mathrm{H}^1(A, B_s)) = s + 2,$$
$$\dim\langle c_1(NS(A)/pNS(A))\rangle = 3,$$
$$\dim\langle c_1^{(s)}(NS(A)/pNS(A)) = 6,$$

we have $2 + 6 = (s + 2) + 3$. Therefore, we have $s = 3$ and we have $\dim \mathrm{H}^1(A, Z_3) \geq 6$. Now, by Lemmas 4.4, 4.5 and Corollary 7.6, we know that $\dim \mathrm{H}^1(A, Z_3) = 6$ and $\langle c_1^{(3)}(NS(A)/pNS(A))\rangle = \mathrm{H}^1(A, Z_3)$. The rest follows from Lemma 4.4. □

Corollary 7.11. *Assume that A is a supersingular abelian surface which is not superspecial. Then* $\dim \mathrm{Im}\ \mathrm{H}^1(A, B_i) = 1$ *for any* $i \geq 1$.

Proof. This follows from Lemma 7.5 and Proposition 7.10. □

Corollary 7.12. *Assume that A is a supersingular abelian surface. Then, we have* $\langle c_1^{(i)}(NS(A)/pNS(A))\rangle = \mathrm{H}^1(A, Z_i)$ *for* $i = 1, 2, 3$. *Moreover,* $\langle c_1(NS(A)/pNS(A))\rangle = \mathrm{Im}\ \mathrm{H}^1(A, Z_i)$ *for any* i.

Proof. Using Lemma 7.1 and Corollary 7.2, we can easily prove the case of a superspecial abelian surface. Assume A is supersingular and not superspecial. The proof of the former part is given in the proof of Proposition 7.10 by Lemma 4.4. Since $\langle c_1(NS(A)/pNS(A))\rangle \subset \mathrm{Im}\ \mathrm{H}^1(A, Z_i)$, we can prove this corollary by Lemmas 4.4, 4.5 and Proposition 7.10. □

We summarize the results on the dimension of $\mathrm{H}^j(Z_i)$ in the following table. The results about $h^0(Z_i)$ and $h^2(Z_i)$ follow from 7.2, 7.10 and the exact sequence $0 \to Z_i \to Z_{i-1} \to d\Omega_A^1 \to 0$.

type	$h^0(Z_i)$	$h^1(Z_i)$	$h^2(Z_i)$
$h = 1, 2$	2	4	2
$h = \infty, a = 1$	2	$3 + i$	$1 + i$
$h = \infty, a = 2$	2	$4 + i$	$2 + i$

8. Degenerate Abelian Surfaces

The moduli space $\mathcal{A}_2$ can be compactified in a canonical way to a compactification $\tilde{\mathcal{A}}_2$. This is a toroidal compactification, called the Delaunay-Voronoi compactification. It is a blow-up of the Satake-compactification due to Igusa, cf. also [3]. Alexeev gives a functorial description of it in [1]. The four strata in codimension 3, 2, 1 and 0 correspond to decompositions of the plane $\mathbf{R}^2$ by triangles, by 4-gons, by infinite strips and by one big cell covering the whole space.

The corresponding degenerations are copies of $\mathbf{P}^2$ glued, copies of $\mathbf{P}^1 \times \mathbf{P}^1$ glued, certain non-normal compactifications of a $\mathbf{G}_m$-bundle over an elliptic curve or honest abelian surfaces. We now first deal with the simplest type of degeneration, the so-called rank 1 degenerations, and these occur in codimension 1.

A rank-1 degeneration of an abelian surface is a non-normal surface obtained as follows. Start with a semi-abelian surface G, i.e. with a $\mathbf{G}_m$-extension of an elliptic curve E

$$1 \to \mathbf{G}_m \to G \to E \to 0 \qquad (*)$$

and consider the associated $\mathbf{P}^1$-bundle $\pi : \tilde{G} \to E$. Then $\tilde{G} - G$ is a union of two sections $\tilde{G}_0 \sqcup \tilde{G}_\infty$ of $\tilde{G}$ over E. The extension class of (*) is an element $e \in \hat{E}$, the dual elliptic curve. The compactification X of G is obtained by glueing the zero section $\tilde{G}_0$ with the infinity section $\tilde{G}_\infty$ by a translation over $e \in E \cong \hat{E}$. We then have $\tilde{G}_0 - \tilde{G}_\infty \equiv \pi^{-1}(p-q)$ with $p - q = e \in E$. The divisor $\tilde{G}_0 + \pi^{-1}(0)$ defines a line bundle $\tilde{L}$ on $\tilde{G}$ which descends to a line bundle L on X with $h^0(L) = 1$. This defines the principal polarization on X, cf. [12]. The surfaces thus constructed are the fibres of the universal family $\tilde{\mathcal{X}}_2$ over $\tilde{\mathcal{A}}_2$ over the codimension 1 stratum in the boundary.

The surface X has $h^1(X, O_X) = 4$, $h^2(X, O_X) = 1$ and we can define a formal Brauer group Φ_X associated to $\mathrm{H}^2(X, O_X)$ as we did for abelian surfaces. This is a 1-dimensional formal group functor. The main result of this section is the following.

Proposition 8.1. *The formal group functor $\Phi_X = \Phi_X^2$ of the rank 1 degeneration X is isomorphic to the formal group functor Φ_E^1 of the elliptic curve E, hence is a formal group.*

Proof. One way of defining the formal Brauer group is as the functor which associates to each nilpotent (commutative quasi-projective) O_X-algebra N the group

$$\Phi_Y(N) = \mathrm{H}^2(Y, (1 + O_X \otimes N)^*),$$

cf. [20]. We now use the Leray spectral sequence for the normalization morphism $f : \tilde{G} \to X$ of X. We have

$$E_2^{ij} = \mathrm{H}^j(X, R^i f_*(F)) \qquad \text{with } F = (1 + O_{\tilde{G}} \otimes N)^*.$$

Because the fibres of f are zero-dimensional $R^1 f_*(F)$ and $R^2 f_*(F)$ vanish. The spectral sequence yields on level 2:

$$\begin{array}{ccc} 0 & 0 & 0 \\ 0 & 0 & 0 \\ \mathrm{H}^0(X, A) & \mathrm{H}^1(X, A) & \mathrm{H}^2(X, A) \end{array}$$

with $A = R^0 f_*(F)$. We have an exact sequence of sheaves on X

$$0 \to F_X \longrightarrow A \longrightarrow B \to 0.$$

Here $F_X = (1 + O_X \otimes N)^*$ and the sheaf B is concentrated on the non-normal locus S of X where the two sections $\tilde{G}_0$ and $\tilde{G}_\infty$ are glued.

Let t_e be the translation over $e \in E$. Then with $F_E = (1 + O_E \otimes N)^*$ there is an exact sequence

$$0 \to F_E \xrightarrow{\alpha} F_E \oplus t_e^* F_E \to B|S \to 0$$

of sheaves on $S \cong E$. Here $\alpha(s) = (s, t_e^* s)$. This implies that $\mathrm{H}^1(S, B|S) \cong \mathrm{H}^1(E, F_E) = \Phi_E^1(N)$. Since $\tilde{G}$ is ruled we have $\mathrm{H}^2(\tilde{G}, F_{\tilde{G}}) = 0$, hence $\mathrm{H}^2(X, A) = 0$. The spectral sequence shows that $\mathrm{H}^1(\tilde{G}, F_{\tilde{G}}) \cong \mathrm{H}^1(X, A)$. In the long exact sequence

$$\to \mathrm{H}^1(F_X) \to \mathrm{H}^1(A) \to \mathrm{H}^1(B) \to \mathrm{H}^2(F_X) \to \mathrm{H}^2(A) = 0$$

the map $\mathrm{H}^1(F_X) \to \mathrm{H}^1(A)$ can be identified with the map $\mathrm{H}^1(X, F_X) \to \mathrm{H}^1(\tilde{G}, F_{\tilde{G}})$. Since the formal group Φ_X^1 fits into an extension

$$1 \to \hat{\mathbf{G}}_m \to \Phi_X^1 \to \Phi_{\tilde{G}}^1 \to 0.$$

we see that $\mathrm{H}^1(X, F_X) \to \mathrm{H}^1(\tilde{G}, F_{\tilde{G}})$ is surjective. This implies that $\mathrm{H}^1(B) \cong \mathrm{H}^2(F_X)$. □

We can do something similar for rank-2 degenerations of abelian surfaces. Then the formal group that we get is a multiplicative group $\hat{\mathbf{G}}_m$. The loci $M^{(h)}$ thus can be extended to the compactification $\tilde{\mathcal{A}}_2$. We can also consider cohomology with coefficients in $W_i(O_X)$ for such degenerate surfaces. The results we obtain are similar. Thus we can extend the loci $M_F^{(h)}$ to the compactification.

The considerations on differential forms using the sheaves Ω^1, B_i and Z_i can be extended to degenerate surfaces X by replacing differential forms by differential forms with log poles along the divisor of non-normal points on X.

9. The Classes of the Loci of Given Height

Let $\mathcal{A}_2$ (or $\mathcal{A}_{2,n}$ with $n \geq 3$, $p \not| n$) be the moduli stack (or fine moduli space) of principally polarized abelian surfaces (with level n-structure) and let $\pi : \mathcal{X} \to \mathcal{A}_2$ be the universal family of principally polarized abelian surfaces. We know that $\mathcal{A}_2$ (resp. $\mathcal{A}_{2,n}$) is a 3-dimensional algebraic stack (resp. variety) which has a natural compactification $\tilde{\mathcal{A}}_2$ (resp. $\mathcal{A}_{2,n}$), see e.g. [1]. It possesses a universal family $\tilde{\pi} : \tilde{\mathcal{X}} \to \tilde{\mathcal{A}}_2$ (resp. $\tilde{\mathcal{A}}_{2,n}$). We set $\tilde{M} = \tilde{\mathcal{A}}_2$ (or $\tilde{M} = \tilde{\mathcal{A}}_{2,n}$) and we denote by v the class of the coherent sheaf $R^0\pi_*\Omega^2_{\tilde{\mathcal{X}}/\tilde{M}}$ in the Chow group $CH^1_{\mathbf{Q}}(\tilde{M})$. We set

$$M^{(h)} := \{s \in \tilde{M} : h(\tilde{\mathcal{X}}_s) \geq h\}.$$

Then, as is well-known, $M^{(h)}$ is an algebraic subvariety (maybe reducible), and is of codimension $h-1$ in $\tilde{\mathcal{A}}_2$ for $h = 1, 2$, and of codimension 2 in $\tilde{\mathcal{A}}_2$ for $h = \infty$. Moreover, $M^{(\infty)}$ is contained in $\mathcal{A}_2$ (or $\mathcal{A}_{2,n}$).

The direct image sheaves $R^2\pi_* W_i(\mathcal{O}_{\tilde{\mathcal{X}}})$ are coherent $W_i(\mathcal{O}_{\tilde{M}})$ sheaves. Note that these sheaves are not $\mathcal{O}_{\tilde{M}}$-modules if $i \geq 2$. We denote by $M_F^{(h)}$

($h = 1, 2, 3$) the scheme-theoretic zero locus of the relative Frobenius map on $R^2\tilde{\pi}_* W_{(h-1)}(\mathcal{O}_{\tilde{\mathcal{X}}})$. We set $M_F^{(\infty)} = M_F^{(3)}$. Then we have $(M_F^{(\infty)})_{\mathrm{red}} = M^{(\infty)}$.

Theorem 9.1. *The classes of $M_F^{(h)}$ for $h = 2, \infty$ in the Chow group $CH_{\mathbf{Q}}^*(\tilde{M})$ are given by*

$$M_F^{(2)} = (p-1)v, \quad M_F^{(\infty)} = (p-1)(p^2-1)v^2.$$

Moreover, $M_F^{(2)} \backslash M_F^{(\infty)}$ is non-singular.

Proof. The locus $M_F^{(2)}$ is given by the support of the cokernel of $F : R^2\tilde{\pi}_*\mathcal{O}_{\tilde{\mathcal{X}}} \to R^2\tilde{\pi}_*\mathcal{O}_{\tilde{\mathcal{X}}}$. Since F is a σ-linear mapping, it follows that $M_F^{(2)} = (p-1)v$. By the same method as we used in van der Geer and Katsura [6] or as in Ogus [14] we see that the tangent space of a point $x = (A_0, D_0) \in M_F^{(2)} \backslash M_F^{(\infty)}$ is given by $\mathrm{Im}\, \mathrm{H}^1(A_0, Z_1) \cap D_0^{\perp}$, which is 2-dimensional. Hence, $M_F^{(2)} \backslash M_F^{(\infty)}$ is non-singular.

Now, let $x = (A_0, D_0)$ be a point in $M_F^{(\infty)}$ such that A_0 is a non-superspecial abelian surface. We denote by S the formal scheme around x in $M_F^{(2)}$, and we denote by $X \to S$ the family which is obtained by the restriction of $\tilde{\pi} : \tilde{\mathcal{X}} \to \tilde{M}$ to S. Then, the Frobenius mapping F is zero on $\mathrm{H}^2(X, \mathcal{O}_{X/S})$. We denote by ∇ the Gauss-Manin connection of $\mathrm{H}^2_{dR}(X/S)$. We consider the Hodge filtration $0 \subset F^2 \subset F^1 \subset \mathrm{H}^2_{dR}(X/S)$, and construct, in the same way as in Section 8 of van der Geer and Katsura [6], the homomorphism

$$\Phi_2 : \mathrm{H}^2(W_2(O_{X/S})) \to \mathrm{H}^2_{dR}(X/S).$$

We take a basis ω of $\mathrm{H}^0(\Omega^1_{X/S})$ and take the dual basis ζ of $\mathrm{H}^2(O_{X/S})$. We take a lifting $\tilde{\zeta} \in \mathrm{H}^2_{dR}(X/S)$ of ζ. Then we have $\langle \tilde{\zeta}, \omega \rangle = 1$. Since we have a surjective homomorphism $R : \mathrm{H}^2(W_2(O_{X/S})) \to \mathrm{H}^2(O_{X/S})$, there exists an element $\alpha \in \mathrm{H}^2(W_2(O_{X/S}))$ such that $R(\alpha) = \zeta$. We set

$$g_h = \langle \Phi_h(\alpha), \omega \rangle.$$

Since $\Phi_h(\alpha) - g_h\tilde{\zeta}$ is orthogonal to ω, the element $\Phi_h(\alpha) - g_h\tilde{\zeta}$ is contained in F^1. Therefore, using the natural isomorphism $\mathrm{H}^2_{dR}/F^1 \cong \mathrm{H}^2(X, \mathcal{O}_X)$, we conclude that

$$\phi_h(\zeta) = g_h\zeta \ \text{ in } \mathrm{H}^2(X, \mathcal{O}_X).$$

This means that $g_h = 0$ gives the scheme theoretic zero-locus $M_F^{(\infty)}$ of ϕ_h. By Theorem 3.9, $(M_F^{(\infty)})_{\mathrm{red}}$ coincides with $M^{(\infty)}$. Since ϕ_h is a σ^2-linear homomorphism, we conclude that the class $M_F^{(\infty)} = (p^2-1)v|_{M_F^{(2)}}$ in the Chow ring $CH^1_{\mathbf{Q}}(M_F^{(2)})$. Hence, as in van der Geer and Katsura [6], using the projection formula, we complete the proof. □

By the result of Ekedahl and van der Geer (cf. [5]), we have $M^{(2)} = (p-1)\lambda_1$ and $M^{(\infty)} = (p-1)(p^2-1)\lambda_2$ in the Chow ring $CH_{\mathbf{Q}}^*(\tilde{M})$. Here λ_i

for $i = 1, 2$ are the Chern classes of the Hodge bundle $E = R^0\pi_*\Omega^1_{\tilde{\mathcal{X}}/\tilde{M}}$, [4, 5]. We know the relation $\lambda_1^2 = 2\lambda_2$ in the Chow ring ([4]) and we have $v = \lambda_1$. Thus we have the following relations between $M_F^{(h)}$ and $M^{(h)}$. Note that we know that all components of $M_F^{(\infty)}$ have multiplicity ≥ 2.

Theorem 9.2. *In the Chow ring $CH^*_{\mathbf{Q}}(\tilde{M})$, we have*

$$M_F^{(2)} = M^{(2)}, \ M_F^{(\infty)} = 2M^{(\infty)}.$$

All components of $M^{(\infty)}$ occur with multiplicity 2 in $M_F^{(\infty)}$.

References

[1] V. Alexeev, Complete moduli in the presence of semi-abelian group action, preprint math.AG/9905103, to appear in J. of Algebraic Geometry.

[2] M. Artin and B. Mazur, Formal groups arising from algebraic varieties, Ann. Scient. Ec. Norm. Sup., 10 (1977), 87–132.

[3] G. Faltings, C.L. Chai, Degenerations of abelian varieties, Ergebnisse der Math. 22, Springer Verlag, 1990.

[4] G. van der Geer, The Chow ring of the moduli space of abelian threefolds, J. of Algebraic Geometry, 7 (1998), 753–770.

[5] G. van der Geer, Cycles on the moduli space of abelian varieties, in: Moduli of Curves and Abelian Varieties, The Dutch Intercity Seminar on Moduli (C. Faber and E. Looijenga, eds.), Aspects of Mathematics E33, Vieweg 1999.

[6] G. van der Geer and T. Katsura, On a stratification of the moduli of K3 surfaces, J. Eur. Math. Soc. 2 (2000) 3, 259–290.

[7] M. Hazewinkel, Formal Groups and Applications, Boston Orlando, Academic Press, 1978.

[8] L. Illusie, Complexe de de Rham-Witt et cohomologie cristalline, Ann. Sci. ENS, 12 (1979), 501–661.

[9] Yu. I. Manin, The theory of commutative formal groups over fields of finite characteristic, Russian Math. Surv., 18 (1963), 1–80 (Usp. Mat. Nauk., 18, (1963), 3–90).

[10] D. Mumford, *Abelian Varieties*, Tata Inst. Fund. Research, Oxford Univ. Press, 1970.

[11] D. Mumford, *Lectures on Curves on an Algebraic Surface*, Annals of Math. Studies 59. Princeton University Press, 1966.

[12] D. Mumford, *The Kodaira dimension of the Siegel modular variety*, Algebraic Geometry–Open Problems, SLNM 997, 348–375.

[13] T. Oda and F. Oort, Supersingular abelian varieties, Proc. Intern. Symp. on Algebraic Geometry, Kyoto, 1977 (M. Nagata, ed.), Kinokuniya Tokyo 1978, 595–621.

[14] A. Ogus, Supersingular K3 crystals, Astérisque 64 (1979), 3–86.

[15] A. Ogus, Singularities of the height strata in the moduli of K3 surfaces, this volume, 325–343.

[16] F. Oort, Which abelian surfaces are products of elliptic curves? Math. Ann., 214 (1975), 35–47.

[17] F. Oort, A stratification of a moduli space of abelian varieties, this volume, 345–416.

[18] A.N. Rudakov, I.R. Shafarevich, T. Zink, The effect of height on degenerations of algebraic $K3$ surfaces. Izv. Akad. Nauk SSSR Ser. Mat. 46 (1982), 117–134.

[19] J.-P. Serre, Sur la topologie des variétés algébriques en caractéristique p, Symposion Internacional de topologia algebraica 1958, 24–53.

[20] T. Shioda, Supersingular K3 surfaces, Algebraic Geometry (Proc. Summer Meeting, Univ. Copenhagen, Copenhagen, 1978), Lecture Notes in Math., 732, 564–591, Springer-Verlag 1979.

Progress in Mathematics, Vol. 195, © 2001 Birkhäuser Verlag Basel/Switzerland

ISOGENY CLASSES OF ABELIAN VARIETIES WITH NO PRINCIPAL POLARIZATIONS

EVERETT W. HOWE

ABSTRACT. We provide a simple method of constructing isogeny classes of abelian varieties over certain fields k such that no variety in the isogeny class has a principal polarization. In particular, given a field k, a Galois extension ℓ of k of odd prime degree p, and an elliptic curve E over k that has no complex multiplication over k and that has no k-defined p-isogenies to another elliptic curve, we construct a simple $(p-1)$-dimensional abelian variety X over k such that every polarization of every abelian variety isogenous to X has degree divisible by p^2. We note that for every odd prime p and every number field k, there exist ℓ and E as above. We also provide a general framework for determining which finite group schemes occur as kernels of polarizations of abelian varieties in a given isogeny class.

Our construction was inspired by a similar construction of Silverberg and Zarhin; their construction requires that the base field k have positive characteristic and that there be a Galois extension of k with a certain non-abelian Galois group.

1. INTRODUCTION

A natural question to ask of an isogeny class $\mathcal{C}$ of abelian varieties over a field k is whether or not it contains a principally polarized variety. If k is algebraically closed then $\mathcal{C}$ will certainly contain a principally polarized variety, and if k is finite then every $\mathcal{C}$ that satisfies some relatively weak conditions will contain a principally polarized variety (see [2]); for example, it is enough for the varieties in $\mathcal{C}$ to be simple and odd-dimensional. In this paper we show that for a large class of fields, including all number fields and function fields over finite fields, it is very easy to construct isogeny classes of abelian varieties that contain no principally polarized varieties. We also provide a framework for considering the more general problem of determining which finite group schemes occur as the kernels of polarizations of varieties in a given isogeny class.

Our construction of isogeny classes containing no principally polarized varieties is very straightforward, but to describe it we must introduce some

1991 *Mathematics Subject Classification.* Primary 14K02; Secondary 11G10, 14K15.
Key words and phrases. Abelian variety, isogeny, polarization, Galois twist.

terminology. If ℓ is a finite extension of a field k and if E is an elliptic curve over ℓ, we let $\operatorname{Res}_{\ell/k} E$ denote the restriction of scalars of E from ℓ to k (see Section 1.3 of [11]). If E is an elliptic curve over k, we define the *reduced restriction of scalars* of E from ℓ to k to be the kernel of the trace map from $\operatorname{Res}_{\ell/k} E$ to E. Let p be an odd prime. We will say that an elliptic curve E over k is *p-isolated* if it has no complex multiplication over k and if it has no p-isogeny to another elliptic curve over k. We will say that a field k is *p-admissible* if there is a p-isolated elliptic curve over k and if there is a Galois extension of k of degree p. One can show, for example, that every number field is p-admissible; that every function field over a finite field of characteristic not p is p-admissible; and that no finite field or algebraically-closed field is p-admissible.

Theorem 1.1. *Let p be an odd prime number and let k be a p-admissible field. Let E be a p-isolated elliptic curve over k, let ℓ be a degree-p Galois extension of k, and let X be the reduced restriction of scalars of E from ℓ to k. Then X is simple, and every polarization of every abelian variety isogenous to X has degree divisible by p^2.*

We provide an elementary proof of this theorem in Part 2 of this paper. In Section 2.1 we prove some basic results about polarizations and endomorphism rings of Galois twists of abelian varieties; we apply these general results in Section 2.2 to prove several results about reduced restrictions of scalars of p-isolated elliptic curves. We use the results of Section 2.2 to prove Theorem 1.1 in Section 2.3.

In Part 3 we prove a very general theorem that sheds additional light on the proof of Theorem 1.1. In Section 3.1 we associate to every isogeny class $\mathcal{C}$ of abelian varieties over a field k a two-torsion group $\mathcal{B}_2(\mathcal{C})$ and a finite set $S_{\mathcal{C}} \subseteq \mathcal{B}_2(\mathcal{C})$, and we prove in Section 3.2 that the set $S_{\mathcal{C}}$ determines the set of kernels of polarizations of varieties in $\mathcal{C}$ up to Jordan-Hölder isomorphism. Then in Section 3.3 we revisit the proof of Theorem 1.1 and show how it can be interpreted in terms of the group $\mathcal{B}_2(\mathcal{C})$ and the set $S_{\mathcal{C}}$.

Theorem 1.1 was inspired by a construction of Silverberg and Zarhin (see [8] and [9]); they too construct twists of a power of an elliptic curve such that every polarization of every abelian variety isogenous to the twist has degree divisible by a given prime. Their original construction is limited to base fields of positive characteristic that have nonabelian Galois extensions of a certain type, but more recently they have produced a new construction that works over an arbitrary number field (see [10]).

Acknowledgments. The author thanks Daniel Goldstein, Bob Guralnick, Alice Silverberg, and Yuri Zarhin for helpful conversations and correspondence.

Conventions and notation. We consider varieties to be schemes over some specified base field; it follows that if U and V are varieties over a field k, then what we call a *morphism* from U to V others might call a k-*morphism* from U to V. If U is a variety over a field k and ℓ is an extension field of k, then we let U_ℓ denote the ℓ-variety $V \times_{\operatorname{Spec} k} \operatorname{Spec} \ell$. If $\alpha\colon U \to V$ is a morphism of varieties over k, we let α_ℓ denote the induced morphism from U_ℓ to V_ℓ. If X is an abelian variety, we let $\widehat{X}$ denote its dual variety, and if $\alpha\colon X \to Y$ is a morphism of abelian varieties, we let $\widehat{\alpha}$ denote the dual morphism $\widehat{Y} \to \widehat{X}$. If G is a group scheme over k and n is an integer, we denote by $G[n]$ the n-torsion subscheme of G.

2. Isogeny classes containing no principally polarized varieties

§2.1. Polarizations and endomorphisms of Galois twists of abelian varieties. In this section we prove some simple general results about Galois twists of abelian varieties that we will need in our proof of Theorem 1.1.

Suppose k is a field, ℓ is a Galois extension of k with Galois group G, and Y is an abelian variety over k. Suppose that X is an ℓ/k-twist of Y and that $f\colon Y_\ell \to X_\ell$ is an isomorphism. Then X corresponds (as in Section III.1.3 of [7]) to the element of $H^1(G, \operatorname{Aut} Y_\ell)$ represented by the cocycle $\sigma \mapsto a_\sigma := f^{-1}f^\sigma$. Our first proposition tells us when an endomorphism α of Y gives rise to an endomorphism of X.

Proposition 2.1. *The endomorphism $f\alpha_\ell f^{-1}$ of X_ℓ descends to an endomorphism of X if and only if we have $a_\sigma\alpha_\ell = \alpha_\ell a_\sigma$ for all σ in G.*

Proof. Let β be the endomorphism $f\alpha_\ell f^{-1}$ of X_ℓ. Then β will descend to X if and only if for all σ in G we have $\beta^\sigma = \beta$, which is

$$(f\alpha_\ell f^{-1})^\sigma = f\alpha_\ell f^{-1}.$$

By multiplying this equality by f^σ on the right and by f^{-1} on the left, and by using the fact that $\alpha_\ell = \alpha_\ell^\sigma$, we see that β descends to X if and only if for all σ we have

$$f^{-1}f^\sigma\alpha_\ell = \alpha_\ell f^{-1}f^\sigma,$$

if and only if $a_\sigma\alpha_\ell = \alpha_\ell a_\sigma$ for all σ in G. □

Now suppose λ is a polarization of Y, and let $x \mapsto x^\dagger$ denote the Rosati involution on $\operatorname{End} Y$ corresponding to λ, so that $x^\dagger = \lambda^{-1}\widehat{x}\lambda$. Our second proposition tells us when the polarization λ gives rise to a polarization of X.

Proposition 2.2. *The polarization $\widehat{f^{-1}}\lambda_\ell f^{-1}$ of X_ℓ descends to a polarization of X if and only if we have $a_\sigma^\dagger a_\sigma = 1$ for all σ in G.*

Proof. Let μ be the polarization $\widehat{f^{-1}}\lambda_\ell f^{-1}$ of X_ℓ. Then μ will descend to X if and only if for all σ in G we have $\mu^\sigma = \mu$, which is

$$(\widehat{f^{-1}}\lambda_\ell f^{-1})^\sigma = \widehat{f^{-1}}\lambda_\ell f^{-1}.$$

By multiplying this equality by f^σ on the right and by $\lambda_\ell^{-1}\widehat{f^\sigma}$ on the left, and by using the fact that $\lambda_\ell = \lambda_\ell^\sigma$, we see that μ descends to X if and only if for all σ we have

$$\lambda_\ell^{-1}(\widehat{f^{-1}f^\sigma})\lambda_\ell(f^{-1}f^\sigma) = 1,$$

if and only if $a_\sigma^\dagger a_\sigma = 1$ for all σ in G. □

§2.2. Reduced restrictions of scalars of p-isolated elliptic curves. Let k be a p-admissible field, let E be a p-isolated elliptic curve over k, and let ℓ be a Galois extension of k of degree p. Let X be the reduced restriction of scalars of E from ℓ to k. In this section we calculate the endomorphism ring of X, a restriction on the degrees of the polarizations of X, and the Galois module structure of the p-torsion of X. These results are the building blocks of our proof of Theorem 1.1.

Lemma 2.3. *The elliptic curve E_ℓ has no complex multiplication.*

Proof. The endomorphism algebra $A = (\operatorname{End} E_\ell) \otimes \mathbf{Q}$ is either $\mathbf{Q}$, an imaginary quadratic field, or a quaternion algebra over $\mathbf{Q}$. The action of $\operatorname{Gal}(\ell/k)$ on $\operatorname{End} E_\ell$ gives us a homomorphism $\operatorname{Gal}(\ell/k) \to \operatorname{Aut}(A/\mathbf{Q})$, and since $\operatorname{End} E = \mathbf{Z}$, this homomorphism must be nontrivial if $A \neq \mathbf{Q}$. Suppose A were a quaternion algebra. Then the image of a generator of $\operatorname{Gal}(\ell/k)$ in $\operatorname{Aut}(A/\mathbf{Q})$ would be a nontrivial automorphism. Since every automorphism of a quaternion algebra is inner, this automorphism would have to be given by conjugation by a non-central element s of A. But then the 2-dimensional sub-algebra $\mathbf{Q}(s)$ would be fixed by the action of $\operatorname{Gal}(\ell/k)$ on A, contradicting the fact that $\operatorname{End} E = \mathbf{Z}$. Therefore A is not a quaternion algebra. Suppose A were a quadratic field. Then $\operatorname{Aut}(A/\mathbf{Q})$ would be a cyclic group of order 2, contradicting the existence of a nontrivial homomorphism $\operatorname{Gal}(\ell/k) \to \operatorname{Aut}(A/\mathbf{Q})$. Thus A must be $\mathbf{Q}$, and $\operatorname{End} E_\ell$ must be $\mathbf{Z}$. □

Let σ be a generator of $\operatorname{Gal}(\ell/k)$ and let R be the restriction of scalars of E from ℓ to k. Then R is the ℓ/k-twist of E^p given by the element of $H^1(\operatorname{Gal}(\ell/k), \operatorname{Aut} E_\ell^p)$ represented by the cocycle that sends σ to the automorphism ξ of E_ℓ^p that cyclically shifts the factors. The kernel S of the trace map $E^p \to E$ is stable under ξ, and the reduced restriction of scalars X of E is the ℓ/k-twist of S given by sending σ to the restriction of ξ to S. The projection map from E^p onto its first $p-1$ factors gives an isomorphism from S to E^{p-1}; under this isomorphism, the restriction of ξ to S is given by the $(p-1)\times(p-1)$ matrix

$$\zeta = \begin{bmatrix} -1 & -1 & \cdots & -1 & -1 \\ 1 & 0 & \cdots & 0 & 0 \\ 0 & 1 & \cdots & 0 & 0 \\ \vdots & \vdots & \ddots & \vdots & \vdots \\ 0 & 0 & \cdots & 1 & 0 \end{bmatrix},$$

where we identify the ring $M_{p-1}(\mathbf{Z})$ of $(p-1)\times(p-1)$ integer matrices with the endomorphism ring of E_ℓ^{p-1}. In other words, X is the ℓ/k-twist of E^{p-1} given by the element of $H^1(\mathrm{Gal}(\ell/k), \mathrm{Aut}\, E_\ell^{p-1})$ represented by the cocycle that sends σ to ζ. Note that the minimal polynomial of the endomorphism ζ is the pth cyclotomic polynomial.

Lemma 2.4. *The abelian variety X is simple over k, and its endomorphism ring is isomorphic to the ring of integers of the pth cyclotomic field.*

Proof. Since $X_\ell \cong E_\ell^{p-1}$ and $\mathrm{End}\, E_\ell = \mathbf{Z} = \mathrm{End}\, E$, every endomorphism of X comes from an endomorphism of E^{p-1}. According to Proposition 2.1, the only endomorphisms of E^{p-1} that give rise to elements of $\mathrm{End}\, X$ are the endomorphisms that commute with the element ζ of $\mathrm{End}\, E_\ell^{p-1}$ defined by the matrix above. Since $\mathbf{Q}(\zeta)$ is a field of degree $p-1$ over $\mathbf{Q}$, it is a maximal commutative subring of the matrix ring $M_{p-1}(\mathbf{Q})$, so the only elements of $\mathrm{End}\, E^{p-1}$ that commute with ζ are those elements that lie in $\mathbf{Q}(\zeta)$. The intersection of $\mathbf{Q}(\zeta)$ with $\mathrm{End}\, E^{p-1}$ is $\mathbf{Z}[\zeta]$, which is the ring of integers of the cyclotomic field $\mathbf{Q}(\zeta)$. Thus $\mathrm{End}\, X$ is isomorphic to the ring of integers of the pth cyclotomic field. And finally, the fact that $(\mathrm{End}\, X)\otimes\mathbf{Q}$ is a field shows that X is simple. □

Lemma 2.5. *If $\alpha \in \mathbf{Q}(\zeta)$ is an endomorphism of X, then the degree of α is the square of the norm of α from $\mathbf{Q}(\zeta)$ to $\mathbf{Q}$.*

Proof. It is easy to see that under the identification of $\mathrm{End}\, E^{p-1}$ with $M_{p-1}(\mathbf{Z})$, the degree function is the square of the determinant function. The lemma then follows from the fact the determinant function from $M_{p-1}(\mathbf{Q})$ to $\mathbf{Q}$, restricted to the maximal subfield $\mathbf{Q}(\zeta)$ of $M_{p-1}(\mathbf{Q})$, is the field norm from $\mathbf{Q}(\zeta)$ to $\mathbf{Q}$. □

Lemma 2.6. *The abelian variety X has a polarization λ of degree p^2.*

Proof. Let μ be the product principal polarization of E^{p-1}, let b be the endomorphism of E^{p-1} defined by the matrix

$$\begin{bmatrix} 2 & 1 & \cdots & 1 \\ 1 & 2 & \cdots & 1 \\ \vdots & \vdots & \ddots & \vdots \\ 1 & 1 & \cdots & 2 \end{bmatrix}$$

with 2's on the diagonal and 1's elsewhere, and let $\overline{\lambda} = \mu b$. The Rosati involution associated to μ is the matrix transpose $x \mapsto x^t$, and the Rosati involution associated to $\overline{\lambda}$ is the matrix transpose conjugated by b — that is, $x \mapsto b^{-1}x^t b$. One checks that $b^{-1}\zeta^t b\zeta$ is the identity matrix, so by Proposition 2.2 the polarization $\overline{\lambda}_\ell$ of E_ℓ^{p-1} descends to give a polarization λ of X.

Since μ is a principal polarization and $\lambda_\ell = \mu_\ell b_\ell$, the degree of λ is the degree of b, which is the square of the determinant of b. An easy calculation shows that $\det b = p$, which proves the lemma. □

Lemma 2.7. *Suppose μ is a polarization of X. Then there is an integer n such that* $\deg \mu = p^2 n^4$.

Proof. Let K be the field $(\operatorname{End} X) \otimes \mathbf{Q}$ and let ζ be the endomorphism of X defined above, so that ζ is a primitive pth root of unity and $K = \mathbf{Q}(\zeta)$. Let $x \mapsto x^\dagger$ be the Rosati involution associated to the polarization λ of Lemma 2.6. Then for every $x \in K$, the element $x^\dagger$ is simply the complex conjugate of x.

Suppose μ is a polarization of X. Then the element $\lambda^{-1}\mu$ of the field K is fixed by the Rosati involution (see §21, Application 3, pp. 208–210 of [3]), and is therefore an element of the maximal real subfield K^+ of K. By Lemma 2.5, the degree of an element of K is equal to the square of its norm to $\mathbf{Q}$, so we find that

$$\deg \mu = \deg \lambda \cdot \deg(\lambda^{-1}\mu) = p^2 \left(N_{K/\mathbf{Q}}(\lambda^{-1}\mu)\right)^2 = p^2 \left(N_{K^+/\mathbf{Q}}(\lambda^{-1}\mu)\right)^4.$$

Let $n = N_{K^+/\mathbf{Q}}(\lambda^{-1}\mu)$. Since $\deg \mu$ is an integer and n is rational, we see that n is an integer, and the lemma is proved. □

We end this section by describing the Galois module structure of $X[p]$. First note that p is not equal to the characteristic of the base field k because E has no p-isogenies. Thus, the group scheme structures of $E[p]$ and $X[p]$ are completely captured by the Galois module structures of the sets of points of these schemes over a separable closure k^{sep} of k. Furthermore, $E[p](k^{\mathrm{sep}})$ is a simple $\operatorname{Gal}(k^{\mathrm{sep}}/k)$-module because E has no p-isogenies to another elliptic curve over k.

Lemma 2.8. *The sequence of modules*

$$0 \subset (\zeta-1)^{p-2}X[p](k^{\mathrm{sep}}) \subset (\zeta-1)^{p-3}X[p](k^{\mathrm{sep}}) \subset \cdots \subset X[p](k^{\mathrm{sep}})$$

is a composition series for the $\operatorname{Gal}(k^{\mathrm{sep}}/k)$*-module* $X[p](k^{\mathrm{sep}})$*, and each composition factor is isomorphic to* $E[p](k^{\mathrm{sep}})$.

Proof. Multiplication by $(\zeta-1)^{p-i-2}$ gives an isomorphism from the quotient $(\zeta-1)^i X[p](k^{\mathrm{sep}})/(\zeta-1)^{i+1}X[p](k^{\mathrm{sep}})$ to $(\zeta-1)^{p-2}X[p](k^{\mathrm{sep}})$, which is the kernel of $\zeta - 1$ acting on $X[p](k^{\mathrm{sep}})$. This kernel is the image of $E[p](k^{\mathrm{sep}})$ under the diagonal embedding of E_ℓ into E_ℓ^{p-1}. □

Lemma 2.9. *If $\varphi\colon X \to Y$ is an isogeny, then the only simple* $\operatorname{Gal}(k^{\mathrm{sep}}/k)$*-module that occurs as a composition factor of the p-power-torsion part of the kernel of φ is* $E[p](k^{\mathrm{sep}})$.

Proof. Immediate from Lemma 2.8. □

§2.3. Proof of Theorem 1.1. For every finite group scheme G over k, let us define the $E[p]$*-rank* of G to be the multiplicity of the simple Galois module $E[p](k^{\text{sep}})$ as a composition factor of the p-power-torsion of G. We will denote the $E[p]$-rank of G by $\text{rank}_{E[p]}(G)$. The $E[p]$-rank is an additive function on exact sequences of finite group schemes.

Suppose $\varphi: X \to Y$ is an isogeny and ν is a polarization of Y. Then the map $\mu: X \to \widehat{X}$ given by $\widehat{\varphi}\nu\varphi$ is a polarization of X, so we have

$$\text{rank}_{E[p]}(\ker \mu) = \text{rank}_{E[p]}(\ker \widehat{\varphi}) + \text{rank}_{E[p]}(\ker \nu) + \text{rank}_{E[p]}(\ker \varphi).$$

Now, $\ker \widehat{\varphi}$ is the Cartier dual of $\ker \varphi$, and $E[p]$ is its own Cartier dual, so $\text{rank}_{E[p]}(\ker \widehat{\varphi}) = \text{rank}_{E[p]}(\ker \varphi)$. Thus the parity of $\text{rank}_{E[p]}(\ker \nu)$ is equal to that of $\text{rank}_{E[p]}(\ker \mu)$, which is odd by Lemmas 2.7 and 2.9. Therefore $E[p]$ appears as a composition factor of the p-power-torsion of $\ker \nu$, so p^2 divides the order of $\ker \nu$, so p^2 divides the degree of ν. □

3. Polarizations up to Jordan-Hölder isomorphism

In this part of the paper we associate to every isogeny class $\mathcal{C}$ of abelian varieties over a field k a two-torsion group $\mathcal{B}_2(\mathcal{C})$ and a finite set $S_\mathcal{C} \subseteq \mathcal{B}_2(\mathcal{C})$, and we show that the set $S_\mathcal{C}$ determines the set of kernels of polarizations of varieties in $\mathcal{C}$ up to Jordan-Hölder isomorphism. Then we revisit the proof of Theorem 1.1 and show how it can be interpreted in terms of the group $\mathcal{B}_2(\mathcal{C})$ and the set $S_\mathcal{C}$.

§3.1. Statement of results. For every isogeny class $\mathcal{C}$ of abelian varieties over a field k, we let $\text{Ker}_\mathcal{C}$ be the category whose objects are finite commutative group schemes over k that can be embedded (as closed sub-group-schemes) in some variety in the isogeny class $\mathcal{C}$ and whose morphisms are morphisms of group schemes. We see that the objects in $\text{Ker}_\mathcal{C}$ are those group schemes that can be written $\ker \varphi$ for some isogeny $\varphi: X \to Y$ of elements of $\mathcal{C}$.

The *Grothendieck group* $G(\text{Ker}_\mathcal{C})$ of $\text{Ker}_\mathcal{C}$ is defined to be the quotient of the free abelian group on the isomorphism classes of objects in $\text{Ker}_\mathcal{C}$ by the subgroup generated by the expressions $X - X' - X''$ for all exact sequences $0 \to X' \to X \to X'' \to 0$ in $\text{Ker}_\mathcal{C}$. If X is an object of $\text{Ker}_\mathcal{C}$ we let $[X]$ denote its class in $G(\text{Ker}_\mathcal{C})$, and we say that two objects X and Y are *Jordan-Hölder isomorphic* to one another if $[X] = [Y]$. The group $G(\text{Ker}_\mathcal{C})$ is a free abelian group on the simple objects of $\text{Ker}_\mathcal{C}$. An element of $G(\text{Ker}_\mathcal{C})$ is said to be *effective* if it is a sum of positive multiples of simple objects. Let us call an element P of $G(\text{Ker}_\mathcal{C})$ *attainable* if there is a polarization λ of a variety in $\mathcal{C}$ such that $P = [\ker \lambda]$. Our goal in this section will be to identify the attainable elements of $G(\text{Ker}_\mathcal{C})$.

To identify the attainable elements we must first define several groups for every isogeny class of abelian varieties. The first two groups will be subgroups of $G(\mathrm{Ker}_{\mathcal{C}})$, and the others will be defined solely in terms of the endomorphism rings of the varieties in the isogeny class.

Let $Z(\mathcal{C})$ denote the subgroup of $G(\mathrm{Ker}_{\mathcal{C}})$ generated by the elements of the form $[G]$, where $G \in \mathrm{Ker}_{\mathcal{C}}$ is a group scheme whose rank is a square and for which there exists a non-degenerate alternating pairing $G \times G \to \mathbf{G}_m$.[1] Cartier duality on $\mathrm{Ker}_{\mathcal{C}}$ defines an involution $P \mapsto \overline{P}$ of $G(\mathrm{Ker}_{\mathcal{C}})$, and $Z(\mathcal{C})$ is stable under this involution. Let $B(\mathcal{C})$ be the subgroup $\{P + \overline{P} : P \in G(\mathrm{Ker}_{\mathcal{C}})\}$ of $G(\mathrm{Ker}_{\mathcal{C}})$; it is not hard to see that $B(\mathcal{C}) \subseteq Z(\mathcal{C})$.

Now let us define the groups that depend on the endomorphism rings of the varieties in $\mathcal{C}$. If X and Y are two varieties in $\mathcal{C}$ then $(\mathrm{End}\, X)\otimes\mathbf{Q} \cong (\mathrm{End}\, Y)\otimes\mathbf{Q}$, so we may define $\mathrm{End}^0 \mathcal{C}$ to be $(\mathrm{End}\, X) \otimes \mathbf{Q}$ for any X in $\mathcal{C}$. We may write $\mathcal{C} \sim \mathcal{C}_1^{n_1} \times \cdots \times \mathcal{C}_r^{n_r}$ for some distinct isogeny classes $\mathcal{C}_i$ of simple varieties and some integers n_i; by this we mean that every X in $\mathcal{C}$ is isogenous to a product $X_1^{n_1} \times \cdots \times X_r^{n_r}$ where $X_i \in \mathcal{C}_i$. Let $A = \mathrm{End}^0 \mathcal{C}$. Then

$$A \cong M_{n_1}(D_1) \times \cdots \times M_{n_r}(D_r),$$

where $D_i = \mathrm{End}^0 \mathcal{C}_i$ and where $M_{n_i}(D_i)$ denotes an n_i-by-n_i matrix algebra over D_i. Each D_i is a division algebra over its center K_i, which is a number field, and the center K of A is the product of the K_i. Only certain kinds of division algebras can occur as the endomorphism algebras of simple isogeny classes, and they are classified into four types (see Theorem 2 (p. 201) of [3]). We will define three subgroups $R_0(A)$, $R_1(A)$, and $R_2(A)$ of K^* by defining these groups first for the four possible types of division algebras D and by then setting

$$R_i(A) = R_i(D_1) \times \cdots \times R_i(D_r).$$

The group $R_1(A)$ will be a subgroup of $R_0(A)$, and $R_2(A)$ will be a subgroup of finite index in $R_1(A)$.

Type I: For this type, $D = K$ is a totally real number field. We let $R_0(D) = K^*$, we let $R_1(D)$ be the multiplicative group of totally positive elements of K, and we let $R_2(D) = R_1(D)$.

Type II: For this type, K is a totally real number field and D is a quaternion algebra over K that is split at every infinite prime of K. We let $R_0(D) = K^*$, we let $R_1(D)$ be the multiplicative group of totally positive elements of K, and we let $R_2(D)$ be the subgroup of $R_1(D)$ consisting of those elements that are squares in $K_{\mathfrak{p}}$ for all primes $\mathfrak{p}$ of K for which the local Brauer invariant $\mathrm{inv}_{\mathfrak{p}} D$ of D at $\mathfrak{p}$ is nonzero.

Type III: For this type, K is a totally real number field and D is a quaternion algebra over K that is ramified at every infinite prime of K. We

[1] The existence of a non-degenerate alternating pairing *implies* that the rank of G is a square, except in characteristic 2; the unique simple local-local group scheme in characteristic 2, which has rank 2, has a non-degenerate alternating pairing.

let $R_0(D)$ be the multiplicative group of totally positive elements of K, we let $R_1(D)$ be the group of squares of elements of $R_0(D)$, and we let $R_2(D) = R_1(D)$.

Type IV: For this type, K is a CM-field with maximal real subfield K^+ and D is a division algebra over K such that if σ is the nontrivial automorphism of K over K^+ then $\mathrm{inv}_{\mathfrak{p}} D = 0$ for every prime of K fixed by σ and $\mathrm{inv}_{\mathfrak{p}} D + \mathrm{inv}_{\mathfrak{p}^\sigma} D = 0$ for every prime $\mathfrak{p}$ of K. We let $R_0(D) = K^*$, we let $R_1(D)$ be the multiplicative group of totally positive elements of K^+, and we let $R_2(D) = R_1(D)$.

An involution $x \mapsto x^\dagger$ of A is *positive* if $\mathrm{Trd}(xx^\dagger)$ is a totally positive element of K for every element x of A^*, where Trd is the reduced trace map from A to K. If $\alpha \in A$ is fixed by a positive involution then $\mathbf{Q}(\alpha)$ is a product of totally real number fields. If $x \mapsto x^\dagger$ is a positive involution of A, we let $A_\dagger$ denote the set of elements of A^* that are fixed by $\dagger$ and that are totally positive.

The following lemma, whose proof we will give in Section 3.2, motivates the definitions of the groups $R_i(A)$.

Lemma 3.1. *Let $A = \mathrm{End}^0 \mathcal{C}$ for an isogeny class $\mathcal{C}$ of abelian varieties over k. Then $R_0(A) = \mathrm{Nrd}(A^*)$, where* Nrd *is the reduced norm map from A to its center K. Suppose $x \mapsto x^\dagger$ is a positive involution of A. Then $R_2(A) \subseteq \mathrm{Nrd}(A_\dagger) \subseteq R_1(A)$.*

Remark. Note that if A is built up out of simple D that are of type I, III, and IV then $R_2(A) = R_1(A)$ and $\mathrm{Nrd}(A_\dagger)$ is a subgroup of K^*, but if a D of type II occurs in A then Lemma 3.1 only allows us to say that $\mathrm{Nrd}(A_\dagger)$ lies between two subgroups of K^*. We cannot expect to do much better than this; one can find examples of D of type II for which $\mathrm{Nrd}(D_\dagger)$ is not a group.

Let $A = \mathrm{End}^0 \mathcal{C}$. We define a homomorphism Prin from the multiplicative group A^* to $G(\mathrm{Ker}_\mathcal{C})$ as follows: Suppose $\alpha \in A^*$ is given. We pick a variety X in $\mathcal{C}$ and choose an endomorphism β of X and an integer n such that $\alpha = \beta/n$. Then we set $\mathrm{Prin}(\alpha) = [\ker \beta] - [\ker n]$. We leave it to the reader to show that the value $[\ker \beta] - [\ker n]$ does not depend on the choice of X, β, and n.

Note that Lemma 3.1 states in part that $R_0(A) = \mathrm{Nrd}(A^*)$. We can use this identity, together with the homomorphism Prin, to define a homomorphism Φ from $R_0(A)$ to $G(\mathrm{Ker}_\mathcal{C})$. Suppose $a \in R_0(A)$ is given, and suppose α and α' are elements of A^* with $\mathrm{Nrd}(\alpha) = \mathrm{Nrd}(\alpha') = a$. Then, by an easy consequence of Wang's Theorem (see [4], Theorem 1.14, p. 38, and the comments on p. 39), we see that $\alpha'\alpha^{-1}$ lies in the commutator subgroup of A^*, so that $\mathrm{Prin}(\alpha'\alpha^{-1}) = 0$. Thus we may define Φ by taking $\Phi(a) = \mathrm{Prin}(\alpha)$ for any choice of $\alpha \in A^*$ such that $\mathrm{Nrd}(\alpha) = a$.

Finally, we let $\mathcal{B}_1(\mathcal{C})$ be the quotient of $Z(\mathcal{C})$ by the subgroup generated by $B(\mathcal{C})$ and $\Phi(R_1(A))$, and we let $\mathcal{B}_2(\mathcal{C})$ be the quotient of $Z(\mathcal{C})$ by the subgroup generated by $B(\mathcal{C})$ and $\Phi(R_2(A))$. Since $2Z(\mathcal{C})$ is contained in $B(\mathcal{C})$ the groups $\mathcal{B}_1(\mathcal{C})$ and $\mathcal{B}_2(\mathcal{C})$ are 2-torsion, and since $R_2(A)$ is a subgroup of finite index in $R_1(A)$, the natural surjection $\mathcal{B}_2(\mathcal{C}) \to \mathcal{B}_1(\mathcal{C})$ has a finite kernel.

Theorem 3.2. *There is a finite subset $S_\mathcal{C}$ of the group $\mathcal{B}_2(\mathcal{C})$ such that an element of $G(\mathrm{Ker}_\mathcal{C})$ is attainable if and only if it is an effective element of $Z(\mathcal{C})$ whose image in $\mathcal{B}_2(\mathcal{C})$ lies in $S_\mathcal{C}$. Furthermore, the image of $S_\mathcal{C}$ in $\mathcal{B}_1(\mathcal{C})$ consists of a single element $I_\mathcal{C}$.*

This theorem must surely seem quite removed from the down-to-earth question of whether there exists a principally-polarized variety in a given isogeny class. However, we will show in Section 3.3 that our proof of Theorem 1.1 can be viewed as an argument showing that there is a homomorphism $\mathcal{B}_1(\mathcal{C}) \to \mathbf{Z}/2\mathbf{Z}$ such that the image of $I_\mathcal{C}$ is nonzero. Also, in certain cases the abstract objects in Theorem 3.2 can be computed. For example, if $\mathcal{C}$ is an isogeny class over a finite field then the group $\mathcal{B}_1(\mathcal{C})$ (which in this case is equal to $\mathcal{B}_2(\mathcal{C})$) can be computed, and one can even compute a subgroup of order at most 2 that contains $I_\mathcal{C}$ (see [2]). If in addition the varieties in $\mathcal{C}$ are ordinary, the element $I_\mathcal{C}$ itself can be calculated (see [1]).

§3.2. Proofs of Lemma 3.1 and Theorem 3.2. In this section we prove the results stated in the preceding section.

Proof of Lemma 3.1. Clearly it will suffice to prove the lemma in the case where $A = M_n(D)$ for some integer n and division algebra D of one of the four types listed in the preceding section. For each of these types of algebras, the Hasse-Schilling-Maass theorem ([5], Theorem 33.15, p. 289) shows that $R_0(A) = \mathrm{Nrd}(A^*)$, so we need only prove the second statement of the lemma.

If D/K is of Type III then the statement we are to prove is Theorem 4.7 of [2], while if D/K is of Type IV then the statement we are to prove is Theorem 4.1 of [2].

Suppose $D = K$ is of Type I. We must show that $\mathrm{Nrd}(A_\dagger)$ is the set of totally positive elements of K. It is clear that $\mathrm{Nrd}(\alpha)$ is totally positive if $\alpha \in A_\dagger$, so all we must show is that every totally positive element of K is the reduced norm of some element of $A_\dagger$.

Let $x \mapsto x^*$ be the transpose on $M_d(K)$. Then by Theorem 8.7.4 (pp. 301–302) and Theorem 7.6.3 (p. 259) of [6], there is an isomorphism $i\colon A \to M_d(K)$ and a diagonal matrix $\alpha \in M_d(K)$ such that the isomorphism i takes the involution $x \mapsto x^\dagger$ to the involution η of $M_d(K)$ defined by $\eta(x) = \alpha x^* \alpha^{-1}$. Now suppose we are given a totally positive b in K. Let $\beta \in M_d(K)$ be the diagonal matrix with b in the upper left corner and 1's elsewhere. Then β is totally positive and fixed by η, and its determinant is b. Therefore $i^{-1}(\beta)$ is an element of $K_\dagger$ with reduced norm equal to b.

Suppose D/K is of Type II. It is clear that the reduced norm of an element of $A_\dagger$ is totally positive, so we have $\operatorname{Nrd}(A_\dagger) \subseteq R_1(A)$, and we must prove that $R_2(A) \subseteq \operatorname{Nrd}(A_\dagger)$.

Let $x \mapsto x^*$ be the conjugate transpose involution on $M_d(D)$, where "conjugation" on D is the standard involution $x \mapsto \operatorname{Trd}_{D/K} x - x$. Then again by Theorem 8.7.4 (pp. 301–302) and Theorem 7.6.3 (p. 259) of [6], there is an isomorphism $i: A \to M_d(D)$ and a diagonal matrix $\alpha \in M_d(D)$ with $\alpha^* = -\alpha$ such that the isomorphism i takes the involution $x \mapsto x^\dagger$ to the involution η of $M_d(D)$ defined by $\eta(x) = \alpha x^* \alpha^{-1}$; furthermore, as is argued on pp. 194–195 of [3], the entries of the diagonal matrix α^2 are totally negative elements of K. Let $\alpha_1, \dots, \alpha_n$ denote the diagonal entries of α and let $c_1 = \alpha_1^2$, so that α_1 satisfies the polynomial $x^2 - c_1$. Since the field $K(\alpha_1)$ splits the quaternion algebra D, the element c_1 of K must be a nonsquare in $K_\mathfrak{p}$ for every prime $\mathfrak{p}$ of K for which $\operatorname{inv}_\mathfrak{p} D \neq 0$.

Suppose b is an element of $R_2(A)$. We claim that there exists an element β of D such that (1) $\beta + \beta^* = 0$ and (2) $\beta\beta^* = -bc_1$ and (3) $\beta\alpha_1^{-1}$ is totally positive. To see that such a β exists, we first note that bc_1 is a nonsquare in $K_\mathfrak{p}$ for every prime $\mathfrak{p}$ of K for which $\operatorname{inv}_\mathfrak{p} D \neq 0$, so the field $K(\sqrt{bc_1})$ splits D. This shows that there is an element β_0 of D such that $\beta_0^2 - bc_1 = 0$, so that β_0 satisfies (1) and (2). If we choose a K-basis for the trace-0 elements of D, then the set of β satisfying (1) and (2) is a level set of a homogeneous ternary quadratic form Q that is indefinite at every infinite prime of K, and we have just shown that there are K-points in this level set. Condition (3) is simply a linear inequality at each of the infinite primes of K, and the inequality can be satisfied locally at each infinite prime by points on the level set because the form Q is indefinite, so by weak approximation we see that there do exist β's satisfying (1), (2), and (3).

Choose such a $\beta \in D$, and let γ be the diagonal matrix with $\beta\alpha_1^{-1}$ in the upper left corner and 1's elsewhere on the diagonal. A computation shows that γ is fixed by the involution η, that the reduced norm of γ is $\operatorname{Nrd}_{D/K}(\beta)/\operatorname{Nrd}_{D/K}(\alpha_1) = b$, and that γ is totally positive. Thus $i^{-1}(\gamma)$ is an element of $A_\dagger$ with reduced norm equal to b. □

Proof of Theorem 3.2. Suppose $\lambda: X \to \widehat{X}$ and $\mu: Y \to \widehat{Y}$ are polarizations of varieties in $\mathcal{C}$. First note that the ranks of $\ker\lambda$ and $\ker\nu$ are squares, and that there are non-degenerate alternating pairings from these groups to the multiplicative group (see §23 of [3]), so that $[\ker\lambda]$ and $[\ker\mu]$ lie in $Z(\mathcal{C})$. Let $\varphi: X \to Y$ be an isogeny, and let ν be the polarization $\widehat{\varphi}\mu\varphi$ of X, where $\widehat{\varphi}: \widehat{Y} \to \widehat{X}$ is the dual isogeny of φ. Let n be any positive integer such that $\ker\lambda \subseteq \ker(n\nu)$ as group schemes. Then there is an isogeny $\alpha: \widehat{X} \to \widehat{X}$ such that $n\nu = \alpha\lambda$. A polarization is equal to its own dual isogeny, so we can equate the right-hand side of the last equality with its dual to get $n\nu = \lambda\widehat{\alpha}$. Using Application III (pp. 208–210) of §21 of [3] (see especially the final paragraph) and the fact that $n\nu$ and λ are polarizations, we find that $\widehat{\alpha} \in \operatorname{End} X$ is fixed by the Rosati involution associated to λ and is totally positive.

The equality $n\widehat{\varphi}\mu\varphi = \lambda\widehat{\alpha}$ translates into the equality

$$[\ker n] + [\ker \widehat{\varphi}] + [\ker \mu] + [\ker \varphi] = [\ker \lambda] + [\ker \widehat{\alpha}]$$

in $G(\mathrm{Ker}_{\mathcal{C}})$. Now, $[\ker \widehat{\varphi}] + [\ker \varphi]$ is an element of $B(\mathcal{C})$, and $\widehat{\alpha}$ and n are totally positive elements of $\operatorname{End}^0 \mathcal{C}$ that are fixed by the Rosati involution, so $[\ker n]$ and $[\ker \widehat{\alpha}]$ lie in $\Phi(R_1(\operatorname{End}^0 \mathcal{C}))$. It follows that the images of $[\ker \mu]$ and $[\ker \lambda]$ in $\mathcal{B}_1(\mathcal{C})$ are equal. Thus, we may define $I_{\mathcal{C}}$ to be the image in $\mathcal{B}_1(\mathcal{C})$ of the kernel of any polarization of any variety in $\mathcal{C}$.

Let $S_{\mathcal{C}}$ be the image in $\mathcal{B}_2(\mathcal{C})$ of the subset

$$\{[\ker \lambda] : \lambda \text{ is a polarization of some } X \text{ in } \mathcal{C}\}$$

of $G(\mathrm{Ker}_{\mathcal{C}})$. We see that $S_{\mathcal{C}}$ is a subset of the preimage of $I_{\mathcal{C}}$ under the natural reduction map $\mathcal{B}_2(\mathcal{C}) \to \mathcal{B}_1(\mathcal{C})$. Since this reduction map has a finite kernel, the set $S_{\mathcal{C}}$ is finite.

To complete the proof of Theorem 3.2, we must show that if P is an effective element of $G(\mathrm{Ker}_{\mathcal{C}})$ whose image in $\mathcal{B}_2(\mathcal{C})$ lies in $S_{\mathcal{C}}$, then P is attainable. So suppose P is an effective element of $G(\mathrm{Ker}_{\mathcal{C}})$ whose image in $\mathcal{B}_2(\mathcal{C})$ is equal to the image of $[\ker \lambda]$ for a polarization $\lambda \colon X \to \widehat{X}$ of a variety in $\mathcal{C}$. Then there is an element Q of $G(\mathrm{Ker}_{\mathcal{C}})$ and an element a of $R_2(\operatorname{End}^0 \mathcal{C})$ such that $P + Q + \overline{Q} = [\ker \lambda] + \Phi(a)$ in $G(\mathrm{Ker}_{\mathcal{C}})$. Lemma 3.1 shows that there is an $\alpha \in \operatorname{End}^0 \mathcal{C}$ that is totally positive and fixed by the Rosati involution associated to λ such that $\operatorname{Nrd} \alpha = a$. Choose an integer n so that $n^2\alpha$ is an actual endomorphism of X and such that $Q + [\ker n]$ is effective. Replacing α with $n^2\alpha$ and Q with $Q + [\ker n]$, we see that we have $P + Q + \overline{Q} = [\ker \lambda] + [\ker \alpha]$. Since α is fixed by the Rosati involution associated to λ and is totally positive, the composite map $\nu = \lambda\alpha$ is also a polarization of X, and we have $P + Q + \overline{Q} = [\ker \nu]$.

Let $G = \ker \nu$ and let $e \colon G \times G \to \mathbf{G}_m$ be the non-degenerate alternating pairing on G whose existence is shown in §23 of [3]. Let H be a simple element of $\mathrm{Ker}_{\mathcal{C}}$ that occurs in Q. Proposition 5.2 of [2] shows that there is an embedding of H into G such that the pairing e restricted to $H \times H$ is the trivial pairing. Let φ be the natural isogeny from X to $Y = X/H$. Then the Corollary to Theorem 2 (p. 231) of §23 of [3] shows that there is a polarization ν' of Y such that $\nu = \widehat{\varphi}\nu'\varphi$. In $G(\mathrm{Ker}_{\mathcal{C}})$ this gives us the equality $[\ker \nu] = [H] + [\ker \nu'] + \overline{[H]}$. If we replace Q by $Q - [H]$ and ν by ν', we will again have the equality $P + Q + \overline{Q} = [\ker \nu]$, but we will have decreased the number of simple group schemes that occur in Q. By applying this argument repeatedly, we can finally obtain the equality $P = [\ker \nu]$ for a polarization ν of a variety in $\mathcal{C}$. This shows that P is attainable. □

§**3.3. Theorem 1.1 revisited.** In this section we show how our proof of Theorem 1.1 can be understood in terms of Theorem 3.2.

Let k be a p-admissible field, let E be a p-isolated elliptic curve over k, let ℓ be a degree-p Galois extension of k, let X be the reduced restriction of scalars of E from ℓ to k, and let $\mathcal{C}$ be the isogeny class of X. Then $A = \operatorname{End}^0 \mathcal{C}$ is the cyclotomic field $K = \mathbf{Q}(\zeta_p)$, so $R_0(A)$ is K^* and $R_1(A)$ and $R_2(A)$ are both equal to the multiplicative group of totally positive elements of the maximal real subfield K^+ of K.

Note that the $E[p]$-rank defines a homomorphism $Z(\mathcal{C})/B(\mathcal{C}) \to \mathbf{Z}/2\mathbf{Z}$. Lemma 2.5 shows that the degree of an element of $R_1(A)$ is equal to the fourth power of its norm from K^+ to $\mathbf{Q}$. Since Lemma 2.8 shows that $E[p]$ is the only simple p-torsion group scheme that occurs in $\operatorname{Ker}_{\mathcal{C}}$, and since the rank of $E[p]$ is p^2, the $E[p]$-rank of every element of $\Phi(R_1(A))$ is even. Thus, the $E[p]$-rank gives us a homomorphism from $\mathcal{B}_1(\mathcal{C})$ to $\mathbf{Z}/2\mathbf{Z}$. Lemma 2.6 shows that the image of $I_{\mathcal{C}}$ under this homomorphism is nonzero, so $I_{\mathcal{C}}$ itself is nonzero. Then Theorem 3.2 shows that the trivial group scheme is not attainable in $\mathcal{C}$, so $\mathcal{C}$ contains no principally polarized varieties.

We leave it to the reader to use the methods of this paper to prove the following generalization of Theorem 1.1:

Theorem 3.3. *Let ℓ/k be a Galois extension of odd prime degree p, and let Y and Z be abelian varieties over k such that*

(1) *$Y[p](k^{\text{sep}})$ is a simple $\operatorname{Gal}(k^{\text{sep}}/k)$-module;*
(2) *the simple module $Y[p](k^{\text{sep}})$ does not occur in $Z[p](k^{\text{sep}})$; and*
(3) *$\operatorname{End} Y_\ell = \mathbf{Z}$.*

Let X be the kernel of the trace map from $\operatorname{Res}_{\ell/k} Y$ to Y. Then X is simple, and every polarization of every abelian variety isogenous to $X \times Z$ has degree divisible by $p^{2 \dim Y}$.

References

1. Everett W. Howe, *Principally polarized ordinary abelian varieties over finite fields*, Trans. Amer. Math. Soc. **347** (1995), 2361–2401.
2. Everett W. Howe, *Kernels of polarizations of abelian varieties over finite fields*, J. Algebraic Geom. **5** (1996), 583–608.
3. David Mumford, *Abelian varieties*, Tata Inst. Fund. Res. Stud. Math. **5**, Oxford University Press, Oxford, 1985.
4. V. Platonov and A. Rapinchuk, *Algebraic Groups and Number Theory*, Pure Appl. Math. **139**, Academic Press, San Diego, California, 1994.
5. I. Reiner, *Maximal Orders*, London Math. Soc. Monographs **5**, Academic Press, London, 1975.
6. W. Scharlau, *Quadratic and Hermitian Forms*, Grundlehren Math. Wiss. **270**, Springer-Verlag, Berlin, 1985.
7. Jean-Pierre Serre, *Cohomologie galoisienne* (*Cinquième édition, revisée et complétée*), Lecture Notes in Math. **5**, Springer Verlag, Berlin, 1994.

8. A. Silverberg and Yu. G. Zarhin, *Self-dual ℓ-adic representations of finite groups*, preprint (available at http://www.math.uiuc.edu/Algebraic-Number-Theory/0151/index.html) (1998).
9. A. Silverberg and Yu. G. Zarhin, *Polarizations on abelian varieties and self-dual ℓ-adic representations of inertia groups*, preprint (available at http://www.math.uiuc.edu/Algebraic-Number-Theory/0171/index.html) (1999).
10. A. Silverberg and Yu. G. Zarhin, *Polarizations on abelian varieties*, preprint (available at http://xxx.lanl.gov/abs/math.AG/0002253) (2000).
11. A. Weil, *Adeles and algebraic groups*, Progr. Math. **23**, Birkhäuser, Boston, 1982.

Progress in Mathematics, Vol. 195, © 2001 Birkhäuser Verlag Basel/Switzerland

IGUSA'S MODULAR FORM AND THE CLASSIFICATION OF SIEGEL MODULAR THREEFOLDS

KLAUS HULEK

0. Introduction

For an integer $d \geq 1$ let

$$E_d = \begin{pmatrix} 1 & 0 \\ 0 & d \end{pmatrix}, \quad \Lambda_d = \begin{pmatrix} 0 & E_d \\ -E_d & 0 \end{pmatrix}.$$

We consider the symplectic group

$$\tilde{\Gamma}_{1,d} := \mathrm{Sp}(\Lambda_d, \mathbb{Z}).$$

For $d = 1$ this is the usual integer symplectic group $\mathrm{Sp}(4, \mathbb{Z})$. The group $\tilde{\Gamma}_{1,d}$ operates on the *Siegel space* of genus 2

$$\mathbb{H}_2 = \left\{ \tau = \begin{pmatrix} \tau_1 & \tau_2 \\ \tau_2 & \tau_3 \end{pmatrix} \in \mathrm{Mat}(2 \times 2, \mathbb{C}); \mathrm{Im}\, \tau > 0 \right\}$$

by

$$\tilde{M} = \begin{pmatrix} \tilde{A} & \tilde{B} \\ \tilde{C} & \tilde{D} \end{pmatrix} : \tau \mapsto (\tilde{A}\tau + \tilde{B}E_d)(\tilde{C}\tau + \tilde{D}E_d)^{-1}E_d.$$

The quotient

$$\mathcal{A}_{1,d} = \tilde{\Gamma}_{1,d} \backslash \mathbb{H}_2$$

is the moduli space of $(1, d)$-polarized abelian surfaces. Alternatively we can consider the following subgroup of the usual rational symplectic group $\mathrm{Sp}(4, \mathbb{Q})$. Let

$$R_d = \mathrm{diag}(1, 1, 1, d)$$

and set

$$\Gamma_{1,d} := R_d^{-1} \tilde{\Gamma}_{1,d} R_d \subset \mathrm{Sp}(4, \mathbb{Q}).$$

Then $\Gamma_{1,d}$ acts in the usual way on $\mathbb{H}_2$ by

$$M = \begin{pmatrix} A & B \\ C & D \end{pmatrix} : \tau \mapsto (A\tau + B)(C\tau + D)^{-1}$$

and

$$\mathcal{A}_{1,d} = \tilde{\Gamma}_{1,d} \backslash \mathbb{H}_2 = \Gamma_{1,d} \backslash \mathbb{H}_2.$$

Let $L = \mathbb{Z}^4$ be the lattice on which Λ_d defines a symplectic form and let $L^\vee$ be the dual lattice of L. We consider the following subgroups of $\tilde{\Gamma}_{1,d}$ defined by

$$\begin{aligned}
\tilde{\Gamma}^{\mathrm{lev}}_{1,d} &:= \left\{M \in \tilde{\Gamma}_{1,d}; M|_{L^\vee/L} = \mathrm{id}\right\}\\
\tilde{\Gamma}_{1,d}(n) &:= \left\{M \in \tilde{\Gamma}_{1,d}; M \equiv \mathbf{1} \bmod n\right\} \quad (n \geq 1)\\
\tilde{\Gamma}^{\mathrm{lev}}_{1,d}(n) &:= \tilde{\Gamma}^{\mathrm{lev}}_{1,d} \cap \tilde{\Gamma}_{1,d}(n).
\end{aligned}$$

This gives rise to subgroups of $\mathrm{Sp}(4, \mathbb{Q})$:

$$\begin{aligned}
\Gamma^{\mathrm{lev}}_{1,d} &:= R_d^{-1}\tilde{\Gamma}^{\mathrm{lev}}_{1,d}R_d\\
\Gamma_{1,d}(n) &:= R_d^{-1}\tilde{\Gamma}_{1,d}(n)R_d\\
\Gamma^{\mathrm{lev}}_{1,d}(n) &:= R_d^{-1}\tilde{\Gamma}^{\mathrm{lev}}_{1,d}(n)R_d = \Gamma^{\mathrm{lev}}_{1,d} \cap \Gamma_{1,d}(n),
\end{aligned}$$

and to the respective moduli spaces

$$\begin{aligned}
\mathcal{A}^{\mathrm{lev}}_{1,d} &= \tilde{\Gamma}^{\mathrm{lev}}_{1,d}\backslash\mathbb{H}_2 &&= \Gamma^{\mathrm{lev}}_{1,d}\backslash\mathbb{H}_2\\
\mathcal{A}_{1,d}(n) &= \tilde{\Gamma}_{1,d}(n)\backslash\mathbb{H}_2 &&= \Gamma_{1,d}(n)\backslash\mathbb{H}_2\\
\mathcal{A}^{\mathrm{lev}}_{1,d}(n) &= \tilde{\Gamma}^{\mathrm{lev}}_{1,d}(n)\backslash\mathbb{H}_2 &&= \Gamma^{\mathrm{lev}}_{1,d}(n)\backslash\mathbb{H}_2.
\end{aligned}$$

The geometric meaning of these moduli spaces is the following:

$$\mathcal{A}^{\mathrm{lev}}_{1,d} = \{(A, H, \alpha); (A, H) \text{ is a } (1, d)\text{-polarized abelian surface, } \alpha \text{ is a canonical level-structure}\}.$$

Here a *canonical* level-structure is a symplectic basis of the kernel of the map $\lambda_H : A \to \hat{A} = \mathrm{Pic}^0 A$. (Note that this kernel is (non-canonically) isomorphic to $\mathbb{Z}/d \times \mathbb{Z}/d$.) Similarly

$$\mathcal{A}_{1,d}(n) = \{(A, H, \beta); (A, H) \text{ is a } (1, d)\text{-polarized abelian surface, } \beta \text{ is a full level-}n\text{ structure}\}.$$

Here a *full level-n structure* is a symplectic basis of the group $A^{(n)}$ of n-torsion points of A. Finally

$$\mathcal{A}^{\mathrm{lev}}_{1,d}(n) = \{(A, H, \alpha, \beta); (A, H) \text{ is a } (1, d)\text{-polarized abelian surface, } \alpha \text{ is a canonical level structure, } \beta \text{ is a full level-}n\text{ structure}\}.$$

Note that

$$\tilde{\Gamma}^{\mathrm{lev}}_{1,d}/\tilde{\Gamma}_{1,d} \cong \Gamma^{\mathrm{lev}}_{1,d}/\Gamma_{1,d} \cong \mathrm{SL}(2, \mathbb{Z}/d)$$

and that we have, therefore, a Galois covering $\mathcal{A}^{\mathrm{lev}}_{1,d} \to \mathcal{A}_{1,d}$ with Galois group $\mathrm{PSL}(2, \mathbb{Z}/d)$.

The aim of this short note is to prove two results about the classification of these Siegel modular varieties. When we say that one of the moduli spaces has some birational property, such as being of general type, we mean that some (and hence any) compactification of the moduli space has this property.

Theorem A. *If* $(d, n) = 1$ *and* $n \geq 4$ *then* $\mathcal{A}_{1,d}(n)$ *is of general type.*

Since $\mathcal{A}_{1,1}(n)$ is rational for $n \leq 3$ this is the best result which one can hope for if one considers all d simultaneously. It was proved by Yamazaki ([Ya]) that $\mathcal{A}_{1,1}(4)$ is of general type. The space $\mathcal{A}_{1,3}(2)$ has a Calabi-Yau model ([BN], [GH]) and hence Kodaira dimension 0, whereas $\mathcal{A}_{1,3}(3)$ is of general type ([GH, Theorem 3.1]). For prime numbers p Sankaran [S] has proved that $\mathcal{A}_{1,p}$ is of general type for $p \geq 173$. A similar result for $\mathcal{A}_{1,d}$, where d is not necessarily prime, is, as far as I know, not known. L. Borisov [Bo] has shown that, up to conjugation, there are only finitely many subgroups Γ of $\mathrm{Sp}(4,\mathbb{Z})$ such that $\mathcal{A}(\Gamma) = \Gamma\backslash\mathbb{H}_2$ is not of general type. Recall however, that the groups $\Gamma_{1,d}(n)$ are not subgroups of $\mathrm{Sp}(4,\mathbb{Z})$ unless d divides n and that, in general, they are also not conjugate to subgroups of $\mathrm{Sp}(4,\mathbb{Z})$. An essential ingredient in the proof of the theorem is Igusa's modular form Δ_{10}.

The above theorem is a result about the birational classification of these varieties. If one wants to ask more precise questions, such as whether K is ample, then one has to specify the compactification with which one wants to work. Here we shall choose the Voronoi compactification for reasons which we shall explain below.

Theorem B. *The Voronoi compactification* $\left(\mathcal{A}_{1,p}^{\mathrm{lev}}(n)\right)^*$ *for a prime number* p *with* $(p,n) = 1$ *is smooth and has ample canonical bundle (i.e. is a canonical model in the sense of Mori theory) if and only if* $n \geq 5$.

When we speak about Voronoi compactification we mean the compactification defined by the second Voronoi decomposition. We choose this compactification because Alexeev [A1] has shown that it appears naturally when one wants to construct a toroidal compactification which represents a geometrically meaningful functor (see [A1] for the principally polarized case). No such compactification of $\mathcal{A}_{1,p}(n)$ is known, and for this reason we introduce the canonical level structure as well. Alexeev [A2] informs us that this is enough to allow a good functorial compactification. Canonical level structures occur very frequently; see for instance [HKW]. Note also that the spaces $(\mathcal{A}_{1,p}(n))^*$ have non-canonical singularities for infinitely many p and n. These singularities come from the toroidal construction, not from fixed points of the group which is neat for $n \geq 3$. Finally the restriction to prime numbers p in Theorem B is done to keep the technical difficulties to an acceptable level. I believe that this restriction is not essential. If we take $p = 1$ then Theorem B can be easily deduced from properties of $\mathcal{M}_2$, but for general p this approach is not available. For this reason Theorem B may be seen as supporting a conjecture made in [H] for moduli spaces of principally polarized abelian varieties.

Acknowledgement: I am grateful to M. Friedland and G. K. Sankaran for useful discussions.

1. General type

In this section we want to prove

Theorem A. *If $(d,n)=1$ and $n\geq 4$ then $\mathcal{A}_{1,d}(n)$ is of general type.*

We shall work with the Voronoi compactification $\mathcal{A}^*_{1,d}(n)$ of $\mathcal{A}_{1,d}(n)$. The proof which follows works for all d including the case $d=1$. In this case (see also Yamazaki's paper [Ya]), however, some of the statements and parts of the construction which follows are trivial. Before we can prove the theorem we need to know something about the coordinates of $\mathcal{A}^*_{1,d}(n)$ near a cusp. Recall that the codimension 1 cusps are given by lines $l\subset\mathbb{Q}^4$ up to the action of the group $\Gamma_{1,d}(n)$ and that the codimension 2 cusps are similarly given by isotropic planes $h\subset\mathbb{Q}^4$. For any such l, resp. h and any group Γ we denote the lattice part of the stabilizer of l, resp. h in Γ by $P'_\Gamma(l)$, resp. $P'_\Gamma(h)$. These lattices have rank 1, resp. 3.

Lemma 1.1. (i) *For every line $l\subset\mathbb{Q}^4$ there is an inclusion $P'_{\mathrm{Sp}(4,\mathbb{Z})}(l)\subset P'_{\Gamma_{1,d}}(l)$ with cokernel $\mathbb{Z}/d_1$ where $d_1|d$.*
(ii) *For every isotropic plane $h\subset\mathbb{Q}^4$ there is an inclusion $P'_{\mathrm{Sp}(4,\mathbb{Z})}(h)\subset P'_{\Gamma_{1,d}}(h)$ with cokernel $\mathbb{Z}/d_1\times\mathbb{Z}/d_2\times\mathbb{Z}/d_3$ where $d_i|d$ for $i=1,2,3$.*

Proof. (i) Let l be a line which corresponds to a given cusp of $\Gamma_{1,d}$. By [FS, Satz 2.1] we may assume that

$$l=\begin{pmatrix}A&0\\C&D\end{pmatrix}l_0\quad,\quad\begin{pmatrix}A&0\\C&D\end{pmatrix}=:M\in\mathrm{Sp}(4,\mathbb{Z})$$

where $l_0=e_3=(0,0,1,0)$. The group $Q'(l_0)=M^{-1}P'_{\Gamma_{1,d}}(l)M$ is a rank 1 lattice which fixes l_0. We want to compare this to the rank 1 lattice

$$P'_{\mathrm{Sp}(4,\mathbb{Z})}(l_0)=\left\{\begin{pmatrix}1&0&s&0\\0&1&0&0\\0&0&1&0\\0&0&0&1\end{pmatrix};s\in\mathbb{Z}\right\}\subset\mathrm{Sp}(4,\mathbb{Z}).$$

Recall from [HKW, Proposition I.1.16] that every element g in $\Gamma_{1,d}$ fulfills the following congruences

$$g-\mathbf{1}\in\begin{pmatrix}\mathbb{Z}&\mathbb{Z}&\mathbb{Z}&d\mathbb{Z}\\d\mathbb{Z}&\mathbb{Z}&d\mathbb{Z}&d\mathbb{Z}\\\mathbb{Z}&\mathbb{Z}&\mathbb{Z}&d\mathbb{Z}\\\mathbb{Z}&\frac{1}{d}\mathbb{Z}&\mathbb{Z}&\mathbb{Z}\end{pmatrix}$$

Hence

$$\begin{pmatrix}{}^tD&0\\-{}^tC&{}^tA\end{pmatrix}g\begin{pmatrix}A&0\\C&D\end{pmatrix}=\begin{pmatrix}*&S\\ *&*\end{pmatrix},\quad S\in\begin{pmatrix}\mathbb{Z}&\mathbb{Z}\\\mathbb{Z}&\mathbb{Z}\end{pmatrix}.$$

In particular $P'_{\mathrm{Sp}(4,\mathbb{Z})}(l_0)$ is contained in $Q'(l_0)$. Hence $P'_{\Gamma_{1,d}}(l)/P'_{\mathrm{Sp}(4,\mathbb{Z})}(l) \cong \mathbb{Z}/d_1$ for some d_1. The claim $d_1|d$ follows since

$$M \begin{pmatrix} 1 & 0 & d & 0 \\ 0 & 1 & 0 & 0 \\ 0 & 0 & 1 & 0 \\ 0 & 0 & 0 & 1 \end{pmatrix} M^{-1} \in P'_{\Gamma_{1,d}}(l).$$

(ii) Again we can choose an element $M \in \mathrm{Sp}(4,\mathbb{Z})$ such that $h = M(h_0)$ where $h_0 = e_3 \wedge e_4$. Then

$$Q'(h_0) = M^{-1} P'_{\Gamma_{1,d}}(h) M$$

consists of elements of the form

$$\begin{pmatrix} 1 & 0 & d_1\mathbb{Z} & d_2\mathbb{Z} \\ 0 & 1 & d_2\mathbb{Z} & d_3\mathbb{Z} \\ 0 & 0 & 1 & 0 \\ 0 & 0 & 0 & 1 \end{pmatrix}.$$

By (i) we can conclude that $d_1, d_3 \in \mathbb{N}$. We claim that also $d_2 \in \mathbb{N}$. To prove this recall that there is a sublattice $L_0 \subset L = \mathbb{Z}^4$ with $L/L_0 \cong \mathbb{Z}/d$ such that $\Gamma_{1,d}(L_0) \subset L$. (This is simply the lattice spanned by e_1, e_2, e_3, de_4). Hence the same statement must be true for $M^{-1}P'_{\Gamma_{1,d}}(h)M$, but this implies that $d_2 \in \mathbb{N}$. The assertions $d_1|d$ and $d_3|d$ follow from (i) and $d_2|d$ follows again since

$$M \begin{pmatrix} 1 & 0 & 0 & d \\ 0 & 1 & d & 0 \\ 0 & 0 & 1 & 0 \\ 0 & 0 & 0 & 1 \end{pmatrix} M^{-1} \in P'_{\Gamma_{1,d}}(h).$$

□

Proof of the theorem. We consider the following maps of moduli spaces

$$\begin{array}{ccccc} & \mathcal{A}_{1,d}^{\mathrm{lev}} & & & \mathcal{A}_{1,d}(n) \\ \swarrow & & \searrow & \swarrow & \\ \mathcal{A}_{1,1} & & \mathcal{A}_{1,d} & & \end{array}$$

The map $\mathcal{A}_{1,d}^{\mathrm{lev}} \to \mathcal{A}_{1,1}$ comes from the inclusion $\Gamma_{1,d}^{\mathrm{lev}} \subset \mathrm{Sp}(4,\mathbb{Z})$. (The argument given in [HKW, Proposition I.1.20] for d prime goes through unchanged for all d.) Note that $\Gamma_{1,d}^{\mathrm{lev}}$ is not normal in $\mathrm{Sp}(4,\mathbb{Z})$ and hence $\mathcal{A}_{1,d}^{\mathrm{lev}} \to \mathcal{A}_{1,1}$ is not Galois. The other maps $\mathcal{A}_{1,d}^{\mathrm{lev}} \to \mathcal{A}_{1,d}$ and $\mathcal{A}_{1,d}(n) \to \mathcal{A}_{1,d}$ are Galois covers.
An essential ingredient in the proof is Igusa's modular form

$$\Delta_{10} = \prod_{m \text{ even}} \Theta_m^2(\tau)$$

given by the product of the squares of all even theta null values. This is a cusp form of weight 10 with respect to $\mathrm{Sp}(4,\mathbb{Z})$. This is, up to scalar, the unique weight 10 cusp form with respect to $\mathrm{Sp}(4,\mathbb{Z})$. Recall that it vanishes exactly along the $\mathrm{Sp}(4,\mathbb{Z})$-translates of the diagonal

$$\mathbb{H}_1 \times \mathbb{H}_1 = \left\{ \begin{pmatrix} \tau_1 & 0 \\ 0 & \tau_3 \end{pmatrix} ; \mathrm{Im}\ \tau_1,\ \mathrm{Im}\ \tau_3 > 0 \right\} \subset \mathbb{H}_2$$

where it vanishes of order 2. Since $\Gamma_{1,d}^{\mathrm{lev}}$ is a subgroup of $\mathrm{Sp}(4,\mathbb{Z})$ we can also consider Δ_{10} as a cusp form with respect to $\Gamma_{1,d}^{\mathrm{lev}}$. Recall that for any modular form G and a matrix M the slash operator is defined by

$$G|_k M := \det(C\tau + D)^{-k}\ G(M\tau) \quad \left(M = \begin{pmatrix} A & B \\ C & D \end{pmatrix} \right).$$

We consider the multiplicative symmetrization

$$F_0 := \prod_{M \in \mathrm{PSL}(2,\mathbb{Z}/d)} \Delta_{10}|_{10} M.$$

It is straightforward to check that F_0 is a cusp form with respect to $\Gamma_{1,d}$ of weight $10\mu(d)$ where

$$\mu(d) = \frac{1}{2}|\mathrm{SL}(2,\mathbb{Z}/d)| = \frac{1}{2} d^3 \prod_{p|d} (1 - \frac{1}{p^2}) \quad (d \geq 3),$$

and $\mu(2) = 6$. Clearly we can also consider F_0 as a cusp form with respect to the smaller group $\Gamma_{1,d}(n)$. Let L be the ($\mathbb{Q}$-)line bundle of modular forms of weight 1. By abuse of notation we shall use the same notation for whatever moduli space we are considering.

Claim 1: For every point P on the boundary of $\mathcal{A}_{1,d}^*(n)$ the modular form F_0 defines an element in $m_P^{n\mu(d)} L^{10\mu(d)}$.

For points on the codimension 1 cusps this follows immediately from Lemma 1.1(i) and $(n,d) = 1$. To prove it in general we consider an isotropic plane h and the lattices $N := P'_{\Gamma_{1,d}}(h)$ and $N' := P'_{\Gamma_{1,d}(n)}(h)$. We denote by Σ_{Vor} the fan defined by the second Voronoi decomposition of the cone of semi-positive definite symmetric real 2×2 matrices (in [HKW] this was also called Legendre decomposition). Let $\sigma \in \Sigma_{\mathrm{Vor}}$ be a 3-dimensional cone and let $T_\sigma(N)$, resp. $T_\sigma(N')$ be the corresponding affine parts in the toric variety $T_{\Sigma_{\mathrm{Vor}}}(N)$, resp. $T_{\Sigma_{\mathrm{Vor}}}(N')$. We claim that Δ_{10} defines a function on the closure of the image of $P'_{\Gamma_{1,d}}(h)\backslash\mathbb{H}_2$ in $T_\sigma(N)$. First of all Δ_{10} is a function on $P'_{\Gamma_{1,d}}\backslash\mathbb{H}_2$ by Lemma 1.1 (ii). Since $T_\sigma(N)$ is normal, it is enough to show that this function extends to the codimension 1 boundary components. This follows from Lemma 1.1 (i). Since Δ_{10} is a cusp form it follows that $\Delta_{10} \in m_P$ for every point P on the boundary. By construction of F_0 this gives claim 1 in the case $n = 1$. Since $(d,n) = 1$ we have $N' = nN$ and this gives the claim for general n.

Let $\mathcal{A}_{1,d}^*(n)$ be the Voronoi compactification of $\mathcal{A}_{1,d}(n)$, i.e. the toroidal compactification given by the second Voronoi decomposition (in [HKW] this was also called Igusa compactification). If $n \geq 3$ the group $\Gamma_{1,d}(n)$ is neat (this follows from a general result of Serre (cf. also [FC, p. 121]) which says that every algebraic integer which is a unit and which is congruent to 1 mod $n (n \geq 3)$ is equal to 1). In particular $\mathcal{A}_{1,d}(n)$ is smooth. The toroidal compactification $\mathcal{A}^*_{1,d}(n)$ will, in general, however have singularities. These arise because the fan given by the Voronoi decomposition is not always basic, i.e. there may be cones which are not spanned by elements of a basis of the lattice. We can always choose a suitable subdivision of the fan given by the Voronoi decomposition and in this way construct a smooth resolution $\psi : \tilde{\mathcal{A}}_{1,d}(n) \to \mathcal{A}^*_{1,d}(n)$ such that the exceptional divisor is a normal crossing divisor.

Let $\omega = d\tau_1 \wedge d\tau_2 \wedge d\tau_3$. It is well known that, if G is a weight $3k$ cusp form which vanishes of order $\geq k$ along all 1-codimensional cusps, then $G\omega^k$ defines a k-fold canonical form on the smooth part of $\mathcal{A}^*_{1,d}(n)$.

Claim 2: The space of k-fold canonical forms which extends to the smooth part of $\mathcal{A}^*_{1,d}(n)$ grows (at least for sufficiently divisible k) as ck^3 for some positive constant c.

To prove this claim recall that $K = 3L - D$ on the smooth part of $\mathcal{A}^*_{1,d}(n)$, where L is the ($\mathbb{Q}$-) line bundle of modular forms of weight 1 and D is the boundary, i.e. the union of all 1-codimensional cusps. By claim 1 the form F_0 gives the equality

$$10\mu(d)L = n\mu(d)D + D_{\text{eff}}$$

for some effective divisor D_{eff} on the smooth part of $\mathcal{A}^*_{1,d}(n)$. Hence we can conclude that

$$-D = -\frac{10}{n}L + \frac{1}{n\mu(d)}D_{\text{eff}}.$$

Combining this equality with the expression for K gives us

$$K = \left(3 - \frac{10}{n}\right) L + \frac{1}{n\mu(d)}D_{\text{eff}}.$$

For $n \geq 4$ the factor in front of L is positive and the claim follows since $h^0(L^k)$ grows as ck^3.

Claim 3: If $F_{3k}\omega^k$ defines a k-fold canonical form on the smooth part of $\mathcal{A}^*_{1,d}(n)$ (i.e. F_{3k} is a modular form of weight $3k$ which vanishes of order k along each boundary component) then $\left(F_0^3\omega^{10\mu(d)}\right)^k \left(F_{3k}\omega^k\right)$ extends to $\tilde{\mathcal{A}}_{1,d}(n)$.

We first notice that, since $n \geq 4$, the form $F_0^3\omega^{10\mu(d)}$ extends to the smooth part of $\mathcal{A}^*_{1,d}(n)$. We shall now give a fairly elementary argument for the extension to a smooth projective model which follows closely the proof of [S, Theorem 6.3]. It is enough to prove that the form in question extends to the generic point of each component of the exceptional divisor. Let E be a component of the exceptional divisor of the resolution $\tilde{\mathcal{A}}_{1,d}(n) \to \mathcal{A}^*_{1,d}(n)$. It is enough to consider points which lie on only one boundary component. We can choose local analytic coordinates z_1, z_2, z_3 on an open set U such that $E = \{z_1 = 0\}$. Recall that U is an open set in some toroidal variety $T_{\tilde{\Sigma}}(N')$ where $N' = P'_{\Gamma_{1,d}(n)}(h)$ for some isotropic plane h and $\tilde{\Sigma}$ is a refinement of the fan Σ_{Vor} defined by the Voronoi decomposition. Moreover the coordinates of the torus are of the form $t_i = e^{2\pi i a_i \tau_i}$ for some rational numbers a_i. A local equation for E is given by $t_1^{b_1} t_2^{b_2} t_3^{b_3}$ for suitable b_i and hence we can set $z_1 = t_1^{b_1} t_2^{b_2} t_3^{b_3}$. Since $\partial z_1/\partial \tau_j = 2\pi i a_j b_j z_1$ we can conclude that the order of $J = \det(\partial \tau_i/\partial z_j)$ along E is $v_E(J) \geq -1$. It follows again from claim 1 that

$$v_E\left(F_0^3 J^{10\mu(d)}\right) \geq (3n-10)\mu(d).$$

Therefore $\left(F_0^3 \omega^{10\mu(d)}\right)$ defines a section of $\mu(d)(10K-(3n-10)E)$. By assumption $F_{3k}\omega^k$ defines a section of $\psi^*\left(kK_{\mathcal{A}^*_{1,d}(n)}\right) = k(K-\alpha E)$ where α is the discrepancy of E. Altogether $\left(F_0^3\omega^{10\mu(d)}\right)^k \left(F_{3k}\omega^k\right)$ defines a section of

$$\begin{aligned}(10\mu(d)kK - k\mu(d)(3n-10)E) + k(K-\alpha E) &= \\ = k\left[(10\mu(d)+1)K - (\mu(d)(3n-10)+\alpha)E\right].\end{aligned}$$

All singularities here are cyclic quotient singularities. This follows from Lemma 1.1 and the fact that $T_{\Sigma_{\mathrm{Vor}}}(P'_{\mathrm{Sp}(4,\mathbb{Z})}(h))$ is smooth. Hence the singularities are log-terminal, i.e. $\alpha > -1$. This implies that $\mu(d)(3n-10)+\alpha > 0$ for $n \geq 4$ and thus the claim follows.

The theorem now follows easily by combining claim 2 and claim 3. □

2. Ample canonical bundle

It is the aim of this section to prove the following

Theorem B. *The Voronoi compactification $\left(\mathcal{A}_{1,p}^{\mathrm{lev}}(n)\right)^*$ for a prime number p with $(p,n)=1$ is smooth and has ample canonical bundle (i.e. is a canonical model in the sense of Mori theory) if and only if $n \geq 5$.*

We remark that this result is also known to be true for $p=1$ (cf. [Bo], [H]). Before we prove this theorem we recall the geometry of the spaces $(\mathcal{A}_{1,p}^{\mathrm{lev}})^*$ which was described in detail in [HKW]. The Tits building of the group $\Gamma_{1,p}^{\mathrm{lev}}$ consists of $1+(p^2-1)/2$ lines and $p+1$ isotropic planes. The lines consist of one so-called central line and $(p^2-1)/2$ peripheral lines. If $D(l_0)$ is the closed boundary surface which belongs to the central line, then there is a map $K(p) \to D(l_0)$ which is an immersion, but not an embedding if $p > 3$. Here $K(p)$ is the Kummer modular surface of level p, i.e. the quotient of Shioda's modular surface $S(p)$ by the involution which acts by $x \mapsto -x$ on every fibre. For each peripheral boundary component $D(l)$ there exists an isomorphism $K(1) \cong D(l)$ where $K(1)$ is the Kummer modular surface of level 1.

If we add a level-n structure clearly the number of inequivalent cusps will increase. We shall, however, still speak about central or peripheral cusps with respect to $\Gamma_{1,p}^{\mathrm{lev}}(n)$ depending on whether this defines a central or peripheral cusp with respect to $\Gamma_{1,p}^{\mathrm{lev}}$. Now assume $n \geq 3$. Then one shows exactly as in the proof of [HKW, Theorem I.3.151] that there are immersions $S(np) \to D(l_{cent})$ if l_{cent} is a central cusp, resp. $S(n) \to D(l_{peri})$ if l_{peri} is a peripheral cusp. The reason why we have Shioda modular surfaces here instead of Kummer modular surfaces is that for $n \geq 3$ the matrix **-1** is not contained in $\Gamma^{\text{lev}}_{1,p}(n)$. It will be immaterial for us whether these maps are immersions or embeddings.

We shall write the boundary as

$$D = \sum_{i \in I} D^i_{cent} + \sum_{j \in J} D^j_{peri}$$

where D^i_{cent} are the central and D^j_{peri} the peripheral boundary components.

We recall the following well known facts about Shioda modular surfaces. For $k \geq 3$ the surface $S(k) \to X(k)$ is the universal elliptic curve with a level-k structure. The base curve $X(k)$ is the modular curve of level k. It has $t(k) = \frac{1}{2}k^2 \prod_{p|k}(1 - \frac{1}{p^2})$ cusps and the fibre of $S(k)$ over the cusps are singular of type I_k, i.e. a k-gon of (-2)-curves. The genus of $X(k)$ equals $1+(k-6)t(k)/12$ and the line bundle $L_{X(k)}$ of modular forms of weight 1 has degree $kt(k)/12$. The elliptic fibration $\pi : S(k) \to X(k)$ has k^2 sections L_{ij} which form a group isomorphic to $\mathbb{Z}/k \times \mathbb{Z}/k$. By F we denote a general fibre of $S(k)$. It is well known (cf. [BH, pp. 78-80]) that

$$\begin{aligned} K_{S(k)} &\equiv \frac{k-4}{4} t(k) F, \\ L^2_{ij} &= \frac{k}{12} t(k) = -\deg L_{X(k)}. \end{aligned}$$

Let $f_{cent} : S(np) \to D_{cent}$, resp. $f_{peri} : S(n) \to D_{peri}$ be the map from $S(np)$ to a central, resp. from $S(n)$ to a peripheral boundary component. Since they are immersions we can consider the normal bundles N_{cent}, resp. N_{peri} of these maps.

Proposition 2.1. (i) $N_{cent} \equiv -\frac{2}{np} L_{X(np)} - \frac{2}{np} \sum_{i,j \in \mathbb{Z}/np \times \mathbb{Z}/np} L_{ij}$

(ii) $N_{peri} \equiv -\frac{2}{n} L_{X(n)} - \frac{2}{n} \sum_{i,j \in \mathbb{Z}/n \times \mathbb{Z}/n} L_{ij}$.

Proof. We shall give the proof for the central boundary components and indicate how it has to be adapted to the peripheral boundary components. There is a natural action of the group $\{L_{ij}\} \cong \mathbb{Z}/np \times \mathbb{Z}/np$ on $S(np)$. It is an easy calculation to check that this action is induced by elements of $\Gamma^{\text{lev}}_{1,p}(n)/\Gamma_{1,p}$. It follows that N_{cent} is invariant under the group $\mathbb{Z}/np \times \mathbb{Z}/np$ and hence

$$N_{cent} \equiv aF + b \sum L_{ij}$$

for some $a, b \in \mathbb{Q}$ (cf. [BH]). To determine a, b we have to compute the degree of the normal bundle N_{cent} on a general fibre of $S(np)$ and on a section, e.g. the zero section L_{00}.

As a representative for a central cusp we can take the line $l_0 = (0, 0, 1, 0)$.

Assume $(n,p)=1$. We set

$$\Gamma_1'(np) := \left\{ \begin{pmatrix} a & b \\ c & d \end{pmatrix} \in \mathrm{SL}(2,\mathbb{Z}); a,d \equiv 1 \bmod np, c \equiv 0 \bmod n, \right.$$
$$\left. b \equiv 0 \bmod np^2 \right\}.$$

Note that by conjugation with $E = \mathrm{diag}(1,p)$ the group $\Gamma_1'(np)$ is conjugate to the principal subgroup $\Gamma_1(np)$. Then by [HKW, Proposition I.3.98] the stabilizer subgroup $P(l_0)$ of $\Gamma_{1,p}^{\mathrm{lev}}(n)$ is given by

$$P(l_0) = \left\{ \begin{pmatrix} 1 & k & s & m \\ 0 & a & m & b \\ 0 & 0 & 1 & 0 \\ 0 & c & -k & d \end{pmatrix}; \begin{pmatrix} a & b \\ c & d \end{pmatrix} \in \Gamma_1'(np), k,s \in n\mathbb{Z}, m \in pn\mathbb{Z} \} \right\}.$$

The action of

$$\begin{pmatrix} 1 & 0 & s & 0 \\ 0 & 1 & 0 & 0 \\ 0 & 0 & 1 & 0 \\ 0 & 0 & 0 & 1 \end{pmatrix} : \begin{pmatrix} \tau_1 & \tau_2 \\ \tau_2 & \tau_3 \end{pmatrix} \mapsto \begin{pmatrix} \tau_1 + s & \tau_2 \\ \tau_2 & \tau_3 \end{pmatrix}$$

gives rise to the partial quotient

$$\begin{array}{ccc} \mathbb{H}_2 & \to & \mathbb{C}^* \times \mathbb{C} \times \mathbb{H}_1 \\ \begin{pmatrix} \tau_1 & \tau_2 \\ \tau_2 & \tau_3 \end{pmatrix} & \mapsto & (t_1 = e^{2\pi i \tau_1/n}, \tau_2, \tau_3). \end{array}$$

To describe the action of the other elements of $P(l_0)$ let

$$j_2\left(\begin{pmatrix} a & b \\ c & d \end{pmatrix}\right) = \begin{pmatrix} 1 & 0 & 0 & 0 \\ 0 & a & 0 & b \\ 0 & 0 & 1 & 0 \\ 0 & c & 0 & d \end{pmatrix}, \quad j_2(k,m) = \begin{pmatrix} 1 & k & 0 & m \\ 0 & 1 & m & 0 \\ 0 & 0 & 1 & 0 \\ 0 & 0 & -k & 1 \end{pmatrix}.$$

Then these elements act as follows

$$j_2\left(\begin{pmatrix} a & b \\ c & d \end{pmatrix}\right) : \begin{pmatrix} \tau_1 & \tau_2 \\ \tau_2 & \tau_3 \end{pmatrix} \mapsto \begin{pmatrix} \tau_1 - \tau_2(c\tau_3+d)^{-1}c\tau_2 & \tau_2(c\tau_3+d)^{-1} \\ \tau_2(c\tau_3+d)^{-1} & (a\tau_3+b)(c\tau_3+d)^{-1} \end{pmatrix}$$

and

$$j_2(k,m) : \begin{pmatrix} \tau_1 & \tau_2 \\ \tau_2 & \tau_3 \end{pmatrix} \mapsto \begin{pmatrix} \tau_1 + k^2\tau_3 + 2k\tau_2 + km & \tau_2 + k\tau_3 + m \\ \tau_2 + k\tau_3 + m & \tau_3 \end{pmatrix}.$$

The group

$$P''(l_0) = \left\{ \begin{pmatrix} 1 & k & m \\ 0 & a & b \\ 0 & c & d \end{pmatrix}; \begin{pmatrix} a & b \\ c & d \end{pmatrix} \in \Gamma_1'(np), k \in n\mathbb{Z}, m \in pn\mathbb{Z} \right\}$$

defines an action on $\mathbb{C} \times \mathbb{H}_1$ by

$$\begin{pmatrix} 1 & 0 & 0 \\ 0 & a & b \\ 0 & c & d \end{pmatrix} : (\tau_2, \tau_3) \mapsto (\tau_2(c\tau_3 + d)^{-1}, (a\tau_3 + b)(c\tau_3 + d)^{-1}),$$

$$\begin{pmatrix} 1 & k & m \\ 0 & 1 & 0 \\ 0 & 0 & 1 \end{pmatrix} : (\tau_2, \tau_3) \mapsto (\tau_2 + k\tau_3 + m, \tau_3).$$

Then $D^0(l_0) = P''(l_0)\backslash\{0\} \times \mathbb{C} \times \mathbb{H}_1$ is the open boundary surface associated to l_0 and conjugation with $E = \operatorname{diag}(1, p)$ shows that $D^0(l_0) \cong S^0(np)$, the open part of $S(np)$ which does not lie over the cusps.

A local equation of $D^0(l_0)$ in $\mathbb{C} \times \mathbb{C} \times \mathbb{H}_1$ is given by $t_1 = 0$ and hence t_1/t_1^2 is a local section of the conormal bundle. Under the action of the group $P(l_0)$ this transforms as follows:

$$(1)\ j_2\left(\begin{pmatrix} a & b \\ c & d \end{pmatrix}\right) : \quad t_1/t_1^2 \mapsto t_1/t_1^2\, e^{2\pi i\left(-\frac{c\tau_2^2}{c\tau_3+d}\right)/n},$$

$$(2)\ j_2(k, m) : \quad t_1/t_1^2 \mapsto t_1/t_1^2\, e^{2\pi i[-k^2\tau_3 - 2k\tau_2]/n}.$$

We can use the formulae (1) and (2) to determine the coefficients a and b. We first determine the degree of N_{cent} on a general fibre F. Since $k \in n\mathbb{Z}$, $m \in pn\mathbb{Z}$ the fibre of $S(np)$ over the point $[\tau_3] \in X(np)$ is given by $E_{[\tau_3]} = \mathbb{C}/(\mathbb{Z}n\tau_3 + \mathbb{Z}np)$. The standard theta function $\vartheta(\tau_3, \tau_2)$ defines a line bundle of degree n^2p on $E_{[\tau_3]}$ and transforms as follows

$$(3)\ \vartheta(\tau_3, \tau_2 + k\tau_3 + m) = \vartheta(\tau_3, \tau_2)e^{2\pi i[-\frac{1}{2}k^2\tau_3 - k\tau_2]}.$$

Comparing formulae (2) and (3) we find that the degree of N_{cent} on F equals $-2np$. Since we have n^2p^2 sections L_{ij} it follows that $b = -2/np$. To determine the coefficient a we have to compute the degree of N_{cent} on the zero section L_{00}. Since $L_{00}^2 = -\deg L_{X(np)}$ we must show that this degree is 0. There are two ways to see this. The first is to use formula (1) and specialise it to $\tau_2 = 0$. One then has to show that this description extends over the cusps which is an easy local calculation. Alternatively one can proceed as follows: The section L_{00} is the transversal intersection of D_{cent} with the closure of the image of the diagonal $\mathbb{H}_1 \times \mathbb{H}_1 \subset \mathbb{H}_2$ which parametrizes products. This closure is isomorphic to $X(n) \times X(np)$ and L_{00} is equal to $\{\text{cusp}\} \times X(np)$. Hence the normal bundle of L_{00} in $X(n) \times X(np)$ is trivial and by adjunction

$$K_{L_{00}} = K|_{L_{00}} + L_{00}|_{L_{00}}$$

where K is the canonical bundle of $(\mathcal{A}_{1,p}^{\text{lev}}(n))^*$. Using the fact that $K = 3L - D$ and pulling this back to $S(np)$ we obtain

$$K_{L_{00}} = \left(3L_{X(np)} - t(np)F - N_c + L_{00}\right)|_{L_{00}}.$$

Since $\deg K_{L_{00}} = t(np)(np-6)/6$ a straightforward calculation shows that the degree of $N_{cent}|_{L_{00}}$ is equal to 0.

The calculation for N_{peri} is essentially the same. The only differences are that $t_3 = e^{2\pi i \tau_3/np^2}$ and that the fibre of $S(n)$ over $[\tau_1] \in X(n)$ is equal to $E_{[\tau_1]} = \mathbb{C}/(\mathbb{Z}np\tau_1 + np\mathbb{Z})$. □

Proof of the theorem: We first remark that $\left(\mathcal{A}_{1,p}^{\text{lev}}(n)\right)^*$ is smooth under the assumptions made. Since $n \geq 4$ the group $\Gamma_{1,p}^{\text{lev}}(n)$ is neat. Therefore it is enough to show that for a given cusp h the toroidal variety $T_{\Sigma_{\text{Vor}}}(P'_{\Gamma_{1,p}^{\text{lev}}(n)}(h))$ is smooth. If n and p are coprime, the lattice $P'_{\Gamma_{1,p}^{\text{lev}}(n)}(h)$ is simply n times the corresponding lattice in $\Gamma_{1,p}^{\text{lev}}$. Hence every $\sigma \in \Sigma_{\text{Vor}}$ is spanned over $\mathbb{R}$ by a basis of the lattice and this implies that T_σ is smooth.
The next observation is that the condition $n \geq 5$ is necessary. We have already remarked that the closure of the diagonal $\mathbb{H}_1 \times \mathbb{H}_1 \subset \mathbb{H}_2$ parametrizing split abelian surfaces is isomorphic to $X(n) \times X(np)$. Consider a curve $C = X(n) \times \{\text{point}\}$. Then

$$K|_C = (3L - D)|_C = 3L_{X(n)} - X_\infty(n)$$

where $X_\infty(n)$ is the divisor of cusps on $X(n)$. Hence

$$K.C = \frac{n}{4}t(n) - t(n)$$

and this is positive if and only if $n \geq 5$.
We shall now assume $n \geq 5$. Let C be an irreducible curve which is not entirely contained in the boundary, i.e. $C \cap \mathcal{A}_{1,p}^{\text{lev}}(n) \neq \emptyset$ and consider a point $[\tau] \in C$. Choose some $\varepsilon > 0$ with $\varepsilon < 3n/5$. By Weissauer's result [We, p. 220] we can find a cusp form F with respect to $\text{Sp}(4,\mathbb{Z})$ such that $F(\tau) \neq 0$ and $o(F) \geq 1/(12+\varepsilon)$. Here $o(F)$ is the order of F, i.e. the vanishing order of F divided by the weight of F. Let m and k be the vanishing order, resp. the weight of F. Since $\Gamma_{1,p}^{\text{lev}}(n) \subset \text{Sp}(4,\mathbb{Z})$ the form F is also a modular form with respect to $\Gamma_{1,p}^{\text{lev}}(n)$. In terms of divisors this gives us

$$kL = mnD + D_{\text{eff}}, \quad C \not\subset D_{\text{eff}}.$$

Here D_{eff} contains in particular multiples of peripheral boundary components since the vanishing order of F along these boundary components is at least np^2. From the above formula we find that

$$\left(\frac{k}{mn}L - D\right).C = \frac{1}{mn}D_{\text{eff}}.C \geq 0.$$

Since $L.C > 0$ we find that $K.C > 0$ provided $3 > k/mn$. But this follows immediately from the inequalities $m/k \geq 1/(12+\varepsilon)$ and $\varepsilon < 3n/5$.
It remains to prove that the restriction of K to every boundary component is ample. Let D_0 be a boundary component and set $D_0' = D - D_0$. We have already observed that there is an immersion $f : \tilde{D} \to D_0$ which is the normalization. The surface $\tilde{D}$ is either isomorphic to $S(np)$ or to $S(n)$ depending on whether we have a central or a peripheral boundary component. The map f embeds every component of a singular fibre. The image of such a component in $\left(\mathcal{A}_{1,p}^{\text{lev}}(n)\right)^*$

is a $\mathbb{P}^1$. Away from $\{0, \infty\}$ this line is either the intersection of 2 different boundary components or 2 branches of D_0 intersecting transversally. In either case we have the formula

$$f^*K = f^*(3L - D_0' - D_0) = 3L_X - F_\infty - N_f.$$

Here L_X is either $L_{X(np)}$ or $L_{X(n)}$ depending on the type of the boundary component, the divisor F_∞ is the union of the singular fibres and N_f is the normal bundle of the immersion f. Let $k = np$ or n. Then

$$\deg(3L_X - F_\infty) = \frac{1}{4}kt(k) - t(k) > 0$$

for $n > 4$. Hence $(3L_X - F_\infty).C \geq 0$ for every curve C and $(3L_X - F_\infty).C > 0$ unless C is contained in a union of fibres. It follows immediately from our proposition that $-N_f.C > 0$ for every curve C which does not contain a section L_{ij}. Since $-N_f.L_{ij} = 0$ we conclude that $f^*K.C > 0$ for every curve C. □

Remark. The above proof can also be adapted to show that K is nef for $n = 4$. We had already seen that K is not ample in this case. In other words $\left(\mathcal{A}_{1,p}^{\mathrm{lev}}(4)\right)^*$ is a minimal, but not a canonical model for $p = 1$ or $p \geq 3$ prime.

References

[A1] V. Alexeev, Complete moduli in the presence of semiabelian group action. Preprint, math.AG/9905103.

[A2] V. Alexeev, Private communication.

[Bo] L. Borisov, A finiteness theorem for subgroups of Sp(4, $\mathbb{Z}$). Algebraic geometry, 9. J. Math. Sci. (New York) **94** (1999), 1073–1099.

[Br] H.J. Brasch, Modulräume abelscher Flächen. Thesis, Erlangen 1994.

[BH] W. Barth and K. Hulek, Projective models of Shioda modular surfaces. Manuscripta Math. **50** (1985), 73–132.

[BN] W. Barth and I. Nieto, Abelian surfaces of type (1, 3) and quartic surfaces with 16 skew lines. J. Algebr. Geom. **3** (1994), 173–222.

[FC] G. Faltings and C.-L. Chai, Degeneration of abelian varieties. Springer Verlag 1990.

[FS] M. Friedland and G. K. Sankaran, Das Titsgebäude einiger arithmetischer Gruppen. Preprint 1999.

[GH] V. Gritsenko and K. Hulek, The modular form of the Barth-Nieto quintic. Intern. Mathematics Research Notices **17** (1999), 915–937.

[H] K. Hulek, Nef divisors on moduli spaces of abelian varieties. In: Complex analysis and algebraic geometry (eds. Th. Peternell and F.-O. Schreyer), de Gruyter 2000, 255–274.

[HKW] K. Hulek, C. Kahn, S. H. Weintraub, Moduli spaces of abelian surfaces: Compactification, degenerations and theta functions. de Gruyter 1993.

[S] G.K. Sankaran, Moduli of polarized abelian surfaces. Math. Nach. **188** (1997), 321–340.

[We] R. Weissauer, Untervarietäten der Siegelschen Modulmannigfaltigkeiten von allgemeinem Typ. Math. Ann. **275** (1986), 207–220.

[Ya] T. Yamazaki, On Siegel modular forms of degree two. Am. J. Math. **98** (1976), 39–53.

Progress in Mathematics, Vol. 195, © 2001 Birkhäuser Verlag Basel/Switzerland

MIRROR SYMMETRY AND QUANTIZATION OF ABELIAN VARIETIES

YU. I. MANIN

0. INTRODUCTION

0.1. Plan of the paper. This paper consists of two sections discussing various aspects of commutative and non-commutative geometry of tori and abelian varieties.

In the first section, we present a new definition of mirror symmetry for abelian varieties and, more generally, complex and p-adic tori, that is, spaces of the form T/B where T is an algebraic group isomorphic to a product of multiplicative groups, K is a complete normed field, and $B \subset T(K)$ is a discrete subgroup of maximal rank in it. We also check its compatibility with other definitions discussed in the literature.

In the second section, we develop an approach to the quantization of abelian varieties first introduced in [Ma1], namely, via theta functions on *non-commutative, or quantum, tori* endowed with a discrete period lattice. These theta functions satisfy a functional equation which is a generalization of the classical one, in particular, they involve a multiplier. Since multipliers cease to be central in the quantum case, one must decide where to put them. In [Ma1] only one-sided multipliers were considered. As a result, the product of two theta functions in general was not a theta function. Here we suggest a partial remedy to this problem by introducing two-sided multipliers. The resulting space of theta functions possesses partial multiplication and has sufficiently rich functorial properties so that rudiments of Mumford's theory ([Mu]) can be developed. The main results of [Ma1] are reproduced here in a generalized form, so that this paper can be read independently.

We will now briefly describe a broader picture into which this work fits.

0.2. On mirror symmetry. In this paper mirror symmetry is understood as a binary relation between (weak) Calabi-Yau manifolds endowed with some additional data. A projective (or compact complex) manifold V is a weak Calabi-Yau, if it admits a nowhere vanishing global volume form. Additional data which are commonly considered are of two types.

A) A symplectic or complexified symplectic structure ω_V on V, which is "sufficiently large" in the case of the complex base field.

B) A cusp c_V in the moduli space (or rather stack) of deformations of V, that is, a neighborhood of a point of "maximal degeneration" to which V belongs.

The mirror partnership relation between (V, c_V, ω_V) and (W, c_W, ω_W) consists in a host of identifications (partly conjectural) of various structures that can be produced starting with such triples. In the case of strict Calabi-Yau's, this includes an identification of two Frobenius manifolds: the quantum cohomology of (V, ω_V) and a germ of the extended moduli space of W with its flat structure determined by c_W, and similarly with the roles of V and W reversed. Generally, one expects also a representation of V and W as dual real Lagrangian torus fibrations over a common base, with a rich structure of Fourier-Mukai transform connecting Lagrangian/complex analytic objects on both sides. For more details on this, see the original works [MirS1], [MirS2], [StYZ], [Ko], [Giv], [Bar], [LYZ], and the report [Ma3].

The most important testing ground for all levels of the mirror correspondence is furnished by the toric mirrors introduced and studied by V. Batyrev. In the simplest version, this construction looks as follows (see [Bat]).

Let T be an n-dimensional algebraic torus, that is, an algebraic group which is isomorphic to a product of n multiplicative groups $\mathbf{G}_m$. It determines (and is functorially determined by any of the) two free abelian groups of rank n: its character group $M_T := \mathrm{Hom}\,(T, \mathbf{G}_m)$ and its group of one-parametric subgroups $N_T := \mathrm{Hom}\,(\mathbf{G}_m, T)$. These groups are naturally dual to each other. Denote by T^t the dual torus, whose character group is $M_{T^t} = N_T$ and respectively $N_{T^t} = M_T$. In the main text of the paper, one of these groups is denoted H and the other H^t.

Among various toric compactifications of T, we are interested in those for which their anticanonical system is ample. Anticanonical divisors on them are n-dimensional Calabi-Yau manifolds.

Such toric compactifications $\overline{T}_\Delta$ are naturally indexed by *reflexive polyhedra* Δ in $N_T \otimes \mathbf{R}$. Via standard convex duality, each such polyhedron determines the dual polyhedron Δ^t in $N_{T^t} \otimes \mathbf{R}$ which is also reflexive. According to Batyrev, families of anticanonical hypersurfaces in the respective toric compactifications are expected to be mirror partners:

$$\overline{T}_\Delta \supset \left|-K_{\overline{T}_\Delta}\right| \iff \left|-K_{\overline{T}^t_{\Delta^t}}\right| \subset \overline{T}^t_{\Delta^t}. \tag{0.1}$$

The relevant maximally degenerate CY's are simply divisors at infinity in these toric compactifications.

Supplementing (T, Δ) with additional combinatorial structure, one can generalize this picture to some Calabi-Yau complete intersections.

The main goal of the first section of this paper is to provide a new definition of mirror symmetry for abelian varieties (and more generally, complex and p-adic tori) similar to (0.1). We use a "multiplicative uniformization" which goes back to Jacobi and which represents $\mathcal{A}$ as a quotient of the n-dimensional algebraic torus T by a multiplicative discrete lattice $B \subset T(K)$. Consider now two dual algebraic tori endowed with period lattices which are explicitly identified, that is a diagram of the form:

$$(i, i^t): \; T \leftarrow B \rightarrow T^t. \tag{0.2}$$

We will say that the pairs $(\mathcal{A} := T(K)/i(B), i^t)$ and $(\mathcal{B} := T^t(K)/i^t(B), i)$ are mirror dual to each other. Over the complex field, we will relate i^t (resp. i) to a structure similar to ω_V (resp. ω_W) above and compare our mirror relation with that of [Gr1], [Gr2], [AP] and [GolLO]. In particular, we will see how this diagram gives rise to two dual fibrations of the relevant abelian varieties (or complex tori) by mutually dual Lagrangian real tori over the same base.

The choice of multiplicative uniformization is not unique, and it provides the environment for a partial compactification of the relevant moduli space and a choice of a maximally degenerate point at the boundary of moduli space to which $\mathcal{A}$ is close, that is, of the relevant cusp of the moduli space. Roughly speaking, at the boundary some generators of the period lattice are forced to vanish so that the rank of the image of B drops. To make it more precise, we can choose a fan Φ in $N_T \otimes \mathbf{R}$, construct the dual fan Φ^t and study the moduli space of the refined diagrams

$$(i, i^t): \; \overline{T}_{\Phi}(K) \leftarrow B \rightarrow \overline{T}^t_{\Phi^t}(K). \tag{0.3}$$

In this context, it turns out that the fibers of the dual mirror fibrations lie in the monodromy invariant homology class τ of middle dimension, as in the strict Calabi-Yau case.

0.3. Commutative and non-commutative tori and theta functions. In the second section of this paper we address the problem of constructing "quantum abelian varieties". As we already mentioned, what we actually construct is a linear space of quantized theta functions (on a noncommutative torus with a period lattice) endowed with a partial multiplication. We regard this semiring as a quantum deformation of the universal multigraded ring $\oplus\Gamma(L)$, $L \in \operatorname{Pic} \mathcal{A}$, $[-1]^*(L) \cong L$, which makes sense for any abelian variety $\mathcal{A}$ (or indeed for any projective variety).

Since the whole construction is algebraic, it can be performed over any complete normed field, for example, a p-adic field, and applied to the classical abelian varieties as well. Only those p-adic abelian varieties admit a multiplicative uniformization which have maximally degenerate stable reduction modulo p. This is the definition of the p-adic cusp. Diagrams of type (0.2) make full sense in this context as well, but i^t admits no straightforward interpretation as anything like a symplectic form. It would be interesting to investigate the meaning of p-adic B-fields for strict Calabi-Yau manifolds.

We will now briefly discuss issues of non-commutative geometry involved in our construction of quantized theta functions. The standard approach is via deformation of classical function rings, and this intuition guided our initial construction in [Ma1] and to a certain degree its extension presented in this paper.

A complementary paradigm, made explicit on many occasions in Connes' papers and the book [Co] is the natural appearance of non-commutative rings as objects encoding commutative spaces "with bad geometric properties", typically quotients of commutative spaces by non-separated equivalence relations.

For example, in the context of multiplicative uniformization of abelian varieties $\mathcal{A}(\mathbf{C}) = T(\mathbf{C})/B$, the multiplicative period lattice B can degenerate also by ceasing to be discrete, although keeping its rank constant, and in this case it becomes natural to interpret $T(\mathbf{C})/i(B)$ in the realm of non-commutative geometry.

To be concrete, consider the case of a one-dimensional commutative torus T. Its maximal toric compactification is $\mathbf{P}^1 = \mathbf{G}_m \cup \{0\} \cup \{\infty\}$. Choose $B = \mathbf{Z}$ so that $\mathcal{A} = \mathbf{C}^*/(q^{\mathbf{Z}})$. On the level of diagram (0.3) we can choose here any $q \in \mathbf{P}^1(\mathbf{C})$. When $|q| \neq 0, 1, \infty$, we get an elliptic curve $\mathcal{A} = E_q$ fibered by images of real tori $|z| = \text{const}$ (where z is the coordinate on $\mathbf{C}^*$) over the circle $\mathbf{R}/\mathbf{Z}\log|q|$. Dualizing this fibration, we will get the mirror dual elliptic curve. The points $0, \infty$ become divisors on a modular curve and provide the familiar degeneration picture in algebraic geometry. Points where q is a root of unity also become visible in algebraic geometry as cusp points of modular curves of higher levels. However, q of infinite order with $|q| = 1$ are not considered in algebraic geometry at all. One way to interpret the space $\mathbf{G}_m/(q^{\mathbf{Z}})$ in this case is to identify it with (one of the versions of) the *two-dimensional* non-commutative torus T_q, whose function ring is generated by $x^{\pm 1}, y^{\pm 1}$ satisfying the commutation relation $xy = qyx$. See the Appendix for an informal explanation of this in the context of noncommutative geometry à la Alain Connes.

One remarkable trace of this origin of T_q for $q = e^{2\pi i\tau}$, $\tau \in \mathbf{R}$, as a limiting elliptic curve is the re-appearance of the modular group $SL(2, Z)$ as a symmetry group in the non-commutative situation: acting on τ, it produces Morita equivalent function rings of T_q's. See [RiS] for a thorough discussion of this in arbitrary dimension.

What I want to stress here, is a somewhat neglected complementary aspect of this picture: namely, that even if the group $q^{\mathbf{Z}}$ is discrete, T_q still can be viewed as a legitimate incarnation of the elliptic curve E_q in the non-commutative world. A systematic treatment of the correspondence between, say, coherent sheaves on E_q and modules over T_q remains a problem for the future, but see the recent preprint [BEG] for some precise facts about this correspondence, and many interesting suggestions are contained in [So].

This remark throws some light on the problems left open with our approach to quantized theta-functions, for example, that of functional equations corresponding to a change of cusp. Some of our non-commutative abelian varieties $T(H,\alpha)/B$ where $T(H,\alpha)$ is a *non-commutative torus* with quantization parameter α (see 2.1) can be understood as a result of taking a quotient of a *commutative torus* $T(H',1)$ of halved rank, by a non-discrete period subgroup with too many generators. This might provide a bridge between our construction and that of Weinstein ([We]) remaining entirely in the realm of commutative geometry.

Acknowledgement. I am grateful to D. Orlov for illuminating correspondence on Abelian mirrors and comments on the paper [GolLO]. I learned from Y. Soibelman the philosophy of treating boundaries of various moduli spaces as bridges to the non-commutative realm. A. Polishchuk has drawn my attention to the paper [Fu] and sent me a copy of it.

1. Mirror symmetry for complex tori and abelian varieties

1.1. Toric formalism. Since we will have to work with several different types of tori which must be carefully distinguished, we start with some terminological conventions.

Let K be a ground field, H a free abelian group of rank n which we will always write additively. An *algebraic torus T with character group H* is the affine spectrum of the group ring of H, and $H = \mathrm{Hom}\,(T, \mathbf{G}_m)$. Put $H^t := \mathrm{Hom}\,(\mathbf{G}_m, T)$. The groups H, H^t are connected by the canonical duality map $H \times H^t \to \mathrm{Aut}\,\mathbf{G}_m = \mathbf{Z}$.

An element $h \in H$ considered as a character of T will be denoted $e(h)$ (e for exponential). We have $e(h+h') = e(h)e(h')$.

In the next section, we will consider noncommutative algebraic tori as well, for which the multiplication rule for $e(h)$ is twisted: see (2.3) below.

1.1.1. Duality of algebraic tori. If T is an algebraic torus with character group H_T, the dual algebraic torus T^t has the character group $H_{T^t} := H_T^t = \mathrm{Hom}\,(H_T, \mathbf{Z})$.

1.1.2. Periods and abstract tori. Let H be the character group of T. Denote by B another free abelian group of the same rank n, and by $i:\ B \to T(K)$ a homomorphism which we will call "period map". Since $T(K) = \mathrm{Hom}(H, K^*)$, to give i is the same as to give a pairing $H \times B \to K^*$ such that (h, b) is the value of $e(h)$ at the point $i(b)$.

We will refer to the quotient space $T/i(B)$ as *an abstract torus*, and to T as *its covering algebraic torus.* One may imagine $T/i(B)$ simply as a functor of points on K-algebras $R \mapsto T(R)/i(B)$. In the case when $K = \mathbf{C}$ and i is an injection with discrete image, $T(\mathbf{C})/i(B)$ is *a complex torus*, which may admit a structure of abelian variety (it is then unique). However, a large part of our elementary formalism will not depend on additional assumptions. Thus, as explained in the Introduction, we will be able to include into our mirror picture p-adic tori and abelian varieties, and eventually non-commutative tori viewed as models of $T(\mathbf{C})/i(B)$ in non-commutative geometry.

In order not to preclude the eventual interpretation of the space $\mathcal{A} = T/i(B)$ we will consistently identify $\mathcal{A}$ with the triple

$$(H_{\mathcal{A}}, B_{\mathcal{A}}, (\,,)_{\mathcal{A}} : H_{\mathcal{A}} \times B_{\mathcal{A}} \to K^*) \tag{1.1}$$

as above.

1.1.3. Poincaré dual abstract tori. By definition, Poincaré duality interchanges characters and periods. More precisely, $\mathcal{A}$ and $\widehat{\mathcal{A}}$ are Poincaré dual if

$$H_{\widehat{\mathcal{A}}} = B_{\mathcal{A}},\ B_{\widehat{\mathcal{A}}} = H_{\mathcal{A}},\ (b,h)_{\widehat{\mathcal{A}}} = (h,b)_{\mathcal{A}}^{-1} \text{ for } h \in H_{\mathcal{A}},\, b \in B_{\mathcal{A}}. \tag{1.2}$$

For abelian varieties this agrees with the classical definition. Notice however, that a choice of the covering algebraic torus is an additional structure, and we explicitly extend Poincaré duality to this context.

1.1.4. Framed tori. A *framing* of the abstract torus (1.1) is a map $i^t : B_{\mathcal{A}} \to T^t_{\mathcal{A}}(K)$. A *framed abstract torus* is a pair $(\mathcal{A}, i^t)$.

A framing of the complex torus or abelian variety consists of its representation as an abstract torus and framing of that abstract torus.

1.1.5. Mirror dual framed abstract tori. Two framed abstract tori $(\mathcal{A}, i^t)$ and $(\mathcal{B}, i)$ are called *mirror dual*, or *mirror partners*, if their covering algebraic tori are dual, and their periods are explicitly identified. More precisely, the relation of mirror partnership is provided by diagrams of the form (0.2). If one thinks about i, i^t in terms of the respective character/period pairings (1.1), the mirror duality is provided by pairings

$$H_{\mathcal{A}} \times B_{\mathcal{A}} \to K^*,\ H^t_{\mathcal{A}} \times B_{\mathcal{A}} \to K^*. \tag{1.3}$$

A framing is called *non-degenerate* if both kernels of the respective pairings are trivial.

This notion of mirror duality is the main definition of this section. We start with studying it for $K = \mathbf{C}$.

1.2. Complex tori. Assume that $i(B)$ is discrete in $T(\mathbf{C})$. Put

$$\Gamma = \Gamma_{\mathbf{Z}} = \pi_1(T(\mathbf{C})/i(B), 0) = H_1(\mathcal{A}, \mathbf{Z}),$$

$\Gamma_{\mathbf{R}} = \Gamma_{\mathbf{Z}} \otimes \mathbf{R}$ and similarly for $\Gamma_{\mathbf{C}}$. If $T(\mathbf{C})/i(B)$ is denoted $\mathcal{A}$, we may write $\Gamma_{\mathcal{A}}$, $\Gamma_{\mathcal{A},\mathbf{R}}$, etc.

The real space $\Gamma_{\mathbf{R}}$ can be identified with the Lie algebra of $\mathcal{A}$, and the exponential map $\exp : \Gamma_{\mathbf{R}} \to \mathcal{A}$ with kernel $\Gamma_{\mathbf{Z}}$ is the universal covering of $\mathcal{A}$.

Let $H = H_{\mathcal{A}}$ be the character group of T. The map $h \mapsto \dfrac{1}{2\pi i}\dfrac{d\,e(h)}{e(h)}$ induces a canonical identification $H = H^1(T(\mathbf{C}), \mathbf{Z})$. Hence we have an exact sequence

$$0 \to H^t_{\mathcal{A}} \to H_1(\mathcal{A}, \mathbf{Z}) \to B_{\mathcal{A}} \to 0, \tag{1.4}$$

where the third arrow is induced by i. Similarly, we have for a mirror dual framed torus $(\mathcal{B} = T^t(\mathbf{C})/i^t(B), i)$

$$0 \to H_{\mathcal{A}} \to H_1(\mathcal{B}, \mathbf{Z}) \to B_{\mathcal{A}} \to 0, \tag{1.5}$$

where the third arrow is now induced by i^t.

1.2.1. Mirror partners as dual real torus fibrations. In this subsection we will show that our definition of mirror partners $(\mathcal{A}, \mathcal{B})$ over $\mathbf{C}$ naturally fits into the general context of Lagrangian/complex duality: see [AP], [StYZ], [Gr2].

We start with a brief description of this context.

Let (X, ω) be a C^∞ symplectic manifold, endowed with a submersion $p_X : X \to U$ whose fibers are Lagrangian tori. We will fix also a Lagrangian section $0_X : U \to X$.

Using ω, we can identify the bundle of Lie algebras of the tori $p_X^{-1}(u)$, $u \in U$, with the cotangent bundle T^*_U. Hence we have a canonical isomorphism $X = T^*_U/H$ where H is a Lagrangian sublattice in T^*_U with respect to the lift of ω which is the standard symplectic form on the cotangent bundle. There exists also a canonical flat symmetric connection on T^*_U for which H is horizontal.

The local system $H^t = \mathcal{H}om\,(H, \mathbf{Z})$ is embedded as a sublattice into T_U, and we can define *the mirror partner* of $(p_X : X \to U, \omega, 0_X)$ as the toric fibration $Y := T_U/H^t$ endowed with the projection to the same base $p_Y : Y \to U$ and the zero section 0_Y.

Passing from X to Y we have lost the symplectic form. To compensate for this loss, we have acquired a complex structure $J : T_Y \to T_Y$ which can be produced from $(p: X \to U, \omega, 0_X)$ in the following way. The flat connection on T_U obtained by the dualization from T_U^* produces a natural splitting $T_Y = p_Y^*(T_U) \oplus p_Y^*(T_U)$. With respect to this splitting, J acts as $(t_1, t_2) \mapsto (-t_2, t_1)$.

Conversely, suppose that we have a complex manifold Y endowed with a fibration by real tori $Y \to U$ with zero section, such that the operator of complex structure along the zero section identifies T_U with the bundle of Lie algebras of fibers. Then we can consecutively construct the lattice $H^t \subset T_U$, the dual fibration $X := T_U^*/H$ and the symplectic form on X coming from the cotangent bundle.

Now we can return to complex tori.

Put $S^1 = \{|z| = 1 \mid z \in \mathbf{C}\}$. We have the Lie group isomorphism $\mathbf{C}^* \to S^1 \times \mathbf{R} : z \mapsto (z/|z|, \log|z|)$. This induces an isomorphism

$$(\alpha, \lambda) : \; T(\mathbf{C}) \to \mathrm{Hom}\,(H, S^1) \times \mathrm{Hom}\,(H, \mathbf{R}). \tag{1.6}$$

If $i(B)$ is discrete of maximal rank, which I will assume, then $\lambda \circ i(B)$ is an additive lattice in the real space $\mathrm{Hom}\,(H, \mathbf{R})$. Thus (1.6) produces a real torus fibration of $T(\mathbf{C})$ over the base which is as well a real torus of the same dimension:

$$0 \to \mathrm{Hom}\,(H, S^1) \to T(\mathbf{C})/i(B) \to \mathrm{Hom}\,(H, \mathbf{R})/\lambda \circ i(B) \to 0\,. \tag{1.7}$$

Similarly, we have

$$0 \to \mathrm{Hom}\,(H^t, S^1) \to T^t(\mathbf{C})/i^t(B) \to \mathrm{Hom}\,(H^t, \mathbf{R})/\lambda^t \circ i^t(B) \to 0 \tag{1.8}$$

where λ^t is defined for T^t in the same way as λ for T. Let us identify the linear real spaces $H_{\mathbf{R}}$ with $H^t_{\mathbf{R}}$ in such a way that lattice points $\lambda \circ i(b)$ and $\lambda^t \circ i^t(b)$ are identified for all $b \in B$. Then (1.7) and (1.8) become dual real torus fibrations over the common base.

The relevant complex structures in our context come from covering tori. We have to introduce symplectic forms. Let us construct, say, $\omega_{\mathcal{A}}$. From (1.8) one sees that $\mathcal{A} = T(\mathbf{C})/i(B)$ can be obtained as the quotient space of the tangent bundle of the base by a lattice. The tangent bundle (and the lattice) is canonically trivialized, and its fiber is $H_{\mathbf{R}}$. Using the two framings, we have identified $H_{\mathbf{R}}$ with $H^t_{\mathbf{R}}$, that is, tangent bundle with cotangent bundle. The canonical symplectic form on the cotangent bundle becomes our $\omega_{\mathcal{A}}$. Clearly, fibers of (1.7) are Lagrangian tori. It remains to check that $\omega_{\mathcal{A}}$ determines $I_{\mathcal{B}}$ as above, but this is quite straightforward.

1.2.2. Maximal degeneration point and monodromy. Let us consider now the situation described by (one half of) the diagram (0.3). More precisely, consider the space of maps of B to a neighbourhood of a point of maximal degeneration in some toric compactification $\overline{T}_\Phi$. Such a point is a zero-dimensional orbit of T, thus it corresponds to a maximal cone in Φ. Assume for simplicity that it is the simplicial cone generated by a basis of H^t. This means that we identify H^t with $\mathbf{Z}^n$, $T(\mathbf{C})$ with $(\mathbf{C}^*)^n$, and choose as partial compactification the embedding $(\mathbf{C}^*)^n \subset \mathbf{C}^n$. Let D_r be the r-th coordinate hyperplane in $\mathbf{C}^n$.

Choosing a basis of B as well, that is, identifying it with $\mathbf{Z}^n$, we see that the region of the partially compactified moduli space of multiplicatively uniformized complex tori that we are interested in can be identified with an open subspace of the matrix space $\mathbf{C}^{n\times n}$ whose columns generate a multiplicative sublattice. The discriminant locus in this region consists of its intersection with $\cup D_{rs}$, where D_{rs} is the r-th coordinate hyperplane in the s-th copy of $\mathbf{C}^n$ times other copies. The origin (intersection of all D_{rs}) is the maximum degeneration point.

Let $q = (q_{rs})$ be a point of the moduli space, $\mathcal{A}_q$ the respective torus. Denote by $\gamma_r \in H_1(\mathcal{A}_q, \mathbf{Z})$ the image of the r-th S^1 in (1.7) with counterclockwise orientation. Let M_{rs} be the monodromy action of a small counterclockwise loop around D_{rs} in the moduli space. All cycles γ_r are monodromy invariant. Let β_s be any lift to $H_1(\mathcal{A}_q, \mathbf{Z})$ of the s-th 1-cycle in the base torus in (1.7). Then M_{rs} transforms β_s into $\beta_s + \gamma_r$ and leaves other 1-cycles invariant.

Thus, the homology class τ of any fiber of (1.7) generates the cyclic group of invariant cycles of middle dimension.

This statement holds independently of the choice of the simplicial fan and a maximal degenerating cone in it. In this sense, the choice of a covering torus alone encodes essential information about the large complex structure.

1.3. Framings and well-becoming pairs. In this subsection we will compare our construction of mirror partners with that of [GolLO]. In that paper, the additional structure on $\mathcal{A}$ is a complex-valued 2-form ω rather than a symplectic form as above. I will show that this as well can be related to an appropriate framing.

Consider first a pair of framed complex tori $(\mathcal{A}, i^t)$ and $(\mathcal{B}, i)$ which are mirror partners as above. In [GolLO], sec. 10, the authors use decompositions

$$\Gamma_{\mathcal{A}} = \Gamma_{1,\mathcal{A}} \oplus \Gamma_{2,\mathcal{A}},\ \Gamma_{\mathcal{B}} = \Gamma_{1,\mathcal{B}} \oplus \Gamma_{2,\mathcal{B}}.$$

We will call such decompositions *compatible* with our choice of covering tori, if $\Gamma_{1,\mathcal{A}} = H^t_{\mathcal{A}}$, $\Gamma_{1,\mathcal{B}} = H_{\mathcal{A}}$ as in (1.4), (1.5). Thus compatible decompositions are simply splittings of (1.4) and (1.5). The spaces $\Gamma_{\mathcal{A},\mathbf{R}}$, $\Gamma_{\mathcal{B},\mathbf{R}}$ are endowed respectively with complex structures $I_{\mathcal{A}}, I_{\mathcal{B}}$ coming from covering tori. Clearly,

$$\Gamma_{\mathcal{A},\mathbf{R}} = H^t_{\mathcal{A},\mathbf{R}} \oplus I_{\mathcal{A}} H^t_{\mathcal{A},\mathbf{R}}$$

so that compatible splittings satisfy the condition 10.3.1 (2) of [GolLO].

Consider now only $\mathcal{A}$, but equipped with a class $\omega \in H^2(\mathcal{A}, \mathbf{C})$ interpreted as an antisymmetric complex-valued form on $\Gamma_{\mathcal{A}}$. Assume that the extension of ω to $\Gamma_{\mathcal{A},\mathbf{R}}$ is $I_{\mathcal{A}}$-invariant (see [GolLO], last lines of 1.4 for an explanation of this condition). Assume moreover that there exists a compatible decomposition of $\Gamma_{\mathcal{A}}$ such that $\Gamma_{1,\mathcal{A}}$ and $\Gamma_{2,\mathcal{A}}$ are ω-isotropic. This means that $(\mathcal{A}, \omega)$ is a well-becoming pair in the sense of [GolLO], 10.3.1. Looking at (1.4), we see that ω can be uniquely reconstructed from its restriction which we also denote ω

$$\omega : H^t_{\mathcal{A}} \times B_{\mathcal{A}} \to \mathbf{C}. \tag{1.9}$$

Exponentiating (1.9) we produce a framing $i^t : B \to T^t(\mathbf{C})$ in the form (1.3), that is

$$e(h)(i^t(b)) = e^{2\pi i\, \omega(h, b+H^t_{\mathcal{A}})},\ h \in H^t_{\mathcal{A}},\ b \in B. \tag{1.10}$$

Clearly, i^t remains the same, if we choose another compatible isotropic splitting. Moreover, it does not change if we add to ω another pairing taking values in $2\pi i\mathbf{Z}$.

In this way we get a map from the set of all well-becoming pairs $(\mathcal{A}, \omega)$ admitting compatible isotropic splittings in the sense of [GolLO] to the set of framed abstract tori $(\mathcal{A}, i^t)$ with non-degenerate framings in our sense.

We can now complete the comparison of our definition of mirror duality with that of [GolLO].

1.3.1. Theorem. *Let $(\mathcal{A}, i^t)$, $(\mathcal{B}, i)$ be a mirror dual pair of framed complex abstract tori, admitting lifts $(\mathcal{A}, \omega_{\mathcal{A}})$, $(\mathcal{B}, \omega_{\mathcal{B}})$ to well-becoming pairs. Then $(\mathcal{A}, \omega_{\mathcal{A}})$, $(\mathcal{B}, \omega_{\mathcal{B}})$ are mirror dual in the sense of [GolLO].*

Proof. We will compare our setting with that of [GolLO], 10.4 and 10.4.1. The groups Γ_1, Γ_2 in [GolLO] are our $H^t_{\mathcal{A}}, B$, the fundamental group of the mirror dual torus is $\Gamma_1^t \oplus \Gamma_2$. This means that the *real* covering torus of their $\mathcal{B}$ is the same as ours, that is T^t. It remains to compare the complex structures. In our case it is simply induced from $T^t(\mathbf{C})$. In [GolLO] it is described with the help of the complex structure operator I acting upon $(\Gamma_1^t \oplus \Gamma_2) \otimes \mathbf{R}$ produced from ω in two steps: via formula (14) and subsequent projection described in 10.4.

In order to check that they coincide, we will reproduce a part of the argument in [GolLO] in a form which hopefully clarifies the meaning of their crucial formula (14).

Since the following construction must be considered in two different situations, described in [GolLO] 9.2 and 10.4 respectively, we slightly change the scope of our notation. From now on, Γ_1, Γ_2 will denote two abstract free abelian groups of the same finite rank, $\Gamma = \Gamma_1 \oplus \Gamma_2$, $\Gamma' = \Gamma_1^t \oplus \Gamma_2$. Introduce two real tori

$$\mathcal{C} = (\Gamma_1 \oplus \Gamma_2)_{\mathbf{R}}/\Gamma_1 \oplus \Gamma_2,\ \mathcal{C}' = (\Gamma_1^t \oplus \Gamma_2)_{\mathbf{R}}/\Gamma_1^t \oplus \Gamma_2. \tag{1.11}$$

Consider data of two types.

(i) Complex structures on $\mathcal{C}'$ described by the operators I on $(\Gamma_1^t \oplus \Gamma_2)_{\mathbf{R}}$ such that their crossover components

$$I_{12} : \Gamma_{2,\mathbf{R}} \to \Gamma_{1,\mathbf{R}}^t,\ I_{21} : \Gamma_{1,\mathbf{R}}^t \to \Gamma_{2,\mathbf{R}}$$

are bijective.

(ii) Forms $\omega \in \wedge^2 \Gamma_{\mathbf{C}}^t$ for which $\Gamma_{1,\mathbf{C}}$ and $\Gamma_{2,\mathbf{C}}$ are maximal isotropic.

We will establish a bijection between them in the following way.

Let us start with a complex structure I in $\mathcal{C}'$. It determines (and is determined by) the space of invariant holomorphic 1-forms on $\mathcal{C}'$. Integrating them over $\Gamma_1^t \subset H_1(\mathcal{C}', \mathbf{Z})$, we will get all additive maps $\Gamma_1^t \to \mathbf{C}$, in particular, all elements of Γ_1. So we have an embedding $\Gamma_1 \to H^0(\mathcal{C}', \Omega^1) : \gamma \to \nu_\gamma$ such that for all $\beta \in \Gamma_1^t, \gamma \in \Gamma_1$,

$$(\beta, \gamma) = \int_\beta \nu_\gamma. \tag{1.12}$$

This allows one to define a non-degenerate scalar product $\langle\,,\rangle : \Gamma_2 \otimes \Gamma_1 \to \mathbf{C}$:

$$\langle \gamma_2, \gamma_1 \rangle = \int_{\gamma_2} \nu_{\gamma_1}. \tag{1.13}$$

Finally, we can extend it to a complex skew-symmetric form ω on Γ declaring Γ_1 and Γ_2 to be isotropic:

$$\omega((\gamma_1, \gamma_2), (\gamma_1', \gamma_2')) = \langle \gamma_2, \gamma_1' \rangle - \langle \gamma_2', \gamma_1 \rangle. \tag{1.14}$$

If we choose a basis of Γ_1^t, Γ_2 and a basis of holomorphic 1-forms whose period matrix over Γ_1^t is the identity E, then the Gram matrix of the pairing $\langle\,,\rangle$ will be just the second half of the total period matrix. Let us denote it τ. Then $e^{2\pi i \tau}$ is the matrix generating the multiplicative period lattice in the covering complex torus $\Gamma_{\mathbf{C}}'/\Gamma_1^t$ which in view of (1.14) agrees with (1.10).

Arguing now in reverse direction, we will show that knowing τ we can reconstruct the operator I in the same basis and get essentially the [GolLO] formula. In fact, I is uniquely determined by the requirement that for all $\gamma \in \Gamma'$ and all holomorphic ν we have $\int_{I\gamma} \nu = i \int_\gamma \nu$. Hence to find I we must solve the matrix system

$$(\operatorname{Re}\tau + i \operatorname{Im}\tau,\ E) \begin{pmatrix} X & Y \\ U & V \end{pmatrix} = (-\operatorname{Im}\tau + i\operatorname{Re}\tau,\ iE)$$

which gives

$$I = \begin{pmatrix} X & Y \\ U & V \end{pmatrix} = \begin{pmatrix} (\operatorname{Im}\tau)^{-1}\operatorname{Re}\tau & (\operatorname{Im}\tau)^{-1} \\ -\operatorname{Im}\tau - \operatorname{Re}\tau\,(\operatorname{Im}\tau)^{-1}\operatorname{Re}\tau & -\operatorname{Re}\tau\,(\operatorname{Im}\tau)^{-1} \end{pmatrix} \tag{1.15}$$

For the first application of (1.15), let us choose a real torus $\mathcal{A}$ with period lattice $\Gamma_{\mathcal{A}}$ and put $\mathcal{C} = \mathcal{A}\oplus\widehat{\mathcal{A}}$ where $\widehat{\mathcal{A}}$ is the Poincaré dual real torus, that is $\Gamma_{\widehat{\mathcal{A}}} = \Gamma^t_{\mathcal{A}}$ (this does not contradict our multiplicative description (1.2) although it may not be immediately obvious). Put $\Gamma_1 = \Gamma_{\mathcal{A}}$, $\Gamma_2 = \Gamma^t_{\mathcal{A}}$. Comparing our formula (1.15) with [GolLO] (14) in this situation, we see that our I coincides with their I_ω. In [GolLO], 8.4, in addition $\mathcal{A}$ possesses a complex structure, which produces the canonical complex structures on $\widehat{\mathcal{A}}$, $\mathcal{C}$ and $\mathcal{C}'$, say, J. Moreover, ω is restricted to lie in the complexified Néron-Severi group, and as a result I commutes with the inherited complex structure J on $\mathcal{C}'$.

The setup which we are discussing in Theorem 1.3.1 is that of [GolLO] 10.4 and 10.5. The relevant torus $\mathcal{C}$ is now $\mathcal{A}$, and its homology lattice is now split by the choice of a compatible decomposition like in 1.3. Extension of this splitting to $\mathcal{A}\oplus\widehat{\mathcal{A}}$ produces the period matrix τ which is block diagonal and consists of two blocks. Formula (1.15) is still valid for the mirror complex structure, when one replaces τ in it by the respective block. Putting everything together we see that our complex structure determined essentially by (1.10) indeed agrees with that of [GolLO].

1.3.2. Remark. [GolLO] contains several tentative descriptions of mirror dual pairs differing mostly by the exact choice of the additional structure that should be added to $\mathcal{A}$. Our Theorem 1.3.1 together with Theorem 10.5 in [GolLO] indicates that the notion of a well-becoming pair endowed with a choice of one half of an isotropic decomposition captures just the right amount of information. Replacing this structure by that of a framing, we make explicit the important aspect of "large complex structure" in the case $K = \mathbf{C}$ and simultaneously extend the definition to abstract tori over arbitrary fields.

2. Quantized theta-functions and abelian varieties

2.1. Category of non-commutative tori. Let H be a free abelian group of finite rank and $\alpha : H\times H \to K^*$ an alternating pairing: for all $h, g \in H$

$$\alpha(h,g) = \alpha(g,h)^{-1}, \quad \alpha(h_1+h_2,g) = \alpha(h_1,g)\alpha(h_2,g). \tag{2.1}$$

A morphism $f : (H_1,\alpha_1) \to (H_2,\alpha_2)$ is a group homomorphism $f : H_1 \to H_2$ such that for all $h, g \in H_1$ we have

$$\alpha_2^2(f(h),f(g)) = \alpha_1^2(h,g). \tag{2.2}$$

The bilinear form

$$\varepsilon_f(h,g) := \alpha_1(h,g)\alpha_2^{-1}(f(h),f(g))$$

with values in $\{\pm 1\}$ is called the *characteristic* of f.

Any such pair (H,α) will be called the *character group* of the non-commutative torus $T(H,\alpha)$ whose *ring of algebraic functions* $Al\,(H,\alpha)$ is the linear space spanned over K by the symbols $e(h)$, $h \in H$, with multiplication rule

$$e(h)e(h') = \alpha(h,h')\,e(h+h'). \tag{2.3}$$

We may write $e_{H,\alpha}(h)$ for $e(h)$ if need be.

Notice that $\varepsilon(h) := \alpha(h,h)$ is a character of H taking values ± 1, and that from (2.3) we get the following formulas:

$$e(h_1)e(h_2)e(h_3) = \alpha(h_1,h_2)\,\alpha(h_1,h_3)\,\alpha(h_2,h_3)\,e(h_1+h_2+h_3), \tag{2.4}$$

$$e(h)^{-1} = \varepsilon(h)\,e(-h). \tag{2.5}$$

We can also consider the two-sided $Al\,(H,\alpha)$-module $Af\,(H,\alpha)$ of *formal functions*, consisting of infinite linear combinations $\sum_h a_h e(h)$, $a_h \in K$, and, in the case of a complete normed field K and a *unitary* quantization parameter α (that is, $|\alpha| = 1$) the ring of *analytic functions* $An\,(H,\alpha)$ consisting of those formal functions for which $|a_h|\,\|h\|^N \to 0$ for any N as $\|h\| \to \infty$, $\|h\|$ being any Euclidean norm on H.

The form α can be called *the quantization parameter.* When $\alpha \equiv 1$, we get the usual notions of commutative geometry, so that $T(H,1)$ is the algebraic torus with character group H.

A *morphism* $F : T(H,\alpha) \to T(H',\alpha')$, by definition, is given by the contravariant K-algebra homomorphism $F^* : Al\,(H',\alpha') \to Al\,(H,\alpha)$.

2.1.1. Proposition. *a) The set of invertible elements of the ring* $Al\,(H,\alpha)$ *is* $\{a\,e(h)\,|\,a \in K^*, h \in H\}$. *If* $F : T(H_2,\alpha_2) \to T(H_1,\alpha_1)$ *is a morphism of non-commutative tori, then the induced map* $f = [F] : H_1 \to H_2$ *determined by* $F^*(e(h)) = a_h e(f(h))$, $a_h \in K^*$, *is additive and satisfies (2.2) and thus is a morphism of character groups.*

b) The set of all morphisms F *with fixed* $f = [F]$ *is either empty, or has a natural structure of principal homogeneous space over the group* $T(H_1,1)(K) = \mathrm{Hom}\,(H_1,K^*)$. *In particular, if the characteristic of* f *is 1, then* $F^* : e(h) \mapsto e(f(h))$ *extends to a uniquely defined morphism of rings of algebraic functions.*

c) Any morphism F^* *extends to* Af *by* $F^*(\sum a_h e(h)) = \sum a_h F^*(e(h))$. *If* K *is normed and* α *unitary, then this extension maps analytic functions to analytic functions.*

Proof. The first statement follows from the fact that H can be endowed with the structure of a well-ordered group. For any such structure, the highest (resp.

lowest) terms of a product are products of the highest (resp. lowest) terms, so that an invertible element coincides with its highest and lowest term.

To prove the second statement, rewrite the equality

$$F^*(e(h)\,e(g)) = F^*(e(h))\,F^*(e(g))$$

using (2.3). Comparing the e-terms, we see that $f(h+g) = f(h) + f(g)$. Comparing the coefficients, we get

$$a_h a_g a_{h+g}^{-1} = \alpha_1(h,g)\alpha_2^{-1}(f(h), f(g)). \tag{2.6}$$

The left hand side is a symmetric form in h, g, whereas the right hand side is alternate. Therefore this form takes values $\{\pm 1\}$. Hence (2.2) holds, and (2.6) is the characteristic $\varepsilon_f(h,g)$ of f.

Finally, let f be a morphism of character groups with characteristic ε. Then ring morphisms F^* with $[F] = f$ correspond bijectively to the solutions $\{a_h \mid h \in H_1\}$ of the equations $a_h a_g a_{h+g}^{-1} = \varepsilon_f(h,g)$. If one such solution exists, then all others are of the form $a_h c(h)$ where $c : H_1 \to K^*$ is an arbitrary homomorphism.

The remaining statements are straightforward.

2.1.2. Direct product. By definition, the ring of algebraic functions of the product $T(H_1,\alpha_1) \times T(H_2,\alpha_2)$ is the tensor product of the respective rings. We can write

$$T(H_1,\alpha_1) \times T(H_2,\alpha_2) = T(H_1 \oplus H_2, \alpha_1 \oplus \alpha_2)$$

by identifying

$$e_{H_1,\alpha_1}(h_1) \otimes e_{H_2,\alpha_2}(h_2) = e_{H_1 \oplus H_2, \alpha_1 \oplus \alpha_2}((h_1,h_2)).$$

2.1.3. Some standard morphisms. *(i) Shifts.* Any point $b \in T(H,1)(K) = \mathrm{Hom}\,(H, K^*)$ determines an automorphism b^* of $T(H,\alpha)$:

$$b^*(e(h)) := h(b)e(h), \tag{2.7}$$

where from now on we denote by $h(b)$ the value of $e_{H,1}(h)$ at the point b.

(ii) Multiplication by n. This is the morphism

$$[n] : T(H,\alpha) \to T(H,\alpha^{n^2})$$

defined by

$$[n]^*(e_{H,\alpha^{n^2}}(h)) = e_{H,\alpha}(nh). \tag{2.8}$$

For $n = -1$ it is an endomorphism of $T(H, \alpha)$. It is also an endomorphism, if α takes values in roots of unity of degree d and $n^2 \equiv 1 \bmod d$.

The commutation rule with shifts is

$$b^* \circ [n]^* = [n]^* \circ (nb)^*. \tag{2.9}$$

(iii) External multiplication. It is the morphism

$$m_{\alpha,\beta} : T(H, \alpha) \times T(H, \beta) \to T(H, \alpha\beta)$$

defined by

$$m^*_{\alpha,\beta}(e_{H,\alpha\beta}(h)) = e_{H,\alpha}(h) \otimes e_{H,\beta}(h). \tag{2.10}$$

(iv) Mumford's morphism. This is the morphism

$$M : T(H \oplus H, \alpha \oplus \alpha) \to T(H \oplus H, \alpha^2 \oplus \alpha^2),$$

$$M^*(e(h, g)) = e(h + g, h - g). \tag{2.11}$$

It is well defined, because

$$(\alpha \oplus \alpha)[(h + g, h - g), (h' + g', h' - g')]$$

$$= \alpha(h + g, h' + g')\alpha(h - g, h' - g') = \alpha^2(h, h')\alpha^2(g, g').$$

2.2. Periods. We choose and fix an abelian group of periods $B \subset T(H, 1)(K)$. The period group is written additively; it acts upon $T(H, \alpha)$ by shifts.

Trying to make sense of the quotient $T(H, \alpha)/B$ we will study formal or analytic functions on $T(H, \alpha)$ with automorphic properties with respect to the group $\{b^* \mid b \in B\}$.

2.3. Definition. *A (two-sided) theta multiplier $\mathcal{L}$ for the non-commutative torus $T(H, \alpha)$ and period group B consists of the data $\mathcal{L} = (h_l, h_r, \psi, (\,,\,))$, where*

(i) $h_l, h_r : B \to H$ are two group homomorphisms.

We also put $h^{\pm} := h_l \pm h_r$ and denote the image of $b \in B$ with respect to h_l (resp. $h_r, h^{\pm}$) as $h_{b,l}$ (resp. $h_{b,r}, h_b^{\pm}$).

(ii) $\psi : B \to K^*$ is also a group homomorphism.

(iii) $(\,,\,) : B \times B \to K^*$ is a symmetric pairing.

These data must satisfy the following condition: for all $b_1, b_2 \in B$

$$h^-_{b_2}(b_1) = (b_1, b_2)^2 \alpha(h_{b_1,l}, h_{b_2,l})\, \alpha(h_{b_1,r}, h_{b_2,r})^{-1}. \tag{2.12}$$

2.3.1. Remark. The moduli space of quotients $T(H,\alpha)/B$ locally splits into a product of the classical moduli space of commutative tori $T(H,1)/B$ and the space of quantization parameters α (which in a sense also are "hidden periods": cf. our discussion of Connes' treatment of bad equivalence relations).

When $K = \mathbf{C}$, existence of sufficiently many theta functions is equivalent to the algebraicity of $T(H,1)/B$ which becomes an abelian manifold. Multipliers of such theta functions satisfy Riemann symmetry and positivity conditions.

The relations (2.12) represent an extension of the symmetry conditions to our enlarged moduli space. For a quantum version of the positivity conditions, see Theorem 2.6.1 b) below.

2.4. Automorphy factors. For any theta multiplier $\mathcal{L}$ and period $b \in B$, the *automorphy factor* $j_{\mathcal{L}}(b)$ is, by definition, the following linear endomorphism of any of the function spaces Al, Af, An of $T(H,\alpha)$:

$$j_{\mathcal{L}}(b):\ \Phi \mapsto \psi(b)\,(b,b)\,e(h_{b,l})\,\Phi\,e(h_{b,r})^{-1}. \tag{2.13}$$

Clearly, it is invertible.

2.5. Proposition. *For $\mathcal{L}$ fixed, the map $b \mapsto j_{\mathcal{L}}(b)^{-1} \circ b^*$ is a group homomorphism.*

It is injective if $\operatorname{Ker} h^- = 0$.

Proof. We must check that

$$j_{\mathcal{L}}(b_1+b_2)^{-1} \circ (b_1+b_2)^* = j_{\mathcal{L}}(b_1)^{-1} \circ b_1^* \circ j_{\mathcal{L}}(b_2)^{-1} \circ b_2^*. \tag{2.14}$$

We have in view of (2.13), (2.4), (2.5):

$$j_{\mathcal{L}}(b)^{-1}(e(h)) = \psi(b)^{-1}\,(b,b)^{-1}\,\varepsilon(h_{b,l})\alpha(h,h_b^+)\alpha(h_{b,r},h_{b,l})\,e(h-h_b^-). \tag{2.15}$$

Now apply both sides of (2.14) to the arbitrary $e(h)$ and then compare them using (2.12). A somewhat lengthy but straightforward calculation gives (2.14).

When h^- is injective, $h_b^- \neq 0$ for non-zero b, so that $j_{\mathcal{L}}(b)^{-1} \circ b^*(e(h)) \neq e(h)$.

2.6. Theta functions and theta types. *A (quantized) theta function with multiplier* $\mathcal{L}$ is a formal series $\theta \in Af\,(T(H,\alpha))$ invariant with respect to the transformation group

$$\{j_{\mathcal{L}}(b)^{-1} \circ b^* \mid b \in B\}.$$

In other words, θ must satisfy the functional equations

$$b^*(\theta) = \psi(b)\,(b,b)\,e(h_{b,l})\,\theta\,e(h_{b,r})^{-1} \tag{2.16}$$

for all $b \in B$. Clearly, theta functions with multiplier $\mathcal{L}$ form a linear space which we denote $\Gamma(\mathcal{L})$. This notation is supposed to remind us of the case of usual abelian varieties where we deal with invertible sheaves and their sections.

Actually, different multipliers may produce the same space of theta functions or even homomorphisms $b \mapsto j_{\mathcal{L}}(b)^{-1} \circ b^*$. Consider, for example, the case of a commutative torus, where $\alpha \equiv 1$. Then $\Gamma(\mathcal{L})$ depends on h_l, h_r only via their difference $h^- = h_l - h_r$. Moreover, if $(\,,)'$ is another symmetric pairing such that $\varphi(b_1, b_2) := (b_1, b_2)'(b_1, b_2)^{-1}$ takes values in $\{\pm 1\}$, then $\varphi(b, b)$ is multiplicative in b, and we may replace $(\psi, (\,,))$ by $(\psi', (\,,)')$ where $\psi'(b) = \phi(b, b)\psi(b)$.

Generally, from (2.15) one sees that if $j_{\mathcal{L}'}(b)^{-1} \circ b^* = j_{\mathcal{L}}(b)^{-1} \circ b^*$ for all b, then $h^- = h'^-$ and moreover,

$$(b, b)^{-1}\alpha(h_{b,r}, h_{b,l}) = \pm(b, b)'^{-1}\alpha(h'_{b,r}, h'_{b,l}),$$

$$\psi(b)^{-1}\alpha(h, h_b^+) = \pm\psi'(b)^{-1}\alpha(h, h_b'^+).$$

We will call two multipliers equivalent, if they have the same space of theta-functions. (This definition is reasonable only for ample multipliers, see below). An equivalence class of multipliers L will be called *a theta type.* The space $\Gamma(\mathcal{L})$ depends only on this class and can be denoted also $\Gamma(L)$.

2.6.1. Theorem. *a) We have*

$$\dim \Gamma(\mathcal{L}) = [H : h^-(B)]. \tag{2.17}$$

b) Assume that K is a normed field. Then all theta functions of type $\mathcal{L}$ are analytic if $[H : h^-(B)] < \infty$ and

$$\log |(b, b)\, \alpha(h_{b,l}, -h_{b,r})| \tag{2.18}$$

is a positive definite quadratic form on B.

In particular, assume that B is free and the quantization parameter α is unitary, $|\alpha(h, h')| \equiv 1$. Then this condition means that $\operatorname{rk} B = \operatorname{rk} H$, B is discrete in $T(H, 1)(K)$ and $\log |(b, b)|$ is positive definite.

Proof. Let $\theta = \sum_{h\in H} a_h e(h)$, $a_h \in K$, $b \in B$. We have

$$b^*(\theta) = \sum_{h\in H} a_h h(b) e(h) = \sum_{h\in H} a_{h+h_b^-}\, (h + h_b^-)(b)\, e(h + h_b^-), \tag{2.19}$$

whereas the right hand side of (2.16) is:

$$\psi(b)(b, b) \sum_{h\in H} a_h\, \varepsilon(h_{b,r})\, \alpha(h_{b,r}, h_{b,i})\, \alpha(h_b^+, h)\, e(h + h_b^-). \tag{2.20}$$

In (2.19) we replace $h_b^-(b)$ by $(b,b)^2\varepsilon(h_{b,l})\varepsilon(h_{b,r})$ (see (2.12)). Comparing coefficients of (2.19) and (2.20), we see that (2.16) is equivalent to

$$a_{h+h_b^-} = a_h\,\psi(b)\,h(b)^{-1}(b,b)^{-1}\varepsilon(h_{b,l})\,\alpha(h_{b,r},h_{b,l})\,\alpha(h_b^+,h) \tag{2.21}$$

for all $h\in H$ and $b\in B$. Thus one can arbitrarily choose values a_h for all h in a system of representatives of $H/h^-(B)$ and then uniquely reconstruct θ. This proves the first statement of the theorem. It also shows that if $\Gamma(\mathcal{L})$ is not finite dimensional, it necessarily contains non-analytic functions.

Assume now that $\Gamma(\mathcal{L})$ is finite dimensional. Then on each coset $h+h^-(B)$ we have

$$\log|a_{h+h_b^-}| = \log|a_h| - \log|(b,b)\,\alpha(h_{b,r},h_{b,l})| + \log|\psi(b)\,h(b)^{-1}\alpha(h_b^+,h)|.$$

The second summand in the right hand side is quadratic in b whereas the third is linear. Hence analyticity follows from the positive definiteness of (2.18).

2.6.2. Multiplication of theta functions. Generally, we can multiply analytic functions, but not formal ones. We will call a theta multiplier $\mathcal{L}$ *analytic*, if $\Gamma(\mathcal{L})$ consists of analytic functions, and *ample*, if $\mathcal{L}$ satisfies the conditions of Theorem 2.6.1 b).

We will call two analytic theta multipliers $\mathcal{L}_i = (h_l^{(i)}, h_r^{(i)}, \psi_i, (\,,\,)_i)$ (with $i=1$ resp. 2), *composable* (in this order) if $h_l^{(2)} = h_r^{(1)}$. Define their *product* as

$$\mathcal{L}_1\otimes\mathcal{L}_2 = \mathcal{L} := (h_l^{(1)},\,h_r^{(2)},\,\psi_1\psi_2,\,(\,,\,)_1(\,,\,)_2).$$

A straightforward calculation shows that if the $\mathcal{L}_i$ are ample, $\mathcal{L}$ is ample as well, and the product of theta functions produces a well defined map

$$\Gamma(\mathcal{L}_1)\otimes\Gamma(\mathcal{L}_2)\to\Gamma(\mathcal{L}):\ \theta_1\otimes\theta_2\mapsto\theta_1\theta_2. \tag{2.22}$$

One can call two ample theta types L_1 composable, if they contain pairs of composable multipliers. The multiplication in (2.22) does not depend on the choice of such a pair, but I did not check that the product type L cannot change.

2.6.3. Quantized abelian varieties. Assume that $(T(H,\alpha),B)$ admits ample theta multipliers. Then we consider $\oplus_L\Gamma(L)$, where L runs over all theta types with $\psi(b)\in\{\pm1\}$, together with the partial multiplication defined above, as a quantized version of the graded coordinate ring of an abelian variety. In the classical case, it is graded by the symmetric elements of Pic lying in the effective cone (see 2.7.2 below).

2.7. Functorial properties of theta functions. Consider a morphism of non-commutative tori $F : T(H_2, \alpha_2) \to T(H_1, \alpha_1)$. As in Proposition 2.1.1, let $F^*(e(h)) = a_h e(f(h))$, $a_h \in K^*$, $f : H_1 \to H_2$. The map f induces a morphism of commutative tori which we also denote F: $T(H_2, 1) \to T(H_1, 1)$. Let $B_i \subset T(H_i, 1)(K)$ be two period lattices such that $F(B_2) \subset B_1$.

Choose a theta multiplier $\mathcal{L}_1 = (h_l, h_r, \psi, (\,,))$ for $T(H_1, \alpha_1)$ and B_1. We will show how to produce from it a new theta multiplier

$$\mathcal{L}_2 = F^*(\mathcal{L}_1) = (h'_l, h'_r, \psi', (\,,)')$$

for $T(H_2, \alpha_2), B_2$, if the following condition holds:

the characteristic of F is 1, that is, $a_{h+g} = a_h a_g$ (cf. end of the proof of Proposition 2.2.1), and f is compatible with α_1 and α_2, and not just their squares.

The map $f : H_1 \to H_2$ induces a map $F :$ Hom $(H_2, K^*) \to$ Hom (H_1, K^*) whose restriction on B_2 sends it to B_1. Put for $b, b_1, b_2 \in B_2$:

$$h'_{l,r} = f \circ h_{l,r} \circ F : \; B_2 \to H_2, \tag{2.23}$$

so that $h'_{b,l} = f(h_{F(b),l})$ and similarly for $h'_r, h'^{\pm}$,

$$\psi'(b) = \psi(F(b))\, a_{h^-_{F(b)}}, \tag{2.24}$$

$$(b_1, b_2)' = (F(b_1), F(b_2)). \tag{2.25}$$

2.7.1. Theorem. *The data (2.23)–(2.25) constitute a theta multiplier for $T(H_2, \alpha_2)$, B_2 such that $F^*(\Gamma(\mathcal{L})) \subset \Gamma(F^*(\mathcal{L}))$.*

Proof. We first have to check that (2.12) holds for $F^*(\mathcal{L})$. For $b_1, b_2 \in B_2$, the left hand side becomes

$$h'^-_{b_2}(b_1) = f(h^-_{F(b_2)})(b_1) = h^-_{F(b_2)}(F(b_1)).$$

Furthermore, if the characteristic of F is 1, the right hand side can be rewritten as

$$(b_1, b_2)'^2 \alpha_2(f(h_{F(b_1),l}), f(h_{F(b_2),l}))\alpha_2(f(h_{F(b_1),r}), f(h_{F(b_2),r}))^{-1}$$
$$= (F(b_1), F(b_2))^2 \alpha_1(h_{F(b_1),l}, h_{F(b_2),l})\alpha_1(h_{F(b_1),r}, h_{F(b_2),r})^{-1}.$$

Applying (2.12) to $F(b_1), F(b_2)$ in lieu of b_1, b_2, we see that both expressions coincide.

Let us now check that $F^*(\Gamma(\mathcal{L})) \subset \Gamma(F^*(\mathcal{L}))$.

Choose a formal function

$$\theta = \sum_{h\in H_1} c_h e(h), \quad e = e_{H_1,\alpha_1}.$$

According to (2.21), it belongs to $\Gamma(\mathcal{L})$ iff the following conditions are satisfied for all $h \in H_1$, $b \in B_1$:

$$c_{h+h_b^-} = c_h\, \psi(b)\, h(b)^{-1}(b,b)^{-1}\varepsilon_1(h_{b,l})\, \alpha_1(h_{b,r}, h_{b,l})\, \alpha_1(h_b^+, h). \tag{2.26}$$

Notice that our notation slightly differs from (2.21): a_h is now reserved for $F^*(e(h)) = a_h e(f(h))$ as in the first lines of 2.7.

Put now

$$F^*(\theta) = \sum_{g\in H_2} C_g e(g), \quad e = e_{H_2,\alpha_2}.$$

We have $C_g = 0$, if $g \notin f(H_1)$; otherwise

$$C_g = \sum_{h\in f^{-1}(g)} c_h a_h. \tag{2.27}$$

To prove that $F^*(\theta) \in \Gamma(F^*(\mathcal{L}))$, we will check that the C_g satisfy the analogs of the relations (2.26) written for all $g \in H_2$ and all $b \in B_2$.

Consider first the case $g \notin f(H_1)$. Then the relevant analog of (2.26) says that $C_{g+h_b'^-}$ must be proportional to C_g, that is, zero. This is indeed true, because in view of (2.23), $h_b'^- = f(h_{F(b)}^-)$ and thus $g + h_b'^- \notin f(H_1)$.

Now assume that $g \in f(H_1)$, fix also $b \in B_2$, and write down separately both sides of the relevant case of (2.26). Since the characteristic of F is 1, the left hand side can be written as

$$C_{g+h_b'^-} = \sum_{h\in f^{-1}(g)} c_{h+h_{F(b)}^-}\, a_h\, a_{h_{F(b)}^-}$$

$$= \sum_{h\in f^{-1}(g)} c_h a_h\, \psi(F(b)) a_{h_{F(b)}^-}\, h(F(b))^{-1}(F(b), F(b))^{-1}$$

$$\times \varepsilon_1(h_{F(b),l})\, \alpha_1(h_{F(b),r}, h_{F(b),l})\, \alpha_1(h_{F(b)}^+, h). \tag{2.28}$$

Here we have rewritten $c_{h+h_{F(b)}^-}$ using (2.26).

The right hand side is

$$\left(\sum_{h\in f^{-1}(g)} c_h a_h\right) \psi'(b)\, g(b)^{-1}(F(b), F(b))^{-1}\, \varepsilon_2(h_{b,l}'^-)\, \alpha_2(h_{b,r}'^-, h_{b,l}'^-)\, \alpha_2(h_b'^+, g). \tag{2.29}$$

We can now compare (2.28) and (2.29) term by term using (2.23)–(2.25), and convince ourselves that they coincide. In particular, we use the identities $g(b) = h(F(b))$ and $g = f(h)$ for any $h \in f^{-1}(g)$.

2.7.2. Examples. All morphisms of tori described in 2.1.3 have characteristic 1. Therefore, complementing these tori by compatible period lattices, we obtain quantized versions of many standard morphisms of abelian varieties.

In particular, $[-1]^*$ acts on the theta multipliers for $T(H,\alpha), B$ by simply inverting ψ. Hence symmetric theta multipliers correspond to $\psi : B \to \{\pm 1\}$.

Mumford's morphism for abelian varieties, induced by the toric morphism (2.11), is the starting point for his study of homogeneous coordinate rings of abelian varieties. Our preparatory work allows us to hope that at least part of this study can be extended to the quantum case.

Appendix. Commutative Geometry as Noncommutative Geometry

The main goal of this Appendix is to illustrate Connes' approach to noncommutative geometry in an algebraic geometric context and to make convincing our claim that an elliptic curve "is" a two-dimensional noncommutative torus.

We take as our starting point Connes' explanations about how to treat as a non-commutative space the quotient of a "commutative space" by an equivalence relation or equivalence groupoid: see [Co], II.2–II.5. This viewpoint is complementary to the deformation paradigm which is more popular in algebraic geometry.

First, a reminder about groupoids.

Let U be a set. Classically, an equivalence relation $\sim$ on U is given by its graph $R \subset U \times U$, $R := \{(a,b) \mid a \sim b\}$ which satisfies the three conditions:

Reflexivity:

$$a \sim a \iff \Delta_U \subset R;$$

Symmetry:

$$(a \sim b \Leftrightarrow b \sim a) \iff s_{12}(R) = R;$$

Transitivity:

$$((a \sim b)\&(b \sim c) \Rightarrow a \sim c) \iff \mathrm{pr}_{13}[(R \times U) \cap (U \times R)] \subset R.$$

All of this can be rephrased as follows: there exists a category with set of objects U and set of morphisms R, such that $R \to U \times U$ is the map $f \mapsto$ (source of f, target of f), and in addition, every morphism is an isomorphism, and all automorphism groups are trivial.

Consider now a diagram $R \to U \times U$ satisfying this description with the last condition deleted so that the automorphism groups can now be arbitrary. We will call such a diagram *an equivalence groupoid* (on the set U). Of course, an equivalence groupoid $R \to U \times U$ comes together with the identity map $U \to R: a \mapsto \mathrm{id}_a$ and the associative multiplication map $R \times_U R \to R$ satisfying the usual categorical axioms which reduce to the reflexivity, symmetry, and transitivity for the usual equivalence relations. Notice that the image of R is in fact an equivalence relation, and the respective quotient is the set of isomorphism classes of objects.

Thus the basic difference between equivalence groupoids and equivalence relations on sets can be demonstrated on one-point sets $U = \{*\}$: in this case R is simply a group. In the framework of homotopy theory, the respective quotient object $\{*\}/R$ is represented by the classifying space BR. Stacks provide a categorical context for constructing such quotients in algebraic geometry.

In fact, the notion of equivalence groupoid was formulated in such a way that it readily generalizes to the case when $R \to U \times U$ is a diagram in an arbitrary category with products, e.g. schemes. One should imagine U as an atlas and R as gluing rules, so that the geometric object we are interested in is symbolically U/R.

Connes' prescription, roughly speaking, consists in reducing the geometric study of $X = U/R$ to the representation theory of a non-commutative ring $\mathcal{R}_X$. The heuristic rule for constructing $\mathcal{R}_X$ can be stated as follows. Let us write elements (points) of R as morphisms $j: u \to u'$ so that $R \to U \times U$ maps such a point to (u, u'). Then $\mathcal{R}_X$ consists of certain functions f on R endowed with a convolution multiplication:

$$(f * g)(k: u \to u'') = \sum_{(i,j):\, ij=k} f(j: u \to u')\, g(i: u' \to u'')\,. \tag{A.1}$$

Of course, the correct choice of the class of functions and making sense of the convolution multiplication may present a problem, but in algebraic geometry it is clear what to start with, at least when U is affine.

Let us illustrate such a setup by several examples.

Example 1. If X is the quotient $\{*\}/G$ where G is a finite group, then $\mathcal{R}_X$ is the group ring of G. Representation theory of $\mathcal{R}_X$ is essentially K-theory of the classifying stack of G, in accordance with common wisdom.

Example 2. If X is the quotient of an affine scheme $U = \operatorname{Spec} A$ by the action of a group G, then $\mathcal{R}_X$ should contain at least the twisted product of A with a version of the group ring of G.

When $U = \mathbf{G}_m$ with coordinate y and $G = \mathbf{Z}$ with generator x acting on y as multiplication by q, then in this twisted product we have $xyx^{-1} = qx$, that is, the basic relation of the noncommutative torus. On the other hand, U/G makes sense as an elliptic curve in analytic geometry, if $|q| \neq 1$.

Example 3. A projective scheme $\operatorname{Proj} A$ is the quotient of its cone $\operatorname{Spec} A$ (with vertex deleted) by the action of $\mathbf{G}_m$ determining the grading. Studying representations of the twisted product of A with the group ring of G_m is equivalent to studying graded A-modules. Deleting the vertex boils down to taking the quotient of the category of graded modules by the subcategory of modules with a finite number of nonvanishing components. This is the familiar Serre picture which is archetypal in the following sense: after finding an appropriate ring $\mathcal{R}_X$, one proceeds to establish an equivalence of categories $\mathcal{R}_X\text{-Mod} \to \mathrm{Coh}_X$, or its localized and/or derived version.

References

[AP] D. Arinkin, A. Polishchuk, *Fukaya category and Fourier transform*, math.AG/9811023.

[Bar] S. Barannikov, *Extended moduli spaces and mirror symmetry in dimensions $n > 3$*, math.AG/9903124.

[BarK] S. Barannikov, M. Kontsevich, *Frobenius manifolds and formality of Lie algebras of polyvector fields*, Int. Math. Res. Notices, 4 (1998), 201–215.

[BEG] V. Baranovsky, S. Evens, V. Ginzburg, *Representations of quantum tori and double-affine Hecke algebras*, preprint math.RT/0005024.

[Bat] V. Batyrev, *Dual polyhedra and the mirror symmetry for Calabi-Yau hypersurfaces in toric varieties*, Journ. Alg. Geom., 3 (1994), 493–535.

[Co] A. Connes, *Noncommutative geometry*, Academic Press, 1994.

[De] P. Deligne, *Local behavior of Hodge structures at infinity*, in: Mirror Symmetry II, ed. by B. Greene and S.-T. Yau, AMS–International Press, 1996, 683–699.

[Fu] K. Fukaya, *Mirror symmetry of abelian variety and multi theta functions*, preprint, 1998.

[Giv] A. Givental, *Equivariant Gromov-Witten invariants*, Int. Math. Res. Notices, 13 (1996), 613–663.

[GolLO] V. Golyshev, V. Lunts, D. Orlov, *Mirror symmetry for abelian varieties*, preprint math.AG/9812003.

[Gr1] M. Gross, *Special Lagrangian fibrations I: Topology*, alg-geom/9710006.

[Gr2] M. Gross, *Special Lagrangian fibrations II: Geometry*, math.AG/9809072.

[Ko] M. Kontsevich, *Homological algebra of Mirror Symmetry*, Proceedings of the ICM (Zürich, 1994), vol. I, Birkhäuser, 1995, 120–139. Preprint alg-geom/9411018.

[LYZ] N. C. Leung, S.-T. Yau, E. Zaslow, *From special Lagrangian to Hermitian-Yang-Mills via Fourier-Mukai transform*, preprint math.DG/0005118.

[Ma1] Yu. Manin, *Quantized theta-functions*, in: Common Trends in Mathematics and Quantum Field Theories (Kyoto, 1990), Progress of Theor. Phys. Supplement, 102 (1990), 219–228.

[Ma2] Yu. Manin, *Frobenius manifolds, quantum cohomology, and moduli spaces*, AMS Colloquium Publications, vol. 47, Providence, RI, 1999, xiii+303 pp.

[Ma3] Yu. Manin, *Moduli, Motives, Mirrors*, Plenary talk at 3rd European Congress of Mathematicians, Barcelona, 2000, preprint.

[MirS1] S.-T. Yau, ed, *Essays on Mirror Manifolds*, International Press Co., Hong Kong, 1992.

[MirS2] B. Greene, S.-T. Yau, eds, *Mirror Symmetry II*, AMS–International Press, 1996.

[Mu] D. Mumford, *On the equations defining abelian varieties I*, Inv. Math. 1 (1966), 355–374.

[Po1] A. Polishchuk, *Massey and Fukaya products on elliptic curve*, math.AG/9803017.

[Po2] A. Polishchuk, *Homological mirror symmetry with higher products*, preprint, math. AG/9901025.

[PoZ] A. Polishchuk, E. Zaslow, *Categorical mirror symmetry: the elliptic curve*, Adv. Theor. Math. Phys., 2 (1998), 443–470. Preprint math.AG/9801119.

[RiS] M. A. Rieffel, A. Schwarz, *Morita equivalence of multidimensional tori*, preprint math.QA/9803057.

[So] Y. Soibelman, *Quantum tori, mirror symmetry and deformation theory*, preprint, 2000.

[StYZ] A. Strominger, S.-T. Yau, E. Zaslow, *Mirror symmetry is T-duality*, Nucl. Phys. B 479 (1996), 243–259.

[We] A. Weinstein, *Classical theta functions and quantum tori*, Publ. RIMS, Kyoto Univ., 30 (1994), 327–333.

Progress in Mathematics, Vol. 195, © 2001 Birkhäuser Verlag Basel/Switzerland

GROUP SCHEMES WITH ADDITIONAL STRUCTURES AND WEYL GROUP COSETS

BEN MOONEN

1. Introduction

Let Y be an abelian variety of dimension g over an algebraically closed field k of characteristic $p > 0$. To Y we can associate its p-kernel $Y[p]$, which is a finite commutative k-group scheme of rank p^{2g}. In the unpublished manuscript [8], Kraft showed that, fixing g, there are only finitely many such group schemes, up to isomorphism. (As we shall discuss later, Kraft also gave a very nice description of all possible types.) About 20 years later, this result was reobtained, independently, by Oort. Together with Ekedahl he used it to define and study a stratification of the moduli space $\mathscr{A}_{g,1} \otimes k$ of principally polarized abelian varieties over k. The strata correspond to the pairs (Y, μ) such that the p-kernel is of a fixed isomorphism type. Their results can be found in [11] and [12]; see also related work by van der Geer in [17].

The present paper is a first step in a programme to develop a similar stratification for (good reductions of) more general Shimura varieties. For Hilbert modular varieties this has been done by Goren and Oort in [4]. In general, such a Shimura variety (i.e., some p-adic model) should come from a $\mathbb{Z}_{(p)}$-version of a Shimura datum, involving a reductive group G over $\mathbb{Z}_{(p)}$ and a homogeneous space X under $G(\mathbb{R})$ such that $(G_{\mathbb{Q}}, X)$ is a Shimura datum in the usual sense. Let us suppose that there is some modular interpretation, say in terms of abelian varieties with certain prescribed Hodge/Tate classes. Now one might hope that from the data (G, X) one can cook up some finite set which is in

Research made possible by a fellowship of the Royal Netherlands Academy of Arts and Sciences.

natural bijection with the set of isomorphism classes of p-kernels with "additional structures" that occur on the special fibre. Further one may hope that this leads to a stratification of the special fibre and that some properties of the strata (say the dimensions, and which strata appear in the boundary of another one) can be described in a group theoretical way, i.e., starting from (G, X).

To give meaning to such speculations, we study Shimura varieties of PEL-type. Here we have a modular interpretation that makes sense in arbitrary characteristics, and we may hope to see a general pattern—if there is one—emerging from this particular case. The programme outlined above leads us to study finite group schemes with an action of a semi-simple $\mathbb{F}_p$-algebra and/or a polarization form. The subject of this paper is the classification of such objects. This is a problem that can be stated without any reference to Shimura varieties, but as motivated above we shall try to obtain a description using a "Shimura-theoretic" language.

The first problem we consider, stated as problem (GE) in section 4, is the classification of BT_1's $\mathscr{G}$ with a given action $\iota\colon D \to \mathrm{End}(\mathscr{G})$ of a semi-simple $\mathbb{F}_p$-algebra D. Here we work over an algebraically closed field k of characteristic p, and by a BT_1 we mean a truncated Barsotti-Tate group of level 1. (See also section 2.)

To explain our result, write (N, F_N, V_N) for the Dieudonné module of such a group scheme $\mathscr{G}$. Consider the reductive k-group $G := \mathrm{GL}_D(N)$. Some of the data involved, such as the structure of N as a D-module, can be viewed as discrete invariants of the pair $(\mathscr{G}, \iota)$, and for the classification problem we may assume them to be fixed. Another discrete invariant, which we call the multiplication type of the pair $(\mathscr{G}, \iota)$, is the structure of $\mathrm{Ker}(F) \subset N$ as a D-module. This multiplication type corresponds to a conjugacy class X of parabolic subgroups of G. Now consider the Weyl group W_G of G. The conjugacy class X gives rise to a subgroup $W_X \subset W_G$; the coset space $W_X\backslash W_G$ is a finite set which can be described explicitly.

Given a pair $(\mathscr{G}, \iota)$ with discrete invariants N and X (formulation to be made more precise) we define an element $\underline{w}(\mathscr{G}, \iota) \in W_X\backslash W_G$. Our first main result is that this element classifies $\mathscr{G}$ with its action by D:

Theorem. — Notations and assumptions as above. Associating $\underline{w}(\mathscr{G}, \iota)$ to the pair $(\mathscr{G}, \iota)$ gives a bijection

$$\left\{\begin{matrix}\text{isomorphism classes of pairs}\\ (\mathscr{G}, \iota) \text{ with invariants } N,\, X\end{matrix}\right\} \xrightarrow{\ \sim\ } W_X\backslash W_G\,.$$

We refer to section 4 for a more precise statement of this result and for further details.

The second classification problem, formulated as problem (GPE) in section 5, is about triples $(\mathscr{G}, \lambda, \iota)$, where $\mathscr{G}$ is a BT_1 over k, where $\lambda\colon \mathscr{G} \xrightarrow{\sim} \mathscr{G}^D$ is a principal quasi-polarization, and where again $\iota\colon D \to \mathrm{End}(\mathscr{G})$ is an action of a semi-simple $\mathbb{F}_p$-algebra. We assume that on D we are given an involution $d \mapsto d^*$, and if $f \mapsto f^\dagger$ is the Rosati involution on $\mathrm{End}(\mathscr{G})$ associated to λ then we require that $\iota(d^*) = \iota(d)^\dagger$ for all $d \in D$.

After some reduction steps, problem (GPE) splits up into three cases, labelled A (unitary), C (symplectic) and D (orthogonal). Each of these cases has its own features, but with suitable notations and definitions our main results can be stated uniformly. Our results are similar in shape to that in case (GE), provided that we work with the group $G := \mathrm{Sp}_D(N, \psi)$ in the place of $\mathrm{GL}_D(N)$. Here ψ is the symplectic form on the Dieudonné module N corresponding to the polarization λ. The chosen labels A, C and D correspond to the root system of G. Again we consider a multiplication type, and similar to what we did in case (GE) we can cook up a finite set $W_X\backslash W_G$. To a triple $(\mathscr{G}, \lambda, \iota)$ we associate an element $\underline{w}(\mathscr{G}, \lambda, \iota) \in W_X\backslash W_G$. Further the inclusion $f\colon \mathrm{Sp}_D(N, \psi) \hookrightarrow \mathrm{GL}_D(N)$ induces an injective map $W(f)\colon W_X\backslash W_G \hookrightarrow W_X\backslash W_{\mathrm{GL}_D(N)}$ which sends $\underline{w}(\mathscr{G}, \lambda, \iota)$ to $\underline{w}(\mathscr{G}, \iota)$. Now we have:

Theorem. — Assume $k = \bar{k}$ *with* $\mathrm{char}(k) > 2$.

(i) *Let $\mathscr{G}$ be a BT_1 over k with an action ι of the semi-simple algebra D. Then there exists a principal quasi-polarization λ such that $(\mathscr{G}, \lambda, \iota)$ is a triple as in* (GPE) *if and only if $\underline{w}(\mathscr{G}, \iota) \in W_X\backslash W_{\mathrm{GL}_D(N)}$ is in the image of the map $W(f)$.*

(ii) *If a form λ as in* (i) *exists then it is unique up to isomorphism (respecting the D-action). In other words: associating $\underline{w}(\mathscr{G}, \lambda, \iota) \in W_X\backslash W_G$ to a triple $(\mathscr{G}, \lambda, \iota)$ gives a bijection*

$$\left\{\begin{matrix}\text{isomorphism classes of triples}\\ (\mathscr{G}, \lambda, \iota) \text{ with invariants } N,\, X\end{matrix}\right\} \xrightarrow{\sim} W_X\backslash W_G\,.$$

Again we refer to the text for a more precise statement of the result.

The present paper only deals with finite group schemes with additional structure. Results about the associated stratifications on PEL-type moduli spaces will appear in [9]. After that it should be investigated whether a similar group-theoretic description also applies to more general Shimura varieties. We hope to return to such questions in the near future. In response to the preprint version of this paper, Torsten Wedhorn informed me that he has also obtained results on the Ekedahl-Oort stratification for PEL-type Shimura varieties; see [18].

Acknowledgements. I have had many pleasant conversations about the subject of this paper with Frans Oort. He explained to me the basic ideas underlying the Ekedahl-Oort stratification; also he pointed out that my original proof of Lemma 4.5 was incomplete. The idea that a good "encoding" could be done using the Weyl group is due to Gerard van der Geer (cf. his paper [17]). On several questions about algebraic groups I received useful suggestions from Wilberd van der Kallen. I thank them all cordially.

Notations and conventions. If S is a set we write $\mathfrak{S}(S)$ for its group of permutations. In case $S = \{1, 2, \ldots, q\}$ we simply write $\mathfrak{S}_q$. We shall write $\mathbb{H}_q \subset \mathfrak{S}_{2q}$ for the subgroup given by

$$\mathbb{H}_q := \{\rho \in \mathfrak{S}_{2q} \mid \rho(j) + \rho(2q+1-j) = 2q+1 \quad \text{for all } j\}.$$

This group, isomorphic to a semi-direct product $\{\pm 1\}^q \rtimes \mathfrak{S}_q$, is known under several names in the literature (e.g., the hyperoctahedral group, or the wreath product of $\mathfrak{S}_q$ with $(\mathbb{Z}/2\mathbb{Z})$). In this paper it will occur as (a realization of) the Weyl group of type C_q.

By a symplectic form on a module N we mean a perfect alternating bilinear pairing. At various points in the paper we shall work with spaces equipped with a symmetric or skew-symmetric pairing. We shall use $\varepsilon \in \{\pm 1\}$ for the sign of the pairing. Given a symplectic or orthogonal space (N, ψ) of even dimension $2q$ we say that an ordered k-basis $\beta = e_1, e_2, \ldots, e_{2q}$ is a symplectic, resp. orthogonal basis, if

$$\psi(e_i, e_j) = \begin{cases} 0 & \text{if } i+j \neq 2q+1; \\ 1 & \text{if } i+j = 2q+1 \text{ and } i < j; \\ \varepsilon & \text{if } i+j = 2q+1 \text{ and } i > j. \end{cases}$$

If $\mathscr{G}$ is a finite commutative group scheme we write $\mathscr{G}^D$ for its Cartier dual. If M is the Dieudonné module of such a group scheme, we write M^D for the dual Dieudonné module, which is the Dieudonné module of $\mathscr{G}^D$. We write $\mathscr{G}[p]$ for the kernel of multiplication by p. Likewise, $\mathscr{G}[F]$ and $\mathscr{G}[V]$ denote the kernels of Frobenius and Verschiebung. Similar notations $M[F]$ and $M[V]$ are used for the Dieudonné module.

If R is a ring of characteristic $p > 0$ then we write $\mathrm{Frob}_R\colon x \mapsto x^p$ for its Frobenius endomorphism. In the particular case that $R = k$ is a perfect field we shall occasionally write σ for Frob_k.

By an involution on a ring we mean what is sometimes called an anti-involution, i.e., an additive homomorphism $x \mapsto x^*$ such that $(xy)^* = y^* x^*$.

2. Review of results of Kraft

2.1. Let k be a perfect field of characteristic $p > 0$. Write $C(1)_k$ for the category of finite commutative k-group schemes which are killed by p. Dieudonné theory tells us that $C(1)_k$ is equivalent to the category of triples (M, F, V), where

— M is a finite dimensional k-vector space,

— $F\colon M \to M$ is a Frob_k-linear map,

— $V\colon M \to M$ is a Frob_k^{-1}-linear map,

such that $F \circ V = 0 = V \circ F$. For the purpose of this paper it is not so important how this equivalence is defined. To fix ideas, let us say we will use contravariant Dieudonné theory. A reference is [3]. In the rest of this paper, whenever we discuss Dieudonné modules we always mean modules of the kind just described, i.e., corresponding to a group scheme in the category $C(1)_k$.

A necessary and sufficient condition for $\mathscr{G} \in C(1)_k$ to be the p-kernel of a Barsotti-Tate group is that the sequence

$$\mathscr{G} \xrightarrow{F_{\mathscr{G}}} \mathscr{G}^{(p)} \xrightarrow{V_{\mathscr{G}}} \mathscr{G}$$

is exact; see [5], Prop. 1.7. On the Dieudonné module this means that

$$\mathrm{Ker}(F) = \mathrm{Im}(V) \qquad \text{and} \qquad \mathrm{Im}(F) = \mathrm{Ker}(V).$$

If these conditions are satisfied we shall say that $\mathscr{G}$ is a BT_1; short for: $\mathscr{G}$ is a truncated Barsotti-Tate group of level 1.

Let us now assume that k is algebraically closed. In the unpublished manuscript [8], Kraft showed that the objects of $C(1)_k$ admit a normal form. In the description of this, one distinguishes two types of group schemes.

(a) *Linear type.* Consider a linear graph Γ, with all edges labelled either by F (in which case we draw it as $\xrightarrow{F}$) or by V (drawn as $\xleftarrow{V}$).

Let $\mathscr{V} = \mathscr{V}(\Gamma)$ be the set of vertices of Γ. We associate to Γ a Dieudonné module (M, F, V) in the following way. As a vector space we take $M = k^{\mathscr{V}}$, that is, for each vertex v we have a base vector e_v. Then define F and V according to the arrows in the graph. So, $F(e_v) = e_w$ if there is an F-arrow from v to w, and $F(e_v) = 0$ if there is not an F-arrow starting at v. Similarly, $V(e_v) = e_w$ if there is a V-arrow from v to w and $V(e_v) = 0$ if v is not the source of a V-arrow.

One readily checks that this defines a Dieudonné module, for which we write $M_\Gamma = (M_\Gamma, F, V)$. Write $\mathscr{G}_\Gamma$ for the corresponding group scheme (well-determined up to isomorphism, or well-determined once we choose a quasi-inverse to the Dieudonné functor $\mathscr{G} \mapsto M(\mathscr{G})$).

(b) *Circular type.* Consider a circular graph

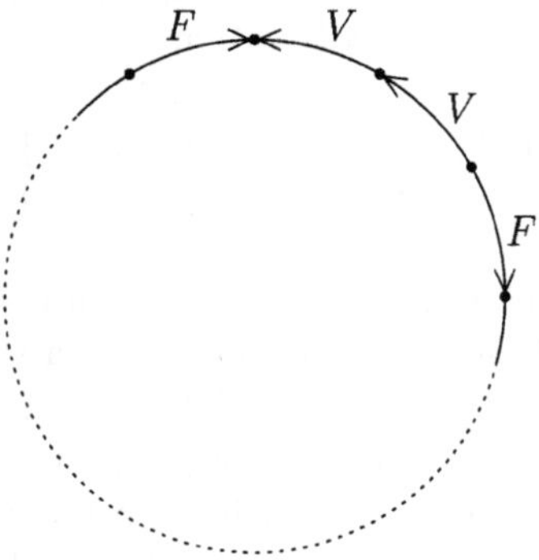

where again all edges are labelled by F or V. We require that the F-V-pattern is not periodic (i.e., the diagram is not invariant under a non-trivial rotation). Two circular diagrams which differ by a rotation will be considered equivalent. By the same rules as in the linear case we obtain a Dieudonné module M_Γ, corresponding to a group scheme $\mathscr{G}_\Gamma$.

2.2. *Examples.* — (i) The category $C(1)_k$ has 3 simple objects: $(\mathbb{Z}/p\mathbb{Z})$, μ_p and α_p. The corresponding diagrams are:

$(\mathbb{Z}/p\mathbb{Z}):$  F $\quad \mu_p:$  V $\quad \alpha_p: \ \bullet$ (no arrows).

(ii) The Cartier dual of $\mathscr{G}_\Gamma$ is isomorphic to $\mathscr{G}_{\check{\Gamma}}$, where $\check{\Gamma}$ is the diagram obtained from Γ by changing all F-arrows into V-arrows and vice versa.

(iii) The group scheme $\mathscr{G}_\Gamma$ is a BT_1 if and only if Γ is a circular diagram.

The category $C(1)_k$ is abelian, and all objects have finite length. It follows that every $\mathscr{G} \in C(1)_k$ is a direct sum of indecomposable objects. Up to isomorphism and permutation of the factors the indecomposables occurring in such a decomposition are uniquely determined. As the following theorem of Kraft [8] shows, the indecomposable objects are precisely the ones corresponding to the diagrams Γ as above.

2.3. *Theorem.* — (i) *If* Γ *is a diagram of the type described above then* $\mathscr{G}_\Gamma$ *is indecomposable. If* $\mathscr{G}_\Gamma \cong \mathscr{G}_{\Gamma'}$ *then* Γ *and* Γ *are equivalent, i.e., either* $\Gamma = \Gamma'$ *of linear type or* Γ *and* Γ' *are of circular type and differ by a rotation.*

(ii) *Every indecomposable object* $\mathscr{G} \in C(1)_k$ *is isomorphic to some* $\mathscr{G}_\Gamma$.

2.4. *Remark.* — Let us relate the above to the description of group schemes given by Raynaud in [13]. We assume that $k = \bar{k}$. Let κ be a finite field with $q = p^r$ elements. Write $\mathscr{I} := \mathrm{Hom}(\kappa, k)$. We view this as a set with a cyclic ordering (see 4.1 below for a precise definition); given $i \in \mathscr{I}$ we write $i+1$ for the embedding $\mathrm{Frob}_k \circ i \colon \kappa \to k$. Consider pairs $(\mathscr{G}, \iota)$, where $\mathscr{G} \in C(1)_k$, and where $\iota \colon \kappa \to \mathrm{End}_k(\mathscr{G})$ gives $\mathscr{G}$ the structure of a scheme in κ-vector spaces. Raynaud studies the pairs $(\mathscr{G}, \iota)$ with the property that the augmentation ideal of $\mathscr{G}$ is free of rank 1 over the group ring $k[\kappa^\times]$. A (very) special case of loc. cit., Cor. 1.5.1. says that there is a bijection

$$\left\{\begin{matrix}\text{sequences of pairs } (\gamma_i, \delta_i)_{i \in \mathscr{I}} \text{ with} \\ \gamma_i, \delta_i \in \{0,1\} \text{ and } \gamma_i \cdot \delta_i = 0 \text{ for all } i\end{matrix}\right\} \xrightarrow{\ \sim\ } \left\{\begin{matrix}\text{pairs } (\mathscr{G}, \iota) \text{ up} \\ \text{to isomorphism}\end{matrix}\right\}.$$

In the language of diagrams Γ, this bijection is obtained as follows. Suppose given a sequence of pairs (γ_i, δ_i). Take $\mathscr{I}$ as a vertex set, and draw

— an arrow $(i) \xleftarrow{V} (i+1)$ if $(\gamma_i, \delta_i) = (1,0)$;

— an arrow $(i) \xrightarrow{F} (i+1)$ if $(\gamma_i, \delta_i) = (0,1)$;

— no arrow between i and $i+1$ if $(\gamma_i, \delta_i) = (0,0)$.

This gives us a disjoint union of diagrams $\Gamma_1, \ldots, \Gamma_n$ of the type described above. The Dieudonné module $M = M_{\Gamma_1} \oplus \cdots \oplus M_{\Gamma_n}$ comes equipped with an action of κ (letting $x \in \kappa$ act as multiplication by $i(x)$ on the base vector corresponding to i), and this gives us the desired scheme in κ-vector spaces. We get a BT_1 if and only if there are no pairs $(\gamma_i, \delta_i) = (0,0)$, in which case $n = 1$ and Γ_1 is circular. (Cf. [13], Rem. 1.5.4.)

2.5. *The canonical filtration.* — An important tool in Kraft's proof of Theorem 2.3 is a natural filtration on the Dieudonné module M of an object

$\mathscr{G} \in C(1)_k$. This filtration is obtained, starting from $(0) \subset M$, by iterated application of the operators F and V^{-1}.

More formally, consider the monoid $\mathscr{M}$ with 1 freely generated by the two symbols F and V^{-1}. On $\mathscr{M}$ we have a total (lexicographic) ordering such that

(i) $FZ_1 < 1 < V^{-1}Z_2$ for all Z_1, $Z_2 \in \mathscr{M}$, and

(ii) if $Z_1 < Z_2$ then $Z_1W < Z_2W$ for all $W \in \mathscr{M}$.

If M is a Dieudonné module, $M' \subset M$ is a k-subspace and $Z \in \mathscr{M}$ then $Z(M')$ is a well-defined k-subspace of M. It is not difficult to show (see [8], p. 70) that for all $Z_1 < Z_2$ we have

$$Z_1F(M) \subseteq Z_2F(M) \subseteq Z_1V^{-1}(0) \subseteq Z_2V^{-1}(0) \,.$$

It follows from this that there is a unique coarsest filtration

$$\mathscr{C}_\bullet\colon \quad (0) = \mathscr{C}_0 \subsetneq \mathscr{C}_1 \subsetneq \mathscr{C}_2 \subsetneq \cdots \subsetneq \mathscr{C}_r = M$$

of M which is stable under all $Z \in \mathscr{M}$, in the sense that for every term $\mathscr{C}_i$ and every $Z \in \mathscr{M}$ there is an index j such that $Z(\mathscr{C}_i) = \mathscr{C}_j$. We call this filtration $\mathscr{C}_\bullet$ the *canonical filtration* of M. Every term $\mathscr{C}_i$ is of the form $\mathscr{C}_i = ZF(M)$ or $\mathscr{C}_i = ZV^{-1}(0)$ for some $Z \in \mathscr{M}$. Note that each term $\mathscr{C}_i$ is in fact a Dieudonné submodule of M.

2.6. *Polarization forms.* — Let $\mathscr{G}$ be a BT_1 over k with Dieudonné module M. As the Cartier dual $\mathscr{G}^D$ represents the functor $\mathscr{H}om(G, \mathbb{G}_{m,k})$, to give a homomorphism $\lambda\colon \mathscr{G} \to \mathscr{G}^D$ is equivalent to giving a bilinear pairing $\varphi\colon \mathscr{G} \times \mathscr{G} \to \mu_p$. On the other hand, λ induces a morphism of Dieudonné modules $\mu\colon M^D \to M$, and this is equivalent to a k-bilinear form $\psi\colon M \times M \to k$ with $\psi(Fm_1, m_2) = \psi(m_1, Vm_2)^p$ for all m_1, $m_2 \in M$.

We define λ (or the corresponding μ) to be a *pq-polarization* (short for: principal quasi-polarization) if the form ψ is a symplectic form, i.e., a perfect alternating pairing. If $\mathrm{char}(k) \neq 2$ then this is equivalent to the requirement that λ is an anti-selfdual isomorphism, i.e., an isomorphism with $\lambda^D = -\lambda \circ \kappa\colon \mathscr{G}^{DD} \to \mathscr{G}^D$, where $\kappa\colon \mathscr{G}^{DD} \xrightarrow{\sim} \mathscr{G}$ is the canonical isomorphism. This, in turn, is equivalent to saying that the composition $\mathscr{G} \xrightarrow{\Delta} \mathscr{G} \times \mathscr{G} \xrightarrow{\varphi} \mu_p$ is trivial, in which case we call φ alternating. Our motivation for using ψ in the definition of a pq-polarization is the fact that for $\mathrm{char}(k) = 2$ we have an implication "ψ is alternating" $\Rightarrow$ "φ is alternating", but as Oort pointed out to us, the converse does *not* hold in general. (A counterexample can be obtained with $\mathscr{G}$ the 2-kernel of the supersingular elliptic curve.) Notice that if $\mathscr{G} = \widetilde{\mathscr{G}}[p]$, where $\widetilde{\mathscr{G}}$ is a Barsotti-Tate group and if λ is induced from a quasi-polarization of $\widetilde{\mathscr{G}}$ then ψ is necessarily alternating.

3. The relative position of two flags

In this section we discuss how the relative position of two flags in a vector space (possibly equipped with a bilinear form) can be measured by a coset in a Weyl group. In the next sections we will use this to associate a Weyl group coset to a BT_1 with additional structures. First we give some abstract definitions; after that we do some concrete examples. The basic case is that of partial flags in a vector space, without further structure (see Example 3.5 below). Once this is understood, the other examples are easy variations.

Not all formulas given in the examples are needed in this paper. However, in our applications to Shimura varieties Weyl group cosets will correspond to strata in certain moduli spaces, and one may ask whether properties of these strata can be read from the corresponding coset. For instance, the dimension of the strata corresponds to the length of the reduced representative of the coset, for which we can give simple explicit formulas. On the other hand, it should be realized that the explicit description of the Weyl groups as permutation groups depends on choices, whereas the actual Weyl group and the elements $w^\sharp(P,Q)$ we consider are canonical.

3.1. We begin by reproducing some terminology and results from Bourbaki [2], Chap. 4, §1, Exercise 3.

Let (W,S) be a Coxeter system. Let X and Y be two subsets of S, and write W_X, resp. W_Y for the subgroups of W they generate. In every double class $(W_X \cdot w \cdot W_Y) \in W_X\backslash W/W_Y$ there is a unique element of minimal length, for which we write $\dot{w}$. Every other element $w' \in W_X \cdot w \cdot W_Y$ can be written as $w' = x\dot{w}y$ with $\ell(w') = \ell(x) + \ell(\dot{w}) + \ell(y)$. An element $w \in W$ is said to be (X,Y)-*reduced* if it is the element of minimal length in its double class $W_X \cdot w \cdot W_Y$. The element w is (X,Y)-reduced if and only if it is both $(X,\emptyset)$-reduced and $(\emptyset,Y)$-reduced.

3.2. Let G be a connected linear algebraic group over an algebraically closed field. Fix a maximal torus and a Borel subgroup $T_0 \subset B_0 \subset G$. Write $W = W_G$ for the Weyl group, and let $S \subset W$ be the set of simple reflections corresponding to the chosen B_0. For $X \subseteq S$, write $P_X \subset G$ for the corresponding standard parabolic subgroup. We call a parabolic P *of type* X if P is conjugate to P_X. Associating to P its type gives a bijection between the set of conjugacy classes of parabolic subgroups and the powerset $\mathfrak{P}(S)$ of S.

The variety $\mathscr{B}$ of Borel subgroups of G is homogeneous under G. The choice of the base point B_0 gives rise to a bijection $\varphi\colon W \xrightarrow{\sim} G\backslash(\mathscr{B} \times \mathscr{B})$, sending $w \in W$ to the orbit of the pair $(B_0, {}^wB_0)$. Thus, given an ordered pair of Borel subgroups (B_1,B_2) we can define $w(B_1,B_2)$ as the unique Weyl group element whose image under φ is the orbit of (B_1,B_2).

If P and Q are parabolic subgroups of types X and Y, respectively, then their relative position is measured by an element

$$w(P,Q) \in W_X\backslash W/W_Y\,.$$

To define this element, choose Borel subgroups $B_1 \subset P$ and $B_2 \subset Q$, and let $w(P,Q)$ be the double class containing $w(B_1, B_2)$. One easily checks that this class is independent of the choice of B_1 and B_2. Alternatively, choose g, $h \in G$ such that ${}^gP = P_X$ and ${}^hQ = P_Y$, and let $v(P,Q) \in P_X\backslash G/P_Y$ be the class of the element gh^{-1}. Notice that $v(P,Q)$ is independent of the chosen elements g and h. Then $w(P,Q)$ is the image of $v(P,Q)$ under the natural bijection $P_X\backslash G/P_Y \xrightarrow{\sim} W_X\backslash W/W_Y$ obtained from Bruhat decomposition.

Given parabolic groups P and $Q_1 \subset Q_2$, the element $w(P,Q_1)$ maps to $w(P,Q_2)$ under the natural map $W_X\backslash W/W_{Y_1} \to W_X\backslash W/W_{Y_2}$. If w represents $w(P,Q)$ then w^{-1} represents $w(Q,P)$.

3.3. *Definition.* — Notations as above. We say that a parabolic subgroup Q is in *optimal position* with respect to a parabolic subgroup P if for any two Borel subgroups B, $B' \subseteq Q$ we have $w(P,B) = w(P,B') \in W_X\backslash W$, that is, if the class $w(P,B)$ is independent of the chosen Borel subgroup $B \subseteq Q$. If Q is in optimal position we write $w^\sharp(P,Q)$ for this class $w(P,B) \in W_X\backslash W$.

3.4. *Remark.* — Consider the (X,Y)-reduced representative $\dot{w}(P,Q)$ of the double class $w(P,Q)$. Then $\dot{w}(P,Q)$ can be represented by a pair (B_1, B_2) with $B_1 \subseteq P$ and $B_2 \subseteq Q$. Hence there exists a Borel subgroup $B \subseteq Q$ (namely $B = B_2$) such that $\dot{w}(P,B) = \dot{w}(P,Q)$. In particular, if Q is in optimal position with respect to P then $\dot{w}(P,B) = \dot{w}(P,Q)$ for *every* $B \subseteq Q$. Conversely, if this last condition holds then Q is evidently in optimal position with respect to P.

3.5. *Example A.* — Let N be a k-vector space of finite dimension d. Set $G := \mathrm{GL}(N)$. Choose a basis $e_1, \dots, e_d$ for N. Let $T_0 \subset G$ be the diagonal torus and $B_0 \subset G$ be the upper triangular Borel. We identify the Weyl group W with $\mathfrak{S}_d$ in the usual way; the simple reflections form the set $S = \{s_1, \dots, s_{d-1}\}$ with s_i corresponding to the transposition $(i \quad i+1)$.

Let

$$C_\bullet\colon \quad (0) = C_0 \subsetneq C_1 \subsetneq \cdots \subsetneq C_{r-1} \subsetneq C_r = N$$

be a partial flag in N. The stabilizer of $C_\bullet$ in $\mathrm{GL}(N)$ is a parabolic subgroup. Associating $\mathrm{Stab}(C_\bullet)$ to $C_\bullet$ gives a bijective correspondence between partial flags in N and parabolic subgroups of G; this allows us to apply the definitions in 3.2 and 3.3 directly to flags.

The type $Y \subseteq S$ of $C_\bullet$ is given by the rule

$$s_i \notin Y \iff \exists j : \dim(C_j) = i\,. \tag{3.5.1}$$

We are interested in the relative position of a flag

$$L\colon \quad (0) \subseteq L \subseteq N$$

and a partial flag $C_\bullet$ as above. If $e := \dim(L)$ then L has type $X = S\backslash\{s_e\}$ and W_X is the subgroup $\mathfrak{S}(\{1,\dots,e\}) \times \mathfrak{S}(\{e+1,\dots,d\})$ of $\mathfrak{S}_d$. (For $e = 0$ or $e = d$ this becomes $X = S$ and $W_X = W$.) The set W^{red} of $(X,\emptyset)$-reduced elements has $\binom{d}{e}$ elements, which can be described as follows. Choose indices $1 \leq j_1 < j_2 < \cdots < j_e \leq d$, and let $1 \leq i_1 < i_2 < \cdots < i_{d-e} \leq d$ be the remaining ones. Given two such sequences $j_\bullet$ and $i_\bullet$, define a permutation

$$w = w(j_\bullet, i_\bullet) \in \mathfrak{S}_d \quad \text{by} \quad w(j_m) = m\,, \quad w(i_m) = e + m\,. \tag{3.5.2}$$

Then $w(j_\bullet, i_\bullet)$ is an element of W^{red} and every element of W^{red} is of this form. The length of $w = w(j_\bullet, i_\bullet)$ is given by

$$\ell(w) = \sum_{m=1}^{e} (j_m - m) = \sum_{n=1}^{d-e} (e + n - i_n)\,.$$

Suppose that $C_\bullet$ is a full flag. Set $\eta(j) := \dim(C_j \cap L)$. The sequence η has the property that for all $j \geq 1$, either $\eta(j) = \eta(j-1)$ or $\eta(j) = \eta(j-1)+1$. Let $1 \leq j_1 < j_2 < \cdots < j_e$ be the indices with $\eta(j) = \eta(j-1)+1$, and let $1 \leq i_1 < i_2 < \cdots < i_{d-e}$ the indices with $\eta(i) = \eta(i-1)$. Then $\dot{w}(L, C_\bullet) \in \mathfrak{S}_d$ is the permutation $w(j_\bullet, i_\bullet)$ as defined above.

Now let $C_\bullet$ be an arbitrary partial flag. To decide whether or not $C_\bullet$ is in optimal position with respect to L we have to refine $C_\bullet$ to a full flag $\tilde{C}_\bullet$ and check whether the class $w(L, \tilde{C}_\bullet)$ depends on the choice of such a refinement. By Remark 3.4 we can check this on the reduced representatives, and combined with the previous computations we find:

$$\begin{array}{c} C_\bullet \text{ is in optimal position} \\ \text{with respect to } L \end{array} \iff \begin{array}{c} \forall j, \text{ either } C_j \subseteq \mathrm{Span}(C_{j-1}, L) \\ \text{or } C_{j-1} \cap L = C_j \cap L. \end{array} \tag{3.5.3}$$

3.6. *Example C.* — Let N be a vector space of dimension $2q$ over k, equipped with a symplectic form $\psi\colon N \times N \to k$. Set $G := \mathrm{Sp}(N, \psi)$. Let $e_1, \ldots, e_{2q}$ be a symplectic basis for N. Let $T_0 \subset G$ be the diagonal torus and $B_0 \subset G$ the upper triangular Borel. The Weyl group W can be identified with the group

$$\mathbb{H}_q := \{\rho \in \mathfrak{S}_{2q} \mid \rho(j) + \rho(2q+1-j) = 2q+1 \quad \text{for all } j\}\,.$$

We have $S = \{s_1, \ldots, s_q\}$ with

$$\begin{cases} s_i = (i \quad i+1) \cdot (2q-i \quad 2q+1-i) & \text{for } i \neq q; \\ s_q = (q \quad q+1)\,. \end{cases} \tag{3.6.1}$$

A partial flag

$$C_\bullet : \qquad (0) = C_0 \subsetneq C_1 \subsetneq \cdots \subsetneq C_{r-1} \subsetneq C_r = N$$

is called a *partial symplectic flag* if $(C_j)^\perp = C_{r-j}$ for all j. Every parabolic subgroup $Q \subset G$ is the stabilizer of a unique partial symplectic flag; as in the previous example we can therefore phrase everything directly in terms of flags. The type $Y \subseteq \{s_1, \ldots, s_q\}$ of a partial symplectic flag $C_\bullet$ is given by (3.5.1).

Let $L \subset N$ be a maximal isotropic subspace. The flag $(0) \subset L \subset N$ has type $X := S \backslash \{s_q\}$. We have

$$W_X = \{\rho \in \mathbb{H}_q \mid \rho(\{1,\ldots,q\}) = \{1,\ldots,q\}\} \cong \mathfrak{S}_q \,.$$

The set $W^{\mathrm{red}} \subset W$ of $(X, \emptyset)$-reduced elements has 2^q elements, which can be described as follows. Of each pair $\{r, 2q+1-r\} \subset \{1, 2, \ldots, 2q\}$, choose one. Let $1 \leq j_1 < j_2 < \cdots < j_q \leq 2q$ be the chosen elements. Let $1 \leq i_1 < i_2 < \cdots < i_q \leq 2q$ be the elements that were not chosen, and notice that $j_m + i_{q+1-m} = 2q + 1$ for all m. Define the permutation $w = w(i_\bullet, j_\bullet)$ by $w(j_m) = m$ and $w(i_m) = q + m$. (Cf. (3.5.2), where we now have $d = 2q$ and $e = q$.) Then w is an element of W^{red}, and every element of W^{red} is of this form. A reduced expression of w as a word in the generators s_i is given by

$$w = t_{i_l} \cdot t_{i_{l-1}} \cdots t_{i_1} \,,$$

where i_l is the largest among the i_m with $i_l \leq q$, and where we write $t_i = s_q \cdot s_{q-1} \cdots s_i$. In particular, we find

$$\ell(w) = \sum_{m=1}^{l} (q + 1 - i_m) \,.$$

Consider a full symplectic flag $C_\bullet$. The reduced element $\dot{w}(L, C_\bullet)$ representing the class $w(L, C_\bullet) \in W_X \backslash W$ is computed in a similar way as in 3.5. More precisely, for $0 \leq i \leq 2q$, set $\eta(i) := \dim(C_i \cap L)$. The sequence η has the property that for every $j \in \{1, \ldots, 2q\}$:

(a) either $\eta(j) = \eta(j-1)$ or $\eta(j) = \eta(j-1) + 1$;

(b) $\eta(j) = \eta(j-1)$ if and only if $\eta(2q+1-j) = \eta(2q-j) + 1$.

Let $j_1 < j_2 < \cdots < j_q$ be the indices where the sequence η jumps and $i_1 < i_2 < \cdots < i_q$ (with $i_m = 2q + 1 - j_{q+1-m}$) the indices where η does not jump. Then $\dot{w}(L, C_\bullet)$ is precisely the permutation $w(i_\bullet, j_\bullet)$ defined above.

To make this concrete, take $q = 7$. Let $L = \langle e_1, e_2, \ldots, e_7 \rangle$. Let $C_\bullet$ be the flag $(0) \subset \langle e_{\pi(1)} \rangle \subset \langle e_{\pi(1)}, e_{\pi(2)} \rangle \subset \cdots$, where $\pi \in \mathbb{H}_7$. Then π is a representative for $w(L, C_\bullet)$, and the element w as defined via the sequence η is the $(X, \emptyset)$-reduced representative for the class $W_X \cdot \pi$. For instance, if π is given by

$$\pi = \begin{bmatrix} 1 & 2 & 3 & 4 & 5 & 6 & 7 & 8 & 9 & 10 & 11 & 12 & 13 & 14 \\ 2 & 11 & 12 & 7 & 9 & 5 & 1 & 14 & 10 & 6 & 8 & 3 & 4 & 13 \end{bmatrix}$$

then the sequence η is given by

$$\begin{array}{rccccccccccccccc} i: & 0 & 1 & 2 & 3 & 4 & 5 & 6 & 7 & 8 & 9 & 10 & 11 & 12 & 13 & 14 \\ \eta(i): & 0 & 1 & 1 & 1 & 2 & 2 & 3 & 4 & 4 & 4 & 5 & 5 & 6 & 7 & 7 \end{array} \,,$$

so that $j_1 = 1$, $j_2 = 4$, $j_3 = 6$, $j_4 = 7$, $j_5 = 10$, $j_6 = 12$, $j_7 = 13$; $i_1 = 2$, $i_2 = 3$, $i_3 = 5$, $i_4 = 8$, $i_5 = 9$, $i_6 = 11$, $i_7 = 14$. This gives

$$w = \dot{w}(L, C_\bullet) = \begin{bmatrix} 1 & 2 & 3 & 4 & 5 & 6 & 7 & 8 & 9 & 10 & 11 & 12 & 13 & 14 \\ 1 & 8 & 9 & 2 & 10 & 3 & 4 & 11 & 12 & 5 & 13 & 6 & 7 & 14 \end{bmatrix},$$

which is indeed a minimal representative (length=14) of the class $W_X \cdot \pi$.

Now let $C_\bullet$ be a partial symplectic flag which has a maximal isotropic term (of dimension q). Combining the previous computations with Remark 3.4 we find that criterion (3.5.3) also holds in this case.

3.7. *Buildings.* — As preparation for the next example we need to extend the definitions in 3.2 and 3.3 to the setting of buildings. All we need from this can be found in [1] or [16]. We do not assume our buildings to be thick, that is, "building" here means "weak building" in the sense of [16].

Let Δ be a building with a strongly transitive, type-preserving action of a group G. Fix an apartment Σ and a chamber $C \in \Sigma$, and let (W, S) be the associated Coxeter system. (So W is the group of type-preserving automorphisms of Σ.) Let $\mathfrak{p}$ and $\mathfrak{q}$ be two elements of Δ, of types X and $Y \subseteq S$, respectively. Let $\mathfrak{p}_X$ and $\mathfrak{p}_Y$ be the faces of C of types X and Y, and write $P_X := \mathrm{Stab}_G(\mathfrak{p}_X)$ and $P_Y := \mathrm{Stab}_G(\mathfrak{p}_Y)$. Bruhat decomposition gives a bijection

$$W_X \backslash W / W_Y \xrightarrow{\sim} P_X \backslash G / P_Y \,. \tag{3.7.1}$$

Choose g, $h \in G$ with $g \cdot \mathfrak{p} = \mathfrak{p}_X$ and $h \cdot \mathfrak{q} = \mathfrak{p}_Y$. Define $w(\mathfrak{p}, \mathfrak{q}) \in W_X \backslash W / W_Y$ to be the element which, via (3.7.1), corresponds to the class of $g \cdot h^{-1}$, noting that this element does not depend on the chosen g and h.

If G is a connected reductive group and Δ is the building of (proper parabolic subgroups of) G then we recover the definition given in 3.2.

The generalization of 3.3 is immediate: we say that $\mathfrak{q}$ is in optimal position with respect to $\mathfrak{p}$ if for any two chambers C_1, C_2 having $\mathfrak{q}$ as a face, $w(\mathfrak{p}, C_1) = w(\mathfrak{p}, C_2)$. If this holds we obtain a well-defined element $w^\sharp(\mathfrak{p}, \mathfrak{q}) \in W_X \backslash W$.

3.8. *Example D.* — Assume $\mathrm{char}(k) \neq 2$. Let N be a vector space of dimension $2q$ over k, equipped with a non-degenerate symmetric k-bilinear form ψ: $N \times N \to k$. Set $G := \mathrm{O}(N, \psi)$.

A partial flag

$$C_\bullet : \qquad (0) = C_0 \subsetneq C_1 \subsetneq \cdots \subsetneq C_{r-1} \subsetneq C_r = N$$

is called a *partial orthogonal flag*, if $(C_j)^\perp = C_{r-j}$ for all j. Let $\Delta := \mathit{Flag}(N, \psi)$ be the building of all such flags, with its natural action of G.

Let $e_1, \ldots, e_{2q}$ be an orthogonal basis for N. Let $\Sigma \subset \mathit{Flag}(N, \psi)$ be the apartment of all flags $C_\bullet$ with the property that each term C_i is of the form $C_i = \langle e_j ; j \in J \rangle$ for some $J \subseteq \{1, 2, \ldots, 2q\}$. The full flag $(0) \subset \langle e_1 \rangle \subset \langle e_1, e_2 \rangle \subset \cdots \subset N$ is a chamber of Σ. The Weyl group $W = W_\Delta$ can be identified with the group $\mathbb{H}_q$ as in 3.6, and the set of simple reflections is the same set $S = \{s_1, \ldots, s_q\}$ as given there. Notice that if $T \subset G$ is a maximal torus (e.g. the diagonal torus) then W_Δ can also be identified with $W_G := \mathscr{N}_G(T)/\mathscr{Z}_G(T)$.

Let $L \subset N$ be a maximal isotropic subspace. We are interested in the relative position of L and a partial flag $C_\bullet$ as above with the property that $r = 2h$ is even; in other words, the flag $C_\bullet$ is assumed to have a maximal isotropic term C_h. The flag $(0) \subset L \subset N$ has type $X := S\backslash\{s_q\}$. The computation of $\dot{w}(L, C_\bullet)$ is now precisely the same as in 3.6, and again we find that criterion (3.5.3) holds.

As G is non-connected, there is an alternative approach to the relative position of L and $C_\bullet$. Namely, set $G^0 := \mathrm{SO}(N, \psi) \subset G$. Let $T_0 \subset G^0$ be the diagonal maximal torus and $B_0 \subset G^0$ the upper triangular Borel (with respect to the chosen orthogonal basis). Define a subgroup $\mathbb{H}_{q,\mathrm{even}}$ and a subset $\mathbb{H}_{q,\mathrm{odd}} \subset \mathfrak{S}_{2q}$ by

$$\mathbb{H}_{q,\mathrm{even/odd}} := \left\{ \rho \in \mathfrak{S}_{2q} \;\middle|\; \begin{array}{l} \rho(j) + \rho(2q+1-j) = 2q+1 \text{ for all } j, \text{ and} \\ \text{the number of } i \leq q \text{ with } \rho(i) > q \text{ is even/odd} \end{array} \right\}.$$

The Weyl group W^0 of G^0 is $\mathbb{H}_{q,\mathrm{even}}$. The set of simple reflections is $S^0 = \{s_1, \ldots, s_{q-1}, \tilde{s}_q\}$, where $s_1, \ldots, s_{q-1}$ are as in (3.6.1) and

$$\tilde{s}_q = (q-1 \quad q+2) \cdot (q \quad q+1)\,.$$

As before, let $L \subset N$ be a totally isotropic subspace. Without loss of generality we may assume that the dimension of $\langle e_1, \ldots, e_q \rangle \cap L$ is congruent to q modulo 2, in which case the stabilizer $P \subset G^0$ of L is a parabolic subgroup of type $X^0 := S\backslash\{\tilde{s}_q\}$. (Otherwise it is of type $S\backslash\{s_{q-1}\}$.) The set $W^{0,\mathrm{red}}$ of $(X^0, \emptyset)$-reduced elements in W^0 has 2^{q-1} elements, which can be described in the same spirit as in the previous examples.

Next consider a partial orthogonal flag

$$C_\bullet\colon \qquad (0) = C_0 \subsetneq C_1 \subsetneq \cdots \subsetneq C_h \subsetneq \cdots \subsetneq C_{2h-1} \subsetneq C_{2h} = N$$

with $\dim(C_h) = q$. Let $Q \subset G^0$ be its stabilizer. Recall that in general an orthogonal flag is not uniquely determined by its stabilizer. (For a parabolic $Q \subset G^0$ there is either a unique partial orthogonal flag with stabilizer Q or there are precisely 3 such flags.) A priori it is therefore not so obvious how the element $w(P, Q)$ is related to $w(L, C_\bullet)$.

To describe this relation, let $\delta = \delta(L, C_\bullet) \in (\mathbb{Z}/2\mathbb{Z})$ be the class of $\dim(L/L \cap C_h)$ modulo 2. The point we want to make is that if we know δ we can compute $w(P, Q) \in W_{X^0}\backslash W_{G^0}$ from $w(L, C_\bullet) \in W_X\backslash W_\Delta$ and vice versa. More precisely, suppose that $C_\bullet$ is in optimal position with respect to L, working inside $\Delta = \mathit{Flag}(N, \psi)$. Notice that $\dot{w}(L, C_\bullet)$ is an element of $\mathbb{H}_{q,\delta} \subset \mathbb{H}_q = W_G$.

Now one can show that also Q is in optimal position with respect to P (in the building of G^0), and that

$$w^\sharp(P,Q) = \begin{cases} w^\sharp(L,C_\bullet) & \text{if } \delta \text{ is even;} \\ w^\sharp(L,C_\bullet)\cdot s_q & \text{if } \delta \text{ is odd.} \end{cases}$$

(Caution: the corresponding formula for reduced representatives $\dot{w}$ does not hold in general.)

In later applications to Shimura varieties the invariant δ will be fixed, and it is the element $w^\sharp(P,Q)$ that we have to consider. With an eye toward such applications, let us describe how the length of the reduced representative $\dot{w}(P,Q)$ can be computed, assuming that $C_\bullet$ is in optimal position with respect to L. Namely, refine $C_\bullet$ to a full orthogonal flag $\Gamma_\bullet$, and set $\eta(i) = \dim(\Gamma_i \cap L)$. As usual, let the j_m's be the indices where the sequence η jumps and let the i_m's be the remaining ones. Then

$$\ell\bigl(\dot{w}(P,Q)\bigr) = \sum_{m=1}^{l} (q - i_m)\,,$$

where $l = \dim(L/L\cap C_h)$ is the largest index such that $i_l \le q$. To illustrate this, take the same data as in the example in 3.6. Then δ is odd, and

$$\dot{w}(P,Q) = \begin{bmatrix} 1 & 2 & 3 & 4 & 5 & 6 & 7 & 8 & 9 & 10 & 11 & 12 & 13 & 14 \\ 1 & 9 & 8 & 2 & 11 & 3 & 10 & 5 & 12 & 4 & 13 & 7 & 6 & 14 \end{bmatrix},$$

which is an element of length 11 in $\mathbb{H}_{7,\mathrm{even}}$.

3.9. *Some functoriality.* — Let $G_1 \subset G_2$ be an inclusion of (possibly non-connected) reductive groups. Suppose G_i $(i=1,2)$ acts in a strongly transitive, type-preserving way on a building Δ_i. Further suppose given an injective map $f\colon \Delta_1 \hookrightarrow \Delta_2$ which realizes Δ_1 as a sub-poset of Δ_2 and which is compatible with the actions of G_1. (The map f will usually *not* be a morphism of complexes.) We assume that there are apartments $\Sigma_i \subset \Delta_i$ and a simplex C which is a chamber of both Σ_1 and Σ_2. Write (W_i, S_i) for the Coxeter system associated to $C \in \Sigma_i \subset \Delta_i$.

Let $\mathfrak{p}_0$ and $\mathfrak{q}_0$ be two faces of C in Σ_1. Write $X_i \subseteq S_i$ for the type of $\mathfrak{p}_0$ and $Y_i \subseteq S_i$ for the type of $\mathfrak{q}_0$, working in Δ_i. Write P_{X_i} and P_{Y_i} for the stabilizer of $\mathfrak{p}_0$ and $\mathfrak{q}_0$, respectively, in G_i. We have $P_{X_1} = G_1 \cap P_{X_2}$, and $P_{Y_1} = G_1 \cap P_{Y_2}$. This gives us a diagram

$$\begin{array}{ccc} P_{X_1}\backslash G_1/P_{Y_1} & \hookrightarrow & P_{X_2}\backslash G_2/P_{Y_2} \\ \wr\downarrow & & \downarrow\wr \\ W_{X_1}\backslash W_1/W_{Y_1} & & W_{X_2}\backslash W_2/W_{Y_2} \end{array}$$

where the vertical bijections are obtained from Bruhat decomposition. Write

$$W(f)\colon W_{X_1}\backslash W_1/W_{Y_1} \hookrightarrow W_{X_2}\backslash W_2/W_{Y_2}$$

for the injective map thus obtained.

Now let $\mathfrak{p}$ be an element in the G_1-orbit of $\mathfrak{p}_0$ and $\mathfrak{q}$ in the G_1-orbit of $\mathfrak{q}_0$. Viewing $\mathfrak{p}$ and $\mathfrak{q}$ as elements of Δ_1 we obtain an element $w_1(\mathfrak{p},\mathfrak{q}) \in W_{X_1}\backslash W_1/W_{Y_1}$; viewing them as elements of Δ_2 gives $w_2(\mathfrak{p},\mathfrak{q}) \in W_{X_2}\backslash W_2/W_{Y_2}$. Now it is immediate from the definitions that $w_2(\mathfrak{p},\mathfrak{q})$ is the image of $w_1(\mathfrak{p},\mathfrak{q})$ under the map $W(f)$.

As an example of this, let $G_1 = \mathrm{Sp}(N,\psi)$ as in 3.6 or $G_1 = \mathrm{O}(N,\psi)$ as in 3.8, and consider the natural inclusion map $G_1 \subset G_2 := \mathrm{GL}(N)$. There is a natural forgetful map $f\colon \Delta := \mathit{Flag}(N,\psi) \hookrightarrow \Delta' := \mathit{Flag}(N)$. Let L and $C_\bullet$ as in 3.6 or 3.8. Write X' for the type of L in Δ' (=the building of $\mathrm{GL}(N)$). Then our computations show that, indeed, the answer for $w(L,C_\bullet)$ is the same whether we compute it in $W_X\backslash W_\Delta$ or in $W_{X'}\backslash W_{\Delta'}$. Even better, in both examples we have a commutative diagram

$$\begin{array}{ccc} W_{G_1} & \hookrightarrow & W_{G_2} \\ \downarrow & & \downarrow \\ W_X\backslash W_\Delta & \underset{W(f)}{\hookrightarrow} & W_{X'}\backslash W_{\Delta'} \end{array}$$

and the reduced representative $\dot{w}(L,C_\bullet)$ is the same whether we compute it in W_{G_1} or in W_{G_2}. (Of course, the length of this element in general does depend on the group in which we work.)

3.10. *Example A (continued).* — Since Type A will occur as a unitary group, let us discuss what happens under duality. Let $\tilde{k} := k\times k$, with involution given by $(x_1,x_2) \mapsto (x_2,x_1)$. Let N be a free $\tilde{k}$-module of rank q, equipped with a non-degenerate alternating hermitian form $\psi\colon N\times N \to \tilde{k}$. We have a natural decomposition $N = N_1 \oplus N_2$, where $(x_1,x_2) \in \tilde{k}$ acts on N_i as multiplication by x_i. The form ψ gives an identification of N_2 with the k-linear dual of N_1. Given a subspace $L_1 \subset N_1$, we write

$$L_1^\perp := \{n_2 \in N_2 \mid \psi(n_1,n_2) = 0 \quad \text{for all } n_1 \in L\} \subset N_2\,.$$

More generally, if $C_{1,\bullet}$ is a partial flag in N_1, we write $C_{1,\bullet}^\perp$ for the partial flag in N_2 whose terms are the spaces $C_{1,i}^\perp$ (with reversed numbering).

Let $G := \mathrm{U}_{\tilde{k}}(N,\psi)$. We have isomorphisms

$$\mathrm{GL}_k(N_1) \xrightarrow{\sim} G \xrightarrow{\sim} \mathrm{GL}_k(N_2)\,.$$

Under the identification $N_2 = N_1^\vee$ the composition of the two maps is given by $g \mapsto (g^\vee)^{-1}$.

Choose k-bases $e_{i,1},\ldots,e_{i,q}$ for N_i $(i=1,2)$ such that $\psi(e_{1,\mu},e_{2,\nu})=0$ whenever $\mu+\nu\neq q+1$. Write $T_i\subset B_i\subset \mathrm{GL}(N_i)$ for the diagonal torus and the upper triangular Borel. As in Example 3.5 we identify the Weyl group W_i of $\mathrm{GL}(N_i)$ with $\mathfrak{S}_q$, and number the set $S_i\subset W_i$ of simple reflections such that (3.5.1) holds.

Let $w_0\in\mathfrak{S}_q$ be the permutation of order 2 given by $w_0(\mu)=q+1-\mu$. For $\pi\in\mathfrak{S}_q$, define $\breve{\pi}:=w_0\circ\pi\circ w_0$. We can describe the Weyl group of G as

$$W=\left\{(\pi_1,\pi_2)\in W_1\times W_2 \mid \pi_2=\breve{\pi}_1\right\},$$

where the simple reflections are the elements $(s_\mu,s_\nu)\in S_1\times S_2$ with $\mu+\nu=q+1$. For all computational purposes we can reduce to the case GL_q and proceed as in 3.5, but up to a duality $\pi\mapsto\breve{\pi}$ the answers obtained depend on which of the two projections $\mathrm{pr}_i\colon G\xrightarrow{\sim}\mathrm{GL}(N_i)$ we use.

More specifically, suppose $L\subset N$ is a maximal isotropic subspace, i.e., $L=L_1\oplus L_2$ with $L_2=L_1^\perp$. Let $C_{1,\bullet}$ be a partial flag in N_1 and $C_{2,\bullet}:=C_{1,\bullet}^\perp$ the "dual" partial flag in N_2. Let $\dim(L_1)=e$, so that the flag $(0)\subset L_1\subset N_1$ has type $X_1:=S_1\backslash\{s_e\}$ and the flag $(0)\subset L_2\subset N_2$ has type $X_2:=S_2\backslash\{s_{q-e}\}$. We have a well-defined bijection $\beta\colon W_{X_1}\backslash W_1\xrightarrow{\sim}W_{X_2}\backslash W_2$ by $\pi\mapsto\breve{\pi}$. Assume that $C_{i,\bullet}$ is in optimal position with respect to L_i; if this holds for one i then also for the other. Then $\dot{w}(L_2,C_{2,\bullet})=\dot{w}(L_1,C_{1,\bullet})^\vee$ and β sends the class $w(L_1,C_{1,\bullet})$ to the class $w(L_2,C_{2,\bullet})$.

4. BT_1's with given endomorphisms

We now start working on the subject proper of this paper: classification of finite group schemes with additional structure. In this section we deal with BT_1's equipped with an action of a finite dimensional semi-simple $\mathbb{F}_p$-algebra—see problem (GE) in 4.2 for a precise statement. Although it does not "correspond" to a Shimura variety (as we do not consider polarization forms), we may view (GE) as the basic classification problem. Understanding this case will make the polarized case (GPE) considered in the next section much more transparent.

After some reduction steps, a solution of problem (GE) is given in Theorem 4.7. The proof will occupy most of the rest of the section.

4.1. *Definition.* — Let $\mathscr{I}$ be a finite set, say of cardinality m. By a *cyclic ordering* of $\mathscr{I}$ we mean an equivalence class of bijections $(\mathbb{Z}/m\mathbb{Z})\to\mathscr{I}$, two such bijections being equivalent if they differ by a translation in $(\mathbb{Z}/m\mathbb{Z})$. If m is even, say $m=2n$ then by a *bi-cyclic ordering* of $\mathscr{I}$ we mean an equivalence class of bijections $(\mathbb{Z}/2\mathbb{Z})\times(\mathbb{Z}/n\mathbb{Z})\to\mathscr{I}$, where two such bijections are equivalent if they differ by a translation over an element $(0,a)$.

If m is even then a cyclic or bi-cyclic ordering of $\mathscr{I}$ gives rise to an involution of $\mathscr{I}$, denoted $\tau\mapsto\tilde{\tau}$. Namely, if $m=2n$ and if $\gamma\colon(\mathbb{Z}/2n\mathbb{Z})\to\mathscr{I}$

defines the cyclic ordering, resp. if $\gamma\colon (\mathbb{Z}/2\mathbb{Z}) \times (\mathbb{Z}/n\mathbb{Z}) \to \mathscr{I}$ defines the bi-cyclic ordering, set

$$\widetilde{\gamma(a)} = \gamma(a+n) \quad \text{(cyclic case)}, \quad \text{resp.} \quad \widetilde{\gamma(r,a)} = \gamma(r+1,a) \quad \text{(bi-cyclic case)}.$$

This defines a map $\tau \mapsto \tilde{\tau}$, independent of the chosen representative γ and compatible with the (bi-)cyclic ordering.

4.2. *The set-up.* — Let k be an algebraically closed field of characteristic $p > 0$. Let D be a finite dimensional semi-simple $\mathbb{F}_p$-algebra. The first problem that we are interested in is the following:

(GE) Classification of pairs $(\mathscr{G}, \iota)$, where $\mathscr{G}$ is a BT_1 over k and $\iota\colon D \to \mathrm{End}_k(\mathscr{G})$ defines an action of D on $\mathscr{G}$ (with $\iota(1) = \mathrm{id}_{\mathscr{G}}$).

As the Brauer group of a finite field is trivial we have $D \cong M_{r_1}(\kappa_1) \times \cdots \times M_{r_n}(\kappa_n)$, where the κ_i are finite fields and $r_1, \ldots, r_n \in \mathbb{Z}_{\geq 1}$. Fixing such an isomorphism, every pair $(\mathscr{G}, \iota)$ as in (GE) is of the form

$$(\mathscr{G}, \iota) = (\mathscr{H}_1^{r_1}, \theta_1^{r_1}) \times \cdots \times (\mathscr{H}_n^{r_n}, \theta_n^{r_n})$$

where $(\mathscr{H}_i, \theta_i)$ is a BT_1 with an action of κ_i, and where we write θ^{r_i} for the induced action of $M_{r_i}(\kappa_i)$ on $\mathscr{H}_i^{r_i}$. Thus, in dealing with problem (GE) we may from now on assume that $D = \kappa$ is a finite field.

Let $\mathscr{I} := \mathrm{Hom}(\kappa, k)$ be the set of embeddings of κ into k. This set comes naturally equipped with a cyclic ordering, where we take the successor of an embedding $i\colon \kappa \to k$ to be $\mathrm{Frob}_k \circ i$. If there is no risk of confusion we simply write $i+1$ for the successor of the element $i \in \mathscr{I}$.

Write

$$k_i := \kappa \underset{\kappa, i}{\otimes} k \qquad \text{and} \qquad K := \kappa \otimes_{\mathbb{F}_p} k = \prod_{i \in \mathscr{I}} k_i\,.$$

Of course, k_i is canonically isomorphic to k; the index i just reminds us of how κ acts. We have a Frob_k-linear isomorphism

$$F_{k_i} := \mathrm{id} \otimes \mathrm{Frob}_k\colon k_i \to k_{i+1}\,.$$

4.3. *Decomposition of the Dieudonné module.* — Let $(\mathscr{H}, \theta)$ be a BT_1 with κ-action. Write N for the (contravariant) Dieudonné module of $\mathscr{H}$. We can view N as a module over K. There is a natural decomposition

$$N = \bigoplus_{i \in \mathscr{I}} N_i\,, \quad \text{with } N_i := \{n \in N \mid a(n) = i(a) \cdot n \quad \text{for all } a \in \kappa\}\,.$$

The Frobenius and Verschiebung F_N and V_N restrict to homomorphisms

$$F_{N_i}\colon N_i \to N_{i+1} \qquad \text{and} \qquad V_{N_i}\colon N_i \leftarrow N_{i+1}\,,$$

equivariant with respect to $F_{k_i}\colon k_i \to k_{i+1}$, respectively its inverse. These maps F_{N_i} and V_{N_i} satisfy

$$\mathrm{Ker}(F_{N_i}) = \mathrm{Im}(V_{N_i}) \qquad \text{and} \qquad \mathrm{Ker}(V_{N_i}) = \mathrm{Im}(F_{N_i}) .$$

This last relation gives (simply writing F for F_N and V for V_N)

$$\begin{aligned}\dim_k(N_i) &= \dim_k\left(\mathrm{Ker}(F_{|N_i})\right) + \dim_k(FN_i)\\ &= \dim_k(VN_{i+1}) + \dim_k\left(\mathrm{Ker}(V_{|N_{i+1}})\right) = \dim_k(N_{i+1}) .\end{aligned}$$

As the index set $\mathscr{I}$ has a cyclic ordering we conclude that $d := \dim(N_i)$ is independent of $i \in \mathscr{I}$, so that N is free of rank d as a K-module.

4.4. *Canonical filtrations.* — Let

$$\mathscr{C}_\bullet\colon \qquad (0) = \mathscr{C}_0 \subsetneq \mathscr{C}_1 \subsetneq \cdots \subsetneq \mathscr{C}_l = N$$

be the canonical filtration of N, introduced in 2.5. As each $\mathscr{C}_\nu \subset N$ is a K-submodule, it is of the form $\mathscr{C}_\nu = \oplus\Gamma_i$, where Γ_i is a k_i-submodule of N_i. Let

$$C_{i,\bullet}\colon \qquad (0) = C_{i,0} \subsetneq C_{i,1} \subsetneq \cdots \subsetneq C_{i,\ell_i} = N_i$$

be the filtration by k_i-submodules of N_i thus obtained. Note that we take the inclusions in this filtration to be strict; hence the projection of $\mathscr{C}_\nu$ to N_i is of the form $C_{i,j}$ with in general $j \neq \nu$.

By construction, for each $i \in \mathscr{I}$ there is an index $h_i \in \{0, 1, \ldots, \ell_i\}$ such that $C_{i,h_i} = \mathrm{Im}(F_{N_{i-1}})$.

4.5. *Lemma.* — (i) *For all $i \in \mathscr{I}$ and $j \in \{0, 1, \ldots, \ell_i\}$ there exist indices $r_i(j) \in \{0, 1, \ldots, h_{i+1}\}$ and $s_i(j) \in \{h_{i+1}, \ldots, \ell_{i+1}\}$ such that*

$$F_{N_i}(C_{i,j}) = C_{i+1,r_i(j)} \qquad \textit{and} \qquad V_{N_i}^{-1}(C_{i,j}) = C_{i+1,s_i(j)} .$$

(ii) *If $j \geq 1$ then either $r_i(j) = r_i(j-1) + 1$ and $s_i(j) = s_i(j-1)$ or $r_i(j) = r_i(j-1)$ and $s_i(j) = s_i(j-1) + 1$. For all i and j we have $r_i(j) + s_i(j) = h_{i+1} + j$.*

(iii) *The maps $\rho_i\colon \{1, \ldots, \ell_i\} \to \{1, \ldots, \ell_{i+1}\}$ given by*

$$\rho_i(j) = \begin{cases} r_i(j) & \text{if } r_i(j) = r_i(j-1) + 1;\\ s_i(j) & \text{if } s_i(j) = s_i(j-1) + 1\end{cases}$$

are bijections. In particular, the integer $\ell := \ell_i$ does not depend on $i \in \mathscr{I}$.

(iv) *If $j \geq 1$ then either*

$$C_{i,j} \cap \mathrm{Ker}(F_{N_i}) = C_{i,j-1} \cap \mathrm{Ker}(F_{N_i}) ,$$

or

$$C_{i,j} \subseteq \mathrm{Span}\big(C_{i,j-1}, \mathrm{Ker}(F_{N_i})\big) .$$

In the first case we have $r_i(j) = r_i(j-1)+1$ *and* $s_i(j) = s_i(j-1)$*; in the second case* $r_i(j) = r_i(j-1)$ *and* $s_i(j) = s_i(j-1)+1$.

Proof. Statement (i) follows from the fact that the canonical filtration $\mathscr{C}_\bullet$ is (by construction) stable under F_N and V_N^{-1}, using that F_N and V_N commute with the action of κ.

Next we prove that either $r_i(j) = r_i(j-1)$ or $r_i(j) = r_i(j-1)+1$. Namely, suppose that $r_i(j) \geq r_i(j-1)+2$. By construction of the canonical filtration, the term $C_{i+1,r_i(j-1)+1}$ arises either as $F(C_{i,\nu})$ or as $V^{-1}(C_{i,\nu})$ for some index ν. But

$$\begin{aligned} F_{N_i}(C_{i,j-1}) = C_{i+1,r_i(j-1)} &\subsetneq C_{i+1,r_i(j-1)+1} \\ &\subsetneq C_{i+1,r_i(j)} = F_{N_i}(C_{i,j}) \subseteq \mathrm{Im}(F_{N_i}) = \mathrm{Ker}(V_{N_i}) , \end{aligned}$$

so that neither of both is possible. A similar argument shows that $s_i(j) = s_i(j-1)$ or $s_i(j) = s_i(j-1)+1$. Further, if $r_i(j) = r_i(j-1)$ then $C_{i,j}$ is contained in the span of $C_{i,j-1}$ and $\mathrm{Ker}(F) = \mathrm{Im}(V)$, so that necessarily $s_i(j) > s_i(j-1)$. Hence we have the implications

$$r_i(j) = r_i(j-1) \Longrightarrow s_i(j) = s_i(j-1)+1$$

and

$$s_i(j) = s_i(j-1) \Longrightarrow r_i(j) = r_i(j-1)+1 .$$

Now we turn to (iii). Define $\rho_i'\colon \{1,\ldots,\ell_i\} \to \{1,\ldots,\ell_{i+1}\}$ by

$$\rho_i'(j) = \begin{cases} r_i(j) & \text{if } r_i(j) = r_i(j-1)+1; \\ s_i(j) & \text{if } r_i(j) = r_i(j-1). \end{cases}$$

We claim that the maps ρ_i' are injective. In fact, this readily follows from the definitions, where we have to note that $\rho_i(j) \leq h_{i+1}$ if $r_i(j) = r_i(j-1)+1$ and $\rho_i(j) > h_{i+1}$ if $r_i(j) = r_i(j-1)$. Using that we have a cyclic ordering on $\mathscr{I}$ it now follows that all ρ_i' are bijections and that ℓ_i is independent of $i \in \mathscr{I}$.

Next remark that we cannot have simultaneously $r_i(j) = r_i(j-1)+1$ and $s_i(j) = s_i(j-1)+1$. Indeed, if this occurs then $s_i(j)$ is not in the image of ρ_i', contradicting what we have just proved. Hence the first claim of (ii) follows, $\rho_i' = \rho_i$, and (iii) is proven. The first statement of (ii) implies that $r_i(j) + s_i(j) = r_i(j-1) + s_i(j-1) + 1$ for all i and j. As $r_i(0) = 0$ and $s_i(j) = h_{i+1}$ this gives $r_i(j) + s_i(j) = h_{i+1} + j$.

For (iv), finally, suppose that $C_{i,j}$ is not contained in the span of $C_{i,j-1}$ and $\mathrm{Ker}(F_{N_i})$. This means that $r_i(j) > r_i(j-1)$, which, by (ii), implies that $s_i(j) = s_i(j-1)$. As $\mathrm{Im}(V_{N_i}) = \mathrm{Ker}(F_{N_i})$ this just means that $C_{i,j} \cap \mathrm{Ker}(F_{N_i}) = C_{i,j-1} \cap \mathrm{Ker}(F_{N_i})$. □

4.6. *Definition of the element* $\underline{w}(\mathscr{H},\theta)$. — For $i \in \mathscr{I}$, let $\mathfrak{f}(i)$ be the k-dimension of $N[F]_i := \mathrm{Ker}(F_{N_i})$. This defines a function

$$\mathfrak{f}\colon \mathscr{I} \to \{0,1,\ldots,d\}$$

which (by analogy to the classical notion of a CM-type) we refer to as the *multiplication type* of the pair $(\mathscr{H},\theta)$.

The integer $d := \mathrm{rk}_K(N)$ and the function $\mathfrak{f}$ should be thought of as "discrete invariants" of the pair $(\mathscr{H},\theta)$, and for problem (GE) we may assume them to be fixed. To state the theorem below we want to rigidify N, i.e., choose an isomorphism $\xi\colon N \xrightarrow{\sim} K^d$. Recall that $K = \prod_{i\in\mathscr{I}} k_i$.

Let $G := \mathrm{GL}_{d,K} = \prod_{i\in\mathscr{I}} G_i$, with $G_i = \mathrm{GL}_{d,k_i}$. The function $\mathfrak{f}$ determines a conjugacy class $X = X_{\mathfrak{f}}$ of parabolic subgroups of G, viz., the conjugacy class containing the stabilizer of $N[F] \subset N \xrightarrow{\sim} K^d$. The type X decomposes as a "product" of types X_i in the factors G_i; identifying the Weyl group of G_i with $\mathfrak{S}_d$ as in 3.5 we have $X_i = S\backslash\{s_{\mathfrak{f}(i)}\}$.

Via the chosen isomorphism ξ we can view $N[F]_i$ and $C_{i,\bullet}$ as flags in k_i^d. Combining (iv) of the lemma with (3.5.3), we find that each $C_{i,\bullet}$ is in optimal position with respect to $N[F]_i$. This gives us a well-defined element $w^\sharp(N[F]_i, C_{i,\bullet}) \in W_{X_i}\backslash W_{G_i}$. Now define

$$\underline{w}(\mathscr{H},\theta) \in W_X\backslash W_G = \prod_{i\in\mathscr{I}} W_{X_i}\backslash W_{G_i}$$

to be the element with $w_i(\mathscr{H},\theta) = w^\sharp(N[F]_i, C_{i,\bullet})$. Alternatively, since each $C_{i,\bullet}$ is in optimal position, it follows that also $Q := \mathrm{Stab}_G(\mathscr{C}_\bullet)$ is in optimal position with respect to $P := \mathrm{Stab}_G(N[F])$, and $\underline{w}(\mathscr{H},\theta)$ is just the element $w^\sharp(P,Q)$. One easily checks that this element does not depend on the choice of ξ. Further it is clear that isomorphic pairs $(\mathscr{H},\theta)$ give the same element $\underline{w}(\mathscr{H},\theta)$.

4.7. *Theorem.* — *Notations and assumptions as in* 4.2 *and* 4.6. *Sending the pair* $(\mathscr{H},\theta)$ *to the element* $\underline{w}(\mathscr{H},\theta)$ *gives a bijection*

$$\left\{\begin{matrix}\text{isomorphism classes of pairs}\\ (\mathscr{H},\theta)\text{ of type }(d,\mathfrak{f})\end{matrix}\right\} \xrightarrow{\sim} W_X\backslash W_G \cong \prod_{i\in\mathscr{I}} \left[\mathfrak{S}_{\mathfrak{f}(i)} \times \mathfrak{S}_{d-\mathfrak{f}(i)}\backslash\mathfrak{S}_d\right].$$

(In the last formula, we have written $\mathfrak{S}_{\mathfrak{f}(i)} \times \mathfrak{S}_{d-\mathfrak{f}(i)}$ as an abbreviation for $\mathfrak{S}(\{1,\ldots,\mathfrak{f}(i)\}) \times \mathfrak{S}(\{\mathfrak{f}(i)+1,\ldots,d\})$.)

The proof of the theorem will occupy the rest of this section.

4.8. *Remark.* — Let $(\mathscr{H},\theta)$ be a pair of type $(d,\mathfrak{f})$. Write θ^D for the induced κ-action on the Cartier dual $\mathscr{H}^D$. Then $(\mathscr{H}^D,\theta^D)$ is of type $(d,\mathfrak{f}')$, where $\mathfrak{f}'\colon \mathscr{I} \to \{0,1,\ldots,d\}$ is the function such that $\mathfrak{f}(i)+\mathfrak{f}'(i) = d$ for all i. Write $X_i' := S\backslash\{s_{\mathfrak{f}'(i)}\}$. Just as in Example 3.10 we have an involution $\pi \mapsto \check{\pi} := w_0\circ\pi\circ w_0$ on $\mathfrak{S}_d$, and this induced a bijection $W_X\backslash W_G \xrightarrow{\sim} W_{X'}\backslash W_G$. This maps $\underline{w}(\mathscr{H},\theta)$ to $\underline{w}(\mathscr{H}^D,\theta^D)$. See also Remark 6.8 below.

4.9. *Standard objects.* — We use the notations introduced in 4.6. Let $\underline{w} = \{w_i\}_{i\in\mathscr{I}} \in W_X\backslash W_G = \prod_{i\in\mathscr{I}} W_{X_i}\backslash W_{G_i}$. We shall define a pair $(\mathscr{H},\theta) = (\mathscr{H}_{\underline{w}},\theta_{\underline{w}})$ of type $(d,\mathfrak{f})$ such that $\underline{w}(\mathscr{H},\theta) = \underline{w}$. We shall refer to this pair as the standard object corresponding to $\underline{w}$.

For $i \in \mathscr{I}$, let $e_{i,1},\ldots,e_{i,d}$ be the standard basis of k_i^d. We use the explicit description of $W_{X_i}\backslash W_{G_i}$ given in 3.5; in particular we identify the Weyl group of G_i with $\mathfrak{S}_d$. Consider the $(X_i,\emptyset)$-reduced representatives $\dot{w}_i \in \mathfrak{S}_d$. Now define $F_i\colon k_i^d \to k_{i+1}^d$ and $V_i\colon k_i^d \leftarrow k_{i+1}^d$ to be the Frob_k-linear, resp. Frob_k^{-1}-linear maps given on the base vectors $e_{i,j}$ by

$$F_i(e_{i,j}) = \begin{cases} 0 & \text{if } \dot{w}_i(j) \le \mathfrak{f}(i); \\ e_{i+1,m} & \text{if } \dot{w}_i(j) = \mathfrak{f}(i)+m. \end{cases}$$

and

$$V_i(e_{i+1,j}) = \begin{cases} 0 & \text{if } j \le d-\mathfrak{f}(i); \\ e_{i,n} & \text{if } j = d-\mathfrak{f}(i)+\dot{w}_i(n). \end{cases}$$

One readily checks that this gives K^d the structure of a Dieudonné module corresponding to a pair $(\mathscr{H}_{\underline{w}},\theta_{\underline{w}})$ of type $(d,\mathfrak{f})$. We shall write $N_{\underline{w}}$ for the Dieudonné module thus constructed. For each i the full flag

$$E_{i,\bullet}\colon \qquad (0) \subset \langle e_{i,1}\rangle \subset \langle e_{i,1},e_{i,2}\rangle \subset \cdots \subset k_i^d$$

is a refinement of the canonical filtration $C_{i,\bullet}$. Hence $w^\sharp(\mathrm{Ker}(F_i),C_{i,\bullet}) = w(\mathrm{Ker}(F_i),E_{i,\bullet})$. Computing $\dot{w}(\mathrm{Ker}(F_i),E_{i,\bullet})$ following the recipe in 3.5 we precisely find back the element $\underline{w}$. Hence $\underline{w}(\mathscr{H}_{\underline{w}},\theta_{\underline{w}}) = \underline{w}$, as desired.

4.10. *The canonical blocks.* — We shall freely use the notations introduced in Lemma 4.5. We refer to the k-vector spaces

$$B_{i,j} := C_{i,j}/C_{i,j-1} \qquad (\text{for } i \in \mathscr{I} \text{ and } j \in \{1,\ldots,\ell\})$$

as the *canonical blocks*. Recall that we have defined permutations $\rho_i \in \mathfrak{S}_\ell$ by

$$\rho_i(j) = \begin{cases} r_i(j) & \text{if } C_{i,j}\cap \mathrm{Ker}(F_{N_i}) = C_{i,j-1}\cap\mathrm{Ker}(F_{N_i}); \\ s_i(j) & \text{if } C_{i,j} \subseteq \mathrm{Span}\big(C_{i,j-1},\mathrm{Ker}(F_{N_i})\big). \end{cases}$$

For each $i \in \mathscr{I}$ and $j \in \{1,2,\ldots,\ell\}$ we obtain a canonical Frob_k-linear isomorphism

$$t_{i,j}\colon B_{i,j} \xrightarrow{\sim} B_{i+1,\rho_i(j)}\,,$$

induced by the Frobenius $F_{N_i}\colon C_{i,j} \to C_{i+1,r_i(j)}$ if $r_i(j) > r_i(j-1)$ or (inversely) by the Verschiebung $V_{N_i}\colon C_{i,j} \leftarrow C_{i+1,s_i(j)}$ if $s_i(j) > s_i(j-1)$. In the first case we refer to $t_{i,j}$ as an *F-arrow*, in the second case we call it a *V-arrow*.

In what follows it will be important to follow how the blocks $B_{i,j}$, thought of as the building blocks of the modules N_i, "migrate" through the circularly ordered system of modules N_i. (For an illustration of this, see 4.12 below.) We shall use the following notations and terminology.

Write $\mathscr{A} := \mathscr{I} \times \{1, \ldots, \ell\}$. Let ρ be the permutation of $\mathscr{A}$ given by $\rho(i,j) = \big(i+1, \rho_i(j)\big)$. By an orbit in $\mathscr{A}$ we simply mean an orbit under the permutation ρ. It will often be convenient to denote an element of $\mathscr{A}$ by a single letter (usually a). Thus, if $a = (i,j) \in \mathscr{A}$, we shall write $t_a \colon B_a \to B_{\rho(a)}$ for the map $t_{i,j}$ above, etc.

4.11. *Lemma. — For every $a \in \mathscr{A}$ we can choose an ordered basis β_a for B_a such that $\beta_{\rho(a)}$ is the image of β_a under $t_a \colon B_a \to B_{\rho(a)}$.*

Proof. Clearly it suffices to find the desired bases for a running through one ρ-orbit $O \subset \mathscr{A}$. Notice that O is a finite set with a natural cyclic ordering. Let $m := \#O$ and choose $a(0) \in O$. Write $a(n) := \rho^n\big(a(0)\big)$, so that $a(m) = a(0)$. The composition

$$\mathscr{T} \colon \big(B_{a(0)} \xrightarrow{t_{a(0)}} B_{a(1)} \xrightarrow{t_{a(1)}} \cdots \xrightarrow{t_{a(m-1)}} B_{a(m)} = B_{a(0)}\big)$$

is a Frob_k^m-linear bijection. Then

$$\mathscr{B}_{a(0)} := \{b \in B_{a(0)} \mid \mathscr{T}(b) = b\}$$

is an $\mathbb{F}_{p^m}$-subspace of $B_{a(0)}$ such that the natural map $\mathscr{B}_{a(0)} \otimes_{\mathbb{F}_{p^m}} k \longrightarrow B_{a(0)}$ is an isomorphism. Now choose an ordered $\mathbb{F}_{p^m}$-basis $\beta_{a(0)}$ for $\mathscr{B}_{a(0)}$, and define $\beta_{a(n)}$ to be the ρ^n-image of $\beta_{a(0)}$. □

4.12. *Remark.* — Consider the K-module $L := \oplus_{(i,j)\in\mathscr{A}} B_{i,j}$, where $c \in \kappa$ acts on a summand $B_{i,j}$ as multiplication by $i(c)$. Let $b \in B_{i,j}$. Define

$$F(b) := \begin{cases} t_{i,j}(b) & \text{if } r_i(j) > r_i(j-1); \\ 0 & \text{if } r_i(j) = r_i(j-1). \end{cases}$$

Set $V_L(b) := 0$ if $C_{i,j} \subseteq \mathrm{Im}(F_{N_{i-1}})$, that is, if $j \leq h_i$. If $j > h_i$ then there is a unique index $\nu \in \{1, \ldots, \ell\}$ such that $j = s_{i-1}(\nu)$, in which case we set $V_L(b) := t_{i-1,\nu}^{-1}(b) \in B_{i-1,\nu}$. With these definitions, L becomes a Dieudonné module of type $(d, \mathfrak{f})$; it should be thought of as the "associated graded" of N with respect to the canonical filtrations $C_{i,\bullet}$.

We claim that a choice of bases β_a as in the lemma gives an isomorphism of L with the standard Dieudonné module $N_{\underline{w}}$. To see this we have to investigate the relation between the permutations ρ_i and the w_i, writing $\underline{w} = \underline{w}(\mathscr{H}, \theta) = \{w_i\}_{i\in\mathscr{I}}$.

In order to describe this relation, we first have to "stretch" the permutations ρ_i to become elements of $\mathfrak{S}_d$; the stretching factors are precisely the dimensions of the blocks $B_{i,j}$. More formally, for $i \in \mathscr{I}$, consider the subset $V_i \subset \{1, \ldots, \ell\} \times \mathbb{Z}_{\geq 1}$ consisting of all pairs (j,n) with $n \leq \dim(B_{i,j})$. Define a total ordering of this set by

$$(j,n) < (j',n') \quad \stackrel{\text{def}}{\Longleftrightarrow} \quad j < j' \quad \text{or} \quad j = j' \text{ and } n < n'.$$

Write $\zeta_i\colon \{1,\ldots,d\} \xrightarrow{\sim} V_i$ for the unique order-preserving bijection. Now we "stretch" ρ_i to a bijection $\rho_i^\sharp\colon V_i \to V_{i+1}$ by $\rho_i^\sharp(j,n) = \bigl(\rho_i(j),n\bigr)$. Via the bijections ζ_i and ζ_{i+1} we can view $\rho_i^\sharp$ as an element of $\mathfrak{S}_d$.

Next define $\pi_i \in \mathfrak{S}_d$ by

$$\pi_i(\nu) = \begin{cases} \mathfrak{f}(i) + \nu & \text{if } \nu \leq d - \mathfrak{f}(i); \\ \nu - \bigl(d - \mathfrak{f}(i)\bigr) & \text{if } \nu > d - \mathfrak{f}(i). \end{cases}$$

Then the permutations $\rho_i^\sharp$ and $\dot{w}_i$ are related by the formula $\dot{w}_i = \pi_i \circ \rho_i^\sharp$. Using this, one now easily verifies that, indeed, a choice of bases β_a as in the lemma gives an isomorphism of Dieudonné modules with κ-action $L \xrightarrow{\sim} N_{\underline{w}}$.

To make this concrete in an example, suppose that κ has p^3 elements. Identify $\mathscr{I} = \mathbb{Z}/3\mathbb{Z}$, take $d = 6$ and $\mathfrak{f}(1) = 2$, $\mathfrak{f}(2) = 1$, $\mathfrak{f}(3) = 5$. Consider the standard object $(\mathscr{H}_{\underline{w}}, \theta_{\underline{w}})$, with $\underline{w}$ given by

$$\dot{w}_1 = \begin{bmatrix} 1 & 2 & 3 & 4 & 5 & 6 \\ 1 & 3 & 4 & 5 & 2 & 6 \end{bmatrix} \quad \dot{w}_2 = \begin{bmatrix} 1 & 2 & 3 & 4 & 5 & 6 \\ 2 & 3 & 1 & 4 & 5 & 6 \end{bmatrix}$$

and

$$\dot{w}_3 = \begin{bmatrix} 1 & 2 & 3 & 4 & 5 & 6 \\ 1 & 2 & 6 & 3 & 4 & 5 \end{bmatrix}.$$

The reader is invited to check that in this case the canonical blocks, and the way they are permuted, look as illustrated in Figure 1.

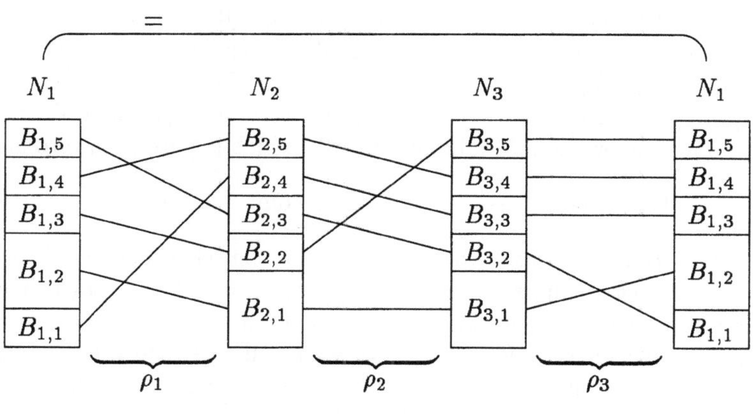

Figure 1.

In each N_i there are four blocks of dimension 1 and one 2-dimensional block. The permutations ρ_i are given by

$$\rho_1 = \begin{bmatrix} 1 & 2 & 3 & 4 & 5 \\ 4 & 1 & 2 & 5 & 3 \end{bmatrix} \quad \rho_2 = \begin{bmatrix} 1 & 2 & 3 & 4 & 5 \\ 1 & 5 & 2 & 3 & 4 \end{bmatrix} \quad \rho_3 = \begin{bmatrix} 1 & 2 & 3 & 4 & 5 \\ 2 & 1 & 3 & 4 & 5 \end{bmatrix};$$

their stretched versions are

$$\rho_1^\sharp = \begin{bmatrix} 1 & 2 & 3 & 4 & 5 & 6 \\ 5 & 1 & 2 & 3 & 6 & 4 \end{bmatrix} \quad \rho_2^\sharp = \begin{bmatrix} 1 & 2 & 3 & 4 & 5 & 6 \\ 1 & 2 & 6 & 3 & 4 & 5 \end{bmatrix}$$

and

$$\rho_3^\sharp = \begin{bmatrix} 1 & 2 & 3 & 4 & 5 & 6 \\ 2 & 3 & 1 & 4 & 5 & 6 \end{bmatrix}.$$

4.13. *F-V-compatible liftings.* — Recall that for $a \in \mathscr{A}$, we call $t_a\colon B_a \to B_{\rho(a)}$ an F-arrow or V-arrow according to whether it is induced by Frobenius or (inversely) by Verschiebung.

Fix bases β_a as in Lemma 4.11. We are going to lift each β_a to an ordered subset $\gamma_a \subset C_a$. (Here $a = (i,j) \in \mathscr{A}$ and $C_a := C_{i,j}$.) For every $i \in \mathscr{I}$ this will give us an ordered basis γ_i of N_i with the property that each $C_{i,j}$ is spanned by the first $\dim(C_{i,j})$ base vectors. (To get the correct indexing for the basis γ_i we use the bijections ζ_i of 4.12.) We want to find liftings γ_a that satisfy the following two conditions:

(i) If t_a is an F-arrow then $\gamma_{\rho(a)}$ is the image of γ_a under $F_N\colon C_a \to C_{\rho(a)}$;

(ii) If t_a is a V-arrow then γ_a is the image of $\gamma_{\rho(a)}$ under $V_N\colon C_a \leftarrow C_{\rho(a)}$.

If these two conditions are satisfied we shall say that $\{\gamma_a\}_{a\in\mathscr{A}}$ is an *F-V-compatible collection of liftings*. If such a collection of liftings exists then it induces an isomorphism of N with the Dieudonné module L as in the above remark, and it follows from that remark that $(\mathscr{H},\theta) \cong (\mathscr{H}_{\underline{w}},\theta_{\underline{w}})$.

4.14. *Tops, bottoms, and chains.* — Let $a(-1)$, $a(0) = \rho\big(a(-1)\big)$ and $a(1) = \rho\big(a(0)\big)$ be three consecutive elements of $\mathscr{A}$. We say that $a(0)$ is a *top* if $t_{a(0)}$ is an F-arrow and $t_{a(-1)}$ is a V-arrow, i.e., if both Frobenius and Verschiebung are injective on $B_{a(0)}$. We say that $a(0)$ is a *bottom* if $t_{a(0)}$ is a V-arrow and $t_{a(-1)}$ is an F-arrow, i.e., if Frobenius and Verschiebung are both zero on $B_{a(0)}$.

$$\text{top:} \quad \xleftarrow{V} \bullet \xrightarrow{F} \qquad\qquad \text{bottom:} \quad \xrightarrow{F} \bullet \xleftarrow{V}$$

Write $\mathscr{A}^t \subset \mathscr{A}$ for the set of tops, $\mathscr{A}^b \subset \mathscr{A}$ for the set of bottom elements.

By an *F-chain* in $\mathscr{A}$ we mean a sequence $a(0), a(1), \ldots, a(n)$ such that $a(0)$ is a top, $a(n)$ is a bottom, and each of the maps $t_{a(\nu)}$ $(0 \leq \nu < n)$ is an F-arrow. Similarly, by a *V-chain* in $\mathscr{A}$ we mean a sequence $a(-n), a(-n+1), \ldots, a(0)$ such that $a(0)$ is a top, $a(-n)$ is a bottom, and each of the maps $t_{a(\nu)}$ $(-n \leq \nu < 0)$ is a V-arrow. With this terminology, the orbits in $\mathscr{A}$ can be pictured as in Figure 2. (Going round, the orbit will in general cross itself; this has not been illustrated.)

4.15. *Pairs of marked orbits.* — For the arguments that follow we need some observations about the mutual position of two orbits. We begin with some terminology. Let $a = (i,j)$ and $a' = (i',j')$ be two elements of $\mathscr{A}$. We say that a and a' are *comparable* if $i = i'$. We write $a' \prec a$ if $i' = i$ and $j' < j$.

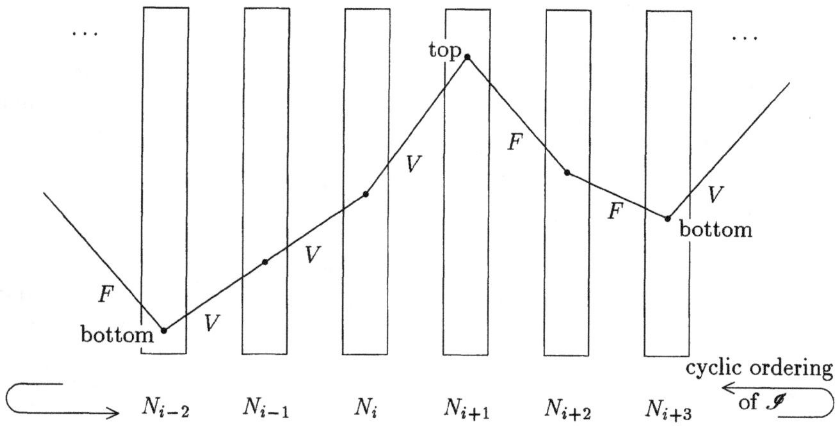

Figure 2.

If $O \subseteq \mathscr{A}$ is an orbit then by a *marking* of O we mean a map $\mathbb{Z} \to O$, written $n \mapsto a(n)$, such that $a(n+1) = \rho\big(a(n)\big)$ for all n. Clearly such a marking is fully determined if we specify one $a(n)$. Let $n \mapsto a(n)$ and $n \mapsto a'(n)$ be markings of orbits O and O', respectively. We assume that $a(n)$ and $a'(n)$ are comparable for all (equivalently: some) n. Given integers $n_1 < n_2$ we say that O and O' are *parallel in the interval* $[n_1, n_2]$ if for all ν with $n_1 \leq \nu < n_2$ the two arrows $t_{a(n)}$ and $t_{a'(n)}$ are of the same kind, i.e., they are either both F-arrows or both V-arrows. We say that the (ordered) pair of orbits O and O' is *up-down* at index n if $t_{a(n)}$ is a V-arrow, $t_{a'(n)}$ is an F-arrow, and that they are *down-up* at n if it is the other way around.

Suppose that $a'(0) = (i'_0, j'_0) \prec a(0) = (i_0, j_0)$. Let us first follow O and O' backwards in time. We know that $t_{a(-1)}\colon B_{a(-1)} \to B_{a(0)}$ is an F-arrow if and only if $j_0 \leq h_{i_0}$. (This is really the definition of h_{i_0}; see 4.4.) In particular, if $t_{a(-1)}$ is an F-arrow then so is $t_{a'(-1)}$, and in this case it is immediate from the definitions that also $a'(-1) \prec a(-1)$. Similarly, if $t_{a'(-1)}$ is a V-arrow then so is $t_{a(-1)}$, in which case we again have $a'(-1) \prec a(-1)$. Thus, if O and O' are parallel in the interval $[n_1, 0]$ and $a'(0) \prec a(0)$ then $a'(\nu) \prec a(\nu)$ for all $\nu \in [n_1, 0]$. Note that if O and O' are no longer parallel in $[n_1 - 1, 0]$ then we may still have that $a'(n_1 - 1) \prec a(n_1 - 1)$. (Configurations c. and d. in Figure 3.)

Still assuming that $a'(0) \prec a(0)$, let us now follow O and O' forward in time. Suppose they are parallel in the interval $[0, n_2]$ but not in $[0, n_2 + 1]$. We find that $a'(\nu) \prec a(\nu)$ for all positive $\nu \leq n$. At index n the pair (O, O') may be either down-up or up-down. In the down-up case we necessarily have $a'(n_2 + 1) \succ a(n_2 + 1)$.

To summarize, if O and O' are marked orbits with $a'(0) \prec a(0)$ then the possible configurations are as illustrated in Figure 3.

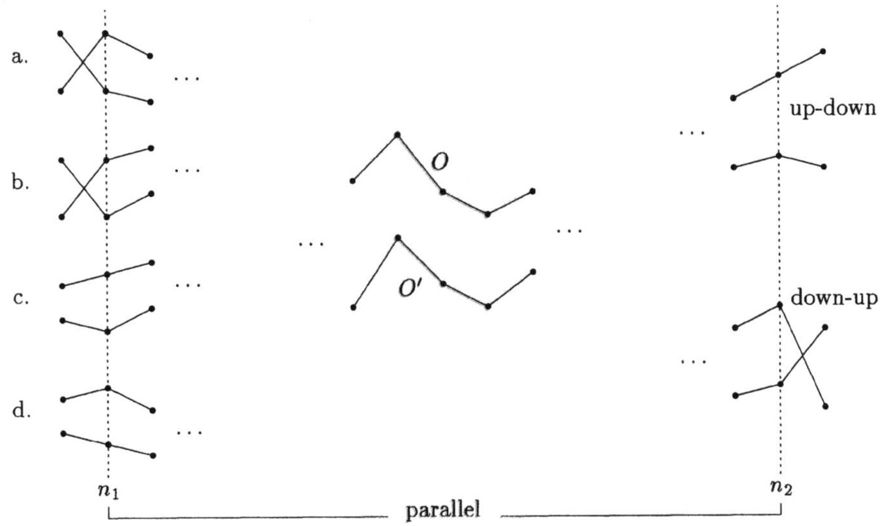

Figure 3.

4.16. *Lemma.* — *Let O and O' be marked orbits as above.*

(i) *If $O \neq O'$ (as marked orbits) then there is no infinite interval in which O and O' are parallel.*

(ii) *Suppose that $a'(0) \prec a(0)$. Let $e := \#O$ and suppose that O and O' are parallel in the interval $[-e, 0]$. Then $a'(-e) \prec a'(0) \prec a(-e) = a(0)$.*

Proof. (i) There is an integer N (the l.c.m. of the lengths of O and O') such that $a(n+N) = a(n)$ and $a'(n+N) = a'(n)$ for all n. Thus, if there is an infinite interval in which O and O' are parallel, then they are actually parallel in all of $\mathbb{Z}$. This implies that for every word Z in the letters F and V^{-1} we have $C_{a(0)} \subseteq Z(N)$ if and only if $C_{a'(0)} \subseteq Z(N)$. (See 2.5 for notations.) By definition of the filtrations $C_{i,\bullet}$ and the assumption that $a(0)$ and $a'(0)$ are comparable, this is possible only if $a(0) = a'(0)$. But then $O = O'$ as marked orbits, contrary to assumption.

(ii) Let $\alpha = (i,j)$ and $\alpha' = (i,j')$ be two comparable elements of $\mathscr{A}$. Suppose that $\rho(\alpha') \prec \rho(\alpha)$; in other words, $\rho_i(j') < \rho_i(j)$. If t_α and $t_{\alpha'}$ are both F-arrows or both V-arrows then $(\rho_i(j) - \rho_i(j')) \leq j - j'$. Iterating this e times we obtain (ii); here we have to remark that by (i) we cannot have $a'(-e) = a'(0)$. □

Now we are ready to prove what we want.

4.17. *Proposition.* — *Let $\{\beta_a\}_{a\in\mathscr{A}}$ be a collection of ordered bases as in 4.11. Then we can lift each β_a to an ordered subset $\gamma_a \subset C_a$ such that the collection $\{\gamma_a\}$ is F-V-compatible.*

As remarked in 4.13 above, if we prove the proposition then Theorem 4.7 follows.

Proof. It suffices to find F-V-compatible liftings γ_a for a running through a fixed orbit $O \subset \mathscr{A}$. In fact, by induction on the integer $r \geq 0$ we shall prove the following assertion:

$L(r)$: If $O \subset \mathscr{A}$ is an orbit and O contains an element (i,j) with $j \leq r$ then we can lift the β_a for $a \in O$ to an F-V-compatible collection $\{\gamma_a\}_{a\in O}$.

Let us first make explicit three remarks that follow directly from the definitions:

(1) Consider an F-chain $a(0), \ldots, a(n)$. Let $\gamma_{a(0)}$ be a lifting of $\beta_{a(0)}$ inside $C_{a(0)}$. For $0 \leq \nu \leq n$, let $\gamma_{a(\nu)}$ be the image (as an ordered set) of $\gamma_{a(0)}$ under F_N^ν. Then $\gamma_{a(\nu)}$ is a lifting of $\beta_{a(\nu)}$ inside $C_{a(\nu)}$.

(2) Consider a V-chain $a(-n), \ldots, a(0)$. Let $\gamma_{a(0)}$ be a lifting of $\beta_{a(0)}$ inside $C_{a(0)}$. For $0 \leq \nu \leq n$, let $\gamma_{a(-\nu)}$ be the image (as an ordered set) of $\gamma_{a(0)}$ under V_N^ν. Then $\gamma_{a(-\nu)}$ is a lifting of $\beta_{a(-\nu)}$ inside $C_{a(-\nu)}$.

(3) Consider an F-chain $a(0), \ldots, a(n)$. Let $\gamma_{a(n)}$ be a lifting of $\beta_{a(n)}$ inside $C_{a(n)}$. Then there is a lifting $\gamma_{a(0)}$ of $\beta_{a(0)}$ inside $C_{a(0)}$ such that $\gamma_{a(n)}$ is the image of $\gamma_{a(0)}$ under F_N^n.

Fix an orbit $O \subset \mathscr{A}$ and a marking $\mathbb{Z} \to O$, denoted $n \mapsto a(n)$. Let $e := \#O$. As a first approximation we shall choose, for n in some subset of $\mathbb{Z}$, liftings $\Gamma(n) \subset C_{a(n)}$ of $\beta_{a(n)}$. These liftings will be compatible under F and V, but in general $\Gamma(n+e)$ will be different from $\Gamma(n)$. After that, we shall modify the $\Gamma(n)$ such that $\Gamma(n+e) = \Gamma(n)$ for all n, in which case we can define the desired liftings γ_a by $\gamma_{a(n)} := \Gamma(n)$.

Suppose there are q top elements and q bottom elements in O. Let

$$\cdots < b_0 < t_0 < b_1 < t_1 < \cdots < b_{q-1} < t_{q-1} < b_q = (b_0+e) < t_q = (t_0+e) < \cdots$$

be the integers such that the $a(b_n)$ are the bottom elements and the $a(t_n)$ are the top elements.

Start by choosing a lifting $\Gamma(b_q)$ of $\beta_{a(b_q)} = \beta_{a(b_0)}$. We apply (3). This tells us that we can choose a lifting $\Gamma(t_{q-1})$ such that $\Gamma(b_q)$ is the image of $\Gamma(t_{q-1})$ under F_N^n, where $n = b_q - t_{q-1}$. By construction, the elements $a(t_{q-1}+\nu)$ with $0 \leq \nu \leq b_q - t_{q-1}$ form an F-chain. The condition that the $\Gamma(n)$ are F-V-compatible forces us to define $\Gamma(t_{q-1}+\nu)$ as the image of $\Gamma(t_{q-1})$ under F_N^ν. Similarly, the elements $a(t_{q-1}-\mu)$ for $0 \leq \mu \leq t_{q-1} - b_{q-1}$ form a V-chain, and we define $\Gamma(t_{q-1}-\mu)$ as the image of $\Gamma(t_{q-1})$ under V_N^μ.

We repeat the previous step, now starting with $\Gamma(b_{q-1})$. This gives us liftings $\Gamma(n)$ for $b_{q-2} \leq n \leq b_{q-1}$. We continue this process until we arrive at b_0. Notice that $\Gamma(b_0)$ and $\Gamma(b_q)$ are both liftings of $\beta_{a(b_0)}$ inside $C_{a(b_0)}$, but that in general they will be different.

Next we want to modify the $\Gamma(n)$, for $b_0 \leq n \leq b_q$, to ordered sets $\Gamma'(n)$, in such a way that the new liftings are still F-V-compatible but also satisfy $\Gamma'(b_0) = \Gamma'(b_q)$. Write $a(b_0) = a(b_q) = (i,j)$. We may assume that b_0 was chosen such that j is minimal among the second coefficients of the elements $a \in O$. By induction we may further assume that statement $L(j-1)$ is already proven.

Here we should remark that for $j = 1$ there is nothing left to prove, as in this case we have $B_{a(b_0)} = C_{a(b_0)}$ and $\Gamma(b_0) = \beta_{a(b_0)} = \Gamma(b_q)$.

Set $\delta(b_q) := \Gamma(b_q) - \Gamma(b_0)$. (Concretely: if $\Gamma(b_q) = x_1, \ldots, x_d$ and $\Gamma(b_0) = y_1, \ldots, y_d$, with $d = \dim(B_{i,j})$, then $\delta(b_q)$ is the ordered set $x_1 - y_1, \ldots, x_d - y_d$.) As $\Gamma(b_q)$ and $\Gamma(b_0)$ are both liftings of $\beta_{a(b_0)}$ we have $\delta(b_q) \subset C_{i,j-1}$. If $\Gamma(b_q) = \Gamma(b_0)$ we are done. If not, let J (with $J \leq j-1$) be the largest index such that $\delta(b_q)$ is not contained in $C_{i,J-1}$. By downward induction on J it suffices to show that we can modify the $\Gamma(n)$ to F-V-compatible liftings $\Gamma'(n)$ such that $\Gamma'(b_q) - \Gamma'(b_0) \subset C_{i,J-1}$.

Let O' be the orbit of (i, J). Let $a'\colon \mathbb{Z} \to O'$ be the marking with $a'(b_q) = (i, J)$. We now apply the induction hypothesis $L(j-1)$. It follows that for $b_0 \leq n \leq b_q$ we can find ordered sets $\varepsilon(n) \subset C_{a'(n)}$, of cardinality $d = \dim(B_{i,j})$, in such a way that

(a) the $\varepsilon(n)$ are F-V-compatible (with respect to the orbit O');
(b) $\varepsilon(b_q)$ and $\delta(b_q)$ have the same image in $B_{i,J}$.

To conclude the proof, there are two cases to consider. First assume that the marked orbits O and O' are not parallel in the interval $[b_0, b_q]$. Let $\nu \in [b_0, b_q]$ be the largest index such that $t_{a'(\nu)}$ and $t_{a(\nu)}$ are of the opposite kind. For $b_0 \leq n \leq b_q$ define $\Gamma'(n)$ by

$$\Gamma'(n) = \begin{cases} \Gamma(n) & \text{if } n \leq \nu; \\ \Gamma(n) - \varepsilon(n) & \text{if } n > \nu. \end{cases}$$

Now the whole point is that, because $t_{a'(\nu)}$ and $t_{a(\nu)}$ are of the opposite kind, these $\Gamma'(n)$ are again F-V-compatible. By construction, $\Gamma'(b_q) - \Gamma'(b_0) = \Gamma(b_q) - \Gamma(b_0) - \varepsilon(b_q)$ is contained in $C_{i,J-1}$.

The other possibility is that O and O' are parallel in the interval $[b_0, b_q]$. In this case, set

$$\Gamma'(n) := \Gamma(n) - \varepsilon(n)\,.$$

These are again F-V-compatible, and it follows from (ii) of Lemma 4.16 that also in this case $\Gamma'(b_q) - \Gamma'(b_0)$ is contained in $C_{i,J-1}$. Proceeding by induction this finally gives us the desired liftings γ_a. □

4.18. We say that a pair $(\mathscr{H}, \theta)$ is *indecomposable* if it is not isomorphic to the product of two non-trivial BT_1's with κ-action, and that it is *isotypic* if it is isomorphic to a power of an indecomposable pair. Viewing the Dieudonné module N as a module over the (non-commutative) ring $\Lambda_\kappa := \kappa \otimes_{\mathbb{F}_p} k_\sigma[F, V]/(FV, VF)$, these notions correspond to the usual notions for Λ_κ-modules. Notice that N is both noetherian and artinian as a Λ_κ-module, since it has finite k-dimension. In particular, N admits a unique decomposition as a direct sum of indecomposables. For the pair $(\mathscr{H}, \theta)$ this means that it can be written as a product of indecomposable pairs, and that the factors in such a decomposition are uniquely determined. The same conclusions can also be derived from the theorem and its proof; in fact, we see that the isotypic factors can be read from the decomposition of the set $\mathscr{A}$ into orbits:

4.19. *Corollary.* — *The pair $(\mathscr{H},\theta)$ is isotypic if and only if the associated index set $\mathscr{A}$ consists of a single orbit. In general, the decomposition of $\mathscr{A}$ into orbits corresponds to a decomposition of $(\mathscr{H},\theta)$ as a product of isotypic pairs. The pair $(\mathscr{H},\theta)$ is indecomposable if and only if $\mathscr{A}$ consists of a single orbit and all canonical blocks are 1-dimensional.*

For instance, the pair $(\mathscr{H},\theta)$ corresponding to the example in 4.12 decomposes as a product of three isotypic factors. Two of the factors are indecomposable; the third one is the square of an indecomposable pair.

5. BT_1's with endomorphisms and a polarization—cases C and D

In this section we state, and solve, a problem (GPE) about BT_1's with an action of a semi-simple $\mathbb{F}_p$-algebra and a pq-polarization. After some reduction steps the main result is given in Theorem 5.5. The problem splits up in three cases, referred to as (C), (D) and (A). Mainly for notational reasons we shall deal with case (A) in the next section. Throughout this section and the next, we assume that k is algebraically closed of characteristic $p > 2$; see also Remark 5.6.

5.1. *The set-up.* — Let k be an algebraically closed field of characteristic $p > 2$. Let $(D, *)$ be a finite dimensional semi-simple $\mathbb{F}_p$-algebra with involution. If $(\mathscr{G},\lambda)$ is a pq-polarized BT_1 over k, write $f \mapsto f^\dagger$ for the Rosati involution on $\mathrm{End}_k(\mathscr{G})$ induced by λ. The problem that we want to consider in this section is:

(GPE) Classification of triples $(\mathscr{G},\lambda,\iota)$, where $(\mathscr{G},\lambda)$ is a pq-polarized BT_1 over k and $\iota\colon D \to \mathrm{End}_k(\mathscr{G})$ is a homomorphism with $\iota(1) = \mathrm{id}_{\mathscr{G}}$ and $\iota(d^*) = \iota(d)^\dagger$ for all $d \in D$.

This problem is easily reduced to the case that $(D, *)$ is a simple algebra with involution. Namely, suppose we have a decomposition of $(D, *)$ as a product of algebras with involution $(D_i, *_i)$. Choose idempotents $e_i \in D$ with $D_i = e_i \cdot D$. Then $\mathscr{G}_i := \mathrm{Im}\big(\iota(e_i)\colon \mathscr{G} \to \mathscr{G}\big)$ is again a BT_1, the pq-polarization λ induces a pq-polarization λ_i on $\mathscr{G}_i$, and $(\mathscr{G},\lambda)$ is the product of the $(\mathscr{G}_i,\lambda_i)$ thus obtained.

From now on we assume that $(D, *)$ is simple as an algebra with involution. Write $\tilde{\kappa}$ for the center of D and put $\kappa = \{a \in \tilde{\kappa} \mid a^* = a\}$. Then κ is a finite field and we can distinguish the following possibilities:

(C) the involution $*$ is orthogonal (of the first kind) and $\tilde{\kappa} = \kappa$;
(D) the involution $*$ is symplectic (of the first kind) and $\tilde{\kappa} = \kappa$;
(A) the involution $*$ is of the second kind; either $\tilde{\kappa} \cong \kappa \times \kappa$ (case (A1)) or $\tilde{\kappa}$ is a quadratic extension of κ (case (A2)).

(The chosen labels refer to the root systems of the algebraic groups that will come into play.) In this section we shall only consider cases (C) and (D). The structures we find in case (A) are not very different but require additional notations; we shall therefore discuss this case in a separate section. Throughout, we freely use the notations and terminology introduced in the previous section.

5.2. *Morita equivalence.* — Dieudonné theory gives an equivalence between the category of triples $(\mathscr{G}, \lambda, \iota)$ as in (GPE) and the category $\mathbf{C}_1$ of 5-tuples $(M, F_M, V_M, \varphi, \iota)$, where

— (M, F_M, V_N) is the Dieudonné module of a BT_1;

— φ: $M \times M \to k$ is a symplectic form such that $\varphi(Fm_1, m_2) = \varphi(m_1, Vm_2)^p$ for all m_1, $m_2 \in M$;

— ι: $D \to \mathrm{End}_{\mathrm{DM}}(M)$ defines an action of D on the Dieudonné module M such that $\varphi(dm_1, m_2) = \varphi(m_1, d^*m_2)$ for all $d \in D$ and m_1, $m_2 \in M$.

As shown in 4.3, M is free as a K-module. (Recall from 4.2 that $K := \kappa \otimes_{\mathbb{F}_p} k$.)

We have a Morita equivalence, which tells us that the category $\mathbf{C}_1$ is equivalent to a category $\mathbf{C}_2$ whose objects are of a simpler kind. The precise description of the category $\mathbf{C}_2$ depends on the type of $(D, *)$. Namely, the objects involve pairs (N, Ψ), where N is a free K-module of finite rank, and where Ψ: $N \times N \to K$ is

(C) a symplectic form if $*$ is orthogonal;

(D) an orthogonal form if $*$ is symplectic.

(We shall see that the K-rank of N is necessarily even; this explains why the orthogonal case is labelled (D) and a type (B) does not occur.) Now define $\mathbf{C}_2$ as the category of 5-tuples $(N, F_N, V_N, \Psi, \theta)$, where

— (N, F_N, V_N) is the Dieudonné module of a BT_1;

— θ: $\kappa \to \mathrm{End}_{\mathrm{DM}}(N)$ defines an action of κ on the Dieudonné module N; via this action we can view N as a module over K, and we require that it is free;

— Ψ: $N \times N \to K$ is a form as just described, such that $\Psi(Fn_1, n_2) = F_K\big(\Psi(n_1, Vn_2)\big)$ for all n_1, $n_2 \in N$.

To describe the equivalence $\mathbf{C}_2 \xrightarrow{\text{eq.}} \mathbf{C}_1$, fix an isomorphism $D \cong M_r(\kappa)$ and set $\mathscr{L} := \kappa^r$, viewed as a left D-module. The involution $*$ on D corresponds to a non-degenerate form χ on $\mathscr{L}$. More precisely, there is a non-degenerate bilinear form χ: $\mathscr{L} \times \mathscr{L} \to \kappa$, unique up to a scalar in $\kappa^\times$, such that

$$\chi(d \cdot l_1, l_2) = \chi(l_1, d^* \cdot l_2) \qquad \text{for all } d \in D \text{ and } l_1,\ l_2 \in \mathscr{L}.$$

The form χ is alternating if $*$ is symplectic and symmetric, non-alternating if $*$ is an orthogonal involution. The equivalence $\mathbf{C}_2 \xrightarrow{\text{eq.}} \mathbf{C}_1$ is obtained by associating to an object $(N, F_N, V_N, \Psi, \theta)$ of $\mathbf{C}_2$ the 5-tuple $(M, F_M, V_M, \varphi, \iota)$ with $M := \mathscr{L} \otimes_\kappa N$, with φ: $M \times M \to k$ the symplectic form given by

$$\varphi(l_1 \otimes n_1, l_2 \otimes n_2) = \mathrm{tr}_{K/k}\big(\chi(l_1, l_2) \cdot \Psi(n_1, n_2)\big)\,,$$

with D acting on M through its natural action on $\mathscr{L}$, and with $F_M := \mathrm{id}_{\mathscr{L}} \otimes F_N$ and $V_M := \mathrm{id}_{\mathscr{L}} \otimes V_N$. That this indeed gives an equivalence of categories is more or less standard; see for instance Knus's book [6], Chap. I, especially Prop. 7.2.4 and Thm. 9.3.5. As remarked above, the M are free K-modules; they therefore indeed correspond to free K-modules N.

Now we translate back to group schemes. Throughout we shall use $\varepsilon \in \{\pm 1\}$ for the sign of the pairings Ψ involved, i.e., $\varepsilon = -1$ in case (C), $\varepsilon = 1$ in case (D). Let $\mathscr{H}$ be a BT_1 equipped with an action of κ. Consider isomorphisms $\mu\colon \mathscr{H} \xrightarrow{\sim} \mathscr{H}^D$ commuting with the κ-actions and such that the corresponding bilinear pairing Ψ on the Dieudonné module is of the type described above; we shall refer to such an isomorphism μ as a *κ-ε-duality*.

In conclusion, we have reduced problem (GPE) to:

(GPE') Classification of triples $(\mathscr{H}, \mu, \theta)$, where $(\mathscr{H}, \theta)$ is a BT_1 with an action of κ, and where $\mu\colon \mathscr{H} \xrightarrow{\sim} \mathscr{H}^D$ is a κ-ε-duality.

5.3. *Decomposition into character spaces.* — Let $(N, F_N, V_N, \Psi, \theta)$ be an object of the category $\mathbf{C}_2$. As in 4.3, the action of κ on N gives rise to a decomposition $N = \oplus_{i\in\mathscr{I}} N_i$. The form Ψ decomposes as a product $\Psi = \prod_{i\in\mathscr{I}} \psi_i$, where $\psi_i\colon N_i \times N_i \to k_i$ is of the same type as Ψ, and

$$\psi_{i+1}\big(F_{N_i}(n_1), n_2\big) = F_{k_i}\big(\psi_i(n_1, V_{N_i}(n_2))\big) \qquad \text{for all } n_1 \in N_i \text{ and } n_2 \in N_{i+1}.$$

If $H_i \subset N_i$ is a k_i-submodule then

$$F_{N_{i-1}}^{-1}(H_i^{\perp_i}) = \big(V_{N_{i-1}}(H_i)\big)^{\perp_{i-1}} \qquad \text{and} \qquad V_{N_i}^{-1}(H_i^{\perp_i}) = \big(F_{N_i}(H_i)\big)^{\perp_{i+1}}, \tag{5.3.1}$$

where $\perp_i$ denotes the perpendicular with respect to the form ψ_i. (In the sequel we shall simply write $\perp$ for $\perp_i$. We trust it will always be clear from the context what is meant.)

Combining (5.3.1) with the fact that $N_i[F] = \mathrm{Im}(V_{N_{i+1}})$ and $N_i[V] = \mathrm{Im}(F_{N_{i-1}})$ we see that $N_i[F] = N_i[F]^\perp$ and $N_i[V] = N_i[V]^\perp$ are maximal totally isotropic subspaces. In particular, the common dimension of the spaces N_i is even, say $\dim(N_i) = 2q$, and $\dim\big(N_i[F]\big) = q$. Repeated further application of (5.3.1) then shows that the canonical filtration $C_{i,\bullet}$ is self-dual with respect to ψ_i. In particular, the common length ℓ of these filtrations is even, say $\ell = 2m$. We have $C_{i,j}^\perp = C_{i,2m-j}$. In particular, if $a = (i,j)$ and $\tilde{a} = (i, 2m+1-j)$ then ψ_i induces a perfect pairing

$$\psi_{(i,j)} = \psi_a\colon B_a \times B_{\tilde{a}} \longrightarrow k_i\,.$$

Summing up, the discrete invariants $(d, \mathfrak{f})$ introduced in 4.6 are given by $d = 2q$ and $\mathfrak{f}(i) = q$ for all i. In case (D) there is another discrete invariant that we need to keep track of. Namely, define a function

$$\delta\colon \mathscr{I} \to (\mathbb{Z}/2\mathbb{Z}) \quad \text{by} \quad \delta(i) = \Big[\dim_k\big(N_i[F]/N_i[F] \cap N_i[V]\big)\Big]\,.$$

This invariant should be thought of as a "mod p" version of the crystalline discriminant, cf. [10], Section 3.

5.4. *Definition of the element* $\underline{w}(\mathscr{H},\mu,\theta)$. — Let $(\mathscr{H},\mu,\theta)$ be a triple as in (GPE'). Write (N,F_N,V_N,Ψ,θ) for the corresponding Dieudonné module. Let $2q$ be the K-rank of N. As explained above, $(\mathscr{H},\theta)$ is of type $(2q,\mathfrak{f})$ where $\mathfrak{f}$ is the constant function q.

Let $\Phi\colon K^{2q}\times K^{2q}\to K$ and $\varphi_i\colon k_i^{2q}\times k_i^{2q}\to k_i$ be the standard symplectic/orthogonal pairings. Write $G=\mathrm{Aut}_K(K^{2q},\Phi)=\prod_{i\in\mathscr{I}}G_i$, with $G_i=\mathrm{Aut}_{k_i}(k_i^{2q},\varphi_i)$. Choose an isometry $\xi\colon(N,\Psi)\xrightarrow{\sim}(K^{2q},\Phi)$; this allows us to view the $N[F]_i$ and $C_{i,\bullet}$ as partial symplectic/orthogonal flags in k_i^{2q}. Consider the building $\Delta:=\mathit{Flag}_K(K^{2q},\Phi)$ of partial symplectic/orthogonal flags of K-modules in K^{2q}; it is the join of the buildings $\Delta_i:=\mathit{Flag}(k_i^{2q},\varphi_i)$. Let X, resp. X_i, be the type of $N[F]\subset K^{2q}$, resp. $N[F]_i\subset k_i^{2q}$. Recall that in the orthogonal case, although Δ is not the building of parabolic subgroups of G, it is still true that $W_\Delta=W_G$.

The definition of $\underline{w}(\mathscr{H},\mu,\theta)$ proceeds exactly as in 4.6. Namely, by (iv) of Lemma 4.5 together with what was found in Examples 3.6 and 3.8, each $C_{i,\bullet}$ is in optimal position with respect to $N[F]_i$. This gives us a well-defined element $w^\sharp(N[F]_i,C_{i,\bullet})\in W_{X_i}\backslash W_{G_i}$. We define

$$\underline{w}(\mathscr{H},\mu,\theta)\in W_X\backslash W_G=\prod_{i\in\mathscr{I}}W_{X_i}\backslash W_{G_i}$$

to be the element with $w_i(\mathscr{H},\mu,\theta)=w^\sharp(N[F]_i,C_{i,\bullet})$. This element is independent of the choice of ξ. It is clear from these definitions that isomorphic triples give the same element $\underline{w}$.

As explained in 3.9, if $X_\mathfrak{f}$ is the type of the stabilizer of $N[F]$ in the group $\mathrm{GL}_{2q,K}$ then there is a natural map

$$W_X\backslash W_G\hookrightarrow W_{X_\mathfrak{f}}\backslash W_{\mathrm{GL}_{2q,K}} \tag{5.4.1}$$

and this maps $\underline{w}(\mathscr{H},\mu,\theta)$ to $\underline{w}(\mathscr{H},\theta)$.

5.5. *Theorem.* — *Notations as above. Assume that* $\mathrm{char}(k)>2$. *Recall that we set* $\varepsilon=-1$ *in case* (C), $\varepsilon=1$ *in case* (D).

(i) *Let* $(\mathscr{H},\theta)$ *be a* BT_1 *with an action of* κ, *of type* $(2q,\mathfrak{f}=q)$. *Let* $\underline{w}(\mathscr{H},\theta)$ *be the associated element of* $W_{X_\mathfrak{f}}\backslash W_{\mathrm{GL}_{2q,K}}$. *Then there exists a* κ-ε-*duality* $\mu\colon\mathscr{H}\to\mathscr{H}^D$ *if and only if* $\underline{w}(\mathscr{H},\theta)$ *is in the image of the map* (5.4.1).

(ii) *If a duality* μ *as in* (i) *exists then it is unique up to isomorphism. In other words: associating to a triple* $(\mathscr{H},\mu,\theta)$ *as in* (GPE') *the element* $\underline{w}(\mathscr{H},\mu,\theta)\in W_X\backslash W_G$ *gives a bijection*

$$\left\{\begin{matrix}\text{isomorphism classes of triples}\\ (\mathscr{H},\mu,\theta)\text{ of type }(2q,\mathfrak{f}=q)\end{matrix}\right\}\xrightarrow{\sim}W_X\backslash W_G\cong\prod_{i\in\mathscr{I}}\mathfrak{S}_q\backslash\mathbb{H}_q\,.$$

5.6. *Remark.* — The "only if" statement of (i) is an immediate consequence of what was explained above. Conversely, if $\underline{w}$ is an element in the image of the map (5.4.1) then we shall see in 5.8 below that there exists a pairing of the desired kind on the Dieudonné module of the standard object $(\mathscr{H}_{\underline{w}}, \theta_{\underline{w}})$. The proof of (ii) is more involved, and will take up a large part of the rest of this section.

The proof that we give only works for $\mathrm{char}(k) \neq 2$. In case (C) the theorem is still true for $\mathrm{char}(k) = 2$. Our proof of this is much more involved, which is why we restrict to the case $p > 2$. For the special case that $\tilde{\kappa} = \mathbb{F}_p$ (no non-trivial endomorphisms) see [12], Section 9.

5.7. *Variant.* — In case (D) we have a variant of the theorem, where we replace G by its identity component; this works provided that we keep the invariant δ fixed.

Concretely, fix q, and set $G^0 := \mathrm{SO}_{2q,K} = \prod_{i\in\mathscr{I}} G_i^0$. If $L \subset K^{2q}$ is a maximal isotropic subspace then $\mathrm{Stab}(L) \subset G^0$ is a maximal parabolic subgroup. The set of all parabolics obtained in this way consists of two G^0-conjugacy classes. Let X^0 be one of these two. Now let $(\mathscr{H}, \mu, \theta)$ be a triple as in (GPE'), of type $(2q, \mathfrak{f} = q)$. Similar to what we did in 5.4 above, we can associate to this triple an element $\underline{w}^0(\mathscr{H}, \mu, \theta) \in W_{X^0}\backslash W_{G^0}$. For this, choose an isometry $\xi\colon (N, \Psi) \xrightarrow{\sim} (K^{2q}, \Phi)$ such that the stabilizer of $N[F]$ is in the chosen class X^0. Let $P_i := \mathrm{Stab}\big(N[F]_i\big)$ and $Q_i := \mathrm{Stab}(C_{i,\bullet}) \subset G_i^0$. Then Q_i is in optimal position with respect to P_i and we can define $\underline{w}^0(\mathscr{H}, \mu, \theta)$ to be the element with $w_i^0(\mathscr{H}, \mu, \theta) = w^\sharp(P_i, Q_i)$.

On the other hand, we have associated to $(\mathscr{H}, \mu, \theta)$ the function δ. As explained in 3.8, this invariant allows us to compute $\underline{w}^0(\mathscr{H}, \mu, \theta)$ from $\underline{w}(\mathscr{H}, \mu, \theta)$ and vice versa. Using this, (ii) of the theorem can be rephrased as follows:

(ii') *Assume we are in case* (D). *Fix a function* $\delta\colon \mathscr{I} \to (\mathbb{Z}/2\mathbb{Z})$. *Then sending a triple* $(\mathscr{H}, \mu, \theta)$ *as in* (GPE') *to the element* $\underline{w}^0(\mathscr{H}, \mu, \theta)$ *gives a bijection*

$$\left\{\begin{matrix}\text{isomorphism classes of triples}\\ (\mathscr{H}, \mu, \theta) \text{ of type } (2q, \mathfrak{f} = q, \delta)\end{matrix}\right\} \xrightarrow{\sim} W_{X^0}\backslash W_{G^0}\,.$$

5.8. *Pairings on standard objects.* — We use the notations of 5.4. Let $\underline{w}$ be an element of $W_X\backslash W_G$. In 4.9 we have associated to $\underline{w}$ a standard Dieudonné module $N_{\underline{w}}$ with an action of κ. The underlying K-module is just $K^{2q} = \oplus_{i\in\mathscr{I}} k_i^{2q}$. We claim that there exists a κ-ε-duality $\mu\colon \mathscr{H} \xrightarrow{\sim} \mathscr{H}^D$, corresponding to a form Ψ on the Dieudonné module, such that $(\mathscr{H}_{\underline{w}}, \mu, \theta_{\underline{w}})$ is a triple as in (GPE'). Since the image of $\underline{w}(\mathscr{H}_{\underline{w}}, \mu, \theta_{\underline{w}})$ under the map (5.4.1) is just $\underline{w}(\mathscr{H}_{\underline{w}}, \theta_{\underline{w}})$ we have $\underline{w}(\mathscr{H}_{\underline{w}}, \mu, \theta_{\underline{w}}) = \underline{w}$ for any such μ.

To prove our claim, let us first recall the structure of $N_{\underline{w}}$; the description we give will also be used later. Let $\ell = 2m$ be the common length of the canonical filtrations $C_{i,\bullet}$. By definition, $\mathscr{A} = \mathscr{I} \times \{1, 2, \ldots, 2m\}$. We have the permutation $\rho\colon \mathscr{A} \to \mathscr{A}$ which sends (i,j) to $\big(i, \rho_i(j)\big)$. If $d_a := \dim(B_a)$ then we can compute the data $\big(\rho, \{d_a\}_{a\in\mathscr{A}}\big)$ from $\underline{w}$ and vice versa. The K-module $N = N_{\underline{w}}$ is given by $N = \oplus_{a\in\mathscr{A}} B_a$ with $x \in \kappa$ acting on $B_{(i,j)}$ as multiplication by $i(x) \in k$. For each $a \in \mathscr{A}$ we have an ordered basis β_a of B_a. Frobenius and Verschiebung are given, writing $a = (i,j)$ by

$$F(\beta_a) = \begin{cases} \beta_{\rho(a)} & \text{if } \rho_i(j) \le m; \\ 0 & \text{if } \rho_i(j) > m \end{cases} \quad \text{and} \quad V(\beta_a) = \begin{cases} \beta_{\rho^{-1}(a)} & \text{if } j > m; \\ 0 & \text{if } j \le m. \end{cases}$$

Given $a = (i,j) \in \mathscr{A}$, we write $\tilde{a} := (i, 2m+1-j)$. Notice that the permutation ρ has the property that $\rho(\tilde{a}) = \widetilde{\rho(a)}$. (If we draw an illustration as in 4.12 this means that the whole picture is symmetric under reflection in a central horizontal axis.) We say that an orbit $O \subset \mathscr{A}$ is self-dual if $a \in O \Leftrightarrow \tilde{a} \in O$.

Assume we are in case (C). In order to describe the desired pairing Ψ, we have to introduce certain constants. If O is an orbit which is not self-dual, set $c(a) := 1$ for all $a \in O$. If O is self-dual then it has even length, say $2s$. Consider the unique non-trivial automorphism $\mathrm{Frob}^s\colon c \mapsto c^{p^s}$ of the field $\mathbb{F}_{p^{2s}}$ over $\mathbb{F}_{p^s}$. Write $\mathbb{F}^-_{p^{2s}} \subset \mathbb{F}_{p^{2s}}$ for the subset of elements c with $c^{p^s} = -c$. Then we choose for every $a \in O$ a constant $c(a) \in \mathbb{F}^-_{p^{2s}}$ in such a way that $c(\rho(a)) = c(a)^p$. In particular, $c(\tilde{a}) = -c(a)$.

For $a \in \mathscr{A}$, let $\psi_a\colon B_a \times B_{\tilde{a}} \to k$ be the perfect bilinear pairing given on the ordered bases β_a and $\beta_{\tilde{a}}$ by the matrix

$$\text{anti-diag}(c(a)) := \begin{pmatrix} & & c(a) \\ & \cdot^{\cdot^{\cdot}} & \\ c(a) & & \end{pmatrix}.$$

For $i \in \mathscr{I}$, let $\psi_i\colon N_i \times N_i \to k_i$ be the orthogonal sum of the pairings $\psi_{(i,j)}$; one checks that this is indeed a symplectic pairing. Further, the construction is such that the form $\Psi := \prod_{i \in \mathscr{I}} \psi_i\colon N \times N \to K$ satisfies

$$\Psi(Fn_1, n_2) = F_K\big(\Psi(n_1, Vn_2)\big) \tag{5.8.1}$$

for all n_1, $n_2 \in N$. In sum, Ψ is a pairing with the required properties.

Next assume we are in case (D). In this case, set $c(a) = 1$ for all $a \in \mathscr{A}$. Applying the same procedure as above gives the standard orthogonal form Ψ on $N \cong K^{2q}$. Again one checks that this is a form as required.

In sum, we have shown that every element $\underline{w} \in W_X \backslash W_G$ is obtained as $\underline{w}(\mathscr{H}, \mu, \theta)$ for some triple $(\mathscr{H}, \mu, \theta)$ as in (GPE').

5.9. *Brief overview of the proof.* — It is clear from the definitions that isomorphic triples $(\mathscr{H}, \mu, \theta)$ give the same element $\underline{w}(\mathscr{H}, \mu, \theta)$. On the other hand, we have just shown that every element of $W_X \backslash W_G$ comes from some triple $(\mathscr{H}, \mu, \theta)$. Thus, the task that remains is to prove unicity of the polarization form: if $(\mathscr{H}, \theta)$ is a BT_1 with κ-action and μ_1, $\mu_2\colon \mathscr{H} \to \mathscr{H}^D$ are two κ-ε-dualities then $(\mathscr{H}, \mu_1, \theta) \cong (\mathscr{H}, \mu_2, \theta)$.

Our proof of this combines two results. First we analyze the group U of automorphisms of a pair $(\mathscr{H},\theta)$ which induce the identity on the canonical blocks. In 5.13 we show that every $f \in U$ with $f^\dagger = f$ can be written as $f = g^\dagger g$ for some $g \in U$. (This uses that $\mathrm{char}(k) \neq 2$.) After that we show that, given a triple $(\mathscr{H},\mu,\theta)$ as in (GPE'), we can choose ordered bases β_a for the canonical blocks B_a which are F-V-compatible and which are also "good" with respect to the pairings $\psi_a\colon B_a \times B_{\tilde{a}} \to k$; this is done in 5.14. Combining everything, the proof of the theorem is completed in 5.15.

5.10. *Lemma.* — *Let k be an algebraically closed field of characteristic $\neq 2$. Let G be a connected unipotent k-group (written multiplicatively), equipped with an anti-involution $g \mapsto g^\dagger$. Then every $f \in G$ with $f^\dagger = f$ can be written as $f = g^\dagger \cdot g$ for some $g \in G$.*

Proof. There is an exact sequence $1 \longrightarrow G_1 \longrightarrow G \xrightarrow{\pi} V \longrightarrow 0$, with V a non-zero vector group (written additively) and $G_1 \subset G$ connected and stable under $\dagger$. Choose $x \in G$ with $\pi(x) = -\frac{1}{2}\pi(g)$. Then $g_1 := x^\dagger g x$ is an element of G_1 with $g_1^\dagger = g_1$. The lemma follows by induction on $\dim(G)$. □

5.11. Let $(\mathscr{H},\theta)$ be a BT_1 with an action of κ. By our results in the previous section, the Dieudonné module N is isomorphic to the standard object associated to $\underline{w} := \underline{w}(\mathscr{H},\theta)$. We shall work with the description of N given in 5.8. As usual we view N (via the given θ) as a module over K. We identify $\mathrm{End}_k(\mathscr{H},\theta)$ with the subring of $\mathrm{End}_K(N) = \prod_{i\in\mathscr{I}} \mathrm{End}_{\tilde{k}_i}(N_i)$ consisting of all endomorphisms which commute with F and V.

Every $f \in \mathrm{End}_k(\mathscr{H},\theta)$ stabilizes the canonical filtrations $C_{i,\bullet}$. For $a \in \mathscr{A}$ we can therefore define $\mathrm{gr}_a(f)\colon B_a \to B_a$ as the homomorphism induced by f. Now consider the subgroup $U \subset \mathrm{Aut}_k(\mathscr{H},\theta)$ given by

$$U = \left\{f \in \mathrm{Aut}_k(\mathscr{H},\theta) \mid \mathrm{gr}_a(f) = \mathrm{id}_{B_a} \quad \text{for all } a \in \mathscr{A}\right\}.$$

Let $\tilde{U}$ be the unipotent radical of the stabilizer of $\mathscr{C}_\bullet$ inside the group $\mathrm{GL}_K(N)$; this $\tilde{U}$ is a connected unipotent k-group. By construction, U is a subgroup of $\tilde{U}$.

Consider an element $f \in U$. Given a, $a' \in \mathscr{A}$, let $P_{a,a'}$ be the matrix of the composition

$$B_a \hookrightarrow N \xrightarrow{f-\mathrm{id}} N \twoheadrightarrow B_{a'}$$

with respect to the bases β_a and $\beta_{a'}$. This is a matrix of size $d_{a'} \times d_a$. Clearly we can have $P_{a,a'} \neq 0$ only if $a' \prec a$ (notation as in 4.15).

It is not so difficult to analyze the relations between the matrices $P_{a,a'}$. To state the result, recall that each $t_a\colon B_a \to B_{\rho(a)}$ is called either an F-arrow or a V-arrow. Further let us write $\sigma := \mathrm{Frob}_k\colon k \to k$. Now suppose that for each pair $(a,a') \in \mathscr{A} \times \mathscr{A}$ we have a matrix $P_{a,a'}$ of size $d_{a'} \times d_a$, with $P_{a,a'} \neq 0$ only if $a' \prec a$. Then a necessary and sufficient condition for this collection of matrices to come from an element $f \in U$ is that for all (a,a') we have

$$\begin{cases} P_{\rho(a),\rho(a')} = P_{a,a'}^{\sigma} & \text{if } t_a \text{ and } t_{a'} \text{ are of the same kind;} \\ P_{a,a'} = 0 & \text{if } t_a \text{ is a } V\text{-arrow, } t_{a'} \text{ is an } F\text{-arrow.} \end{cases} \tag{5.11.1}$$

(As we have already assumed that $P_{a,a'} = 0$ whenever $a' \not\prec a$, the only case where we get a new relation $P_{a,a'} = 0$ is when $a' \prec a$ and $\rho(a') \prec \rho(a)$, with t_a a V-arrow, $t_{a'}$ an F-arrow. In the terminology of 4.15 this is the up-down case.)

At this point we use what was found in 4.15. Write $\mathscr{V}$ for the set of pairs $(a, a') \in \mathscr{A} \times \mathscr{A}$ such that $a' \prec a$. Given such a pair (a, a'), consider the corresponding orbits O and O'. We mark these orbits such that $a(0) = a$ and $a'(0) = a'$. Let $[n_1, n_2]$ (with $n_1 \leq 0 \leq n_2$) be the largest interval around 0 in which the marked orbits O and O' are parallel. We say that $(b, b') \in \mathscr{V}$ is equivalent to (a, a') if there exists an index $n \in [n_1, n_2]$ with $(b, b') = \big(a(n), a'(n)\big)$. This defines an equivalence relation on $\mathscr{V}$. We say that the class of (a, a') is a *nil-class* if the pair of orbits (O, O') is up-down at index n_2. (See the illustration in 4.15.) Finally, we call the pair $\big(a(n_1), a'(n_1)\big)$ the distinguished representative for the equivalence class of (a, a').

With this terminology, (5.11.1) can be rephrased as follows. If the class of (a, a') is a nil-class then we must have $P_{a,a'} = 0$. If the class is not a nil-class, with distinguished representative $\big(a(n_1), a'(n_1)\big)$, then $P = P_{a(n_1),a'(n_1)}$ can be chosen arbitrarily and

$$P_{a,a'} = P^{\sigma^{-n_1}} .$$

In particular, this implies the following result.

5.12. *Lemma.* — *The group U is a connected unipotent subgroup of $\tilde{U}$.*

5.13. *Corollary.* — *Assume that* $\mathrm{char}(k) \neq 2$. *Let $(\mathscr{H}, \mu, \theta)$ be a triple as in* (GPE'). *Write $f \mapsto f^\dagger$ for the Rosati involution on* $\mathrm{End}_k(\mathscr{H}, \theta)$ *induced by μ, and notice that U is stable under $\dagger$. Then every $f \in U$ with $f^\dagger = f$ can be written as $f = g^\dagger \cdot g$ for some $g \in U$.*

Now we come to the final main ingredient of the proof of Theorem 5.5.

5.14. *Lemma.* — *Let $(\mathscr{H}, \mu, \theta)$ be a triple as in* (GPE'). *Choose constants $c(a)$ $(a \in \mathscr{A})$ as in* 5.8. *Then for every $a \in \mathscr{A}$ we can choose an ordered basis β_a for B_a such that:* (a) *$\beta_{\rho(a)}$ is the image of β_a under $t_a\colon B_a \to B_{\rho(a)}$ and* (b) *the form $\psi_a\colon B_a \times B_{\tilde{a}} \to k$ is given on the bases β_a and $\beta_{\tilde{a}}$ by the matrix* $\text{anti-diag}\big(c(a)\big)$.

Proof. Recall that for $a = (i, j) \in \mathscr{A}$ we write $\tilde{a} := (i, 2m + 1 - j)$. Given an orbit $O \subset \mathscr{A}$, write $\tilde{O} := \{\tilde{a} \mid a \in O\}$, which is also an orbit. It suffices to find the bases β_a for a running through $O \cup \tilde{O}$. If $O \neq \tilde{O}$ this is easy: for $a \in O$ choose ordered bases β_a satisfying condition (a). (See Lemma 4.11.) For $\tilde{a} \in \tilde{O}$ let $\beta_{\tilde{a}}$ then be the basis defined by condition (b). These again satisfy (a), and we are done.

Next suppose that $O = \tilde{O}$. Then O has even length, say $2s$. Fix a marking $n \mapsto a(n)$ of O. For $r \geq 1$ consider the Frob_k^r-linear isomorphisms

$$T^{(r)} := \big(B_{a(0)} \xrightarrow{t_{a(0)}} B_{a(1)} \xrightarrow{t_{a(1)}} \cdots \xrightarrow{t_{a(r-1)}} B_{a(r)}\big)$$

and

$$U^{(r)} := \left(B_{a(s)} \xrightarrow{t_{a(s)}} B_{a(s+1)} \xrightarrow{t_{a(s+1)}} \cdots \xrightarrow{t_{a(s+r-1)}} B_{a(s+r)}\right).$$

Using the relation (5.8.1) we see that the diagram

$$\begin{array}{ccc} B_{a(0)} \times B_{a(s)} & \xrightarrow{\psi_{a(0)}} & k \\ {\scriptstyle T^{(r)}\times U^{(r)}}\downarrow & & \downarrow{\scriptstyle \mathrm{Frob}_k^r} \\ B_{a(r)} \times B_{a(s+r)} & \xrightarrow[\psi_{a(r)}]{} & k \end{array} \tag{5.14.1}$$

is commutative.

Consider the $\mathbb{F}_{p^{2s}}$-vector spaces

$$\mathscr{B}_{a(0)} := \{b \in B_{a(0)} \mid T^{(2s)}(b) = b\} \quad \text{and} \quad \mathscr{B}_{a(s)} := \{b \in B_{a(s)} \mid U^{(2s)}(b) = b\}.$$

We have $\mathscr{B}_{a(0)} \otimes_{\mathbb{F}_{p^{2s}}} k \xrightarrow{\sim} B_{a(0)}$ and $\mathscr{B}_{a(s)} \otimes_{\mathbb{F}_{p^{2s}}} k \xrightarrow{\sim} B_{a(s)}$. The maps $T^{(s)}$ and $U^{(s)}$ restrict to Frob^s-linear isomorphisms

$$\mathscr{T}^{(s)}\colon \mathscr{B}_{a(0)} \xrightarrow{\sim} \mathscr{B}_{a(s)} \qquad \text{and} \qquad \mathscr{U}^{(s)}\colon \mathscr{B}_{a(s)} \xrightarrow{\sim} \mathscr{B}_{a(0)},$$

inverse to each other. Taking $r = 2s$ in diagram (5.14.1) we find that $\psi_{a(0)}$ restricts to a perfect pairing $\mathscr{B}_{a(0)} \times \mathscr{B}_{a(s)} \longrightarrow \mathbb{F}_{p^{2s}}$. Next taking $r = s$ in (5.14.1) we find that

$$H\colon \mathscr{B}_{a(0)} \times \mathscr{B}_{a(0)} \longrightarrow \mathbb{F}_{p^{2s}} \qquad \text{defined by} \quad H(b, b') = \psi_{a(0)}\big(b, \mathscr{T}^{(s)}(b')\big)$$

is an ε-symmetric hermitian form with respect to the involution Frob^s of $\mathbb{F}_{p^{2s}}$ over $\mathbb{F}_{p^s}$. Notice that the discriminant of such a form is in $\mathbb{F}_{p^s}^\times/\mathrm{Norm}\big(\mathbb{F}_{p^{2s}}^\times\big)$, which is the trivial group.

Essentially, all we have to do now is to bring H in standard form. Note that the forms H are classified by a cohomology group $H^1\big(\mathbb{F}_{p^s}, \mathrm{SU}(\mathscr{B}_{a(0)}, H)\big)$, and by Lang's theorem this group is trivial. (See Knus et al., [7], Section 29.19, and Serre, [14], Chap. III, §2.) Hence we can choose a basis $\beta_{a(0)}$ on which the form H is given by the matrix anti-diag$\big(c(a(0))\big)$. Now define $\beta_{a(n)}$ to be the image of $\beta_{a(0)}$ under $T^{(n)}$. This gives bases β_a as required. □

5.15. *End of the proof.* — Combining all previous results, let us complete the proof of Theorem 5.5. We take a triple $(\mathscr{H}, \mu, \theta)$ as in (GPE'). Write (N, F, V, Ψ, θ) for the corresponding Dieudonné module with additional structure. Let $f \mapsto f^\dagger$ be the Rosati involution on $\mathrm{End}_k(\mathscr{H}, \theta)$.

Write $\underline{w} := \underline{w}(\mathscr{H}, \mu, \theta)$. Choose constants $c(a)$, for $a \in \mathscr{A}$ as in 5.8. Write $N_{\underline{w}} = \oplus_{a\in\mathscr{A}} B_a$ for the standard Dieudonné module corresponding to $\underline{w}$; we use the description given in 5.8, but in order to avoid confusion with the module N we now do write the subscript "$\underline{w}$". Let $\Psi_{\underline{w}}$ be the pairing on $N_{\underline{w}}$ constructed in loc. cit.

Choose bases β_a as in Lemma 5.14. This gives us an isomorphism of Dieudonné modules with κ-action $h\colon N \xrightarrow{\sim} N_{\underline{w}}$. Further, taking h as an identification, the two pairings Ψ and $\Psi_{\underline{w}}$ induce the same pairings $\psi_a\colon B_a \times B_{\bar{a}} \to k$ on the canonical blocks. It follows that there is an element $f \in U$ with $f^\dagger = f$ such that $\Psi_{\underline{w}}(n_1, n_2) = \Psi(fn_1, n_2)$ for all $n_1, n_2 \in N$. By Corollary 5.13 this f can be written as $f = g^\dagger \cdot g$ for some $g \in U$. Then $h \circ g^{-1}\colon N \xrightarrow{\sim} N_{\underline{w}}$ is an isomorphism which is compatible both with the κ-actions and the pairings. The existence of such an isomorphism is what we had to prove. □

6. BT$_1$'s with endomorphisms and a polarization—case A

In this section we deal with problem (GPE) stated in 5.1 in case the algebra with involution $(D, *)$ is simple of type (A). There are some new features, but once we have set up notations most steps in the proof of Theorem 6.7 are the same as in the previous section. Therefore we shall explain the new features and state the main result, but only outline the proof.

6.1. *The set-up.* — We consider a semi-simple algebra D with involution $*$ of the second kind; as explained in 5.1 we may assume that $(D, *)$ is simple as an algebra with involution. Write $\tilde{\kappa}$ for the center of D and $\kappa := \{a \in \tilde{\kappa} \mid a^* = a\}$. Then κ is a finite field and either $\tilde{\kappa} = \kappa \times \kappa$, which we call case (A1), or $\tilde{\kappa}$ is a quadratic field extension of κ, called case (A2).

Set $\mathscr{I} := \mathrm{Hom}(\kappa, k)$ and $\tilde{\mathscr{I}} := \mathrm{Hom}(\tilde{\kappa}, k)$. We shall typically use the letter i for an element of $\mathscr{I}$ and τ for an element of $\tilde{\mathscr{I}}$. For $i \in \mathscr{I}$, set $i + 1 := \mathrm{Frob}_k \circ i$. For $\tau \in \tilde{\mathscr{I}}$, set $\tau + 1 = \mathrm{Frob}_k \circ \tau$ and $\bar{\tau} = \tau \circ *$. In case (A2), taking $\tau + 1$ as the successor of τ gives $\tilde{\mathscr{I}}$ a cyclic ordering. We have $\bar{\tau} = \tilde{\tau}$, using the notation of 4.1. Similarly, if in case (A1) we impose the condition that $\bar{\tau} = \tilde{\tau}$ then we have a natural bi-cyclic ordering on $\tilde{\mathscr{I}}$. In both cases the set $\mathscr{I}$ has a cyclic ordering. We have a natural $2:1$ map $\tilde{\mathscr{I}} \to \mathscr{I}$; if this sends $\tau \in \tilde{\mathscr{I}}$ to $i \in \mathscr{I}$ then we refer to τ and $\bar{\tau}$ as the elements of $\tilde{\mathscr{I}}$ lying over i.

For $i \in \mathscr{I}$ and $\tau \in \tilde{\mathscr{I}}$ we write

$$k_\tau := \tilde{\kappa} \underset{\tilde{\kappa},\tau}{\otimes} k\,, \quad k_i := \kappa \underset{\kappa,i}{\otimes} k\,, \quad \text{and} \quad \tilde{k}_i := \tilde{\kappa} \underset{\kappa,i}{\otimes} k\,.$$

Let

$$K := \kappa \otimes_{\mathbb{F}_p} k = \prod_{i \in \mathscr{I}} k_i \qquad \text{and} \qquad \tilde{K} := \tilde{\kappa} \otimes_{\mathbb{F}_p} k = \prod_{i \in \mathscr{I}} \tilde{k}_i = \prod_{\tau \in \tilde{\mathscr{I}}} k_\tau\,.$$

The maps $\mathrm{id} \otimes \mathrm{Frob}_k$ give Frob_k-linear isomorphisms

$$F_{k_\tau}\colon k_\tau \to k_{\tau+1}\,, \quad F_{k_i}\colon k_i \to k_{i+1}\,, \quad F_{\tilde{k}_i}\colon \tilde{k}_i \to \tilde{k}_{i+1}\,,$$

and

$$F_K\colon K \to K\,, \quad F_{\tilde{K}}\colon \tilde{K} \to \tilde{K}\,.$$

6.2. *Morita equivalence.* — Similar to what was done in 5.2, we can reduce problem (GPE) to a problem with $D = \tilde{\kappa}$. This time, the category $\mathbf{C}_1$ is equivalent to the category $\mathbf{C}_2$ of 5-tuples $(N, F_N, V_N, \Psi, \theta)$, where

— (N, F_N, V_N) is the Dieudonné module of a BT_1;

— $\theta\colon \tilde{\kappa} \to \mathrm{End}_{\mathrm{DM}}(N)$ defines an action of $\tilde{\kappa}$ on N, making it a free module over $\tilde{K}$;

— $\Psi\colon N \times N \to \tilde{K}$ is an alternating hermitian pairing with respect to the non-trivial automorphism of $\tilde{K}$ over K.

Note that we only consider modules N which are free over $\tilde{K}$. To explain why, recall from 4.3 that our modules N are free over K. As we shall see below, the existence of a hermitian pairing with values in $\tilde{K}$ implies that they are even free over $\tilde{K}$.

Translating back to group schemes the problem becomes the following.

(GPE') Classification of triples $(\mathscr{H}, \mu, \theta)$, where $(\mathscr{H}, \theta)$ is a BT_1 with an action of $\tilde{\kappa}$, and where $\mu\colon \mathscr{H} \xrightarrow{\sim} \mathscr{H}^D$ is a $\tilde{\kappa}$-(-1)-duality.

Here we use the ad hoc term "$\tilde{\kappa}$-(-1)-duality" for a duality μ corresponding to an alternating hermitian form Ψ on the Dieudonné module.

6.3. Let $(N, F_N, V_N, \Psi, \theta)$ be an object of the category $\mathbf{C}_2$. As in 4.3, the action of κ gives rise to a decomposition $N = \oplus_{i \in \mathscr{I}} N_i$. Using the full action of $\tilde{\kappa}$ this is refined to a decomposition $N = \oplus_{\tau \in \tilde{\mathscr{I}}} N_\tau$. If τ and $\bar{\tau}$ are the two elements of $\tilde{\mathscr{I}}$ lying over $i \in \mathscr{I}$ then $N_i = N_\tau \oplus N_{\bar{\tau}}$.

Frobenius and Verschiebung become homomorphisms $F_{N_\tau}\colon N_\tau \to N_{\tau+1}$ and $V_{N_\tau}\colon N_\tau \leftarrow N_{\tau+1}$, equivariant with respect to $F_{k_\tau}\colon k_\tau \to k_{\tau+1}$, respectively its inverse.

By a similar procedure as in 4.4, we obtain canonical filtrations

$$C_{\tau,\bullet}\colon \qquad (0) = C_{\tau,0} \subsetneq C_{\tau,1} \subsetneq \cdots \subsetneq C_{\tau,\ell_\tau} = N_\tau$$

By construction, for each $\tau \in \tilde{\mathscr{I}}$ there is an index $h_\tau \in \{0, 1, \ldots, \ell_\tau\}$ such that $C_{\tau,h_\tau} = \mathrm{Im}(F_{N_{\tau-1}})$. The analogue of Lemma 4.5 in this setting is as follows.

6.4. *Lemma.* — (0) *The integer $\ell := \ell_\tau$ is independent of $\tau \in \tilde{\mathscr{I}}$.*

(i) *For each $\tau \in \tilde{\mathscr{I}}$ and $j \in \{0, 1, \ldots, \ell\}$ there exist indices $r_\tau(j) \in \{0, 1, \ldots, h_{\tau+1}\}$ and $s_\tau(j) \in \{h_{\tau+1}, \ldots, \ell\}$ such that*

$$F_{N_\tau}(C_{\tau,j}) = C_{\tau+1,r_\tau(j)} \qquad \textit{and} \qquad V_{N_\tau}^{-1}(C_{\tau,j}) = C_{\tau+1,s_\tau(j)}\,.$$

(ii) *For $j \geq 1$, either*

$$C_{\tau,j} \cap \mathrm{Ker}(F_{N_\tau}) = C_{\tau,j-1} \cap \mathrm{Ker}(F_{N_\tau})\,,$$

or

$$C_{\tau,j} \subset \mathrm{Span}\big(C_{\tau,j-1}, \mathrm{Ker}(F_{N_\tau})\big)\,.$$

In the first case we have $r_\tau(j) = r_\tau(j-1)+1$ *and* $s_\tau(j) = s_\tau(j-1)$*; in the second case* $r_\tau(j) = r_\tau(j-1)$ *and* $s_\tau(j) = s_\tau(j-1)+1$*. Further,*

$$C_{\tau,j} \cap \mathrm{Ker}(F_{N_\tau}) = C_{\tau,j-1} \cap \mathrm{Ker}(F_{N_\tau}) \iff C_{\bar\tau,\ell+1-j} \subset \mathrm{Span}\big(C_{\bar\tau,\ell-j}, \mathrm{Ker}(F_{N_{\bar\tau}})\big)\,.$$

(iii) *For all* τ *and* j *we have* $r_\tau(j)+s_\tau(j) = \ell+j$ *and* $r_\tau(j)+s_{\bar\tau}(\ell-j) = \ell$.

6.5. We define the canonical blocks by

$$B_{\tau,j} := C_{\tau,j}/C_{\tau,j-1} \qquad (\text{for } \tau \in \widetilde{\mathscr{I}} \text{ and } j \in \{1,\ldots,\ell_\tau\}).$$

Further we have permutations $\rho_\tau \in \mathfrak{S}_\ell$, given by

$$\rho_\tau(j) = \begin{cases} r_\tau(j) & \text{if } C_{\tau,j} \cap \mathrm{Ker}(F_{N_\tau}) = C_{\tau,j-1} \cap \mathrm{Ker}(F_{N_\tau}); \\ s_\tau(j) & \text{if } C_{\tau,j} \subset \mathrm{Span}\big(C_{\tau,j-1}, \mathrm{Ker}(F_{N_\tau})\big). \end{cases}$$

As in 4.10 we obtain, for $\tau \in \widetilde{\mathscr{I}}$ and $j \in \{1,2,\ldots,\ell\}$, a Frob_k-linear isomorphism

$$t_{\tau,j}\colon B_{\tau,j} \xrightarrow{\ \sim\ } B_{\tau+1,\rho_\tau(j)}\,,$$

induced either by Frobenius or (inversely) by Verschiebung.

We have an involution $(a,b) \mapsto (b,a)$ of $\tilde{k}_i \cong k_\tau \times k_{\bar\tau} = k \times k$ over $k_i = k$. We shall write this automorphism as $x \mapsto \bar{x}$. Note that if we restrict this automorphism to one of the two factors we get an isomorphism $k_\tau \to k_{\bar\tau}$ which, via the identifications $k_\tau = k = k_{\bar\tau}$, is just the identity on the field k; in other words: if $a \in k_\tau$ then $\bar{a}$ is the same element of k, but now viewed of an element of $k_{\bar\tau}$. The map $k_\tau \to k_{\bar\tau}$ given by $a \mapsto \bar{a}$ should be thought of as "complex conjugation"; in particular, if C_τ is a k_τ-module and $C_{\bar\tau}$ is a $k_{\bar\tau}$-module then it makes sense to say that a pairing $\psi\colon C_\tau \times C_{\bar\tau} \to k_\tau$ is anti-linear in the second variable.

With this terminology, we have perfect pairings, linear in the first variable, anti-linear in the second variable,

$$\psi_\tau\colon N_\tau \times N_{\bar\tau} \to k_\tau \qquad \text{with} \quad \psi_{\bar\tau}(y,x) = -\overline{\psi_\tau(x,y)}\,.$$

In particular, $\dim(N_\tau) = \dim(N_{\bar\tau})$, and it follows that the $\tilde{K}$-module N is free, as claimed in 6.2.

The canonical filtrations $C_{\tau,\bullet}$ and $C_{\bar\tau,\bullet}$ are dual with respect to the pairings ψ_τ and $\psi_{\bar\tau}$. In particular, on the canonical blocks we get induced perfect pairings

$$\bar\psi_{\tau,j}\colon B_{\tau,j} \times B_{\bar\tau,\ell+1-j} \longrightarrow k_\tau\,.$$

6.6. *Definition of the element* $\underline{w}(\mathscr{H},\mu,\theta)$. — Let $(\mathscr{H},\mu,\theta)$ be a triple as in (GPE'). Write $(N, F_N, V_N, \Psi, \theta)$ for the corresponding Dieudonné module. Let q be the $\tilde{K}$-rank of N. For $\tau \in \widetilde{\mathscr{I}}$, let $\mathfrak{f}(\tau)$ be the k-dimension of $N[F]_\tau :=$

$\mathrm{Ker}(F_{N_\tau})$. This defines a function $\mathfrak{f}\colon \tilde{\mathscr{I}} \to \{0,1,\ldots,q\}$ with the property that $\mathfrak{f}(\tau)+\mathfrak{f}(\bar\tau)=q$ for all τ.

Let $\Phi\colon \tilde{K}^q \times \tilde{K}^q \to \tilde{K}$ be an alternating hermitian form, $\tilde{K}$-linear in the first variable, $\tilde{K}$-antilinear in the second. Write $G = \mathrm{U}_{\tilde{K}}(\tilde{K}^q, \Phi)$ for the group of $\tilde{K}$-linear unitary automorphisms. The pairing Φ is the product of pairings

$$\varphi_i = (\varphi_\tau, \varphi_{\bar\tau})\colon \tilde{k}_i^q \times \tilde{k}_i^q = (k_\tau^q \oplus k_{\bar\tau}^q) \times (k_\tau^q \oplus k_{\bar\tau}^q) \longrightarrow \tilde{k}_i = k_\tau \times k_{\bar\tau}$$

(where φ_τ pairs k_τ^q with $k_{\bar\tau}^q$), and writing $G_i = \mathrm{Aut}_{\tilde{k}_i}(k_i^{2q}, \varphi_i)$ we have $G = \prod_{i\in\mathscr{I}} G_i$. If τ and $\bar\tau$ are the two elements of $\tilde{\mathscr{I}}$ lying over $i \in \mathscr{I}$ then we have isomorphisms

$$\mathrm{GL}_{q,k_\tau} \xrightarrow{\ \sim\ } G_i \xrightarrow{\ \sim\ } \mathrm{GL}_{q,k_{\bar\tau}}\,.$$

Choose an isometry $\xi\colon (N,\Psi) \xrightarrow{\ \sim\ } (\tilde{K}^q, \Phi)$ of hermitian $\tilde{K}$-modules; this allows us to view the $N[F]_\tau$ and $C_{\tau,\bullet}$ as partial flags in k_τ^q. Viewing $k_{\bar\tau}^q$ as the dual of k_τ^q via the perfect pairing ψ_τ we have $N[F]_{\bar\tau} = N[F]_\tau^\perp$ and $C_{\bar\tau,\bullet} = C_{\tau,\bullet}^\perp$.

Write $P := \mathrm{Stab}_G(N[F])$ and $Q := \mathrm{Stab}_G(\mathscr{C}_\bullet)$. Let $W = W_G$, and write X for the type of P. We have a natural decomposition $W_X\backslash W = \prod_{i\in\mathscr{I}} W_{X_i}\backslash W_i$. The inclusion map $G \hookrightarrow \mathrm{GL}_{q,\tilde{K}}$ gives rise to an injective map

$$W_X\backslash W \hookrightarrow W_{X_\mathfrak{f}}\backslash W_{\mathrm{GL}_{q,\tilde{K}}}\,, \tag{6.6.1}$$

writing $X_\mathfrak{f}$ for the type of the stabilizer of $N[F]$ in $\mathrm{GL}_{q,\tilde{K}}$ (which is equivalent to knowing the function $\mathfrak{f}$).

Working as in 4.6 and 5.4, using (ii) of Lemma 6.4 and what was explained in Examples 3.5 and 3.10, we obtain a well-defined element

$$\underline{w}(\mathscr{H}, \mu, \theta) := w^\sharp(P, Q) \in W_X\backslash W\,,$$

independent of the choice of ξ.

6.7. *Theorem.* — *Notations as above. Assume that* $\mathrm{char}(k) \neq 2$.

(i) *Let* $(\mathscr{H}, \theta)$ *be a* BT_1 *with an action of* $\tilde\kappa$, *of type* $(q, \mathfrak{f})$. *Let* $\underline{w}(\mathscr{H}, \theta)$ *be the associated element of* $W_{X_\mathfrak{f}}\backslash W_{\mathrm{GL}_{q,\tilde{K}}}$. *Then there exists a* $\tilde\kappa$-(-1)-*duality* $\mu\colon \mathscr{H} \to \mathscr{H}^\vee$ *if and only if* $\underline{w}(\mathscr{H}, \theta)$ *is in the image of the map* (6.6.1).

(ii) *If a duality* μ *as in* (i) *exists then it is unique up to isomorphism. In other words: associating to a triple* $(\mathscr{H}, \mu, \theta)$ *as in* (GPE') *the element* $\underline{w}(\mathscr{H}, \mu, \theta) \in W_X\backslash W_G$ *gives a bijection*

$$\left\{\begin{matrix}\text{isomorphism classes of triples}\\ (\mathscr{H}, \mu, \theta) \text{ of type } (q, \mathfrak{f})\end{matrix}\right\} \xrightarrow{\ \sim\ } W_X\backslash W_G\,.$$

As explained in 3.10, if W_τ is the Weyl group of GL_{q,k_τ} and X_τ is the type of the stabilizer of $N[F]_\tau$ then we have natural bijections

$$W_{X_\tau}\backslash W_\tau \xrightarrow{\ \sim\ } W_{X_{\bar\tau}}\backslash W_{\bar\tau} \quad \text{denoted } \pi \mapsto \check\pi.$$

Then the set $W_X\backslash W_G$ can be described as

$$W_X\backslash W_G = \left\{(\pi_\tau)_{\tau\in\widetilde{\mathscr{I}}} \mid \pi_\tau \in W_{X_\tau}\backslash W_\tau \text{ and } \pi_{\bar\tau} = \check\pi_\tau \text{ for all } \tau.\right\}$$

6.8. *Remark.* — Suppose we are in case (A1), meaning that $\tilde\kappa = \kappa \times \kappa$ with involution $(x_1, x_2) \mapsto (x_2, x_1)$. Let us number the two factors κ as $\kappa^{(1)}$ and $\kappa^{(2)}$. The set $\widetilde{\mathscr{I}}$ naturally decomposes, as a set with bi-cyclic ordering, as $\widetilde{\mathscr{I}} = \mathscr{I}^{(1)} \amalg \mathscr{I}^{(2)}$. Write $\mathfrak{f}^{(\nu)}\colon \mathscr{I} = \mathscr{I}^{(\nu)} \longrightarrow \{1, \ldots, q\}$ for the restriction of $\mathfrak{f}$ to $\mathscr{I}^{(\nu)}$.

For $\nu = 1$, 2, let $(\mathscr{H}^{(\nu)}, \theta^{(\nu)})$ be a BT_1 with κ-action, of type $(q, \mathfrak{f}^{(\nu)})$, with $(\mathscr{H}^{(\nu)}, \theta^{(\nu)})$ dual to $(\mathscr{H}^{(\nu+1)}, \theta^{(\nu+1)})$. Then take $(\mathscr{H}, \theta) := (\mathscr{H}^{(1)}, \theta^{(1)}) \times (\mathscr{H}^{(2)}, \theta^{(2)})$, which comes with a natural action of $\tilde\kappa$. Take

$$\mu = \begin{pmatrix} 0 & \zeta \\ -\zeta^D & 0 \end{pmatrix}, \text{ where } \zeta\colon (\mathscr{H}^{(2)}, \theta^{(2)}) \xrightarrow{\ \sim\ } (\mathscr{H}^{(1)}, \theta^{(1)})^D .$$

This gives us a triple $(\mathscr{H}, \mu, \theta)$ as in (GPE'), and every such triple is of this form. One verifies that Theorems 4.7 and 6.7 are consistent. (Compare also with Remark 4.8.)

6.9. *Outline of the proof.* — The proof of 6.7 is essentially the same as that of Theorem 5.5. By the previous remark we may assume that $\tilde\kappa$ is a field.

The first step is to show that if $\underline{w}$ is an element in the image of the map (6.6.1), there exists a $\tilde\kappa$-(-1)-duality on the standard object $(\mathscr{H}_{\underline{w}}, \theta_{\underline{w}})$. For the set $\mathscr{A}$ we now take $\mathscr{A} := \widetilde{\mathscr{I}} \times \{1, \ldots, \ell\}$, with permutation ρ given by $\rho(\tau, j) = \bigl(\tau + 1, \rho_\tau(j)\bigr)$. For $a = (\tau, j) \in \mathscr{A}$ we set $\tilde a := (\bar\tau, \ell + 1 - j)$. Then we proceed exactly as in 5.8, noting that the desired pairings are now hermitian pairings $\psi_a\colon B_a \times B_{\tilde a} \to k$.

After that we need to prove that the $\tilde\kappa$-(-1)-duality is unique (up to isomorphism). The arguments in 5.11 do not use any polarization, so that Corollary 5.13 is still valid. Thus only Lemma 5.14 remains. Our notations have been set up in such a way that we can copy the proof in 5.14 verbatim.

6.10. *An example: the CM-case.* — To illustrate the use of the "multiplication type" $\mathfrak{f}$, and to give a brief preview of what will be done in [9], let us show how to compute the p-kernel of a reduction of an abelian variety of CM-type. This is very similar to the recipe for computing the Newton polygon of such a reduction, which we shall recall.

Let F be a CM-field of degree $2g$ over $\mathbb{Q}$, i.e., a totally imaginary quadratic extension of a totally real field F_0. Given an embedding $\sigma\colon F \to \mathbb{C}$, write $\bar\sigma$ for the conjugate embedding. By a *CM-type* for F we mean a subset $\mathfrak{f} \subset \mathrm{Hom}(F, \mathbb{C})$ such that precisely one of each pair $(\sigma, \bar\sigma)$ is in $\mathfrak{f}$.

Let Y be an abelian variety of dimension g over a number field $E \subset \mathbb{C}$. Suppose Y is equipped with an action by an order $\mathscr{O} \subset F$; this gives us a homomorphism $F \to \mathrm{End}(Y) \otimes \mathbb{Q}$. The first homology $H = H_1(Y_{\mathbb{C}}, \mathbb{C})$ is free of rank 1 as a module over

$$F \otimes_{\mathbb{Q}} \mathbb{C} = \prod_{\sigma \in \mathrm{Hom}(F,\mathbb{C})} \mathbb{C}^{(\sigma)} .$$

Hodge theory gives us a canonical $(F \otimes_{\mathbb{Q}} \mathbb{C})$-submodule $F^0H \subset H$, and we define $\mathfrak{f} \subset \mathrm{Hom}(F, \mathbb{C})$ to be the set of weights of this submodule. Then $\mathfrak{f}$ is a CM-type and we say that Y is of CM-type $(F, \mathfrak{f})$.

Choose embeddings $\overline{\mathbb{Q}} \hookrightarrow \overline{\mathbb{Q}}_p \hookrightarrow \mathbb{C}$. Let $\mathfrak{p}$ be the corresponding place of E above p. Possibly after replacing E by a finite extension, Y has good reduction Y_0 over $k(\mathfrak{p})$. Via the chosen field embeddings we can identify

$$\mathrm{Hom}(F, \mathbb{C}) = \mathrm{Hom}(F, \overline{\mathbb{Q}}_p) = \mathrm{Hom}(F \otimes_{\mathbb{Q}} \mathbb{Q}_p, \overline{\mathbb{Q}}_p) = \amalg_\pi \mathrm{Hom}(F_\pi, \overline{\mathbb{Q}}_p),$$

where π runs through the set of places of F above p and we write F_π for the π-adic completion of F. The set $\mathfrak{f}$ is then a disjoint union of sets $\mathfrak{f}_\pi \subset \mathrm{Hom}(F_\pi, \overline{\mathbb{Q}}_p)$.

In Tate's paper [15] it is explained how to compute the Newton polygon of Y_0. Namely, the slopes that occur are just the numbers $\#\mathfrak{f}_\pi/[F_\pi : \mathbb{Q}_p]$.

Write k for the residue field of the maximal unramified subfield $\mathbb{Q}_p^{\mathrm{nr}} \subset \overline{\mathbb{Q}}_p$; this k is an algebraically closed field containing $k(\mathfrak{p})$. Assume that the order $\mathcal{O}$ is maximal at p and that p does not ramify in F. (This is what the usual conditions for being a prime of good reduction mean in the present case.) We can also compute the isomorphism type of the p-kernel $Y_0[p]$ over k. Note that $\mathcal{O}/p\mathcal{O}$ is the product of the residue fields $k(\pi)$, and that the Dieudonné module N of $Y_0[p]$ is free of rank 1 over $\mathcal{O}/p\mathcal{O}$. In particular, the structure of $Y_0[p]_k$ with its action of $\mathcal{O}/p\mathcal{O}$ is completely determined by the multiplication type $\mathfrak{f}_0 \subset \mathrm{Hom}(\mathcal{O}/p\mathcal{O}, k)$. (According to the definition in 4.6, the multiplication type is a function $\mathfrak{f}_0\colon \mathrm{Hom}(\mathcal{O}/p\mathcal{O}, k) \to \{0,1\}$. Such a function is completely determined by the subset $\mathfrak{f}_0^{-1}(1) \subset \mathrm{Hom}(\mathcal{O}/p\mathcal{O}, k)$. In this example we work with the subset rather than the function.) The assumption that p is unramified in F gives us a natural identification

$$\mathrm{Hom}(F \otimes \mathbb{Q}_p, \overline{\mathbb{Q}}_p) = \mathrm{Hom}(F \otimes \mathbb{Q}_p, \mathbb{Q}_p^{\mathrm{nr}}) \xrightarrow{\sim} \mathrm{Hom}(\mathcal{O}/p\mathcal{O}, k),$$

and by looking at the Dieudonné module of the full p-divisible group $Y[p^\infty]$ we readily find that $\mathfrak{f}_0$ is just the image of $\mathfrak{f}$ under this map.

In concrete terms, write $\mathcal{I} := \mathrm{Hom}(\mathcal{O}/p\mathcal{O}, k)$ and compute the multiplication type $\mathfrak{f}_0 \subset \mathcal{I}$ as the image of the type $\mathfrak{f}$. Given $i \in \mathcal{I}$, draw an arrow $(i) \overset{V}{\longleftarrow} (i+1)$ if $i \in \mathfrak{f}_0$, and arrow $(i) \overset{F}{\longrightarrow} (i+1)$ if not. This gives us a union of circular diagrams Γ_π as described in 2.1 (one diagram, of size $[F_\pi : \mathbb{Q}_p]$, for each prime π), with a natural action of $\mathcal{O}/p\mathcal{O}$ (letting x act as multiplication by $i(x)$ on the base vector corresponding to i.) This describes the p-kernel of Y_0.

References

[1] K.S. Brown, *Buildings*, Springer-Verlag, New York, 1989.

[2] N. Bourbaki, *Groupes et Algèbres de Lie*, Chap. 4–6, Masson, Paris, 1981.

[3] M. Demazure, *Lectures on p-divisible groups*, Lecture Notes in Math. **302**, Springer-Verlag, Berlin, 1972.
[4] E.Z. Goren and F. Oort, *Stratifications of Hilbert Modular Varieties*, J. Alg. Geom. **9** (2000), 111–154.
[5] L. Illusie, *Déformations de Groupes de Barsotti-Tate (d'après A. Grothendieck)*, in: Séminaire sur les Pinceaux Arithmétiques: la Conjecture de Mordell, L. Szpiro, ed., Astérisque **127** (1985), pp. 151–198.
[6] M.-A. Knus, *Quadratic and Hermitian Forms over Rings*, Grundlehren der mathematischen Wissenschaften **294**, Springer-Verlag, Berlin, 1991.
[7] M.-A. Knus et al., *The Book of Involutions*, Colloquium Publ. **44**, A.M.S., Providence, R.I., 1998.
[8] H. Kraft, *Kommutative algebraische p-Gruppen (mit Anwendungen auf p-divisible Gruppen und abelsche Varietäten)*, manuscript, University of Bonn, September 1975, 86 pp.
[9] B.J.J. Moonen, *The Ekedahl-Oort Stratification for Moduli Spaces of PEL-Type*, in preparation.
[10] A. Ogus, *Hodge Cycles and Crystalline Cohomology*, in: Hodge Cycles, Motives, and Shimura Varieties, P. Deligne et al., Lecture Notes in Math. **900**, Springer-Verlag, Berlin, 2nd printing, 1989, pp. 357–414.
[11] F. Oort, *A Stratification of a Moduli Space of Polarized Abelian Varieties in Positive Characteristic*, in: Moduli of Curves and Abelian Varieties, C. Faber and E. Looijenga, eds., Aspects of Mathematics E 33, Vieweg, Braunschweig/Wiesbaden, 1999, pp. 47–64.
[12] F. Oort, *A Stratification of a Moduli Space of Abelian Varieties*, this volume, pp. 345–416.
[13] M. Raynaud, *Schémas en groupes de type* $(p, \ldots, p)$, Bull. Soc. math. France **102** (1974), 241–280.
[14] J.-P. Serre, *Cohomologie Galoisienne*, Lecture Notes in Math. **5**, Springer-Verlag, Berlin, 5th ed., 1994.
[15] J. Tate, *Classes d'Isogenie des Variétés Abéliennes sur un Corps Fini*, Sém. Bourbaki, exp. 352, Novembre 1968, in: Lecture Notes in Math. **179**, Springer-Verlag, Berlin, 1971.
[16] J. Tits, *Buildings of Spherical Type and Finite BN-Pairs*, Lecture Notes in Math. **386**, Springer-Verlag, Berlin, 2nd printing, 1986.
[17] G. van der Geer, *Cycles on the Moduli Space of Abelian Varieties*, in: Moduli of Curves and Abelian Varieties, C. Faber and E. Looijenga, eds., Aspects of Mathematics E 33, Vieweg, Braunschweig/Wiesbaden, 1999, pp. 65–89.
[18] T. Wedhorn, *The Dimension of Oort Strata of Shimura Varieties of PEL-Type*, this volume, pp. 441–471.

Progress in Mathematics, Vol. 195, © 2001 Birkhäuser Verlag Basel/Switzerland

MODULI SPACE OF ELLIPTIC CURVES WITH HEISENBERG LEVEL STRUCTURE

IKU NAKAMURA AND TOMOHIDE TERASOMA

ABSTRACT. A Heisenberg level structure on an elliptic curve is a kind of marking of the linear action by the *noncommutative* Heisenberg group on the space of global sections of a line bundle. The moduli scheme of polarized elliptic curves with Heisenberg level structures is known to be a quasi projective curve over $\mathbf{Z}[\zeta_d, 1/d]$ if the degree of polarization is equal to $d \geq 3$. We prove that it is isomorphic to the (open subscheme of the) moduli scheme $X(d)$ of elliptic curves with level d-structures in the classical sense. Moreover, we also prove that the universal curve over the moduli is the same as that over $X(d)$ if d is odd, while they are different for d even. In particular, the universal elliptic curve over $SQ^0(d)$ has no global sections for $d \geq 4$ and even.

§0 INTRODUCTION

In [N2] level $G(K)$-structures or rigid $G(K)$-structures on polarized abelian schemes have been introduced as well as certain singular degenerate abelian schemes. Among other things the existence of the complete fine moduli scheme $SQ_{g,K}$ in the category of reduced base schemes was established in [N2] in some major cases. Over an algebraically closed field of good characteristic, the first notion is the same as the level K-structure in the classical sense. However in the case of nonclosed fields or in the case of families over certain base schemes the notions can become different. Moreover, the difference becomes critical when one considers the problem of compactifying the moduli scheme of polarized abelian schemes.

The purpose of the present article is to study the structure of the complete moduli schemes $SQ_{1,K}$ and their universal curves in the one dimensional case. For notational convenience we denote it by $SQ^0(d)$ when $K = (\mathbf{Z}/d\mathbf{Z})\oplus(\mathbf{Z}/d\mathbf{Z})$. We also refer to a family Z over S of smooth elliptic curves with level $G(K)$-structures simply as an elliptic curve over S with Heisenberg level structure.

We prove that $SQ^0(d)$ is isomorphic to the moduli scheme $X(d)$ of generalized elliptic curves with level d-structures in the classical sense. Moreover, we also prove that the universal curve over $SQ^0(d)$ is the same as that over $X(d)$

Research of N (resp. of T) was supported in part by the Grants-in-aid No. 12304001, No. 11874001 (resp. No. 11640012) for Scientific Research, the Ministry of Education.

if d is odd, while they are different for d even. In particular, the universal (generalized) elliptic curve over $SQ^0(d)$ has no rational sections for $d \geq 4$ and even. We remark that since the universal (generalized) elliptic curve over $SQ^0(d)$ or $X(d)$ is two-dimensional, and there is a unique minimal surface extending the universal elliptic curve, it is enough for the proof to discuss only the loci of smooth elliptic curves.

In section 1 we recall the definitions of both new and classical level structures. Moreover, we introduce the third moduli space $SQX(d)$ to compare the two moduli spaces $SQ^0(d)$ and $X(d)$. In sections 2, 3, and 4, we prove that the natural covering $SQX(d) \to X(d)$ is a union of double coverings (resp. trivial coverings) if d is even (resp. odd). In section 5, we prove that the natural morphism $SQ^0(d) \to X(d)$ given by the Jacobian fibration is an isomorphism for any d. In section 6, we show that the universal elliptic curve $E_{SQ^0(d)}$ over $SQ^0(d)$ is obtained by twisting the universal elliptic curve $S(d)$ over $X(d)$ by an order two element if d is even. As a consequence, we see that $E_{SQ^0(d)}$ has no section over $SQ^0(d)$ if d is even. In section 7, we study the relatively minimal model of $E_{SQ^0(d)}$. An example for $d = 4$ is given in the last section.

Acknowledgement. The present article was completed during the authors' stay in 1998 at the Max-Planck-Institut für Mathematik in Bonn. We would like to thank all the people in the institute for their hospitality and the referee for pointing out a mistake in the first manuscript.

§1 Moduli problems and Heisenberg group

Let d be an integer greater than or equal to 3 and $\mathcal{O}$ be the ring $\mathbf{Z}[\zeta_d, \frac{1}{d}]$, where ζ_d is a primitive root of unity. We put $H = H^\vee = \mathbf{Z}/d\mathbf{Z}$ and define a bilinear form $\langle , \rangle$ by

$$H \times H^\vee \to \mu_d \quad : \quad (a, b) \to \langle a, b \rangle = \zeta_d^{ab}.$$

We define a group structure on the set $\mu_d \times H \times H^\vee$ by

$$(a, b, c)(a', b', c') = (aa'\langle b, c' \rangle, b + b', c + c').$$

It is called the Heisenberg group $G(d)$ of level d. The natural projection $\pi : G(d) \to H \oplus H^\vee$ is a surjective group homomorphism. The inclusion $H \to G(d)$ defined by $b \mapsto (1, b, 0)$ is a group homomorphism and H is identified as a subgroup of $G(d)$. Let S be a scheme over $Spec(\mathcal{O})$ and define the Schroedinger representation $U(d)$ of $G(d)$ on $\mathcal{O}_S[H^\vee]$ by

$$(U(d)(a, b, c))v(\chi) = a\langle b, \chi \rangle v(\chi + c),$$

where $v(\chi)$ denotes the base of $\mathcal{O}_S[H^\vee]$ corresponding to $\chi \in H^\vee$ regarded as a function on H. For any point s in S, the fiber of the representation

$(U(d), \mathcal{O}_S[H^\vee] \otimes k(s))$ is an irreducible representation over the residue field $k(s)$.

Now we introduce a $G(d)$-structure on a genus 1 curve on S. Let $E \to S$ be a proper smooth geometrically connected scheme of relative dimension 1. It is called a genus 1 curve on S if the genus of every fiber is 1. If in addition $E \to S$ has a specified section e, it is called an elliptic curve on S. If E is a genus 1 curve on S, then the group of translations $Tr(E/S)$ is an elliptic curve on S. Set $K = H \oplus H^\vee$. An inclusion T of K into $Tr(E/S)$ compatible with the Weil pairing is called a translation structure on E under K. The translation of E corresponding to the image of $g \in K$ is denoted by T_g. We fix a translation structure on E under K. A line bundle of E is called degree d if the degree of the restriction to each fiber of E at $s \in S$ is d. Let $\mathcal{L}$ be a line bundle of degree d on E. A $G(d)$-linearization is a set of sheaf isomorphisms $\varphi = \{\varphi_g : T^*_{\pi(g)}\mathcal{L} \to \mathcal{L}\}_{g \in G(d)}$ indexed by $g \in G(d)$ with the 1-cocycle condition, i.e. the following diagram commutes:

$$\begin{array}{ccc} T^*_{\pi(h)}T^*_{\pi(g)}\mathcal{L} & \xrightarrow{T^*_{\pi(h)}\varphi_g} & T^*_{\pi(h)}\mathcal{L} \\ \downarrow & & \downarrow{\scriptstyle \varphi_h} \\ T^*_{\pi(gh)}\mathcal{L} & \xrightarrow[\varphi_{gh}]{} & \mathcal{L}, \end{array}$$

and the central condition, i.e. $\varphi_{(a,0,0)} =$ (multiplication by a). Let $\phi : E \to S$ be the structure homomorphism. If $\mathcal{L}$ is $G(d)$-linearized, then for $g \in G(d)$ we define a homomorphism $\rho(g)$ on the locally free $\mathcal{O}_S$-module $\phi_*\mathcal{L}$ by

$$\rho(g) : \phi_*\mathcal{L} \overset{\text{natural}}{\longrightarrow} \phi_*(T^*_{\pi(g^{-1})}\mathcal{L}) \overset{\phi_*(\varphi_{g^{-1}})}{\longrightarrow} \phi_*\mathcal{L}.$$

By the 1-cocycle condition, $\rho(g)$ gives rise to a representation ρ of $G(d)$ on $\phi_*\mathcal{L}$. The restriction of ρ to H is denoted by ρ_H.

Definition (admissible linearization). Let (E,T) be a translation structure on a genus 1 curve E over S and let $(\mathcal{L}, \varphi)$ be a $G(d)$-linearized line bundle on (E,T). If the fiber of the representation $(\rho \otimes k(s), \phi_*\mathcal{L} \otimes k(s))$ is isomorphic to the Schroedinger representation $U(d)$ over $k(s)$ for all $s \in S$, the pair $((E,T),(\mathcal{L},\varphi))$ is called an admissible pair and φ is called an admissible $G(d)$-linearization.

Using the notion of admissible linearization, we define a contravariant functor $SQ^0(d)$ from $(Sch/Spec(\mathcal{O}))$ to $(Sets)$. For a $Spec(\mathcal{O})$ scheme S, $SQ^0(d)(S)$ is the set of equivalence classes of admissible pairs $((E,T),(\mathcal{L},\varphi))$. Here the equivalence class is defined as follows. Let $((E_1,T_1),(\mathcal{L}_1,\varphi_1))$ and $((E_2,T_2),(\mathcal{L}_2,\varphi_2))$ be two admissible pairs over S. They are equivalent if and only if (1) there exists an S-isomorphism $h : E_1 \to E_2$ compatible with the translation structures, and (2) there exists an invertible sheaf $\mathcal{M}$ on S and an isomorphism of sheaves $\tilde{h} : h^*\mathcal{L}_2 \to \mathcal{L}_1 \otimes \phi_1^*\mathcal{M}$ compatible with the $G(d)$-linearizations, where $\phi_1 : E_1 \to S$ is the structure morphism. By [N2], the functor $SQ^0(d)$ is representable by a scheme $SQ^0(d)$.

Next we recall the classical level d structure on elliptic curves on S. Let $E \to S$ be an elliptic curve on S. A level d structure on E is a group injection $\iota : K \to E(S)$ from $K = H \oplus H^\vee$ to the group $E(S)$ of S-valued sections of E compatible with the symplectic structure of K and the Weil pairing of E. Note that a level d structure provides a translation structure of E under K. Since the group of translations $Tr(E/S)$ is isomorphic to the Picard scheme $Pic^0(E)$, we have that $Pic^0(E)$ has a natural level d structure $\iota : K \to Pic^0(E)$ for an object $((E,T),(\mathcal{L},\varphi))$ of $SQ^0(d)(S)$. By attaching $(Pic^0(E),\iota)$ to $((E,T),(\mathcal{L},\varphi))$, we get a functor $J : SQ^0(d) \to X(d)$. Note that by the existence of a $G(d)$-linearized line bundle, the inclusion ι is compatible with the symplectic structure of K and the Weil pairing of E. For an elliptic curve E with a level structure ι, the image of the subgroup H of K is a smooth divisor on E. The sheaf of relative 1-forms on S with logarithmic poles along H is denoted by $\Omega^1(\log H) = \Omega^1_{E/S}(\log H)$. For $h \in H$, the sheaf $T_h^*\Omega^1(\log H)$ is canonically isomorphic to $\Omega^1(\log H)$ and the identification is denoted by $N_h : T_h^*\Omega^1(\log H) \to \Omega^1(\log H)$.

Lemma 1.1. *Let s be a point of S. As a representation of H, $(\phi_*\Omega^1(\log H))_s$ is isomorphic to the restriction $U(d)_H$ to H of the Schroedinger representation $U(d)$ over $k(s)$.*

Proof. By using the residue map $\Omega^1(\log H) \to \mathcal{O}_H$, we have the following exact sequence

$$0 \to \phi_*\Omega^1_{E/S} \to \phi_*\Omega^1_{E/S}(\log H) \to \phi_*\mathcal{O}_H \overset{\text{sum}}{\to} R^1\phi_*\Omega^1_{E/S} \simeq \mathcal{O}_S.$$

On the other hand, $(\phi_*\Omega^1_{E/S})_s \simeq H^0(E_s,\Omega^1)$ is the trivial representation of H, which gives the lemma.

Definition (H-admissible linearization). A $G(d)$-linearization $(\Omega^1(\log H),\varphi)$ on $\Omega^1(\log H)$ is called H-admissible if $\varphi_h = N_h$ for $h \in H$.

Proposition 1.2. *An H-admissible linearization $(\Omega^1(\log H),\varphi)$ on $\Omega^1(\log H)$ is admissible.*

Proof. We may assume that S is the spectrum of a field k. Then the restriction of ρ on $H^0(E,\Omega^1(\log H))$ to H is isomorphic to $U(d)_H$. Therefore it is enough to prove that any representation ρ of $G(K)$ with the central condition, whose restriction to H is isomorphic to $U(d)_H$, is isomorphic to $U(d)$. Let $\rho_H = \oplus_{\chi\in H^\vee}\rho_H(\chi)$ and $U(d)_H = \oplus_{\chi\in H^\vee}U(d)_H(\chi)$ be the decomposition of ρ_H and $U(d)_H$ respectively. Then $\rho_H(\chi)$ and $U(d)_H(\chi)$ are 1-dimensional vector spaces. We fix an isomorphism $\eta(1) : \rho_H(1) \simeq U(d)_H(1)$ and define $\eta(\chi) : \rho_H(\chi) \simeq U_H(\chi)$ by $(U(d)(1,0,\chi)) \circ \eta(1) \circ \rho(1,0,-\chi)$. Then $\oplus_{\chi\in H^\vee}\eta(\chi)$ gives an isomorphism between ρ and $U(d)$ as $G(d)$-modules. See also [M, p.295].

The set of isomorphism classes of level d structures on S is denoted by $X(d)(S)$. Then it is known that the functor $X(d)$ is represented by a connected smooth scheme over $Spec(\mathcal{O})$. The pair $((E,\iota),(\Omega^1(\log H),\varphi))$, where (1) the map $\iota : K \to E$ is a level d structure, and (2) the $G(d)$-linearization φ of $\Omega^1(\log H)$ is H-admissible, is called a $(G(d),H)$-elliptic curve. The set of isomorphism classes of $(G(d),H)$-elliptic curves on S is denoted by $SQX(d)(S)$. The functor $SQX(d)$ is a contravariant functor from $(Sch/Spec(\mathcal{O}))$ to $(Sets)$.

Lemma 1.3. *The functor $SQX(d)$ is representable by a scheme $SQX(d)$ over $\mathcal{O}$.*

Proof. Later, in Proposition 2.2, we prove that the natural forgetful functor $SQX(d) \to X(d)$ is relatively representable. The lemma is a consequence of the representability of $X(d)$.

§2 The morphism $SQX(d) \to X(d)$

In this section, we investigate the natural forgetful functor $SQX(d) \to X(d)$ attaching (E,ι) to $((E,\iota),(\Omega^1(\log H),\varphi))$. Let $h^\vee$ be the generator of $H^\vee$ corresponding to 1.

Lemma 2.1. *Set $T^*_{h^\vee}\Omega^1(\log H) = \Omega^1(\log H-h^\vee)$. Then $\Omega^1(\log H-h^\vee)\otimes\phi^*\mathcal{M}$ is isomorphic to $\Omega^1(\log H)$ on E, where $\mathcal{M}$ is an invertible sheaf on the base scheme S such that $\mathcal{M}^{\otimes d} \simeq \mathcal{O}_S$. Here $\phi : E \to S$ is the structure morphism and $H - h^\vee$ denotes the divisor $T^{-1}_{h^\vee}(H)$.*

Proof. Since the degree of $\Omega^1(\log H)$ is d and $h^\vee$ is a d-torsion section of E, by Abel's theorem, the restriction of $\Omega^1_{E/S}(\log H - h^\vee) \otimes \Omega^1_{E/S}(\log H)^{-1}$ to each fiber E_s at $s \in S$ is trivial. Therefore there exists an invertible sheaf $\mathcal{M}$ on S such that $\Omega^1_{E/S}(\log H - h^\vee) \otimes \phi^*\mathcal{M} \simeq \Omega^1_{E/S}(\log H)$. By applying $T^*_{kh^\vee}$ to both sides of this isomorphism, we have $\Omega^1_{E/S}(\log H - (k+1)h^\vee) \otimes \phi^*\mathcal{M} \simeq \Omega^1_{E/S}(\log H - kh^\vee)$ and as a consequence, we have $\mathcal{M}^{\otimes d} \simeq \mathcal{O}_S$.

Using Lemma 2.1, we fix an isomorphism $f : \Omega^1_E(\log H - h^\vee) \otimes \mathcal{M} \to \Omega^1_E(\log H)$. Note that if S is the spectrum of a field, the morphism f corresponds to the multiplication of a rational function f on E such that $(f) = (H - h^\vee) - (H)$. For i with $0 \le i \le d-1$, we have that f induces the isomorphism

$$T^*_{ih^\vee} f \otimes 1^{\otimes i} : \Omega^1(\log H - (i+1)h^\vee) \otimes \mathcal{M}^{\otimes(i+1)} \to \Omega^1(\log H - ih^\vee) \otimes \mathcal{M}^{\otimes i}.$$

By composing these morphisms, we have the norm $Nm(f)$ of f defined by

$$Nm(f) = (T^*_{(d-1)h^\vee} f \otimes 1^{\otimes(d-1)}) \circ \cdots \circ (T^*_{h^\vee} f \otimes 1) \circ f :$$
$$\Omega^1(\log H) \otimes \mathcal{M}^{\otimes d} \xrightarrow{\simeq} \Omega^1(\log H).$$

Since E is proper and geometrically connected, the morphism $Nm(f)$ corresponds to an isomorphism $\mathcal{M}^{\otimes d} \to \mathcal{O}_S$ and it is also denoted by $Nm(f)$.

Proposition 2.2. *Let $\phi : E \to S$ be an element of $X(d)(S)$. The fiber product $S \times_{X(d)} SQX(d)$ defined by*

$$\begin{array}{ccc} S \times_{X(d)} SQX(d) & \longrightarrow & SQX(d) \\ \downarrow & & \downarrow \\ S & \xrightarrow[(E,\phi)]{} & X(d) \end{array}$$

is representable by the scheme $Spec(\oplus_{i=0}^{d-1} \mathcal{M}^{\otimes i})$ where the ring structure on the sheaf $\oplus_{i=0}^{d-1} \mathcal{M}^{\otimes i}$ is given by the morphism $Nm(f) : \mathcal{M}^{\otimes d} \simeq \mathcal{O}_S$. As a consequence, the morphism $SQX(d) \to X(d)$ is relatively representable.

Proof. To give a morphism $U \to S \times_{X(d)} SQX(d)$ is equivalent to giving an H-admissible $G(d)$-linearization on $E_U = E \times_S U$. That is equivalent to giving a set of isomorphisms

$$\varphi_g : T_g^* \Omega^1_{E_U/U}(\log H) \to \Omega^1_{E_U/U}(\log H)$$

satisfying (1) $\varphi_h = N_h$ for $h \in H$, and (2) the 1-cocycle condition and the central condition for $\{\varphi_g\}_{g \in G(d)}$. Suppose that $U \to S \times_{X(d)} SQX(d)$ is given. Then by the 1-cocycle condition, we have

$$T^*_{(d-1)h^\vee}(\varphi_{h^\vee}) \circ \cdots \circ T^*_{h^\vee}(\varphi_{h^\vee}) \circ \varphi_{h^\vee} = 1.$$

Therefore $f \circ \varphi_{h^\vee}^{-1} : \Omega^1(\log H) \otimes \mathcal{M} \otimes \mathcal{O}_U \to \Omega^1(\log H)$ corresponds to an element $g : \mathcal{M} \otimes \mathcal{O}_U \to \mathcal{O}_U$ such that $Nm(g) = Nm(f)$. Therefore we have $U \to Spec(\oplus_{i=0}^{d-1} \mathcal{M}^{\otimes i})$ by sending $g : \mathcal{M} \to \mathcal{O}_U$. On the other hand, if we have a morphism $U \to Spec(\oplus_{i=0}^{d-1} \mathcal{M}^{\otimes i})$, then the restriction $\mathcal{M} \to \mathcal{O}_U$ gives an isomorphism $g^{-1} : \Omega^1_{E_U/U}(\log(H)) \otimes \mathcal{M}^{-1} \to \Omega^1_{E_U/U}(\log(H))$. We define $\varphi_{h^\vee}$ as

$$\varphi_{h^\vee} = (g^{-1}) \cdot f : \Omega^1_{E_U/U}(\log(H - h^\vee)) \to \Omega^1_{E_U/U}(\log(H)).$$

We define $\varphi_{ch^\vee}$ and $\varphi_{(a,b,c)}$ as

$$\varphi_{ch^\vee} = T^*_{(c-1)h^\vee}(\varphi_{h^\vee}) \circ \cdots \circ T^*_{h^\vee}(\varphi_{h^\vee}) \circ \varphi_{h^\vee},$$
$$\varphi_{(a,b,c)} = a \cdot T^*_{bh}(\varphi_{ch^\vee}).$$

Note that this is the unique way to define φ_g satisfying the 1-cocycle condition and the central condition regarding the following relations

$$(a,b,c) = (1,0,c)(1,b,0)(a,0,0)$$
$$\varphi_{g_1g_2g_3} = \varphi_{g_3}T^*_{g_3}(\varphi_{g_2})T^*_{g_2g_3}(\varphi_{g_1}).$$

We check the 1-cocycle condition. Since the Weil pairing is compatible with the symplectic structure of $K = H \oplus H^\vee$, we have $\varphi_{c'h^\vee} = \langle b, c'\rangle T^*_{-bh}(\varphi_{c'h^\vee})$. Therefore we have

$$\begin{aligned}
&\varphi_{(a',b',c')}T^*_{b'h+c'h^\vee}(\varphi_{(a,b,c)})\\
&= a'T^*_{b'h}(\varphi_{c'h^\vee}) \circ T^*_{b'h+c'h^\vee}(aT^*_{bh}(\varphi_{ch^\vee}))\\
&= aa'T^*_{(b+b')h}(T^*_{-bh}(\varphi_{c'h^\vee}) \circ T^*_{c'h^\vee}(\varphi_{ch^\vee}))\\
&= aa'T^*_{(b+b')h}(T^*_{-bh}(\varphi_{c'h^\vee}) \circ \varphi^{-1}_{c'h^\vee} \circ \varphi_{c'h^\vee} \circ T^*_{c'h^\vee}(\varphi_{ch^\vee}))\\
&= aa'\langle b,c'\rangle T^*_{(b+b')h}(\varphi_{(c+c')h^\vee})\\
&= \varphi_{(aa'\langle b,c'\rangle, b+b', c+c')}\\
&= \varphi_{(a,b,c)(a',b',c')}
\end{aligned}$$

Thus we get an H-admissible $G(d)$-linearization on $\Omega^1_{E_T/T}(\log H)$.

§3 Field extension and linearization

In this section, we investigate the minimal field extension on which a given elliptic curve with a level d structure admits an H-admissible linearization. Let $S = Spec(k)$, where k is a field over the ring $\mathcal{O} = \mathbf{Z}[\zeta_d, \frac{1}{d}]$, $E \to S$ be an elliptic curve and $\iota : K \to E$ be a level d-structure. An isomorphism $T^*_{h^\vee}\Omega^1(\log H) \to \Omega^1(\log H)$ corresponds to a rational function f on E such that $(f) = (H + h^\vee) - (H)$. Then $Nm(f) = \prod_{i=0}^{d-1} T^*_{ih^\vee}(f)$ is an element of $k^\times$ and the class $[Nm(f)]$ in $k^\times/k^{\times d}$ does not depend on the choice of f. We prove the next theorem in this section.

Proposition 3.1. *Put $d' = d$ if d is odd and $d' = d/2$ if d is even. With notation as above, we have $[Nm(f)] \in k^{\times d'}/k^{\times d}$.*

Proof. In this proof the divisor on E defined by a rational function f is denoted by $(f)_E$ to avoid confusion. Let θ and δ be rational functions on E such that

$$(\theta)_E = (-h - h^\vee) - (-h) - (-h^\vee) + (0), \quad (\delta)_E = (-2h^\vee) - 2(-h^\vee) + (0).$$

There exists such rational functions on E by Abel's theorem and the existence of a level d structure. For $w = \sum_{g\in K} w_g T^*_g \in \oplus_{g\in K} T^*_g \mathbf{Z}$, and a rational function h on E, we define $h^w = \prod_{g\in K} T^*_g(h)^{w_g}$. If we put

$$A_i = T^*_{(i-1)h} + \cdots + T^*_h + 1, A_1 = 1, B = T_{h^\vee} + 2T^*_{2h^\vee} + \cdots + (d-1)T_{(d-1)h^\vee},$$

then we have

$$(\theta^{A_i})_E = (-ih - h^\vee) - (-ih) - (-h^\vee) + (0), \quad (\delta^B)_E = d((-h^\vee) - (0)).$$

Therefore, we have

$$(\prod_{i=1}^{d-1} \theta^{A_i} \cdot \delta^B)_E = \sum_{i=0}^{d-1}((-ih - h^\vee) - (-ih)) = (H - h^\vee) - (H).$$

Therefore, to compute $[Nm(f)]$, we may assume $f = \prod_{i=1}^{d-1} \theta^{A_i} \cdot \delta^B$. Put $N = 1 + T_{h^\vee} + \cdots + T^*_{(d-1)h^\vee}$. Since the T^*_g $(g \in K)$ commute with each other, we have

$$(3.1) \qquad Nm(f) = f^N = (\prod_{i=1}^{d-1} \theta^{A_i} \cdot \delta^B)^N$$
$$= (\prod_{i=1}^{d-1} \theta^{A_i})^N \cdot (\delta^B)^N = \prod_{i=1}^{d-1} (\theta^N)^{A_i} \cdot (\delta^N)^B.$$

Since $(\theta^N)_E = 0$ and $(\delta^N)_E = 0$, we have that $\Theta = \theta^N$ and $\Delta = \delta^N$ are non-zero constants. Therefore (3.1) is equal to $\prod_{i=1}^{d-1} \Theta^{A_i} \cdot \Delta^B = \Theta^{d(d-1)/2} \Delta^{d(d-1)/2}$. This proves the proposition.

Corollary 3.2. *If d is odd, each component of $SQX(d)$ is isomorphic to $X(d)$ via the natural morphism J of §1. If d is even, either every component of $SQX(d)$ is a double covering of $X(d)$ or every component is isomorphic to $X(d)$.*

Remark. In the next section, we show that the former case always occurs if d is even.

Proof. Let η be the generic point of $X(d)$ and $E_\eta \to \eta$ be the generic fiber of the universal elliptic curve. If d is odd, $Nm(f) \in k_\eta^{\times d}$ by Proposition 3.1. Therefore $\eta \times_{X(d)} SQX(d) = Spec(k_\eta[t]/(t^d - Nm(f))$ is isomorphic to the union of $Spec(k)$. Since the morphism $SQX(d) \to X(d)$ is finite etale, we get the first assertion of the corollary. If d is even, $Nm(f) \in k_\eta^{\times d'}$. Therefore $k_\eta[t]/(t^d - Nm(f))$ is either a product of quadratic extensions of k_η or a product of copies of k_η.

§4 Divisors on elliptic surfaces $\bar{S}(d)$

In this section, we assume that d is even. We prove the following theorem.

Theorem 4.1. *If d is even, every connected component of $SQX(d)$ is a finite étale double covering of $X(d)$.*

To prove this theorem, it is enough to prove it after base change to $Spec(\mathbf{C})$. Therefore in this section, $X(d)$ and $SQX(d)$ are considered as schemes over $Spec(\mathbf{C})$.

Let $\bar{X}(d)$ be the smooth compactification of $X(d)$ and $\bar{S}(d)$ be the minimal model of a compactification of the universal elliptic curve on $X(d)$. For details on $\bar{S}(d)$, we refer to [S]. Let η be the generic point of $X(d)$ and $S(d)_\eta$ the generic fiber of $S(d) \to X(d)$. The closure of the universal section $g \in K$ in $\bar{S}(d)$ is denoted by S_g. Then S_g is a section from $\bar{X}(d)$ to $\bar{S}(d)$. Let f be a rational function on $S(d)$ such that $(f)_{S(d)_\eta} = (H - h^\vee) - (H)$. Let $(f)_{\bar{S}(d)}$ be the divisor of f on $\bar{S}(d)$. Then

$$(f)_{\bar{S}(d)} = \sum_{h \in H} (S_{h-h^\vee} - S_h) + \Phi_r + \Phi_s,$$

where Φ_r (resp. Φ_s) is a divisor whose support is contained in a finite union of regular fibers (resp. singular fibers). Let $\Sigma = \bar{X}(d) - X(d)$, $\mu = \#\Sigma$ and $\Sigma = \{p_1, \ldots, p_\mu\}$. It is known that the singular fiber at $p \in \Sigma$ is type I_d. Let $D_{0,j}, \ldots, D_{d-1,j}$ be the components of the fiber at p_j. Suppose that $D_{0,j}$ intersects with the zero section S_0 and we set the numbering of $D_{i,j}$ as in the following figure.

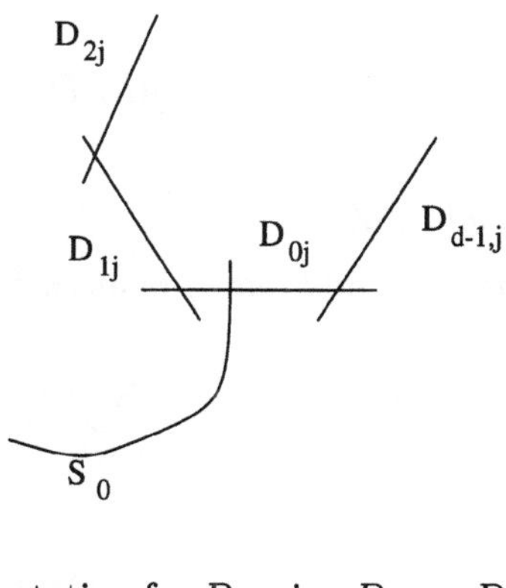

We use the cyclic notation for $D_{i,j}$, i.e. $D_{i,j} = D_{i+d,j}$. Let F be a smooth fiber of $\bar{S}(d)$. The intersection numbers are given as follows:

$$D_{i,j} \cdot D_{i',j'} = 0 \qquad \text{if } j \neq j' \text{ or } \mid i - i' \mid \geq 2,$$
$$D_{i,j} \cdot D_{i+1,j} = 1, \quad D_{i,j} \cdot D_{i,j} = -2, \quad D_{i,j} \cdot F = 0.$$

Let $T_{\mathbf{Q}}$ be the $\mathbf{Q}$-subspace of $NS(\bar{S}(d))_{\mathbf{Q}} = NS(\bar{S}(d)) \otimes \mathbf{Q}$ generated by S_0, F and $D_{i,j}$ $(1 \leq i \leq d-1, 1 \leq j \leq \mu)$. Then they are linearly independent and the intersection form on $T_{\mathbf{Q}}$ is non-degenerate. Let $k(g,j)$ be the index k $(1 \le k \le d-1)$ such that $D_{k,j} \cdot S_g = 1$. The numerical class of D in $NS(\bar{S}(d))_{\mathbf{Q}}$ is denoted by $[D]$.

Lemma 4.2. *If we put*

$$[S_g] = [S_0] + a[F] + \sum_{i=1}^{d-1} a_{i,j}[D_{i,j}],$$

in $T_{\mathbf{Q}}$, *and* $k_j = k(g,j)$, *then*

$$a_{i,j} = \begin{cases} -\frac{i(d-k_j)}{d} & (1 \le i \le k) \\ -\frac{k_j(d-i)}{d} & (k < i), \end{cases}$$

if $k_j \neq 0$. *If* $k_j = 0$, *then* $a_{i,j} = 0$ *for all* $1 \le i \le d-1$.

Proof. We get the result by solving the equations $[S_g] \cdot [D_{i,j}] = \delta_{i,k_j} = a_{i-1} - 2a_{i,j} + a_{i+1,j}$ and $[S_g] \cdot [F] = 1$. Here formally we put $a_{0,j} = 0$. (To get the coefficient a, it is sufficient to know $[S_0] \cdot [S_0]$, since $[S_g] \cdot [S_0] = 0$ for $g \neq 0$. But we don't need this coefficient for the moment.)

The coefficient $a_{i,j}$ of $[D_{i,j}]$ in $[S_g]$ is denoted by $a_{i,j}(g)$. As a corollary to Lemma 4.2, we get the following lemma.

Lemma 4.3. *The component of* Φ_s *contained in the fiber at* p_j *is equal to*

$$-\sum_{i=1}^{d-1} \sum_{h \in H} (a_{i,j}(h - h^{\vee}) - a_{i,j}(h)) D_{i,j} + b_j F_j,$$

where $F_j = \sum_{i=0}^{d-1} D_{i,j}$ *and* $b_j \in \mathbf{Z}$.

Corollary 4.4. *The order of* $Nm(f)$ *at* p_j *is given by*

$$-\sum_{i=1}^{d-1} (\sum_{h \in H} (a_{i,j}(h - h^{\vee}) - a_{i,j}(h))) + db_j = -\sum_{h \in H} (A_j(h - h^{\vee}) - A_j(h)) + db_j.$$

Here we put $A_j(g) = \sum_{i=1}^{d-1} a_{i,j}(g)$.

Proof. By taking the average over $h \in H$, and using the fact that $Nm(f)$ is a constant function, we get the lemma.

By a simple calculation, we get $A_j(g) = \mathcal{A}(k(g,j))$, where

$$\begin{aligned} \mathcal{A}(k) &= \sum_{1 \le i \le k} -\frac{i(d-k)}{2} + \sum_{k+1 \le i \le d-1} -\frac{k(d-i)}{2} \\ &= -k(d-k)/2. \end{aligned}$$

Proof of Theorem. If

$$k(h,j) = 0 \text{ and } k(h-h^\vee, j) = 1 \text{ for all } h \in H, \tag{4.1}$$

then $\mathrm{ord}_{p_j} Nm(f) = -d(d-1)/2 + db_j \not\equiv 0 \pmod d$. Therefore $Nm(f) \in k_\eta^{\times d'} - k_\eta^{\times d}$. By the description of $\bar{S}(d)$ (see [S, p.39]), there exists j $(1 \le j \le \mu)$ satisfying condition (4.1). Therefore every component of $SQX(d)$ is a double covering of $X(d)$ branching at p_j.

We can investigate the branching points of $\overline{SQX(d)} \to \bar{X}(d)$ using Corollary 4.4, where $\overline{SQX(d)}$ is the compactification of $SQX(d)$.

Corollary 4.5. *The covering $\overline{SQX(d)} \to \bar{X}(d)$ branches at p_j if and only if*

$$\sum_{h \in H} (\mathcal{A}(k(h-h^\vee, j)) - \mathcal{A}(k(h,j))) \not\equiv 0 \pmod d$$

Using this corollary, we get the following proposition.

Proposition 4.6. *The point p_j is a ramified point if and only if (1) $k(-h^\vee, j)$ is odd and (2) $(\#I - 1)\dfrac{d}{\#I}$ is even, where $I = \{k(h,j) \mid h \in H\}$.*

Proof. We denote $k(g) = k(g,j)$ for short. Since $\mathcal{A}(k+l) - \mathcal{A}(k) - \mathcal{A}(l) = kl$ and $\mathcal{A}(d) = 0$, we have $\mathcal{A}(k) \equiv \mathcal{A}(k') \pmod d$ if $k \equiv k' \pmod d$. By the numbering of $D_{i,j}$, we have $k(h-h^\vee) \equiv k(h) + k(-h^\vee) \pmod d$. Therefore we have

$$\begin{aligned} \mathcal{A}(k(h-h^\vee)) - \mathcal{A}(k(h)) &\equiv \mathcal{A}(k(h) + k(-h^\vee)) - \mathcal{A}(k(h)) \pmod d \\ &\equiv \mathcal{A}(k(-h^\vee)) + k(h)k(-h^\vee) \pmod d. \end{aligned}$$

Let $I = \{k(h) \mid h \in H\}$. Then we have

(4.2)

$$\begin{aligned} \sum_{h \in H} (\mathcal{A}(h-h^\vee) - \mathcal{A}(h)) &\equiv d\mathcal{A}(k(-h^\vee)) + k(-h^\vee) \cdot \frac{d}{\#I} \sum_{k \in I} k \pmod d \\ &\equiv \frac{dk(-h^\vee)}{2} + k(-h^\vee) \cdot \frac{d^2(\#I-1)}{2\#I} \pmod d. \end{aligned}$$

Therefore if $k(-h^\vee)$ is even, (4.2) $\equiv 0 \pmod d$. If k is odd, (4.2) $\not\equiv 0 \pmod d$ if and only if $\dfrac{d(\#I-1)}{\#I}$ is even. This proves the proposition.

The following corollary is useful to prove the reducedness of the minimal model of the descent family in §7.

Corollary 4.7. *If p_j is ramified in $\overline{SQX(d)} \to \bar{X}(d)$, then T_v acts trivially on the components of $S(d)_{p_j}$, where $v = \sum_{h\in H} h$.*

Proof. Suppose that p_j is ramified. Let I be the image of H in the automorphism group of components of the singular fiber. Since $\dfrac{d(\#I-1)}{\#I}$ is even, $\dfrac{d}{\#I}$ is even. Since T_v is an involution, its image in I is trivial.

We finish this section by proving the non-existence of an admissible linearized line bundle on $S(d)$, if the base field is $\mathbf{C}$.

Proposition 4.8. *Assume that d is even. On $S(d)$, there exists no line bundle $\mathcal{L}$ with an admissible $G(d)$-linearization if the base field is* $\mathbf{C}$.

Proof. Let η be a generic point of $X(d)$, $S(d)_\eta$ the generic fiber of $S(d) \to X(d)$ and $\mathcal{L}_\eta$ a degree d line bundle on $S(d)_\eta$. Suppose that $\mathcal{L}_\eta$ is equipped with an admissible $G(d)$-linearization $\varphi = \{\varphi_g\}$. Let O be the zero section of $S(d)_\eta$. Then $\mathcal{L}_\eta \otimes \mathcal{O}(dO)^{-1}_{S(d)_\eta}$ is a degree 0 divisor on $S(d)_\eta$. Therefore it is written as $\mathcal{O}_{S(d)_\eta}((g)-(O))$ with an element $g \in K$ by the result of Theorem 5.5 in [S], if the base field is $\mathbf{C}$. Therefore we may assume that $\mathcal{L}_\eta = \mathcal{O}(\sum_{i=1}^d \alpha_i)$, where $\alpha_i \in K$. Let h and $h^\vee$ be the generator of H and $H^\vee$ corresponding to 1 respectively. Then $T_h^*\mathcal{L}_\eta \to \mathcal{L}_\eta$ and $T_{h^\vee}^*\mathcal{L}_\eta \to \mathcal{L}_\eta$ correspond to the multiplication of rational functions f and $f^\vee$ such that

$$(f) = \sum_{i=1}^d ((\alpha_i - h) - (\alpha_i)), \quad (f^\vee) = \sum_{i=1}^d ((\alpha_i - h^\vee) - (\alpha_i)).$$

Let $p \in \Sigma$ and k be the surjective homomorphism $K \to \mathbf{Z}/d\mathbf{Z}$ corresponding to the rotation of the cycle at p, i.e. the substitution of the irreducible components of the singular fiber $\bar{S}(d)_p$ at p. We put

$$Nm^\vee(f) = f\cdot T_h^* f \cdots T_{(d-1)h}^* f, \quad Nm(f^\vee) = f^\vee \cdot T_{h^\vee}^* f^\vee \cdots T_{(d-1)h^\vee}^* f^\vee.$$

Then as in the proof of Proposition 4.6, we have

$$\begin{aligned}\mathrm{ord}_p(Nm^\vee(f)) &\equiv (\mathcal{A}(k(\alpha_i - h)) - \mathcal{A}(k(\alpha_i)))\\ &\equiv d\mathcal{A}(k(-h)) + (\sum_{i=1}^d k(\alpha_i))k(-h),\\ \mathrm{ord}_p(Nm(f^\vee)) &\equiv d\mathcal{A}(k(-h^\vee)) + (\sum_{i=1}^d k(\alpha_i))k(-h^\vee).\end{aligned}$$

Since $h^d = (h^\vee)^d = 1$, $\mathrm{ord}_p(Nm^\vee(f)) \equiv \mathrm{ord}_p(Nm(f^\vee)) \equiv 0 \pmod d$. Therefore we have

$$(E_k): \begin{cases} d\mathcal{A}(k(-h)) + (\sum_{i=1}^d k(\alpha_i))k(-h) \equiv 0 (\mathrm{mod}\ d) \\ d\mathcal{A}(k(-h^\vee)) + (\sum_{i=1}^d k(\alpha_i))k(-h^\vee) \equiv 0 (\mathrm{mod}\ d). \end{cases}$$

Since any surjective homomorphism $k : K \to \mathbf{Z}/d\mathbf{Z}$ is realized by the rotation of cycles at some p, the following lemma leads us to a contradiction to the existence of φ.

Lemma 4.9. *There is no common solution for (E_k) for all surjective homomorphisms $k : K \to \mathbf{Z}/d\mathbf{Z}$.*

Proof. Let $\alpha_i = a_i h + b_i h^\vee$ and k_1, k_2, k_3 be surjective homomorphisms $K \to \mathbf{Z}/d\mathbf{Z}$ such that

$$k_1(-h) = k_2(-h) = k_2(-h^\vee) = k_3(-h^\vee) = 1, \quad k_1(-h^\vee) = k_3(-h) = 0.$$

Then we have

(4.3)

$$d\mathcal{A}(k_1(-h)) + (\sum_{i=1}^d k_1(\alpha_i))k_1(-h) \equiv \frac{d}{2} + \sum_{i=1}^d a_i \equiv 0 \pmod d,$$

$$d\mathcal{A}(k_2(-h)) + (\sum_{i=1}^d k_2(\alpha_i))k_2(-h) \equiv \frac{d}{2} + \sum_{i=1}^d (a_i + b_i) \equiv 0 \pmod d,$$

$$d\mathcal{A}(k_3(-h^\vee)) + (\sum_{i=1}^d k_3(\alpha_i))k_3(-h^\vee) \equiv \frac{d}{2} + \sum_{i=1}^d b_i \equiv 0 \pmod d.$$

It is obvious that there is no common solution for (4.3).

§5 The morphism $SQX(d) \to SQ^0(d)$

In this section, we investigate the forgetful functor $SQX(d) \to SQ^0(d)$.

Proposition 5.1. *The morphism $SQX(d) \to SQ^0(d)$ is a finite etale Galois covering with Galois group H.*

Proof. Let $((E,T),(\mathcal{L},\varphi))$ be an object of $SQ^0(d)(S)$. By changing $\mathcal{L}$ to $\mathcal{L} \otimes \phi^*\mathcal{M}$, where $\mathcal{M} = \mathcal{H}om_{G(d)}(\mathcal{O}_S[H^\vee], \phi_*\mathcal{L})$, we may assume that $\phi_*\mathcal{L}$ is isomorphic to $\mathcal{O}_S[H^\vee]$ as $\mathcal{O}_S$-module with a $G(d)$-action. For an S-scheme $\theta : U \to S$, $E \times_S U$, $\theta^*\mathcal{L}$ and $\phi \times_T U$ are denoted by E_U, $\mathcal{L}_U$ and ϕ_U respectively. The functor $F = S \times_{SQ^0(d)} SQX(d)$ is given by

$$F : (U/S) \mapsto \{\text{sections } p : U \to E_U \text{ such that there exists an isomorphism } \Omega^1_{E_U/U}(\log Hp) \overset{\alpha}{\simeq} \mathcal{L}_U \text{ compatible with } H\text{-linearization}\}.$$

Let $s_{\mathcal{L}} : S \to Pic^d(E)$ be the section of $Pic^d(E) \to S$ corresponding to $\mathcal{L}$ and $\mu : E \to Pic^d(E)$ the morphism obtained by attaching $p \mapsto \Omega^1_{E/S}(\log Hp)$. Let $M = \mu^{-1}(s_{\mathcal{L}}(S)) \subset E$. By Abel's theorem, M is an étale K-torsor over S.

$$\begin{array}{ccc} M & \longrightarrow & s_{\mathcal{L}}(S) \\ \cap\downarrow & & \downarrow\cap \\ E & \xrightarrow[\mu]{} & Pic^d(E) \end{array}$$

By the definition of μ and $Pic^d(E)$, to give a morphism U to M is equivalent to giving a morphism $p : U \to E_U$ such that $\Omega^1_{E_U/U}(\log Hp)$ is isomorphic to $\mathcal{L}_U \otimes \phi_U^* \mathcal{M}$ for some invertible sheaf $\mathcal{M}$ on U. We compare the natural H-linearization N_h of $\Omega^1_{E_U/U}(\log Hp)$ and the given H-linearization of $\mathcal{L}_U$ via the isomorphism α. Let h be a generator of H.

$$(5.1) \qquad \begin{array}{ccc} T_h^* \mathcal{L}_U \otimes \phi_U^* \mathcal{M} & \xrightarrow{\varphi_h} & \mathcal{L}_U \otimes \phi_U^* \mathcal{M} \\ T_h^*\alpha \uparrow & & \uparrow \alpha \\ T_h^* \Omega^1_{E_U/U}(\log Hp) & \xrightarrow[N_h]{} & \Omega^1_{E_U/U}(\log Hp) \end{array}$$

Then there exists $g \in \mathcal{O}_U^*$ such that $\alpha \circ N_h = g \cdot \varphi_h \circ T_h^* \alpha$. By taking the norms, we conclude $g^d = 1$. Therefore g is locally constant, a d-th root of unity on each connected component. If we apply $U \simeq M$, then M can be decomposed as $M = \coprod_{\xi \in \mu_d} M(\xi)$, where $M(\xi) = \{m \in M \mid g(m) = \xi\}$. Note that the diagram (5.1) commutes if and only if $T \to M$ factors through $M(1)$. Therefore the functor F is represented by $M(1)$. Therefore to prove the theorem, it is enough to prove the following proposition.

Proposition 5.2. *For $g \in K$, $gM(\xi) = M(\langle g, h\rangle \xi)$.*

Proof. Since g is a locally constant function on M, it is enough to show that $T_g p \in M(\langle g, h\rangle \xi)$ where $p : Spec(k) \to M(\xi)$ is a k-valued point of $M(\xi)$. (We may assume that k is a field.) Let q be the composite $Spec(k) \to M(\xi) \to SQ^0(d)$ and let $((E_q, T), (\mathcal{L}, \varphi))$ be the corresponding genus 1 curve with an admissible $G(d)$-linearization. Then by the definition of $M(\xi)$, there is an isomorphism $\alpha : \Omega^1_{E_q}(\log Hp) \to \mathcal{L}$ such that the following diagram commutes.

$$(5.2) \qquad \begin{array}{ccc} T_h^* \mathcal{L} & \xrightarrow{\varphi_h} & \mathcal{L} \\ T_h^*\alpha \uparrow & & \uparrow \alpha \\ T_h^* \Omega^1_{E_q}(\log Hp) & \xrightarrow[\xi^{-1} N_h]{} & \Omega^1_{E_q}(\log Hp) \end{array}$$

Let $g \in K$ and $T_g(p) = p + g$, then there exists an isomorphism

$$\beta : \Omega^1_{E_q}(\log H(p+k)) \to \Omega^1_{E_q}(\log Hp).$$

Then β corresponds to a rational function f on E_q such that $(f) = (H(p+g)) - (Hp)$. Therefore the following diagram commutes.

$$\begin{array}{ccc} T_h^*\Omega^1_{E_q}(\log Hp) & \xrightarrow[\langle g,h\rangle N_h]{} & \Omega^1_{E_q}(\log Hp) \\ {\scriptstyle T_h^*\beta}\uparrow & & \uparrow{\scriptstyle\beta} \\ T_h^*\Omega^1_{E_q}(\log H(p+g)) & \xrightarrow[N_h]{} & \Omega^1_{E_q}(\log H(p+g)) \end{array} \tag{5.3}$$

By the commutative diagrams (5.2) and (5.3), we get the proposition.

Theorem 5.3. *The morphism $J : SQ^0(d) \to X(d)$ obtained by attaching the Picard scheme $Pic^d(E)$ to E is an isomorphism.*

Moreover, combining with Proposition 4.8 we get the following theorem.

Theorem 5.4. *Assume that d is even. The universal family $E_{SQ}(d) \to SQ^0(d)$ has no section, if the base field is $\mathbf{C}$.*

Proof. Assume that $E_{SQ}(d) \otimes \mathbf{C} \to SQ(d) \otimes \mathbf{C} \simeq X(d) \otimes \mathbf{C}$ has a section. Then $E_{SQ}(d) \otimes \mathbf{C} \simeq S(d) \otimes \mathbf{C}$. Since $E_{SQ}(d) \otimes \mathbf{C}$ has a $G(d)$-linearized line bundle, this contradicts Proposition 4.8.

Remark 5.5. *If d is odd, then $E_{SQ}(d) \to SQ^0(d)$ is isomorphic to $S(d) \to X(d)$ via the isomorphism J in Theorem 5.3.*

§6 Descent of Heisenberg structure

In this section, we construct the universal family $E_{SQ}(d)$ of genus 1 curves on $SQ^0(d)$ and the universal $G(d)$-structure from the classical universal family $S(d) \to X(d)$ with level d structure by using descent theory. As an application, we show that $E_{SQ}(d)/(\mathbf{Z}/2\mathbf{Z} \oplus \mathbf{Z}/2\mathbf{Z}) \simeq S(d)$. From now on we assume that d is even in this section. We choose a connected component $\tilde{X}(d)$ of $SQX(d)$, which is a double covering of $X(d)$. (See Theorem 4.1.) Let $\tilde{S}(d)$ be the base change $S(d) \times_{X(d)} \tilde{X}(d)$ of the universal curve $S(d)$. Let ι and $\tilde{\iota}$ be the involution of $\tilde{X}(d)$ and $\tilde{S}(d)$ over $X(d)$ and $S(d)$ respectively. The $SQX(d)$-scheme structure of $\tilde{X}(d)$ gives an H-admissible $G(d)$-linearization $\tilde{\varphi}$ for the sheaf $\tilde{\mathcal{L}} = \Omega^1_{\tilde{S}(d)/\tilde{X}(d)}(\log H)$ on $\tilde{S}(d)$: $\tilde{\varphi}_g : T_g^*\tilde{\mathcal{L}} \to \tilde{\mathcal{L}}$. Since $\tilde{\mathcal{L}}$ is a pull-back of $\mathcal{L} = \Omega^1_{S(d)/X(d)}(\log H)$, there is a natural descent data $\nu : \tilde{\iota}^*\tilde{\mathcal{L}} \to \tilde{\mathcal{L}}$ for $\tilde{\mathcal{L}}$. The

1-cocycle condition for ν is $\tilde{\iota}^*\nu \circ \nu = id$. For $g \in K$, we have the following diagram, which might be not commutative.

$$(D_g) \qquad \begin{array}{ccc} \tilde{\iota}^* T_g^* \tilde{\mathcal{L}} = T_g^* \tilde{\iota}^* \tilde{\mathcal{L}} & \xrightarrow{T_g^*\nu} & T_g^* \tilde{\mathcal{L}} \\ {\scriptstyle \tilde{\iota}^*\tilde{\varphi}_g}\downarrow & & \downarrow{\scriptstyle \tilde{\varphi}_g} \\ \tilde{\iota}^*\tilde{\mathcal{L}} & \xrightarrow[\nu]{} & \tilde{\mathcal{L}} \end{array}$$

The morphism $\tilde{\varphi}_g$ descends to a morphism $\varphi_g : T_g^*\mathcal{L} \to \mathcal{L}$ if and only if the diagram (D_g) commutes. For $h \in H$, H-admissibility implies that $\tilde{\varphi}_h$ is a base change of $\varphi_h = N_h$. Therefore (D_h) commutes for $h \in H$. As for the diagram $(D_{h^\vee})$, we have the following identity.

Lemma 6.1.

$$\tilde{\varphi}_{h^\vee} \circ T_g^*\nu = -\nu \circ \tilde{\iota}^* \tilde{\varphi}_{h^\vee}.$$

Proof. To choose a connected component $\tilde{X}(d)$ is equivalent to choosing a d-th root τ of $Nm(f)$. Then $\tau \in \mathcal{O}_{\tilde{X}(d)}^\times - \mathcal{O}_{X(d)}$ and $\iota^*\tau = -\tau$. By the explicit description of the isomorphism $Spec(\mathcal{O}_{X(d)}[t]/(t^d - Nm(f))) \simeq SQX(d)$ in Proposition 2.2, we have $\varphi_{h^\vee} = \tau^{-1} f : T_{h^\vee}^*\mathcal{L} \to \mathcal{L}$. This implies the lemma.

Now we construct a twist $S(d)^\#$ of $S(d)$ on $X(d)$ with a K-translation structure $T^\#$, an invertible sheaf $\mathcal{L}^\#$ and $G(d)$-linearization $\varphi^\# = \{\varphi_g^\#\}_g$. To make a twist of $S(d)$, it is sufficient to give descent data $\mu^\# : \iota^*\tilde{S}(d) \to \tilde{S}(d)$ and $\nu^\# : \iota_1^*\tilde{\mathcal{L}} \to (\mu^\#)^*\tilde{\mathcal{L}}$ satisfying the following conditions. Here, $\iota^*\tilde{S}(d)$ and ι_1 are defined by the fiber product:

$$\begin{array}{ccc} \iota^*\tilde{S}(d) & \xrightarrow{\iota_1} & \tilde{S}(d) \\ \downarrow & & \downarrow \\ \tilde{X}(d) & \xrightarrow[\iota]{} & \tilde{X}(d) \end{array}$$

(1) The morphism $\mu^\#$ is an isomorphism as $\tilde{X}(d)$-schemes and $\nu^\#$ is a sheaf isomorphism on $i^*\tilde{S}(d)$.

We define $\iota^*\mu^\#$ and ι_2 by the following fiber product.

$$\begin{array}{ccccc} \tilde{X}(d) & \longleftarrow & \iota^*\tilde{S}(d) & \xleftarrow{\iota^*\mu^\#} & \tilde{S}(d) = \iota^*\iota^*\tilde{S}(d) \\ {\scriptstyle \iota}\downarrow & & {\scriptstyle \iota_1}\downarrow & & \downarrow{\scriptstyle \iota_2} \\ \tilde{X}(d) & \longleftarrow & \tilde{S}(d) & \xleftarrow{\mu^\#} & \iota^*\tilde{S}(d) \end{array}$$

(2) 1-cocycle conditions are satisfied, i.e. $\mu^{\#} \circ i^*\mu^{\#} = id$ and the composite

$$\tilde{\mathcal{L}} = \iota_2^*\iota_1^*\tilde{\mathcal{L}} \overset{\iota_2\nu^{\#}}{\to} \iota_2^*(\mu^{\#})^*\tilde{\mathcal{L}} = (\iota^*\mu^{\#})^*\iota_1^*\tilde{\mathcal{L}}$$
$$\overset{(\iota^*\mu^{\#})^*\nu^{\#}}{\to} (\iota^*\mu^{\#})^*(\mu^{\#})^*\tilde{\mathcal{L}} = \tilde{\mathcal{L}}$$

is the identity.

(3) Compatibility for the K-action.

$$\begin{array}{ccc} \iota^*\tilde{S}(d) & \xrightarrow{\mu^{\#}} & \tilde{S}(d) \\ {\scriptstyle \iota^*T_g}\downarrow & & \downarrow{\scriptstyle T_g} \\ \iota^*\tilde{S}(d) & \xrightarrow[\mu^{\#}]{} & \tilde{S}(d) \end{array}$$

(4) Compatibility for the $G(d)$-linearization

$$\begin{array}{ccccc} \iota_1\tilde{\mathcal{L}} & \longrightarrow & (\mu^{\#})^*\tilde{\mathcal{L}} & \xleftarrow[(\mu^{\#})^*\tilde{\varphi}_g]{} & (\mu^{\#})^*T_g^*\tilde{\mathcal{L}} \\ {\scriptstyle \iota_1^*\tilde{\varphi}_g}\uparrow & & & & \uparrow\simeq \\ \iota_1^*T_g^*\tilde{\mathcal{L}} & \xrightarrow[\simeq]{} & (\iota^*T_g)^*\iota_1^*\tilde{\mathcal{L}} & \xrightarrow[(\iota^*T_g)^*\nu]{} & (\iota^*T_g)^*(\mu^{\#})^*\tilde{\mathcal{L}} \end{array} .$$

By condition (2), we obtain a genus 1 curve $S(d)^{\#}$ on $X(d)$ and a line bundle $\mathcal{L}^{\#}$ on $S(d)^{\#}$. Condition (3) assures that the K-action descends to $S(d)^{\#}$ and by condition (4), the $G(d)$-linearization $\varphi = \{\tilde{\varphi}_g\}$ on $\tilde{\mathcal{L}}$ descends to that on $\mathcal{L}^{\#}$, say $\varphi^{\#} = \{\varphi_g^{\#}\}$. Note that the original descent data for $S(d)$, $\mathcal{L}$ satisfies all the conditions except (4). For a better understanding, see the example for $d = 4$ in §8.

We define $\mu^{\#}$ and $\nu^{\#}$. Since the morphism T_g is defined on $S(d)$, the following diagram is commutative.

$$\begin{array}{ccc} \tilde{S}(d) & \xleftarrow[\mu]{} & \iota^*\tilde{S}(d) \\ {\scriptstyle T_g}\downarrow & & \downarrow{\scriptstyle \iota^*T_g} \\ \tilde{S}(d) & \xleftarrow[\mu]{} & \iota^*\tilde{S}(d). \end{array}$$

Now we define $\mu^{\#} = T_v \circ \mu$. Next consider the following diagram:

$$\begin{array}{ccc} (\mu^{\#})^*\tilde{\mathcal{L}} & \xrightarrow[\simeq]{} & \iota_1^*\tilde{\mathcal{L}} \\ \simeq\downarrow & & \uparrow{\scriptstyle \nu} \\ \mu^*T_v^*\tilde{\mathcal{L}} & \xrightarrow[\mu^*\tilde{\varphi}_v]{} & \mu^*\tilde{\mathcal{L}}. \end{array}$$

The sheaf isomorphism $\nu^{\#}$ is defined by $\nu^{\#} = \nu \circ \mu^*\tilde{\varphi}_v$.

Proposition 6.2. *The isomorphisms $\mu^{\#}$ and $\nu^{\#}$ satisfy the conditions (1) - (4) of descent data.*

Proof. The proof is done by diagram chasing. Here, we give the diagrams; the diagram chasings are left to the readers.

Condition (2). First, the 1-cocycle condition for $\mu^{\#}$.

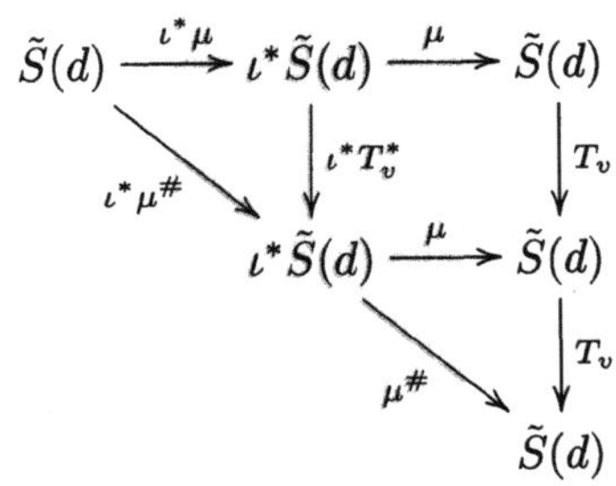

To show the 1-cocycle condition for $\nu^{\#}$, we give a diagram of varieties for diagram chasing invertible sheaves on $\tilde{S}(d)$.

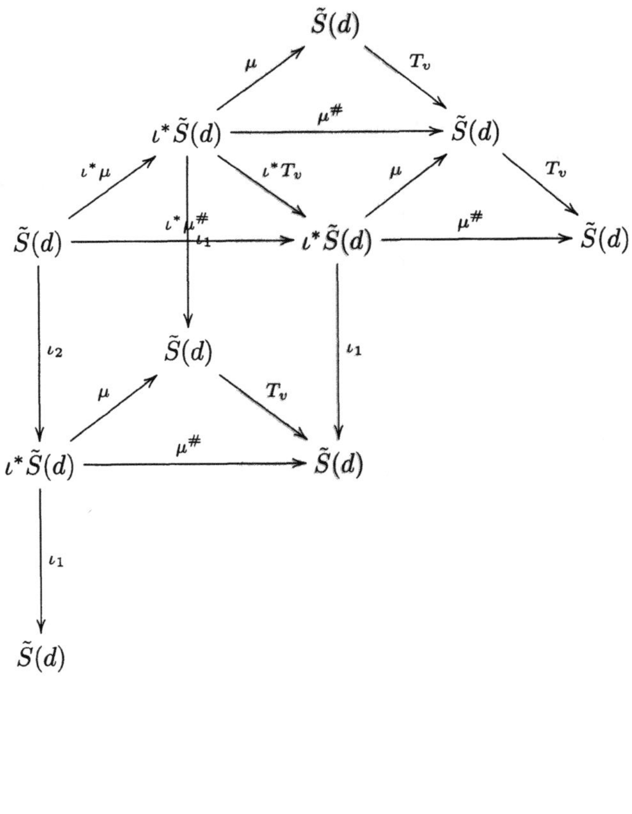

To check the 1-cocycle condition for $\nu^\#$, chase the following diagram.

$$\begin{array}{ccccccc}
 & (\iota^*\mu^\#)^*(\mu^\#)^*\tilde{\mathcal{L}} & \xrightarrow{=} & (\iota^*\mu^\#)^*\mu^*T_v^*\tilde{\mathcal{L}} & & & \\
 \swarrow{\scriptstyle (\iota^*\mu^\#)^*\nu^\#} & & \swarrow{\scriptstyle (\iota^*\mu^\#)^*\mu^*\tilde\varphi_v} & & \searrow{\scriptstyle id} & & \\
(\iota^*\mu^\#)^*\iota_1^*\tilde{\mathcal{L}} & \xleftarrow{(\iota^*\mu^\#)^*\nu} & (\iota^*\mu^\#)^*\mu^*\tilde{\mathcal{L}} & \xrightarrow{=} & (\iota^*\mu)^*\mu^*T_v^*\tilde{\mathcal{L}} & & \\
\downarrow{\scriptstyle =} & & & & \searrow{\scriptstyle (\iota^*\mu)^*\mu^*\tilde\varphi_v} & \searrow & \\
\iota_2^*\mu^*T_v^*\tilde{\mathcal{L}} & \xrightarrow{\iota_2^*\mu^*\tilde\varphi_v} & \iota_2^*\mu^*\tilde{\mathcal{L}} & \xleftarrow{=} & (\iota^*\mu)^*\iota_1^*\tilde{\mathcal{L}} & \xleftarrow{(\iota^*\mu)^*\nu} & (\iota^*\mu)^*\mu^*\tilde{\mathcal{L}} \\
\searrow{\scriptstyle \iota_2^*\nu^\#} & & \downarrow{\scriptstyle \iota_2^*\nu} & & \swarrow{\scriptstyle id} & & \\
 & & \iota_2^*\iota_1^*\tilde{\mathcal{L}} & & & &
\end{array}$$

Condition (3).

$$\begin{array}{ccccc}
\tilde S(d) & \xleftarrow{T_v} & \tilde S(d) & \xleftarrow{\mu} & \iota^*\tilde S(d) \\
\uparrow{\scriptstyle T_g} & & \uparrow{\scriptstyle T_g} & & \uparrow{\scriptstyle \iota^*T_g} \\
\tilde S(d) & \xleftarrow{T_v} & \tilde S(d) & \xleftarrow{\mu} & \iota^*\tilde S(d).
\end{array}$$

Condition (4). First note that the following diagram is commutative by the definition of Weil pairing.

$$\begin{array}{ccccc}
T_v^*T_g^*\tilde{\mathcal{L}} & \xrightarrow{=} & T_g^*T_v^*\tilde{\mathcal{L}} & \xrightarrow{T_g^*\tilde\varphi_g} & T_g^*\tilde{\mathcal{L}} \\
\downarrow{\scriptstyle T_v\tilde\varphi_g} & & & & \downarrow{\scriptstyle \langle g,v\rangle\tilde\varphi_g} \\
T_v^*\tilde{\mathcal{L}} & & \xrightarrow[\tilde\varphi_v]{} & & \tilde{\mathcal{L}}
\end{array}$$

Therefore the following diagram is commutative.

$$\begin{array}{ccccccc}
(\mu^\#)^*T_g^*\tilde{\mathcal{L}} & \xrightarrow{=} & \mu^*T_v^*T_g^*\tilde{\mathcal{L}} & & T_g^*\mu^*\tilde{\mathcal{L}} & \xrightarrow{T_g^*\nu} & T_g^*\iota_1^*\tilde{\mathcal{L}} \\
\Big\downarrow & & \downarrow{\scriptstyle =} & & \downarrow{\scriptstyle =} & & \downarrow{\scriptstyle =} \\
\Big\downarrow & & \mu^*T_g^*T_v^*\tilde{\mathcal{L}} & \xrightarrow[\mu^*T_g^*\tilde\varphi_v]{} & \mu^*T_g^*\tilde{\mathcal{L}} & & \iota_1^*T_g^*\tilde{\mathcal{L}} \\
\Big\downarrow & & & & \downarrow{\scriptstyle \langle g,v\rangle\mu^*\tilde\varphi_g} & & \downarrow{\scriptstyle \iota_1^*\tilde\varphi_g} \\
(\mu^\#)^*\tilde{\mathcal{L}} & \xrightarrow{=} & \mu^*T_v^*\tilde{\mathcal{L}} & \xrightarrow[\mu^*\tilde\varphi_v]{} & \mu^*\tilde{\mathcal{L}} & \xrightarrow[\nu]{} & \iota_1^*\tilde{\mathcal{L}}
\end{array}$$

Definition. The descent genus 1 curve on $X(d)$ with a translation structure and an admissible $G(d)$-linearized invertible sheaf are denoted by

$$((S(d)^{\#}, T^{\#}), (\mathcal{L}^{\#}, \varphi^{\#})).$$

Proposition 6.3. *Via the isomorphism $S(d) \to X(d)$, $S(d)^{\#}$ is isomorphic to $E_{SQ}(d)$.*

Proof. Since $((S(d)^{\#}, T^{\#}), (\mathcal{L}^{\#}, \varphi^{\#}))$ is a genus 1 curve on $X(d)$ with an admissible $G(d)$-linearized line bundle $\mathcal{L}^{\#}$, we get a morphism $\alpha : X(d) \to SQ^0(d)$ such that

$$\begin{array}{ccc} S(d)^{\#} & \longrightarrow & E_{SQ}(d) \\ \downarrow & & \downarrow \\ X(d) & \underset{\alpha}{\longrightarrow} & SQ^0(d) \end{array}$$

is a fiber product. By taking the Jacobian fibration, the following diagram is a fiber product

$$\begin{array}{ccc} Pic^0(S(d)^{\#}) & \longrightarrow & S(d) \\ \downarrow & & \downarrow \\ X(d) & \underset{J\circ\alpha}{\longrightarrow} & X(d). \end{array}$$

On the other hand, by the construction of $S(d)^{\#}$, $Pic^0(S(d)^{\#}) \simeq S(d)$ as elliptic curves with level structures. Therefore $J \circ \alpha$ is an isomorphism. Since the natural morphism $J : SQ^0(d) \to X(d)$ is an isomorphism by Theorem 5.3, α is also an isomorphism and $S(d)^{\#} \to E_{SQ}(d)$ is an isomorphism.

Lemma 6.4. *Let V be the subgroup $V = \langle h^{d/2}, (h^{\vee})^{d/2} \rangle$ of the translation group K. The quotient $S(d)^{\#}/V$ of $S(d)^{\#}$ by V is isomorphic to $S(d)$ over $X(d)$.*

Proof. This is a consequence of the following commutative diagram:

$$\begin{array}{ccc} \iota^* S(d) & \xrightarrow{\mu^{\#}} & S(d) \\ \downarrow & & \downarrow \\ \iota^* S(d)/V & \xrightarrow{\mu^{\#}} & S(d)/V \\ \simeq\downarrow & & \downarrow\simeq \\ \iota^* S(d) & \xrightarrow{\mu} & S(d). \end{array}$$

By combining Proposition 6.3 and Lemma 6.4, we obtain the following theorem.

Theorem 6.5.

$$E_{SQ}(d)/V \simeq S(d)$$

§7 Minimal model for E_{SQ}

In this section, we investigate the relatively minimal model $\bar{E}_{SQ}(d)$ of a compactification of $E_{SQ}(d)$ over $\bar{X}(d)$. We assume that d is even in this section.

Proposition 7.1. *The type of all the singular fibers of $\bar{E}_{SQ}(d)$ is I_d.*

Proof. Let $\overline{\tilde{X}(d)}$ be the compactification of $\tilde{X}(d)$ defined in §6. Let $p \in \overline{\tilde{X}(d)} - \tilde{X}(d)$ be a unramified point over $X(d)$, then by the construction of $S(d)^{\#}$, the singular fiber is isomorphic to that of $\bar{S}(d) \to \bar{X}(d)$. Therefore the singular fiber is isomorphic to I_d (see [K], [S]). Let $p \in \overline{\tilde{X}(d)} - \tilde{X}(d)$ be a ramified point over $X(d)$. Then by Corollary 4.7 the translation T_v stabilizes all the irreducible components of $S(d)_p$. Let $U \ni p$ be a sufficiently small neighbourhood of p, $U^* = U - \{p\}$ and $\tilde{U}$ (resp. $\tilde{U}^*$) be the double covering of U (resp. U^*). Then $S(d)^{\#} |_{U^*}$ is isomorphic to the quotient of $\tilde{S}(d) |_{\tilde{U}^*}$ by the involution $T_v \circ \tilde{\iota}$, where $\tilde{\iota}$ is the non-zero element of $\mathrm{Gal}(\tilde{S}(d)/S(d))$. Now we compute the relatively minimal model of $\overline{\tilde{S}(d)} |_{\tilde{U}} /\langle T_v \circ \tilde{\iota}\rangle$, where $\overline{\tilde{S}(d)}$ is the pull back of $\bar{S}(d)$ by $\overline{\tilde{X}(d)}$. The variety $\bar{S}(d) |_U$ is denoted by X. Then the fiber $\bar{S}(d)_p \subset X$ is a cycle of rational curves as in §4. The fiber product $X \times_U \tilde{U}$ is branching at the normal crossing divisor $S(d)_p$. To get a smooth branched covering, we blow up X at the singular point of $S(d)_p$ to get $\hat{X}$. The natural morphism $\hat{X} \to X$ is denoted by b. Then the proper transform of $\bar{S}(d)_p$ is a disjoint union of $\mathbf{P}^1$'s and the normalization $\tilde{X} \to \hat{X}$ of $X \times_U \tilde{U}$ over $\hat{X}$ is unramified at the exceptional divisors. Let E_i be the exceptional divisor over the intersection of D_i and D_{i+1}, and $\hat{D}_i$ the proper transform of D_i. The multiplicities $\mu(\hat{D}_i)$, $\mu(E_i)$ of $\hat{D}_i$ and E_i in $b^*(\bar{S}(d)_p)$ and their self intersections $(E_i)^2$, $(\hat{D}_i)^2$ are as follows:

$$\mu(\hat{D}_i) = 1, \quad \mu(E_i) = 2, \quad (\hat{D}_i)^2 = -4, \quad (E_i)^2 = -1.$$

By taking a double covering, the special fiber for $\tilde{X} \to \tilde{U}$ is isomorphic to I_{2d}. Here $\tilde{D}_i$ is the reduced part of the inverse image of $\hat{D}_i$ and $\tilde{E}_i$ is the inverse image of E_i. The involution T_v induces an involution $\tilde{T}_v$ on $\tilde{X}$ and it stabilizes all the components $\tilde{D}_i$ and $\tilde{E}_i$ and acts on them effectively. The involution $\tilde{\iota}$ induces an involution on $\tilde{X}$ and it is also denoted by $\tilde{\iota}$. Then $\tilde{\iota}$ fixes $\tilde{D}_i$ and acts effectively on $\tilde{E}_i$. Therefore the involution $T_v \circ \tilde{\iota}$ fixes $\tilde{E}_i$ and acts on $\tilde{D}_i$ effectively. Then the quotient $\hat{X}^{\#} = \tilde{X}/\langle T_v \circ \tilde{\iota}\rangle$ is a smooth variety over U. The natural map $\tilde{X} \to \hat{X}^{\#}$ is denoted by $\pi^{\#}$ and we put $\pi^{\#}(\tilde{E}_i) = \hat{E}_i^{\#}$ and $\pi^{\#}(\tilde{D}_i) = D_i^{\#}$. The multiplicities of the components in the fiber of $\hat{X}^{\#} \to U$

and self intersections of $\hat{E}_i^\#$ and $D_i^\#$ are given as follows:

$$\mu(\hat{E}_i^\#) = 1, \quad \mu(D_i^\#) = 2, \quad (\hat{E}_i^\#)^2 = -4, \quad (D_i^\#)^2 = -1.$$

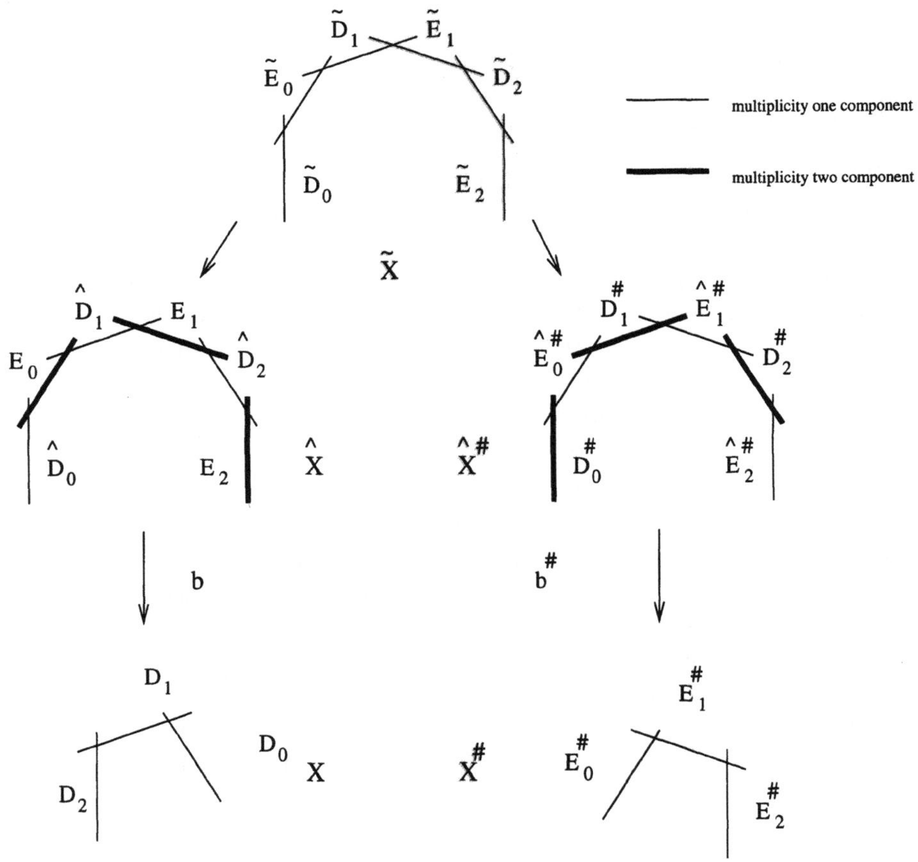

By blowing down $D_i^\#$, we get $X^\#$ with I_d type singularities.

Proposition 7.2. *The varieties $\bar{S}(d)$ and $\bar{S}(d)^\#$ have the same Euler characteristic.*

§8 Example for $n = 4$

In this section, we illustrate an example for $n = 4$. For simplicity, we assume that the base field is $\mathbf{C}$, but almost all arguments are valid for any characteristic.

In this case, the function fields of $X(4)$ and $SQ(4)$ are rational function fields $\mathbf{C}(\sigma)$ and $\mathbf{C}(\lambda)$, respectively. The generic fiber $S(4)_\eta$ of $S(4) \to X(4)$ is given by the famous Jacobi quartic (see [S]):

$$S(4)_\eta : y^2 = (1 - \sigma^2 x^2)(1 - \sigma^{-2} x^2).$$

The group of $\mathbf{C}(\sigma)$-valued points of $S(4)_\eta$ consists of 4-torsion points and they are given by $x = \pm\sigma, \pm\sigma^{-1}, \pm 1, \pm\sqrt{-1}, 0, \infty$. The equation of the generic fiber $E_{SQ}(4)_\eta$ of $E_{SQ}(4) \to SQ(4)$ is given by the $(2,2)$-complete intersection in $\mathbf{P}^3$ (see [N2]):

$$E_{SQ}(4)_\eta : \begin{cases} X_0^2 + X_2^2 = \lambda X_1 X_3, \\ X_1^2 + X_3^2 = \lambda X_0 X_2. \end{cases}$$

The K-action of $E_{SQ}(4)_\eta$ is given by

$$h^\vee : X_i \mapsto X_{i+1}, \quad h : X_i \mapsto (\sqrt{-1})^i X_i.$$

Here we used the cyclic notation, $X_{i+4} = X_i$. If we consider the following finite Galois morphism π,

$$\begin{aligned} \pi : \mathbf{P}^3 &\to \mathbf{P}^3 \\ (X_i)_i &\mapsto (T_0 : T_1 : T_2 : T_3) \\ &= ((X_0 + X_2)^2, (X_0 - X_2)^2, (X_1 + X_3)^2, (X_1 - X_3)^2), \end{aligned}$$

the variety $E_{SQ}(4)_\eta$ is nothing but the inverse image of a line L_λ under the morphism π, where

$$L_\lambda : \begin{cases} 2(T_0 + T_1) = \lambda(T_2 - T_3) \\ 2(T_2 + T_3) = \lambda(T_0 - T_1). \end{cases}$$

Therefore $E_{SQ}(4)_\eta$ is a finite Galois covering with group $(\mathbf{Z}/2)^{\oplus 4}/\Delta(\mathbf{Z}/2)$, where $\Delta(\mathbf{Z}/2)$ is the diagonal subgroup $\{(a,a,a,a) \mid a \in \mathbf{Z}/2\}$ of $(\mathbf{Z}/2)^{\oplus 4}$. The subgroup $V = \langle h^2, (h^\vee)^2 \rangle$ corresponds to the subgroup $N/\Delta(\mathbf{Z}/2)$, where $N = \{(a_0, \ldots, a_3) \mid a_0 \cdots a_3 = 1\}$. Therefore the quotient variety $E_{SQ}(4)/V$ corresponds to a double covering of L_λ corresponding to the group $N/\Delta(\mathbf{Z}/2)$. Therefore the rational function field of $E_{SQ}(4)_\eta$ is generated by

$$q = \frac{T_2 - T_3}{T_0 - T_1}, \quad p = \frac{X_0 - X_2}{X_0 + X_2} \cdot \frac{X_1 + X_3}{X_0 + X_2} \cdot \frac{X_1 - X_3}{X_0 + X_2}.$$

Let $k = \mathbf{C}(\lambda)$. Since

$$\frac{T_1}{T_0} = \frac{\lambda q - 2}{\lambda q + 2}, \quad \frac{T_2}{T_0} = \frac{\lambda + 2q}{\lambda q + 2}, \quad \frac{T_3}{T_0} = \frac{\lambda - 2q}{\lambda q + 2},$$

the quadratic extension $k(p,q)$ over the rational function field $k(q)$ is given by

$$\begin{aligned} p^2 &= \frac{(X_0 - X_2)^2(X_1 + X_3)^2(X_1 - X_3)^2}{(X_0 + X_2)^6} = \frac{T_1T_2T_3}{T_0^3} \\ &= \frac{(\lambda q - 2)(\lambda^2 - 4q^2)}{(\lambda q + 2)^3} \\ &= \frac{-\sigma^2}{(\sigma q + 1)^4}(1 - \sigma^2 q^2)(1 - \sigma^{-2}q^2), \end{aligned}$$

where $\lambda = 2\sigma$. Therefore if we put $y = \sqrt{-1}\frac{(\sigma q + 1)^2}{\sigma}p$ and $t = q$, then $E_{SQ}(4)_\eta/V$ is isomorphic to $S(4)_\eta$ under the isomorphism $\mathbf{C}(\sigma) \simeq \mathbf{C}(\lambda)$ given by $\lambda = 2\sigma$.

Now we will show that $E_{SQ}(4)_\eta$ is obtained from $S(4)_\eta$ by twisting using the translation T_v over the torsion point v of order two. By a suitable choice of origin of $S(4)_\eta$, we may assume that the automorphism T_v is given as follows:

$$T_v : y \mapsto -y, \quad x \mapsto -x.$$

There are exactly two surjective homomorphisms $k : \mathbf{Z}/d\mathbf{Z} \oplus \mathbf{Z}/d\mathbf{Z} \to \mathbf{Z}/d\mathbf{Z}$ up to signature which satisfy the ramification condition of Proposition 4.6. If they are denoted by k_1 and k_2, then up to permutation we have

$$k_1(h) = 0, \quad k_1(h^\vee) = 1, \quad k_2(h) = 2, \quad k_2(h^\vee) = 1.$$

The singular points of $S(4)_\eta$ are given by $\sigma = \pm 1, \pm\sqrt{-1}, 0, \infty$. We claim that $\{k_1, k_2\}$ corresponds to $\sigma \in \{0, \infty\}$. Express the family of elliptic curves as a double covering of $(x, \sigma - 1)$ near $\sigma = 1$. Then the branch divisors are given as $D_1 = \{x = \sigma^{-1}\}$, $D_2 = \{x = \sigma\}$, $D_1 = \{x = -\sigma\}$ and $D_1 = \{x = -\sigma^{-1}\}$. By blowing up and taking the double covering, we have the following configuration of divisors.

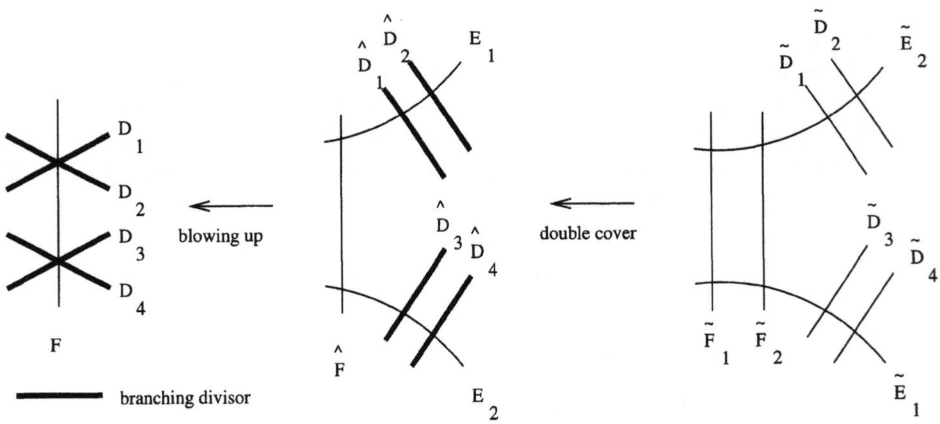

The fiber of the minimal model of the elliptic fibration is equal to $\tilde{F}_1 + \tilde{F}_2 + \tilde{E}_1 + \tilde{E}_2$. Since T_v interchanges components $D_1 \leftrightarrow D_4$ and $D_2 \leftrightarrow D_3$, we have that T_v interchanges components $\tilde{E}_1$ and $\tilde{E}_2$. Therefore T_v does not fix components of fibers. In the same way, at $\sigma = -1, \pm i$, T_v does not fix components of fibers. Therefore $\tilde{X}(4)$ is given by the double covering defined by $\sqrt{2\sqrt{-1}\sigma}$. Therefore the twist $S(4)^{\#}$ of $S(4)$ is obtained by the fixed part of $\mathbf{C}(\sqrt{1\sqrt{-1}\sigma}, x, y)$ under the action

$$\tilde{\iota}^{\#} = T_v \circ \tilde{\iota} : \quad x \mapsto -x, \quad y \mapsto -y, \quad \sqrt{2\sqrt{-1}\sigma} \mapsto -\sqrt{2\sqrt{-1}\sigma}.$$

Actually it is easy to see that $\mathbf{C}(\sqrt{2\sqrt{-1}\sigma}, x, y)^{\tilde{\iota}^{\#}} = \mathbf{C}(\sigma, X, Y)$, where $X = \sqrt{-2\sqrt{-1}\sigma}x, Y = \sqrt{2\sqrt{-1}\sigma}y$. As an elliptic curve on $\mathbf{C}(\sigma)$, it is expressed as

$$S(4)^{\#}_{\eta} : Y^2 = 2\sqrt{-1}\sigma(1 - \frac{\sqrt{-1}\sigma X^2}{2})(1 - \frac{\sqrt{-1}\sigma^{-3}X^2}{2}). \tag{8.1}$$

To show that $S(4)^{\#}_{\eta}$ is isomorphic to $ES_{SQ}(4)_{\eta}$, we consider an involution ι' on $E_{SQ}(4)$ using inhomogeneous coordinates $x_1 = \frac{X_1}{X_0}$, $x_2 = \frac{X_2}{X_0}$ and $x_3 = \frac{X_3}{X_0}$ as follows:

$$\iota' \begin{cases} x_2 \mapsto -x_2 \\ x_1 \mapsto -\sqrt{-1}x_3 \\ x_3 \mapsto \sqrt{-1}x_1 \end{cases}$$

By the coordinate change $p = x_1 + \sqrt{-1}x_3$, $q = x_1 - \sqrt{-1}x_3$, the equation of $E_{SQ}(4)_{\eta}$ is given by

$$1 + x_2^2 = \frac{\lambda}{4\sqrt{-1}}(p^2 - q^2), \quad pq = \lambda x_2 .$$

Lemma 8.1.

$$k(x_1, x_2, x_3)^{\iota'} = k(q).$$

Proof. Since $1 + \frac{pq}{\lambda}^2 = \frac{\lambda}{4\sqrt{-1}}(p^2 - q^2)$, we have

$$p^2 = \frac{(4\sqrt{-1} + 2\sigma q^2)\sigma^2}{2\sigma^3 - \sqrt{-1}q^2}. \tag{8.2}$$

As a consequence, we have $k(q) \supset k(q, p^2) \supset k(p^2, q, x_2)$. On the other hand, $k(p^2, q, x_2^2)(p) = k(p, q, x_2) = k(p, q, x_2)$. Therefore $k(p, q, x_2) \supset k(q) \supset k(p^2, q, x_2)$ and $[k(p, q, x_2) : k(p^2, q, x_2^2)] = 2$ and we get the lemma.

The minimal polynomial of p over $k(q)$ is given by the equation (8.2). Therefore (8.2) is the equation of the elliptic curve $E_{SQ}(4)_\eta$. By substituting

$$Y = \frac{p(\sigma^3 - \frac{\sqrt{-1}q^2}{2})}{\sigma^2}, \quad X = q,$$

(8.2) is equivalent to the equation (8.1) of $S(4)^\#$.

Remark. Both $S(4)^\#$ and $S(4)$ are singular K3 surfaces. And by Theorem 6.5, they are isogenous.

References

[K] Kodaira, K., *On compact analytic surfaces II*, Ann. Math. **77** (1963), 563–626.

[M] Mumford, D., *On the equations defining Abelian varieties, I*, Inv. Math. **3** (1966), 287–354.

[N1] Nakamura, I., *Compactification of moduli of abelian varieties over* $\mathbf{Z}[\zeta_N, 1/N]$, C. R. Acad. Sci. Paris **327** (1998), 875–880.

[N2] Nakamura, I., *Stability of degenerate abelian varieties*, Inv. Math. **136** (1999), 659–715.

[S] Shioda, T., *On elliptic modular surfaces*, J. Math. Soc. Japan **24** (1972), 20–59.

Progress in Mathematics, Vol. 195, © 2001 Birkhäuser Verlag Basel/Switzerland

SINGULARITIES OF THE HEIGHT STRATA IN THE MODULI OF K3 SURFACES

ARTHUR OGUS

ABSTRACT. Moduli spaces of varieties in characteristic p are stratified by Newton polygons. For example, the moduli space of polarized K3 surfaces of degree d admits a stratification by the height h. Using the classification of supersingular K3 crystals, we determine the dimension and singular locus of each stratum. For example, if d is prime to p, the singular locus of the set where the height is at least h coincides with the set of points where the Artin invariant σ_0 is $\leq h-1$.

Let k be a perfect field of characteristic $p > 0$, let W be its Witt ring, and let F_k (resp. F_W) denote the absolute Frobenius automorphism of k (resp. of W). Fix a positive integer d. A *polarized K3 surface of degree d* over a k-scheme S is a smooth projective morphism $f: \mathcal{X} \to S$ and a relatively ample invertible sheaf $\mathcal{L}$ on $\mathcal{X}$ such that for every $s \in S$, the fiber $\mathcal{X}_s$ is a K3 surface and the self-intersection of $\mathcal{L}_s$ on X_s is $2d$. We shall also assume that $\mathcal{L}$ is *p-primitive*, *i.e.*, that $\mathcal{L}_s$ is not divisible by p for any $s \in S$. In his seminal paper [1], Artin used the formal Brauer group to define a height stratification S_h on the base S of a versal family $\mathcal{X}/S$ of polarized K3 surfaces. He found a sequence of closed subschemes

$$S = S_1 \supseteq S_2 \supseteq \cdots S_i \supseteq \cdots$$

and showed that each $S_{i+1} \subseteq S_i$ is defined by a single equation. For example, the "nonordinary locus" S_2 is defined by the vanishing of the action of Frobenius on $R^2 f_*(\mathcal{O}_X)$. For $i \geq 11$, all the S_i have the same reduced structure, and their intersection is the "supersingular locus S_∞," which (at least set-theoretically) admits a further stratification

$$S_{\infty,1} \subseteq S_{\infty,2} \subseteq \cdots S_{\infty,10} = S_\infty$$

by the "Artin invariant" σ_0. Artin proved that, in a neighborhood of a certain point of $S_{\infty,1}$, S_i has codimension $i-1$ if $i \leq 11$. Recently Katsura and van der Geer have looked again at these strata, and proved, among other things, that each S_i is equidimensional and that its singular locus is contained in the supersingular stratum S_{11} [6].

Katsura and van der Geer use the formal group methods of Artin, so to continue the propaganda campaign of [4], in this paper we use crystalline methods to define and analyze the strata S_h and their singularities in more detail. In our attempt to convince the reader that it is not only possible, but even easy, to extract geometric information from crystalline periods, we will try to make our argument self-contained, without reference to the theory of formal groups. This attempt is not meant to hide our gratitude to the aforementioned authors for their inspiration to return to the beautiful subject of K3 surfaces. It is also a pleasure to thank the referee for a very thorough job which resulted in many improvements to our exposition.

Our main new result is easiest to state when the polarization degree is prime to p. In this case the singular locus of S_i is, set-theoretically, $S_{\infty,i-1}$. It follows that S_i is normal if $i \leq 10$ and is generically nonreduced if $i \geq 11$. We also show that $S_{11} = S_\infty$ scheme-theoretically in this case.

The crystalline cohomology of a K3 surface X/k over k is a free W-module $H := H_{cris}(X/W)$ of rank 22. The absolute Frobenius endomorphism F_X of X induces an F_W-linear endomorphism Φ of H, which we sometimes view as a W-linear map $F_W^*H \to H$. The cup-product pairing and trace map endow H with a symmetric bilinear form

$$(\,|\,)\colon Sym^2 H \to W$$

which defines an isomorphism from H to its dual $H^\vee$. It is known (and trivial if p is odd) that this form is *even*: $(x|x)$ is divisible by 2 for every $x \in H$ [5]. Writing $Q(x) := 1/2(x|x)$, we find that Q is a *quadratic form* on H, *i.e.*, $Q \in Sym^2(H^\vee)$, and $(x|y) = Q(x+y) - Q(x) - Q(y)$ for $x, y \in H$. In fact, the triple (H, Φ, Q) forms a *K3-crystal*, as defined below.

Definition 1. *A K3 crystal over k of rank r is a finitely generated free module over W of rank r equipped with a nondegenerate quadratic form Q and an F_W-linear endomorphism $\Phi\colon H \to H$ satisfying the following conditions:*

1. *The reduction of Φ modulo p has rank one.*
2. *$Q(\Phi(x)) = p^2 F_W^*(Q(x))$ for any $x \in H$.*

The nondegeneracy of Q means that the associated bilinear form $(x, y) := Q(x+y) - Q(x) - Q(y)$ defines an isomorphism from H to its dual $H^\vee$.

If L is a line bundle on X, its crystalline Chern class $\xi := c_{cris}(L) \in H^2_{cris}(X/W)$ lies in $T_H := \{x \in H : \Phi(x) = px\}$, and it is p-primitive if and only if ξ is not divisible by p in T_H [4, 1.5]. We therefore define a *polarized K3 crystal* to be a pair (H, ξ), where H is a K3-crystal and ξ is an element of T_H which is not divisible by p. The degree of (H, ξ) is by definition $Q(\xi) \in \mathbf{Z}_p$. If the degree is prime to p and p is odd, then the orthogonal complement of ξ in H is again a K3-crystal. Many of the arguments we shall give can be simplified somewhat in this case by working with this orthogonal complement, and we invite the reader to adopt this point of view.

The structure of K3 crystals over an algebraically closed field k can be described fairly explicitly. Suppose that H is a K3 crystal of rank 22 and that k is algebraically closed. We refer to [3] for the definition of the Hodge and Newton polygons of an F-crystal. The fundamental theorem of Mazur tells us that the Newton polygon of (H, Φ) lies on or above the Hodge polygon, has integral break-points, and, because of the duality coming from Q, is symmetric. We shall call the x-coordinate of its first breakpoint the *Newton height* (or just *height*) of H, and we say that H is *supersingular* if all its Newton slopes are 1; in this case we say that the height is ∞. Mazur's theorem implies that $\mathrm{ht}(H) \geq 1$, and the symmetry implies that $22-2\mathrm{ht}(H)$ is the horizontal length of the Newton segment of slope 1. Since $\Phi(\xi) = p\xi$, this length is at least 1, so $\mathrm{ht}(H) \in \{1, 2 \ldots 10, \infty\}$. If $\mathrm{ht}(H) < \infty$, then according to a theorem of Katz [3], H can be expressed as a direct sum $H = H_h \bigoplus H_1 \bigoplus H'_h$, where H_h is an F-crystal with Hodge numbers $(1, h-1, 0)$, H'_h is the twisted dual of H_h, with Hodge numbers $(0, h-1, 2)$, and $H_1 \cong W \otimes T_H$ has rank $22-2h$ and Hodge numbers $(0, 22-h, 0)$. If $h(H) = \infty$, T has rank 22 and the p-adic ordinal of the discriminant of the restriction of Q to T is $p^{2\sigma_0}$, where σ_0 is an integer with $1 \leq \sigma_0 \leq 11$, called the *Artin invariant* of H. Furthermore, it is shown in [4] that H admits an orthogonal direct sum decomposition $H \cong H_0 \oplus H_1$, where H_0 is a supersingular K3 crystal of rank $2\sigma_0$ and H_1 is a twist of a unit root crystal. The quadratic form restricted to H_1 is nondegenerate, and its restriction to H_0 is p times a nondegenerate form Q'. It is apparent that $\sigma_0 \leq 10$ if H admits a polarization of degree prime to p, and in fact it is always true for the K3 crystals arising from K3 surfaces, as we shall see later.

Suppose that $H := (H, \Phi, Q)$ is a K3 crystal over an algebraically closed field k and $h \geq 1$ is an integer such that Φ^h is divisible by p^{h-1}; write $\Phi^h = p^{1-h}\Phi_h$. Note that $Q(\Phi_h(x)) = p^2 F_W^* Q(x)$ for $x \in H$. The abstract *Hodge and conjugate filtrations* M_h and N_h of (H, Φ_h) are defined as follows:

$$\begin{aligned} M_h^i H &:= \Phi_h^{-1}(p^i H) \\ N_h^i H &:= p^{i-2}\Phi_h(M_h^{2-i} H) \end{aligned}$$

We shall especially consider the filtrations induced by M_h and N_h on H_{DR}, the reduction modulo p of H. It follows from the compatibility of Φ_h and Q that $M_h^1 H_{DR}$ (resp. $N_h^1 H_{DR}$) is the annihilator of $M_h^2 H_{DR}$ (resp. $N_h^2 H_{DR}$). Mazur's theorem implies that, if X/k is a K3 surface, the filtration M_1 on $H_{DR} \cong H_{DR}(X/k)$ is just the Hodge filtration F_{Hdg}, and the filtration N_1 on $H_{DR}(X/k)$ is the conjugate filtration F_{con}. Consequently we shall write F_{Hdg} and F_{con} for the filtrations induced by M_1 and N_1 on H_{DR} in the abstract case as well. It follows that Φ induces an injective map

$$\psi \colon F_k^* \operatorname{Gr}^0_{Hdg} H_{DR} \to H_{DR}$$

whose image is exactly $F^2_{con} H_{DR}$, and an isomorphism

$$\gamma \colon F_k^* \operatorname{Gr}^1_{Hdg} H_{DR} \to \operatorname{Gr}^1_{con} H_{DR}.$$

Note that $F^2_{Hdg} H_{DR}$ and $F^2_{con} H_{DR}$ are isotropic, so that Q induces quadratic forms on $\operatorname{Gr}^1_{Hdg} H_{DR}$ and $\operatorname{Gr}^1_{con} H_{DR}$, and γ is compatible with these forms.

It is possible to compute the Newton slopes of a K3 crystal from the purely mod p data consisting of H_{DR}, F_{Hdg}, F_{con} and γ. We shall define, by induction on h, a class $\mathbf{K3}_h$ of K3 crystals over k, and for each $H \in \mathbf{K3}_h$ an increasing sequence of submodules

$$E_0 \subseteq E_1 \subseteq \cdots E_h$$

of H_{DR}, as follows. For $h = 1$, $\mathbf{K3}_1$ is the class of all K3-crystals over k, and E_1 is the submodule $N_1^2 H_{DR} = F_{con}^2 H_{DR}$. If $\mathbf{K3}_h$ has been defined and $H \in \mathbf{K3}_h$, then by definition $H \in \mathbf{K3}_{h+1}$ if and only if $E_h \subseteq F_{Hdg}^1 H_{DR}$. Then there is a map:

$$F_S^* E_h \to F_S^* F_{Hdg}^1 H_{DR} \to F_S^* \operatorname{Gr}_{Hdg}^1 H_{DR} \xrightarrow{\gamma} \operatorname{Gr}_{con}^1 H_{DR}$$

and E_{h+1} is defined to be the inverse image in $F_{con}^1 H_{DR} \subseteq H_{DR}$ of the image of this map. It follows by induction that each E_{h+1}/E_h has dimension at most one, and, by the compatibility of Q with γ, that each E_h is totally isotropic.

Lemma 2. *Let H be a K3 crystal over k and h a positive integer such that Φ^i is divisible by p^{i-1} for all $i \leq h$. Then $H \in \mathbf{K3}_h$, and*

$$E_h = N_h^2 + E_{h-1} = N_h^2 + N_{h-1}^2 + \cdots N_1^2 \subseteq H_{DR}.$$

Proof: It is convenient to set $E_0 := 0$; then the lemma is true for $h = 1$ by the definition of E_1. Suppose that $h > 1$ and that Φ^i is divisible by p^{i-1} for all $i \leq h$. By the induction assumption, $H \in \mathbf{K3}_{h-1}$, $E_{h-1} = E_{h-2} + N_{h-1}^2 H_{DR}$, and $E_{h-2} \subseteq F_{Hdg}^1 H_{DR}$. By definition, E_{h-1} is the inverse image of $\gamma F_k^*(E_{h-2})$ under the projection $\pi\colon F_{con}^1 H_{DR} \to \operatorname{Gr}_{con}^1 H_{DR}$. Since $\Phi\Phi_{h-1}$ is divisible by p, $N_{h-1}^2 H \subseteq M_1^1 H$, so $E_{h-1} \subseteq F_{Hdg}^1 H_{DR}$ and $H \in \mathbf{K3}_h$. Furthermore, $N_h^2 H$ is the image of $p^{-1}\Phi\Phi_{h-1}H$, *i.e.*, the image of $N_{h-1}^2 H$ under $p^{-1}\Phi$. Modulo $F_{con}^2 H_{DR}$, this is the image of $N_{h-1}^2 H$ under the map

$$N_{h-1}^2 H_{DR} \to F_k^* F_{Hdg}^1 H_{DR} \to \operatorname{Gr}_{Hdg}^1 H_{DR} \xrightarrow{\gamma} \operatorname{Gr}_{con}^1 H_{DR}.$$

which we just denote by $\gamma F_k^*(N_{h-1}^2 H_{DR})$. We conclude that

$$\pi(E_h) := \gamma F_k^*(E_{h-1}) = \gamma F_k^*(E_{h-2} + N_{h-1}^2) = \pi(E_{h-1}) + \pi(N_h^2).$$

Since E_h and $E_{h-1} + N_h^2$ both contain F_{con}^2 and have the same image modulo F_{con}^2, they must be equal. □

Proposition 3. *Let H be a K3 crystal over k and let h be a positive integer. Then the following are equivalent:*

1. *H has Newton height at least h.*
2. *For all $i \leq h$, Φ^i is divisible by p^{i-1}.*
3. *$H \in \mathbf{K3}_h$.*
4. *For all $i < h$, Φ^i is divisible by p^{i-1} and $N_i^2 H_{DR} \subseteq F_{Hdg}^1 H_{DR}$.*

Proof: The proof is by induction on h. The case $h = 1$ is trivial, and we assume that the proposition is true for h and that conditions (1)–(4) all hold, and we will prove their equivalence with $h+1$ in place of h. In particular, E_i is defined for $i \leq h$ and contained in $F^1_{Hdg}H_{DR}$ for $i < h$. It follows immediately from Lemma 2 that $E_h \subseteq F^1_{Hdg}H_{DR}$ if and only if $N^2_h H_{DR} \subseteq F^1_{Hdg}H_{DR}$, so conditions (3) and (4) are equivalent. Since $\Phi^{h+1} = p^{h-1}\Phi\Phi_h$, Φ^{h+1} is divisible by p^h if and only if $N^2_h H_{DR} \subseteq F^1_{Hdg}H_{DR}$. Thus (3) and (4) are also equivalent to (2). If Φ^{h+1} is divisible by p^h, then its Newton slopes are at least h, and hence the Newton slopes of Φ are at least $h/(h+1)$ and Φ has Newton height at least $h+1$. Conversely, suppose that H has Newton height at least $h+1$. The Newton slopes of Φ are at least $h/(h+1)$, and so the Newton slopes of $\Phi_h = p^{1-h}\Phi^h$ are at least $h^2/(h+1) - (h-1) = 1/(h+1) > 0$. Then Φ_h has no unit root part, and its action on H_{DR} is nilpotent. Since Φ_h and Φ commute, the image $F^2_{con}H_{DR}$ of Φ on H_{DR} is invariant under Φ_h, and since it is one-dimensional, Φ_h acts as zero on $F^2_{con}H_{DR}$. This says that $\Phi_h\Phi$ is divisible by p, *i.e.*, that Φ^{h+1} is divisible by p^h. □

Lemma 4. *Suppose that $H \in \mathbf{K3}_h$ and $N^2_h H_{DR} \subseteq E_{h-1} + F^2_{Hdg}H_{DR}$. Then H is supersingular.*

Proof: It follows from the hypothesis and Lemma 2 that for any $x \in H$, $\Phi_h(x) \in N^2_{h-1}H + \cdots N^2_1 H + pH + M^2 H$. Thus, there exist elements $x_1, \ldots x_h \in H$ and $z \in M^2 H$ such that

$$\Phi_h(x) = \Phi_{h-1}(x_1) + \Phi_{h-2}(x_2) + \cdots \Phi(x_{h-1}) + px_h + z.$$

Multiplying by p^{h-1} and applying Φ, we find that

$$\Phi^{h+1}(x) = p\Phi^h(x_1) + p^2\Phi^{h-1}(x_2) + \cdots p^h\Phi(x_h) + p^{h+1}x_{h+1},$$

where $x_{h+1} := p^{-2}\Phi(z)$. Then it follows by induction on m that for any $m \geq 1$ and $x \in H$, there exist $x_i \in H$ such that

$$\Phi^{h+m}(x) = p^m\Phi^h(x_1) + p^{m+1}\Phi^{h-1}(x_2) + \cdots p^{m+h}\Phi(x_h) + p^{h+1+m}x_{h+1}.$$

In particular Φ^{h+m} is divisible by p^m, and all its slopes are at least m. Then all the slopes of Φ are at least $m/(h+m)$, and taking the limit with m, we see that the slopes of Φ are all 1, *i.e.*, H is supersingular. □

The following is the key semi-linear algebra result that enables us to determine the singularities of the height strata.

Lemma 5. *Let H be a K3 crystal over k.*

1. *If $\mathrm{ht}(H)$ is finite, and $i \leq \mathrm{ht}(H)$, then $E_i = N^2_i H_{DR}$ and has dimension i.*
2. *If H is supersingular and $i \leq \sigma_0(H)$, then $E_i = N^2_i H_{DR}$ and has dimension i. Moreover, $E_{\sigma_0} = E_{\sigma_0 - 1} + F^2_{Hdg}H_{DR}$, and for $i \geq \sigma_0$, $E_i = E_{\sigma_0}$ and is the image of the annihilator of Q on $k \otimes T_H$ and also the annihilator of the image of $k \otimes T_H \to H_{DR}$.*

Proof: Suppose that $\mathrm{ht}(H)$ is finite and $i \leq \mathrm{ht}(H)$. Then by Lemmas 2 and 4, E_{i-1} is strictly contained in E_i. On the other hand, $\dim E_i \leq i$ for all i, so we can conclude that the dimension of E_i is i. Consider the Katz decomposition $H \cong H_h \oplus H_1 \oplus H'_h$, and restrict attention to the factor H_h. Since H_h has only 0 and 1 as Hodge slopes, there is an endomorphism V of H_h such that $\Phi V = p$. Then $\Phi_i V^i := p^{1-i}\Phi^i V^i = p$, so that Φ_i also has only 0 and 1 as Hodge slopes. Moreover, Φ has $(h-1)/h$ as its unique Newton slope, and so Φ_i has $i(h-1)/h-(i-1) = (h-i)/h$ as its unique Newton slope. Since the Hodge and Newton polygons of Φ_i have the same endpoint, the Hodge numbers of Φ_i must be $(i, h-i)$, and this proves that N_i^2 has dimension i. Up to this point we have been considering only the factor H_h, but we can conclude that N_i^2 of the entire crystal has dimension at least i. Lemma 2 implies that $N_i^2 H_{DR} \subseteq E_i$ for all i, and it follows that $N_i^2 H_{DR} = E_i$.

The proof of (2) uses the classification of supersingular K3 crystals from [4], which we briefly recall. Let

$$\begin{aligned} E &:= \bigcap_{h=1}^{\infty} M_h^1 H = \bigcap_{h=1}^{\infty} (\Phi^h)^{-1}(p^h H) \\ E'' &:= \sum_{h=1}^{\infty} N_h^2 H = \sum_{h=1}^{\infty} p^{1-h}\Phi^h(H) \end{aligned}$$

Then E is the largest submodule of H which is invariant under $\phi := p^{-1}\Phi$, $E'' \subseteq E$, and E'' is the set of all $h \in H$ such that $(h|e) \in pW$ for all $e \in E$. Better, let $E' := p^{-1}E \subseteq \mathbf{Q} \otimes H$, which is then the smallest submodule of $\mathbf{Q} \otimes H$ containing H which is invariant under ϕ. Then E' is the set of $h \in \mathbf{Q} \otimes H$ such that $(h|e) \in W$ for all $e \in E$, and $pE' \subseteq E \subseteq H \subseteq E'$. The k-vector spaces H/E and E'/H are naturally dual, and so have the same dimension σ_0. Then $2\sigma_0$ is the dimension of E'/E, and is also the p-adic ordinal of the discriminant of Q on E. Let $T := \{e \in E : \phi(e) = e\}$, and $T' := \{e' \in E' : \phi(e') = e'\}$. Then $W \otimes_{\mathbf{Z_p}} T \cong E$, $W \otimes_{\mathbf{Z_p}} T' \cong E'$, and these isomorphisms identify ϕ with $F_W \otimes \mathrm{id}$. The image of H in $E'/E \cong k \otimes T'/T$ is a k-linear subspace not contained in any proper ϕ-stable subspace. If e'_1 and e'_2 are elements of E', then $pe'_1 \in E$, hence $(e'_1|e'_2)' := (pe'_1|e'_2) \in W$ and $(\ |\ )'$ is a symmetric bilinear form on E' which is compatible with ϕ and hence descends to T'. The annihilator of this form modulo p is just the image of E in E', and so we can view it as a symmetric bilinear form on E'/E. When $p = 2$, the set of all $v' \in E'/E$ such that $(v'|v')' = 0$ is a linear subspace containing the image of H. Since this space is ϕ-invariant, it must be all of E'/E, so in fact $(\ |\ )'$ is even and we can define $Q'(e') := 1/2(e'|e')$ for $e' \in E'$. Then Q' is a nondegenerate quadratic form on T'/T, and the image K' of H in $k \otimes T'/T$ is a *strictly characteristic subspace* [4, 3.19]: it is totally isotropic of dimension σ_0, $(K' + \phi(K'))/K'$ has dimension one, and K' is not contained in any proper $\mathbf{F}_p$-rational subspace. The automorphism group of k over $\mathbf{F}_p$ acts on the set of all strictly characteristic subspaces, and it is convenient (elsewhere) to set $K := \phi^{-1}(K')$. As explained on page 33 of [4], there is a (unique) line ℓ in V such that

$$\phi(K) = \ell \oplus \phi(\ell) \oplus \cdots \phi^{\sigma_0 - 1}(\ell),$$

and in fact

$$V = \ell \oplus \phi(\ell) \oplus \cdots \phi^{2\sigma_0 - 1}(\ell).$$

There is an exact sequence

$$k \otimes T \to H_{DR} \to K' \to 0.$$

The subspaces $F^1_{Hdg}H_{DR}$ and $N^1_h H_{DR}$ contain the image of $k \otimes T$, and hence are determined by their images $F^1_{Hdg}K'$ and $N^1_h K'$ in K', and it is easy to check that $F^1_{Hdg}K' = K \cap \phi(K)$ and $N^1_h K' = \phi(K) \cap \phi^{h+1}(K)$. Suppose $h \leq \sigma_0$. Then because the spaces $\{\phi^i(\ell) : 0 \leq i \leq 2\sigma_0 - 1\}$ form a direct sum decomposition of V,

$$\begin{aligned} F^1_{Hdg}K' &= K \cap \phi(K) = \ell \oplus \cdots \phi^{\sigma_0 - 2}(\ell) \\ N^1_h K' &= \phi(K) \cap \phi^{h+1}(K) = \phi^h(\ell) \oplus \phi^{h+1}(\ell) \oplus \cdots \phi^{\sigma_0 - 1}(\ell) \end{aligned}$$

Thus in fact $0 = N^1_{\sigma_0}K' \subseteq N^1_{\sigma_0 - 1}K' \subseteq K'$, and each $N^1_i K'$ has codimension i if $i \leq \sigma_0$. Hence $N^2_i H_{DR}$ has dimension i if $i \leq \sigma_0$, and $N^2_{\sigma_0} H_{DR}$ is exactly the image of $k \otimes T \to H_{DR}$. By Lemma 2, E_i has dimension i if $i \leq \sigma_0$, $E_i = E_{\sigma_0}$ if $i \geq \sigma_0$, and E_{σ_0} is the annihilator of the image of $k \otimes T \to H_{DR}$. This annihilator is $pE^\vee/E$, and hence is also the image of the annihilator of $k \otimes T$. Moreover,

$$N^1_h K' \cap F^1_{Hdg}K' = \phi^h(\ell) \oplus \phi^{h+1}(\ell) \oplus \cdots \phi^{\sigma_0 - 2}(\ell),$$

so $N^1_{\sigma_0}K' = N^1_{\sigma_0 - 1}K' \cap F^1_{Hdg}K' = 0$. Again by duality,

$$E_{\sigma_0} = E_{\sigma_0 - 1} + F^2_{Hdg}H_{DR}.$$

□

To summarize:

Proposition 6. *Let H be a K3 crystal over an algebraically closed field k and let $h \leq \mathrm{ht}(H)$ be a positive integer. Then the following are equivalent:*

1. $E_h^\perp \cap F^1_{Hdg}H_{DR} = E_{h-1}^\perp \cap F^1_{Hdg}H_{DR}$.
2. $E_h + F^2_{Hdg}H_{DR} = E_{h-1} + F^2_{Hdg}H_{DR}$.
3. *H is supersingular, and $\sigma_0(H) \leq h$.*
4. $F^2_{Hdg}H_{DR} \subseteq E_h$.
5. *$E_{h-1} + F^2_{Hdg}H_{DR}$ contains the null space of Q on $\mathbf{F}_p \otimes T$.*
6. *$E_{h-1} + F^2_{Hdg}H_{DR}$ contains a nonzero point of $\mathbf{F}_p \otimes T$.*

Proof: Since $F^1_{Hdg}H_{DR}$ is the annihilator of $F^2_{Hdg}H_{DR}$, conditions (1) and (2) are equivalent. If (2) holds, then by Lemma 2,

$$N^2_h H_{DR} \subseteq E_{h-1}H_{DR} + F^2_{Hdg}H_{DR},$$

and by Lemma 4, H is supersingular. Then Lemma 5 shows that (2) implies (3) and that (3) implies (4) and (5). Suppose that H has height $h \leq 10$. Consider the Katz decomposition $H \cong H_h \oplus H_1 \oplus H'_h$. Since Φ is divisible by p on $H_1 \oplus H'_h$, the same is true of Φ_i for all $i \leq h$, hence $N_i^2 H_{DR}$ and E_i are contained in H_{hDR} for $i \leq h$. On the other hand, $F^2 H_{DR} \subseteq H'_{hDR}$. Thus condition (4) implies that H is supersingular, and by Lemma 5 $h \leq \sigma_0$. Evidently (5) implies (6). Similarly if $h < \infty$, then in the Katz decomposition the image of $T \to H$ is contained in H_1, and hence modulo p cannot intersect $E_{h-1} + F^2_{Hdg} H_{DR}$. If (6) holds, it follows that H is supersingular, and $E_{h-1} + F^2_{Hdg} H_{DR}$ contains a ϕ-invariant line. Then by duality $E^{\perp}_{h-1} K \cap F^1_{Hdg} K$ is contained in a proper ϕ-invariant subspace, so in fact $h \geq \sigma_0$. □

If H is the K3 crystal of a K3 surface, then the rank of H is 22 and its crystalline discriminant is $-1 \in \mathbf{Z}_p/\mathbf{Z}_p^2$ [5, 4.9, 4.10]. We can use this fact to prove that the Artin invariant of a supersingular K3 crystal coming from a surface is at most 10, without assuming the Tate conjecture.

Proposition 7. *The Artin invariant σ_0 of a K3 crystal associated to a supersingular K3 surface X/k is at most* 10.

Proof: This is a consequence of the Tate conjecture, as explained by Artin [1]. To prove it unconditionally, use the direct sum decomposition $H \cong H_0 \oplus H_1$. As shown in [4, 3.15], the crystalline discriminant $(\frac{H_0}{p}) \in \{\pm 1\} \cong \mathbf{F}_p^*/\mathbf{F}_p^{*2}$ is $-(\frac{-1}{p})^{\sigma_0}$ if p is odd. Since the discriminant of H is $(\frac{-1}{p})$, we cannot have $\sigma_0 = 11$. If $p = 2$, it is easy to see that T_{H_0} has a dual basis (x_i, y_i), $i = 1, \ldots \sigma_0$ with respect to which the quadratic form is $x_1^2 + y_1^2 + x_1 y_1 + \sum_{i>1} x_i y_i$, and that its discriminant is $3(-1)^{\sigma_0} \in \mathbf{Q}_2/\mathbf{Q}_2^{*2}$. Since the discriminant is -1, it cannot be the case that $\sigma_0 = 11$, since in that case H_1 would be zero. □

It is convenient to introduce a variation of the above construction for polarized K3 crystals. If (H, ξ) is a polarized K3 crystal, it follows from the fact that $\Phi(\xi) = p\xi$ that the image ξ_{DR} of ξ in H_{DR} lies in $F^1_{Hdg} \cap F^1_{con} H_{DR}$ and is fixed by γF_k^*. If $H \in \mathbf{K3}_h(k)$ and $i \leq h$ we define $E_{i,\xi}$ to be the span in H_{DR} of E_i and the image ξ_{DR} of ξ in H_{DR}. If H is supersingular we define $\tau(H, L)$ to be $\sigma_0(H) - 1$ if $\xi_{DR} \in Ann(T/pT)$ and to be $\sigma_0(H)$ otherwise. For example, if the degree of ξ is prime to p, $\tau(H, L) = \sigma_0(H)$. It is also convenient to define $\tau(H, L)$ and $\sigma_0(H)$ to be ∞ if H is not supersingular. If (X, L) is a polarized K3 surface over k, we write $\tau(X, L)$ and $\sigma_0(X)$ for $\tau(H, \xi)$ and $\sigma_0(H)$, where H is the crystalline cohomology of X/W and ξ is the crystalline Chern class of L.

Corollary 8. *Let (H, ξ) be a polarized K3 crystal with $H \in \mathbf{K3}_h(k)$, let $i \leq h - 1$, and let $\overline{E}_{i,\xi}$ be the image of $E_{i,\xi}$ in $\mathrm{Gr}^1_{Hdg} H_{DR}$, i.e.,*

$$\overline{E}_{i,\xi} = (E_{i,\xi} + F^2_{Hdg})/F^2_{Hdg}.$$

Then

$$\dim(\overline{E}_{i,\xi}) = \min(i+1, \tau(H,L)).$$

Furthermore, if H is supersingular, $E_{\tau,\xi} = E_{\tau-1,\xi} + F^2_{Hdg}H_{DR}$.

Proof: Evidently $\dim(\overline{E}_{i,\xi}) \leq \dim(E_{i,\xi}) \leq \dim(E_i) + 1 \leq i+1$. If H is supersingular with Artin invariant σ_0, by Lemma 5 $\overline{E}_i \subseteq \overline{E}_{\sigma_0} = \overline{E}_{\sigma_0 - 1}$, so $\dim(\overline{E}_{i,\xi}) \leq \sigma_0$. If $\xi \in Ann(T/pT)$, then in fact $E_{\sigma_0,\xi} = E_{\sigma_0}$, and $\dim(\overline{E}_{i,\xi}) \leq \sigma_0 - 1 = \tau$. Thus in any case

$$\dim(\overline{E}_{i,\xi}) \leq \min(i+1, \tau(H,L)).$$

Suppose that H is not supersingular. Then by Lemma 5, E_i has dimension i and by condition (4) of Proposition 6, E_i does not meet $F^2_{Hdg}H_{DR}$. Furthermore, the Katz decomposition shows that $E_i + F^2_{Hdg}H_{DR}$ does not meet the image of T/pT. Hence $\dim(\overline{E}_{i,\xi}) = i+1$. Suppose that H is supersingular. By Lemma 5, the dimension of the image $\overline{E}_i$ of E_i in $\operatorname{Gr}^1_{Hdg} H_{DR}(X/k)$ is $\min(i, \sigma_0 - 1)$. Hence if ξ_{DR} does not belong to $E_i + F^2_{Hdg}H_{DR}$, the dimension of $\overline{E}_{i,\xi}$ is $\min(i+1, \sigma_0) \geq \min(i+1, \tau)$. On the other hand, if $\xi_{DR} \in E_i + F^2_{Hdg}H_{DR}$, then condition (6) of Proposition 6 holds with $h = i+1$, so that $\sigma_0 \leq i+1$, and by Lemma 5, ξ_{DR} belongs to the annihilator of T/pT. Then $\overline{E}_{i,\tau} = \overline{E}_i = \overline{E}_{\sigma_0 - 1}$ has dimension $\sigma_0 - 1 = \tau$. □

We are now ready to investigate the moduli of K3 surfaces. If $f \colon \mathcal{X} \to S$ is a family of K3 surfaces over a k-scheme S with polarization $\mathcal{L}$ of degree d, the relative De Rham cohomology $H_{DR}(\mathcal{X}/S) := R^2 f_* \Omega^{\cdot}_{\mathcal{X}/S}$ is locally free of rank 22. It is equipped with its Hodge and conjugate filtrations F_{Hdg} and F_{con}, and the associated graded sheaves $\operatorname{Gr}_{Hdg} H_{DR}$ and $\operatorname{Gr}_{con} H_{DR}$ are also locally free. If $\mathcal{X}/S$ is obtained by pullback from a smooth k-scheme, $H_{DR}(\mathcal{X}/S)$ has a canonical Gauss-Manin connection $\nabla \colon H_{DR} \to \Omega^1_{X/S} \otimes H_{DR}$. The Hodge filtration is moved by at most one step by ∇, and the induced map

$$F^2_{Hdg}H_{DR}(\mathcal{X}/S) \to \operatorname{Gr}^1_{Hdg} H_{DR}(\mathcal{X}/S) \otimes \Omega^1_{S/k}$$

is $\mathcal{O}_S$-linear and is equivalent to a map

$$\kappa' \colon F^2_{Hdg}H_{DR}(\mathcal{X}/S) \otimes \operatorname{Gr}^1_{Hdg} H_{DR}(\mathcal{X}/S)^{\vee} \to \Omega^1_{S/k}.$$

Since $\Omega^2_{X/S}$ is trivial along the fibers and the Hodge spectral sequence of $\mathcal{X}/S$ degenerates, there is a natural isomorphism

$$F^2_{Hdg} \otimes \operatorname{Gr}^1_{Hdg} H_{DR}(\mathcal{X}/S)^{\vee} \cong f_*(\Omega^2_{X/S}) \otimes R^1 f_*(\Omega^1_{\mathcal{X}/S})^{\vee} \cong R^1 f_*(T_{X/S})^{\vee}.$$

Thus the map κ' can be interpreted as the differential of the map from S into the moduli stack. These facts, as well as the basic deformation theory of K3 surfaces, are very clearly explained in [2].

The absolute Frobenius morphism of X induces an $\mathcal{O}_S$-linear map

$$\phi \colon F^*_S H_{DR}(\mathcal{X}/S) \to H_{DR}(\mathcal{X}/S),$$

whose kernel is exactly $F_S^*(F^1_{Hdg}H_{DR})$ and whose image is exactly $F^2_{con}H_{DR}$, and the inverse Cartier isomorphism induces an isomorphism:

$$\gamma\colon F_S^* \operatorname{Gr}^1_{Hdg} H_{DR} \to \operatorname{Gr}^1_{con} H_{DR}. \tag{1}$$

The filtration F_{con} is stable under ∇, and if $F_S^* \operatorname{Gr}^1_{Hdg} H_{DR}$ is endowed with the connection killing the elements of $\operatorname{Gr}^1_{Hdg} H_{DR}$, the map γ is horizontal. The quadratic form Q induces nondegenerate forms on $\operatorname{Gr}^1_{Hdg} H_{DR}$ and $\operatorname{Gr}^1_{con} H_{DR}$ compatible with γ. The Chern class $\xi_{DR} := c_{DR}(\mathcal{L})$ of $\mathcal{L}$ is a global horizontal section of $H_{DR}(\mathcal{X}/S)$ contained in $F^1_{Hdg} \cap F^1_{con} H_{DR}(\mathcal{X}/S)$.

If S is a k-scheme, let $\mathcal{F}(S)$ denote the category of polarized K3 surfaces of degree d, with morphisms the isomorphisms preserving the polarizations. A morphism $S' \to S$ defines a functor $\mathcal{F}(S) \to \mathcal{F}(S')$, and $\mathcal{F}$ forms a fibered category with effective descent. It should be easy to verify that $\mathcal{F}$ is in fact a Deligne-Mumford stack, but since we shall be only interested in local properties of families of K3 surfaces, we shall not attempt to carry this out here. If d is prime to p, $\mathcal{F}$ is formally smooth of dimension 19 over k, and there exist formally étale morphisms $S \to \mathcal{F}$ with S a smooth k-scheme of dimension 19. Formation of H_{DR} is compatible with base change $S' \to S$, and the same is true of its filtrations and their associated graded sheaves.

We define a sequence of stacks and closed immersions $\mathcal{F}_{h+1} \subseteq \mathcal{F}_h \subseteq \cdots \mathcal{F}_1 = \mathcal{F}$. That is, each $\mathcal{F}_i(S)$ will be a strictly full subcategory of $\mathcal{F}_{i-1}(S)$, and for each object $(\mathcal{X}/S, \mathcal{L})$ of $\mathcal{F}(S)$, the map $S_i := \mathcal{F}_i \times_{\mathcal{F}} S \to S$ will be a closed immersion. Furthermore, associated to each object $(\mathcal{X}/S, \mathcal{L})$ of $\mathcal{F}_i(S)$, will be a quasi-coherent subsheaf $E_i(\mathcal{X}/S) \subseteq H_{DR}(\mathcal{X}/S)$. These objects are defined inductively as follows: For any object $(\mathcal{X}/S, \mathcal{L})$ of $\mathcal{F}_1(S) = \mathcal{F}(S)$, let

$$E_1(\mathcal{X}/S) := F^2_{con} H_{DR}(\mathcal{X}/S) \subseteq H_{DR}(\mathcal{X}/S).$$

If $h \geq 1$ and $(\mathcal{X}/S, \mathcal{L})$ is an object of $\mathcal{F}_h(\mathcal{X}/S)$, then $E_h(\mathcal{X}/S) \subseteq H_{DR}(\mathcal{X}/S)$ is defined, and $\mathcal{F}_{h+1}(S)$ is defined to be the full subcategory of $\mathcal{F}_h(S)$ whose objects are the elements $(\mathcal{X}/S, \mathcal{L})$ such that $E_h(\mathcal{X}/S) \subseteq F^1_{Hdg} H_{DR}(\mathcal{X}/S)$. If $(\mathcal{X}/S, \mathcal{L})$ is an object of $\mathcal{F}_{h+1}(S)$, then $E_{h+1}(\mathcal{X}/S)$ is by definition the inverse image in $F^1_{con} H_{DR}(\mathcal{X}/S) \subseteq H_{DR}(\mathcal{X}/S)$ of the image of $F_S^* E_h$ under the map

$$F_S^* F^1_{Hdg} H_{DR}(X/S) \longrightarrow F_S^* \operatorname{Gr}^1_{Hdg} H_{DR}(X/S) \xrightarrow{\gamma} \operatorname{Gr}^1_{con} H_{DR}(X/S). \tag{2}$$

Since γ is horizontal, it follows that E_h is stable under ∇. Let $E_{h,\mathcal{L}}(\mathcal{X}/S, \mathcal{L})$ denote the $\mathcal{O}_S$-submodule of H_{DR} generated by $E_h(\mathcal{X}/S)$ and $c_{DR}(\mathcal{L})$, which again is contained in $F^1_{Hdg} \cap F^1_{con} H_{DR}(\mathcal{X}/S)$ and invariant under ∇ and γF_S^*. We also define $E_0(\mathcal{X}/S) := 0$, so that $E_{0,\mathcal{L}}(\mathcal{X}/S)$ is the $\mathcal{O}_S$-submodule of S generated by $c_{DR}(\mathcal{L})$. Note that $E_{h,\mathcal{L}}$ can be defined inductively in the same way as E_h, and that an object $(\mathcal{X}/S, \mathcal{L})$ of $\mathcal{F}_h(S)$ lies in $\mathcal{F}_{h+1}(S)$ if and only if $E_{h,\mathcal{L}}(\mathcal{X}/S, \mathcal{L})$ is contained in $F^1_{Hdg} H_{DR}(\mathcal{X}/S)$.

It follows from Proposition 3 that $\mathcal{F}_h(k)$ is the category of polarized K3 surfaces of degree d with Newton height at least h.

Lemma 9. *Let $(\mathcal{X}/S, \mathcal{L})$ be an object of $\mathcal{F}_h(S)$ and let $i \leq h$ be a positive integer.*

1. *Each $E_i(\mathcal{X}/S)$ is a coherent totally isotropic subsheaf of $H_{DR}(\mathcal{X}/S)$ locally generated by i elements, and*
$$E_0(\mathcal{X}/S) \subseteq E_1(\mathcal{X}/S) \subseteq \cdots E_h(\mathcal{X}/S) \subseteq H_{DR}(\mathcal{X}/S).$$
2. *If $g\colon S' \to S$ is a morphism of k-schemes and $(\mathcal{X}'/S', \mathcal{L})$ is obtained from $(\mathcal{X}/S, \mathcal{L})$ by pullback, then the natural isomorphism*
$$g^* H_{DR}(\mathcal{X}/S) \to H_{DR}(\mathcal{X}'/S')$$
induces a surjection $g^* E_i(\mathcal{X}/S) \to E_i(\mathcal{X}'/S')$ and an isomorphism
$$g^*(Q_i(\mathcal{X}/S) \to Q_i(X'/S),$$
where $Q_i(\mathcal{X}/S) := H_{DR}(\mathcal{X}/S)/E_i(\mathcal{X}/S)$.
3. *Suppose that S is of finite type over k and that no closed fiber of $\mathcal{X}/S$ is supersingular with Artin invariant less than i. Then $E_i(\mathcal{X}/S)$ is a local direct factor of $H_{DR}(\mathcal{X}/S)$ of rank i, and its formation commutes with any base change $S' \to S$.*

Proof: The statements are true if $h = 1$, and, proceeding by induction on h, we assume that the lemma holds for h and that $h \geq 1$. Furthermore, it suffices to treat the case $i = h$. Then $E_{h-1} \subseteq E_h$ and the quotient is locally monogenic. Let $\tilde{E}_i \subseteq \mathrm{Gr}^1_{con} H_{DR}(\mathcal{X}/S)$ denote the image of of $F_S^*(E_i)$ under the map (2). Then there is an exact sequence
$$0 \to \tilde{E}_{h-1} \to \tilde{E}_h \to \tilde{E}_h/\tilde{E}_{h-1} \to 0,$$
and the quotient is locally monogenic. If π is the projection $F^1_{con} H_{DR} \to \mathrm{Gr}^1_{con} H_{DR}(X/S)$, E_{i+1} is by definition $\pi^{-1}(E_i)$, and so $E_h \subseteq E_{h+1}$ and $E_{h+1}/E_h \cong \pi(E_{h+1})/\pi(E_h) \cong \tilde{E}_h/\tilde{E}_{h-1}$ is locally monogenic. Since E_h is totally isotropic, the same is true of $\gamma F_k^*(E_h)$ and E_{h+1}. Statements (2) and (3) can be verified locally on S, so we may assume that there exist splittings of F_{Hdg} and F_{con}, and hence that there exists a map $\tilde{\gamma}\colon F_S^* H_{DR} \to H_{DR}$ extending the map γ defined in (1). Then E_h is the sum of the images of $(\tilde{\gamma} F_S *)^i F^2_{con} H_{DR}$ for $i = 0 \cdots h-1$, and is independent of the choice of $\tilde{\gamma}$, and since g is compatible with Frobenius, the base change map sends $g^* E_h(\mathcal{X}/S)$ onto $E_h(\mathcal{X}/S')$. It follows that formation of the quotient $Q_i(\mathcal{X}/S)$ is compatible with any base change $S' \to S$.

Suppose now that S is of finite type and that no closed fiber is supersingular with Artin invariant less than h. For any closed point s of S, the natural map from the fiber $E_h(s)$ of E_h at s to $E_h(X_s)$ is a surjective map of k-vector spaces. Its source $E_h(s)$ is generated by h elements, and by Lemma 5, the target $E_h(X_s)$ has dimension h. It follows that the map
$$E_h(s) \to E_h(X_s)$$

is an isomorphism. Since H_{DR} is locally free, there is an exact sequence

$$0 \to \mathrm{Tor}_1(k(s), Q_h) \to E_h(s) \to H_{DR}(s) \to Q_h(s) \to 0,$$

and since the map $E_h(s) \to H_{DR}(s)$ is injective, $\mathrm{Tor}_1(k(s), Q_h) = 0$ and Q_h is locally free in a neighborhood of s. Hence E_h is a local direct summand and its formation commutes with base change. □

Corollary 10. *If $(\mathcal{X}/S, \mathcal{L})$ is an object of $\mathcal{F}_{11}(S)$ and $j \in \mathbf{N}$, then the set of closed points of S for which $\sigma_0(X_s) \leq j$ is a Zariski closed subset of S_{red}, and similarly for $\tau(X_s, \mathcal{L}_s)$.*

Proof: By Lemma 5 and Proposition 7, $\sigma_0(X_s) \leq j$ if and only if the dimension of $E_{10}(X_s) \leq j$, *i.e.*, if and only if the dimension of $Q_{10}(X_s) \geq 22 - j$. Since $Q_{10}(\mathcal{X}/S)$ is a coherent sheaf on S whose formation commutes with base change, the corollary follows from the semi-continuity theorem. The same proof works with τ in place of σ, using Corollary 8. □

Proposition 11. *Each $\mathcal{F}_{h+1} \subseteq \mathcal{F}_h$ is a closed immersion, locally defined by a single equation, and the ideal of $\mathcal{F}_{h+1}$ in $\mathcal{F}_h$ is the image of the pairing $(\ |\ ) \colon F^2_{Hdg} \otimes E_h/E_{h-1} \to \mathcal{O}_{\mathcal{F}_h}$.*

Proof: We argue by induction on h, and, assuming the proposition true for $h-1$, we may work locally on the base S of a polarized family $(\mathcal{X}/S, \mathcal{L})$ defining an object of $\mathcal{F}_{h-1}(S)$. Then $E_{h-1} \subseteq F^1_{Hdg}H_{DR}$, and hence is orthogonal to F^2_{Hdg}, and the map $F^2_{Hdg} \otimes E_h \to \mathcal{O}_{\mathcal{F}_h}$ factors through $F^2_{Hdg} \otimes E_h/E_{h-1}$. Its image is an ideal I of $\mathcal{O}_{\mathcal{F}_h}$, and since E_h/E_{h-1} is locally monogenic, so is I. If $g \colon S' \to S$ is any morphism, the pullback of $\mathcal{X}/S$ to S' lies in $\mathcal{F}_{h+1}$ if and only if $g^*E_h \to g^* \mathrm{Gr}^0_{Hdg} H_{DR}$ is zero, *i.e.*, if and only if g^*E_h is orthogonal to $g^*F^2_{Hdg}$. Thus the morphism $S' \to S \to \mathcal{F}$ factors through $\mathcal{F}_{h+1}$ if and only if $S' \to S$ factors through the closed subscheme of S defined by I. □

If $(\mathcal{X}/S, \mathcal{L})$ is a polarized K3 surface over S, there is a commutative diagram

$$\begin{array}{ccc}
F^2_{Hdg}H_{DR}(\mathcal{X}/S) \otimes \mathrm{Gr}^1_{Hdg} H_{DR}(\mathcal{X}/S) & & \\
\downarrow \delta & \searrow \kappa & \\
F^2_{Hdg}H_{DR}(\mathcal{X}/S) \otimes \mathrm{Gr}^1_{Hdg} H_{DR}(\mathcal{X}/S)^\vee & \xrightarrow{-\kappa'} & \Omega^1_{S/k}
\end{array}$$

Here δ comes from the canonical pairing and κ and κ' are the (Kodaira-Spencer) maps induced by the Gauss-Manin connection. That is, if ω is a section of $F^2_{Hdg}H_{DR}$ and η (resp. ϕ) is a section of $\mathrm{Gr}^1_{Hdg} H_{DR}$ (resp. of its dual), then $\kappa(\omega \otimes \eta) = (\omega | \nabla(\eta'))$ for any $\eta' \in F^1_{Hdg}H_{DR}$ lifting η, $\kappa'(\omega \otimes \phi) = \phi(\nabla(\omega))$, and $\kappa'(\omega \otimes \delta(\eta)) = (\nabla\omega | \eta')$.

Suppose now that $(\mathcal{X}/S, \mathcal{L})$ is a polarized K3 surface over S defining an object of $\mathcal{F}_h(S)$, so that $E_i \subseteq H_{DR}(\mathcal{X}/S)$ is defined for $i \leq h$. As we saw in the construction, these spaces are invariant under the Gauss-Manin connection. Furthermore, $E_{h-1} \subseteq F^1_{Hdg}H_{DR}$, and since it is horizontal, it is annihilated by the Kodaira-Spencer mapping κ of $\mathcal{X}/S$. Let S' be the closed subscheme $\mathcal{F}_{h+1} \times_{\mathcal{F}_h} S$ of S. Note that E_h does not lie in $F^1_{Hdg}H_{DR}$ until we restrict to S'. In fact the following proposition shows that the conormal to S' in S is precisely measured by this fact.

Proposition 12. *Let $(\mathcal{X}/S, \mathcal{L})$ be an object of $\mathcal{F}_h(S)$ and let $S' := \mathcal{F}_{h+1} \times_{\mathcal{F}_h} S$. Then there is a commutative diagram:*

$$\begin{array}{ccccc}
F^2_{Hdg} \otimes E_h & \longrightarrow & F^2_{Hdg} \otimes \mathrm{Gr}^1_{Hdg} H_{DR}(\mathcal{X}/S)_{|S'} & & \\
\Big\downarrow (\,|\,) & & \Big\downarrow -\kappa & \searrow & \\
I_{S'}/I^2_{S'} & \xrightarrow{\quad d \quad} & \Omega^1_{S/k}{}_{|S'} & \longrightarrow & \Omega^1_{S'/k}
\end{array}$$

In particular, the image of the conormal to S' by the differential d coincides with the image of $F^2_{Hdg} \otimes E_h/E_{h-1}$ under the Kodaira-Spencer mapping κ.

Proof: Choose local sections ω of $F^2_{Hdg}H_{DR}$ and η of E_h over S. Then by Proposition 11, $f := (\omega|\eta)$ is a local generator for the ideal $I_{S'}$ of S' in S. Since E_h is horizontal, there exists $\theta \in \Omega^1_{S/k}$ such that $\nabla(\eta) = \theta \otimes \eta$ modulo E_{h-1}. Hence

$$df = (\nabla(\omega)|\eta) + (\omega|\nabla(\eta)) = (\nabla(\omega)|\eta) + \theta(\omega|\eta)$$

Taking the images in $\Omega^1_{S/k}{}_{|S'}$, we see that $df = \kappa'(\omega \otimes \delta(\eta)) = -\kappa(\omega \otimes \eta)$. □

If $(X/k, L)$ is a polarized K3 surface, by standard deformation theory (and the fact that X admits no vector fields!), the unpolarized surface X/k admits a versal formal k-deformation $\mathcal{X}/\mathcal{S}$, with $\mathcal{S} \cong \operatorname{Spec} k[[t_1, \ldots t_{20}]]$. As explained in [2], the versal formal k-deformation $(\mathcal{X}/\mathcal{S}_L)$ embeds as a closed formal subscheme of $\mathcal{S}$, and its ideal I_L is monogenic (and nonzero). If $\mathfrak{m}$ is the (maximal) ideal of definition of $\mathcal{S}$, the Kodaira-Spencer mapping defines an isomorphism. Using this fact and Proposition 12 we can easily calculate the following first-order description of $\mathcal{F}_h$.

Corollary 13. *Let $(\mathcal{X}/\mathcal{S})$ be the unpolarized versal formal deformation of a polarized K3 surface $(X/k, L)$, let i be an integer with $i \leq \mathrm{ht}(X)$, and let $I_{i,L} \subseteq \mathfrak{m}$ be the ideal of $\mathcal{S}_{i,L}$ in $\mathcal{S}$.*

1. *The ideal $I_{i,L}$ of $\mathcal{S}_{i,L}$ in $\mathcal{S}$ can be generated by i elements, and its image in $\mathfrak{m}/\mathfrak{m}^2$ is the image of $F^2_{Hdg} \otimes \overline{E}_{i-1,L}$ under the Kodaira-Spencer isomorphism κ:*

$$F^2_{Hdg} \otimes \overline{E}_{i-1,L} \subseteq F^2_{Hdg} \otimes \mathrm{Gr}^1_{Hdg} H_{DR}(X/k) \xrightarrow{\kappa} \mathfrak{m}/\mathfrak{m}^2.$$

2. *The Zariski tangent space of $\mathcal{F}_i$ at the k-valued point corresponding to $(X/k, L)$ is naturally isomorphic to*

$$\mathrm{Gr}^0_{Hdg} \otimes \overline{E}^{\perp}_{i-1,L} \subseteq \mathrm{Gr}^0_{Hdg} \otimes H^1(X/k, \Omega^1_{X/k}).$$

In particular, its dimension is $\max(20-i, 20-\tau(X,L))$.

Proof: We have already seen in Proposition 11 that each $\mathcal{S}_{i,\mathcal{L}} \subseteq \mathcal{S}_{i-1,\mathcal{L}}$ is a closed immersion defined by a single equation. It follows by induction that $\mathcal{S}_{i,\mathcal{L}} \subseteq \mathcal{S}_{\mathcal{L}}$ is defined by $i-1$ equations and hence that $\mathcal{S}_{i,\mathcal{L}} \subseteq \mathcal{S}$ is defined by i equations. Let $T \subseteq \mathcal{S}$ be the closed subscheme defined by $\mathfrak{m}^2$. Then there is a surjective map $\Omega^1_{T/k} \to \mathfrak{m}/\mathfrak{m}^2$, which is in fact an isomorphism if p is odd. As is well-known, the ideal of $T_{\mathcal{L}} := T \cap \mathcal{S}_{\mathcal{L}}$ is the image of $F^2_{Hdg} \otimes c_{Hdg}(L) \in F^2_{Hdg} \otimes H^1(X/k, \Omega^1_{X/k})$ under the Kodaira-Spencer mapping [4, 1.14]. This says exactly that the ideal $I_{1,\mathcal{L}}$ of $S_{1,\mathcal{L}}$ in T is the image of $F^2_{Hdg} \otimes \overline{E}_{0,L}$ under the Kodaira-Spencer mapping, proving (1) when $i = 1$. On the closed subscheme $S := T_{i,\mathcal{L}}$ of T, $(\mathcal{X}, \mathcal{L})$ defines an object of $\mathcal{F}_i$. Proposition 12 says that the image of $I_{i+1,\mathcal{L}}/I_{i,\mathcal{L}}$ in $\Omega^1_{S/k}$ coincides with the image of $F^2_{Hdg} \otimes E_i$. Taking images in $\mathfrak{m}/\mathfrak{m}^2 + I_{i,\mathcal{L}}$, we see that the image of $I_{i+1,\mathcal{L}}$ coincides with the image of $F^2_{Hdg} \otimes E_i$. Assuming (1) for i, it follows that the image of $I_{i+1,\mathcal{L}}$ mod $\mathfrak{m}^2$ coincides with the image of $\overline{E}_i$, and so (1) follows by induction. The Zariski tangent space of $\mathcal{F}_i$ is therefore the dual of the quotient of $\mathfrak{m}/\mathfrak{m}^2$ by the image of $F^2_{Hdg} \otimes E_{i-1,L}$, which can be identified with $\mathrm{Gr}^0_{Hdg} \otimes \overline{E}^{\perp}_{i-1,L}$ using the duality map. Thus by Corollary 8, its dimension is $\max(20-i, 20-\tau(X,L))$. □

Proposition 14. *Let $(\mathcal{X}/S, \mathcal{L})$ be a family of polarized K3 surfaces over a finite type k-scheme S such that the resulting map $S \to \mathcal{F}$ is unramified.*

1. *If i is an integer such that every fiber X_s is supersingular with Artin invariant $\leq i$, then the dimension of S is at most $i-1$.*
2. *If every fiber X_s is supersingular, then the dimension of S is at most* 9.
3. *If $h \leq 11$ is an integer such that every fiber X_s has height at least h, then the dimension of S is at most* $20-h$.

Proof: Without loss of generality we may suppose that S is reduced, and even smooth, and that every fiber X_s has height at least i. Then $E_i(\mathcal{X}/S)$ is defined, and by Lemma 9, formation of

$$Q_i(\mathcal{X}/S) := H_{DR}(\mathcal{X}/S)/E_i(\mathcal{X}/S)$$

commutes with any base change $S' \to S$. Since S is reduced and Q_i is coherent, there is a dense open subset of S on which Q_i is locally free, and we may assume that this true on all of S. Then $E_i(\mathcal{X}/S)$ is locally free, and its formation commutes with any base change $S' \to S$.

Now suppose that every fiber has Artin invariant at most i. Then by Lemma (5), $F^2_{Hdg}(X_s) \subseteq E_i(s)$ for every s, and since F^2_{Hdg} and E_i are local direct factors and S is reduced and of finite type over k, $F^2_{Hdg}(\mathcal{X}/S)H_{DR} \subseteq E_i(\mathcal{X}/S)$. Since E_i is horizontal, the Kodaira-Spencer mapping factors:

$$\begin{array}{ccc} F^2_{Hdg}H_{DR} \otimes \mathrm{Gr}^1_{Hdg} H_{DR}(\mathcal{X}/S)^\vee & & \\ \downarrow & \searrow & \\ F^2_{Hdg}H_{DR} \otimes \overline{E}_i^\vee & \longrightarrow & \Omega^1_{S/k} \end{array}$$

Since $S \to \mathcal{F}$ is unramified, for each s in S the map $H^1(X_s, \Omega^1_{X_s/k}) \to \mathfrak{m}_s/\mathfrak{m}_s^2$ is surjective, and hence the dimension of $\mathfrak{m}_s/\mathfrak{m}_s^2$ is at most the dimension of $\overline{E}_i(X_s/k)$, which by Lemma 5 is at most $i-1$. This proves (1), and (2) follows, since the Artin invariant σ_0 is at most 10, by Proposition 7

To prove (3), suppose that $h \leq 11$ and that every fiber has height at least h. If $h = 11$, every fiber is supersingular, so by (2), the dimension of S is at most $9 = 20-11$. Suppose that $h \leq 10$, and that the dimension of S is at least $20-h$. Since $20 - h \geq 10$, not all the points of S are supersingular, and in particular we may restrict to an open set on which every point has finite height. Then by Lemma 5, every $\overline{E}_{h-1,\mathcal{L}}(X_s/k)$ has dimension h, and we may assume that $\overline{E}_{h-1,\mathcal{L}}(\mathcal{X}/S, \mathcal{L})$ is locally free of rank h and that its formation commutes with base change. Since $E_{h-1,\mathcal{L}}$ is a horizontal subspace of $H_{DR}(\mathcal{X}/S)$, its image $\overline{E}_{h-1,\mathcal{L}}$ in $\mathrm{Gr}^1_{Hdg} H_{DR}(\mathcal{X}/S)$ is annihilated by the Kodaira-Spencer mapping κ. Then the image of κ has dimension at most $20-h$, and hence S has dimension at most $20 - h$. □

We can assemble what we have proved as follows:

Theorem 15. *For $1 \leq h \leq 11$, $\mathcal{F}_h \subseteq \mathcal{F}$ is a local complete intersection of codimension $h-1$ and dimension $20-h$. For $11 \geq h \geq 2$, the ideal of $\mathcal{F}_h$ in $\mathcal{F}_{h-1}$ is invertible and isomorphic to $(F^2_{Hdg})^{1-p^h}$. If $(X/k, L)$ is a polarized K3 surface of height at least h, then $(X/k, L)$ defines a singular point of $\mathcal{F}_h$ if and only if $\tau(X/k, L) < h$.*

Proof: We saw above, and it has long been known, that a versal deformation of a polarized K3 surface $(X/k, L)$ is a local complete intersection of dimension 19 (and is even smooth unless $\tau(X/k, L) = 0$). Since $S_h \subseteq S_1$ is locally defined by $h-1$ equations its dimension is everywhere at at least $20 - i$. By Proposition 14, the dimension is exactly $20 - i$, so S_h is a local complete intersection. It follows that I_h/I_{h-1} is invertible, and by Proposition 12, it is isomorphic to E_h/E_{h-1}, which is locally generated by, and hence isomorphic to, $(F^2_{Hdg})^{1-p^h}$. Furthermore, if $(X/k, L)$ has height at least h, it defines a singular point of $\mathcal{F}_h$ if and only if the tangent space of $\mathcal{F}_h$ at $(X/k, L)$ has dimension larger than the dimension $20 - h$. By Corollary 13, this is true if and only if $\tau < h$. □

Corollary 16. *Suppose d is prime to p and $1 \leq h \leq 11$ and $(\mathcal{X}/S, \mathcal{L})$ is an object of $\mathcal{F}(S)$ defining an étale morphism $S \to \mathcal{F}$. Then the singular locus of S_h coincides with the set $S_{\infty,h-1}$. Its codimension is everywhere at least $22-2h$ in S_h, and S_h is normal if $h \leq 10$. On the other hand, S_{11} is generically nonreduced.*

Proof: If d is prime to p, $\tau(X, L) = \sigma_0(X)$, and so by Theorem 15, the singular set of S_h consists exactly of $S_{\infty,h-1}$. If $1 \leq h \leq 10$, Proposition 14 says that this set has dimension at most $h-2$. Thus the codimension of the singular set is at least $20 - h - (h-2) = 22 - 2h \geq 2$. Since S_h is a local complete intersection, it is normal by Serre's criterion. The singular set of S_{11} is $S_{\infty,10}$, which by Proposition 7 is, set-theoretically, all of S_{11}. Thus S_{11} is generically singular, hence generically nonreduced, since the ground field k is algebraically closed. □

The following result gives a more precise description of the generic singularities of $\mathcal{F}_h$, which occur at those points at which $\tau(X, L) = \tau + 1$.

Theorem 17. *Let $(\mathcal{X}/S, \mathcal{L})$ be a family of polarized K3 surfaces over S and let s be a k-rational point of S at which the fiber X/k is supersingular and let $\tau := \tau(X, L)$. Let $\mathfrak{m}$ be the maximal ideal of s in S_τ, so that the ideal I of $\mathcal{F}_{\tau+1}$ in $\mathcal{F}_\tau$ is contained in $\mathfrak{m}^2$. Use any basis of $H^0(X/k, \Omega^2_{X/k})$ to identify $F^2_{Hdg} H_{DR}(X/k)$ with k and let*

$$\kappa' \colon H^1(X/k, \Omega^1_{X/k})^\vee \to \mathfrak{m}/\mathfrak{m}^2$$

be the resulting Kodaira-Spencer mapping. Then the image of Q via the map

$$\kappa^* \colon Sym^2(H^1(X/k, \Omega^1_{X/k})^\vee) \to Sym^2(\mathfrak{m}/\mathfrak{m}^2) \to \mathfrak{m}^2/\mathfrak{m}^3$$

generates I (modulo $\mathfrak{m}^3$).

Proof: We may assume without loss of generality that $(\mathcal{X}/S, \mathcal{L})$ is versal. Then s is a smooth point of S_τ and a singular point of $S_{\tau+1}$. We may work locally around s, and let S' denote an affine open neighborhood of s in S_τ, which we may assume is smooth.

Lemma 18. *Locally on S', if $i \leq \tau$, $E_i \subseteq H_{DR}(\mathcal{X}/S')$ has a basis annihilated by the Gauss-Manin connection.*

Proof: Let T be the p-adic completion of a smooth lifting of S' Then the crystalline cohomology of X/T is a lifting H_T of $H_{DR}(\mathcal{X}/S')$ to T, *i.e.*, a locally free sheaf of $\mathcal{O}_T$-modules whose reduction modulo p is canonically isomorphic to $H_{DR}(\mathcal{X}/S')$. Locally we may also choose a lifting F_T of the absolute Frobenius endomorphism of S'. Then the Frobenius endomorphism of F_X induces a map $\Phi_T \colon F_T^* H_T \to H_T$. Each point $s \in S'(k)$ admits a unique Teichmüller lifting $t \in T(W)$ (compatible with F_T), and $t^*(\Phi_T)$ becomes identified with the action of Frobenius Φ on the crystalline cohomology of $\mathcal{X}_s/W$.

For $i \geq 1$, the ith iterate $\Phi_T^i\colon F_T^{i*}H_T \to H_T$ is defined in the obvious way. For every closed point s with corresponding Teichmüller point t, $t^*\Phi_T$ is divisible by p^{i-1}, by Proposition (3), and it follows easily by induction that the same is true of Φ_T^i. Write $\Phi_T^i := p^{1-i}\Phi_{i,T}$ and let $N_i^2 H_T$ denote the image of $\Phi_{i,T}$. For each Teichmüller point $t \in T(W)$, the image of $\tau^*(N_i^2 H_T)$ in $\tau^* H_T$ is the image $N_i^2(t)H_{T,t}$ of Φ_i acting on the crystalline cohomology of the corresponding $\mathcal{X}_s$. By Lemma 5, the fiber $E_i(s)$ identifies with the image of $N_i^2(t)$ in the de Rham cohomology at s. Since $E_i(\mathcal{X}/S')$ is a local direct factor of $H_{DR}(\mathcal{X}/S)$, it follows that in fact $E_i(\mathcal{X}/S') = N_i^2 H_{DR}(\mathcal{X}/S')$, *i.e.*, the image of the map $\Phi_{i,S'}: F_S^{*i}H_{DR} \to H_{DR}$. Since the map $\Phi_{i,T}$ is horizontal, the same is true of $\Phi_{i,S'}$. But $F_S^* H_{DR}$ has a basis of horizontal sections, and consequently so does its image $E_i(\mathcal{X}/S')$ under $\Phi_{i,S'}$. □

Localizing as necessary, we may choose a basis (x, y_i, z) for $H_{DR}(\mathcal{X}/S')$, compatible with the Hodge filtration. We may choose z so that $(x|z) = 1$ and $(y_i|z) = 0$ for all i. Then the expression for Q as an element of $Sym^2(H^\vee)$ is

$$Q = \sum_{i<j}(y_i|y_j)y_i'y_j' + \sum_i Q(y_i){y_i'}^2 + x'z' + Q(z){z'}^2.$$

If $\eta \in E_{\tau,\mathcal{L}}$ lifts a basis for $E_{\tau,\mathcal{L}}/E_{\tau-1,\mathcal{L}}$, we can write $\eta = fx + \sum g_i y_i + hz$, and I is generated by $(x|\eta) = h$. By Lemma 5, x is a basis for $E_{\tau,L}(s)/E_{\tau-1,L}(s)$. Hence η can be chosen so that $\eta = x$ modulo $\mathfrak{m}$, *i.e.*, so that g_i and h belong to the maximal ideal $\mathfrak{m}$ and f is a unit. Since η and x are isotropic,

$$0 = Q(\eta) = \sum_{i<j} g_i g_j (y_i|y_j) + \sum_i g_i^2 Q(y_i) + fh + h^2 Q(z).$$

Then

$$-fh = \sum_{i<j} g_i g_j (y_i|y_j) + \sum_i g_i^2 Q(y_i) + h^2 Q(z) \in \mathfrak{m}^2$$

also generates I, and $h \in \mathfrak{m}^2$.

According to the previous lemma, we can choose η to be horizontal, so that

$$-\nabla(x) = df \otimes x + \sum dg_i \otimes y_i + g_i \nabla(y_i) + dh \otimes z + h\nabla(z).$$

But g_i and h annihilate $\Omega^1_{S'/k}(s)$, and since $h \in \mathfrak{m}^2$, dh maps to zero in $\Omega^1_{S'/k}(s)$. Thus $\kappa'(y_j') := y_j'(\nabla(x))$ maps to the class of $-dg_j$ in $\Omega^1_{S'/k}(s)$, that is, to the class of $-g_j$ in $\mathfrak{m}/\mathfrak{m}^2$. Consequently

$$\kappa^*(Q) = \sum_{i<j} g_i g_j (y_i|y_j) + \sum_i g_i^2 Q(y_i) = -fh \pmod{\mathfrak{m}^3}$$

□

Artin has asked if $\mathcal{F}_\infty := \cap \mathcal{F}_i$ is reduced. For simplicity we restrict to the case in which d is prime to p and p is odd.

Theorem 19. *Let $(\mathcal{X}/S, \mathcal{L})$ be a versal deformation of a polarized K3 surface $(X/k, L)$ of degree d. If d is prime to p and p is odd, then $S_\infty = S_{11}$, and in particular is generically nonreduced. If X is supersingular with $\sigma_0 = 10$, then étale locally in a neighborhood of the point of S corresponding to X, $\Sigma := S_{11,red}$ is smooth and $\mathcal{O}_{S_{11}} \cong \mathcal{O}_\Sigma[T]/(T^2)$.*

Proof: Let S' denote the open subset of S_{10} obtained by omitting the set of supersingular points with Artin invariant σ_0 at most 9. By Lemma 5, $E_{10}(\mathcal{X}/S')$ is a local direct factor of $H_{DR}(\mathcal{X}/S')$ of rank 10, and it is contained in the orthogonal complement H_{prim} of $c_{DR}(\mathcal{L})$. Consequently its orthogonal complement $E_{10}^\perp$ in H_{prim} is also a local direct summand. In fact E_{10} is, fiber by fiber, a maximal totally isotropic subspace of H_{prim}, hence $E_{10} = E_{10}^\perp$ fiber by fiber, and hence $E_{10} = E_{10}^\perp$. Now over S'_{11}, E_{11} is a totally isotropic subspace containing E_{10}, hence contained in $E_{10}^\perp$, and hence $E_{10} = E_{11}$. Then it follows from the definition of S'_{12} that $S'_{12} = S'_{11}$, and then that $S'_i = S'_{11}$ for every $i \geq 11$. To extend this result to all of S, we use the fact that S_{11} is a local complete intersection of dimension 9, and hence has depth 9. The ideal I of S_{12} in S_{11} vanishes outside the set of points with Artin invariant at most 9, and hence outside a closed set of dimension at most 8. Consequently $I = 0$. Continuing by induction, we conclude that $S_\infty = S_{11}$.

In fact, it is easy and instructive to calculate S'_{11} explicitly. Let us work locally around a point s corresponding to a supersingular K3 surface with $\sigma_0 = 10$. Then S'_{10} is smooth, and E_i is a local direct factor of $H_{DR}(\mathcal{X}/S_{10})$ for $i \leq 10$. On S_{10}, $E_9 \subseteq F^1_{Hdg}H_{DR}$, and E_9 is horizontal and totally isotropic. Furthermore, $E_9 \subseteq E_{10} \subseteq E_{9,\mathcal{L}}^\perp$, and all are local direct factors of H_{DR}. Let $V := E_{9,\mathcal{L}}^\perp/E_9$, a locally free sheaf of rank 3. The induced quadratic form on V is nondegenerate, and by Lemma 5, the image of $F^2_{Hdg}H_{DR}$ in V is a local direct factor of rank one and is isotropic. In some étale neighborhood of s there exists a basis (x, y, z) for V compatible with the Hodge filtration and such that $(x|x) = (x|y) = (z|z) = 0$ and $(x|z) = Q(y) = 1$. Let η be a basis for E_{10}/E_9, and write $\eta = fx + gy + hz$, with f, g, h sections of $\mathcal{O}_{S_{10}}$. At each point of S_{11}, $F^2_{Hdg}H_{DR} \subseteq E_{10}$, so that f is a unit in a neighborhood of s. Changing η if necessary, we may assume that $f = 1$. Then since η is isotropic, $h = -g^2$, and h generates the ideal of S_{11}. Since E_{10} is horizontal, we can write $\nabla(x) + dg \otimes y = \theta\eta = \theta x$ modulo g, so that $\nabla(x) = -dg \otimes y$ modulo $F^1_{Hdg}H_{DR}$ and g. Then $\kappa(y) = (\nabla(x)|y) = -dg$ modulo g. Since κ induces an isomorphism $H^1(X, \Omega^1_{X/k})/E_9 \cong \Omega^1_{S_{10}/k}(s)$, it follows that $dg(s)$ is not zero in $\Omega^1_{S_{10}/k}(s)$. Thus the closed subscheme S'' of S_{10} defined by g is smooth at s, and S_{11} is defined by g^2. In $\mathcal{O}_{S_{11}}$, $g^2 = 0$, and $\eta = x + gy$. Hence $F^*(\eta) = F^*(x) + g^p y = F^*(x)$, which maps to zero in Gr^1_{Hdg} and hence is killed by γ. This shows again that $S_{12} = S_{11}$. □

Remark 20. If $\mathcal{X}/S$ is a smooth proper morphism of schemes in characteristic p, in general it is not clear how to define the maximal subscheme of S over which $\mathcal{X}$ is supersingular. This difficulty is illuminated by the fact that S_∞ is not reduced. Especially, note that if S is a versal deformation of a supersingular K3 surface X with $\sigma_0 = 10$, then $NS(X)$ prolongs over $S_{11,red}$, not to all of S_{11}.

Remark 21. If $(\mathcal{X}/S, \mathcal{L})$ is a versal family of polarized K3 surfaces, the tangent and cotangent spaces of S inherit from the quadratic form on $H^1(\Omega^1_{\mathcal{X}/S})$ quadratic forms with values in $(F^2_{Hdg})^{\otimes 2}$; if d is prime to p and p is odd, these forms are nondegenerate. Furthermore, the conormal sheaves to each S_i are totally isotropic, as are the cotangent spaces to each S_{∞,σ_0}. It is this fact that underlies the proof of Theorem 17, and we wonder if it has further depth.

References

[1] Michael Artin. Supersingular K3 surfaces. *Annales Scientifiques de l'Ecole Normale Supérieure*, 7(4):543–568, 1974.

[2] Pierre Deligne and Luc Illusie. Relèvement des surfaces K3 en caractéristique nulle. In *Surfaces Algébriques*, volume 868 of *Lecture Notes in Mathematics*. Springer Verlag, 1981.

[3] Nicholas Katz. Slope filtration of F-crystals. *Astérisque*, 63:113–164, 1979.

[4] Arthur Ogus. Supersingular K3 crystals. *Astérisque*, 64:3–86, 1979.

[5] Arthur Ogus. A crystalline Torelli theorem for supersingular K3 surfaces. In Michael Artin and John Tate, editors, *Arithmetic and Geometry*, volume II, pages 361–394. Birkhäuser, 1983.

[6] G. van der Geer and T. Katsura. On a stratification of the moduli of K3 surfaces. *J. Eur. Math. Soc.*, 2:259–290, 2000.

Progress in Mathematics, Vol. 195, © 2001 Birkhäuser Verlag Basel/Switzerland

A STRATIFICATION OF A MODULI SPACE OF ABELIAN VARIETIES

FRANS OORT

CONTENTS

INTRODUCTION

In this paper we study the moduli space $\mathcal{A}_g \otimes \mathbb{F}_p$ of polarized abelian varieties of dimension g in positive characteristic. We construct a stratification of this space. The strata are indexed by isomorphism classes of group schemes killed by p; a polarized abelian variety (X, λ) has its moduli point in a certain stratum if $X[p]$ belongs to the isomorphism class given by a certain discrete invariant. We define these invariants by a numerical property of a filtration of $N = X[p]$.

Passing from one stratum to a stratum in its boundary feels like "degenerating the p-structure". The fact that these strata are all quasi-affine allows us to keep going in this process until we arrive at the unique zero-dimensional stratum, the superspecial locus. One can formulate this idea by saying that the ordinary locus has several "boundaries", one where the abelian variety degenerates, one where the p-structure "becomes more special" (and an analogous idea for all non-zero-dimensional strata). This phenomenon, non-present in this form in characteristic zero, but available and powerful in positive characteristic, is expected to have many applications.

We feel that the strata S_φ and the closed subsets they define inside $\mathcal{A}_{g,1} \otimes \mathbb{F}_p$ merit further study, see [11], where another description for our strata is given and where the cycle classes of the closures of the strata are computed. This is a natural way of producing Chow classes, or cohomology classes on $\mathcal{A}_{g,1}$ which have a geometric interpretation when going over to characteristic p. It is clear that these classes depend (on φ but also) on the characteristic p chosen. Further study might show how useful they are.

Remark. A polarization $\lambda : X \to X^t$ on an abelian variety X induces a homomorphism $\zeta : N \to N^D$ on $N := X[p]$. The morphism λ is symmetric with respect to the duality $X \mapsto X^t$, and ζ is *anti-symmetric* with respect to the duality $N \mapsto N^D$. Going back from ζ to some λ, and following these and their relation under deformation theory is possible, and in fact easy, if the characteristic of the base field is $\neq 2$. However, in case $p = 2$, then $+\zeta = -\zeta$ on $N = X[2]$, and in general, there are (anti-)symmetric morphisms $N \to N^D$ which do not come from a polarization. This causes technical complications. Sections 9 and 12 can be simplified considerably under the extra assumption $p > 2$; we have chosen to treat all cases uniformly. For a proof of (9.4) in case $p > 2$, see Moonen, [33]. For a proof of (12.5) in case $p > 2$, see Wedhorn, [61].

Sections 2 and 4 follow closely the classification in [27]; as we need also a description of the form on $X[p]$ obtained from a principal polarization on X we describe "standard types" in Section 9; this gives a classification of group schemes annihilated by p with an alternating form on their Dieudonné modules over an algebraically closed field.

A short survey of the paper. The basic idea of the paper is: over an algebraically closed field there are only a finite number of finite group schemes annihilated by p of a given rank. Clearly this gives subsets of $\mathcal{A}$ by considering for every abelian variety X the group scheme $X[p]$. We prove, using an adaptation of an idea by Raynaud, that these strata are quasi-affine: Sections 1 – 4.

Using a principal polarization on X we can make this more precise, and study these strata also at the boundary, Sections 5 – 6.

There is exactly one stratum of dimension one. This is contained in the supersingular locus; purely algebraic arguments show that the closure of this stratum is *connected*: Sections 7 – 8.

Then we study in which way strata fit together. To this end we use deformation theory, in the disguise of "displays", see (15.6); we study how we can move inside a stratum, how we can move transversally out of a stratum, and we show that strata fit together as is required in a stratification: Sections 10 – 12. In order to be able to describe these deformations we show that every "polarized BT_1" over an algebraically closed field can be put in standard form, see Section 9; this proves the number of strata is finite.

Logical order in some arguments in this paper:

- an idea by Raynaud, generalized to our strata, proves that these strata in $\mathcal{A}^*$ are quasi-affine, see (6.5); using boundary behavior of strata, see (6.3), this shows that positive dimensional strata in $\mathcal{A}$ contain in their closure points of another stratum;
- using standard types we show that the boundary of a stratum is a union of lower dimensional strata, see (9.4) and (12.5);
- in Section 8 we describe the unique one-dimensional stratum and in Section 7 we show it is connected; hence all positive dimensional strata are connected.

Remark. These methods use techniques in positive characteristic. One of the corollaries of the main result of this paper is: *Let k be a field, fix a positive integer n not divisible by* char(k) *and consider the moduli space $\mathcal{A}_{g,1,n} \otimes k$ of principally polarized abelian varieties of dimension g with a symplectic level-n-structure defined over a field containing k; this space is irreducible.*

This result is well-known: for fields of characteristic zero this is classical. For fields of positive characteristic this was proved by C.-L. Chai and by G. Faltings around 1985. Their proof followed the same line as set out by Zariski-Grothendieck-Deligne-Mumford for the moduli space of algebraic curves: first prove the result in characteristic zero, then construct "good" compactifications of moduli spaces in all characteristics, i.e., "over" Spec$(\mathbb{Z})$; then conclude by Zariski's connectedness theorem that the compactified moduli scheme in positive characteristic is connected, and derive the result. In this paper we prove the irreducibility of $\mathcal{A}_{g,1,n} \otimes k$ for fields of positive characteristic, and hence it also shows the irreducibility in characteristic zero; as all constructions are algebraic (also the construction of the minimal compactification $\mathcal{A}^*$ is), this gives an algebraic proof of this fact.

The research for this paper started as joint work with Torsten Ekedahl. In an early stage we had an idea how to apply the "Raynaud trick" to non-ordinary strata; this is described in Section 4. The construction of L, as in Section 7, and the proof it is connected is due to Ekedahl. Several technical details still had to be supplied; in particular the definition of the stratification, and methods like "standard types" became clear much later. Finally we decided that this author would publish the results. - I thank Torsten Ekedahl for sharing his ideas with me; the first idea of the paper is joint work, but I take full responsibility for correctness of the final result. - I apologize to several authors, who used already this structure, e.g. see [2], see [61], and [13], for the delay in publication of this paper.

Acknowledgments: I thank Johan de Jong, who made a suggestion which simplified the original definition of the stratification. I am grateful to Ben Moonen who spotted a mistake in an earlier version, who had an essential suggestion in one of our many stimulating discussions, and who contributed in writing down results in Section 7. I thank Fabrizio Andreatta for useful suggestions. I thank Ching-Li Chai, Eyal Goren, Ke-Zheng Li, and the referee for helpful critical remarks.

Notations. We fix a prime number p. All base fields, all base schemes will be in characteristic p. All group schemes considered will be commutative. Polarizations on abelian varieties are supposed to be principal polarizations starting from Section 5.

For $n \in \mathbb{Z}_{>0}$ we consider the ring $\mathbb{Z}[\zeta_n, 1/n]$, where ζ_n is a choice of a primitive n-th root of unity. If p is a prime number and $n \in \mathbb{Z}_{>0}$ not divisible by p, we write $\mathbb{F}_{p,n}$ for the smallest field containing $\mathbb{F}_p$ and containing (a fixed choice of) a primitive n-th root of unity. We write $\mathcal{A}_{g,d,n} \to \mathrm{Spec}(\mathbb{Z}[\zeta_n, 1/n])$ for the moduli scheme of abelian varieties with a polarization of degree d^2 and a symplectic level-n-structure. When p and n are chosen, typically, we write $\mathcal{A} = \mathcal{A}_{g,1,n} \otimes \mathbb{F}_{p,n}$.

1. Results

For every elementary sequence φ, see (5.6), there is a locally closed subset $S_\varphi \subset \mathcal{A} = \mathcal{A}_{g,1,n} \otimes \mathbb{F}_{p,n}$. The unique zero-dimensional stratum $\Sigma := S_{\{0,\cdots,0\}}$ is closed. A point $[(X,\lambda)]$ is in Σ iff it is *superspecial*, i.e., for an algebraically closed field k, we have $X \otimes k \cong E^g$, where E is a supersingular elliptic curve. The set Φ_g of all elementary sequences of length g has cardinality $\#(\Phi_g) = 2^g$.

(1.1). **Theorem.** *The unique one-dimensional stratum $S_{\{0,\cdots,0,1\}}$ equals L^0, and its Zariski closure is*

$$L \quad = \quad (S_{\{0,\cdots,0,1\}})^c,$$

where $L^0 \subset L$ are as constructed in (7.1). *The closed curve $L \subset \mathcal{A}$ is* connected.

(1.2). **Theorem.** *Every stratum $S_\varphi \subset \mathcal{A}$ is* non-empty and is quasi-affine (i.e., dense-open in an affine scheme). *For every elementary sequence φ all components of S_φ have the same dimension and we have:*

$$\dim(S_\varphi) \quad = \quad |\varphi| \quad := \quad \sum_{i=1}^{i=g} \varphi(i).$$

(1.3). **Theorem.** *The disjoint union*

$$\mathcal{A} \quad = \quad \bigsqcup_{\varphi} S_\varphi$$

is a finite stratification: the boundary $\partial_{\mathcal{A}}(S_\varphi) := (S_\varphi)^c - S_\varphi$ is the union of all strata meeting that boundary,

$$\partial_{\mathcal{A}}(S_\varphi) = \bigsqcup_{\varphi'} S_{\varphi'}, \quad \text{the union taken over} \quad \forall\ (S_\varphi)^c \cap S_{\varphi'} \neq \emptyset;$$

i.e., either $S_{\varphi'}$ does not meet $(S_\varphi)^c$, or it is contained in $(S_\varphi)^c$.

For every $\varphi \in \Phi$, with φ not superspecial, and every irreducible component $S \subset S_\varphi$, there exists an irreducible component $L' \subset L$ contained in S^c. Hence, *every stratum* S_φ *with* $\varphi \neq \{0, \cdots, 0\}$, *i.e., which is not the superspecial locus, has the property that* $(S_\varphi)^c$ *is connected.*

(1.4). **Corollary** (Faltings, Chai). *Let* k *be an algebraically closed field, and let* $n \in \mathbb{Z}_{>0}$, *such that* $\mathrm{char}(k)$ *does not divide* n. *The moduli space* $\mathcal{A}_{g,1,n} \otimes k$ *is* irreducible.

(1.5). **Corollary.** *Consider* $V_{g-1} \subset \mathcal{A}_{g,1,n} \otimes \mathbb{F}_{p,n}$, *the locus of the non-ordinary principally polarized abelian varieties. Suppose* $g > 1$. *This locus is geometrically irreducible.*

(1.6). **"Polarized BT$_1$ truncated group schemes".** For a complete classification of all $(N, <,>) = (X, \lambda)[p]$, where $N = X[p]$ over an algebraically closed field k, see Section 9.

(1.7). For a complete classification of all principally polarized abelian varieties (X, λ) in characteristic p with $a(X) = g - 1$, see Section 8.

2. Filtrations on finite group schemes

In this section we define the canonical filtration on certain finite group schemes. We derive some elementary properties.

(2.1). All base schemes in this paper will be in characteristic p. All finite group schemes considered will be commutative. If we consider $N \to S$, a group scheme over a base, it is supposed to be finite and flat over S.

We write $F : N \to N^{(p)}$ for the Frobenius homomorphism and $V : N^{(p)} \to N$ for the Verschiebung homomorphism. All finite group schemes considered in this section will be annihilated by p.

For a group scheme $N \to S$, commutative and finite and flat over S, we will write $N[F] := \mathrm{Ker}(F : N \to N^{(p)})$, we write $V(N) = \mathrm{Im}(V : N^{(p)} \to N)$, if these exist as finite flat group schemes over S; if this is the case we say that N/S is a *Barsotti-Tate truncated level one group scheme* if it is annihilated by p, i.e., $[p]_N = 0$, and

$$\mathrm{Im}(V : N^{(p)} \to N) = \mathrm{Ker}(F : N \to N^{(p)}),$$
$$\mathrm{Im}(F : N \to N^{(p)}) = \mathrm{Ker}(V : N^{(p)} \to N).$$

Such a group scheme will be called a BT$_1$.

Suppose $T \subset N$ a subgroup scheme. We will write $F^{-1}(T) \subset N$ for the finite flat subgroup scheme (if it exists as a finite flat group scheme over S) which is the pull back by $F : N \to N^{(p)}$ of $T^{(p)} \subset N^{(p)}$.

(2.2). **Construction: the canonical filtration.** We suppose N is a BT_1 over a field $K \supset \mathbb{F}_p$. We construct a filtration:

$$0 = N_0 \subset \cdots \subset N_r = V(N) \subset \cdots \subset N_s = N;$$

here

$$N_r = H := V(N) = N[F],$$

and $0 \leq r \leq s$ are integers. The construction is as follows: in the first step one considers all images $V^i(N)$; this gives a filtration of N; in the second step we take all $F^{-j}(N')$, $j \in \mathbb{Z}_{>0}$, for every N' in the previous filtration; then we go on by induction: in all odd numbered steps we take all $V^i(N')$, for all N' in the previous filtration; in all even numbered steps we take all $F^{-j}(N')$. Each step gives a filtration which is a refinement of the previous one. Note that odd numbered steps add at most new steps in the filtration of $0 \subset H$, and even numbered steps add at most new steps in the filtration of $H \subset N$. After a finite number of steps the process stabilizes, in fact after at most $2(q-1)$ steps if $\mathrm{rk}(N) = p^q$; the filtration reached will be called *the canonical filtration of* N.

Let E be the set of all finite words in the symbols V and F^{-1}. The set of subgroup schemes $\{W(N) \mid W \in E\}$ is the canonical filtration of N.

Note that we performed the construction of the canonical filtration for a BT_1 over a field. For a group scheme over an arbitrary base the canonical filtration need not exist (steps might produce non-flat group schemes). We will come back to this.

(2.3). **Notation.** A BT_1 over a field K has a canonical filtration, and we derive a triple $\tau(N) = \tau = \{v, f, \rho\}$, where:

$$\rho : \{0, \cdots, s\} \to \mathbb{Z}_{\geq 0}, \quad v : \{0, \cdots, s\} \to \{0, \cdots, r\}, \quad f : \{0, \cdots, s\} \to \{r, \cdots, s\},$$

are defined by:

$$\mathrm{rk}(N_i) = p^{\rho(i)}, \quad V(N_i) = N_{v(i)}, \quad F^{-1}(N_i) = N_{f(i)}.$$

The triple $\{v, f, \rho\}$ will be called *the canonical type* of N; it will be denoted by $\tau(N)$. We describe some properties. We write $\Gamma = \Gamma_s = \{0, \cdots, s-1\}$. We define $\pi_\tau = \pi : \Gamma \to \Gamma$ by:

$$v(i+1) > v(i) \Rightarrow \pi(i) := v(i); \quad v(i+1) = v(i) \Rightarrow \pi(i) := f(i).$$

We write $B_i = N_{i+1}/N_i$ for every $i \in \Gamma$. Note that the maps v and f are monotone, i.e., $v(i+1) \geq v(i)$ and $f(i+1) \geq f(i)$ for all i.

(2.4). **Lemma.** *Let N be a BT_1 over a field K, and let $\{v, f, \rho\}$ be its canonical type. Then:*
(i) the maps $v : \{0, \cdots, s\} \to \{0, \cdots, r\}$ and $f : \{0, \cdots, s\} \to \{r, \cdots, s\}$ are surjective; the map $\pi : \Gamma \to \Gamma$ is bijective.
(ii)

$$v(i+1) > v(i) \Longleftrightarrow f(i+1) = f(i); \quad \textit{in this case} \quad V : B_i^{(p)} \xrightarrow{\sim} B_{\pi(i)}$$

is an isomorphism;
(iii)

$$v(i+1) = v(i) \Longleftrightarrow f(i+1) > f(i); \quad \textit{in this case} \quad F : B_{\pi(i)} \xrightarrow{\sim} B_i^{(p)}$$

is an isomorphism;
(iv)

$$f(i) + v(i) = s + i.$$

Remark. Consider the set E of all finite words in the letters V and F^{-1}. For $w \in E$ and a BT_1 over a field as above we consider the rank of $w(N)$. The group scheme N determines a function $E \to \mathbb{Z}$ by $w \mapsto \mathrm{rk}(w(N))$; we write this function as indexed set $\{\mathrm{rk}(w(N)) \mid w \in E\}$; if these functions are equal for N_1 and N_2, then $\tau(N_1) = \tau(N_2)$.

(2.5). **Notation: Cycles.** The bijection π splits up Γ into cycles, orbits under the group generated by π:

$$\Gamma = \Gamma_1 \sqcup \cdots \sqcup \Gamma_m.$$

For an element $i \in \Gamma$ we say $n \in \mathbb{Z}_{>0}$ is the order if n is the smallest positive number such that $\pi^n(i) = i$, i.e., if $i \in \Gamma_b$ then $n = \#(\Gamma_b)$.

Proof of (2.4). In the canonical filtration every $0 \subset N_i \subset H = V(N)$ is the image under V of a group scheme in the filtration (this is the way these N_i are constructed); hence $v : \{0, \cdots, s\} \to \{0, \cdots, r\}$ is surjective. The same arguments for $H \subset N_i \subset N$ show that f is surjective. We show that $\pi : \Gamma \to \Gamma$ is injective; indeed, suppose $\pi(i_1) = \pi(i_2) < r$; then $\pi(i_1) = v(i_1) = b = v(i_2) = \pi(i_2)$; as v is monotone, the fact $v(i_1+1) > v(i_1) = b = v(i_2) < v(i_2+1)$ implies $i_1 = i_2$ (there is precisely one i at which $v(i)$ jumps from b to $b+1$); note that $\pi(i) \geq r$ implies $v(i+1) = v(i)$; if $\pi(i_1) = \pi(i_2) \geq r$ the same arguments involving f show that $i_1 = i_2$; this proves that $\pi : \Gamma \to \Gamma$ is bijective.

As π is bijective, we obtain the cycle structure on Γ as indicated above.

Claim. *For every $i \in \Gamma$ we have* $\mathrm{rk}(B_i) \geq \mathrm{rk}(B_{\pi(i)})$. For every $i \in \Gamma$ the homomorphism $V : B_i^{(p)} \to N_{v(i+1)}/N_{v(i)}$ is *surjective*, and $F : N_{f(i+1)}/N_{f(i)} \to (N_{i+1}/N_i)^{(p)}$ is *injective*. This proves the claim.

Following the elements in a cycle under π, we see that

$$\mathrm{rk}(B_i) \geq \mathrm{rk}(B_{\pi(i)}) \geq \cdots \geq \mathrm{rk}(B_{\pi^n(i)}) = \mathrm{rk}(B_i).$$

We conclude that $v(i+1) > v(i)$ implies that

$$V : B_i^{(p)} \xrightarrow{\sim} N_{v(i+1)}/N_{v(i)} = B_{v(i)}$$

is an isomorphism. This shows that $(\mathrm{Ker}(V) \cap N_{i+1}) \subset N_i$; hence we conclude that $v(i+1) > v(i) \Rightarrow f(i+1) = f(i)$.

In the same way we show that $f(i+1) > f(i)$ implies that

$$F : N_{f(i+1)}/N_{f(i)} = B_{f(i)} \xrightarrow{\sim} B_i^{(p)}$$

is an isomorphism and that $v(i+1) = v(i)$.

Clearly $f(0) + v(0) = r + 0$; by (ii) and (iii) the last claim follows. This concludes the proof of Lemma (2.4). □

(2.6). **Corollary.** *Let $i \in \Gamma$, and let $n \in \mathbb{Z}_{>0}$ be the order of i under π. The maps in (ii) and the inverses of the maps in (ii) above give a (canonical) isomorphism*

$$B_i^{(p^n)} \xrightarrow{\sim} B_i.$$

□

(2.7). **Definition.** *A triple $\tau = \{v, f, \rho\}$ is a* canonical type *if $0 \leq r \leq s$ are integers, v, f and ρ are maps as in (2.3), such that:*

- *$\rho : \{0, \cdots, s\} \to \mathbb{Z}_{\geq 0}$ is strict monotone with $\rho(0) = 0$,*
- *the maps v and f are monotone and surjective, with*

$$v(i+1) > v(i) \Longleftrightarrow f(i+1) = f(i), \quad v(i+1) = v(i) \Longleftrightarrow f(i+1) > f(i);$$

  this defines a permutation π of $\Gamma := \{0, \cdots, s-1\}$;
- *$\rho(i+1) - \rho(i) = \rho(\pi(i+1)) - \rho(\pi(i))$ for every $i \in \Gamma$.*

We have seen that a BT_1 over a field defines a canonical type. Conversely:

(2.8). **Remark.** For every canonical type, there exists a BT_1 over a field (even over $\mathbb{F}_p$) having this canonical type. We will not use this; this simple observation can be easily proved with methods indicated later.

The methods of this section can be easily derived from [27]. In Section 5 we will give more details on filtrations, and we give a classification of symmetric BT_1 group schemes. All these methods will be easy, and in fact almost equivalent to [27]. Then, in Section 9 we will classify "polarized BT_1 truncated group schemes"; this will be the backbone of the deformation theory in later sections.

(2.9). **Remark.** In the case of symmetric BT_1's we will study the possible canonical types in more detail in Section 5.

The cycles in Γ studied in this section will be called (V, F^{-1})-cycles, in contrast with "$(V, \perp)$-cycles" studied later.

3. Strata: the canonical stratification of the moduli space of abelian varieties

(3.1). **Lemma.** *Let $h : M_1 \to M_2$ be a homomorphism of finite flat group schemes over a noetherian scheme S; let R be a positive integer. Consider all points $s \in S$ such that $\mathrm{Im}(h_s : M_{1,s} \to M_{2,s})$ has rank equal to R. These points give a locally closed set $S(R) \subset S$; this is a locally closed, reduced subscheme, and $\mathrm{Im}(h : M_1 \to M_2)|_{S(R)} \subset M_2|_{S(R)}$ exists as a finite flat group scheme.*

Let $h : M_1 \to M_2$ be a flat surjective homomorphism of finite flat group schemes over a noetherian scheme S, and let $N_2 \subset M_2$ be a flat subgroup scheme; let R be a positive integer. Consider all points $s \in S$ such that $h^{-1}((N_2)_s)$ has rank equal to R. These points give a locally closed set $T(R) \subset S$; this is a locally closed, reduced subscheme, and $h^{-1}(N_2)|_{T(R)} \subset M_1|_{T(R)}$ exists as a finite flat group scheme.

Proof. The rank of the fibers of the finite group scheme $\mathrm{Ker}(h : M_1 \to M_2)$ is upper-semicontinuous. Hence $S(R) \subset S$ is locally closed. Over this scheme $\mathrm{Ker}(h)|_{S(R)}$ is finite with geometric fibers of constant rank; hence it is flat by [4], Lemma (1.13) on page 240. Hence the quotient $M_1|_{S(R)}/\mathrm{Ker}(h)|_{S(R)}$ exists as finite, flat group scheme, see [14], page 212-17, Th. 7.2; this is a closed subscheme of $M_2|_{S(R)}$, the image of $h|_{S(R)}$. The same arguments show the second part of the claim. □

Notation. Let τ be a canonical type, and let $N \to S$ be a BT_1 over S. We denote by $\mathcal{D}_\tau(S)$ the set, considered as reduced subscheme, of points $s \in S$ such that the fiber N_s has canonical type $\tau(N_s) = \tau$.

(3.2). **Proposition.** *Let τ be given, and let $N \to S$ be a BT_1. Then*

$$\mathcal{D}_\tau(S) \subset S$$

is locally closed.

Let a canonical type τ be given, and let $N \to T$ be a BT_1, with T reduced, such that for every $s \in T$ we have $\tau(N_s) = \tau$, i.e., "the type in the fibers of $N \to T$ is constant". Then the canonical filtration of N/T exists, i.e., there exist finite flat subgroup schemes $N_i \subset N \to S$, which are obtained by the construction of the canonical filtration, and which in every fiber N_s give the canonical filtration.

Proof. As we have seen, $\tau(N_s)$ is determined by the set $\{\mathrm{rk}(N) \mid w \in \mathcal{W}\}$. It suffices to consider only words of length at most $2q(q-1)$, where $q = \mathrm{rk}(N)$. Consider the collection of all such words; for a given type τ fix the ranks of the images of each word belonging to the given type τ. Start with $S = S_0$. Suppose $b \in \mathbb{Z}_{>0}$, suppose we work over a base S_{b-1}, and suppose that all words of length at most $b-1$ have "images" which are flat group schemes over S_{b-1} of the given ranks. The set $S_b \subset S_{b-1}$ is the reduced locally closed subscheme where the image of every word of length b has a given rank (determined by τ). This exists: use the previous lemma. Repeating this argument for all words and all ranks we have obtained from τ a locally closed set $S_{2q(q-1)}$ in the base where all "images" have constant ranks, and give flat group schemes. This is the set we are looking for.

If $\tau(-)$ is constant on T, in each step of the proof of the previous lemma we obtain subgroup schemes of constant rank in N, hence flat subgroup schemes by [4], Lemma (1.13) on page 240. This shows that the construction of the canonical filtration under the condition $\tau(-)$ is constant on T produces a filtration by flat subgroup schemes, and the proposition is proved. □

Obviously the canonical filtration on a group scheme over an arbitrary base does not exist if the type changes.

(3.3). **Notation.** Let τ be a canonical type. For a fixed $n \in \mathbb{Z}_{>0}$ consider the moduli space

$$\mathcal{A}' := \mathcal{A}_{g,*,n} \otimes \mathbb{F}_{p,n} = \cup_d \ \mathcal{A}_{g,d,n} \otimes \mathbb{F}_{p,n}.$$

We write

$$D_\tau := \mathcal{D}_\tau(\mathcal{A}') \subset \mathcal{A}'$$

for the locally closed reduced subscheme consisting of moduli points $s \in \mathcal{A}'$, where s comes over some field from an abelian variety X_s, such that $\tau(X_s[p]) = \tau$. This is a locally closed subset. Note that the rank of $X_s[p]$ equals $2g$. Note that only a finite number of canonical types exist for a given rank.

(3.4). **Definition.** We have obtained a finite disjoint union into locally closed subsets:

$$\mathcal{A}' \quad = \quad \bigsqcup_\tau D_\tau.$$

This is called the *canonical stratification* of $\mathcal{A}'$. In some papers this is called the EO-stratification, or the Oort stratification.

(3.5). **Remark.** We study some properties of this stratification on $\mathcal{A}'$. Then, in order to describe more precise results we will restrict later to the case of

$$\mathcal{A} = \mathcal{A}_{g,1,n} \otimes \mathbb{F}_{p,n},$$

the moduli space of *principally* polarized abelian varieties in characteristic p.

4. THE RAYNAUD TRICK ON $\mathcal{A}_g \otimes \mathbb{F}_p$

(4.1). **Theorem.** *For every canonical type τ every irreducible component of the stratum $D_\tau \subset \mathcal{A}' := \cup_d \ \mathcal{A}_{g,d,n} \otimes \mathbb{F}_{p,n}$ is a* quasi-affine *subscheme.*

Proof. It suffices to show this in case of arbitrary high level structure. Hence we suppose $\mathcal{A}' = \cup_d \mathcal{A}_{g,d,n} \otimes \mathbb{F}_{p,n}$, with $n \in \mathbb{Z}_{\geq 3}$ not divisible by p. We choose a component $T \subset D_\tau \subset \mathcal{A}'$.

We are going to show that the theorem follows from ampleness of the determinant line bundle associated with the sheaf of invariant differential forms on the "universal semi-abelian scheme", as proved by Moret-Bailly, see [35], also see [10], combined with the properties we proved on cycles in the p-kernel constructed by the canonical filtration, see (2.4).

Let $\mathcal{X} \to \mathcal{A}'$ be the universal abelian scheme. We write $\mathfrak{t}_{\mathcal{X}/\mathcal{A}'}$ for the tangential sheaf of $\mathcal{X} \to \mathcal{A}'$ along the zero-section, and we write $\omega_{\mathcal{X}/\mathcal{A}'}$ for the determinant line bundle of its dual:

$$\omega_{\mathcal{X}/\mathcal{A}'} = \mathrm{Det}(\mathfrak{t}^{\mathrm{dual}}_{\mathcal{X}/\mathcal{A}'}) = \mathrm{Det}(\Omega_{\mathcal{X}/\mathcal{A}'}),$$

see [10], page 24 and page 137.

Claims. 1) *Under these conditions, the line bundle $\omega_{\mathcal{X}/T}$ on T is a torsion line bundle.*
2) *From this it follows that T is quasi-affine.*
By [35], Chap. IX, and [10], I.5 and V.2, Th. 2.3, we conclude that $\omega|_T$ is ample; if we assume (1) it is also torsion; hence T is quasi-affine. This proves: (1) $\Rightarrow$ (2).
We show that (1) is true. We write $N := \mathcal{X}[p]\,|_T$. By (3.2) we see that the canonical filtration $\{N_i \mid 0 \leq i \leq s\}$ by finite flat subgroup schemes of N exists. We have:

$$0 = N_0 \subset N_1 \subset \cdots \subset N_r = N[F] = V(N);$$

we see that all tangential sheaves

$$\mathfrak{t}_{N_1/T} \subset \cdots \subset \mathfrak{t}_{N_r/T} = \mathfrak{t}_{\mathcal{X}/T}$$

are well-defined, of constant rank. We write $\mathcal{L}_i := \mathrm{Det}(\mathfrak{t}_{N_{i+1}/T}/\mathfrak{t}_{N_i/T})$, for $0 \leq i < r$, and we are going to show that these are torsion line bundles. By (2.4) we know that if $i \in \Gamma$ has order $n(i) = n > 0$ under $\pi : \Gamma \to \Gamma$, then for $B_i := N_{i+1}/N_i$ we have

$$(B_i)^{(p^n)} \cong B_i; \quad \text{hence} \quad (\mathcal{L}_i)^{p^n} \cong \mathcal{L}_i;$$

this shows that $\mathrm{Det}(\mathcal{L}_i)^{p^n-1} = 1$. We conclude that $\mathrm{Det}(\mathfrak{t}_{\mathcal{X}/T}) = \mathrm{Det}(\mathcal{L}_1) \otimes \cdots \otimes \mathrm{Det}(\mathcal{L}_r)$ is torsion; hence its inverse $\omega|_T$ is torsion. This shows (1), and it finishes the proof of Theorem (4.1). □

(4.2). **Remark.** A special case of this proposition is due to Raynaud: if S is irreducible and complete, in characteristic p, and $\mathcal{X} \to S$ is a family of ordinary abelian varieties (i.e., all fibers have maximal p-rank), then the natural map $S \to \mathcal{A}_g \otimes \mathbb{F}_p$ is constant, i.e., the family $\mathcal{X} \to S$ is isotrivial, cf. [60], Th. 5, page 62, cf. [35], XI.5, page 237. This beautiful idea by Raynaud stimulated us to look for some of the basic notions as exposed in this paper.

The example of p-rank equal to $f = g - 1$ is worked out in [45], 6.2; that case can be seen as a motivating example illustrating the proof of the "Raynaud trick" in the general setting, as given above.

In the case of a family of ordinary abelian varieties the tangent bundle can be trivialized by a finite covering, because "μ_p is rigid"; that was the original idea by Raynaud. In case of lower p-rank, the fact that α_p does not admit a canonical base seems to spoil the case; however then we consider the p-kernel $\mathcal{X}[p]$, "go up and down" by V and by F^{-1} and we prove that the determinant of the tangent bundle is torsion, as is done above.
From now on abelian varieties considered will be *principally* polarized.

5. Filtrations on finite symmetric group schemes

In this section we consider *symmetric* BT_1's, and we describe and characterize filtrations on such finite group schemes.

Finite group schemes will be considered, usually over a field, sometimes (finite and flat) over a more general base. We will write N for a finite (commutative) group scheme. All finite group schemes considered in this section will be annihilated by p; when over an arbitrary base we suppose the group scheme is finite and flat over that base. In all cases we assume that we work with a BT_1.

(5.1). **Definition.** We say that ζ is a *symmetry* on a finite group scheme N if

$$\zeta : N \xrightarrow{\sim} N^D$$

is an isomorphism onto its Cartier dual group scheme. A finite group scheme which admits a symmetry is called a symmetric finite group scheme.
Remark. Later we will be more precise whether this symmetry has extra properties (like being alternating or anti-symmetric).
For commutative group schemes we have

$$(F_N : N \to N^{(p)})^D \quad = \quad (V_{N^D} : N^{(p)D} = N^{D(p)} \to N^D);$$

as V commutes with base change we derive

$$\zeta{\cdot}V_N = V_{N^D}{\cdot}\zeta^{(p)} = (F_N)^D{\cdot}\zeta^{(p)}.$$

Notation. Suppose $\zeta : N \to N^D$ is a symmetry, and $i : T \hookrightarrow N$ is a finite subgroup scheme; we write:

$$\perp_\zeta(T) = \perp(T) := \mathrm{Ker}(N \xrightarrow{\zeta} N^D \xrightarrow{i^D} T^D).$$

Earlier we have considered the set E of all (finite) words in the symbols V and F^{-1}. For a BT_1 group scheme N over a field the filtration $\{W(N) \mid W \in E\}$ was called the *canonical filtration.*

Consider the set e of all (finite) words in the symbols V and $\perp$. We consider the set of finite subgroup schemes $\{w(N) \mid w \in e\}$ of N.

(5.2). **Lemma.** *Let N be a finite group scheme over a field K. Assume this is a BT_1. Assume there is a symmetry $\zeta : N \xrightarrow{\sim} N^D$ on N. Then the following properties hold:*
1) $\perp(\perp(T)) = T$;
2) $\perp(V(N)) = V(N)$;
3) the set $\{w(N) \mid w \in e\}$ is a finite filtration on N;
4) $\perp(V(T)) = F^{-1}(\perp(T))$ for every subgroup scheme $T \subset N$;
5) $\{w(N) \mid w \in e\} = \{W(N) \mid W \in E\}$.
Remark. The last property shows that the canonical filtration does not depend on a particular choice of the symmetry.
Proof. 1) The first property follows by dualizing

$$\perp(T) \to N \xrightarrow{\zeta} N^D \to T^D.$$

2) The second follows from: $\perp(V(N)) = \mathrm{Ker}(F) = V(N)$.
3) Note that $\perp(N) = 0$; hence for every word of the form $w = w'\cdot(\perp\cdot V^i\cdot\perp)$ we have $w(N) = w'(N)$. Hence we need only consider words ending with a positive power of V. The set $\{V^i(N) \mid i \in \mathbb{Z}_{>0}\}$ is a filtration of $V(N)$ which stabilizes after a finite number of steps. The set $\{\perp(V^i(N)) \mid i \in \mathbb{Z}_{>0}\}$ is a filtration of $V(N) \subset N$. Hence $\{(V^j\cdot\perp\cdot V^i)(N) \mid i,j \in \mathbb{Z}_{\geq 0}\}$ is a filtration, etc.: we carry on by induction on the number of letters in the words in consideration; this shows that we obtain a filtration; because N is a finite group scheme, this is a finite filtration.
4) This follows because

$$(F : N \to N^{(p)})^D = (V : (N^{(p)})^D = (N^D)^{(p)} \to N^D).$$

5) As we have seen, we need only consider words w ending with V, and the same for words W. If $w = V \in e$, then also $w \in E$. If $w = w'\cdot\perp\cdot V$, then $w(N) = (w'\cdot V)(N)$. Suppose w ends in V^2; note that if $i, j \in \mathbb{Z}_{\geq 0}$ then

$$(V^i\cdot\perp\cdot V^j\cdot\perp)(T) = (V^i\cdot F^{-j})(T).$$

We write either $w = w'\cdot V^b$ (in case $\perp$ appears an even number of times in w), or $w = w'\cdot V^a\cdot\perp\cdot V^b$ (in the odd case), where $a \in \mathbb{Z}_{>0}$, and $b \in \mathbb{Z}_{>1}$, and where w' is a finite product of factors of the form $V^i\cdot\perp\cdot V^j\cdot\perp$; in the second case, note that

$$w(N) = (w'\cdot V^a\cdot\perp\cdot V^b)(N) = (w'\cdot V^a\cdot F^{b-1}\cdot V)(N);$$

hence in both cases there exists a word $W \in E$ such that $w(N) = W(N)$. Hence

$$\{w(N) \mid w \in e\} \subset \{W(N) \mid W \in E\}.$$

Conversely, considering all $V^i\cdot F^{-j}$, with $i \geq 0$, in a word W ending on V shows the opposite inclusion. This ends the proof of the lemma. □

(5.3). **Definition.** Let (N, ζ) be a symmetric BT_1 over a base S. We say that

$$0 = N_0 \subset \cdots \subset N_i \subset \cdots \subset N_r = H = VN = N[F] \subset \cdots \subset N_{2r} = N$$

is a *good filtration* if the following conditions are satisfied:

1): Each $N_j \to S$ is a flat subgroup scheme, $N_i \neq N_{i+1}$.

2): For every $0 \leq j \leq 2r$ duality gives an equality

$$\perp(N_j) = N_{2r-j}.$$

3): For every $0 \leq j \leq 2r$, the image $\mathrm{Im}(V : N_j^{(p)} \to N)$ exists as a finite flat group scheme over S, and it equals one of the group schemes in the filtration, and

4): for every $0 \leq i \leq r$, the group scheme N_i is such an image.

We say that a good filtration is a *final filtration* if moreover its length equals $2g$ with $p^{2g} := \mathrm{rk}(N/S)$:

$$0 = N_0 \subset \cdots \subset N_g = V(N) \subset \cdots \subset N_{2g} = N$$

by finite, flat group schemes over S, such that:

5): the rank of N_i over S equals p^i.

(5.4). **Proposition.** *Let N be a symmetric* BT_1 *over a field. Its canonical filtration,* as constructed in Section 2, *is a good filtration.*

Proof. In Lemma (5.2) we have seen that the canonical filtration constructed with the help of V and F^{-1} is the same as the one constructed with V and $\perp$. Hence the canonical filtration is stable under $\perp$; hence it satisfies (2). Conditions (3) and (4) follow by construction. □

We see that the canonical filtration (if it exists) is the "minimal good filtration" and a final filtration (if it exists) a "maximal good filtration".

(5.5). **Proposition.** *Suppose that N is a symmetric* BT_1 *over a base $S \to \mathbb{F}_p$ with a good filtration*

$$0 = N_0 \subset \cdots \subset N_i \subset \cdots \subset N_r = H \subset \cdots \subset N_{2r} = N$$

on it. For every j with $0 \leq j \leq 2r$ we write $\nu(j)$ for the integer with the property

$$\mathrm{Im}(V : N_j^{(p)} \to N) = N_{\nu(j)}.$$

1) For every $0 \leq j \leq 2r$ the inverse image of $N_j^{(p)}$ under $F : N \to N^{(p)}$ equals

$$F^{-1}(N_j^{(p)}) = N_{2r-\nu(2r-j)}.$$

2) For every $0 \leq j < 2r$ there is an exact sequence

$$0 \to N_{2r-\nu(2r-j-1)}/N_{2r-\nu(2r-j)} \xrightarrow{F} N_{j+1}^{(p)}/N_j^{(p)} \xrightarrow{V} N_{\nu(j+1)}/N_{\nu(j)} \to 0.$$

3) For every $0 \leq j < 2r$

$$\nu(j) = \nu(j+1) \iff \nu(2r-j) = \nu(2r-j-1)+1.$$

4) If $0 \leq j < 2r$ and $\nu(j) < \nu(j+1)$, then $\nu(j)+1 = \nu(j+1)$, and

$$V : N_{j+1}^{(p)}/G_j^{(p)} \xrightarrow{\sim} N_{\nu(j+1)}/N_{\nu(j)}.$$

Proof. From

$$V(N_{2r-j}^{(p)}) \subset N_{\nu(2r-j)}$$

we deduce

$$F(N_{2r-\nu(2r-j)}) \subset N_j^{(p)}.$$

Hence all arrows as claimed in the proposition exist. In order to show the statements it suffices to show these for every fiber $N_s \to \mathrm{Spec}(K) \subset S$, where K is a perfect field. We write $A = \mathbb{D}(N_s)$ for its Dieudonné module.

1) In order to show $\mathcal{V}^{-1}(A_j^{(p)}) = A_{2r-\nu(2r-j)}$ we observe:

$$x \in \mathcal{V}^{-1}(A_j^{(p)}) \iff \mathcal{V}(x) \in A_j^{(p)} \iff$$

$$\iff \forall y \in A_{2r-j}, \quad <\mathcal{V}x, y>= 0 \iff \forall y \in A_{2r-j}, \quad <x, \mathcal{F}y>= 0 \iff$$

$$\Longleftrightarrow \forall z \in \mathcal{F}(A^{(p)}_{2r-j}) = A_{\nu(2r-j)}, \quad < x, z >= 0 \Longleftrightarrow$$
$$\Longleftrightarrow x \in \perp(A_{\nu(2r-j)}) = A_{2r-\nu(2r-j)}.$$

2) The sequence as in (2) exists, surjectivity on the right follows from the definition of ν in a good filtration, and injectivity on the left follows from (1). In order to show exactness in the middle, we choose $x \in A^{(p)}_{j+1}$ such that

$$\mathcal{F}x \in A_{\nu(j)} = \mathcal{F}(N^{(p)}_j).$$

Let $y \in A^{(p)}_j$ with $\mathcal{F}x = \mathcal{F}y$. Then $\mathcal{F}(x-y) = 0$, hence there exists z with $\mathcal{V}z = x - y \in A^{(p)}_{j+1}$. By (1) this implies $z \in A_{2r-\nu(2r-j-1)}$. As $\mathcal{V}z$ and x are congruent mod A_j this proves exactness in the middle.
3) For any index $j < 2r$ we have $\nu(j+1) \leq \nu(j)+1$. If $\nu(j) = \nu(j+1)$ then by (2) we conclude $\nu(2r-j) = \nu(2r-j-1)+1$. Note that $\nu(2r) = r$. Hence by the previous argument, it is excluded that for any j we have $\nu(j) < \nu(j+1)$ ánd $\nu(2r-j) > \nu(2r-j-1)$. This proves (3).
4) This follows from (2) and (3). □

(5.6). **Definition: Final types and elementary sequences.** A *final sequence* is a map $\psi : \{1, \cdots, 2g\} \to \mathbb{Z}_{\geq 0}$ with $\psi(0) = 0$, and $\psi(2g) = g$, such that

$$\psi(i) \leq \psi(i+1) \leq \psi(i) + 1, \quad 0 \leq i < 2g,$$

and

$$\psi(i) + 1 = \psi(i+1) \Longleftrightarrow \psi(2g-i) = \psi(2g-i-1).$$

Note that this implies:

$$\psi(2g-i) = \psi(i) + g - i.$$

An *elementary sequence* is a map $\varphi : \{1, \cdots, g\} \to \mathbb{Z}_{\geq 0}$ with $\varphi(0) = 0$ and

$$\varphi(i) \leq \varphi(i+1) \leq \varphi(i) + 1, \quad 0 \leq i < g.$$

A *symmetric canonical type* is a canonical type as in (2.3) with $s = 2r$ and the extra conditions

$$f(j) = 2r - v(2r-j), \quad \forall 0 \leq j < r,$$

and

$$\rho(j+1) - \rho(j) = \rho(2r-j) - \rho(2r-j-1), \quad \forall 0 \leq j < r.$$

Note that in a symmetric canonical type we have $f(j) = 2r - v(2r-j) = r + j - v(j)$ hence $v(2r-j) - v(j) = r - j$ for all j. If N is a symmetric BT_1 over a field, its canonical filtration, given by V and F^{-1}, defines a canonical type, and this is symmetric.

We could define a *good type* as a pair of maps $\nu : \{0, \cdots, 2r\} \to \{0, \cdots, r\}$ and $\rho : \{0, \cdots, 2r\} \to \mathbb{Z}_{\geq 0}$ satisfying obvious rules generalizing the definition of a symmetric canonical type. Here f is given by $f(j) = 2r - \nu(2r-j)$. We will not need this definition.

The set of all final sequences for a given value of g is denoted by $\Psi = \Psi_g$. The set of all elementary sequences for a given value of g is denoted by $\Phi = \Phi_g$. Clearly $\#(\Phi_g) = 2^g$ (in each step the sequence φ either jumps by one, or does not jump).

We define a partial ordering on Φ by:

$$\varphi' \quad \prec \quad \varphi \quad \stackrel{\text{def}}{\Longleftrightarrow} \quad \varphi'(i) \leq \varphi(i) \qquad \forall i \leq g.$$

Observe that any maximal totally ordered subset in Φ has length equal to $g(g+1)/2$.

We define a "dimension function" $|\cdot| : \Phi \to \mathbb{Z}$ on Φ, by:

$$|\varphi| := \sum_{i=1}^{i=g} \varphi(i).$$

Note that for a final sequence we have $\psi(2g-i) - \psi(i) = g - i$ for $0 \leq i < g$; we call this "symmetry". A final sequence ψ defines an elementary sequence φ by $\varphi(i) := \psi(i), \quad \forall\, 0 \leq i \leq g$; we call this "truncation" of the final sequence. Conversely, an elementary sequence φ defines a final sequence ψ by: $\psi(i) = \varphi(i) = \psi(2g-i) - g + i$ for $0 \leq i \leq g$; this is inverse to the "truncation" process.

Combinatorial constructions. Suppose given a symmetric canonical type $\tau = \{v, f, \rho\}$. We associate to this a final sequence and an elementary sequence $\varphi = \text{stretch}(\tau)$ by "stretching": suppose $\{\varphi(0), \varphi(1), \cdots, \varphi(\rho(i))\}$ is defined, for $0 \leq i$; then define $\{\varphi(0), \varphi(1), \cdots, \varphi(\rho(i+1))\}$ by $\varphi(\rho(i)) = \varphi(\rho(i)+1) = \cdots = \varphi(\rho(i+1))$ if $v(i+1) = v(i)$; $\{\varphi(0), \varphi(1), \cdots, \varphi(\rho(i+1))\}$ by $\varphi(\rho(i)) < \varphi(\rho(i)+1) < \cdots < \varphi(\rho(i+1))$ if $v(i+1) > v(i)$; clearly this procedure produces an elementary sequence.

Suppose given a final type. Single out those indices which can be reached by starting with $2g$ and applying $j \to \psi(j)$ and $i \to 2g - i$ as many times until no further indices appear (this we call the canonical construction); then delete all numbers not appearing, and construct a canonical type τ from these data in the obvious way.

Remark. We shall see that for a "polarized BT_1" over an algebraically closed field the canonical filtration can be refined to a final filtration, see (9.6). The final type associated to a canonical type by stretching is the type of any final filtration associated with the canonical filtration. A refinement from a canonical filtration to a final filtration is not unique in general. However the final type and the elementary type are uniquely determined by the canonical type, and are the ones obtained by stretching and truncation.

(5.7). **Lemma.** *Suppose given $g \in \mathbb{Z}_{\geq 0}$. There is a natural bijection of sets:*

(CT) *all symmetric canonical types $\tau = \{v, f, \rho\}$ with $\rho(2r) = 2g$,*

(FT) *all final sequences $\psi \in \Psi_g$,*

(ES) *all elementary sequences* $\varphi \in \Phi_g$,
where (CT) $\to \Phi$ *is given by stretching, and* $\Psi \to \Phi$ *is given by truncation.*

Proof. It is clear that stretching indeed produces an element in Φ. From an elementary sequence φ we define a final sequence by "symmetry"; this map $\Phi \to \Psi$ is left and right inverse to truncation. Given a final sequence (or a final filtration) we produce a canonical type (or a canonical filtration) by applying the construction of the canonical type (or construction of the canonical filtration). It is straightforward that all maps involved are bijective (and in fact inverse to each other where applicable). □

(5.8). **Notation.** Let N be a symmetric BT_1 over a field; its canonical type defines an elementary sequence; this will be denoted by $\mathrm{ES}(N) \in \Phi$. For an abelian variety X over a field, which admits a *principal polarization*, the isomorphism $X \xrightarrow{\sim} X^t$, using the duality theorem, see [42], 19.1, shows that $X[p] \cong X^t[p] \cong X[p]^D$, and it follows that $N = X[p]$ is symmetric; we write $\mathrm{ES}(X) := \mathrm{ES}(X[p]) \in \Phi$.

(5.9). **Cycles.** Suppose given a canonical type (or a canonical filtration on a BT_1 over a base). We have defined a (V, F^{-1})-cycle in Section 2 for a canonical type by

$$\Gamma = \{0, \cdots, 2r-1\}, \quad \pi : \Gamma \to \Gamma$$

and we obtain a disjoint union

$$\Gamma = \Gamma_1 \sqcup \cdots \sqcup \Gamma_m$$

of (V, F^{-1})-cycles.

Suppose we have a *symmetric* canonical type. We generate an equivalence relation, the $(V, \perp)$-relation, by iterating:

$$\nu(j) > \nu(j-1) \quad \Longrightarrow \quad j-1 \;\sim\; \nu(j-1)$$

and

$$j-1 \;\sim\; 2r-j+1.$$

Explanation: in a canonical filtration, we declare the steps, where

$$V : N_j^{(p)}/N_{j-1}^{(p)} \to N_{\nu(j)}/N_{\nu(j-1)}$$

is an isomorphism, to be equivalent steps, and we declare dual steps under $\perp$ to be equivalent.

Remark. We can achieve the (V, F^{-1})-equivalence and the $(V, \perp)$-equivalence as follows. Connect vertices in Γ by ordered edges V and $F^{(-1)}$ and by unordered edges $\perp$. Deleting the $\perp$-edges gives the (V, F^{-1})-equivalence and the (V, F^{-1})-cycles. Deleting the $F^{(-1)}$-edges gives the $(V, \perp)$-equivalence and the $(V, \perp)$-cycles. We easily see that a $(V, \perp)$-equivalence class consists of a graph which contains exactly one cycle and in each vertex j in the cycle:
1) Either in the cycle an edge $\perp$ ends in j, and an edge V starts; no V ends

in j.
2) Or an edge V ends, an edge V starts in j, and there is an edge $\perp$ attached to j, not included in the cycle;
3) Or an edge V ends in j, an edge $\perp$ is attached to j, no V starts in j; the cycle passing through j reads:

$$\cdots \xrightarrow{V} j \xleftrightarrow{\perp} 2r-j+1 \xrightarrow{V} \cdots ;$$

only in the second case there is is an edge attached to j not contained in the cycle ("a loose edge").

We study the relation between the (V,F^{-1})-equivalence relation and the $(V,\perp)$–equivalence relation.

(5.10). **Lemma and definition.** *Suppose given a symmetric canonical type τ. Let $\Gamma = \Gamma_1 \sqcup \cdots \sqcup \Gamma_m$ be the disjoint union of (V,F^{-1})-cycles. Then:*

(i) *The operation $\perp : \Gamma \to \Gamma$ respects the disjoint union into (V,F^{-1})-cycles.*

Suppose $\Gamma' = \Gamma_i$ for some index i; then

(ii) **odd** *either $\perp(\Gamma') = \Gamma'$; in this case $\#(\Gamma')$ is an even number; in this case there is precisely one $(V,\perp)$-cycle in Γ' and Γ' is a $(V,\perp)$-equivalence class; this $(V,\perp)$-cycle in Γ' contains an odd number of symbols $\perp$; this will be called an* odd cycle; *in this case the length of the $(V,\perp)$-cycle is this odd number plus $\#(\Gamma')/2$,*

(iii) **even** *or $\perp(\Gamma') = \Gamma'' \neq \Gamma'$; in this case $\#(\Gamma' \cup \Gamma'')$ is a $(V,\perp)$-equivalence class; in this case the $(V,\perp)$-cycle in $\Gamma' \cup \Gamma''$ contains an even number of symbols $\perp$; this will be called an* even cycle.

Proof. By construction we have $\pi{\cdot}\perp = \perp{\cdot}\pi$. This proves (i).

Suppose $\Gamma' = \Gamma_i = \perp(\Gamma')$ for some index i. Let $j \in \Gamma'$. Then $\perp(j) \in \perp(\Gamma') = \Gamma'$. Let w be the shortest word in V and F^{-1} with $w(j) = \perp(j)$. Then $\overline{w}(\perp(j)) = j$, where $\overline{w}$ is the word obtained from w by replacing V by F^{-1} and F^{-1} by V. Hence $\#(\Gamma')$ is even. In the circular graph Γ' every vertex is connected to its next vertex either by V or by F^{-1}, and for opposite vertices one is V, and the opposite is F^{-1}. From this combinatorial picture we see: deleting all edges named F^{-1} in Γ' we obtain a connected graph, it contains exactly one cycle containing an odd number of symbols $\perp$, and it contains all symbols V in w and in $\overline{w}$. This proves (ii).

Suppose $\Gamma' = \Gamma_i \neq \perp(\Gamma') = \Gamma''$ for some index i. Let w be the shortest word in V and F^{-1} needed to go through Γ' once. Then $\overline{w}$ is the shortest word needed to go through Γ'' once. Every vertex of Γ' is connected to a unique vertex in Γ'' by $\perp$; every V-edge in Γ' corresponds to a unique F^{-1}-edge in Γ''. This proves that the $(V,\perp)$-equivalence class of any element in $\Gamma' \cup \Gamma''$ equals $\Gamma' \cup \Gamma''$, and it shows that there is precisely one $(V,\perp)$-cycle; this cycle has an even number of symbols $\perp$ in the cycle. The lemma is proved. □

(5.11). **Notation.** Let φ be an elementary sequence, with associated canonical type τ. We write $\mathcal{C}_\varphi() = \mathcal{D}_\tau()$. We write

$$S_\varphi := \mathcal{D}_\tau(\mathcal{A}) \subset \mathcal{A} := \mathcal{A}_{g,1,n} \otimes \mathbb{F}_{p,n},$$

see (3.3), for the locally closed subset where the elementary sequence is constant and equal to φ.

6. Strata at the boundary, strata in $\mathcal{A}^*_{g,1} \otimes \mathbb{F}_p$

(6.1). **The elementary sequence of a semi-abelian variety.** For the definition of a semi-abelian scheme $X \to S$ we refer to [10], Def. 2.3 on page 8. In particular, every geometric fiber X_s is an extension over an abelian variety with kernel $T_s \cong (\mathbb{G}_m)^\beta$ for some non-negative integer β.

For a semi-abelian variety X over a field K of characteristic p we define its p-rank $f = f(X)$ by:

$$\mathrm{Hom}(\mu_p, X \otimes k) \cong (\mathbb{Z}/p)^f,$$

where k is an algebraically closed field containing K. Note that if X is an abelian variety, then $\#(X[p](k)) = p^{f(X)}$.

Let X be a semi-abelian variety over a field K; suppose that $L \subset X$ is the maximal connected linear subgroup; we write $Y = X/L$. We define the *elementary sequence* $\mathrm{ES}(X) = \varphi \in \Phi_g$ as follows. We consider $G := X[p]$. This finite group has a filtration $0 \subset L[F] = G_1 \subset X[F] \subset X[p]$; then $G_1 \otimes k \cong (\mu_p)^\beta$, where $\beta := \dim(L)$ and $X[p]/G_1 \cong Y[p]$. We have a canonical filtration on $Y[p]$; this defines an elementary sequence $\theta := \mathrm{ES}(Y) \in \Phi_{g-\beta}$. We define $\varphi = \mathrm{ES}(X)$ by: $\varphi(i) = i$ for all $0 \leq i \leq \beta$ and $\varphi(i) = \theta(i) + \beta$ for all $\beta \leq i \leq g$; in this way $\mathrm{ES}(X)$ is defined:

$$\varphi = \{1, \cdots, \beta, \cdots, \theta(i) + \beta, \cdots\}.$$

Here is another way of defining this sequence. Consider, as in (2.2), the set E of all (finite) words in the symbols V and F^{-1}. The set of such words defines on $N = X[p]$ a set of finite subgroup schemes $\{W(N) \mid W \in E\}$; this is a filtration; to this we associate a "canonical filtration" of N, a canonical type, and by truncation an elementary sequence. This is the one defined earlier in case $\beta = 0$, i.e., X is an abelian variety; it coincides with the elementary sequence just defined in all cases.

Suppose that $X_1/L_1 = Y_1$ and $X_2/L_2 = Y_2$ are semi-abelian varieties with $Y_1 \cong Y_2$. Then $\mathrm{ES}(X_1) = \mathrm{ES}(X_2)$; this follows from the construction.

By $\mathcal{A}^*$ we denote the minimal compactification of $\mathcal{A} = \mathcal{A}_{g,1,n} \otimes \mathbb{F}_{p,n}$, see [10], Chap. V, in particular see pp. 150, 152, Theorem 2.3 and Theorem 2.5. Note that a geometric point of $x' \in \mathcal{A}^*$ defines an abelian variety Y (with a principal polarization, with some level structure); hence the point x' defines an elementary sequence (by the data Y and g). Let $\overline{\mathcal{A}} \to \mathcal{A}^*$ be a toroidal compactification, let x be a geometric point of $\overline{\mathcal{A}}$, and $x \mapsto x'$. The elementary sequence defined by x only depends on $x' \in \mathcal{A}^*$. Essential feature: $\mathrm{ES}(X)$ only depends on $Y = X/L$, and on $X[F]$ we can give a filtration (see below).

Let φ be an elementary sequence. We write $T_\varphi \subset \mathcal{A}^*$ for the set of points $s \in \mathcal{A}^*$ with elementary sequence equal to φ.

(6.2). **Proposition.** *The set $T_\varphi \subset \mathcal{A}^*$ is locally closed.*
Proof. Let $\psi : \overline{\mathcal{A}} \to \mathcal{A}^*$ be a toroidal compactification and $T'_\varphi \subset \overline{\mathcal{A}}$ be the set of points $x \in \overline{\mathcal{A}}$ with elementary sequence equal to φ. Then $\psi^{-1}(T_\varphi) = T'_\varphi$; moreover ψ is a proper morphism. Hence it suffices to prove that $T'_\varphi \subset \overline{\mathcal{A}}$ is locally closed.

It suffices to prove the proposition locally on $\overline{\mathcal{A}}$. For points $s \in T_\varphi \cap \mathcal{A} = T'_\varphi \cap \mathcal{A} = S_\varphi \subset \mathcal{A}$ this has already been proved. Let $s \in T'_\varphi \cap \partial(\overline{\mathcal{A}})$ with $\partial(\overline{\mathcal{A}}) = \overline{\mathcal{A}} - \mathcal{A}$. We use results about "Mumford's uniformization theorem", and about "Raynaud extensions", see [10], in particular pp. 35/52, pp. 102 - 112 and pp. 149/150. By these theories there exist the following data:

- a complete local normal domain R, a semi-abelian scheme $\mathcal{X} \to S := \operatorname{Spec}(R)$ with generic fiber X_η which is an abelian variety with a principal polarization λ_η, and a "Kodaira-Spencer" morphism $f : \operatorname{Spec}(R) \to \overline{\mathcal{A}}$ defined by this family such that the closed point $0 \in \operatorname{Spec}(R)$ maps to $f(0) = s$ and such that $R \hookleftarrow \mathcal{O}_{\overline{\mathcal{A}},s}$ is injective;
- an exact sequence $0 \to \mathcal{L} \to \mathcal{Z} \to \mathcal{Y} \to 0$ of group schemes over S, where $\mathcal{Z} \to S$ is a semi-abelian scheme, where $\mathcal{Y} \to S$ is an abelian scheme, with a principal polarization λ' on $\mathcal{Y} \to S$, and there exists an isomorphism $\mathcal{L} \cong (\mathbb{G}_m)^\beta \to S$ (actually there is more structure here involved, which however we do not need);
- these are related by the fact that the formal semi-abelian schemes derived from these two sets of data are isomorphic: $\mathcal{X}^\wedge \cong_{\operatorname{Spf}(R)} \mathcal{Z}^\wedge$ and the isomorphism transforms the principal polarization induced by λ into the abelian part of $\mathcal{X}_0$ with λ'_0 on $Y_0 = \mathcal{Y} \times_S \{0\}$.

Suppose $\varphi \in \Phi_g$ and $\theta \in \Phi_{g-\beta}$ are as above. Using [10], III.7, we see that the set of points $T = f^{-1}(T'_\varphi) \subset S$ equals $\mathcal{C}_\theta(\mathcal{Y} \to S)$; by Proposition (3.2) this is locally closed in S. By assumption we have $0 \in T$. By $R \hookleftarrow \mathcal{O}_{\overline{\mathcal{A}},s}$ and $s \in T_\varphi \cap \partial(\overline{\mathcal{A}})$ we conclude that $T'_\varphi \subset \overline{\mathcal{A}}$ is closed at s. This finishes the proof of the proposition. □

(6.3). **Proposition.** *Let $\varphi \in \Phi_g$. Let Z be an irreducible component of $S_\varphi = \mathcal{C}_\varphi(\mathcal{A})$, and let Z^* be the closure of $Z \subset \mathcal{A}^*$. Suppose there is an elementary sequence $\varphi' \in \Phi_g$ with $Z^* \cap T_{\varphi'} \neq \emptyset$. Then the Zariski closure Z^c of $Z \subset \mathcal{A}$ meets $S_{\varphi'}$:*

$$Z^c \cap S_{\varphi'} \neq \emptyset.$$

Let $\varphi' \in \Phi$, and let W be an irreducible component of $T_{\varphi'}$. Then

$$W \cap \mathcal{A} \neq \emptyset.$$

In particular, every irreducible component of T_φ *meets* $\mathcal{A}$*, and intersects this interior into a component of* S_φ*.*

Proof. The second and the third statement follow from the first.

As in the proof of the previous proposition it suffices to prove this result for a toroidal compactification $\psi : \overline{\mathcal{A}} \to \mathcal{A}^*$.

Let $s \in \overline{Z} \cap \partial(\overline{\mathcal{A}})$ and $s \in T'_{\varphi'}$. Using results on toroidal compactifications and on "Mumford's uniformization theorem", and about "Raynaud extensions", we conclude that we have the following data:

- a noetherian complete domain R, which is complete for an ideal $I \subset R$, we write $S = \mathrm{Spec}(R)$, and $S_0 = \mathrm{Spec}(R_0)$, we write $S^\wedge$ for the completion of S along S_0 by the ideal I, a semi-abelian scheme $\mathcal{X} \to S$, with generic fiber X_η which is an abelian variety with a principal polarization λ_η, and a "Kodaira-Spencer" morphism $f : \mathrm{Spec}(R) \to \overline{\mathcal{A}}$ defined by this family such that the closed point $0 \in \mathrm{Spec}(R)$ maps to $f(0) = s$, such that $R \hookleftarrow \mathcal{O}_{\overline{\mathcal{A}},s}$, and a morphism $\pi : S \to \mathcal{A}^*$ obtained as the composition $S \to \mathcal{A}_{g-\beta,1,n} \otimes \mathbb{F}_{p,n} =: \mathcal{A}' \subset \partial(\mathcal{A}) \subset \mathcal{A}^*$; N.B. if $\beta > 0$, we will see that the morphism π in general does not factor f;
- an exact sequence $0 \to \mathcal{L} \to \mathcal{Z} \to \mathcal{Y} \to 0$ of group schemes over S, where $\mathcal{Z} \to S$ is a semi-abelian scheme, where $\mathcal{Y} \to S$ is an abelian scheme, with a principal polarization λ' on $\mathcal{Y} \to S$, and there exists an isomorphism $\mathcal{L} \cong (\mathbb{G}_m)^\beta \to S$, such that the "Kodaira-Spencer" morphism given by $(\mathcal{Y}, \lambda') \to S$ defines $\pi : S \to \mathcal{A}$;
- these are related by the fact that the formal semi-abelian schemes derived from these two sets of data are isomorphic: $\mathcal{X}^\wedge \cong_{S_0} \mathcal{Z}^\wedge$, and the isomorphism transforms the principal polarization on the generic fiber of $\mathcal{X}$ into λ' on $\mathcal{Y} \to S$.

From this the proposition follows: consider $\overline{Z} \to \overline{\mathcal{A}} \xleftarrow{f} S$, and the fiber product P of this diagram; there exists an irreducible component $U \subset \mathcal{C}_\varphi(S) \subset S$ such that its closure $Z^{**} = U^c \subset (\mathcal{C}_\varphi(S))^c \subset S$ is an irreducible component of P containing $0 \in S$. Let θ be as before. We see that $\pi(Z^{**}) \subset \pi((\mathcal{C}_\varphi(S))^c) \subset (\mathcal{C}_\theta(\mathcal{A}'))^c \subset \mathcal{A}'$, the last closures taken in $\mathcal{A}'$. We see that there exists a complete discrete valuation ring Λ, with $\Gamma = \mathrm{Spec}(\Lambda)$ and a morphism $h : \Gamma \to S$ such that $h(\Gamma) \subset Z^{**} \cap \pi^{-1}(s)$ and $f(h(\Gamma)) \not\subset \partial(\mathcal{A})$. The generic point $a \in \Gamma$ has the property $\pi(h(a)) = s \in T'_{\varphi'}$; because $h(\Gamma) \subset \pi^{-1}(s)$ we conclude that $\mathrm{ES}(\mathcal{X}_{h(a)}) = \mathrm{ES}(X_0) = \varphi'$. Moreover $h(a)$ is in the closure of U, hence $f(h(a)) \in \overline{Z}$. Hence $\overline{Z} \cap S_{\varphi'} \neq \emptyset$. □

(6.4). **Proposition.** *Let S be an irreducible, reduced, normal, noetherian scheme over $\mathbb{F}_p$, and let $X \to S$ be a semi-abelian scheme, with generic fiber an abelian variety with a principal polarization. Suppose all fibers have the same elementary sequence $\varphi \in \Phi_g$, i.e., the moduli map $S \to \mathcal{A}^*$ maps S into T_φ. Consider $X[p] \to S$. There exists a filtration $0 \subset U \subset M \subset X[p]$ with finite* *group schemes U and M flat over S, such that $U \to S$ is of multiplicative type over S in every fiber, and such that for every point $s \in S$, the fiber $N_s := M_s/U_s$ is of local-local type, and every fiber $X[p]_s/M_s$ is étale-local. The group schemes $U \to S$ and $M \to S$ are flat over S. The finite group scheme $M/U =: N \to S$ is flat over S, and it is symmetric; it admits a canonical filtration and it satisfies the properties for the canonical filtration as described in* Corollary (2.6).

Remark. In general, the rank of $X[p]_s/M_s$ is not constant on S.

Proof. The finite group schemes U and M defined by the properties mentioned have constant rank (all geometric fibers have the same rank) because of the fact that the elementary sequence is constant on S. Moreover note that for every $j \in \mathbb{Z}_{>0}$ the finite group scheme $X[F^j] \to S$ is flat, and $U \subset M \subset X[F^N]$ are closed subgroup schemes, hence finite, over S for large N (e.g. $N = g$). As the elementary sequence of $X \to S$ is constant on S, we conclude that the finite group schemes $U \to S$ and $M \to S$ have constant rank, and hence these are flat by [4], Lemma 1.13 on page 240. The principal polarization on the generic fiber extends to a symmetry on $N = M/U$, e.g. use [10], Proposition I.2.7. The rest of the proof is as in Section 2, especially as in the proof of (2.6). □

(6.5). **Theorem.** *For every $\varphi \in \Phi_g$ the stratum $T_\varphi \subset \mathcal{A}^*$ is* quasi-affine.

Proof. It suffices to show this in case of arbitrary high level structure. Hence we suppose $\mathcal{A}^* = \mathcal{A}^*_{g,1,n} \otimes \mathbb{F}_{p,n}$, with $n \in \mathbb{Z}_{\geq 3}$ not divisible by p. It suffices to show that every irreducible component $V \subset T_\varphi \subset \mathcal{A}^*$ is quasi-affine. We choose such a component V. Note that we know that V meets the interior of $\mathcal{A}^*$, see (6.3). We proceed as in the proof of (4.1).

By results in [10], V.2, there exist an irreducible, normal, reduced scheme $S \to \operatorname{Spec}(\mathbb{F}_p)$, a semi-abelian scheme $\mathcal{X} \to S$, a principal polarization on its generic fiber which is an abelian variety, and a symplectic level-n-structure, such that the moduli morphism $S \to \mathcal{A}^*$ defined by these data factors through V, and such that $f : S \to V$ is proper and surjective. We write $\mathfrak{t}_{\mathcal{X}/S}$ for the tangential sheaf of $\mathcal{X} \to S$ along the zero-section, and we write $\omega_{\mathcal{X}/S}$ for the determinant line bundle of the dual of $\mathfrak{t}_{\mathcal{X}/S}$, see [10], page 137. We know, see [10], V.2.5, that the moduli morphism $S \to \mathcal{A}^*$ is defined by the line bundle $\omega_{\mathcal{X}/S}$.

Claims. 1) *Under these conditions, the line bundle $\omega_{\mathcal{X}/S}$ on S is a torsion line bundle.*

2) *From this it follows that V is quasi-affine.*

Indeed, suppose (1) is proven and let us see that the result is established; consider the Stein factorization $S \to V' \to V$, as in EGA II, 1.3.1; then $f' : S \to V'$ is proper with connected fibers, and $V' \to V$ is finite. By EGA III, 4.3.1 we know that $f'_*(\mathcal{O}_S) \cong \mathcal{O}_{V'}$. Because $V' \to V$ is finite, assuming (1), it follows that $f_*(\omega_{\mathcal{X}/S})$ is torsion on V. Using [35], Chap. IX, and [10], V.2, Th. 2.3, we conclude that $\omega|_V$ is ample and torsion; hence V is quasi-affine. This proves (1) $\Rightarrow$ (2).

We show that (1) is true. Let γ be the canonical type associated with φ, say $\gamma = (\nu, \rho)$. We write $f = f(\gamma)$, i.e., the largest index with $\varphi(f) = f$, e.g. $f = 0$ if $\nu(1) = 0$, and $f = \rho(1)$ if $\nu(1) = 1$. We write $U = 0$ if $f = 0$, and we write $U = G_1$ if $f > 0$. We use Proposition (5.9), we use that the elementary sequence on $\mathcal{X} \to S$ is constant, we use [4], Lemma 1.13 on page 240, and we conclude that we have finite flat subgroup schemes

$$0 = G_0 \subset G_1 = U \subset G_2 \subset \cdots \subset G_r = G[F] = V(G[p]),$$

where $G_1 \times_S S' \cong (\mu_p)^f \times S'$ for some finite cover $S' \to S$, and where the other steps in the filtration come from the canonical filtration on $M/U = N$, with the notation as in (5.9). Note that the partial quotients G_i/G_{i-1} are annihilated by F and by V for $1 < i \leq r$ (in the terminology of [29], 2.4, these are α-groups). We see that all tangential sheaves

$$\mathfrak{t}_{G_1/S} \subset \cdots \subset \mathfrak{t}_{G_r/S} = \mathfrak{t}_{\mathcal{X}/S}$$

are well-defined, of constant rank. We write $\mathcal{L}_i := \mathrm{Det}(\mathfrak{t}_{G_i/S}/\mathfrak{t}_{G_{i-1}/S})$, for $0 < i \leq r$, and we are going to show that these are torsion line bundles.

This is clear for the first step in the filtration: over S' as above the pull-back of $\mathcal{L}_1$ is trivial, hence its norm is trivial, hence $(\mathcal{L}_1)^{\mathrm{degree}(S'/S)} = 1$.

Consider one of the other steps G_j/G_{j-1} with $j > 1$. We apply Corollary (2.6). In case

$$(G_j/G_{j-1})^{(p^n)} \xrightarrow{\sim} G_j/G_{j-1}$$

we get

$$(\mathcal{L}_j)^{p^n} = (\mathcal{L}_j);$$

in this case this shows the class to be torsion (note that $n > 0$).

We conclude that $\mathrm{Det}(\mathfrak{t}_{\mathcal{X}/S}) = \mathrm{Det}(\mathcal{L}_1) \otimes \cdots \otimes \mathrm{Det}(\mathcal{L}_r)$ is torsion; hence its inverse $\omega_{\mathcal{X}/S}$ is torsion. This shows (1), and it finishes the proof of Theorem (6.5). □

7. Connecting the superspecial locus inside the supersingular locus

In this section we construct a complete algebraic curve $L \subset \mathcal{A}$, with $\Sigma \subset L$. We write $L^o = L - \Sigma$. We show that $L^o \subset S_{\{0,\cdots 0,1\}}$, and in Section 8 we show equality. An essential tool in the rest of the paper will be the property that L *is connected*, see Proposition (7.3). Actually, the curve L is contained in the supersingular locus of $\mathcal{A}$ and the proof of (7.3) uses pure algebra. The basic idea of this section is due to T. Ekedahl. The present form of this section is mainly due to Ben Moonen who helped with most details in this section.

In case $g = 1$ we will write $L = \mathcal{A}$ and $L^o = L - \Sigma$. In this section we consider abelian varieties of dimension $g \geq 2$.

(7.1). **Construction of L.** Let k be an algebraically closed field of characteristic p. Choose a point $x \in \Sigma(k)$; let (X, μ, η) be the corresponding principally polarized abelian variety with symplectic level-n-structure.

For the construction of the locus L we are interested in coverings $f\colon Y \to X$ such that

(i): $\mathrm{Ker}(f) \cong \alpha_p$,
(ii): Y is superspecial,
(iii): $\mathrm{Ker}(f^*\mu) \cong \alpha_p \times \alpha_p$.

Coverings f satisfying (i) correspond (by duality) to subgroup schemes of X isomorphic to α_p. Such subgroup schemes are parameterized by the projective space $\mathbb{P}(\mathrm{Lie}(X)) \cong \mathbb{P}^{g-1}$. One can show, see [34], pp. 138/139, [21], Prop. 5.2, that the coverings f satisfying (i) and (ii) are in bijection with $\mathbb{P}^{g-1}(\mathbb{F}_{p^2})$, and that the coverings satisfying (i)–(iii) correspond to those vectors in $\mathbb{P}^{g-1}(\mathbb{F}_{p^2})$, "very good directions", which are isotropic with respect to a certain hermitian form.

Given a covering $f : Y \to X$ satisfying (i)–(iii), write $\mathfrak{t}$ for the tangent space of $\mathrm{Ker}(f^*\mu)$; this is a 2-dimensional k-vector space. A point $s \in \mathbb{P}(\mathfrak{t}) \cong \mathbb{P}^1$ corresponds uniquely to a subgroup scheme $\alpha_p \cong N = N_s \subset \mathrm{Ker}(f^*\mu) \subset Y[p]$, and conversely every such subgroup scheme gives a point in $\mathbb{P}(\mathfrak{t})$. Moreover every such subgroup scheme is totally isotropic for the pairing induced on $\mathrm{Ker}(f^*\mu)$; it follows from condition (iii) that for every such N, the polarization $f^*\mu$ descends to a principal polarization μ_s on the quotient $X_s := Y/N_s$. Further, as we have natural isomorphisms $X_s[n] \stackrel{\sim}{\leftarrow} Y[n] \stackrel{\sim}{\rightarrow} X[n]$, the symplectic level-$n$-structure η of (X, μ) induces a symplectic level-n-structure η_s of (X_s, μ_s). Taking s to be the point corresponding to $\mathrm{Ker}(f) \subset \mathrm{Ker}(f^*\mu)$ gives $(X_s, \mu_s, \eta_s) = (X, \mu, \eta)$.

In total, the choice of $x \in \Sigma(k)$ and the covering f gives rise to a non-constant morphism $\mathbb{P}^1_k \cong \mathbb{P}(\mathfrak{t}) \longrightarrow \mathcal{A}_k = \mathcal{A}_{g,1,n} \otimes k$; the image is denoted by $L^{(f)}$. We define $L_k \subset \mathcal{A}_k$ to be the reduced closed subscheme which is the union of the images of all morphisms thus obtained (varying x and f), $L_k = \cup_f L^{(f)}$. One can show that the closed subscheme L_k is defined over $\mathbb{F}_p$, i.e., there is a closed subscheme $L \subset \mathcal{A}_{g,1,n}$ such that $L_k = L \otimes k$.

We indicate a different construction of L, and we derive some properties. We write $H := G_{1,1}$ for the formal p-divisible group of dimension one and height 2. For a supersingular elliptic curve E over k we have $E[p^\infty] \cong_k H \otimes k$.

(7.2). **Lemma.** *A homomorphism $f : Y \to X$ as above can be constructed by a choice $X[p^\infty] \cong H^2 \oplus H^{g-2}$ such that μ is in block form $\mu_2 \oplus (\mu_1)^{g-2}$, correspondingly $Y[p^\infty] \cong H^2 \oplus H^{g-2}$ and*

$$\mathrm{Ker}(f^*(\mu)) \subset H^2 \subset H^2 \oplus H^{g-2}.$$

We have $\Sigma \subset L \subset (S_{\{0,\dots,0,1\}})^c$.

Proof. For the superspecial Y we have $Y[p^\infty] \cong H^g$. By [29], 6.1 we see that the polarization $f^*(\mu)$ on Y gives a quasi-polarization on $Y[p^\infty] \cong H^g \cong M_1 \oplus M_2 \oplus \cdots \oplus M_{g-1}$ in block-form with $M_1 \cong H^2$, with $\mathrm{Ker}(f^*(\mu)) \subset M_1 \cong H^2$ and $M_j \cong H$ for $2 \leq j \leq g-1$. This proves the first claim.

Suppose $s \in \mathbb{P}(\mathrm{t})$, with $X_s := Y/N_s$. Choose coordinates on $\mathbb{P}(\mathrm{t}) \cong \mathbb{P}^1$ by $M_1 \cong H^2$. We know that $a(X_s) = 1$ iff $s \notin \mathbb{P}^1(\mathbb{F}_{p^2})$; suppose this is the case; then $X_s[p] := N \cong N_1 \oplus N_2 \oplus \cdots \oplus N_{g-1}$ with $\mathrm{rk}(N_1) = p^4$ and $\mathrm{ES}(N_1) = \{0,1\}$ and for $2 \leq j \leq g-1$ we have $\mathrm{rk}(N_j) = p^2$ and $\mathrm{ES}(N_j) = \{0\}$. Direct verification shows that in this case $\mathrm{ES}(X_s) = \{0, \ldots, 0, 1\}$. As we see that $\Sigma \subset L$, all components of L have dimension one, and $L^o := L - \Sigma \subset S_{\{0,\ldots,0,1\}}$ the second claim follows. □

Conclusion. Every component $L^{(f)} \subset L_k$ as in (7.1) can be given by choosing $[(X,\mu)] \in \Sigma(k)$, by choosing an isomorphism $X[p^\infty] \cong H^g$ on which μ is in diagonal form $\mathrm{diag}(\lambda, \cdots, \lambda)$, see [29], 6.1, by choosing two factors: $H^g = H^2 \oplus H^{g-2}$, and by performing a construction of type [34], pp. 138/138, on $(H^2, \mathrm{diag}(\lambda, \lambda))$.

Remark. The fact that $S_{\{0,\cdots,0\}} = \Sigma \subset \left(S_{\{0,\cdots,0,1\}}\right)^c$ is a special case of (11.1).

(7.3). **Proposition.** *The locus $L \subset \mathcal{A}$ is 1-dimensional and connected.*

(7.4). **Remark.** In Section 8 we shall show that in fact $L = (S_{\{0,\ldots,0,1\}})^c$. In this section we shall prove the proposition as stated here. For this we need a number of preparations about hermitian forms, which will be given in (7.5) – (7.11). The proof of (7.3) is given in (7.18) below.

(7.5). Let R be a ring equipped with an (anti-)involution $r \mapsto r^\dagger$. Let M be a finitely generated projective right R-module. By a hermitian form on M (with respect to the involution $\dagger$) we mean a bi-additive map

$$\psi \colon M \times M \longrightarrow R$$

such that

(a): ψ is sesquilinear, meaning that $\psi(m_1 r_1, m_2 r_2) = r_1^\dagger \cdot \psi(m_1, m_2) \cdot r_2$ for all r_1, $r_2 \in R$ and m_1, $m_2 \in M$;

(b): $\psi(m', m) = \psi(m, m')^\dagger$ for all m, $m' \in M$.

We say that a form ψ as above is skew-hermitian if it satisfies (a) and:

(b'): $\psi(m', m) = -\psi(m, m')^\dagger$ for all m, $m' \in M$.

The dual $M^\vee := \mathrm{Hom}_R(M, R)$ of M has a natural structure of a left R-module. Let $M^\dagger := M^\vee$ as an additive group, and give it the structure of a right R-module by $fr := r^\dagger \cdot f$. A form ψ satisfying (a) gives rise to a homomorphism $M \to M^\dagger$ by $m \mapsto \psi(m, -)$. If this map is an isomorphism we say that ψ is perfect.

By a hermitian space over R we mean a pair (M, ψ) consisting of a projective right R-module of finite type equipped with a perfect hermitian form ψ.

(7.6). Let P be a field. Write $A \mapsto A^\dagger = \mathrm{Trd}(A) - A$ for the canonical involution on the algebra $M_2(P)$ of 2×2 matrices over P. Consider $V_1 := P^2$ with its natural structure of a right $M_2(P)$-module. Up to a scalar in $P^\times$ there is a unique perfect skew-hermitian form on V_1 with respect to $\dagger$, namely the form $\Psi_1 \colon V_1 \times V_1 \longrightarrow M_2(P)$ given by

$$\Psi_1\big((x,y),(x',y')\big) = \begin{pmatrix} yx' & yy' \\ -xx' & -xy' \end{pmatrix}.$$

Let (V_2, Ψ_2) be a symplectic space over P, by which we mean a finite dimensional P-vector space V_2 with a non-degenerate alternating bilinear form $\Psi_2 \colon V_2 \times V_2 \longrightarrow P$. The dimension of V_2 is even, say $\dim_P(V_2) = 2g$. Then $V := V_1 \otimes_P V_2$ becomes a free right $M_2(P)$-module of rank g via the action of $M_2(P)$ on V_1, and the tensor product $\Psi_1 \otimes_P \Psi_2$ is a perfect hermitian form on $M_1 \otimes_P M_2$.

Morita equivalence tells us that every hermitian space (V, Ψ) of rank g over $M_2(P)$ (with respect to the canonical involution) is isometric to such a tensor product, see [25], I.9. In particular, since any two symplectic spaces of the same dimension over P are isometric, also all hermitian spaces of the same rank over $M_2(P)$ are isometric (also see: [8], Section 3).

(7.7). Let D be a quaternion algebra over $\mathbb{Q}$. Write $d \mapsto d^\dagger$ for its canonical involution. Fix a positive integer g. Let (V, Ψ) be a hermitian space of rank g over D. Then

$$G := \mathrm{Aut}_D(V, \Psi)$$

is a linear algebraic group over $\mathbb{Q}$. It is immediate from what was explained in (4.4) that if $\mathbb{Q} \subset L$ is a field extension which splits D then $G \otimes_{\mathbb{Q}} L \cong \mathrm{Sp}_{2g,L}$. In particular, G is absolutely simple and simply connected. Assume that $D_{\mathbb{R}} := D \otimes_{\mathbb{Q}} \mathbb{R}$ is non-split. Write $\Psi_{\mathbb{R}}$ for the $\mathbb{R}$-bilinear extension of Ψ to a hermitian form on $V_{\mathbb{R}}$. Define $q \colon V_{\mathbb{R}} \to \mathbb{R}$ by $q(v) = \Psi_{\mathbb{R}}(v,v)$; note that it follows from (b) in (7.5) that indeed $\Psi_{\mathbb{R}}(v,v) \in \mathbb{R} \subset D_{\mathbb{R}}$. Then q is a quadratic form on $V_{\mathbb{R}}$. The signature of Ψ is defined to be $\mathrm{sign}(\Psi) := \mathrm{sign}(q)/4$. Two hermitian spaces (V, Ψ) and (V', Ψ') over D are isometric if and only if $\dim(V) = \dim(V)$ and $\mathrm{sign}(\Psi) = \mathrm{sign}(\Psi')$, see for instance [53], Chapter 10, §1. We say that Ψ is positive definite if q is; this is equivalent to $\mathrm{sign}(\Psi) = g$.

(7.8). With $(D, \dagger)$ as in (7.7), fix a hermitian space (V, Ψ) of rank g over D. Let $R \subset D$ be an order; note that R is stable under the involution $\dagger$. Let (M, ψ) be a hermitian space of rank g over R such that there exists an isometry

$$\alpha \colon (M_{\mathbb{Q}}, \psi_{\mathbb{Q}}) \ \overset{\sim}{\to}\ (V, \Psi).$$

Then $\alpha(M) \subset V$ is an R-lattice in V on which Ψ induces a perfect hermitian form; we shall refer to such lattices as perfect hermitian R-lattices in V. Note that R is a hereditary ring; hence every R-lattice $M \subset V$ is projective, and if $\Psi_{|M}$ is perfect then $(M, \Psi_{|M})$ is a hermitian space.

Let n be a positive integer. Let us write H_n for the standard hermitian form on $(R/nR)^g$, i.e., the form given by

$$H_n\big((x_1,\ldots,x_g),(y_1,\ldots,y_g)\big) = x_1^\dagger y_1 + \cdots + x_g^\dagger y_g\,.$$

If (M,ψ) is a hermitian space of rank g over R then by a level-n-marking of (M,ψ) we shall mean an isometry

$$\eta\colon (M/nM,\bar{\psi}) \ \xrightarrow{\sim}\ \big((R/nR)^g, H_n\big)\,,$$

where we write $\bar{\psi}$ for the hermitian form on M/nM induced by ψ.

(7.9). **Lemma.** *Let R be an order in a quaternion algebra D over $\mathbb{Q}$. Let p be a prime number, n a positive integer prime to p. Let g be an integer ≥ 2, and fix a hermitian space (V,Ψ) of rank g over D (with respect to the canonical involution on D).*

Let (M_1,ψ_1,η_1) and (M_2,ψ_2,η_2) be hermitian spaces of rank g over R equipped with level-n-markings. Assume that $(M_{j,\mathbb{Q}},\psi_{j,\mathbb{Q}})$ is isometric to (V,Ψ), for $j=1$, 2. Then there exist isometries

$$\alpha_j\colon (M_{j,\mathbb{Q}},\psi_{j,\mathbb{Q}}) \ \xrightarrow{\sim}\ (V,\Psi)$$

such that $\alpha_1(M_1)\cap\alpha_2(M_2)$ has a p-power index in both $\alpha_1(M_1)$ and $\alpha_2(M_2)$, and such that via the isomorphism $M_1/nM_1 \xrightarrow{\sim} M_2/nM_2$ induced by $\alpha_2^{-1}\circ\alpha_1$ the level-n-markings η_1 and η_2 correspond.

Note that if α_1 and α_2 satisfy the first condition then indeed $\alpha_2^{-1}\circ\alpha_1$ induces an isomorphism $M_1/nM_1 \xrightarrow{\sim} M_2/nM_2$.

Proof. As above, write $G := \mathrm{Aut}_D(V,\Psi)$. We claim that $G(\mathbb{Q}_p)$ is non-compact; this is equivalent to saying that $V\otimes\mathbb{Q}_p$ has an isotropic vector with respect to Ψ. If $D\otimes_{\mathbb{Q}}\mathbb{Q}_p \cong M_2(\mathbb{Q}_p)$ then the discussion in (7.6) shows that $G_{\mathbb{Q}_p}$ is split, and the claim is clear. Next suppose that $D_p := D\otimes_{\mathbb{Q}}\mathbb{Q}_p$ is the (unique) quaternion algebra over $\mathbb{Q}_p$. Because $g\geq 2$ there exist two non-zero vectors v_1, $v_2\in V\otimes\mathbb{Q}_p$ with $\Psi(v_1,v_2)=0$. Now we use that the reduced norm map $D_p^\times\to\mathbb{Q}_p^\times$ is surjective; using this it easily follows that some linear combination $v_1d_1+v_2d_2$ is isotropic.

Write $\hat{\mathbb{Z}}^{(p)} := \prod_{\ell\neq p}\mathbb{Z}_\ell$, where the product runs over all prime numbers ℓ different from p. Write $\mathbb{A}_f^{(p)} = \hat{\mathbb{Z}}^{(p)}\otimes_{\mathbb{Z}}\mathbb{Q}$ for the ring of finite adèles of $\mathbb{Q}$ without component at p. As $G(\mathbb{Q}_p)$ is non-compact and G is absolutely simple and simply connected, we can apply strong approximation, see for example [49], Th. 7.12 on p. 427, or see [24], see [51]. This tells us that the diagonal image of $G(\mathbb{Q})$ in $G(\mathbb{A}_f^{(p)})$ is dense in the adèlic topology.

For $j=1,2$, write $\hat{M}_j^{(p)} := M_j\otimes_{\mathbb{Z}}\hat{\mathbb{Z}}^{(p)}$. Choose isometries

$$\beta_j\colon (M_{j,\mathbb{Q}},\psi_{j,\mathbb{Q}}) \ \xrightarrow{\sim}\ (V,\Psi).$$

Write

$$\tilde{\beta}_j\colon \hat{M}_j^{(p)}\otimes_{\hat{\mathbb{Z}}^{(p)}}\mathbb{A}_f^{(p)} \ \xrightarrow{\sim}\ V\otimes_{\mathbb{Q}}\mathbb{A}_f^{(p)}$$

for the $\mathbb{A}_f^{(p)}$-linear extension of β_j. Consider the set Γ of all elements $\gamma \in G(\mathbb{A}_f^{(p)})$ such that

$$(\gamma \circ \tilde{\beta}_1)(\hat{M}_1^{(p)}) = \tilde{\beta}_2(\hat{M}_2^{(p)})$$

inside $V \otimes \mathbb{A}_f^{(p)}$. For $\gamma \in \Gamma$, write $t_\gamma \colon (M_1/nM_1) \xrightarrow{\sim} (M_2/nM_2)$ for the isomorphism induced by $\tilde{\beta}_2^{-1} \circ (\gamma \circ \tilde{\beta}_1)$. Let $\Gamma' \subset \Gamma$ be the subset of all γ which are compatible with the given level-n-markings, i.e.,

$$\Gamma' := \{g \in \Gamma \mid \eta_1 = \eta_2 \circ t_\gamma\}.$$

Then Γ' is a non-empty open subset of $G(\mathbb{A}_f^{(p)})$. Hence it follows from strong approximation that there exists an element $g \in G(\mathbb{Q})$ which maps into Γ under the diagonal embedding $G(\mathbb{Q}) \hookrightarrow G(\mathbb{A}_f^{(p)})$. Then $\alpha_1 := g \circ \beta_1$ and $\alpha_2 := \beta_2$ are isometries with the required property. $\square$

(7.10). Keep the notations of (7.9). Assume that D is inert at p. There is a unique 2-sided prime ideal $\mathfrak{p} \subset R$ above p, which is in fact a maximal ideal (both left and right). The assumption that D is inert at p implies that $k(\mathfrak{p}) := R/\mathfrak{p}$ is a field with p^2 elements, see [52], Th. 22.4 and Th. 22.14. The involution $\dagger$ on R induces the unique non-trivial automorphism $\mathrm{Frob}_{k(\mathfrak{p})}$ of $k(\mathfrak{p})$.

Given an R-submodule $L \subset V$, write

$$\check{L} := \{v \in V \mid \Psi(v, y) \in R \text{ for all } y \in L\}.$$

In particular, if $M \subset V$ is an R-lattice then $M = \check{M}$ if and only if Ψ induces a perfect form on M.

(7.11). **Lemma** *Let M and M' be perfect hermitian R-lattices in V such that $M \cap M'$ has p-power index in both M and M'. Assume that $\mathfrak{p} \subset R$ is a principal ideal. Then there exist perfect hermitian R-lattices*

$$M = L_0, L_1, \cdots, L_m = M'$$

such that $L_j/(L_j \cap L_{j+1}) \cong k(\mathfrak{p}) \cong L_{j+1}/(L_j \cap L_{j+1})$ as R-modules for $0 \leq j < m$.

Proof. Write $N := M \cap M'$. We claim that $\mathrm{length}_R(M/N) = \mathrm{length}_R(M'/N)$. To see this, note that $\mathrm{Hom}_R(M/N, R) = 0$ and that $\mathrm{Ext}^1_R(M, R) = 0$. Hence we obtain an exact sequence

$$0 \to M^\dagger \longrightarrow N^\dagger \longrightarrow \mathrm{Ext}^1_R(M/N, R) \to 0.$$

Now $M/N \cong R/\mathfrak{p}^{d_1} \oplus \cdots \oplus R/\mathfrak{p}^{d_r}$ as an R-module, from which

$$\mathrm{length}_R(M/N) = d_1 + \cdots + d_r.$$

But if $\mathfrak{p} = (\pi)$ then we easily find that $\mathrm{Ext}^1_R(R/\mathfrak{p}^d, R) \cong R/\mathfrak{p}^d$. Hence $\check{N}/\check{M} = \check{N}/M \cong N^\dagger/M^\dagger$ has the same length as M/N. Doing the same with M' instead of M we find that

$$\mathrm{length}_R(\check{N}/N) = \mathrm{length}_R(M/N)^2 = \mathrm{length}_R(M'/N)^2,$$

which proves our claim.

We prove the lemma by induction on $\mathrm{length}_R(M/N)$. By what we have just shown, $\mathrm{length}_R(M/N) = 0$ corresponds to the case $M = M'$. Note that $M + M' = \check{N}$. On $\check{N}/N$ we have a hermitian form

$$\chi\colon \check{N}/N \times \check{N}/N \longrightarrow D/R$$

induced by Ψ. The subspaces M/N and M'/N are maximal isotropic with respect to χ.

Let $T \subset M/N$ be an R-submodule with $M/T \cong k(\mathfrak{p})$. Let $T^\perp$ be the orthogonal complement of T with respect to χ, set $T' := T^\perp \cap (M'/N)$, and let

$$S := T + T' \subset \check{N}/N\,.$$

An easy calculation shows that S is maximal isotropic for χ. Hence the pre-image $L \subset \check{N}$ of S is a perfect hermitian lattice in V. Now note that, by construction,

$$\mathrm{length}_R\bigl(L/(L \cap M')\bigr) = \mathrm{length}_R(T) < \mathrm{length}_R(M/N)\,.$$

Hence we can take $L_1 = L$ and proceed by induction. □

(7.12). Let k be an algebraically closed field of characteristic p. Fix a supersingular elliptic curve E over k. Then E has a unique principal polarization, which we call λ. The endomorphism algebra $\mathrm{End}^0(E)$ is the quaternion algebra $Q_{p,\infty}$ over $\mathbb{Q}$ which is inert exactly at p and ∞. The endomorphism ring $R := \mathrm{End}(E)$ is a maximal order of this quaternion algebra. The Rosati involution $r \mapsto r^\dagger$ on $\mathrm{End}^0(E)$ associated to λ (or to any other polarization) is the principal involution, i.e., $r^\dagger = \mathrm{Trd}(r) - r$.

Let X be a superspecial abelian variety of dimension $g \geq 2$ over k. This means that X is isomorphic to E^g as an abelian variety. Write

$$M := \mathrm{Hom}(E, X),$$

which has a natural structure of a free right R-module of rank g. We have a canonical isomorphism $M \otimes_R E \cong X$.

Note that E^t has a natural structure of a right R-module. Similar to the definition of $M^\dagger$ in (4.3), let us write $E^\dagger$ for the elliptic curve E^t with it structure of a left R-module given by $Pr := r^\dagger \cdot P$. Then X^t is naturally isomorphic to $M^\dagger \otimes_R E^\dagger$.

(7.13). **Remark.** For any abelian variety X over k, we can form the right R-module $\mathrm{Hom}(E, X)$. Conversely, for any right R-module N of finite type, $N \otimes_R E$ is a supersingular abelian variety; to see this, take a finite presentation of N; we shall see that in this case X is superspecial. We have natural homomorphisms

$$\alpha\colon \mathrm{Hom}(E, X) \otimes_R X \longrightarrow X \quad \text{and} \quad \beta\colon N \longrightarrow \mathrm{Hom}(E, N \otimes_R E)\,.$$

In order to understand these functors $\mathrm{Hom}(E,-)$ and $-\otimes_R E$, the following two observations are useful.

(i) Given an abelian variety X over k, there exists a superspecial abelian variety $S^0(X)$ and a homomorphism $\rho\colon S^0(X)\to X$ with the following universal property: if Y is a superspecial abelian variety, any homomorphism $f\colon Y\to X$ factors through ρ, see [29], 1.8. Further, ρ is an isogeny if and only if X is supersingular.

(ii) The class number of $Q_{p,\infty}$ equals 1. This means that for a maximal order R as above, every torsion-free right R-module of finite type is free, see [57].

We find that $\mathrm{Hom}(E,X)=\mathrm{Hom}\big(E,S^0(X)\big)$ combining these two remarks, so that $\mathrm{Hom}(E,X)\otimes_R X\cong S^0(X)$, and α is just the isogeny ρ. In the other direction, if N is a right R-module of finite type, consider the natural map $\pi\colon N\to N':=N/\mathrm{Tors}(N)$. Then $N\otimes_R E \ \stackrel{\sim}{\to}\ N'\otimes_R E$. We have $\mathrm{Hom}(E,N\otimes_R E)\cong N'$, and via this isomorphism β is just the map π.

(7.14). **Lemma.** *Choose notation as in* (7.12); *in particular X is superspecial. There is a natural bijection*

$$\left\{\begin{matrix}\text{symmetric isomorphisms}\\ X\ \stackrel{\sim}{\to}\ X^t\end{matrix}\right\}\ \stackrel{\sim}{\to}\ \left\{\begin{matrix}\text{perfect hermitian}\\ \text{forms on } M\end{matrix}\right\} \tag{1}$$

which restricts to a bijection

$$\left\{\begin{matrix}\text{principal polarizations}\\ X\ \stackrel{\sim}{\to}\ X^t\end{matrix}\right\}\ \stackrel{\sim}{\to}\ \left\{\begin{matrix}\text{positive definite perfect}\\ \text{hermitian forms on } M\end{matrix}\right\}.$$

Proof. Given a homomorphism $\mu\colon X\to X^t$, let $\psi=\psi_\mu$ be the hermitian form on M given by

$$\psi(h,h')=\big(E\xrightarrow{h'}X\xrightarrow{\mu}X^t\xrightarrow{h^t}E^t\xrightarrow{\lambda^{-1}}E\big).$$

In the opposite direction, given a hermitian form ψ on M, consider the associated homomorphism $\tilde{\psi}\colon M\to M^\dagger$, and define $\mu=\mu_\psi\colon X\to X^t$ by

$$\mu_\psi=\tilde{\psi}\otimes\lambda\colon X=M\otimes_R E\longrightarrow X^t=M^\dagger\otimes_R E^\dagger.$$

One checks that $\mu\mapsto\psi_\mu$ and $\psi\mapsto\mu_\psi$ are inverse to each other, and that a perfect form ψ correspond to an isomorphism μ. This gives the first bijection. Under this bijection the principal polarizations correspond to the positive definite perfect hermitian forms: this follows from what is explained in [37], § 21, Application III.

(7.15). We keep the notations introduced in (7.12). Let ℓ be a prime number different from p. Write $R_\ell:=R\otimes_{\mathbb{Z}}\mathbb{Z}_\ell$; then there exists an isomorphism $R_\ell\cong M_2(\mathbb{Z}_\ell)$.

The canonical pairing

$$e_\ell\colon T_\ell E \times T_\ell E^\dagger \longrightarrow \mathbb{Z}_\ell(1)$$

has the property that $e_\ell(r\xi, \xi') = e_\ell(\xi, r^\dagger\xi')$ for all $r \in R$, $\xi \in T_\ell E$ and $\xi' \in T_\ell E^\dagger$. Using this, one can show that there exists a unique bi-additive form

$$\tilde{e}_E\colon T_\ell E \times T_\ell E^\dagger \longrightarrow R_\ell(1)$$

such that

$$e_E(\xi, \xi') = \mathrm{Trd}\big(\tilde{e}_E(\xi, \xi')\big)\,, \quad \text{and} \tag{1}$$

$$\tilde{e}_E(r_1\xi, r_2\xi') = r_1 \cdot \tilde{e}_E(\xi, \xi') \cdot r_2^\dagger \tag{2}$$

for all $\xi \in T_\ell E$ and $\xi' \in T_\ell E^\dagger$, and r_1, $r_2 \in R$.

Next we remark that the canonical "evaluation" pairing $\mathrm{ev}\colon M \times M^\dagger \longrightarrow R$ has the property that

$$\mathrm{ev}(mr_1, m'r_2) = r_2^\dagger \cdot \mathrm{ev}(m, m') \cdot r_1\,. \tag{3}$$

We have canonical isomorphisms $T_\ell X \cong M \otimes_R T_\ell E$ and $T_\ell X^t \cong M^\dagger \otimes_R T_\ell E^\dagger$. Taking these as identifications, the canonical pairing $e_X : T_\ell X \times T_\ell X^\dagger \longrightarrow \mathbb{Z}_\ell(1)$ is given by

$$e_X(m \otimes \xi, m' \otimes \xi') = \mathrm{Trd}\big(\mathrm{ev}(m, m') \cdot \tilde{e}_E(\xi, \xi')\big)\,. \tag{4}$$

Let us note here that the form

$$\big((m, \xi), (m', \xi')\big) \mapsto \mathrm{ev}(m, m') \cdot \tilde{e}_E(\xi, \xi')$$

does *not* descend to a well-defined form on $T_\ell X \times T_\ell X^t$. However, using (1–3) together with the fact that Trd is a trace form (so $\mathrm{Trd}(\alpha\beta) = \mathrm{Trd}(\beta\alpha)$) we find that the right hand side of (4) *is* well-defined on $T_\ell X \times T_\ell X^t$.

(7.16). Let a be a positive integer, and let $n := \ell^a$. Recall that we write H_n for the standard hermitian pairing on $(R/nR)^g$. Assume given a primitive n-th root of unity in k, via which we identify $\mathbb{Z}/n\mathbb{Z}$ and $\mathbb{Z}/n\mathbb{Z}(1) = \mu_n(k)$. Using what was explained in (7.15), we find that the Weil pairing $e_n^\lambda\colon E[n] \times E[n] \longrightarrow \mathbb{Z}/n\mathbb{Z}$ lifts to a pairing

$$\tilde{e}_n^\lambda\colon E[n] \times E[n] \longrightarrow R/nR.$$

Consider the pairing

$$\theta\colon \big((R/nR)^g \otimes_R E[n]\big) \times \big((R/nR)^g \otimes_R E[n]\big) \longrightarrow \mathbb{Z}/n\mathbb{Z}$$

given by $\theta\big((x \otimes P), (x' \otimes P')\big) = \mathrm{Trd}\big(H_n(x, x') \cdot \tilde{e}_n^\lambda(P, P')\big)$. One checks that this is a well-defined symplectic pairing. In other words, if Φ_n is the standard symplectic pairing on $(\mathbb{Z}/n\mathbb{Z})^{2g}$ then there exists an isometry

$$\big((R/nR)^g \otimes_R E[n], \theta\big) \quad \xrightarrow{\sim} \quad \big((\mathbb{Z}/n\mathbb{Z})^{2g}, \Phi_n\big).$$

More generally, if n is an arbitrary positive integer prime to p then we find that there exist isometries as above by writing $n = \ell_1^{a_1} \cdots \ell_r^{a_r}$ and applying the preceding to each of the factors $\ell_i^{a_i}$.

(7.17). **Lemma** *Let n be a positive integer, not divisible by p. Assume chosen a primitive n-th root of unity in k. Fix an isometry as in (7.16). Let μ be a principal polarization of X which, as in (7.14), corresponds to a perfect hermitian form ψ on M. Then there is a natural bijection*

$$\left\{\begin{matrix}\text{symplectic level-}n\\ \text{structures on }(X,\lambda)\end{matrix}\right\} \overset{\sim}{\leftarrow} \left\{\begin{matrix}\text{level-}n\text{-markings}\\ \text{of }(M,\psi)\end{matrix}\right\} \qquad (1)$$

Proof. For simplicity, let us assume that n is a power of a prime number ℓ. (The general case is easily reduced to this.) Suppose given a level-n-marking of (M,ψ). Using what was explained in (7.15) we find that the induced isomorphism

$$X[n] = (M/nM) \otimes_R E[n] \overset{\sim}{\to} (R/nR)^g \otimes_R E[n]$$

gives an isometry

$$\left(X[n], e_n^{\mu}\right) \overset{\sim}{\to} \left((R/nR)^g \otimes_R E[n], \theta\right),$$

where θ is as in (7.16). Composing with the fixed isometry (7.16) this gives a symplectic level-n-structure of (X,μ). This defines the map in (7.17).

By an easy "integral" variant of (7.6) we see that $\mathrm{Aut}\left((R/nR)^g, H_n\right) \cong \mathrm{Sp}_{2g}(\mathbb{Z}/n\mathbb{Z})$. In particular, both sides of (7.17)(1) are principal homogeneous under $\mathrm{Sp}_{2g}(\mathbb{Z}/n\mathbb{Z})$. Further, one easily sees that the map (7.17)(1) just defined is compatible with these structures of homogeneous spaces. Hence our map is a bijection. □

(7.18). **Proof of Proposition** (7.3). Let x and x' be points of $\Sigma(k)$, corresponding to triples (X,μ,η) and (X',μ',η'), respectively. By Lemma (7.9) we can embed the corresponding R-modules $M := \mathrm{Hom}(E,X)$ and $M' := \mathrm{Hom}(E,X')$ as perfect hermitian R-lattices in (V,Ψ) in such a way that $M\cap M'$ has p-power index in both M and M' and that the given level-n-markings of M and M' agree. As $\mathfrak{p} \subset R$ is a principal ideal, see (ii) of Remark (7.13), we can apply Lemma (7.11). This gives us perfect hermitian lattices

$$M = L_0, L_1, \ldots, L_m = M'$$

with $L_j/(L_j\cap L_{j+1}) \cong k(\mathfrak{p}) \cong L_{j+1}/(L_j\cap L_{j+1})$. Fix $j \in \{0,\ldots,m-1\}$. Write $N := L_j \cap L_{j+1}$, and consider the homomorphisms

$$X_j := L_j \otimes_R E \overset{f}{\longleftarrow} Y := N \otimes_R E \overset{h}{\longrightarrow} X_{j+1} = L_{j+1} \otimes_R E$$

induced by the inclusions $N \hookrightarrow L_j$ and $N \hookrightarrow L_{j+1}$, respectively. Via the correspondence (7.14), we have principal polarizations μ_j on X_j and μ_{j+1} on X_{j+1}; these satisfy $f^*\mu_j = h^*\mu_{j+1}$. The given level-n-markings of M and M' induce level-n-markings on each of the lattices L_j; we write η_j for the corresponding symplectic level-n-structure of (X_j,μ_j).

We are done if we show that f is a covering satisfying conditions (i)–(iii) of (7.1). Indeed, by symmetry the same then holds for h, and we conclude that

(X_j, μ_j, η_j) and $(X_{j+1}, \mu_{j+1}, \eta_{j+1})$ lie on the same irreducible component of the locus L, as constructed in (7.1).

By construction, f is an isogeny of superspecial abelian varieties, and $\deg(f)$ is a p-power. (It is easily seen that f induces an isomorphism $T_\ell Y \overset{\sim}{\to} T_\ell X_j$ for all primes $\ell \neq p$.) Further, f is minimal, in the sense that it does not admit a non-trivial factorization. Hence $\mathrm{Ker}(f) \cong \alpha_p$. This implies that $\mathrm{Ker}(f^*\mu_j)$ is either isomorphic to $\alpha_p \times \alpha_p$, or it is a non-trivial extension of α_p by α_p. But in the latter case $\mathrm{Ker}(f^*\mu)$ has a unique subgroup scheme of rank p, so that necessarily $\mathrm{Ker}(f) = \mathrm{Ker}(h)$. This is in contradiction with our construction.

□(7.18)

8. Abelian varieties with $a = g - 1$

In this section we work over an algebraically closed field $k \supset \mathbb{F}_p$. We consider *principally* polarized abelian varieties and *principally* quasi-polarized p-divisible groups. We study strata defined by an elementary sequence φ with $\varphi(g) = 1$. These describe abelian varieties of dimension g with $a(\varphi) = a(X) = g - 1$. Results of this section are rather elementary once you know the Dieudonné-Manin classification; we do not use results of the previous sections. We use a result in [29], and we use some notations as explained in [19]. For the notation $G_{m,n}$ we refer to (15.5).

(8.1). For every pair $m, n \in \mathbb{Z}_{\geq 0}$ of coprime integers there exists a p-divisible group $H_{m,n}$, defined over $\mathbb{F}_p$, isogenous with $G_{m,n}$, which is characterized by the fact that its endomorphism ring over an algebraically closed field is maximal within the isogeny class; see [19], Section 5. This p-divisible group can be defined by the Dieudonné module given by the semimodule $[0, \infty)$, in the notation as explained in [19]. If $m > 1$ and $n > 1$ there exist p-divisible groups isogenous with $G_{m,n}$ not isomorphic with $H_{m,n}$.

We know that $a(H_{m,n}) = \min(m, n)$ (as is easy to see form the defining semi-module. However this property is in general not sufficient to characterize $H_{m,n}$ within its isogeny class, but in the cases used in this section it will be the case (see the lemma below). One can feel the object $H_{m,n}$ as the (unique) "minimal" one in the isogeny class of $G_{m,n}$, and groups with $a(G) \leq 1$ as the most general ones; this can be made precise, see [19], Section 5, especially see (5.7).

(8.2). *If G is a formal group isogenous with $G_{m,n}$ then*

$$a(G) \leq m, \quad a(G) \leq n.$$

Moreover we know that the maximum $a(G) = \min(m, n)$ is achieved within this isogeny class.

See [42], page II.15-10, see [50].

A proof is not difficult; e.g. using notation as introduced in [19] we see: let $G \sim G_{m,n}$ over an algebraically closed field, and let $A = \mathrm{Type}(G)$ be the semi-module associated with G. Consider

$$A\backslash(m + A \cup n + A) \quad \to \quad \mathbb{Z} \to \mathbb{Z}/m.$$

This map is injective. Hence

$$\#(A\backslash(m + A \cup n + A)) = a(A) = a(G) \leq m.$$

The same holds for the natural mapping to $\mathbb{Z}/n$. This proves the desired inequality. As moreover $a(H_{m,n}) = \min(m, n)$ we are done. □

Remark. If the isogeny type of a p-divisible group G is given, consider "an isogeny factor", i.e., the image of some $G_{m,n}$ appearing in the isogeny type under an isogeny into G. In general these "isogeny factors" are not direct summands, and it is usually difficult to decide which member of the isogeny class of $G_{m,n}$ appears in this way. However in the situation studied in this section all these questions turn out to have a unique and simple answer.

In this paper we use $V - U$ to denote the complement of U in V; however in the proof above we use $V\backslash U$ for this notion, in order to avoid confusion with subtraction.

(8.3). **Theorem.** *Let X be an abelian variety of dimension g over an algebraically closed field of characteristic p which admits a principal polarization. Suppose* $\mathrm{ES}(X) = \varphi$ *with* $\varphi(g) = 1$, *i.e.,* $a(X) = g - 1$. *Then:*

(I): *either there exists an integer b with $0 \leq b \leq [\frac{g-1}{2}]$ and an isomorphism*

$$X[p^\infty] \quad \cong \quad H_{b+1,b} \quad \bigoplus \quad (g - 2b - 1)\cdot G_{1,1} \quad \bigoplus \quad H_{b,b+1},$$

and in this case

$$\varphi = (0, \cdots, 0, \overbrace{1, \cdots, 1}^{g-b}), \quad |\varphi| = g - b,$$

(II): *or there exists an integer r with $0 < r \leq [\frac{g}{2}]$, there exists an isogeny $X \sim E^g$, where E is a supersingular elliptic curve* (i.e., X is supersingular), *and*

$$\varphi = (0, \cdots, 0, \overbrace{1, \cdots, 1}^{r}), \quad |\varphi| = r$$

(and the structure of $X[p^\infty]$ will be explained in the proof). Note that for all b as in (I) and all r as in (II) we have $g - b > r$ and $0 \leq r \leq [\frac{g-1}{2}] \leq g - b$; the value of $|\varphi|$ determines in which case we are.

As a corollary of this theorem and its proof we deduce:

(8.4). **Corollary.** *Let X be an abelian variety with*

$$\mathrm{ES}(X) = \varphi := (0, \cdots, 0, 1).$$

Then X is as described in the construction (7.1). *In particular*

$$L^o = S_{(0,\cdots,0,1)}, \quad L = S_{(0,\cdots,0,1)} \cup \Sigma = \left(S_{(0,\cdots,0,1)}\right)^c.$$

From the theorem we conclude that $S_{(0,\cdots,0,1)} \subset L$; hence the corollary follows from (7.2) and the previous theorem. $\square$(8.3) $\Rightarrow$ (8.4)

The proof of this theorem will be given in various steps. In the first step we analyze the isogeny type of $X[p^\infty]$.

First step. Suppose $a(X) = g - 1$. From (8.2) it follow that we are in one of the following cases:

(I.0): *either $f(X) = 1$; we claim: in this case*

$$X[p^\infty] \;\cong\; G_{1,0} \;\oplus\; (g-1) \cdot G_{1,1} \;\oplus\; G_{0,1},$$

and we conclude $\varphi(1) = 1$; hence $\varphi = (1, \cdots, 1)$; this is case I.0;

(I.b): *or $f(X) = 0$ and there exists an integer b with $0 < b \leq [\frac{g-1}{2}]$ and an isogeny*

$$X[p^\infty] \;\sim\; G_{b+1,b} \;\bigoplus\; (2g-2b-1) \cdot G_{1,1} \;\bigoplus\; G_{b,b+1};$$

we shall analyze this case;

(II): *or*

$$X[p^\infty] \sim g \cdot G_{1,1};$$

we shall describe all supersingular cases with $a(X) = g - 1$.

Indeed, the fact that $a(X) = g - 1$ limits the possibilities of the formal isogeny types to the cases as indicated. If $f(X) = 1$ and $a(X) = g - 1$ we see (note that the base field is algebraically closed) that $X[p^\infty] \cong G_{1,0} \oplus G_{0,1} \oplus G$ with $a(G) = \dim(G)$. This proves the first statement. The remaining possibilities have been listed. $\square$

(8.5). **Lemma.** *Let $b \in \mathbb{Z}_{\geq 0}$, let G be a p-divisible group over an algebraically closed field with $G \sim G_{b+1,b}$, and $a(G) = b$; then $G \cong H_{b+1,b}$.*

Proof. The case $b = 0$ is well-known; we suppose $b > 0$. In [19] we have seen that to every $G \sim G_{b+1,b}$ we can attach a semi-module $A = \mathrm{Type}(G) \subset \mathbb{Z}$. This is a subset of $\mathbb{Z}$ bounded from below, and stable under $+m = +(b+1)$ and $+n = +b$; let $t \in A$ be the smallest element.

We show that (under all conditions mentioned) from $a(A) = b$ it follows that $A = [t, \infty)$; in fact, let $C = A \backslash ((b+A) \cup ((b+1)+A))$, hence $\#(C) = a(A)$; if there exist $x, y \in \mathbb{Z}$ such that $t + x + 1, \cdots, t + y - 1 \notin A$, then

$$\#((\cup_j [t + x + 1 + j \cdot (b+1), t + y + b - 1 + j \cdot b]) \cap C) < y - x - 1;$$

we see that such gaps in $[t, t+B-1] \cap A$ do not appear if $a(A) = b$; we conclude that $A = [t, \infty)$. This implies $G \cong H_{b+1,b}$, which proves the lemma. $\square$

(8.6). **Remark.** Let $m, n \in \mathbb{Z}$ with $m > n+1 > n > 1$. The semi-module $A = \{0\} \cup [2, \infty]$ has $a(A) = n$. We see that in this case there exists $G \cong G_{m,n}$ with $a(G) = n = \min(m, n)$ and $G \not\cong H_{m,n}$.

(8.7). **Second step. Lemma.** *Let G be a p-divisible group over an algebraically closed field, which admits a principal polarization, such that $G \sim H_{b+1,b} \oplus (g - 2b-1){\cdot}G_{1,1} \oplus H_{b,b+1}$ and $a(G) = g-1$. Then*

$$G \quad \cong \quad H_{b+1,b} \quad \bigoplus \quad (2g-2b-1){\cdot}G_{1,1} \quad \bigoplus \quad H_{b,b+1},$$

and we have

$$\mathrm{ES}(G) = \varphi = (0, \cdots, 0, 1, \cdots 1), \quad |\varphi| = g - b.$$

Proof. Let $I, S, J \subset G$ be the images of the three factors under an isogeny $H_{b+1,b} \oplus (2g-2b-1){\cdot}G_{1,1} \oplus H_{b,b+1} \to G$. We are going to use that in the isogeny classes of I, S, J there is a unique (up to isomorphism) group with maximal a-number. We claim: $a(I) = b = a(J)$ and $a(S) = 2g-2b-1$ and hence $I \cong H_{b+1,b}$, and $S \cong (2g-2b-1){\cdot}G_{1,1}$, and $J \cong H_{b,b+1}$; for a p-divisible group Y over a perfect field we write $\alpha(Y) \subset Y$ for the smallest subgroup scheme containing all $\alpha_p \subset Y$ (denoted by $A(Y)$ in [29], 1.5, but we have used that notation already for a different purpose). The exact sequence $\alpha(I) \subset Y \to Y/(\alpha(I))$ together with the fact that $a(Y)$ and $a(Y/(\alpha(I)))$ are "maximal" show that $a(I) = b$; analogous statements for S and for J; these imply the existence of the isomorphisms indicated, hence the claim. Next we show that the natural map $I \oplus S \oplus J \to I + S + J = G$ is an isomorphism. We show this for $S \oplus J \to S + J$; because the a-numbers are "maximal" we see that

$$0 \to \alpha(S) \to \alpha(S+J) \to \alpha((S+J)/S) \to 0$$

is exact; it follows that the

$$\mathrm{Ker}(S \oplus J \to S + J) \cap \alpha(S \oplus J) = 0,$$

which proves $S \oplus J \cong S + J$. An analogous statement holds for $I + S$ and for $I + J$. From these the result follows. □

(8.8). **Remark.** We see cases where a p-divisible group G is determined up to isomorphism by $G[p]$: the case $f(G) = \dim(G)$, the case $a(G) = \dim(G)$, and the case as in **(I)** of the theorem. However in general $G[p]$ does not determine G. We have a general conjecture which illustrates what should be (im)possible, see (14.7), and (14.8).

(8.9). **Third step.** Suppose we are in the situation as in (II), i.e., X is a supersingular abelian variety with a principal polarization with $a(X) = g-1$. We use the description of the locus $\mathcal{S}_{g-1}(a \geq g-1)$ as given in [29], Section

9, in particular in (9.10) and (9.11). From that description we see that in this case there exists an integer r with $0 < r \leq [\frac{g}{2}]$, an isomorphism $\xi_p(X) \cong (g-2r){\cdot}G_{1,1} \oplus G$, and an exact sequence

$$0 \to (\alpha_p)^r \to (2r){\cdot}G_{1,1} \to G \to 0,$$

with $a(G) = 2r - 1$. We assume the integer r is chosen to be the minimal one allowing this structure. The polarization in that case gives a form on the Dieudonné-module, with $< e_i, \mathcal{F}(e_i) >$ non-zero for $1 \leq i < 2g - 2r >$ and $< e_{2g-2j}, e_{2g-2j+1} >$ non-zero for $r \geq j \geq 1$. An explicit computation shows that in this case $\varphi = (0, \cdots, 0, 1, \cdots 1)$, $|\varphi| = r$. This ends the proof of the theorem. $\square$(8.3)

9. Standard types

In this section we attach to every elementary sequence φ a finite group scheme N_φ plus a non-degenerate alternating pairing $<,>=<,>_{\mathrm{st}}$ on $A_\varphi := \mathbb{D}(N_\varphi)$ such that the pair $(N_\varphi, <,>)$ is defined over $\mathbb{F}_p$, and such that $\mathrm{ES}(N_\varphi) = \varphi$.

Then we show that a "polarized BT_1 truncated group scheme" over an algebraically closed field k is isomorphic (over k) with the standard type given by its elementary sequence. This result will be used in (12.5), where we use that two polarized BT_1 group schemes over an algebraically closed field with equal elementary sequences are isomorphic.

Remark. It might seem strange that we define $<,>$ on the Dieudonné module (and not on the group scheme itself). If we would avoid characteristic 2, in fact we could work with anti-symmetric morphisms on group schemes, prove the results we want to achieve, and develop a mechanism which works over any base in characteristic $p \neq 2$. However in characteristic $= 2$ we do not see how this simplification can be carried out with success. We have chosen for a unified treatment applicable in all characteristics. For a further discussion we refer to Section 12.

I do not know how to characterize the fact that $<,>: A \times A \to K$ is *alternating* on $A := \mathbb{D}(N)$ in terms of the related symmetric $\zeta : N \to N^D$ in chase the characteristic of the perfect base field K equals $p = 2$.

Reminder: $\mathbb{D}(V : N^{(p)} \to N) = (\mathcal{F} : A^{(p)} \to A)$. The constructions involving V and F^{-1} on N we find back as construction involving $\mathcal{F}$ and $\mathcal{V}^{-1}$ on $A = \mathbb{D}(N)$.

Remark. We use in this paper the following notation, also see Section 12: let (X, λ) be a principally polarized abelian variety over a perfect field K of characteristic p; write $N := X[p]$, with Dieudonné module $A = \mathbb{D}(N)$; the polarization induces a symmetry on N, and it defines a pairing $<,>: A \times A \to K$. We write $(N, <,>) =: (X, \lambda)[p]$ for this "polarized BT_1 truncated group scheme".

Remark. In what follows we use a classification of a BT_1 over an algebraically closed field found by Kraft, see [27], and rediscovered independently much later by the present author. That structure is not very difficult to describe. However the classification of "polarized BT_1 truncated group schemes" turns out to be more involved, as will be seen below.

(9.1). **Construction.** Suppose given an elementary sequence

$$\{\varphi(1), \cdots, \varphi(g)\} = \varphi \in \Phi, \quad \varphi(0) = 0,$$

and let ψ be the related final sequence:

$$\psi(i) = \varphi(i) \quad \text{for} \quad 0 \leq i \leq g, \quad \text{and} \quad \psi(g-i) = \varphi(i) + g - i \quad \text{for} \quad 0 \leq i \leq g.$$

Consider the set of indices where the ψ-sequence jumps:

$$1 \leq m_1 < m_2 < \cdots < m_g \leq 2g;$$

this is the set of all integer i such that $\psi(i-1) < \psi(i)$. We denote by

$$1 \leq n_g < n_{g-1} < \cdots < n_1 \leq 2g$$

the complimentary set, i.e., the set of indices where the ψ-sequence does not jump. Note that:

$$m_i + n_i = 2g + 1 \quad \text{for} \quad 0 \leq i \leq g.$$

For convenience we write $m_0 = 0, \;\; 2g+1 = m_{g+1}, \;\; n_{g+1} = 0, \;\; n_0 = 2g+1$. Given $\varphi \in \Phi$ we are going to construct a covariant Dieudonné-module A_φ with a non-degenerate alternating pairing over $\mathbb{F}_p$. The related group scheme with pairing will be a polarized BT_1, which will be denoted by $(N_\varphi, <,>)$. It will be called the *standard type* given by φ.

We construct $A = A_\varphi$ as the vector space of dimension $2g$ over $\mathbb{F}_p$ generated by a base $\{Z_1, \cdots, Z_{2g}\}$. We write

$$Z_{m_i} = X_i \quad \text{and} \quad Z_{n_i} = Y_i \quad \text{for} \quad 1 \leq i \leq g;$$

note that the numbering of the Y_i is decreasing, and the numbering of the X_i is increasing in $\{Z_j \mid j\}$. We define

$$\mathcal{F} : A \to A \quad \text{by} \quad \mathcal{F}(X_i) := Z_i, \quad \mathcal{F}(Y_i) = 0 \quad \text{for} \quad 0 \leq i \leq g.$$

We define the alternating pairing by

$$< X_i, Y_j >= \delta_{i,j}, \quad < X_i, X_j >= 0, \quad < Y_i, Y_j >= 0 \quad \text{for} \quad 0 \leq i,j \leq g.$$

Note that these determine the action of $\mathcal{V}$; in fact $\mathcal{V} : M \to M$ is given by

$$\mathcal{V}(Z_i) = 0, \quad \mathcal{V}(Z_{2g-i+1}) = \pm Y_i \quad \text{for} \quad 0 \leq i \leq g,$$

where $\mathcal{V}(Z_{2g-i+1}) = +Y_i$ if $Z_{2g-i+1} \in \{Y_g, \cdots, Y_1\}$ and $\mathcal{V}(Z_{2g-i+1}) = -Y_i$ if $Z_{2g-i+1} \in \{X_1, \cdots, X_g\}$.

It follows that $\mathcal{F}, \mathcal{V}$ and $<,>$ are connected by the rules they should satisfy:

$$< \mathcal{F}(a), b >=< a, \mathcal{V}(b) > \quad \text{for all} \quad a, b \in A;$$

note that the base field is $\mathbb{F}_p$; hence the Dieudonné modules are over the ring $\mathbb{F}_p[\mathcal{F}, \mathcal{V}]$, which is commutative in this case. Note that

$$\text{Im}(\mathcal{F}) = \mathbb{F}_p{\cdot}Z_1 \oplus \cdots \oplus \mathbb{F}_p{\cdot}Z_g = \text{Ker}(\mathcal{V}),$$

and

$$\mathrm{Ker}(\mathcal{F}) = \mathbb{F}_p{\cdot}Y_g \oplus \cdots \oplus \mathbb{F}_p{\cdot}Y_1 = \mathrm{Im}(\mathcal{V}).$$

We denote by $(N_\varphi, <,>_{\mathrm{st}})$ the associated group scheme with this non-degenerate pairing on $A_\varphi = \mathbb{D}(N_a) = \oplus\mathbb{F}_p{\cdot}Z_i$, all defined over $\mathbb{F}_p$. From the formulas just derived we see that this is a "polarized BT_1 of rank p^{2g} defined over $\mathbb{F}_p$", see below. Moreover we see that the submodules $A_i = \oplus_{j=1}^{j=i}\mathbb{F}_p{\cdot}Z_j$ define a final filtration on $N = N_\varphi$, and we see that the final type of this final filtration equals φ.

(9.2). **Definition.**Let K be a perfect field. We say that $(N, <>)$ is a *polarized* BT_1 *truncated group scheme*, or: a polarized BT_1, defined over K, if N is a BT_1 over K, with Dieudonné module $A := \mathbb{D}(N)$ and $<,>: A \times A \to K$ is an alternating, non-degenerate pairing such that $< \mathcal{F}x, y > = < x, \mathcal{V}y >^p$ for all $x, y \in A$.

(9.3). **Conclusion.** *For every elementary sequence $\varphi \in \Phi$ we have constructed a polarized BT_1 $(N_\varphi, <,>)$ defined over $\mathbb{F}_p$ with (covariant) Dieudonné module $(A_\varphi, <,>)$, possessing a final filtration given by a symplectic $\mathbb{F}_p$-base $\{Y_g, \cdots, Y_1, X_1, \cdots, X_g\}$ such that $Y_i \in \mathrm{Ker}(\mathcal{F})$ for every i and such that every base element under $\mathcal{F}$ is either mapped to zero or to a base vector, and such that* $\mathrm{ES}(N_\varphi) = \varphi$. It will be called the **standard type** defined by φ. The base $\{Z_1, \cdots, Z_{2g}\} = \{Y_g, \cdots, Y_1, X_1, \cdots, X_g\}$ will be called a **standard base** for A_φ. In case confusion is possible, we will write $<,>_{\mathrm{st}}$ for the "standard pairing" on the "standard module" A_φ. A final filtration is given by $A_i = \oplus_{j=1}^{j=i}\mathbb{F}_p{\cdot}Z_j$.

(9.4). **Theorem.** *Suppose k is an algebraically closed field of characteristic p. Let $(N, [,])$ be a polarized BT_1 of rank p^{2g} defined over k. Suppose $\varphi = \mathrm{ES}(N)$ is the elementary sequence determined by N. Then*

$$(N, [,]) \;\cong\; (N_\varphi, <,>_{\mathrm{st}}) \otimes_{\mathbb{F}_p} k.$$

Remark. This isomorphism is not unique in general.

As Ben Moonen shows, see [33], a different, easier proof of this theorem can be given in case the characteristic of the base field is not equal to two. We could not generalize that proof to include the case $\mathrm{char}(k) = p = 2$. Below I give a proof which covers all cases uniformly. It seems desirable to simplify the methods of this section.

Strategy of the proof. We study on the one hand all pairings on $A = \mathbb{D}(N)$ such that a given $(\mathcal{F}, \mathcal{V})$-base is in standard form on the associated grade module; on the other hand we study all transformations changing a $(\mathcal{F}, \mathcal{V})$-base into another such base. We show that the group of these transformations acts transitively on the set of all such pairings, see (9.20). In order to carry out this programme we study the combinatorics related with the values of a pairing on base vectors, and the combinatorics related to the coefficients needed in such transformations.

The proof of the theorem will be given in several steps.

(9.5). Let (X, λ) be a principally polarized abelian variety. Then $N := X[p]$ is a symmetric BT_1. In fact, more is true. The pairing on $X[p^\infty]$ is symmetric, and this implies that the map $\zeta : X[p] =: N \to N^D$ is *anti-symmetric*, see Section 12 for more details. However more is true: *if (X, λ) is an abelian variety over a perfect field $K \supset \mathbb{F}_p$, the symmetry $\zeta : N \to N^D$ obtained defines a pairing $<,>: A \times A \to K$ on $A =: \mathbb{D}(N)$ which is alternating.*
Indeed, the pairing on $\mathbb{D}(X[p^\infty]$ with values in the characteristic zero domain $W_\infty(K)$ is symmetric, and the result follows (for details, see (12.2)).

(9.6). **Proposition.** *Let k be an algebraically closed field. Let $(N, <,>)$ be a polarized BT_1 truncated group scheme defined over k, and let $\zeta : N \to N^D$ be the symmetry defined by $<,>$. The canonical filtration on N can be refined to a final filtration of (N, ζ).*
Remark. For a given $(N, <,>)$ the number of final filtrations on $(N, <,>)$ is finite. We will give a reformulation of this proposition (in a special case) in (9.13).
Proof. We consider steps in the canonical filtration of $A := \mathbb{D}(N)$. In (5.10) we have constructed $(\mathcal{F}, \perp)$-cycles in the set of partial quotients in the Dieudonné module of the canonical filtration; such a cycle consists of vertices, each is a quotient in the canonical filtration of the Dieudonné module, edges of the cycle are given by bijective maps $\mathcal{F}$ or $\perp$, and if $\mathcal{F}$ and $\mathcal{F}$ are consecutive edges, there is a "loose edge" $\perp$. We construct a refinement to a final filtration for each of the cycles separately. Let P be one of the k-vector spaces $P = \mathbb{D}(N_j/N_{j-1})$ as in (5.10). Using a $(\mathcal{F}, \perp)$-cycle in A according to $<,>$, we see that composition of the maps $\mathcal{F}$ and isomorphism $\perp$ give a linear mapping

$$f : (P^{(p^d)})^{(D^\epsilon)} \xrightarrow{\sim} P;$$

here d is the number of steps involving $\mathcal{F}$, and ϵ is the parity of the number of steps involving $\perp$. Note that $d > 0$. Note that in the same cycle the values d and ϵ are the same for all edges in a cycle. We distinguish two cases:
(even): suppose the number of edges named $\perp$ in the loop of the cycle is even; in this case the parity of ϵ is even, i.e.,

$$f : (P^{(p^d)}) \xrightarrow{\sim} P;$$

(odd): suppose the number of edges $\perp$ in the loop of the cycle is odd; in this case the parity of ϵ is odd.

Suppose we are in the case of an **(even)** cycle. This is case (iii) of Lemma (5.10). We use the "Jordan reduction" for the map f as exposed in [16], or in [6], pp. 232 - 234, [56], page 38. Hence there exists a k-base $P = k{\cdot}e_1 \oplus \cdots \oplus k{\cdot}e_c$

such that $f(e_i) = e_i$. We transport this base through every step of the cycle, pushing is forward by $\mathcal{F}$, or going to the dual base under $\perp$. As ϵ is even we come back with the same ordered base. This defines in every step of the filtration connected with the cycle a filtration by steps of dimension one; these are related under $\mathcal{F}$ or dual under $\perp$ whenever applicable. This gives a final filtration for all quotients in the cycle studied, hence a final filtration for N_γ, the direct sum of all N_{j+1}/N_j for all quotients appearing in this cycle.

Suppose we are in the case of an **(odd)** cycle. This is case (ii) of Lemma (5.10). Let us consider one odd $(\mathcal{F},\perp)$-cycle γ. As we have seen this is constructed from a word w in $\mathcal{F}$ and $\mathcal{V}^{-1}$, describing the combinatorics of the related $(\mathcal{F},\mathcal{V}^{-1})$-cycle on A, i.e., the word in V and F^{-1} in the (V,F^{-1})-cycle on N. As γ is odd, the word w is self-dual, the number $|w|$ of letters is even; we write $|w| = 2m$. Let $P = P_0$ be one of the subquotients of $\mathbb{D}(N)$ appearing in γ; consider $P_0, P_1, \cdots, P_{2m-1}, P_m = P$, the subquotients appearing in the $(\mathcal{F},\mathcal{V}^{-1})$-cycle; consider $g_i : P = P_0 \to P_i$, which is obtained by going clockwise i steps through the $(\mathcal{F},\mathcal{V}^{-1})$-cycle given by w, with the related maps, using either $\mathcal{F}$ or $\mathcal{V}^{-1}$; it is a σ^i-linear map. We write $g = g_{2m} : P_0 = P \to P = P_{2m}$. We see that this is a σ^{2m}-linear endomorphism of the vector space P. Let us write $c := \dim_k P$.

We use the "Jordan reduction" for the map g as exposed in [16], or in [6], pp.232 - 234, [56], page 38. This shows: define $\mathbb{F} := \mathbb{F}_{p^{2m}}$; there exists a $\mathbb{F}$-base $\mathcal{P} := \mathbb{F}{\cdot}e_1 \oplus \cdots \oplus \mathbb{F}{\cdot}e_c$, and $g : \mathcal{P} \to \mathcal{P}$ such that $g(e_i) = e_i$, and such that $(\mathcal{P}, g) \otimes_{\mathbb{F}} k \cong (P, g)$. From now on we work with vector spaces over $\mathbb{F}$, in particular we write $g_i(\mathcal{P}) = \mathcal{P}_i$.

We write $\iota := F^m$, i.e., the map $x \mapsto x^{p^m}$ when restricted to $\mathbb{F}$ we have $\iota \in \mathrm{Gal}(\mathbb{F}/\mathbb{F}_{p^m})$. Let us write $g_m =: h : P_0 = P \to Q = P_m$, and analogously $H : \mathcal{P} \to \mathcal{P}$. We define

$$\psi : P \times P \to \mathbb{F} \quad \text{by} \quad \psi(x,y) = < x, hy > .$$

Remark that $g : P \to P$ is ι-linear over k, hence $g : \mathcal{P} \to \mathcal{P}$ is linear over $\mathbb{F}$; hence every base change over $\mathbb{F}$ leaves g in diagonal-**1**-form.

Note that ψ is:

1) non-degenerate,
2) linear in the first variable,
3) ι-linear in the second variable, and
4) $\psi(x,y) = -\psi(y,x)^\iota\ \forall x,y \in P$, i.e., "$\psi$ is hermitian with respect to ι".

From this last equality we see that for all $x, y \in \mathcal{P}$ we have $\psi(x,y) = \psi(x,y)^{\iota\cdot\iota}$; this shows that ψ factors as $\psi : \mathcal{P} \to \mathcal{P}$:

5) for all $x, y \in \mathcal{P}$ we have $\psi(x,y) \in \mathbb{F}$.

Choose an element $a \in \mathbb{F}^*$ such that $b^\iota = -b$, i.e., an element such that $b^{p^m-1} = -1$. *We are going to show that we can choose a $\mathbb{F}$-base $\mathcal{P} = \mathbb{F}{\cdot}f_1 \oplus \cdots \oplus \mathbb{F}{\cdot}f_c$ such that on this $\mathbb{F}$-base the form ψ is anti-diagonal with b on every anti-diagonal place, i.e., such that $\psi(f_i, f_j) = b{\cdot}\delta_{j,c-i+1}$ (Kronecker-delta) for every* $0 \le i, j \le c$; let us call this the *standard anti-diagonal form over* $\mathbb{F}$.

Suppose we have proven that ψ can be put standard anti-diagonal form over $\mathbb{F}$; then we are done with this step: the image of the base $\{f_j \mid 1 \leq j \leq c\}$ under g_i gives an ordered base for $\mathcal{P}_i$, for $1 \leq i \leq 2m$; note that the matrix of g on this $\mathbb{F}$-base equals $\mathbf{1}$; thus for every $\mathcal{P}_i$ we obtain a filtration by subvectorspaces with relative quotients of dimension one, and the filtrations are carried into each other under $\mathcal{F}$ respectively $\mathcal{V}$; the equality $< x, z = \mathcal{F}(y) >=< \mathcal{V}(x), y >^{\sigma}$ shows this gives a final filtration on all subquotients in A_γ, compatible with $\perp$.

We are going to use methods exposed in [26], 29.19. Over $\overline{\mathbb{F}}$ the form ψ can be put in standard anti-diagonal form; let us write $P = \mathcal{P} \otimes_{\mathbb{F}} \overline{\mathbb{F}}$; indeed, if $\dim P > 1$, there is an isotropic vector $x \in P$; we can write $P = k{\cdot}x \oplus k{\cdot}z \oplus (k{\cdot}x \oplus k{\cdot}z)^{\perp}$ with $\psi(x, z) \neq 0$, and we proceed by induction. Moreover the discriminant of this bilinear form is in

$$\mathbb{F}^*/N_{\mathbb{F}/\mathbb{F}_{p^m}}(\mathbb{F}^*) = \{1\}.$$

Let us write $\mathbb{U}$ for the group of linear transformations of P respecting the data (1) $\sim$ (4) above; this is a connected linear group. By [26], page 403 the set of isometry classes of forms isometric with ψ is given by $\mathrm{H}^1(\mathbb{F}, \mathbb{U})$; by [55], page 139 we know this cohomology set is trivial; this shows there exists a transformation in $\mathbb{U}(\mathbb{F})$ transforming ψ into the standard anti-diagonal form over $\mathbb{F}$. We have seen that this is enough to construct a final refinement. This ends the construction of a final refinement for N_γ in the case of an odd cycle γ.

Performing the constructions described above for every cycle, we refine the original filtration to a good filtration with relative steps of dimension one, i.e., to a final filtration. This proves Proposition (9.6). □

We start with the proof of Theorem (9.4).

(9.7). **First step.** As k is perfect we can write $N = N_{loc,et} \oplus N_{loc,loc} \oplus N_{et,loc} = N' \oplus N_{loc,loc}$. Because k is algebraically closed, $N_{loc,et} \cong (\mu_{p,k})^f$, and $N_{et,loc} \cong ((\mathbb{Z}/p)_k)^f$, and we choose a standard base for $N' = N_{loc,et} \oplus N_{et,loc}$. Note that $\mathrm{ES}((G, <, >)$ starts as follows: $\{1, 2, \cdots, f, f, \cdots\}$. It suffices to choose a standard base for $(M_{loc,loc}, <, >)$. From now on we suppose that N is purely local-local, and hence $\mathcal{F}$ and $\mathcal{V}$ act nilpotently on A.

(9.8). We fix some notation, and we recall a result from [27]. In this section a *word* will be a finite sequence of the letters $\mathcal{F}$ and $\mathcal{V}$. A word will be used in a cyclic way, i.e., putting the first letter as last we will consider the two words as the same. A word is called *simple* if it is not periodic of period larger than one. We only consider words in which both $\mathcal{V}$ and $\mathcal{F}$ appear. The number of letters in a words will be called the length of the word, indicated by $|w|$.

For a word w we construct, as in [27], a finite group scheme Z_w with Dieudonné module $B_w = \mathbb{D}(Z_w)$ over $\mathbb{F}_p$. If $w = L_1 \cdots L_d$ we choose $z_1, \cdots, z_d$ plus the convention $z_{d+1} = z_1$; if $L_i = \mathcal{F}$ we write $\mathcal{F}(z_i) = z_{i+1}$ and $\mathcal{V}(z_{i+1}) =$

0; if $L_i = \mathcal{V}$ we write $\mathcal{V}(z_{i+1}) = z_i$ and $\mathcal{F}(z_i) = 0$ (the maps $\mathcal{F}$ are written "clockwise" and $\mathcal{V}$ "anti-clockwise"). This defines the structure of a Dieudonné module on $B_w := \oplus_i \ \mathbb{F}_p{\cdot}z_i$ over $\mathbb{F}_p[\mathcal{F},\mathcal{V}]$.

Structure theorem (see [27], Section 5): *Let $k \supset \mathbb{F}_p$ be an algebraically closed field. Let N be a BT_1 over k. Then there exists a finite set of mutually different simple words w_i, and integers $n_i \in \mathbb{Z}_{>0}$ such that*

$$N \ \cong \ \bigoplus_i \ (\ Z_{w_i})^{n_i}.$$

This will be called *a decomposition into isotypic summands*; note that the decomposition as in right hand side above is uniquely determined by N, but the isomorphism is far from unique.

Moreover: *If w and w' are different simple words, a homomorphism $N_w \to N_{w'}$ has non-zero kernel and non-zero cokernel.*

From a word w we construct its dual w^D by replacing $\mathcal{F}$ in w by $\mathcal{V}$ in w^D and by replacing $\mathcal{V}$ by $\mathcal{F}$; a word w is called *symmetric* if $w = w^D$ (equality in the cyclic way). Note that $(N_w)^D = N_{w^D}$ (a canonical isomorphism over $\mathbb{F}_p$).

(9.9). **Second step: Lemma.** *Let k be an algebraically closed field, an let $(N, <,>)$ a a polarized* BT_1 *over k; i.e., N is a* BT_1 *truncated group scheme, with Dieudonné module $A = \mathbb{D}(N)$, and $<,>$ is a non-degenerate, bilinear, alternating form on A satisfying $< x, \mathcal{F}(z) >=< \mathcal{V}(x), z >^{\sigma}$. Then there exist symmetric simple symmetric words w_i, integers n_i, and pairs of non-symmetric words v_j, v_j^D and integers m_j and pairings such that*

$$(N,<,>) \ \cong \ \bigoplus_i \ (\ Z_{w_i}, <,>_i)^{n_i} \ \oplus \ \bigoplus_j \ \left(Z_{v_j} \oplus Z_{v_j^D}, <,>_j \right)^{m_i}.$$

Proof. Suppose we write $N = M \oplus P$ in such a way that M and P are sums of isotypic summands of N and such a that for a non-symmetric word v_j the summands Z_{v_j} and $Z_{v_j^D}$ both appear either in M or in P. The pairing gives a morphism

$$\begin{pmatrix} \beta & \gamma \\ \delta & \varepsilon \end{pmatrix} = \zeta : \ \ M \oplus P \longrightarrow (M \oplus P)^D.$$

Because $<,>$ was supposed to be non-degenerate, ζ is an isomorphism; because of the assumptions on the decomposition $N = M \oplus P$ we see that β is an isomorphism (and ε as well). We define $\varphi := \beta^{-1}{\cdot}(-\gamma)$; we define $P' = (1 + \varphi)P \subset N$; we see that on $N = M \oplus P'$ the form is in diagonal shape. As $<,>$ is alternating on $\mathbb{D}(M)$, it is alternating on the summands $\mathbb{D}(M)$ and $\mathbb{D}(P')$. Induction on the number of isotypic summands in N finishes the proof. ☐

Remark. The lemma just proved is strictly speaking not necessary in the proof below, but it simplifies considerably the notation in the last steps of our proof.

(9.10). In order to apply the structure theorem for BT_1's to "standard types" we are going to change notation a little bit. The group schemes Z_w appearing in the classification by Kraft we are going to replace by group schemes C_w, more suited to our purpose (and over an algebraically closed field the same up to isomorphism). Let w be a (simple, cyclic) word as above. We define a BT_1 E_w and its Dieudonné module $C_w = \mathbb{D}(E_w)$ almost analogous as before: if $w = L_1 \cdots L_d$ we choose $z_1, \cdots, z_d$ plus the convention $z_{d+1} = z_1$; we write $\{z_1, \cdots, z_d\} = \{x_1, \cdots, x_b, y_1, \cdots, y_c\}$ with the convention that "the y's are images under $\mathcal{V}$", i.e., if $L_j = \mathcal{V}$, then z_j will be baptized y, and otherwise x; we define $C_w := \oplus_i \ \mathbb{F}_p \cdot z_i$. If $L_i = \mathcal{F}$ we write $\mathcal{F}(z_i) = z_{i+1}$ and $\mathcal{V}(z_{i+1}) = 0$; if $L_i = \mathcal{V}$ we write $\mathcal{V}(z_{i+1}) = \pm z_i$ and $\mathcal{F}(z_i) = 0$; in fact we write $\mathcal{V}(z_{i+1}) = +z_i$ if $z_{i+1} \in \{y_1, \cdots, y_c\}$, and $\mathcal{V}(z_{i+1}) = -z_i$ if $z_{i+1} \in \{x_1, \cdots, x_b\}$ (in short: $\mathcal{F} : x \mapsto x, \mathcal{F} : x \mapsto y, \mathcal{V} : y \mapsto +y, \mathcal{V} : x \mapsto -y$). This constructs the structure of a Dieudonné module on C_w over $\mathbb{F}_p[\mathcal{F}, \mathcal{V}]$; hence we have constructed a BT_1 E_w.

For a word w we define $a(w)$ to be the number of pairs $\mathcal{FV}$ in the word.

Third step: Lemma. *Let k be an algebraically closed field, let w be a word. Then $Z_w \otimes k \cong E_w \otimes k$.*

Proof. Suppose $\mathbb{D}(Z_w) = \oplus_i \mathbb{F}_p \cdot z'_i$ and $\mathbb{D}(E_w) = C_w = \oplus_i \mathbb{F}_p \cdot z_i$. Let $|w| = d$ and $a(w) = a$. Consider $z'_i \mapsto z'_{i+1}$ by $\mathcal{F}$, if $L_i = \mathcal{F}$ or by $\mathcal{V}^{-1}$ if $L_i = \mathcal{V}$; composition of these maps gives $z'_1 \mapsto z'_1$, and $b \cdot z'_1 \mapsto b^{\sigma^d} \cdot z'_1$ for every $b \in k$. The same process yields $z_1 \mapsto (-1)^a \cdot z'_1$ in C_w. Choose $b \in k$ with $b^{\sigma^d} = (-1)^a$. We define a k-linear map

$$\mathbb{D}(C_w) \otimes k \to \mathbb{D}(Z_w) \otimes k \quad \text{by} \quad z'_i \mapsto \beta_i \cdot z_i,$$

where $\beta_1 = b$ and, inductively, $\beta_{i+1} = \beta_i^\sigma$ if $L_i = \mathcal{F}$, and $\mathcal{V}(\beta_{i+1} \cdot z_{i+1}) = \beta_i \cdot z_i$ if $L_i = \mathcal{V}$. This yields $\beta_{d+1} = b^{\sigma^d} \cdot (-1)^a$. Hence the k-linear maps commute with the actions of $\mathcal{F}$ and $\mathcal{V}$, hence we have constructed a Dieudonné module isomorphism $\mathbb{D}(C_w) \otimes k \to \mathbb{D}(Z_w) \otimes k$. This proves the lemma. □

Using the structure theorem, and the two lemmas just proved, in order to prove (9.4) it suffices to show this for isotypic summands:

(9.11). *Suppose k is an algebraically closed field of characteristic p. Let $(N, [,])$ be a polarized BT_1 of rank p^{2g} defined over k. Suppose either $N = ((E_v \oplus E_{v^D}) \otimes k)^n$, where v is a simple, non-symmetric word, and $n \in \mathbb{Z}_{>0}$ or $N = (E_w \otimes k)^n$, where w is a simple symmetric word and $n \in \mathbb{Z}_{>0}$. Suppose $\varphi = \mathrm{ES}(N)$. Then*

$$(N, [,]) \quad \cong \quad (N_\varphi, <,>_{\mathrm{st}}) \otimes_{\mathbb{F}_p} k.$$

For the rest of the section we keep notations as in (9.11).

We write $A = \mathbb{D}(N)$. A k-base for a Dieudonné module over $k[\mathcal{F}, \mathcal{V}]$ will be called a $(\mathcal{F}, \mathcal{V})$-base if it equals $\{z_1, \cdots, z_d\} = \{x_1, \cdots, x_b, y_1, \cdots, y_c\}$ with the conventions as in the definition of C_w, in particular "$\mathcal{V} : x \mapsto -y$"; in the situation to be studied we will have $d = 2m$ and $b = m = c$. By the structure theorem and by the lemma in (9.10) we have a $(\mathcal{F}, \mathcal{V})$-base for A. On A we have the form $[,]$. We say that that a base for A is in standard form (with respect to $[,]$) if it is a $(\mathcal{F}, \mathcal{V})$-base, and if the form $[,]$ is the standard symplectic form on that base. We study all possible $(\mathcal{F}, \mathcal{V})$-bases for A and we show there is at least one which is a standard base for $[,]$.

We will write $z_i < z_j$ iff $i < j$, i.e., the base vectors in z_i are in a strictly lower piece of the canonical stratification

(9.12). Suppose we are in case $N = ((E_v \oplus E_{v^D}) \otimes k)^n$ as in (9.11). Then $a(N)$ is *even*, and we refer to this case as the even case. We shall write $|v|+|v^D| = 2m$

Suppose w is a simple symmetric word; then $a(w)$ is odd; this can be seen as follows; choose a permutation in such a way that w starts with $\mathcal{F}$ and ends with $\mathcal{V}$. From the fact that w is simple and symmetric it follows that $|w|$ is even, say $= 2m$, and that $L_i \neq L_{i+m}$; hence the first m letters are in the form $\mathcal{F} \cdots \mathcal{F}$ and the last m are $\mathcal{V} \cdots \mathcal{V}$; we see that the number of times that $\mathcal{F}\mathcal{V}$ appears in w is odd. The case $N = (E_w \otimes k)^n$ as in (9.11) is called the odd case; here we have $a(N) = a(w) \cdot n$.

On A we have the canonical filtration

$$0 \subset \cdots \subset A^{(j)} \subset \cdots \subset A.$$

We call $\oplus_j A^{(j)}/A^{(j-1)}$ the associated graded of A. Note that the canonical filtration has exactly $|v| + |v^D|$, respectively $|w|$ steps. Note that the vectors in z_i are in $A^{(j)}$ iff $i \leq j$. The notions of even and odd just explained are the same as the ones used in Section 2.

(9.13). **Fourth step: Lemma.** *We use the notations introduced above, in particular as in* (9.11). *There exists a base* $\{z_1, \cdots, z_{2g}\} = \{x_1, \cdots, x_g, y_1, \cdots, y_g\}$ *which is a* $(\mathcal{F}, \mathcal{V})$*-base for* A, *which defines a final filtration, and which is in standard form on the associated graded module for the given pairing* $[,]$.

Proof. We start with a $(\mathcal{F}, \mathcal{V})$-base for A as given in the structure theorem in (9.7) plus the lemma in (9.10). The linear substitutions on P as in Proposition (9.6) can be carried over to the whole $(\mathcal{F}, \mathcal{V})$-base, and we see that we can derive a new $(\mathcal{F}, \mathcal{V})$-base for A which is in standard form on the associated graded module. □

From now on we are allowing only substitutions (choices of another base), which leave the residue classes of the base vectors in the associated graded invariant; i.e., a new base vector Z_i' will be of the form $Z_i+?$ with $? \in A^{(j-1)}$ if $Z_i \in A^{(j)}$ and $Z_i \notin A^{(j-1)}$.

(9.14). We study the canonical filtration of N as in (9.11). We denote by z_j, x_i, y_i the steps in this filtration, as well as the ordered sets of n base vectors in this step of the filtration (the ones which give non-zero residue classes in that step of the associated graded). As we work with $(\mathcal{F}, \mathcal{V})$-bases, it is clear what is meant by a notation like $\mathcal{F}(x_i) = z_i$: it maps the ordered set of base vectors encoded by x_i to the analogous ordered set of base vectors given by z_i; by $\mathcal{V}(x_i) = y_j$ we mean combinatorially that $\mathcal{V}$ is bijective on these steps in the canonical filtration of A, and that for base vectors $x_i = \{x_i^{(1)}, \cdots, x_i^{(n)}\}$, analogously for y_j, we have that $\mathcal{V}(x_i^{(t)}) = -y_j^{(t)}$. When we write $[z_i, z_j]$ we mean a $n \times n$-matrix given by the pairings on the pairs of vectors in these sets.

(9.15). Sets of base vectors as assembled in the canonical filtration on A in consideration will be distributed in 4 subsets. We write

$$\mathrm{TX} = \{x_i \mid x_i \notin \mathcal{F}A\}$$

(and remember, no base vector in x_i is $\mathcal{V}A$), and

$$\mathrm{BX} = \{x_i \mid x_i \in \mathcal{F}A\}, \quad \mathrm{X} = \mathrm{TX} \cup \mathrm{BX},$$

(where T stand for Top and B stands for Bottom). Further

$$\mathrm{UY} = \{y_i \mid y_i \notin \mathcal{F}A\} \quad \text{and} \quad \mathrm{LY} = \{y_i \mid y_i \in \mathcal{F}A\}, \quad \mathrm{Y} = \mathrm{UY} \cup \mathrm{LY}$$

(Upper and Lower). Note that for $a(w) = a$, respectively $a(v) + a(v^D) = a$ we have $\mathrm{TX} = \{x_1, \cdots, x_{g-a+1}\}$ and $\mathrm{LY} = \{y_a, \cdots, y_1\}$. Note that a Top-X-vector is not an image of any base vector, and a Lower-Y-vector is an image under $\mathcal{F}$ and under $\mathcal{V}$. All other base vectors are an image either under $\mathcal{F}$ (when in BX), or under $\mathcal{V}$ (when in UY). Every base vector is an image under a power of $\mathcal{F}$ or a power of $\mathcal{V}$ of a $\pm$ Top-X-vector.

For every index j, with $1 \leq j \leq a$ we have uniquely determined indices $\xi(j), \eta(j)$, with $1 \leq \xi(j), \quad \eta(j) \leq g - a + 1$ (they correspond to Top-X-coordinates) and positive integers $f(j), v(j)$ such that for base vectors we have:

$$\mathcal{F}^{f(j)} : x_{\xi(j)} \longmapsto y_j, \qquad \mathcal{V}^{v(j)} : x_{\eta(j)} \longmapsto -y_j, \qquad y_j \in \mathrm{LY} \qquad (*_j).$$

Note that for every $i \leq g - a + 1$ there exist j and j' such that $\xi(j) = i$ and $\eta(j') = i$.

Claim. *A base $\{z_1, \cdots, z_{2g}\} = \{x_1, \cdots, x_g, y_1, \cdots, y_g\}$ is a $(\mathcal{F}, \mathcal{V})$-base for A iff and only if: the Top-X-base vectors satisfy $(*_j)$ for all $1 \leq j \leq a$ and all base vectors are obtained as a $\mathcal{F}^?$-image or a $\mathcal{V}^?$-image of a Top-X-base vector.*

The proof of this claim follows directly from the definition of a $(\mathcal{F}, \mathcal{V})$-base. Again, note that z_j stands for an ordered sequence of n elements; equations should be understood in that sense.

(9.16). We study all possible alternating pairings on A, and we are interested to have a base on which the pairings $[z_i, z_j]$ are in standard form. We are considering the following combinatorics. Consider the set $\Pi = (\{1, \cdots, 2m\})^{(2)} = (\{1, \cdots, 2m\} \times \{1, \cdots, 2m\})/(\mathbb{Z}/2)$ of unordered pairs (z_i, z_j). We will define $\Pi_0 \subset \Pi$, the subset of "zero-classes", and an equivalence relation on Π respecting $\Pi_0 \subset \Pi$, and we consider "1-elements". To this end we write: $(p, q = \mathcal{V}(r)) \prec (\mathcal{F}(p), r)$ if $\mathcal{F}p \neq 0 \neq \mathcal{V}r$, and $\sim$ is the equivalence relation generated by the steps in this partial ordering. We write $(p, q = \mathcal{V}(r)) \in \Pi_0$ if $\mathcal{F}(p) = 0$; we write $(\mathcal{F}(p), r) \in \Pi_0$ if $\mathcal{V}(r) = 0$; we write $(z_i, z_j) \in \Pi_0$ if $i + j \leq 2m$; we denote by Π_0 the set of all classes equivalent with any of the previously mentioned zero classes; in particular we see that $(\mathcal{V}(p), \mathcal{V}(r)) \in \Pi_0$ and $(\mathcal{F}(p), \mathcal{F}(r)) \in \Pi_0$; note that $A^{(i)}$ and $A^{(j)}$ pair to zero if $i + j \leq 2m$, hence:

Observation. *If* $[,]$ *is an alternating pairing on* A*, and* $(z_i, z_j) \in \Pi_0$ *then* $[z_i, z_j] = 0$. This follows from $[p, q = \mathcal{V}(r)]^\sigma = [\mathcal{F}(p), r]$.

Observation. *Suppose we have non-zero classes* $(z_i, z_j) \prec (p, q)$*. Then either* $z_i \in \mathrm{X}$ *and* $z_j \in \mathrm{Y}$ *or conversely.* Indeed, we can only "move on" if at least one of the elements is in the image of $\mathcal{V}$, and as the element is non-zero, the other is not in the image of $\mathcal{V}$.

An element of the form (x_j, y_j) is called a 1-element.

Observation. *If* $[,]$ *is a non-degenerate alternating pairing which is in standard form on the associated graded, then* $[x_j, y_j] = 1$ (i.e., the diagonal matrix of size $n \times n$) $\forall j$. □

Basic observation. *If in* Π *we have equivalent pairs* $(p, q) \sim (r, s)$ *then* $[p, q] = 0 \Leftrightarrow [r, s] = 0$.

Claim. (i) *All 1-elements form one equivalence class.*
(ii) *Let* $\{\pi_1 \prec \cdots \prec \pi_e\} \subset \Pi$ *be a longest chain of non-zero, non-1 equivalent elements,* $\pi_j = (z'_j, z''_j)$*; then* z'_1 *does not occur in* $\pi_2, \cdots, \pi_e$ (i.e., there are no "cyclic" equivalence classes in the case of non-zero and non-1-classes).
(iii) *Let* $\{\pi_1 \prec \cdots \prec \pi_e\} \subset \Pi$ *be a longest chain of non-zero and non-1 equivalent elements. Then* $\pi_1 = (x_i, z_j)$ *with* $x_i \in \mathrm{TX}$ *and* $z_j \in \mathrm{TX} \cup \mathrm{UY}$.
(iv) *Let* $\delta + 1 = \#((\Pi - \Pi_0)/\sim)$ *be the number of equivalence classes of non-zero elements. Consider* A *with a* $(\mathcal{F}, \mathcal{V})$*-base for* A*. The set* $\mathcal{Q}$ *of alternating non-degenerate pairings on* A *which are in standard form on the associated graded equals* $k^{\delta \cdot n^2} \cong \mathcal{Q}$ (and the bijection will be indicated in the proof).

Proof. (i) In the situation (9.11) all 1-elements form a single equivalence class: consider separately the symmetric word w or the pair of words v and v^D. □

(ii) Suppose there would exist a cyclic equivalence class which consists of non-zero and non-1 elements. In case of w there would be an integer q, say $0 < q < 2m$, such that $L_i \neq L_{i+q}$ for all i. As w is simple this can only happen if $q = m$. The case v and v^D follows in the same way (because v is simple and non-symmetric). □

(iii) There is an element "before" (p, q) if one of the p, q is an image under $\mathcal{F}$. So the left end (p, q) of a non-zero, non-1 equivalence class has elements not in the image of $\mathcal{F}$, not both in the image of $\mathcal{V}$. □

(iv) Let $(x^{(1)}, z^{(1)}), \cdots, (x^{(\delta)}, z^{(\delta)})$ be the left-ends of the non-zero and non-1 equivalence classes. Any $[,]$ as in (iv) defines $n \times n$-matrices $M_1, \cdots, M_\delta \in k^{n^2}$ by $[x^{(j)}, z^{(j)}] =: M_j$; for this choice we take once and for all an ordering for each pair $(x^{(j)}, z^{(j)})$. Conversely, given such matrices, we define $[x_i, y_i] = 1$, and

$[x^{(j)}, z^{(j)}] = M_j$; this defines all pairings between base-vectors, and we check that this gives a $[,]$ as in (iv). This finishes the proof of the claim. □

We write $\Pi^0 := \Pi - \Pi_0 - \Pi_1$ for the set of non-zero and non-1-elements. Conclusion: The evaluation map $\mathcal{Q} \to k^{\delta \cdot n^2}$, given by computing $[,]$ on left ends of equivalence classes in Π^0, is bijective. We will consider the set $\mathcal{Q}$ as $\mathbb{A}^{\delta \cdot n^2}(k)$ and treat it as a variety over k.

(9.17). We consider the group $\mathcal{S}$ of all k-linear transformations of A which are Dieudonné module isomorphisms, and which map a given $(\mathcal{F}, \mathcal{V})$-base to a $(\mathcal{F}, \mathcal{V})$-base. It acts on the set $\mathcal{Q}$ of non-degenerate alternating forms on A (by conjugation): $\mathcal{S} \times \mathcal{Q} \to \mathcal{Q}$; this will be considered as the action of a linear group over k onto a variety over k (identifying a linear group over k with the group of its k-rational points, etc.). We are going to prove that the orbit of every element of $\mathcal{Q}$ under $\mathcal{S}$ is the whole of $\mathcal{Q}$ (and hence we can change $[,]$ into $<,>_{\mathrm{st}}$). We see that in general the stabilizer of an element of $\mathcal{Q}$ is non-trivial. We give the proof by considering an algebraic subvariety $L \subset \mathcal{S}$ containing the identity of this algebraic group, and showing that the differential of the action parameterized by L maps surjectively onto the tangent space at $[,] \in \mathcal{Q}$.
Remark: The dimension of $\mathcal{S}$ in general is much larger that the dimension of $\mathcal{Q}$, but that information is not of much help, because in general the stabilizer in $\mathcal{S}$ of an element of $\mathcal{Q}$ is non-trivial.

(9.18). *Description of $\mathcal{S}$ and of L.* We introduce "variables" $\beta_{s,j}$ and $\gamma_{s,t}$, each is a $n \times n$-matrix (indices will be specified below). Given A as above with a $(\mathcal{F}, \mathcal{V})$-base we write:

$$x'_s = x_s + \sum_{j<s} \beta_{s,j} \cdot x_j + \sum_{y_t < x_s} \gamma_{s,t} \cdot y_t, \qquad \forall x_s \in \mathrm{TX} \qquad (**);$$

these are subject to the relations $(*_j)$ as explained in (9.15); we write $(*)$ for the set $(*_j)$ $\forall j$. We define vectors z'_i by: taking all images under powers of $\mathcal{F}$ and powers of $\mathcal{V}$, and deleting the ones which are zero. The new base vectors in the sequel will be denoted by $\{\rho(z_i)\}$ once all $\beta_{s,j}$ and $\gamma_{s,t}$ satisfying $(*)$ (and perhaps several conditions).

Claim. *The substitutions given by $(**)$ subject to $(*)$ form a group which are all substitution which are the identity on the associated graded, and which map the given $(\mathcal{F}, \mathcal{V})$-base for A to a $(\mathcal{F}, \mathcal{V})$-base. This defines a group scheme.* □
Remark. In fact, this group scheme in general is non-reduced. This will play no role later in our considerations. We are going to define a subvariety $L \subset \mathcal{S}$ by imposing extra conditions on the variables.

(9.19). For a set z_j of base vectors we define $d(z_j) = z_{2m+1-j}$; note that $d(x_i) = y_i$ and $d(y_i) = x_i$. We consider the set Λ of ordered pairs $(z_i \& z_j)$, such that $z_j < z_i$:

$$\Lambda = \{(z_i \& z_j) \in (\{1, \cdots, 2m\})^2 \mid z_j < z_i\}.$$

We write $\Lambda_1 = \{(z\&d(z))\}$, and we call elements of this set "1-elements". We define

$$D : \Lambda \to \Pi \quad \text{by} \quad D(z_i \& z_j) := (z_i, d(z_j)).$$

We write $\Lambda_0 := D^{-1}(\Pi_0)$.

Claim. *The map $\Lambda - \Lambda_1 \to \Pi - \Pi_1$ is 2-1-surjective.*

In fact, (z', z'') is the image of $(z', d(z''))$ if $d(z'') < z'$ respectively of $(z'', d(z'))$ if $d(z') < z''$; hence $D : \Lambda \to \Pi$ is surjective. On $\Lambda - \Lambda_1$ the map $(z'\&z'') \mapsto (d(z'')\&d(z'))$ is fixed point free with quotient $\Pi - \Pi_1$. □

On Λ we define an equivalence relation, indicated by $\approx$; we write $(z'\&z'') \prec (\mathcal{F}z'\&\mathcal{F}z'')$ if $\mathcal{F}z' \neq 0 \neq \mathcal{F}z''$; we write $(\mathcal{V}z'\&\mathcal{V}z'') \prec (z'\&z'')$ if $\mathcal{V}z' \neq 0 \neq \mathcal{V}z''$; a partial ordering, and the equivalence relation given by this is the one generated by iterating these definitions.

Claim. *We have:*

$$(z'\&z'') \prec (u'\&u'') \quad \Longleftrightarrow \quad D(z'\&z'') \prec D(u'\&u'').$$

In fact, $d(z'') = \pm\mathcal{V}(d(\mathcal{F}z''))$, hence $(z'\&z'') \prec (\mathcal{F}z'\&\mathcal{F}z'')$ implies that $(z', \mathcal{V}(d(\mathcal{F}z'') \prec (\mathcal{F}z', d(\mathcal{F}z''))$, and conversely. An analogous statement with $(\mathcal{V}z'\&\mathcal{V}z'') \prec (z'\&z'')$ proves the claim. □

We number the $\sim$-equivalence classes $E_1, \cdots, E_\delta \subset \Pi - \Pi_0 - \Pi_1$. We define variables $T_1, \cdots, T_\delta$, each is a $n \times n$-matrix of variables, and we are going to define a morphism

$$\mathrm{Spec}(k[T_1, \cdots, T_\delta]) \to \mathcal{S};$$

the image of this maps will be a closed, reduced subvariety $L \subset \mathcal{S}$; we study the action

$$L \times \mathcal{Q} \to \mathcal{Q}.$$

Note that for every $\sim$-equivalence class $E \subset \Pi^0$ there are exactly two disjoint $\approx$-equivalence classes $E', E'' \subset \Lambda^0$ such that

$$D : E' \xrightarrow{\sim} E, \quad D : E'' \xrightarrow{\sim} E.$$

We consider an $\sim$-equivalence class $E \subset \Pi - \Pi_0$. There are three possibilities:

- **(1):** The class contains (x_i, y_i) for some index i. In this case there is only one $\sim$-equivalence class, it contains all (x_j, y_j). This is what we called the 1-equivalence-class.
- **(2):** The left end of E equals (x_s, x_t) with $x_s, x_t \in$ TX. In this case E consists of one element: $E = \{(x_s, x_t)\}$.
- **(3):** The left end of E equals (x_s, z) with $z \in$ UY.

For *zero classes* we define:

$$D(x_s\&z) \in \Pi_0 \Longrightarrow \beta_{s,j} = 0, \text{ if } z = x_j, \quad \text{respectively } \gamma_{s,t} = 0, \text{ if } z = y_t \quad \text{(L0)}.$$

Suppose we are in case **(2)**. Then $D(x_s\&y_t) = (x_s, x_t) = D(x_t\&y_s)$. Assume that $t < s$ (otherwise interchange); then $y_s < y_t < x_s$. We write:

$$x_s \in \mathrm{TX}, \quad y_t < y_s \quad \Longrightarrow \quad \gamma_{t,s} = 0 \qquad \text{(L1)}.$$

This condition will remove certain ambiguities. If $E_i = E = \{(x_s, x_t)\} = \{x_{s(i)}, x_{t(i)}\}$ with $x_s, x_t \in \mathrm{TX}$ and $x_t < x_s$, hence $y_s < y_t < x_s$, we define:

$$T_i =: \gamma_{s(i),t(i)} = \gamma_{s,t} \qquad \text{(L2)}.$$

Suppose we are in case **(3)**. We write $(x, z) \in E \subset \Pi^0$ for its left end; hence $x \in \mathrm{TX}$, and $z \in \mathrm{UY}$. Note that $d(z) < x$ and $d(x) < z$ hence $D(x\&d(z)) = (x, z) = D(z\&d(x))$. We choose the notation for E', E'' mapping onto E in such a way that $(x\&d(z) \in E'$ and $(z\&d(x)) \in E''$.

If $E = E_i$ with left end $(x_{s(i)}, z_{u(i)})$ we define

$$\beta_{s(i),j} = T_i, \text{ if } d(z_{u(i)}) = x_j \quad \text{respectively} \quad \gamma_{s,t} = T_i, \text{ if } d(z_{u(i)}) = y_t \qquad \text{(L3)};$$

we follow the ordered set E'_i, and we define $\beta_{s,j} = \pm T_i^{p^c}$ respectively $\gamma_{s,t} = \pm T_i^{p^c}$ with the appropriate sign, and with the power of p according to the number of steps c in E'_i that is $(x_s\&x_j) \in E'_i$ respectively $(x_s\&y_t) \in E'_i$ from the left end side of E'_i.

All elements appearing in E'' give rise to variables which we equate to zero:

$$(x_s\&x_j) \in E'' \Longrightarrow \beta_{s,j} = 0, \qquad (x_s\&y_t) \in E'' \Longrightarrow \gamma_{s,t} = 0 \qquad \text{(L4)}.$$

The relations $(*)$, (L0), (L1), (L2), (L3) and (L4) define the value of all $\beta_{s,j}$ and $\gamma_{s,t}$ introduced in (9.18); under these conditions we have defined

$$\mathrm{Spec}(k[T_1, \cdots, T_\delta) \xrightarrow{\sim} L \subset \mathcal{S}.$$

We see that indeed this defines a closed subvariety of $\mathcal{S}$.

(9.20). **Fifth step: Proposition.** *Equating all variables T_i to zero gives $1 \in L \subset \mathcal{S}$, the identity transformation. Choose any $[,] \in \mathcal{Q}$ as in (9.11). The action $L \times \mathcal{Q} \to \mathcal{Q}$ at $[,]$ gives a morphism $L \to L\cdot[,]$. The differential of the action $L \times \mathcal{Q} \to \mathcal{Q}$ at $[,]$ is a surjective linear map $\mathfrak{t}_{L,1} \to \mathfrak{t}_{\mathcal{Q},[,]}$. It follows that the orbit of $[,]$ under the action $\mathcal{S} \times \mathcal{Q} \to \mathcal{Q}$ equals $\mathcal{S}\cdot[,] = \mathcal{Q}$.*

Proof. We write $I = (T_i \mid i) \subset k[T_1, \cdots, T_\delta] =: k[T]$ for the ideal generated by these variables. We compute the action of $\mathrm{Spec}(k[T]/I^2)$ on $\mathcal{Q}$. Let us write $\tau_i = T_i \bmod I^2$. We see that all elements $\beta_{s,t}$ and $\gamma_{s,t}$ not defined by a left end are in I^2.

Claim. Under the action of $\mathrm{Spec}(k[T]/I^2)$ on $\mathcal{Q}$ the matrices $M_j = [z'_j, z''_j]$ are transformed into $M_j + \pm\tau_j$; in fact, let us write $\{\rho(z_i)\}$ as in (9.15); then we compute $\rho\cdot M_j = [\rho(z'_j), \rho(z''_j)]$; in case **(2)**, using (L0) and (L2) and using $I^2 = 0$ this gives:

$$[\rho(x_s), \rho(x_t)] = [x_s + T_i\cdot y_t + \sum_{j<s} \beta_{s,j}\cdot x_j \ + \sum_{y_u<x_s, u\neq t} \gamma_{s,u}\cdot y_u, \quad x_t + 0\cdot y_s + \cdots]$$

$$= [x_s, x_t] + \pm\tau_i;$$

in case **(3)**, if $(x,z) \in E_i$ is the left end, using (L3) and (L4) and $I^2 = 0$ in case $(x,z) = (x_s, y_j)$ we obtain:

$$[\rho(x), \rho(z)] = [x_s + T_i{\cdot}x_j + \sum_{u<s, u\neq j} \beta_{s,u}{\cdot}x_u + \sum_{y_t<x_s} \gamma_{s,t}{\cdot}y_t, \quad y_j + 0{\cdot}y_s + \cdots],$$

respectively if $(x,z) = (x_s, x_t)$ we obtain:

$$= [x_s + T_i{\cdot}y_t + \cdots, \quad x_t + 0{\cdot}y_s + \cdots].$$

This proves the claim.

Hence it proves that $\mathfrak{t}_{L,1} \to \mathfrak{t}_{\mathcal{Q},[,]}$ is surjective. Hence for every $[,] \in \mathcal{Q}$ the set $L{\cdot}[,] \subset \mathcal{Q}$ is dense. We know that $\mathcal{Q}$ is irreducible. This proves that $\mathcal{S}$ acts transitively on $\mathcal{Q}$. □

In the basic observation in (9.16) we see that pairings between base vectors (non dual ones) are zero iff these parings are zero on all left ends of equivalence classes. The previous proposition shows that, starting from (9.11) there is a transformation $\{x_s \mapsto x'_s \mid x_s \in \mathrm{TX}\} = \rho \in \mathcal{S}(k)$ such that $\rho{\cdot}[,] = [\rho(), \rho()]$ is zero on left ends of non-1-$\sim$-equivalence classes. Hence $\rho{\cdot}[,] = <,>_{\mathrm{st}}$. This proves the claim as in (9.11). By what has been said and proven in the first three steps, this proves Theorem (9.4). □

(9.21). **Remark.** By [59], Proposition 2.4.14 we know that every orbit of a unipotent group $\mathcal{S}_{\mathrm{red}}$ acting on an affine variety $\mathcal{Q}$ is closed; hence we see that in the proof of (9.20) it suffices to compute the tangent action of L at $<,>_{\mathrm{st}}$ in order to prove that every $[,]$ is in the orbit $\mathcal{S}{\cdot} <,>_{\mathrm{st}}$.

(9.22). **Remark.** Suppose a final filtration on a polarized BT_1 truncated group scheme is given. We note there are in general many choices for a standard base matching this filtration. here is one example. Choose indices i and j with $n_j < m_i$; then $n_i = 2g+1-m_i < 2g+1-n_j = m_j$; we choose some β in the base field k, we change X_i to $X'_i = X_i + \beta{\cdot}Y_j$ and X_j to $X'_j = X_j + \beta{\cdot}Y_i$, and we leave all other base vectors the same. Clearly the new set is a standard base matching the same filtration. Examples of a different nature are easy to give, as we have seen in the proof above: in general there are many isomorphisms on A_φ which by conjugation do not change $<,>_{\mathrm{st}}$.

10. Moving in a stratum

In this section we fix an elementary sequence $\varphi \in \Phi$, and we prove:

(10.1). **Proposition.** *Every component of S_φ has dimension at least $\mid \varphi \mid$.* Later we will show in fact that:

$$\dim(S_\varphi) = \mid \varphi \mid.$$

Let $x \in S_\varphi(k)$, where k is an algebraically closed field. We are going to study deformation theory around $[(X_0,\lambda_0)] = x_0 \in \mathcal{A}$, especially a part which "stays inside S_φ". We use notations introduced earlier, especially we use a standard base for $\mathbb{D}((X_0,\lambda_0)[p])$ as described in Section 9. We write ψ for the final sequence associated with φ, and we suppose we have chosen, see (9.6), a final filtration

$$A_0 \subset A_1 \subset \cdots \subset A_g = \mathcal{F}(A) \subset \cdots \subset A_{2g} = A = \mathbb{D}(N_0).$$

The standard base $\{Z_1,\cdots,Z_g\} = \{X_1,\cdots,X_g,Y_g,\cdots,Y_1\}$ for A is lifted to a symplectic base $\{X'_1,\cdots,X'_g,Y'_g,\cdots,Y'_1\}$ for $M = \mathbb{D}(X_0[p^\infty])$ such that $Y'_i \in \mathcal{F}(M)$.

We use the integers $\{n_g,\cdots,n_1\}$ and $\{m_1,\cdots,m_g\}$ as introduced in Section 9. We write

$$I_\varphi = I = \{n_g,\cdots,n_{g-a-1}\} \quad \text{and} \quad J_\varphi = J = \{m_1,\cdots,m_{g-a}\};$$

this corresponds with the Y_j, respectively X_i, contained in $A_g = \mathcal{F}(A)$, where $a = a(X_0) = a(\varphi) = g - \varphi(g)$. We introduce variables $t_{u,v}$, some of which are put to zero, some of which are considered as variables, and in this way we are going to construct a ring R_φ. To this end we take subset $D_\varphi \subset \{1,\cdots,g\}\times\{1,\cdots,g\}$ defined as follows:

(J) it consists of all (u,v) such that $u \geq v > a$ and

(I) it consists of all (u,v) such that $1 \leq v \leq a$ and $m_{g-u+1} \leq n_{g-v+1}$.

(10.2). **Observation.**

$$\#(D_\varphi) = \mid \varphi \mid.$$

Indeed, $\sum_{j\in J} \varphi(j)$ equals the number of pairs in (J) above, and $\sum_{i\in I} \varphi(i)$ equals the number of pairs in (I) above. □

We define

$$R_\varphi = k[[t_{u,v} \mid (u,v) \in D \ \text{ or } \ (v,u) \in D]]/(t_{u,v} - t_{v,u}).$$

Note that if (u,v) in (I) above then it is not in (J), and (v,u) is not in (J). Note that the Krull dimension of this ring equals $\mid \varphi \mid$. We write $T_{(u,v)} \in W_\infty(R_\varphi)$ for the Teichmüller lift of $t_{u,v} \in R_\varphi$, and we write $T = (T_{u,v} \mid 1 \leq u, v \leq v)$ for the corresponding matrix.

The display of M on the symplectic W-base $\{X'_1,\cdots,X'_g,Y'_g,\cdots,Y'_1\}$ for M we denote by

$$\begin{pmatrix} A & B \\ C & D \end{pmatrix}$$

(with apologies that we use the symbol A with two different meanings; we expect there will be no confusion). With the matrix (T) defined above we obtain a display

$$\begin{pmatrix} A+TC & B+TD \\ C & D \end{pmatrix}.$$

This defines a quasi-polarized p-divisible group over $\mathrm{Spec}(R_\varphi)$ that deforms $(X_0[p^\infty], \lambda_0)$; hence by the theorem of Serre and Tate, see [30], [22], [32], we obtain a principally polarized abelian scheme $(X^{(T)}, \lambda) \to \mathrm{Spec}(R_\varphi)$; we write $[(X^{(T)}, \lambda)] = x \in \mathcal{A}(R_\varphi)$ for the corresponding point.

(10.3). **Claim.** *We have* $[(X^{(T)}, \lambda)] = x \in S_\varphi(R_\varphi)$.
In fact, if $\psi(m_i) = n_s$, then

$$\mathcal{F}^{(T)}(X_i') = \sum T_{j,g-s+1} \cdot X'_{g-j+1} \quad + Y_s'.$$

The variables coming from (J) above show up only multiplied by a factor p. These two fact show that this bases induces a filtration on $A^{(T)}$ on which the elementary sequence φ is realized, say over a perfect closure of the field of fractions of R_φ. This proves the claim. □

Hence we have constructed $x_0 \in \mathrm{Spec}(R_\varphi) \subset S_\varphi$. By the observation above this proves the proposition. □

11. Moving out of a stratum

In this section we consider elementary sequences φ and φ' such that

$$\varphi \prec \varphi' \quad \text{and} \mid \varphi \mid +1 = \mid \varphi' \mid .$$

This means there is an index b with $0 < b \leq g$ and

$$\varphi(i) = \varphi'(i) \quad \text{for} \quad i \neq b, \;\; 0 < i \leq g, \quad \text{and} \quad \varphi(b) + 1 = \varphi'(b).$$

Note that this is only possible when $\varphi(b-1) = \varphi(b) < \psi(b+1)$.

(11.1). **Proposition.** *Under these conditions*

$$S_{\varphi'} \quad \subset \quad (S_\varphi)^c.$$

Proof. It suffices to prove the case that $\varphi(1) = 0$, which we suppose from now on. As in the previous section we write: $[(X_0, \lambda_0)] = x_0 \in S_\varphi \subset \mathcal{A}$ and $\{Z_1, \cdots, Z_{2g}\} = \{X_1, \cdots, X_g, Y_g, \cdots, Y_1\}$, a standard basis for A is lifted to a symplectic base $\{X_1', \cdots, X_g', Y_g', \cdots, Y_1'\}$ for $M = \mathbb{D}(X_0[p^\infty])$ such that $Y_i' \in \mathcal{F}(M)$. Note that $\varphi(b-1) = \varphi(b) < \psi(b+1)$; hence Z_b is one of the base vectors $Y_?$, in fact $Z_b = Y_y$ with $y = g-b+\varphi(b)$; moreover Z_{b+1} is one of the base vectors $X_?$, in fact $Z_{b+1} = X_x$ with $x = \varphi(b+1)$. On the module A we have $\mathcal{F}(X_b) = Y_y$. We define $u = g + 1 - \varphi(b+1)$, and $v = b + 1 - \varphi(b)$; we write $t_{u,v} = q$, a variable over k, and $t_{v,u} = q$, and we put all other $t_{i,j} = 0$; the Teichmüller lift we denote by $Q \in W_\infty(k[[q]])$ (and all others equal to 0), thus obtaining a matrix T. From this we obtain a deformation $(X^{(T)}, \lambda) \to \mathrm{Spec}(k[[q]])$.
Claim. *For the generic fiber, we have* $\mathrm{ES}(X_\eta^{(T)}, \lambda) = \varphi'$.
Observe that $Y_y \in \mathcal{F}(A)$, hence $y \geq g - a + 1$, hence $v \leq a$. In the deformation we see that $T_{v,u}$ is multiplied by a factor p. We conclude that:
$\mathcal{F}^{(T)}(X_b) = Y_y + T \cdot X_x$, and

$\mathcal{F}^{(T)}(Z_i) = \mathcal{F}(Z_i)$ for all $i \neq m_b$, i.e., $Z_i \neq X_b$.
Let K be a perfect field containing $k[[q]]$, write $(X, \lambda) = (X^{(T)}, \lambda)$, and let us show that $\mathrm{ES}(X, \lambda) = \varphi'$.

In case $1 \leq b < g$ the old standard base is

$$\{Z_1, \cdots, Z_{b-1}, Y_y, X_x, Z_{b+2}, \cdots, Z_{g-b-2}, X_y, Y_x, Z_{g-b+1}, \cdots, Z_{2g}\}.$$

We choose a new base for the module on which the display is defined by choosing

$$\{Z_1, \cdots, Z_{b-1}, Y_y + T{\cdot}X_x, Y_y, Z_{b+2}, \cdots, Z_{g-b-2}, TX_y + Y_x, Y_x, Z_{g-b+1}, \cdots, Z_{2g}\}.$$

This ordered set defines a filtration; with the σ-linear map given by $\mathcal{F}^{(T)}(X_b) = Y_y + T{\cdot}X_x$, etc. as above, we obtain a final filtration that has elementary sequence φ'.

In case $b = g$, we have $x = g - a + 1 = y$; the old standard base is $\{Z_1, \cdots, Z_{b-1}, Y_y, X_x, Z_{b+2}, \cdots, Z_{2g}\}$, and on the new base given by the ordered set $\{Z_1, \cdots, Z_{b-1}, Y_y + T{\cdot}X_x, Y_y, Z_{b+2}, \cdots, Z_{2g}\}$ we obtain a final filtration with elementary sequence φ'. This proves the claim, and it finishes the proof of the proposition. □

(11.2). **Corollary.** *For every $\varphi \in \Phi$ we have:*

$$\dim(S_\varphi) \quad = \quad |\, \varphi \,| \, .$$

In fact, if $(g(g+1)/2) - |\, \varphi \,| = c$, there is a sequence

$$\varphi = \varphi_0 \quad \supsetneqq \quad \varphi_1 \quad \supsetneqq \quad \cdots \quad \supsetneqq \quad \varphi_c = \{1, \cdots, g\}.$$

By the previous corollary this shows that every component of S_φ has dimension at most $|\, \varphi \,|$. By the main result of the proceeding section the opposite inequality holds, and we are done. □

Remark. It seems strange that we obtain a regular deformation space starting at a point in the boundary of $S_{\varphi'}$, although we know that in general a stratum is "very singular" at its boundary. However note that we consider deformations respecting a final filtration. For example, if we start in a point of Σ, and and we want to move into the interior of L, there are many ways of doing that. But what we proved, implies that giving a final filtration for the closed fiber, this singles out a "direction of deformation" into L.

12. Transport of structure along the boundary of a stratum

In this section we prove that *the boundary of a stratum S_φ is a union of (lower dimensional) strata.* We use Lemma (12.4), and give the proof in Proposition (12.5). In (12.1) we explain some notation. The results (12.2) and (12.3) will not be used in the sequel; these are included in order to explain the complexity of the situation; we expect that (12.3) can be of independent interest. The main result of this section is not difficult if we work over fields of characteristic $p > 2$; in such a case an anti-symmetric pairing on N gives an alternating pairing on $A := \mathbb{D}(N)$; in that case a proof of Proposition (12.5) is not so difficult. We have chosen to include also the case $p = 2$ in the general proof, leaving simplifications in other cases to the reader, but indicating how biextensions can be brought in.

(12.1). **Polarizations and biextensions.**
(AV) Let $(\mathcal{X}, \lambda) \to S$ be a polarized abelian scheme. By [54], 1.4 on page 224, we know:

$$\mathrm{Biext}^1(X_1, X_2; \mathbb{G}_m) \xrightarrow{\sim} \mathrm{Hom}(X_1, X_2)$$

(we will write $X_1 = X = X_2$, numbering the copies of X involved in order to keep track what is what). The polarization is an isogeny $\lambda : X \to X^t$. We know that λ is symmetric in the sense that

$$(\lambda : X_1 \to X_2^t) \quad = \quad (\lambda^t : (X_2)^{tt} = X_2 \to X_1^t);$$

here we write $X^{tt} = X$ meaning we have a canonical, functorial isomorphism, which is denoted by

$$\kappa_X : X \xrightarrow{\sim} X^{tt}$$

for abelian schemes, and for p-divisible groups.
(Biext) Let $N \to S$ be a finite flat group scheme. By [54], 1.1.4 and 3.3.2 , see page 258, we know that

$$\mathrm{Ext}(N_1, N_2^D) \xrightarrow{\sim} \mathrm{Biext}^1(N_1, N_2; \mathbb{G}_m).$$

We write $\epsilon \mapsto E$ for the extension and the biextension corresponding under this canonical identification. If (X, λ) is as above, and q is a positive integer, $N := X[q]$, we write $(X, \lambda) \,|\, _N$ for the restriction of the biextension defined by (X, λ)

$$|_N : \mathrm{Biext}^1(X_1, X_2; \mathbb{G}_m) \longrightarrow \mathrm{Biext}^1(N_1, N_2; \mathbb{G}_m).$$

Suppose P and Q are BT_1. We say that

$$(\epsilon) \qquad 0 \to Q \to T \to P \to 0$$

is a 2-extension if $Q = T[p]$ and $[p] : T \to \mathrm{Im}([p] : T \to T) = T/T[p] = P$ under the natural identifications. We say that ϵ is *anti-symmetric* if $\epsilon^{DD} = -\epsilon$; we intend to say: the exact sequence ϵ defines

$$(\epsilon^D) \qquad 0 \to P^D \to T^D \to Q^D \to 0;$$

dualizing again we obtain ϵ^{DD}; for finite flat group schemes we have a duality isomorphism

$$\iota_P : P \xrightarrow{\sim} P^{DD}$$

(denoted by κ in [42], I.2-2, but we use a different symbol because we have used already κ for abelian schemes); we intend to say:

$$(\iota_P)^*(\epsilon^{DD}) = -(\iota_Q)_*(\epsilon).$$

We write $\mathrm{Ext}(N_1, N_2^D; \mathbb{G}_m)^{(2,a)}$ for the *set* of 2-extension which are anti-symmetric.

(Pairings) Let $K \supset \mathbb{F}_p$ be a perfect field, and let N be a finite group scheme over K. We consider $<,>: A \times A \to W$, a pairing on the Dieudonné module $A := \mathbb{D}(N)$. If (X, λ) is as above, and $N := X[p]$, we write $(X, \lambda)[p] = (N, <,>)$ for the pairing induced by λ on $A = \mathbb{D}(A)$.

Note the different notations. If (X, λ) is a polarized abelian scheme, and $N := X[p]$, we write $(N, E) = (X, \lambda) \mid_N$ for the pair obtained by the restriction of the biextension, and we write $(X, \lambda)[p] = (N, <,>)$ for the pair obtained by restricting the polarization morphism to N.

Between these concepts there are some relations. Most of these are well-known, and we list them for convenience.

(12.2). **Lemma** (1) *Let (X, λ) be a polarized abelian scheme. Let q be a positive integer, and $N := \mathcal{X}[q]$. The restriction of $\lambda \in$* Biext$^1(X_1, X_2; \mathbb{G}_m)$ *gives a* symmetric *biextension $E \in$* Biext$(N_1, N_2; \mathbb{G}_m)$, *corresponding with an* anti-symmetric *extension $\epsilon \in$* Ext$^1(N_1, N_2^D)$.
(2) *If moreover $q = p$, and the polarization λ is principal, the related extension ϵ is an* anti-symmetric 2-extension:

$$\lambda \mid_N = \epsilon \in \operatorname{Ext}(N_1, N_2^D; \mathbb{G}_m)^{(2,a)}.$$

(3) *Let $K \supset \mathbb{F}_p$ be a perfect field, and $N := X[p]$. A principal polarization restricts to a non-degenerate pairing $(X, \lambda)[p] = (N, <,>)$ on $A := \mathbb{D}(N)$ which is* alternating.
(4) *An anti-symmetric 2-extension on a BT_1 over a perfect field restricts to a non-degenerate,* alternating *pairing $(N, <,>)$ on $A := \mathbb{D}(N)$.*
(5) *Let $k \supset \mathbb{F}_p$ be an algebraically closed field. The restriction map from* $\operatorname{Ext}(N_1, N_2^D; \mathbb{G}_m)^{(2,a)}$ *to the set of pairs $(N, <,>)$ of BT_1 group schemes with a non-degenerate alternating pairing on its Dieudonné module is* surjective. [Remark: in general this surjection is far from being injective.]
Proof. (1). This follows basically from [40], Coroll. (1.3.), pp. 69/70. In fact, if $\psi : X \to Y$ is an isogeny of abelian schemes, the duality theorem gives a canonical identification

$$\nu : \operatorname{Ker}(\psi^t) \xrightarrow{\sim} (\operatorname{Ker}(\psi))^D.$$

The last diagram in the corollary cited reads: $-\nu{\cdot}\iota_N = \nu{\cdot}(\kappa_X \mid_N)$. This proves (1).
(2). The extension is obtained as $\epsilon = (N_1 \to N_2^D)^*((\mathcal{X}^t[p^2]/\mathcal{X}^t[p]) \cong \mathcal{X}^t[p])$. Hence it is a 2-extension. This proves (2).
(3) and (4). The pairing on $A := \mathbb{D}(N)$, where $N = X[p]$ in (3) can be defined as follows: consider a 2-extension

$$0 \to N_2^D \longrightarrow T \longrightarrow N_1 \to 0.$$

Then multiplication by p on each member of this exact sequence, and applying the snake lemma we obtain an isomorphism $N_1 \to N_2^D$; this is the pairing we are looking for. Let us make it explicit in terms of Dieudonné modules. Let

$B := \mathbb{D}(T)$. This is a free module over $W_2 = W_\infty(k)/(p^2)$. The isomorphism $T \to T^D$, anti-symmetric by (2), is written out by a non-degenerate, anti-symmetric pairing $[,] : B \times B \to W_2$, and the pairing $<,>$ on A obtained form the extension can be described as: for $x \in A_1, y \in A_2$, we choose $x' \in B$ such that x' mod $A_2^D = x$; note that $A_2^D \subset B$; with these notations:

$$< x, y >= (px')(y),$$

and using $B \cong B^D$ by $[,]$ and $(A_2^D \to B)^D = B^B \to A_2)$ we can rewrite this as

$$< x, y >= p{\cdot}[x', y'].$$

We remark that $-1 \neq +1 \in W_2$. As the pairing $[,]$ is anti-symmetric this implies that for every $z \in B$, with $z \notin pB$ we have $[z, z] = -[z, z]$, hence $p{\cdot}[z, z] = 0$. We obtain $< x, x >= p{\cdot}[x', x'] = 0$ for every $x \in A$. This proves (3) and (4).

Remark. We do not have (in case the characteristic of the base field equals two) a characterization of those pairings on $N = \mathcal{X}[p]$ which come from a principal polarization on $\mathcal{X}$ only in terms of group schemes, not using $\mathbb{D}(N)$.

(12.3). **Construction.** *Given a pair $(N, <,>)$ of a BT_1 with a non-degenerate alternating pairing on $A = \mathbb{D}(N)$ over k, we construct a p-divisible group G, and a principal quasi-polarization λ on G such that the restriction $(G, \lambda : G \to G^t)[p]$ gives $(N, \zeta : N = G[p] \to N^D = G^t[p])$.* - From this it clearly follows that the restriction map in (5) is surjective. One could indicate the polarized p-divisible group obtained below by something like Can(φ). Note that in general $\mathrm{Can}(\mathrm{ES}(G, \lambda)) \ncong (G, \lambda)$.

Indeed, we use the structure theorem of standard types as in Section 9. Hence we can choose a basis $\{X_1, \cdots, X_g, Y_1, \cdots, Y_g\}$ with the properties specified above. We choose $M := \bigoplus W{\cdot}X_i' \oplus \bigoplus W{\cdot}Y_j'$, and we construct the structure of a Dieudonné module on this. We analyze the construction (9.1). We denote the base vectors in the standard type describing $A = \mathbb{D}(N)$ by $\{Z_1, \cdots, Z_{2g}\} = \{X_1, \cdots, Y_g\}$. We have seen that $\{Z_1, \cdots, Z_g\} = \{X_1, \cdots, X_{g-a}, Y_g, \cdots, Y_{g-a+1}\}$ is a k-base for $\mathcal{F}(A) = A[\mathcal{V}]$, and every of these base vectors is an image under $\mathcal{F}$ of one of the base vectors, in fact $\mathcal{F}(X_i) = Z_i$. We have also seen that each of the elements of $\{Z_g, \cdots, Z_{2g}\} = \{Y_{g-a}, \cdots, Y_1, X_{g-a+1}, \cdots, X_g\}$ is mapped under $\mathcal{V}$ onto either $-Y_i$ or onto $+Y_i$. For every $1 \leq i \leq g$ the image $\mathcal{F}(X_i) = Z_j \in A = M/pM$ is defined, and we require $\mathcal{F}(X_i') = Z_j' \in M$. For every $g < i \leq 2g$ the image $\mathcal{V}(Z_i) = Z_j$ is defined, and we require $\mathcal{V}(Z_i') = Z_j'$.

If $1 \leq i \leq a$ and $X_i = \mathcal{F}(X_j)$, we require $\mathcal{V}(X_i') = p{\cdot}X_j'$.

If $g \geq i > g - a$ and $Y_i = \mathcal{F}(X_j)$ we require $\mathcal{V}(Y_i') = p{\cdot}X_j'$.

In this way we have defined $\mathcal{V}$ for all base vectors Z_i'.

If $a \geq j \geq 1$ and $Y_i = \mathcal{V}(Y_j)$ we require $\mathcal{F}(Y_i') = p{\cdot}Y_j'$.

If $a \geq j \geq 1$ and $-Y_i = \mathcal{V}(X_j)$ we require $\mathcal{F}(Y_i') = -p{\cdot}X_j'$.

In this way we have defined the image under $\mathcal{F}$ and under $\mathcal{V}$ of every base vector (and every image is either $\pm$ a base vector, or p times $\pm$ a base vector). We extend the map $\mathcal{F}$ in a σ-linear way and $\mathcal{V}$ in a τ-linear way on M, where $\tau = \sigma^{-1}$. Direct verification shows that this defines the structure of a Dieudonné module of a p-divisible group with a symplectic W-base $\{Z_i' \mid 1 \leq i \leq 2g\}$. By construction, modulo p, this gives back the standard type which is k-isomorphic with A with its symplectic form; this ends the construction.

Hence the proof of the lemma is complete. $\square$(12.2)

(12.4). **Lemma.** *Let K be a perfect field. Let (N_1, ζ_1) and (N_2, ζ_2) be finite group schemes with an anti-symmetric symmetry over K. Let $A_i := \mathbb{D}(N_i)$ be their Dieudonné modules. Let $f : A_1 \to A_2$ be a W-linear, bijective map, where $W = W_\infty(K)$. Suppose that f commutes with Frobenius and with the symmetries, i.e., the diagrams*

$$\begin{array}{ccc} A_1^{(p)} & \xrightarrow{f^{(p)}} & A_2^{(p)} \\ \mathcal{F}\downarrow & & \downarrow\mathcal{F} \\ A_1 & \xrightarrow{f} & A_2 \end{array} \qquad \textit{and} \qquad \begin{array}{ccc} A_1 & \xrightarrow{\zeta_1} & A_1^D \\ f\downarrow & & \uparrow f^D \\ A_2 & \xrightarrow{\zeta_2} & A_2^D \end{array}$$

are commutative. Then f is an isomorphism of Dieudonné modules, hence

$$(N_1, \zeta_1) \xrightarrow{\sim} (N_2, \zeta_2).$$

Proof. On A_1 and on A_2 we have the Verschiebung, $\mathcal{V} : A_i \to A_i^{(p)}$. We have to show that f also commutes with these maps in order to conclude that f is an an isomorphism of Dieudonné modules. We show this by pushing $\mathcal{F}$ with the help of the ζ_i onto the duals, and then by duality transforming $\mathcal{F}$ on the duals into $\mathcal{V}$ on the A_i.

At first we show:

$$(A_2^{(p)D} \to A_2^D \to A_1^D) = f^D{\cdot}\mathcal{F} \stackrel{?}{=} \mathcal{F}{\cdot}f^{(p)D} = (A_2^{(p)D} \to A_1^{(p)D} \to A_1^D).$$

We should draw a large diagram, and chasing leads to:

$$(A_1^{(p)} \to A_2 \to A_2^{(p)D} \to A_2^D \to A_1^D) = (f^D{\cdot}\mathcal{F}{\cdot}(\zeta_2^{(p)}{\cdot}f^{(p)})) =$$

$$= -f^D{\cdot}\mathcal{F}{\cdot}\zeta_2^{(p)DD}{\cdot}\kappa{\cdot}f^{(p)} = f^D{\cdot}\zeta_2{\cdot}(f{\cdot}\mathcal{F}) =$$

$$f^D{\cdot}\zeta_2{\cdot}(f{\cdot}\mathcal{F}) = \zeta_1{\cdot}\mathcal{F} = \mathcal{F}{\cdot}f^{DD}{\cdot}(\zeta_2^{(p)}{\cdot}f^{(p)});$$

hence

$$(f^D{\cdot}\mathcal{F}) = (\mathcal{F}{\cdot}f^{(p)D}) : A_2^{(p)D} \to A_1^D.$$

We use:

$$(A_i^{(p)D} \xrightarrow{\mathcal{F}} A_i^D)^D = (A_i^{(p)D} = A_i^{(p)} \xleftarrow{\mathcal{V}} A_i = A_i^{DD})$$

under the identification $\iota : A_i \to A_i^{DD}$. This proves that f "commutes" with $\mathcal{V}$, and the lemma is proved. $\square$

We show that the boundary of a stratum is a union of (lower dimensional) strata:

(12.5). **Proposition.** *Let $\varphi \in \Phi$ be an elementary sequence, let $[(X_0, \lambda_0)] = 0 \in (S_\varphi)^c$, and let $[(Y_s, \mu_s)] = s \in \mathcal{A}$ such that* $\mathrm{ES}(X_0, \lambda_0) = \mathrm{ES}(Y_s, \mu_s)$. *Then $s \in (S_\varphi)^c$. This implies:*

$$S_{\varphi'} \cap ((S_\varphi)^c) \neq \emptyset \quad \Longrightarrow \quad S_{\varphi'} \subset (S_\varphi)^c.$$

Proof. It suffices to show this under the extra condition that $X_0[p]$ is a local-local group scheme, i.e., $\varphi'(1) = 0$. We suppose k is an algebraically closed field, with $(X_0, \lambda_0) \in \mathcal{A}(k)$, over which (Y_s, μ_s) is also defined. As $0 \in (S_\varphi)^c$ we can choose a complete local domain R of characteristic p, with residue class field k, and an abelian scheme $(X, \lambda) \to \Delta = \mathrm{Spec}(R)$ such that $(X, \lambda) \otimes_R k = (X_0, \lambda_0)$ and for the generic point $\eta \in \Delta$ we have $\mathrm{ES}(X_\eta, \lambda_\eta) = \varphi$. As $\mathrm{ES}(X_0, \lambda_0) = \varphi' = \mathrm{ES}(Y_s, \mu_s)$, using (9.4), (!), we can choose an isomorphism (which we will write as an identification):

$$(X_0, \lambda_0)[p] = (N_0, \zeta_0) = (Y_s, \mu_s)[p].$$

Let $\{Z_i\}$ be a symplectic W-base for the Dieudonné module $M_0 := \mathbb{D}(X_0[p^\infty])$ such that $Z_{g+1}, \cdots, Z_{2g} \in \mathcal{V}M$. On this basis the matrix of $\mathcal{F}$ on M_0 is given by the display

$$\begin{pmatrix} A & B \\ C & D \end{pmatrix} \quad \text{and} \quad \begin{pmatrix} A+TC & B+TD \\ C & D \end{pmatrix}$$

is the display of the deformation $(X_0, \lambda_0) \subset (X, \lambda)$, where $T \in \mathrm{Mat}(g \times g, W_\infty(R))$ is a symmetric matrix. We choose a symplectic basis for $\mathbb{D}(Y_s[p^\infty])$ such that this modulo p coincides with $\{Z_i \bmod p\}$ on $\mathbb{D}(X_0[p]) = \mathbb{D}(N_0) = \mathbb{D}(Y_s[p])$. We write

$$\begin{pmatrix} a & b \\ c & d \end{pmatrix}$$

for the display on this basis on $\mathbb{D}(Y_s[p^\infty])$. Note that

$$\begin{pmatrix} A \bmod p & 0 \\ C \bmod p & 0 \end{pmatrix} = \begin{pmatrix} a \bmod p & 0 \\ c \bmod p & 0 \end{pmatrix}.$$

We define a deformation of $(Y_s[p^\infty], \mu_s)$, and hence by the theorem of Serre and Tate, a deformation (Y, μ) of (Y_s, μ_s), by the display

$$\begin{pmatrix} a+Tc & b+Td \\ c & d \end{pmatrix}.$$

We claim: $\mathrm{ES}(Y_\eta, \mu_\eta) = \varphi$. Indeed, let K be a perfect field containing $Q(R)$. We see that $\mathcal{F}$ on $\mathbb{D}(X_\eta)[p] \otimes K$ is given by the same display

$$\begin{pmatrix} (A+TC) \bmod p & 0 \\ C \bmod p & 0 \end{pmatrix} = \begin{pmatrix} (a+Tc) \bmod p & 0 \\ c \bmod p & 0 \end{pmatrix}$$

as the one giving $\mathcal{F}$ on $\mathbb{D}(Y_\eta)[p] \otimes K$, and the pairings on these finite group schemes coincide under this identification. Hence the conditions of the previous lemma are fulfilled, and we conclude that $(X_\eta, \lambda)[p] \otimes K \cong (Y_\eta, \mu)[p] \otimes K$. Hence $\mathrm{ES}(X_\eta, \lambda_\eta) = \mathrm{ES}(Y_\eta, \mu_\eta) = \varphi$, and the proof of the proposition is concluded. □

We expect that a proof of the previous proposition can be given by local deformation theory, by using [17], and giving an interpretation of the image of biextension on finite group schemes to a certain set of anti-symmetric pairings over arbitrary base schemes. This was our original approach, which works in characteristic $p > 2$, also see [61]. We have not carried out this programme for $p = 2$.

(12.6). **Remark.** It seems plausible (using the notations in the previous proof) that $(X, \lambda)[p] \cong (Y, \mu)[p] \to \Delta$. However, in general there is no $(Y, \mu) \to \Delta$ extending (Y_s, μ_s) such that $(X, \lambda) \mid_N$ and $(Y, \mu) \mid_N$ are isomorphic over Δ, where $X[p] = N = Y[p]$; in other terms: the finite group scheme with the symmetry (N, ζ) can be included in a new $(Y, \mu) \supset (Y_s, \mu_s)$, but in general the finite group scheme with biextension (N, E) does not admit such a construction.

13. Proof of the results

The set of superspecial points $\Sigma \subset \mathcal{A}$ is a finite set of points. It is clear that

$$\mathrm{ES}(X) = g \iff a(X) = g \iff X \quad \text{is superspecial,}$$

i.e., there is a supersingular elliptic curve E and an isomorphism $X \otimes k \cong E^g$.

(13.1). In Section 7 we have seen the construction of L, in (7.3) we have proved that L is connected. In Section 8 we have proved that $L = (S_{\{0,\cdots,0,1\}})^c$. □ Theorem (1.1)

(13.2). In Section 11 we have seen how to produce new abelian varieties with an elementary sequence φ' (see the notation in that section). Hence it follows by induction from the superspecial locus that every elementary sequence does appear, i.e., for every $\varphi \in \Phi$ the locus S_φ is non-empty. Using Section 10 and Section 11, in particular (11.2) it follows that $\dim(S_\varphi) = \mid \varphi \mid$.

□ Theorem (1.2).

(13.3). In (12.5) we have proved that a locus $S_{\varphi'}$ either does not meet $(S_\varphi)^c$ or is contained in $(S_\varphi)^c$: this is the first claim in Theorem (1.3).

Let $\varphi \in \Phi$, with $\varphi \neq \{0, \cdots, 0\}$, i.e., $S_\varphi \neq \Sigma$. By (6.5) we know that every irreducible component $T \subset T_\varphi \subset \mathcal{A}^*$ is quasi-affine; because its dimension is positive, this implies that its boundary $\partial_{\mathcal{A}^*}(T) \neq \emptyset$. By (6.3) we conclude that $\partial_{\mathcal{A}^*}(T) \cap \mathcal{A}$ is non-empty; hence we see that every irreducible component $S \subset S_\varphi$ is not complete. Moreover using (12.5) this proves that every irreducible component $S \subset S_\varphi$ has a non-empty boundary in $\mathcal{A}$, which is of codimension one by (4.1), and we see it is a union of components of lower dimensional strata. Hence by induction we see that every component of S_φ contains in its boundary a component of the (unique) zero-dimensional stratum $S_{\{0,\cdots,0,1\}}$; hence by (8.4) we know that $\partial(S)$ contains an irreducible component of L. Hence if $\varphi \neq \{0,\cdots,0\}$, i.e., $S_\varphi \neq \Sigma$, then $(S_\varphi)^c$ is connected.

□ Theorem (1.3)

Remark. Instead of using (6.5) in the proof of (1.3), we can use (6.3), induction on g, and (12.5).

As $\mathcal{A} = (S_{\{1,2,\cdots,g\}})^c$, we know by Theorem (1.3) that it is connected. Moreover for $n \geq 3$ it is non-singular. Hence $\mathcal{A}_{g,1,n} \otimes \mathbb{F}_{p,n}$ is geometrically irreducible for every g and every n (not divisible by p). □ Corollary (1.4)

(13.4). It suffices to show (1.5) in case we moreover assume $n \geq 3$. We note that the locus $V_{g-1} \subset \mathcal{A}$ of non-ordinary polarized abelian varieties equals $V_{g-1} = (S_{\{0,1,2,\cdots,g-1\}})^c$. Let us work over an algebraically closed field k. Every component of this closed locus contains at least one irreducible component of L, and it contains at least one point of Σ.

Let $[(X_0 = E^g, \lambda)] = x \in \Sigma \subset V_{g-1}$. By [29], 6.1 we can put a principal polarization on E^g in diagonal form. We write out the display of the universal deformation of $(X_0 = E^g, \lambda)$ on this basis. The Hasse-Witt matrix of X_0 is zero, and we see that a formal neighborhood of x in $\mathcal{A}$ is given by $\mathrm{Spf}(k[[T_{i,j} \mid 1 \leq i,j \leq g]]/(T_{i,j} - T_{j,i}))$; moreover

$$V_{g-1} \otimes k \text{ at } x \text{ is given by } \mathrm{Det}(T_{i,j}) = 0.$$

Claim. *This equation is given by a non-zero, irreducible polynomial.* This we prove by induction on g. For $g = 1$ this is clear. Let $T = T_{g,g}$, and write

$$\mathrm{Det}(T_{i,j} | 1 \leq i,j \leq g) =: D_g = A \cdot T + B,$$

where $A = D_{g-1}$ and B are polynomials in the other variables. The polynomial A is non-zero, and for $g > 1$ it is irreducible by the induction assumption. For $g > 1$ we substitute $T_{j,j} = 1$ for all $j < g-1$, and $T_{g-1,g} = T_{g,g-1} = 1$, and all other variables $= 0$; this gives $B \mapsto 1$ and $A \mapsto 0$; because A is irreducible we see that A and B are coprime. Hence $AT + B$ is irreducible.

Thus we see that $V_{g-1} \otimes k$ is locally irreducible at all $x \in \Sigma$. Hence for every irreducible component $L' \subset L_k$ there is exactly one component of $V_{g-1} \otimes k$ containing L'. Hence there is precisely one component of $V_{g-1} \otimes k$ having non-empty intersection with L_k. By what has been said before this implies that V_{g-1} is geometrically irreducible. □ Corollary (1.5)

14. Some questions

(14.1). We have seen that some strata are reducible (like Σ, like L, say for large p). Some strata are irreducible (like $\mathcal{A}$, like V_1 for $g > 1$). What can be said in general? We expect:

Conjecture. *Let $\varphi \in \Phi$ such that S_φ is not contained in the supersingular locus. Then we expect that $(S_\varphi)^c$ is geometrically irreducible.*

This is in agreement with the examples just mentioned; it was proved in certain cases by Van der Geer, see [11], Theorem (9.9): take $\varphi = \{1, 2, \cdots, g-a, \cdots, g-a\}$, with $a < g$.

(14.2). In (4.1) we have seen that $T_\varphi \subset \mathcal{A}^*$ is a quasi-affine set, and the same for S_φ. *We expect that T_φ is affine subscheme of $\mathcal{A}^*$ for every φ. We expect that there exists φ such that $S_\varphi \subset \mathcal{A}$ is not affine.*

(14.3). Given two elementary sequences $\varphi, \varphi' \in \Phi$ we have defined the relation $\varphi' \prec \varphi$ by $\varphi'(i) \leq \varphi(i)$ for all i. We define

$$\varphi' \subset \varphi \quad \overset{\text{def}}{\Longleftrightarrow} \quad S_{\varphi'} \subset (S_\varphi)^c.$$

It follows from (11.1) that $\varphi' \prec \varphi \Rightarrow \varphi' \subset \varphi$. It is not so difficult to give examples that the converse does not hold:

Example. *We choose $\varphi', \varphi \in \Phi_7$, and a deformation which shows that*

$$\varphi' \subset \varphi, \quad \textit{while} \quad \varphi' \not\prec \varphi.$$

We choose $\varphi' = \{0,0,1,1,2,2,2\}$, hence $\psi' = \{0,0,1,1,2,2,2;3,4,4,5,5,6,7\}$; we choose $\varphi = \{0,0,0,1,2,3,3\}$, hence $\psi = \{0,0,0,1,2,3,3;4,4,4,4,5,6,7\}$; clearly $\varphi' \not\prec \varphi$; we prove that $\varphi' \subset \varphi$. The standard type for φ' has a Dieudonné module with basis:

$$\{Y_7, X_1, Y_6, X_2, Y_5, Y_4; X_3, X_4, Y_2, X_5, Y_1, X_6, X_7\}.$$

We write out a display, where $T_{4,5} = T = T_{5,4}$, and all other variables are put to zero. Explicitly, on the basis given above, we have $\mathcal{F} : X_6 \mapsto Y_4 + TX_4 =: Z$, and all other images under $\mathcal{F}$ of base vectors are the same as before: under $\mathcal{F}$, all Y_i are mapped to zero, and we have $X_7 \mapsto Y_3$ and $X_6 \mapsto Z \mapsto T^\sigma \cdot Y_5$, and $X_5 \mapsto X_2 \mapsto Y_6$ and $X_4 \mapsto Y_5$ and $X_3 \mapsto X_1 \mapsto Y_7$. Over the field of fractions $k((T))$ we choose a basis by:

$$\{Y_7, Y_6, Y_5, X_1, X_2, \ \xi_3 = Z = Y_4 + TX_4, \ \eta_4 = Y_3;$$
$$\xi_4 = X_3, \ \eta_3 = (1/T)Y_4, \ Y_2, Y_1, X_5, X_6, X_7\};$$

this puts the generic fiber of the deformed module in standard form, with elementary sequence φ. □

We note that φ' is canonical, and it belongs to the cycles

$$\{\mathcal{FVV}; \mathcal{VFF}; \mathcal{FVFFVFVV}\};$$

the final sequence belonging to φ is not canonical; it belongs to the cycles $\{\mathcal{FVV}; \mathcal{VFF}; (\mathcal{FFVV})^2\}$; we have seen that $\varphi' \subset \varphi$. However, consider the cycles $\{\mathcal{FVFFVFVV}\}$ and $\{(\mathcal{FFVV})^2\}$; these define $\beta', \beta \in \Phi_4$, and they correspond with elementary sequences $\{0,1,1,1\}$ and $\{0,0,1,2\}$; hence, although $\varphi' \subset \varphi$, "subtracting equal parts", gives $\beta' \subsetneq \beta$ and $\beta' \supsetneq \beta$!?

We see: in general

$$\varphi \subset \varphi' \not\Rightarrow \varphi \prec \varphi'.$$

Question. *Is there an easy algorithm, using the combinatorics of sequences, describing the relation $\varphi \subset \varphi'$?*

(14.4). On $\mathcal{A}$ we have defined a stratification $\mathcal{A} = \sqcup_{\varphi\in\Phi} S_\varphi$. For every symmetric Newton polygon β there is a stratum $W_\beta \subset \mathcal{A}$; this is the (closed) set of points where the Newton polygon is either equal or above β; we write W^0_β for the locally closed set of points where the Newton polygon is equal to β; we have $\mathcal{A} = \sqcup_\beta W^0_\beta$; for this structure, see [47]. Both stratifications are reasonably well understood by now. However intersections are not so easy to understand. For example, I have some information, but no proven complete results, which tell us exactly which elementary sequences appear on the set W^0_β.

(14.5). Even more difficult is the question to describe intersections of all NP-strata, and of all EO-strata with the Torelli locus $j(\mathcal{M}_g) = \mathcal{T}^0_g \subset (\mathcal{T}^0_g)^c =: \mathcal{T}_g \subset \mathcal{A}$; in these case I have no reasonable conjecture to offer. Here is an example:
Question. *Does every symmetric Newton polygon appear on $\mathcal{T}_g$?*

It is known that even for large values of g there are supersingular curves of genus g, contrary what might be thought after dimension considerations, see [12]. I expect the answer to this question to be negative for large g, but I have no reasonable theory, no method, no examples to support this expectation.

I do not know which elementary sequences appear on $\mathcal{T}_g$.

Also I have no idea to describe the intersection of $\mathcal{T}_g$ with either of the two foliations of W^0_β which will be described in [48].

(14.6). **Remark.** Suppose that the answer to the question in (14.5) is negative, i.e., suppose there exists a symmetric Newton polygon which does not appear on the corresponding $\mathcal{T}_g$. Then it follows that there exists an abelian variety over $\overline{\mathbb{Q}}$ which is not isogenous with a Jacobian; this would answer a question asked by N. Katz (we expect a negative answer):
Question. *Is every abelian variety over $\overline{\mathbb{Q}}$, or over $\overline{\mathbb{F}_p}$ isogenous with a Jacobian?* Note that over $\mathbb{C}$ there exists for every $g \geq 4$ an abelian variety not isogenous with a Jacobian.

(14.7). *In general a p-divisible group G is not determined by the structure of its p-kernel G[p].* For example, Choose $g = 3$, and consider all principally polarized abelian varieties (X, λ) with $f(X) = 0$ and $a(X) = 1$. We have seen that in such a case we have $(X[p], <,>) \cong N_{\{0,1,2\}}$, in the notation of Section 9; but we know that the formal isogeny type of $X[p^\infty]$ can be either $G_{2,1} \oplus G_{1,2}$ or $3{\cdot}G_{1,1}$. Moreover, even if the formal isogeny type is given, then within one of these classes there are infinitely many mutually non-isomorphic possibilities. Many more examples along these line can be given.

However there are cases in which the p-kernel does determine the p-divisible group. For example, if $f(X)$ is maximal (the ordinary case), or if $a(X)$ is maximal (the superspecial case): over an algebraically closed field there is a unique isomorphism class for the p-divisible group having this p-kernel.

Also, in (8.3), case (I), we have seen other cases where the p-kernel determines the p-divisible group.

It would be nice to have a general criterion which tells us when it should be true that the p-kernel of a p-divisible group already determines the structure of the p-divisible group. To that end we formulate a conjecture, which includes the cases recorded above. We hope you can soon consult [48] for a more extensive discussion.

We work over an algebraically closed field k. Let $\beta = \sum(m_i, n_i)$ be a symmetric Newton polygon. We write $H(m,n)$ for the "minimal" p-divisible group in the isogeny class of $G_{m,n}$, see [19], (5.3), see (8.1) above. We write $H(\beta) = \bigoplus H(M_i, n_i)$. There is an (obvious) principal quasi-polarization μ on $H(\beta)$ (note that β is symmetric). For "minimal isogeny types" we could hope the following to be true:

(14.8). **Conjecture.** *Suppose (G, λ) is a p-divisible group with a principal quasi-polarization over an algebraically closed field k; suppose that there exists a symmetric Newton polygon β and an isomorphism*

$$(G,\lambda)\,|_{G[p]} \cong (H(\beta),\mu)\,|_{H(\beta)[p]}; \quad \textit{then we expect:} \quad (G,\lambda) \cong (H(\beta),\mu).$$

(14.9). **Remark.** We have seen that every elementary sequence appears in $\mathcal{A}_{g,1} \otimes \mathbb{F}_p$. We have constructed for every *symmetric* canonical type τ a non-empty, locally closed subscheme $D_\tau \subset \mathcal{A}' = \mathcal{A}_g \otimes \mathbb{F}_p$; however we do not know which finite group schemes $N = X[p]$ appear on a given irreducible component of $\mathcal{A}_g \otimes \mathbb{F}_p$, i.e., we do not know which strata D_τ for an arbitrary canonical type τ are non-empty, and how they appear in components of $\mathcal{A}'$.

(14.10). The following is true: *Let S be a complete, irreducible scheme over $\mathbb{F}_p$. Let $G \to S$ be a p-divisible group, such that any two geometric fibers of $G[p] \to S$ are isomorphic*; i.e., if $s, t \in S(k)$, with $k = \overline{k}$, then there exists an isomorphism $G[p]_s \cong G[p]_t$; equivalently: over some finite covering $T \to S$ the group scheme $G[p]_T$ is constant. *Then it follows that any two geometric fibers of $G \to S$ are isomorphic.*

We hope to publish a proof soon, and we expect to come back to this and related questions. Note that in general G does not become constant over any finite cover. Note that in general the conclusion does not hold if we delete the condition "S is complete".

15. APPENDIX. SOME NOTATIONS

(15.1). Throughout this paper all base schemes, and all base fields, will be in characteristic p. Let X be an abelian variety over a field K. The p-rank of X, denoted by $f = f(X)$ is defined by:

$$X[p](k) \cong (\mathbb{Z}/p)^f,$$

where k denotes an algebraically closed field containing K, and $X[p]$ is the group scheme $\mathrm{Ker}(\times p : X \to X)$. An abelian variety X of dimension g such that $f(X) = g$ (i.e., the case where the p-rank is maximal) is called *ordinary*.

We fix a positive integer g, a prime number p and a positive integer n not divisible by p, and we denote by

$$\mathcal{A} = \mathcal{A}_{g,1,n} \otimes \mathbb{F}_{p,n},$$

the moduli space of principally polarized abelian varieties with a symplectic level-n-structure in characteristic p.

Let G be a group scheme over a field K. We write $a(G) := \dim_L \mathrm{Hom}(\alpha_p, G_L)$, where $L \supset K$ is a perfect field containing K. Here $\alpha_p = \mathbb{G}_a[F]$, the kernel of the Frobenius morphism on the additive linear group of dimension one; this group scheme is defined over $\mathbb{F}_p$, and we will consider this over any base without further indicating over which base it is considered. An abelian variety is ordinary iff $a(X) = 0$.

Let E be an elliptic curve over a field K (of characteristic p). Then either E has a point of order exactly p over an algebraically closed field $k \supset K$, and in this case the elliptic curve E is ordinary, or

$$f(E) = 0, \quad \text{and the curve is called } \textbf{supersingular}.$$

In short: an elliptic curve is supersingular iff its p-rank is zero, iff $a(E) = 1$. We say that an abelian variety X is *supersingular* if it isogenous to a product of supersingular elliptic curves over some extension field. It is *superspecial* if it isomorphic to a product of supersingular elliptic curves over some extension field.

For a commutative group scheme G and $q \in \mathbb{Z}_{>0}$ we denote by $G[q]$ the kernel of multiplication by q on G. In this paper the p-kernel $X[p]$ of abelian varieties in consideration will play a central role. We write $X[p^\infty]$ for the union of the finite group schemes $X[p^i]$ for all $1 \leq i$; note that it contains the formal group $\hat{X}$ of X and $\hat{X} = X[p^\infty]$ iff $f(X) = 0$.

(15.2). Here is a survey of some of the notions about supersingular and *superspecial* abelian varieties:

Theorem/Definition: *Let k be an algebraically closed field of characteristic p, and let E be a supersingular elliptic curve over k. Let X be an abelian variety of dimension g over a field $K \subset k$. Let $g \geq 2$.*

a)

$$\hat{X}_k \sim \hat{E}^g \Longleftrightarrow X_k \sim E^g \quad \stackrel{\text{def}}{\Longleftrightarrow} \quad X \ \ \textit{is} \ \ \textbf{supersingular}.$$

b) *If* $E_1, \cdots, E_{2g}$ *are supersingular elliptic curves, then*

$$E_1 \times \cdots \times E_g \cong E_{g+1} \times \cdots \times E_{2g}.$$

c) *A supersingular elliptic curve is superspecial;*

$$\hat{X}_k \cong \hat{E}^g \Longleftrightarrow a(X) = \text{dim}X \Longleftrightarrow X \cong E^g \quad \stackrel{\text{def}}{\Longleftrightarrow} \quad X \ \ \textit{is} \ \ \textbf{superspecial}.$$

The first statement was proved in [43], Th. 4.2. Statement (b) is a theorem by Deligne [58], Th. 3.5 on page 580, which uses a class number computation by Eichler, cf. [7]; also see [41]. For proving (c), note that if $\hat{X} \cong \hat{E}^g$, then $a(X) = \text{dim}X$, and by [44] it follows that X is isomorphic with a product of supersingular elliptic curves, $X \cong E_1 \times \cdots \times E_g$; such a product is isomorphic with E^g by (b). For further information, see [29].

(15.3). **Covariant Dieudonné modules** (with apologies to the Contravarianists.) Over a perfect field (of positive characteristic) finite group schemes and p-divisible groups can be classified by the theory of Dieudonné modules. This can be done in a contravariant way as in [40], [31]. One can also use the covariant theory. And it is not important which one we choose as long as we work over a perfect field: these two theories are anti-equivalent.

However as soon as we are working over a more general base ring (e.g. when studying deformation theory) things are different. For finite group schemes the theory is much more difficult, but see [17], [18]. For formal groups Cartier developed a theory of covariant Dieudonné modules which works well. Later this was developed and used in the disguise of *displays*, invented by Mumford, see [36], and see [38], [39], [63], [64]. **Therefore, in all cases we shall work with the covariant theory.**

For a domain R of positive characteristic p we write $W = W(R) = W_\infty(R)$ for the ring of infinite Witt vectors. The "Frobenius" $x \mapsto x^p$ on R lifts to a ring homomorphism $\sigma : W \to W$. From now on we use a perfect field K. In that case $\sigma : W(K) \to W(K)$ is an isomorphism. We write E for the ring "power series" in the variables F and V with coefficients in W with the relations $FV = p = VF$ and $Fa = a^\sigma F$ and $aV = Va^\sigma$ for $a \in W$. Note that E is commutative iff $K = \mathbb{F}_p$.

Finite group schemes over a perfect field. Denote by $\mathcal{N} = \mathcal{N}_{loc,loc,K}$ the category of finite group schemes over the perfect field K which are of local-local type. Dieudonné module theory tells us:

There is a covariant equivalence: $\mathbb{D}$ *between* $\mathcal{N}$ *and the category of* E*-modules of finite length on which* F *and* V *act nilpotently. This equivalence has the following properties:*

$$\text{rk}(N) = p^t \quad \Longleftrightarrow \quad \ell(\mathbb{D}(N)) = t.$$

Moreover (with apologies...):

$$\mathbb{D}(F : N \to N^{(p)}) = (\mathcal{V} : A \to A^{(p)}),$$

$$\mathbb{D}(V : N^{(p)} \to N) = (\mathcal{F} : A^{(p)} \to A);$$

notation: the module $A^{(p)}$ is the one obtained from A obtained by the base change σ. We have distinguished the action of Frobenius on group schemes, denoted by F, and the action on Dieudonné modules given by Frobenius, which we indicate by $\mathcal{F}$. In the central part of this paper, we consider an abelian variety X, we write $N := X[p]$, we write $A := \mathbb{D}(N)$ for the Dieudonné module of this p-kernel. We use the operation V on N, which corresponds with $\mathcal{F}$ on A.

(15.4). **Serre duality.** For a p-divisible group $G = \text{ind.lim.}_{i\to\infty} G_i$ (over an arbitrary base) we follow Serre in defining the dual p-divisible group:

$$G^t = \text{ind.lim.}_{i\to\infty}\ G_i^D;$$

here $-^D$ is the Cartier dual, and $\times p : G_{i+1} \to G_i$ dualizes to inclusions $G_i^D \subset G_{i+1}^D$ which we use to define G^t.

Note that the duality theorem on abelian schemes, see [42], Th. 19.1, implies that for an abelian scheme $X \to S$ we have:

$$X[p^\infty]^t \quad = \quad X^t[p^\infty].$$

(15.5). **Classification by Dieudonné and Manin, notation.** In [31] we find a classification for isogeny classes of p-divisible groups over an algebraically closed field $k \supset \mathbb{F}_p$. For a pair of non-negative integers $m, n \in \mathbb{Z}_{\geq 0}$ a p-divisible group $G_{n,m}$ is defined. We write $G_{1,0} = \mathbb{G}_m^\wedge = \mathbb{G}_m[p^\infty]$. We write $G_{0,1}$ for its dual, which in fact is the "constant group scheme" $\underline{\mathbb{Q}_p/\mathbb{Z}_p}$.

For $m > 0$ and $n > 0$ which are relatively prime positive integers we write $G_{m,n}$ for a formal group of dimension m, whose Serre dual has dimension n, and which is "isosimple". It is defined by:

$$\mathbb{D}(G_{m,n}) = W[[F,V]]/W[[F,V]]{\cdot}(F^m - V^n).$$

Over $\mathbb{F}_p$ already we have the isomorphism:

$$(G_{m,n})^t \quad = \quad G_{n,m}.$$

The Dieudonné - Manin classification theorem reads:

Let $k \supset \mathbb{F}_p$ be an algebraically closed field. For a p-divisible group G over k, of dimension d and height h, there exist pairs $\{(m_i, n_i) \mid 1 \leq i \leq s\}$ and an isogeny:

$$G \sim \sum_i\ G_{(m_i,n_i)},\ \sum m_i = d,\ \sum (m_i + n_i) = h.$$

The set of these pairs is called the *formal isogeny type* of G. In case $G \cong X[p^\infty]$, the p-divisible group of an abelian variety, the formal isogeny type is "symmetric" and we can change the notation above into:

$$X[p^\infty] \sim f\cdot(G_{1,0} \oplus G_{0,1}) \bigoplus s\cdot G_{1,1} \bigoplus \sum_i^{<\infty} (G_{m_i,n_i} \oplus G_{n_i,m_i}),$$

where $f \in \mathbb{Z}_{\geq 0}$, and $s \in \mathbb{Z}_{\geq 0}$, and the integers m_i, and n_i are pairwise coprime with $m_i > n_i > 0$.

A formal isogeny type can be encoded via the notion of a **Newton polygon** abbreviated NP: for a p-divisible group G of height h and dimension d a Newton polygon is a polygon in $\mathbb{R}^2$ (or in $\mathbb{Q}^2$, if you prefer that), starting at $(0,0)$, ending at $(h, h-d)$, which is lower convex and which has break points in $\mathbb{Z}^2$. The p-divisible group $G_{m,n}$ gives the slope $n/(n+m)$ with multiplicity $n+m$. A direct sum as above gives a Newton polygon by ordering the slopes in increasing order. Note that the local-étale part gives all slopes equal to 0, the local-local summands give slopes strictly between 0 and 1, and the étale-local part gives all slopes 1.

(15.6). **Displays.** This tool was invented by Mumford, see [36], developed in [38], [39], and by Zink in various manuscripts, see [63], [64], [65]; this describes deformation theory by giving the Frobenius mapping on what we would like to consider as the "Dieudonné module of the deformed p-divisible group". In this paper we do not use the full strength of that method. What we need is the following (for more details on the notations, see [46], 1.9 - 1.12). Suppose G_0 is a p-divisible group over a perfect field $K \supset \mathbb{F}_p$, and let $M_0 = \mathbb{D}(G_0)$ be its covariant Dieudonné module. Let $\{X'_1, \cdots, X'_d, Y'_1, \cdots Y'_c\}$ be a W-basis for M_0, such that $Y'_1, \cdots Y'_c \in \mathcal{V}(M)$; here we write X'_i etc. because in Section 9 the notation X_i, and Y_j has been used for elements of a standard base for $\mathbb{D}(G_0[p])$.

Displays of a p-divisible group over a field. In this situation we define the display of this p-divisible group over a field. The σ-linear map $F : M_0 \to M_0$ can be written out on the W-basis $\{e_1 = X'_1, \cdots, e_d = X'_d, e_{d+1} = Y'_1, \cdots, e_{d+c} = Y'_c\}$ as:

$$\begin{aligned} Fe_j &= \sum_{i=1}^{h} a_{ij}e_i & & 1 \leq j \leq d, \\ e_j &= V\left(\sum_{i=1}^{h} a_{ij}e_i\right) & & d < j \leq h = c+d; \end{aligned}$$

we say that we have written the module in *displayed form*. We shall write

$$(a_{i,j} \mid 1 \leq i,j \leq h) = \begin{pmatrix} A & B \\ C & D \end{pmatrix};$$

This matrix, denoted by (a), will be called the matrix of the display. Note that in this case the σ-linear map F is given on this base by the matrix

$$\begin{pmatrix} A & pB \\ C & pD \end{pmatrix},$$

where

$$A = (a_{ij} \mid 1 \le i,j \le d), \quad B = (a_{ij} \mid 1 \le i \le d < j \le h),$$

$$C = (a_{ij} \mid 1 \le j \le d < i \le h,), \quad D = (a_{ij} \mid d < i,j \le h).$$

Where the display-matrix is symbolically denoted by (a), we write (pa) symbolically for the associated F-matrix (it is clear what is meant as soon as d is given).

Suppose R is a complete Noetherian local ring with perfect residue class field K. We assume $p{\cdot}1 = 0 \in R$. Let $t_{r,s}$ be elements in the maximal ideal of R, with $1 \le r \le d < s \le h$, and let

$$T_{r,s} = (t_{r,s}, 0, \cdots) \in W(R)$$

be their Teichmüller lifts. Write

$$T = \begin{pmatrix} T_{1,d+1} & \cdots & T_{1,h} \\ \vdots & & \vdots \\ T_{d,d+1} & \cdots & T_{d,h} \end{pmatrix}, \qquad \begin{pmatrix} A+TC & pB+pTD \\ C & pD \end{pmatrix}.$$

We can ask whether this is the matrix of a display connected with a deformation of G_0. In fact:

Let K be a perfect field, let R be a complete Noetherian local ring with residue class field K, and let G_0 be a formal p-divisible group over K. The formulas above define a display over the ring W of Witt vectors over R (in the sense of [64]). *Hence these formulas define a deformation $G \to \mathrm{Spec}(R)$ of G_0.*

Remark. One can show that the deformation just given is the universal deformation of G_0 in equal characteristic p, by taking the elements $t_{r,s}$ as parameters,

$$R := K[[t_{r,s} \mid 1 \le r \le d < s \le h]].$$

Suppose moreover the p-divisible group G_0 has a principal quasi-polarization λ_0 and let the base $\{X'_1, \cdots, Y'_d\}$ be symplectic (in this case $c = h - d = d$). *Then assume moreover that*

$$t_{r,s} = t_{s-d,r+d} \in R;$$

the displayed form above defines a deformation (G, λ) as quasi-polarized formal p-divisible group of (G_0, λ_0).

In this case we renumber the elements as $x_{i,j} = t_{i,j+d}$, with $d = g = c = h/2$. In this case the formal power series ring

$$R := K[[x_{i,j} \mid 1 \le i,j \le g]]/(x_{i,j} - x_{j,i} \mid 1 \le i,j \le g)$$

defines the universal deformation space of (G_0, λ_0).

This gives an explicit description of the deformation theory of local-local p-divisible groups over complete local domains in characteristic p; also the deformation theory of such groups with a principal quasi-polarization, and hence, by the theorem of Serre and Tate, of principally polarized abelian varieties (the theory can be applied to much more general situations, but we will not need that here).

References

[1] C.-L. Chai - *Compactification of Siegel moduli schemes.* London Math. Soc. Lect. Notes Ser. 107, Cambridge Univ. Press, 1985.

[2] C.-L. Chai - *Every ordinary symplectic isogeny class in positive characteristic is dense in the moduli.* Invent. Math. **121** (1995), 439 - 479.

[3] C.-L. Chai & F. Oort - *Density of Hecke orbits.* [In preparation.]

[4] P. Deligne & M. Rapoport - *Les schémas de modules de courbes elliptiques.* In: Modular functions of one variable, II (Ed. P. Deligne, W. Kuyk), Lect. Notes Math. 349, Springer - Verlag 1973; pp. 143 - 316 (pp. DeRa 1 - 174).

[5] M. Demazure - *Lectures on p-divisible groups.* Lect. Notes Math. 302, Springer-Verlag, 1972.

[6] J. Dieudonné - *Lie groups and Lie hyperalgebras over a field of characteristic p.* II. Amer. Journ. Math. **77** (1955), 218 - 244.

[7] M. Eichler - *Zur Zahlentheorie der Quaternionen-Algebren.* J. Reine Angew. Math. **195** (1955), 127 - 151.

[8] T. Ekedahl - *On supersingular curves and abelian varieties.* Math. Scand. **60** (1987), 151 - 178.

[9] G. Faltings - *Arithmetische Kompaktifizierung des Modulraums der abelschen Varietäten.* Arbeitstagung Bonn 1984 (Ed. F. Hirzebruch et al.), Lect. Notes Math. 1111, Springer-Verlag, 1985.

[10] G. Faltings & C.-L. Chai - *Degeneration of abelian varieties.* Ergebn. Math. 3. Folge, Bd. 22, Springer - Verlag, 1990.

[11] G. van der Geer - *Cycles on the moduli space of abelian varieties.* In: Moduli of curves and abelian varieties (the Dutch intercity seminar on moduli) (Eds C. Faber, E. Looijenga). Aspects of Math. E 33, Vieweg 1999; pp. 65 - 89.

[12] G. van der Geer & M. van der Vlugt – *On the existence of supersingular curves of given genus.* Journ. reine angew. Math. **45** (1995), 53 - 61.

[13] E. Z. Goren & F. Oort - *Stratification of Hilbert modular varieties I.* Journ. Algebraic Geom. **9** (2000), 111 - 154.

[14] A. Grothendieck - *Fondements de la géométrie algébrique.* [Extraits du Séminaire Bourbaki 1957 - 1962.] Exp. 212, Sém. Bourbaki **13** (1960/1961): *Préschémas quotients.*

[15] A. Grothendieck & J. Dieudonné - *Éléments de géométrie algébrique.* Vol. II: Étude globale élémentaire de quelques classes de morphismes. Vol. III^1: Étude cohomologique des faisceaux cohérents (Première Partie). Publ. Math. **8**, **11**, Inst. Hautes Ét. Sc. 1961, 1961 [Cited as EGA II, EGA III].

[16] H. Hasse & E. Witt - *Zyklische unverzweigte Erweiterungskörper vom Primzahlgrade p über einem algebraischen Funktionenkörper der Charakteristik p.* Monatsh. Math. Phys. **43** (1963), 477 - 492.

[17] L. Illusie - *Déformations de groupes de Barsotti-Tate.* Exp.VI in: Séminaire sur les pinceaux arithmétiques: la conjecture de Mordell (Ed. L. Szpiro), Astérisque 127, Soc. Math. France, 1985.

[18] A. J. de Jong – *Finite locally free group schemes in characteristic p and Dieudonné modules.* Invent. Math. **114** (1993), 89 - 137.

[19] A. J. de Jong & F. Oort - *Purity of the stratification by Newton polygons.* Journ. A.M.S. **13** (2000), 209 - 241. See: http://www.ams.org/jams

[20] T. Katsura & F. Oort - *Families of supersingular abelian surfaces.* Compos. Mat. **62** (1987), 107 - 167.

[21] T. Katsura & F. Oort - *Supersingular abelian varieties of dimension two or three and class numbers.* Algebraic Geometry, Sendai 1985 (ed. T. Oda); Adv. St. Pure Math. **10** (1987), Kinokuniya & North-Holland, 1987.

[22] N. Katz – *Serre-Tate local moduli.* Exp. V^{bis} in: Surfaces algébriques, Sém. Géom. Alg. Orsay 1976-78 (Ed. J. Giraud, L. Illusie and M. Raynaud), Lect. Notes Math. 868, Springer - Verlag 1981, pp. 138 - 202.

[23] N. M. Katz - *Slope filtrations of F-crystals.* Journ. Géom. Algébr. de Rennes, Vol. I; Astérisque 63, Soc. Math. France, 1979; pp. 113 - 163.

[24] M. Kneser - *Strong approximation.* Proceed. Sympos. Pure Math., Vol. 9: Algebraic groups and discontinuous subgroups. A.M.S. 1966.

[25] M.-A. Knus - *Quadratic and hermitian forms over rings.* Grundl. math. Wissensch. 294, Springer - Verlag 1991.

[26] M.-A. Knus, A. Merkurjev, M. Rost & J.-P. Tignol - *The book of involutions.* AMS Colloq. Publ. 44, Amer. Math. Soc. 1998.

[27] H.-P. Kraft - *Kommutative algebraische p-Gruppen (mit anwendungen auf p-divisible Gruppen und abelsche Varietäten).* Sonderforsch. Bereich Bonn, September 1975. Ms. 86 pp.

[28] H.-P. Kraft & F. Oort - *Finite group schemes annihilated by p.* [In preparation.]

[29] K.-Z. Li & F. Oort - *Moduli of supersingular abelian varieties.* Lecture Notes Math. 1680, Springer - Verlag 1998.

[30] J. Lubin, J-P. Serre & J. Tate – *Elliptic curves and formal groups.* In: Lect. Notes, Summer Inst. Algebraic Geometry, Woods Hole, July 1964; 9 pp.

[31] Yu. I. Manin - *The theory of commutative formal groups over fields of finite characteristic.* Usp. Math. **18** (1963), 3-90; Russ. Math. Surveys **18** (1963), 1 - 80.

[32] W. Messing – *The crystals associated to Barsotti-Tate groups: with applications to abelian schemes.* Lect. Notes Math. 264, Springer-Verlag 1972.

[33] B. Moonen – *Group schemes with additional structures and Weyl group cosets.* This volume, pp. 255–298.

[34] L. Moret-Bailly - *Familles de courbes et de variétés abéliennes sur* $\mathbb{P}^1$. *Familles de courbes et de variétés abéliennes.* Sém. sur les pinceaux de courbes de genre aux moins deux. Ed. L. Szpiro. Astérisque 86, Soc. Math. France, 1981; pp. 109 - 140.

[35] L. Moret-Bailly - *Pinceaux de variétés abéliennes.* Astérisque 129, Soc. Math. France, 1985.

[36] D. Mumford – *Constructions of deformation of Dieudonné modules.* Private communication, 1968.

[37] D. Mumford - *Abelian varieties.* Tata Inst. Fund. Res. & Oxford Univ. Press, 1970 (2nd print. 1974).

[38] P. Norman – *An algorithm for computing local moduli of abelian varieties.* Ann. Math. **101** (1975), 499-509.

[39] P. Norman & F. Oort - *Moduli of abelian varieties.* Ann. Math. **112** (1980), 413 - 439.

[40] T. Oda - *The first de Rham cohomology group and Dieudonné modules.* Ann. Sc. Ecole Norm. Sup. **2** (1969), 63 - 135.

[41] A. Ogus - *Supersingular K3 crystals.* In: Journ. Géom. Algébr. de Rennes, Vol II; Astérisque 64, Soc. Math. France, 1979; pp. 3 - 86.

[42] F. Oort - *Commutative group schemes.* Lect. Notes Math. 15, Springer-Verlag 1966.

[43] F. Oort - *Subvarieties of moduli spaces.* Invent. Math. **24** (1974), 95 - 119.

[44] F. Oort - *Which abelian surfaces are products of elliptic curves?* Math. Ann. **214** (1975), 35 - 47.

[45] F. Oort - *A stratification of moduli spaces of polarized abelian varieties in positive characteristic.* In: Moduli of curves and abelian varieties (the Dutch intercity seminar on moduli) (Eds C. Faber, E. Looijenga). Aspects of Math. E 33, Vieweg 1999; pp. 47 - 64.

[46] F. Oort - *Newton polygons and formal groups: conjectures by Manin and Grothendieck.* [To appear: Ann. Math. **152** (2000).]

[47] F. Oort -*Newton polygon strata in the moduli space of abelian varieties.* This volume, pp. 417–440.

[48] F. Oort - *Foliations of moduli spaces.* [In preparation.]

[49] V. Platonov & A. Rapinchuk - *Algebraic groups and number theory.* Pure and appl. math. Vol. 139, Academic Press 1994.

[50] M. Poletti - *Differenziali esatti di prima specie su varietà abeliane.* Ann. Scuola Norm. Sup. Pisa **21** (1967), 107 - 110.

[51] G. Prasad - *Strong approximation for semi-simple groups over function fields.* Ann. Math. **105** (1977), 553 - 572.

[52] I. Reiner - *Maximal orders.* Academis Press 1975.

[53] W. Scharlau - *Quadratic and hermitian forms.* Grundl. math. Wissensch. 270, Springer - Verlag 1985.

[54] *Séminaire de géométrie algébrique*, 1967-1969, SGA 7 : *Groupes de monodromie en géométrie algébrique.* (A. Grothendieck, M. Raynaud, D. S. Rim.) Vol. I, Lect. Notes Math. 288, Springer - Verlag, 1972.

[55] J-P. Serre - *Cohomologie galoisienne.* Cinquième édition, révisée et complétée. Lect. Notes Math. 5, Springer - Verlag 1994.

[56] J-P. Serre - *Sur la topologie des variétés algébriques en caractéristique p.* Symp. Int. Top. Alg. Mexico (1985), pp. 24-53 (see Collected works, Vol. I, pp. 501-530).

[57] G. Shimura - *Arithmetic of alternating forms and quaternion hermitian forms.* Journ. Math. Soc. Japan **15** (1963), 33 - 65.

[58] T. Shioda - *Supersingular K3 surfaces.* In: Algebraic Geometry, Copenhagen 1978 (Ed. K. Lønsted). Lect. Notes Math. 732, Springer-Verlag, 1979; pp. 564-591.

[59] T. A. Springer – *Linear algebraic groups.* Second edition. Progress Math. 9, Birkhäuser, 1998.

[60] L. Szpiro - *Propriétés numériques du faisceau dualisant relatif.* Sém. sur les pinceaux de courbes de genre aux moins deux. Ed. L. Szpiro. Astérisque 86, Soc. Math. France, 1981; pp. 44-78.

[61] T. Wedhorn - *The dimension of Oort strata of Shimura varieties of PEL-type.* This volume, pp. 441–471.

[62] Yu. G. Zarhin - *Endomorphisms of abelian varieties and points of finite order in characteristic p.* Math. Notes Acad. Sci. USSR **21** (1977), 734 - 744.

[63] T. Zink - *Cartiertheorie kommutativer formaler Gruppen.* Teubner-Texte Math. 68, Teubner, Leipzig, 1984.

[64] T. Zink - *The display of a formal p-divisible group,* University of Bielefeld, Preprint 98–017, February 1998.

[65] T. Zink - *A Dieudonné theory for p-divisible groups.* Manuscript, 23 pp., September 1998.

Progress in Mathematics, Vol. 195, © 2001 Birkhäuser Verlag Basel/Switzerland

NEWTON POLYGON STRATA IN THE MODULI SPACE OF ABELIAN VARIETIES

FRANS OORT

INTRODUCTION

We consider p-divisible groups (also called Barsotti-Tate groups) in characteristic p, abelian varieties, their deformations, and we draw some conclusions.

For a p-divisible group (in characteristic p) we can define its Newton polygon. This is invariant under isogeny. For an abelian variety the Newton polygon of its p-divisible group is "symmetric". We are interested in the strata defined by Newton polygons in local deformation spaces, or in the moduli space of polarized abelian varieties.

In deformation theory of a p-divisible group it is difficult to keep track of behavior of its Newton polygon. Isogeny correspondences between components of the moduli space of polarized abelian varieties in characteristic zero are finite-to-finite; however in general such correspondences blow up and down in characteristic p. Hence an isogeny invariant seems difficult to follow in local deformation theory. That is the true origin of fascinating aspects of the problem we are considering.

Grothendieck showed that Newton polygons "go up" under specialization, see [5], page 149, see [10], Th. 2.3.1 on page 143; we obtain Newton polygon strata as closed subsets in the deformation space of a p-divisible group or in the moduli space of polarized abelian varieties in positive characteristic.

In 1970 Grothendieck conjectured the converse. In [5], the appendix, we find a letter of Grothendieck to Barsotti, and on page 150 we read: "$\cdots$ *The wishful conjecture I have in mind now is the following: the necessary conditions* $\cdots$ *that* G' *be a specialization of* G *are also sufficient. In other words, starting with a BT group* $G_0 = G'$*, taking its formal modular deformation* $\cdots$ *we want to know if every sequence of rational numbers satisfying* $\cdots$ *these numbers occur as the sequence of slopes of a fiber of* G *at some point of* S*.*"

In this paper we show that this conjecture for p-divisible groups by Grothendieck indeed is true, see (2.1). We show that the analogue for *principally* (quasi) polarized formal groups is true, see (3.1). We show that the analogue is true for *principally* polarized abelian varieties, see (3.2); this is a particular case of the theorems we prove about Newton polygon strata. As a result we can give precise information on these strata in Section 3 (their dimension can be computed from combinatorial data; we show that generically on such a stratum the a-number equals one) in case of *principally* polarized abelian varieties.

We study deformations keeping track of information about the Newton polygon. The proof relies on two rather different aspects of deformation theory of p-divisible groups.

- On the one hand we have studied deformations of simple p-divisible groups, *keeping the Newton polygon constant.* We use methods and results derived from "Purity" as obtained in [8]. This works fine for simple groups. However the use of "catalogues" for non-isoclinic groups does not seem to give what we want; it is not even clear that nice catalogues exist in general.
- On the other hand we studied deformation theory with $a(G_0) = 1$ as closed fiber; here we study deformations where *the Newton polygon jumps.* This works by using a non-commutative version of the theorem of Cayley-Hamilton from linear algebra, see [18]. Note however that the Cayley-Hamilton approach breaks down essentially for $a(G_0) \geq 2$.

In this paper a combination of the two methods gives what we want. Hence, in spirit, the proof of a straight statement is not uniform. We have not been able to unify these methods in one straightforward method. We wonder what Grothendieck would have substituted for our proof.

Here is a survey of results on Newton polygon strata inside $\mathcal{A} := \mathcal{A}_{g,1} \otimes \mathbb{F}_p$.

- *For every symmetric Newton polygon ξ the locus $\mathcal{W}_\xi(\mathcal{A}) = W_\xi \subset \mathcal{A}$ is closed* (Grothendieck-Katz); the definition of $\mathcal{W}_\xi(\mathcal{A})$ is in (1.2).
- *For every ξ and every irreducible component $W \subset W_\xi$ its dimension can be computed easily:* $\dim(W) = \mathrm{sdim}(\xi)$; for the notation of this combinatorial invariant, see (1.9).
- *Generically on W the Newton polygon equals ξ and generically the a-number is at most one,* see (4.1).
- Note that the supersingular locus $\mathcal{S}_{g,1}$ has "many components" (for $p >> 0$; Deuring-Eichler for $g = 1$, Katsura-Oort for $g \leq 3$, Li-Oort for all g), but we expect:
- *for every $\beta \neq \sigma$ the locus W_β is geometrically irreducible* (?), see (5.1).

We remark that in case of quasi-polarized p-divisible groups, or polarized abelian varieties, *without supposing that the polarization is principal,* there are counterexamples to the existence of deformations with given closed fiber, and expected Newton polygon for the generic fiber.

The interesting phenomenon that Hecke correspondences (finite-to-finite in characteristic zero) cause blowing up and down between various strata in positive characteristic is the origin of these interesting facts. It seems a miracle that for the principally polarized case we obtain such coherent, aesthetically beautiful theorems.

Many people have patiently listened to me in discussions about this topic. Especially I mention, and I thank: Ching-Li Chai, Johan de Jong, Ke-Zheng Li, Shinichi Mochizuki and Ben Moonen for sharing their time and interests with me.

1. Some definitions, notations, and results we are going to use

Throughout the paper we fix a prime number p. We apply notions as defined and used in [18], and in [8]. For a p-divisible group G, or an abelian variety X, over a field of positive characteristic we use its Newton polygon, abbreviated by NP, denoted by $\mathcal{N}(G)$, respectively $\mathcal{N}(X)$. For dimension d and height $h = d+c$ of G (respectively dimension $g = d = c$ of X) this is a lower convex polygon in $\mathbb{R} \times \mathbb{R}$ starting at $(0,0)$, ending at (h,c), with integral break points, such that every slope is non-negative and at most equal to one. *We write $\gamma \prec \beta$ if every point of β is on or below* γ (the locus defined by β contains the one defined by γ). For further details we refer to [18].

(1.1). Suppose given a formal p-divisible group G_0 over k. We write $\mathcal{D} = \mathrm{Def}(G_0)$ for the universal deformation space in equal characteristic p. By this we mean the following. Formal deformation theory of G_0 is prorepresentable; we obtain a formal scheme $\mathrm{Spf}(A)$ and a prorepresenting family $\mathcal{G}' \to \mathrm{Spf}(A)$. By [7], Lemma 2.4.4 on page 23, we know that there is an equivalence of categories of p-divisible groups over $\mathrm{Spf}(A)$ respectively over Spec(A). We will say that $\mathcal{G} \to \mathrm{Spec}(A) = \mathcal{D} = \mathrm{Def}(G_0)$ is the universal deformation of G_0 if the corresponding $\mathcal{G}' \to \mathrm{Spf}(A) = \mathcal{D}^\wedge$ prorepresents the deformation functor.

(1.2). A theorem by Grothendieck and Katz, see [10], Th. 2.3.1 on page 143, says that for any family $\mathcal{G} \to S$ of p-divisible groups over a scheme S in characteristic p, and for any Newton polygon β there is a unique maximal closed, reduced subscheme $W \subset S$ containing all points s at which the fiber has a Newton polygon equal to or lying above β:

$$s \in W \iff \mathcal{N}(\mathcal{G}_s) \prec \beta.$$

This set will be denoted by

$$\mathcal{W}_\beta(\mathcal{G} \to S) \subset S.$$

For every Newton polygon β with $\beta \succ \gamma = \mathcal{N}(G_0)$ we define $V_\beta \subset \mathcal{D} = \mathrm{Def}(G_0)$ as the maximal closed, reduced subscheme carrying all fibers with Newton polygon equal to or above β in the universal deformation space of G_0.

In case of a family $(\mathcal{G}, \lambda) \to S = \mathrm{Spec}(R)$ of quasi-polarized p-divisible groups we have an analogous notation. In case of symmetric Newton polygons we write

$$\mathcal{W}_\xi(\mathcal{A}_g \otimes \mathbb{F}_p) = W_\xi$$

for the Newton polygon stratum given in the moduli space of polarized abelian varieties in characteristic p; this is obtained by choosing some level structure, studying the universal family, taking its p-divisible group, constructing the Newton polygon stratum, and considering the image in $\mathcal{A}_g \otimes \mathbb{F}_p$. We will study this mainly inside $\mathcal{A} := \mathcal{A}_{g,1} \otimes \mathbb{F}_p$, the moduli space of *principally* polarized abelian varieties in characteristic p.

(1.3). In "Purity" we obtained, see [8], Th. 4.1:
If in a family of p-divisible groups (say, over an irreducible, noetherian scheme) the Newton polygon jumps, then it already jumps in codimension one.

(1.4). For a commutative group scheme G over a field K (in characteristic p), we write $a(G)$ for the dimension of the L-vector space $\mathrm{Hom}(\alpha_p, G_L)$, where $L \supset K$ is any perfect field containing K.

Note that there exist examples in which

$$\dim_K(\mathrm{Hom}(\alpha_p, G)) < \dim_L(\mathrm{Hom}(\alpha_p, G_L)).$$

However, if $a(G) = 1$, then $\dim_K(\mathrm{Hom}(\alpha_p, G)) = 1 = \dim_L(\mathrm{Hom}(\alpha_p, G_L))$.

Note that $\mathrm{Hom}(\alpha_p, G) \neq 0$ iff the local-local part of G is non-trivial, i.e. iff G is not ordinary. Hence if we write $a(G) \leq 1$ we intend to say: either G is ordinary, or $a(G) = 1$.

(1.5). We recall one of the results obtained by the method of "Purity" as described in [8], Corollary 5.12:
Let G_0 be an absolutely simple p-divisible group over a field K of characteristic p. Then there exists a deformation G_η with $\mathcal{N}(G_\eta) = \mathcal{N}(G_0)$ and $a(G_\eta) \leq 1$.

(1.6). We use the following notation: we fix integers $h \geq d \geq 0$, and we write $c := h - d$. We consider Newton polygons ending at (h, c). For a point $(x, y) \in \mathbb{Z} \times \mathbb{Z}$ we write $(x, y) \prec \gamma$ for the property "*the point (x, y) is on or above the Newton polygon γ*". For a Newton polygon β we write:

$$\Diamond(\beta) = \{(x, y) \in \mathbb{Z} \times \mathbb{Z} \mid y < c, \quad y < x, \quad (x, y) \prec \beta\},$$

and we define

$$\dim(\beta) := \#(\Diamond(\beta)).$$

Note that for the "ordinary" Newton polygon $\rho := d{\cdot}(1,0) + c{\cdot}(0,1)$ the set of points $\Diamond = \Diamond(\rho)$ is a parallelogram; this explains our notation. Note that $\#(\Diamond(\rho)) = d{\cdot}c$.

(1.7). We recall one of the results obtained by the method of "Cayley-Hamilton" as described in [18], Theorem 3.2; from that we see, using the notation as in (1.6):
Let G_0 be a p-divisible group over an algebraically closed field $k \supset \mathbb{F}_p$ with $a(G_0) \leq 1$. In $\mathcal{D} := \mathrm{Def}(X_0)$ there exists a coordinate system $\{t_j \mid j \in \Diamond(\rho)\}$

and an isomorphism $\mathcal{D} \cong \mathrm{Spec}(k[[t_j \mid j \in \Diamond(\rho)]])$ *such that for any* $\beta \succ \mathcal{N}(X_0)$ *we have*

$$\mathcal{W}_\beta(\mathcal{D}) = \mathrm{Spec}(R_\beta),$$

with $R_\beta := k[[t_j \mid j \in \Diamond(\beta)]] = k[[t_j \mid j \in \Diamond(\rho)]]/(t_j \mid j \notin \Diamond(\beta))$.

(1.8). **Corollary.** *Let* G_0 *be a p-divisible group over a field* K *with* $a(G_0) \leq 1$. *In* $\mathrm{Def}(G_0)$ *every Newton polygon* $\beta \succ \mathcal{N}(G_0)$ *is realized.*
In fact, in $\mathrm{Def}(G_0)$ the NP-strata are closed subsets, and we have seen that the finite union of strata belonging to all Newton polygons above β and not equal to β is properly contained in the stratum given by β. □

(1.9). We fix an integer g. For every *symmetric* Newton polygon ξ of height $2g$ we define:

$$\triangle(\xi) = \{(x,y) \in \mathbb{Z} \times \mathbb{Z} \mid y < x \leq g, \quad (x,y) \prec \xi\},$$

and we write

$$\mathrm{sdim}(\xi) := \#(\triangle(\xi)).$$

For the ordinary symmetric Newton polygon $\rho = g{\cdot}((1,0)+(0,1))$ indeed $\triangle = \triangle(\rho)$ is a triangle; this explains our notation. But you can rightfully complain that the "triangle" $\triangle(\beta)$ in general is not a triangle.

For symmetric Newton polygons we have an analogous statement: for a symmetric Newton polygon $\xi \succ \gamma = \mathcal{N}(G_0)$ and $\mathcal{D} = \mathcal{D}(G_0, \lambda_0)$ we have:

$$\mathcal{W}_\xi(\mathcal{D}) = \mathrm{Spec}(R_\xi),$$

with $R_\xi := k[[t_j \mid j \in \triangle(\xi)]] = k[[t_j \mid j \in \triangle(\rho)]]/(t_j \mid j \notin \triangle(\xi))$.

(1.10). **Displays.** Given a Dieudonné module M of a p-divisible group, and a W-basis for the W-free module, the map $F: M \to M$ is given by a matrix, called a display. Mumford showed that deformations of certain p-divisible groups can be given by writing out a display over a more general base ring. What we need is contained in [19], [20]; also see [14], [15]. Below we construct deformations of local-local p-divisible groups. We shall write out the display, and use several times (without further mention) that this defines a deformation, see [19], Chapter 3, in particular Corollary 3.16. Deformations of quasi-polarized formal p-divisible groups can be described with the help of displays, see [15], Section 1.

Notations and results as described in [18] and in [8] will be used below.

We write K for an arbitrary field in characteristic p, and k for an algebraically closed field.

(1.11). **Remark.** We use *covariant* Dieudonné module theory. In that theory we have $\mathbb{D}(V : N^{(p)} \to N) = (F : M^{(p)} \to M)$, etc. The Newton polygon of a p-divisible group G over a perfect field K of characteristic p is defined as the Newton polygon of F on $M = \mathbb{D}(G)$, hence as the Newton polygon of V on G. This has the advantage that the natural filtration on group schemes (μ_p first, then α_p, finally $\mathbb{Z}/p$) gives rise on p-divisible groups to increasing slopes, as they appear naturally in the Newton polygon of a p-divisible group.

2. Deformations of p-divisible groups

(2.1). **Theorem** (conjectured by Grothendieck, Montreal 1970). *Let K be a field of characteristic p, and let G_0 be a p-divisible group over K. We write $\mathcal{N}(\mathcal{G}_0) =: \gamma$ for its Newton polygon. Suppose given a Newton polygon β "below" γ, i.e. $\gamma \prec \beta$. Then there exists a deformation G_η of G_0 such that $\mathcal{N}(\mathcal{G}_\eta) = \beta$.*

Deformations of p-divisible groups. For p-divisible groups there exists a solution to the universal deformation problem in equi-characteristic: given G_0 over a field K, there exists a formal p-divisible group $\mathcal{G} \to \mathrm{Spec}(A)$ which is universal for this problem, and $A \cong K[[t_1, \cdots, t_{dc}]]$. Here we use the passage from formal p-divisible groups over $\mathrm{Spf}(A)$ to p-divisible groups over $\mathrm{Spec}(A)$ as explained in (1.1) without further comments.

We say that $\mathcal{H}_\eta$ is a *deformation* of G_0, if there exists a complete local domain B of characteristic p with residue class field K and field of quotients $Q(B)$,

$$Q(B) \supset B \to K \quad \text{and a } p\text{-divisible group} \quad \mathcal{H} \to \mathrm{Spec}(B)$$

with special fiber $\mathcal{H} \otimes K \cong G_0$ and generic fiber $\mathcal{H} \otimes Q(B) = \mathcal{H}_\eta$.

Reduction of the problem. Inside $\mathcal{D} := \mathrm{Spec}(A)$, the base of the universal deformation space of G_0, for a given Newton polygon γ there is a closed $V_\gamma \subset \mathcal{D}$ carrying all p-divisible groups with Newton polygon equal or "above" γ, see [10], 2.3.2. In order to prove Theorem (2.1) we have to show that there is an irreducible component of V_γ with generic point parameterizing a p-divisible group having Newton polygon exactly equal to γ. Hence it suffices to show (2.1) in case $K = k$, an algebraically closed field, and in case G_0 is a local-local p-divisible group. From now on we make these assumptions, and we write h for the height of G_0, and d for its dimension, and $h = d + c$.

(2.2). *Filtered groups.* Consider a p-divisible group G_0 over an algebraically closed field k. Suppose its Newton polygon $\beta = \mathcal{N}(G_0)$ has $m+1$ points with integral coordinates

$$\#(\beta \cap (\mathbb{Z} \times \mathbb{Z})) = m + 1;$$

then there exist simple groups Z_i, with $1 \leq i \leq m$, corresponding with the slopes between integral points in γ (we arrange these slopes in some order), and an isogeny $\sum Z_i \to G_0$. We define a filtration

$$0 = G_0^{(0)} \subset \cdots \subset G_0^{(i)} \subset \cdots \subset G_0^{(m)} = G_0$$

by taking the image

$$\sum_{i \leq j} Z_i \quad \longrightarrow \quad G_0^{(j)} \subset G_0.$$

Such a filtration has the following property:

- The successive quotients $G_0^{(i)}/G_0^{(i-1)}$, with $0 < i \leq m$ are *simple p-divisible groups.*

This is called a *maximal filtration* of G_0. We will consider deformations of filtered groups.

Remark. We could have taken the filtration for example such that the successive quotients have non-decreasing slopes; however we do not need such a condition for our construction below.

(2.3). *A basis adapted to a filtration.* Suppose given a maximal filtration of G_0 over k as above. We write d_i for the dimension of $G_0^{(i)}$ and c_i for the dimension of its dual. Let $M_0 = \mathbb{D}(G_0)$ be the covariant Dieudonné module. This has a filtration by $\mathbb{D}(G_0^{(i)}) =: M_0^{(i)} \subset M_0^{(m)} = M_0$. We write $W = W_\infty(k)$. We say that $\{x_1, \cdots, x_d;\ y_1, \cdots, y_c\} \subset M_0$ is a *basis adapted to the filtration* if it is a W-basis for M_0 with $y_j \in VM_0$, and moreover $x_j \in M_0^{(i)}$ iff $j \leq d_i$ and $y_j \in M_0^{(i)}$ iff $j \leq c_i$.

Claim: *Such a basis exists.* In fact, note that if $H \subset G$ is a sub-p-divisible group, then $\mathbb{D}(H) \subset \mathbb{D}(G)$ is a submodule such that the quotient has no W-torsion; hence it is a W-direct summand. Using this, we conclude the existence of a W-basis of the given form.

(2.4). **Lemma.** *There exists a deformation $\{\mathcal{G}^{(i)} \mid 0 \leq i \leq m\}$ of filtered p-divisible groups such that every sub-quotient*

$$\mathcal{Y}^{(i)} := \mathcal{G}^{(i)}/\mathcal{G}^{(i-1)}$$

is an absolutely simple p-divisible group of constant slope, with $a(\mathcal{Y}_\eta^{(i)}) \leq 1$ for all $0 < i \leq m$.

Proof. We show that this is a consequence of [8], Corollary (5.12). It suffices to prove the lemma in case G_0 is local-local. We describe deformations of p-divisible groups (in equal characteristics), and in particular of filtered p-divisible groups with the help of displays. Consider $M_0 = \mathbb{D}(G_0)$ and a basis adapted to the filtration as above. We consider the display of M_0, i.e. the matrix of the σ-linear map $F : M_0 \to M_0$ on this W-basis can be given in block form

$$\begin{pmatrix} A & pB \\ C & pD \end{pmatrix},$$

and its display is

$$\begin{pmatrix} A & B \\ C & D \end{pmatrix} = \begin{pmatrix} (A_{a,b}) & (B_{a,f}) \\ (C_{e,b}) & (D_{e,f}) \end{pmatrix}, \qquad 1 \leq\ a,\ b,\ e,\ f\ \leq m.$$

Here, for fixed $a = b = e = f = i$ the four "diagonal" blocks give the display of the induced display on $Y_0^{(i)} = G_0^{(i)}/G_0^{(i-1)}$. As this is the display of a filtered group we have $A_{a,b} = 0$ for $a > b$, $\cdots$, $D_{e,f} = 0$ for $e > f$.

All formal deformations of G_0 are given by a $d \times c$ matrix T, with $T_{i,j} = (t_{i,j}, 0, \cdots)$, the Teichmüller lifts of the variables $t_{i,j}$. According to the mumbers d_i and c_i this matrix is written in blockform $T = (\tau_{a,b} \mid 1 \leq a, b \leq m)$, and the universal deformation given by T is the display

$$\begin{pmatrix} A+TC & B+TD \\ C & D \end{pmatrix}.$$

For every simple factor $Y_0^{(i)}$ there is a matrix $\tau_{i,i}$ giving a deformation of $Y_0^{(i)}$ to $Y_\tau^{(i)}$ with generic fiber having the same Newton polygon as $Y_0^{(i)}$ and a-number equal to one: this is precisely [8], Corollary (5.12). Using the matrix T in block form with these blocks as diagonal elements, and $\tau_{i,j} = 0$ for all $i \neq j$ we achieve a deformation of filtered groups of the filtered G_0 which satisfies the requirements of the lemma. This finishes the proof. □

(2.5). **Remark.** In case we are in the polarized case, the basis will be slightly different from the choice in the proof of (2.4); however it seemed unnecessary to take already here precautions needed in that case.

(2.6). We recall the notion of a display *in normal form*; in this paper we need only the "modulo p version", which is slightly different from the one considered in [18], 2.1. Let Y be a local-local p-divisible group of dimension u and codimension v over a field. Suppose Y is given by a display. Here, a W-basis for the display is said to be in *normal form* if we have a basis $\{x_1, \cdots, x_u;\ y_1, \cdots, y_v\}$ with $F(x_i) = x_{i+1}$ for $i < u$ and $F(x_u) = y_1$ (i.e. the left hand upper block has entries equal to one just below the diagonal, rest zero, and the left hand lower block has one entry one in the right hand upper corner and rest zero). [Note that such a basis can exist only if $a(Y) = 1$.]

Suppose a filtration $\{H_0^{(i)} \mid 0 \leq i \leq m\}$ of a local-local G_0 considered as above moreover has the property that $a(H_0^{(i)}/H_0^{(i-1)}) = 1$. We say a basis adapted to the filtration is in *normal form adapted to the filtration* if the basis can be induced on each of the subfactors $\mathbb{D}(H_0^{(i)}/H_0^{(i-1)})$ and if on each of these factors it is in normal form.

Claim. *A display associated with a p-divisible group over a field K with a filtration with the extra condition* $a(H_0^{(i)}/H_0^{(i-1)}) = 1$ *for all i, allows the choice of a basis in normal form adapted to the filtration.* Indeed, for $N = \mathbb{D}(Y)$ as above, with $a(Y) = 1$, there exists a vector $x = x_1 \in N$ with $F^{u-1}x \notin VN + pN$; this follows by methods of p-linear algebra; note that the rank of the map induced by F on N/VN equals $d-1$; note that this is the same as the rank of the matrix on any basis; then $F^u x \notin pN$; hence for this module N a basis in normal form exists. We can lift these, chosen for each of the relative factors, to a basis in normal form for $\mathbb{D}(G_0)$.

(2.7). **Lemma.** *Suppose $\{H_0^{(i)} \mid 0 \leq i \leq m\}$ is a filtered local-local p-divisible group over a field K, with $a(H_0^{(i)}/H_0^{(i-1)}) = 1$ for $1 \leq i \leq m$. Then there exists a deformation $\{\mathcal{H}^{(i)} \mid 0 \leq i \leq m\}$ of filtered p-divisible groups such that $\mathcal{H}_\eta^{(i)}/\mathcal{H}_\eta^{(i-1)}$ and $H_0^{(i)}/H_0^{(i-1)}$ have the same slopes for $1 \leq i \leq n$, and such that $a(\mathcal{H}_\eta^{(m)}) = 1$.*

Proof. It suffices to prove this lemma in case we work over an algebraically closed field k. We use a basis $\{x_1, \cdots, x_d;\ y_1, \cdots, y_c\}$ for $M_0 = \mathbb{D}(G_0)$ in normal form adapted to the filtration; its existence was shown above. Next we choose the deformation matrix T. As above, this is in block form $T = (\ \tau_{i,j} \mid 1 \leq\ i,\ j \leq m)$. We choose variables $s_1, \cdots, s_{m-1}$, and their Teichmüller lifts $S_i = (s_i, 0, \cdots) \in W(k[[s_j \mid 1 \leq j \leq m-1]])$; we define: $\tau_{i,j} = 0$ if $i+1 \neq j$; we define $\tau_{i,i+1}$, with $1 \leq i < m$ as the matrix consisting of zero elements, except in the left hand upper corner, and there the element equals S_i; i.e. $T_{d_{i-1}+1, c_i+1} = S_i,\ \ 1 \leq i < m$, and all other elements in the matrix T are equal to zero (we write $d_0 = 0$). This matrix T defines a display; hence it defines a p-divisible group $\mathcal{H}_S \to \operatorname{Spec}(k[[s_1, \cdots, s_{m-1}]])$. By construction the deformation respects the filtration; moreover it leaves the successive quotients undeformed; hence the Newton polygon of its generic fiber equals $\beta := \mathcal{N}(G_0)$. Next we compute the Hasse-Witt matrix of the deformed group. This is given by the matrix $(A + TC) \bmod p$. In A we have diagonal blocks with zeros everywhere, except on the subdiagonal in the blocks, where the elements are one (this is because we have chosen the basis such that on partial quotients the matrix is in normal form). The contribution of the deformation is changing (by the variables s_i) the right hand upper corner of the blocks in $(A+TC)$ mod p immediately above the diagonal. Note that the element $(A+TC)_{c_i+1, d_{i+1}-1}$ is non-zero modulo p, for $1 \leq i \leq m-1$. We claim: *the rank of the Hasse-Witt matrix equals $d-1$*, i.e. $a(\mathcal{H}_\eta) = 1$; in fact, in the Hasse-Witt matrix of the display, delete the d_1-column and the (c_m+1)-th row; then permute columns and rows in such a way that the elements 1 which were just below the diagonal in the same order come on the diagonal, and then such that the elements $S_1, \cdots, S_{m-1}$ in this order are on the diagonal ; in the new matrix (modulo p), we have non-zero elements on the diagonal, and zeros below the diagonal; hence its determinant is non-zero; hence the original matrix has rank equal to $d-1$, which proves the claim. This finishes the proof. □

(2.8). **Proposition.** *Suppose G_0 is a p-divisible group over a field K. There is a deformation G_η of G_0 such that $a(G_\eta) \leq 1$ and $\mathcal{N}(G_0) = \mathcal{N}(G_\eta)$.*

Proof. We show that this follows, using (2.4) and (2.7). It suffices to give the proof in case of a local-local p-divisible group G_0 over an algebraically closed field k; we suppose this to be the case.

Let $\mathcal{G} \to \mathrm{Def}(G_0) = \mathrm{Spec}(R)$ be the universal deformation space. We choose a filtration on G_0, and let $\mathrm{Spec}(R_1) \subset \mathrm{Spec}(R)$ be a closed subset defined by an integral domain R_1 over which there is a deformation $Y \to \mathrm{Spec}(R_1)$ of filtered p-divisible groups as explained in (2.4). This defines (and is defined) by a matrix $(\mathcal{F})$ in displayed form over $W(R_1)$ in block form. Let $d = \dim(G_0)$ and $c = \dim(G_0^t)$. We write t for the matrix of variables $(t_{i,j} \mid 1 \leq i \leq d,\ 1 \leq j \leq c)$, we write T for the matrix of Teichmüller lifts to $W_\infty(R_1[[t]])$, and we write $(\mathcal{F})_T$ for the "deformation matrix" as in (2.4). This defines a filtered p-divisible group $\mathcal{H} \to \mathrm{Spec}(R_1[[t]])$. By the universal property of $\mathcal{G} \to \mathrm{Def}(G_0) = \mathrm{Spec}(R)$ there exists a morphism $\varphi : \mathrm{Spec}(R_1[t]]) \to \mathrm{Spec}(R)$ and $\mathcal{H} \to \mathcal{G}$ which is the pull back. We write $R_1 \subset Q(R_1) = L_1$ for its field of fractions, and $G_1 = \mathcal{G} \otimes_R L_1 = Y \otimes L_1$. We define

$$(\mathcal{H} \to \mathrm{Spec}(R_1[[t]])) \otimes_{R_1[[t]]} L_1[[t]] =: (\mathcal{K} \to \mathrm{Spec}(L_1[[t]]))\,.$$

Let L_2 be a perfect field containing L_1. We know that $G_2 := G_1 \otimes_{L_1} L_2$ is given by the matrix $(\mathcal{F})_T$. Hence $(\mathcal{K} \to \mathrm{Spec}(L_1[[t]])) \otimes_{L_1} L_2$ is the universal deformation space of $\mathcal{D}_2 = (\mathcal{L} \to \mathrm{Def}(G_2) = \mathrm{Spec}(L_2[[t]])$.

Let $\gamma = \mathcal{N}(G_0)$. By (2.7) the subscheme $W_\gamma(\mathcal{L} \to \mathrm{Def}(G_2) = \mathrm{Spec}(L_2[[t]])$ contains an irreducible component over which the generic fiber has $a(\mathcal{L}_\eta) = 1$. Hence the same holds for $W_\gamma(\mathcal{K} \to \mathrm{Spec}(L_1[[t]]))$. The morphisms

$$(\mathcal{K} \to \mathrm{Spec}(L_1[[t]])) \to (\mathcal{H} \to \mathrm{Spec}(R_1[[t]])) \to (\mathcal{G} \to \mathrm{Def}(G_0))$$

show that the same holds for the universal deformation space of G_0. This proves (2.8). □

(2.9). **Proof of** (2.1). We show that this follows, using (2.8) and [18]. Indeed, by (2.8) we have a deformation of G_0, keeping the Newton polygon constant, and achieving $a(G_\eta) \leq 1$ for the generic fiber over $L = K(\eta)$. Using the corollary in (1.7) we conclude that G_η over L admits a deformation to a fiber having Newton polygon equal to $\beta \succ \mathcal{N}(G_\eta) = \mathcal{N}(G_0)$. As in the previous proof we conclude that this shows that in $\mathrm{Def}(G_0)$ the Newton polygon β is realized. This finishes the proof of (2.1). □

(2.10). **Theorem.** *Suppose given a p-divisible group G_0 over a field K. Let β be a Newton polygon with $\beta \succ \mathcal{N}(G_0) =: \gamma$. Consider the closed reduced subscheme $\mathcal{W}_\beta(\mathcal{D}(G_0)) =: V_\beta \subset \mathcal{D}(G_0)$, see* (1.2). *The dimension of every component of V_β equals* $\dim(\beta) = \#(\Diamond(\beta))$*; generically on such a closed set the Newton polygon is β; on V_β the a-number generically is at most one.* (In fact on V_β the a-number generically is equal to one iff $\beta \neq \rho := d{\cdot}(1,0) + c{\cdot}(0,1)$.)

Proof. It suffices to prove the theorem in case G_0 is defined over an algebraically closed field k. In (2.8) we have seen that every p-divisible group can be deformed to one with the same Newton polygon, and having a-number at most one; as in the proof of (2.8) "transitivity of methods" shows that every irreducible component of V_β has at its generic point these properties. In [18] we find a description of the deformation theory of p-divisible groups with $a \leq 1$, see (1.7). Hence we know that deforming p-divisible groups with $a = 1$ every Newton polygon below β can be achieved. In [18] the dimension of the locus $V_\gamma(a = 1) \subset \mathrm{Def}(G_0)$ is computed to be purely equal to $\diamondsuit(\gamma)$. We see that the theorem follows from (2.8) and [18]. □

3. Deformations of principally quasi-polarized formal groups and of principally polarized abelian varieties

(3.1). **Theorem** (the principally polarized variant of the conjecture by Grothendieck, Montreal 1970). *Let K be a field of characteristic p, and let (G_0, λ_0) be a* principally *quasi-polarized p-divisible group over K. We write $\mathcal{N}(G_0) = \gamma$ for its Newton polygon. Suppose given a symmetric Newton polygon ξ "below" γ, i.e. $\gamma \prec \xi$. Then there exists a deformation (G_η, λ) of (G_0, λ_0) such that $\mathcal{N}(\mathcal{G}_\eta) = \xi$.*

(3.2). **Corollary.** *Let K be a field of characteristic p, and let (X_0, λ_0) be a* principally *polarized abelian variety over K. We write $\mathcal{N}(X_0) = \gamma$ for its Newton polygon. Suppose given a symmetric Newton polygon ξ "below" γ, i.e. $\gamma \prec \xi$. Then there exists a deformation $(\mathcal{X}_\eta, \lambda)$ of (G_0, λ_0) such that $\mathcal{N}(\mathcal{X}_\eta) = \xi$.*

Proof. The existence of a formal polarized abelian scheme as wanted follows from (3.1) using the Serre-Tate theorem on the equivalence between formal deformations of polarized abelian schemes and the corresponding quasi-polarized p-divisible groups, see [9], Th. 1.2.1. By the Chow-Grothendieck algebraization method for polarized formal schemes ("formal GAGA"), see [4], III^1.5.4, it follows that we obtain an actual abelian scheme. Its Newton polygon can be read off from its p-divisible group. Hence we see that (3.2) follows from (3.1). □

In order to give a proof of (3.1) we first make the reduction to the case that (G_0, λ_0) is defined over an algebraically closed field, and that G_0 is of local-local type (for polarized formal groups this is the same as saying that it is a local p-divisible group): if we assume the theorem proved for this special case, then it follows in general. From now on we assume that G_0 is a local p-divisible group with a principal quasi-polarization over an algebraically closed field; we write h for the height of G_0, and d for its dimension; then $h = 2d$, and G_0 is a local-local p-divisible group.

(3.3). Analogous to the definitions as in (2.2) we consider a filtration related to (G_0, λ), a *principally quasi-polarized filtered formal group* over an algebraically closed field.

Suppose the number of points with integral coordinates on the Newton polygon $\gamma = \mathcal{N}(G)$ is $m + 1$:

$$\#(\gamma \cap (\mathbb{Z} \times \mathbb{Z})) = m + 1.$$

We write s for the number of $G_{1,1}$-factors in the isogeny type of G_0; i.e. the number of slopes equal to $1/2$ is $2s$; we speak of the *even* case, respectively the *odd* case, in case s is even, respectively odd; we write $s = 2t$, respectively $s = 2t+1$. We write $m = 2n+s$. A *maximal symplectic filtration* of a *principally* quasi-polarized p-divisible group (G_0, λ) is a filtration

$$0 = G_0^{(0)} \subset \cdots \subset G_0^{(i)} \subset \cdots \subset G_0^{(m)} = G_0$$

- such that $G_0^{(i)}/G_0^{(i-1)}$ is simple for every $0 < i \leq m$, such that
- moreover $\lambda_0 : G_0 \to (G_0)^t$ induces an isomorphism on the subquotients

$$G_0^{(i)}/G_0^{(i-1)} \to (G_0^{(m-i+1)}/G_0^{(m-i)})^t$$

for every $0 < i \leq (m+1)/2$.

Let us explain the last condition. Note that an exact sequence of p-divisible groups

$$0 \to X \longrightarrow Y \longrightarrow Z = Y/X \to 0$$

yields an exact sequence

$$0 \to Z^t \longrightarrow Y^t \longrightarrow X^t \to 0.$$

The last condition in the definition of a (maximal) symplectic filtration says that for $0 < i \leq (m+1)/2$ we have a natural commutative diagram:

$$\begin{array}{ccccccc}
G_0 & \hookleftarrow & G_0^{(m-i+1)} & \longrightarrow & \left(G_0^{(m-i+1)}/G_0^{(i-1)}\right) & \hookleftarrow & \left(G_0^{(i)}/G_0^{(i-1)}\right) \\
\lambda_0 \downarrow\cong & & & & \downarrow\cong & & \downarrow\cong \\
G_0^t & \longrightarrow & \left(G_0^{(m-i+1)}\right)^t & \longleftarrow & \left(G_0^{(m-i+1)}/G_0^{(i-1)}\right)^t & & \\
 & & & & \| \cong & & \\
 & & & & G_0^{(m-i+1)}/G_0^{(i-1)} & \to & \left(G_0^{(m-i+1)}/G_0^{(m-i)}\right).
\end{array}$$

Such diagrams follow if the composite map $G_0^{(i-1)} \to G_0 \to (G_0)^t \to (G_0^{(m-i+1)})^t$ is zero for every $0 < i \leq (m+1)/2$.

Note that in the "odd case", i.e. m is odd, there is a "middle" step in the filtration, and it is required to be self dual; in case m is even, there is no middle step, and we need only consider $i \leq m/2$.

The conditions necessary for a filtration to be maximal and symplectic can also be expressed on the Dieudonné module $\mathbb{D}(G_0) = M_0$. One has to assume that $N_0^{(i)} \subset M_0$ has a torsion-free quotient, in order to obtain $G_0^{(i)} \subset G_0$. The property of being symplectic can be expressed with the help of the bilinear, non-degenerate, skew pairing on M_0 induced by the principal quasi-polarization.

In this way we see: if $X \subset G_0$ is a p-divisible subgroup, and λ_0 is a principal quasi-polarization, such that $(X \to G_0 \to (G_0)^t \to X^t) = 0$, then there exists a unique $X^\perp \subset G_0$ such that λ_0 induces a principal quasi-polarization on $X^\perp/X$.

(3.4). We recall a lemma which was proved in [11], 6.1. Over $\mathbb{F}_{p^2}$ we consider two series of quasi-polarized supersingular p-divisible groups.

$$\mathbf{I}_r \qquad (S,\tau_r),\ \ r\in\mathbb{Z}_{\geq 0},\quad S=G_{1,1},\quad \deg(\tau_r)=p^{2r},$$

$$\mathbf{II}_r \qquad (T,\nu_r),\ \ r\in\mathbb{Z}_{\geq 0},\quad T=(G_{1,1})^2,\quad \deg(\nu_r)=p^{2r}.$$

These are defined as follows. On the Dieudonné modules we take in the first case:

$$M=W{\cdot}x\oplus W{\cdot}Fx,\quad Fx=Vx,\quad \epsilon\in W-pW,\quad \epsilon^{\sigma}=-\epsilon,\quad <x,Fx>=p^r{\cdot}\epsilon.$$

In the second case:

$$M=W{\cdot}x\oplus W{\cdot}Fx\oplus W{\cdot}y\oplus W{\cdot}Fy,\quad Fx=Vx,\quad Fy=Vy,$$

and

$$<x,y>=p^r,\quad <x,Fx>=0=<y,Fy>=<x,Fy>=<y,Fx>.$$

With these notations: *For every supersingular quasi-polarized p-divisible group (H,ς) over an algebraically closed field k there is an isomorphism:*

$$(H,\varsigma)\quad\cong\quad\bigoplus\nolimits_j\ (H_j,\varsigma_j),$$

where each of the summands (H_j,ς_j) is of type **(I)** *or of type* **(II)**, *with some degree for the quasi-polarization.*

Remark. Suppose (S,τ_r) is of type $(\mathbf{I}_r)$ and u is an integer, with $u\geq r$. Then there is an isogeny $\psi: S\to S$ such that $(S,\psi^*\tau_r)\cong(S,\tau_u)$.

(3.5). **Lemma**. *Given a principally quasi-polarized p-divisible group (G_0,λ_0) over an algebraically closed field k, there exists a maximal symplectic filtration $(\{G_0^{(i)}\mid 0\leq i\leq m\},\lambda_0)$.*

Proof. A maximal symplectic filtration which will be constructed in the proof will depend on choices: the subgroups appearing in the first half of the filtration and their order will depend on the choice which factors in the isogeny type are chosen to appear on which place. We shall prove the lemma for local-local p-divisible groups (and, if necessary, one can insert later the local-étale factors and their duals).

The lemma will be proved by induction on the height of G_0. So we assume that for all principally polarized p-divisible groups of smaller height the lemma has been established.

In the induction step we write (Y,λ) for the pair (G_0,λ) in the lemma. We choose a direct sum of simple p-divisible groups H, an isogeny $\varphi: H\to Y$ and we construct $\varsigma:=\varphi^*(\lambda)$, arriving at a quasi-polarized p-divisible group (H,ς). We choose this isogeny φ in such a way that on all supersingular factors of type **(I)** the degree of the induced quasi-polarization is equal to some fixed number p^{2r}; by the remark above this choice can be made.

We perform the induction step by choosing a simple direct summand $D \overset{\leftarrow}{\hookrightarrow} H$ such that the induced map

$$(\varsigma|_D \colon D \to H \xrightarrow{\varsigma} H^t \to D^t) = 0.$$

After we make such a choice, we define $(X \subset Y) = \mathrm{Im}(\varphi) = (\varphi(X) = D \subset H))$. We conclude that

$$(X \hookrightarrow Y \xrightarrow{\lambda} Y^t \to X^t) = 0;$$

we see that in this case λ induces a principal quasi-polarization on $X^{\perp}/X$, and we carry on by induction.

As for the choice of $D \subset Y$ as above there are in general four basically different choices possible.

3) In case Y admits a simple factor of slope not equal to $1/2$, we can choose for $D \subset H$ such a factor. In this case $\mathrm{Hom}(D, D^t) = 0$, and we are done.

2) In case (H, ς) contains a supersingular direct summand T of Type (**II**), we choose $D \subset H$ such that $\mathbb{D}(D) = W{\cdot}x \oplus W{\cdot}Fx \subset \mathbb{D}(T) \subset \mathbb{D}(Y)$, using the notation explained above. This is a totally isotropic subspace, and a choice $X \subset Y$ as desired follows.

1) Suppose (H, ς) contains at least two direct summands of Type (**I**) (by construction with quasi-polarizations of the same degree)

$$W{\cdot}e \oplus W{\cdot}Fe \oplus W{\cdot}f \oplus W{\cdot}Ff = N \subset \mathbb{D}(H),$$

with $< e, Fe >= p^r{\cdot}\epsilon =< f, Ff >$, etc. We choose $\xi \in W(k)$ with $\xi \cdot \xi^{\sigma} = -1$; note that $\xi^{\sigma^2} = \xi$; in this case $z := e + \xi{\cdot}f$ generates a Dieudonné submodule $N' = W{\cdot}z \oplus W{\cdot}Fz \subset N$ which is isotropic for the form induced by ς; it is the Dieudonné module $N' = \mathbb{D}(D)$ of a p-divisible group $D \subset H$ of height 2, and its image $X \subset Y$ satisfies the condition $(X \hookrightarrow Y \to Y^t \to X^t) = 0$. In this case a choice for the induction step can be made.

0) Suppose that (H, ς) satisfies none of the possibilities in the previously considered three cases. Then either $Y = 0 = H$, and we are done. Or Y is supersingular of height 2; in this case it is of Type ($\mathbf{I}_0$), and it is filtered by $0 \subset Y$, which is of the form wanted in the lemma. Hence induction finishes the proof of (3.5). □

(3.6). **Definition.** *A symplectic basis adapted to a symplectic filtration.* Suppose given a principally quasi-polarized p-divisible group (G_0, λ) over an algebraically closed field k. We write $M := \mathbb{D}(G_0)$. Suppose $\{G_0^{(i)} \mid 0 \leq i \leq m\}$ is a symplectic filtration as in (3.3). Suppose $Y^{(i)} := G_0^{(i)}/G_0^{(i-1)}$ has dimension d_i and codimension c_i. Note that in this case $d_i = c_{m-i+1}$. Let us write $D_i := d_1 + \cdots + d_i$, i.e. this is the dimension of $G_0^{(i)}$.

We say that $\{x_1, \cdots, x_d;\ y_1, \cdots, y_d\}$ is a symplectic basis adapted to this symplectic filtration if

$$x_1, \cdots, x_{D_i}, y_{D_{m-i}+1}, \cdots, y_d \in M_i := \mathbb{D}(G_0^{(i)})$$

is a W-basis for $M := \mathbb{D}(G_0)$ (note this definition, in the symplectic case, is slightly different from (2.3)), and if it is symplectic i.e. $< x_i, y_i >= 1, \quad 0 < i \leq h$, and all other products between basis vectors are zero. We say moreover this basis adapted to a filtration is *in normal form* if it is in normal form modulo p, in the sense of (2.6), i.e. normal modulo p on all simple subfactors.

(3.7). **Lemma.** *Suppose given a principally polarized p-divisible group* (G, λ) *over an algebraically closed field* k*, with a maximal symplectic filtration* $\{X_i \mid 0 \leq i \leq m\}$*. We write* $\mathbb{D}(G) = M$*, and* $\{N_i = \mathbb{D}(X_i) \mid 0 \leq i \leq m\}$ *for its filtered Dieudonné module.*
(a) *There exists a symplectic basis adapted to this symplectic filtration.*
(b) *If moreover* $a(X_i/X_{i-1}) = 1$*, for all* $0 < i \leq m$*, there exists a normal symplectic basis adapted to this symplectic filtration.*
Proof. In case the number m of steps in the filtration is even we change slightly the notation; in that case we introduce an extra trivial step in the middle. In case m is odd, we do not change the filtration. In both cases we obtain $m = 2n + 2t + 1$. We achieve a filtration

$$0 = N_0 \subset N_1 \subset \cdots \subset N_n \subset \cdots \subset N_{n+t} \subset N_{n+t+1} \subset \cdots \subset N_{n+2t+1} \subset \cdots \subset N_m.$$

On this module there is an alternating non-degenerate bilinear form, denoted by $< -, - >$; we have $N_i^{\perp} = N_{m-i}$. We will choose the basis on M inductively by considering $P_j = N_{n+t+j+1}/N_{n+t-j}$. Note that the pairing on M induces a pairing with the same properties on each $P_j, \quad 0 \leq j \leq n+t$. Induction starts at $j = 0$: if $P_0 = 0$ (the case the original filtration had an even number of steps) we are done; if $P_0 \neq 0$, then $X_{n+t+j+1}/X_{n+t-j} \cong G_{1,1}$, the module P_0 has rank 2, and the choice of a symplectic basis in normal form is easy: $P_0 = W{\cdot}e \oplus W{\cdot}Fe$.

Induction hypothesis: On P_j *for* $0 \leq j < n+t$ *there exits a symplectic basis adapted to the filtration, in normal form in case* **(b)**. Induction step: We are going to construct the same for P_{j+1}.

In order to simplify, we choose a different notation, which will only be used in the induction step. We fix j as before, and write $P := P_j$, and $Q := P_{j+1}$. We assume we have already a basis $\{B_1'', \cdots, B_d''; \ b_1'', \cdots, b_d''\}$ in the required form for P. We write $N = N_{n+t-j}/N_{n+t-j-1} \subset Q = N_{n+t+j+2}/N_{n+t-j-1}$, and $N^{\perp} \subset Q$; note that $N^{\perp} = N_{n+t+j+1}$,

$$0 \subset N \subset N^{\perp} \subset Q \quad \text{and} \quad N^{\perp}/N = P.$$

We choose a basis $\{A_1, \cdots, A_u; \ a_1, \cdots, a_v\}$ for $N \subset Q$, and a basis $\{C_1'', \cdots, C_v''; \ c_1'', \cdots, c_u''\}$ for $Q/N^{\perp}$; these are chosen such that the duality $N \xrightarrow{\sim} (Q/N^{\perp})^D$ agrees with the choices made, and in case **(b)**, such that these bases are in normal form; these choices follow in case **(a)** from the structure of the problem: compare the various possibilities in the proof of (3.5); in case **(b)** we consider the finite length Dieudonné modules $U := N/p{\cdot}N$ and

$W := (Q/N^{\perp})/p{\cdot}(Q/N^{\perp})$; these are in perfect duality; both are generated by one element, as Dieudonné modules, in case **(b)**; a generator $A_1^{(0)}$ for U not in $FU+VU$ has the property that $F^u(A_1^{(0)})$ generates V^vU; using this we choose dual bases for U and W in normal form; then we lift these to the required bases for N and $N^D = Q/N^{\perp}$. By these preparations we have chosen bases:

$$\text{for} \quad N: \quad \{A_1, \cdots, A_u;\ a_1, \cdots, a_v\}, \quad N \subset Q,$$
$$\text{for} \quad P: \quad \{B_1'', \cdots, B_d'';\ b_1'', \cdots, b_d''\}, \quad P = N^{\perp}/N,$$
$$\text{for} \quad Q/N^{\perp}: \quad \{C_1'', \cdots, C_v'';\ c_1'', \cdots, c_u''\},$$

which satisfy the conditions for being symplectic on P and on $N \oplus (Q/N^{\perp})$, and being normal in **(b)** as explained:

$$\begin{array}{ccccc} A_i, a_j \in N & \subset & N^{\perp} & \subset & Q \\ & & \downarrow & & \downarrow \\ & & B_i'', b_j'' \in N^{\perp}/N & \subset & Q/N \\ & & & & \downarrow \\ & & & C_i'', c_j'' \in & Q/N^{\perp}. \end{array}$$

Note that Q is a successive extension with three quotients as given above. In order to perform the induction step, we lift the given bases for the three (sub)quotients to a W-basis

$$\{A_1, \cdots, A_u;\ B_1', \cdots, B_d';\ C_1', \cdots, C_v';\ c_1', \cdots, c_u';\ b_1, \cdots, b_d;\ a_1, \cdots, a_v\}$$

for Q, using $Q \to Q/N \to Q/N^{\perp}$ and $N^{\perp} \to N^{\perp}/N$, such that this basis respects the filtration, and such that the last $u+d+v$ basis vectors are in VQ. We are going to show it can be changed into a symplectic basis adapted to the filtration (which will be normal by the choices already made in case of **(b)**). Note that the following inner products already satisfy the conditions in the condition of being symplectic: $< A, - >$, $< a, - >$, $< B', B' >$, $< B', b' >$; by this we mean that $< A_i, z >= 0$ for all vectors in the given basis, unless $z = c_i'$, in which case it equals one, etc. We are going to change some of the basis vectors, such that the basis still is adapted to the filtration, that it becomes symplectic (and the condition of being normal will remain unchanged).

Compute: $< B_i', C_j' >= \beta_{i,j}$, and $< B_i', c_j' >= \gamma_{i,j}$; write $B_i := B_i' - \sum_j (\beta_{i,j}{\cdot}a_j + \gamma_{i,j}{\cdot}A_j)$. After this step all products of the form $< B_i, - >$ satisfy the conditions required.

Compute: $< C_i', C_j' >= \xi_{i,j}$, and $< C_i', c_j' >= \delta_{i,j} + \varepsilon_{i,j}$ (here $\delta_{i,j}$ is the Kronecker delta) and $< C_i', b_j >= \zeta_{i,j}$; write $C_i := C_i' - \sum_j (\xi_{i,j}{\cdot}a_j + \varepsilon_{i,j}{\cdot}A_j) - \sum_j \zeta_{i,j}{\cdot}B_j$. Compute $< c_i', c_j' >= p{\cdot}\eta_{i,j}$ and $< c_i.b_j >= p{\cdot}\theta_{i,j}$; write $c_i = c_i' - \sum_j \eta_{i,j}{\cdot}F{\cdot}VA_j - \sum_j \theta_{i,j}{\cdot}F{\cdot}VB_j'$. In this way we obtain a basis for Q

which satisfies the requirements: it is symplectic, it is adapted to the filtration in the sense of (3.3), and in case **(b)** it is in normal form. This ends the induction step. Performing all steps we find a basis for M satisfying the conditions of the lemma. □

(3.8). **Lemma.** *Start with data and the notation as in* (3.3). *There exists a deformation* $(\{\mathcal{G}^{(i)} \mid 0 \leq i \leq m\}, \lambda)$ *of quasi-polarized filtered p-divisible groups such that all slopes of*

$$Y^{(i)} := \mathcal{G}^{(i)}/\mathcal{G}^{(i-1)}$$

are constant, and $a(Y_\eta^{(i)}) \leq 1$ *for all* $0 < i \leq m$.

Proof. We adapt the proof of (2.4) with small changes. It suffices to prove the lemma in case we have a local-local p-divisible group. We choose a symplectic basis $\{x_1, \cdots, x_h;\ y_1, \cdots, y_h\}$ for $\mathbb{D}(G_0)$ adapted to the filtration as constructed in the previous lemma. The display matrix is written in block form and the blocks $A_{i,i}, B_{i,m-i}, C_{m-i,i}, D_{m-i,m-i}$ describe the display of $N_0 = \mathbb{D}(G_0^{(i)}/G_0^{(i+1)})$; here the choice of the basis is slightly different from the one made in (2.4), as we have to adapt to the polarized case. By [8], Corollary (5.15) we can choose $\tau_{i,m-i}$ (for $0 < i \leq (m+1)/2$) which deforms $G_0^{(i)}/G_0^{(i-1)}$ keeping the Newton polygon constant, and achieving $a = 1$. In case $G_0^{(i)}/G_0^{(i-1)}$ is supersingular of height 2, we already have $a = 1$, and we choose $\tau_{i,m-i} = 0$. We complete the matrix T in a symmetric way; this means that the block $\tau_{m-i,i}$ is obtained from $\tau_{i,m-i}$ (for $1 \leq i \leq n$) by symmetry in the main diagonal of T. All block matrices $T_{i,j}$ outside the anti-diagonal are required to be zero. Hence T is defined, and the deformation $\mathcal{G}_\tau$ thus obtained satisfies the requirements of the lemma. This proves Lemma (3.8). □

(3.9). **Lemma.** *Suppose* $(\{H_0^{(i)} \mid 0 \leq i \leq m\}, \lambda)$ *is a local principally quasi-polarized filtered p-divisible local group over an algebraically closed field k, with all* $a(H_0^{(i)}/H_0^{(i-1)}) = 1$. *Then there exists a deformation* $\{\mathcal{H}^{(i)} \mid 0 \leq i \leq m\}$ *of filtered p-divisible groups with the same slopes, such that* $a(\mathcal{H}_\eta^{(m)}) = 1$.

Proof. We adapt the proof of (2.7) with small changes. We reduce to the local-local case. For the case of the Dieudonné module of principally quasi-polarized formal p-divisible groups we choose a symplectic basis in normal form adapted to the symplectic filtration. We study deformation displays on this basis. We choose variables $s_1, s_2, \cdots, s_{m-1}$, with $s_j = s_{m-j}$ and their Teichmüller lifts $S_i \in W(k[[s_i \mid 1 \leq i \leq m-1]]/(\cdots s_j = s_{m-j} \cdots))$; we define the block matrix $\tau_{i,m-i}$ with S_i, $1 \leq i < m$, in the left hand upper corner; we define $\tau_{i,j} = 0$ if $j \neq m - i$ (i.e. equal to zero if not in the upper-antidiagonal). The display matrix in block form $T = (\tau_{i,j} \mid 1 \leq i, j \leq m)$ defines a deformation $\mathcal{H}_s$ which satisfies the requirements of (3.9). □

(3.10). **Corollary.** *Let (G_0, λ_0) be a principally quasi-polarized p-divisible group over a field K, and let ξ be a symmetric Newton polygon, $\xi \succ \mathcal{N}(G_0)$. There is a deformation (G_η, λ_η) of (G_0, λ_0) such that $\xi = \mathcal{N}(G_\eta)$ and $a(G_\eta) \leq 1$.*

Proof of (3.1) and of (3.10). From (3.8) and (3.9) we conclude that a principally quasi-polarized p-divisible group (G_0, λ_0) defined over a field K can be deformed to a principally quasi-polarized p-divisible group G_η with the same Newton polygon and with $a(G_\eta) = 1$: as in the proof of (2.8) use the universal deformation space of (G_0, λ_0). For this (over some field) it follows from [18] that it can be deformed to the situation where the generic fiber has a given Newton polygon, see (1.7). Hence using the same methods as in the proof of (2.8) we see that Theorem (3.1) and Corollary (3.10) follow from (3.8), (3.9) and [18]. □

(3.11). **Corollary.** *Let (X_0, λ_0) be a principally polarized abelian variety over a field K and let ξ be a symmetric Newton polygon, $\xi \succ \mathcal{N}(G_0)$. There is a deformation (G_η, λ_η) of (G_0, λ_0) such that $\xi = \mathcal{N}(G_\eta)$ and $a(G_\eta) \leq 1$.*
Proof. This follows from (3.10) and [18], using Serre-Tate and Chow-Grothendieck. In fact, the p-divisible group $X_0[p^\infty]$ plus principal polarization can be deformed as in (3.10). The rest follows as in the proof of (3.2). □

(3.12). Here is an easy example which might explain the esssence of what we are doing. Consider a supersingular elliptic curve E, with its unique principal polarization, and let $X_0 = E \times E$ be equiped with the "diagonal" principal polarization λ_0. Study deformations which stay inside the supersingular locus. In this case this is still not so difficult. The display of the universal deformation over $\mathrm{Spec}(k[[s,t,u]])$ gives a Hasse-Witt matrix equal to

$$H = \begin{pmatrix} s & t \\ t & u \end{pmatrix}.$$

For abelian surfaces, being supersingular amounts to having p-rank equal to zero. Hence the supersingular locus here is given by $H \cdot H^{(p)} = 0$. One easily solves the four equations thus obtained. Choose $\zeta \in W$ with $\zeta^\sigma \cdot \zeta = -1$, and consider all deformations given by $s = \zeta \cdot t$ and $u = \zeta^{-1} \cdot t$ over $\mathrm{Spec}(k[[t]])$. Clearly there are $p+1$ such choices; these are exactly the components of the local deformation space of the supersingular locus in the deformation space of principally polarized abelian surfaces. We can also argue as follows (and obtain the same result). We see that (X_0, λ_0) admits exactly $p+1$ maximal symplectic filtrations; each of these defines a regular supersingular deformation space (keeping the filtration); in this way we find back the $p+1$ branches of the local supersingular moduli space at (X_0, λ_0).

(3.13). **Remark.** Here is another example. Suppose (G_0, λ) is a principally quasi-polarized p-divisible group, such that $G_0 = H_0' \oplus H_0''$, say with $a(H_0') = 1 = a(H_0'')$, where H' is a simple formal group with slope s, with $0 < s < 1/2$. We can choose a deformation of the filtered group $0 \subset H_0' \subset G_0$, and arrive at a filtered p-divisible group $0 \subset \mathcal{H}' \subset \mathcal{X}$ with a principal quasi-polarization. However we can also arrive at $0 \subset \mathcal{H}'' \subset \mathcal{Y}$. This last case gives an example of a filtered p-divisible group over a base, where the steps in the filtration have *decreasing* slopes. It is easy to show that the p-divisible groups $\mathcal{X}$ and $\mathcal{Y}$ (over some common base) are not isomorphic. This phenomenon deserves further study, especially in case of more isogeny factors.

In case of a p-divisible group over a complete local base such that the Newton polygon is constant we find in [10], Th. 2.4.2 the existence of a filtration by isoclinic groups of increasing slopes if certain conditions on Hodge polygons are satisfied. We see that such a filtration in general does not exist.

(3.14). We work this out in a special case. Suppose given $H := G_{2,1} \oplus G_{1,2}$ and a quasi-polarization μ on it with $\mathrm{Ker}(\mu) = G_{2,1}[V] \oplus G_{1,2}[F] = (\alpha_p)^2$. We construct $(\mathcal{G}, \lambda) \to \mathbb{P}^1$ in the usual way of "taking the quotient by a variable α_p", i.e. above the point $(x, y) \in \mathbb{P}^1$ the fiber $(\mathcal{G}, \lambda)_{(x,y)} = H/(x, y)(\alpha_p)$, and the polarization μ descends to a principal polarization on $\mathcal{G}_{(x,y)}$. Around the point $0 = (0, 1)$ we have a filtration $\mathcal{G}'$ of $(\mathcal{G}, \lambda)$, and every fiber of $\mathcal{G}$ has Newton polygon equal to $(2, 1)$; on can show that the factor group does not come from a subgroup (and here the filtration is by increasing slopes). Around the point $\infty = (1, 0)$ we have a filtration $\mathcal{G}'' \subset \mathcal{G}$; here the filtration is by decreasing slopes and around the point $\infty \in \mathbb{P}^1$ no isoclinic filtration with increasing slopes is possible.

Conversely, both local situations can be reconstructed by a deformation of filtered groups.

(3.15). We have seen deformations obtained as deformations of filtered groups. However, in general an isoclinic, non-constant $\mathcal{G} \to B$ over an integral base need not be derived from a deformation of a filtered group. A component of the Newton polygon stratum locally at a point need not be given by a deformation of filtered groups.

In case $X_0[p^\infty] \cong G_{2,1} \oplus G_{1,2}$ the local deformation space of (X_0, λ) keeping the Newton polygon constant has two irreducible branches of dimension equal to 2. One can be given by a filtered deformation (the one with the increasing slopes), the other does contain a one-dimensional filtered deformation (the one with the decreasing slopes). Here we see an example that we can move to the interior of a branch, but not reach the full branch by filtered deformations. Here is another example:

Consider an irreducible component $W \subset \mathcal{S}_{3,1,n}$ of the supersingular locus (with some level structure given by an integer $n \geq 3$ not divisible by p). This has a very singular point $x \in W$ obtained by any flag type quotient $(E^3 \to E^3/(\alpha_p)^2 \to X_0) = (F : E^3 \to E^3/E^3[F])$; this is the point described in [11], (9.4.16) on page 59. The universal family over W, locally at x, does not come from the deformation of any maximal symplectic filtration of X_0; in fact, no subspace on which the $a = 1$ case on W is realized can be obtained by a filtered deformation.

4. Newton polygon strata

We recall some notations. We fix $g \in \mathbb{Z}_{>0}$, and we write $\mathcal{A} = \mathcal{A}_{g,1} \otimes \mathbb{F}_p$ for the moduli space of *principally* polarized abelian varieties in characteristic p.

For a symmetric Newton polygon ξ belonging to g we write $W_\xi \subset \mathcal{A}$ for the subset of all $[(X, \lambda)] \in \mathcal{A}$ with $\mathcal{N}(X) \prec \xi$, which is a closed subset by [10], Th. 2.3.1 on page 143.

For an irreducible subset $W \subset \mathcal{A}$ we write $\mathcal{N}(-, W \subset \mathcal{A})$ for the Newton polygon of the generic fiber, and $a(-, W \subset \mathcal{A})$ for the a-number of the generic fiber above W.

(4.1). **Theorem**, see [17]. *Suppose given p, and g as as above. For every Newton polygon ξ we have:*

(a) *For every irreducible component W of $W_\xi := \mathcal{W}_\xi(\mathcal{A}) \subset \mathcal{A}_{g,1} \otimes \mathbb{F}_p$ we have*

$$\mathcal{N}(-, W \subset \mathcal{A}) = \xi \quad \textit{and} \quad a(-, W \subset \mathcal{A}) \leq 1,$$

i.e. generically on W the Newton polygon of $\mathcal{X} \to W$ equals ξ, and generically the a-number is at most one.

(b) *The dimension of every irreducible component W of W_ξ equals* $\mathrm{sdim}(\xi) = \#(\triangle(\xi))$.

This follows from (3.10) and [18]; compare with (2.9) and with (1.7). □

(4.2). **Remark.** In particular a supersingular principally polarized abelian variety can be deformed (keeping polarization and Newton polygon) to one with $a = 1$. Hence we have a new proof for [11], Theorem (4.9). In particular we obtain the dimension of the supersingular locus, and the number of components of the supersingular locus, expressed as a class number. The difference in methods between [11] and this paper is as follows. In [11] *polarized flag type quotients*, and several variants were considered; with the correct definitions (especially once having found the notion of "rigid" quotients) the fact that every component of the supersingular locus generically has $a = 1$ results; also the dimension of the supersingular locus can be computed by methods of flag type quotients. In the proof in this paper we avoid such considerations, but instead we consider deformations of *filtered p-divisible groups*; it works for all Newton polygons, and it shows that $a \leq 1$ generically on Newton polygon strata; for those the method of Cayley-Hamilton works, as in [18]; it gives a computation for the dimension of the various strata. Of course, in both cases we obtain the same answer for the dimension of the supersingular locus: $\dim(\mathcal{S}_{g,1}) = [g^2/4]$, as conjectured by Tadao Oda and the present author, see [16], pp. 615/616. By the method of flag type quotients we obtain a global description of every component of the supersingular locus $\mathcal{S}_{g,1}$, as to be found in [11].

(4.3). **Remark.** We have studied deformations of *principally* polarized abelian varieties. However if we would consider deformations of abelian varieties with a non-separable polarization things are not as uniform as above. In case of polarized p-divisible groups, or polarized abelian varieties, *without supposing the polarization is principal*, there are counterexamples to the existence of deformations with given closed fiber, and expected Newton polygon for the generic fiber, see [18], Section 6: there exist supersingular polarized abelian varieties of dimension $g = 3$, i.e. its Newton polygon equals $3{\cdot}(1,1)$, which cannot be deformed to a polarized abelian variety with Newton polygon equal to $(2,1) + (1,2)$. In the non-principally polarized case dimensions of strata may be different from what we compute in the principally polarized case. In [11], Section 12 some examples have been worked out. The phenomenon that Hecke correspondences blow up and down is present for Newton polygons which allow $a \geq 2$. Also see (5.8).

5. Some questions and conjectures

(5.1). For every symmetric Newton polygon ξ (and every g and every p) we obtain $W_\xi \subset \mathcal{A} = \mathcal{A}_{g,1} \otimes \mathbb{F}_p$. For $\xi = \sigma$, the supersingular Newton polygon, this locus has "many" components (for $p >> 0$; in fact this number is a class number, asymptotically going to ∞ with $p \to \infty$).
Conjecture. *For given p, g, and $\xi \neq \sigma$, we conjecture that this locus W_ξ is* geometrically irreducible.

(5.2). We consider complete subvarieties of moduli spaces. It is known that for any field K, and any *complete* subvariety $W \subset \mathcal{A}_g \otimes K$, the dimension of W is at most $(g(g+1)/2) - g$. We wonder is this maximum ever achieved? If yes, in which cases?
Conjecture. *Let $g \geq 3$. Suppose W is a complete subvariety $W \subset \mathcal{A} = \mathcal{A}_{g,1} \otimes \mathbb{F}_p$ of dimension equal to* $(g(g+1)/2) - g$ (the maximal possible dimension for complete subvarieties; see [3], Coroll. 2.7). *We expect that under these conditions W is equal to the locus V_0 of principally polarized abelian varieties with p-rank equal to zero.* (This locus is complete and has the right dimension.)
If this is true, probably we have a proof for:

(5.3). **Conjecture.** *Let $g \geq 3$. Let $W \subset \mathcal{A}_g \otimes \mathbb{C}$ be a* complete *subvariety. We expect that under these conditions:*

$$\dim(W) < (g(g+1)/2) - g.$$

Hecke orbits are dense in $\mathcal{A}_g \otimes \mathbb{C}$. Chai proved the same for Hecke orbits of *ordinary* polarized abelian varieties in positive characteristic, see [1]. In his case only ℓ-power isogenies need to be considered for one prime $\ell \neq p$.

(5.4). **Conjecture.** *Fix a polarized abelian variety* $[(X,\lambda)] = x \in \mathcal{A} \otimes \mathbb{F}_p$. *Choose any prime number* $\ell \neq p$. *Consider the Hecke orbit of* x *admitting isogenies of degree equal to a product of a power of* ℓ *and of a succession of isogenies with kernel* α_p. *We conjecture that this Hecke orbit is everywhere dense in the Newton polygon stratum* W_β *with* $\beta := \mathcal{N}(X)$.
This will be studied in [2].

(5.5). As in the case of simple p-divisible groups, for any p-divisible group G there is a catalogue for all groups isogenous with G. Probably there exists an irreducible one (constructed using the methods of Section 2). It could be useful to study that case.

(5.6). **Conjecture; Foliations.** We expect the following facts to be true. For every Newton polygon β there should exist integers $i(\beta),\ \ c(\beta) \in \mathbb{Z}_{\geq 0}$, and for every point $[(X,\lambda)] = x \in \mathcal{A} = \mathcal{A} \otimes \mathbb{F}_p$ with $\mathcal{N}(X) = \beta$ there should exist a closed subset $x \in \mathcal{I}(x) = \mathcal{I}_\beta(x) \subset W_\beta \subset \mathcal{A}$, and a locally closed subset $x \in \mathcal{C}(x) = \mathcal{C}_\beta(x) \subset W_\beta \subset \mathcal{A}$ such that:

- $\dim(\mathcal{I}(x)) = i(\beta)$ and $\dim(\mathcal{C}(x)) = c(\beta)$.
- For every geometric point $[(Z,\zeta)] = z \in \mathcal{C}(x)(k)$ there is an isomorphism $(Z[p^\infty],\zeta) \cong (X[p^\infty],\lambda)$. All irreducible components of the locally closed set $\mathcal{C}(x)$ contain x, and it is the maximal locally closed set with the property just mentioned.
- For every geometric point $[(Y,\mu)] = y \in \mathcal{I}(x)$ there is a Hecke-correspondence using only iterates of α_p-isogenies relating $[(X,\lambda)]$ and $[(Y,\mu)]$. All irreducible components of the closed set $\mathcal{I}(x)$ contain x, and it is the maximal closed set with the property just mentioned.
- The dimensions are complementary: $i(\beta) + c(\beta) = \mathrm{sdim}(\beta)$, and locally at x their intersection is zero dimensional.
- If moreover $a(X) \leq 1$, the (locally) closed sets $\mathcal{I}(x)$ and $\mathcal{C}(x)$ are regular at $x \in \mathcal{A}$, intersect transversally at x, and together their tangent spaces span the tangent space of $x \in W_\beta$.
- Examples:
  for the supersingular locus we have $i(\sigma) = \mathrm{sdim}(\sigma) = [g^2/4]$ and $c(\sigma) = 0$;
  for the ordinary locus we have $i(\rho) = 0$, and $c(\rho) = \mathrm{sdim}(\sigma) = (g(g+1))/2$;
  for the case the p-corank equals one, i.e. $\beta = (g-1)\cdot(1,0) + (1,1) + (g-1)\cdot(0,1)$ we have $i(\beta) = 0$, and $c(\beta) = \mathrm{sdim}(\sigma) = ((g(g+1))/2) - 1$.
  We have: p-rank $f(\beta) < g-1$ iff $i(\beta) > 0$.
  We have: $\beta \neq \sigma$ iff $c(\beta) > 0$.

- There is an easy combinatorial argument by which the numbers $i(\beta)$ and $c(\beta)$ can be read off from the Newton polygon diagram of β.

(5.7). In general $G[p]$ does not determine a p-divisible group G. But in some cases it does. Let β be a symmetric Newton polygon. For a pair of relatively prime integers (m,n) we have defined in [8], Section 5 a p-divisible group $H_{m,n}$; it is characterized by: $H_{m,n} \sim G_{m,n}$, and for an algebraically closed field $k \supset \mathbb{F}_p$, the ring $\mathrm{End}(H_{m,n} \otimes k)$ is a maximal order in $\mathrm{End}^0(G_{m,n} \otimes k)$. We define H_β to be the direct sum of all $H_{m,n}$ ranging over all slopes of β. We expect:
Conjecture. *Suppose (G, λ) is a principally quasi-polarized p-divisible group over an algebraically closed field k, such that $G[p] \cong H_\beta[p]$; then (?) we should conclude $G \cong H_\beta$.*

Note that in the special cases $\beta = \rho$ (the ordinary case), and $\beta = \sigma$ (supersingular) this conjecture is true; the conjectural statement above seems the natural generalization of this. Special cases have been proved.

(5.8). **Conjecture.** (Newton polygon strata, the non-principally polarized case). Consider $\mathcal{A}_g$, the moduli space of all polarized abelian varieties, where the polarization need not be principal. Let ξ be a symmetric Newton polygon. This gives a stratum $\mathcal{W}_\xi(\mathcal{A}_g \otimes \mathbb{F}_p)$. Let $f = f(\xi)$ be the p-rank of ξ, i.e. this Newton polygon has exactly f slopes equal to zero. We expect: *under these conditions, there is an irreducible component*

$$W \subset \mathcal{W}_\xi(\mathcal{A}_g \otimes \mathbb{F}_p) \quad \text{with} \quad \dim(W) = (g(g+1)/2) - (g-f),$$

i.e. we expect that there is a component of every Newton polygon stratum which is a whole component of its p-rank stratum.

If this is the case, we see that there are "many" pairs of polarized abelian variety (X,λ) and a Newton polygon $\gamma \succ \mathcal{N}(X)$ such that there exists no deformation of (X,λ) to a polarized abelian variety with Newton polygon equal to γ, namely consider $\beta \prec \gamma$ with $\beta \neq \gamma$ and $f(\beta) = f(\gamma)$.

References

[1] C.-L. Chai - *Every ordinary symplectic isogeny class in positive characteristic is dense in the moduli.* Invent. Math. **121** (1995), 439 - 479.

[2] C.-L. Chai & F. Oort - [Still in preparation.]

[3] G. van der Geer - *Cycles on the moduli space of abelian varieties.* In: Moduli of curves and abelian varieties (The Dutch intercity seminar on moduli) (Eds. C. Faber & E. Looijenga), Asp. Math. E 33, Vieweg 1999; pp. 65 - 89.

[4] A. Grothendieck & J. Dieudonné - *Eléments de géométrie algébrique. Ch. III^1: Etude cohomologique des faisceaux cohérents.* Publ. Math. 11, IHES 1961.

[5] A. Grothendieck - *Groupes de Barsotti-Tate et cristaux de Dieudonné.* Sém. Math. Sup. Univ. Montréal. Presses Univ. Montréal, 1974.

[6] A. J. de Jong - *Homomorphisms of Barsotti-Tate groups and crystals in positive characteristic.* Invent. Math. **134** (1998), pp. 301 - 333.

[7] A. J. de Jong - *Crystalline Dieudonné module theory via formal rigid geometry.* Publ. Math. IHES **82** (1995), 5 - 96.

[8] A. J. de Jong & F. Oort - *Purity of the stratification by Newton polygons.* Journ. A.M.S. **13** (2000), 209 - 241.
[9] N. M. Katz - *Serre-Tate local moduli.* In: Surfaces algébriques (Sém. de géom. algebr. d'Orsay 1976-78). Lect. Notes Math. 868, Springer - Verlag 1981; Exp. V-bis, pp. 138 - 202.
[10] N. M. Katz - *Slope filtrations of F-crystals.* Journ. Géom. Alg. Rennes, Vol. I, Astérisque 63, Soc.Math. France, 1979.
[11] K.-Z. Li & F. Oort - *Moduli of supersingular abelian varieties.* Lect. Notes Math. 1680, Springer - Verlag, 1998.
[12] Yu. I. Manin - *The theory of commutative formal groups over fields of finite characteristic.* Usp. Math. **18** (1963), 3-90; Russ. Math. Surveys **18** (1963), 1-80.
[13] D. Mumford - *Geometric invariant theory.* Ergebn. Math., Neue Folge, Bd. 34, Springer - Verlag, 1965.
[14] P. Norman - *An algorithm for computing local moduli of abelian varieties.* Ann. Math. **101** (1975), 499-509.
[15] P. Norman & F. Oort - *Moduli of abelian varieties.* Ann. Math. **112** (1980), 413-439.
[16] T. Oda & F. Oort - *Supersingular abelian varieties.* Proceed. Intl. Symp. on Algebraic Geometry (Ed. M. Nagata), Kyoto 1977, Kinokuniya, 1978; pp. 595-621.
[17] F. Oort - *Moduli of abelian varieties and Newton polygons.* C. R. Acad. Sci. Paris **312** (1991), 385-389.
[18] F. Oort - *Newton polygons and formal groups: conjectures by Manin and Grothendieck.* [To appear Ann. Math.]
[19] T. Zink - *The display of a formal p-divisible group,* University of Bielefeld, Preprint 98–017, February 1998.
[20] T. Zink - *A Dieudonné theory for p-divisible groups.* Manuscript, 23 pp., September 1998.

Progress in Mathematics, Vol. 195, © 2001 Birkhäuser Verlag Basel/Switzerland

THE DIMENSION OF OORT STRATA OF SHIMURA VARIETIES OF PEL-TYPE

TORSTEN WEDHORN

Introduction

The moduli space of principally polarized abelian varieties of dimension g admits in positive characteristic p a stratification by the p-rank of the abelian variety, i.e. by the rank of the p-torsion. In the case of the moduli space of elliptic curves this is simply the stratification given by the (open and dense) ordinary locus and the zero-dimensional supersingular locus. But for $g > 2$ this stratification is too coarse for many purposes. A refinement is provided by stratifying the moduli space according to the Newton polygon of an abelian variety. This stratification has been studied thoroughly by Oort ([Oo1], [Oo2], see also [dJO]). He also suggested to study another stratification given by the isomorphism type of the p-torsion which we will call the Oort stratification in the sequel (often it is also called Ekedahl-Oort stratification). This stratification is also a refinement of the p-rank stratification.

Both stratifications can also be considered for models of Shimura varieties of PEL-type which classify abelian varieties with certain additional structures. The precise definition of the Newton polygon stratification in this case together with a description of the strata which can occur and the specialization property of the stratification is due to Rapoport and Richartz [RR]. It follows from results of de Jong and Oort [dJO] that the codimension of a stratum in the closure of any non-empty stratum lying directly above can be only one. Combining these results one gets lower bounds for the dimension of the strata, but in general it is not known whether they are empty or not except for some special cases: In the case of the moduli space of principally polarized abelian varieties Oort [Oo2] shows that all strata are nonempty. In fact he shows a much more precise statement, namely that each point x with Newton polygon ν lies in the closure of the locus of a Newton polygon stratum corresponding to a Newton polygon lying below ν. For general Shimura varieties of PEL-type there is in good reduction the much weaker statement that the ordinary stratum (in the sense of PEL-moduli spaces) is open and dense ([We1]).

The Oort stratification has been examined in the case of Hilbert-Blumenthal varieties by Goren and Oort [GO] where they give a precise description of the occurring isomorphism classes and the dimension of the strata. Further they prove that the closure of a stratum is a union of strata and that all strata are quasi-affine. In the case of the moduli space of principally polarized abelian varieties, Oort [Oo3] has obtained analogous results. Further, van der Geer [vGe] computed the Chow classes of the Oort strata. For arbitrary good reductions of Shimura varieties of PEL-type, B. Moonen [Mo] has given a description of the possible occurring strata in terms of the Shimura datum. More precisely, he proved that the set of these strata is in canonical bijection to the quotient of the Weyl group of the Shimura group by the Weyl group of the Levi component corresponding to the minuscule coweight given by the Shimura datum. In particular, it turns out that there are only finitely many Oort strata.

The main goal of this article is the definition and the calculation of the dimension of the Oort strata for good reductions of arbitrary Shimura varieties of PEL-type. We show that the closure of a stratum is the union of strata. To formulate the main result, let $\underline{X} = (X, \lambda, \iota)$ be the p-torsion of an abelian variety together with a polarization and an action of a $\mathbb{Z}_{(p)}$-algebra, unramified at p, and denote by ξ its isomorphism class. If we denote by $\mathtt{A}_\xi$ the corresponding Oort stratum, we show

Theorem: *For $p > 2$, $\mathtt{A}_\xi$ is locally closed and its codimension is equal to* $\dim(\underline{\mathrm{Aut}}(\underline{X}))$.

To do this, we follow an idea of J. de Jong which can be formulated roughly as follows: Define the moduli space $\mathtt{B}$ of truncated Barsotti-Tate groups with PEL-structure and show that the canonical morphism from the moduli space $\mathtt{A}$ of abelian varieties with PEL-structure into this moduli space is smooth. Define a smooth cover $\tilde{\mathtt{B}}$ of $\mathtt{B}$ as the moduli space of truncated Barsotti-Tate groups with PEL-structure and with a trivialisation of its structure sheaf as a module over the base and denote by $\tilde{\mathtt{A}}$ the fibre product of $\mathtt{A}$ with $\tilde{\mathtt{B}}$ over $\mathtt{B}$. Then the inverse image of $\mathtt{A}_\xi$ in $\tilde{\mathtt{A}}$ is the inverse image of a locally closed subscheme of $\tilde{\mathtt{B}}$ of codimension $\dim(\underline{\mathrm{Aut}}(\underline{X}))$ which implies the claim.

In order to calculate the dimension of $\underline{\mathrm{Aut}}(\underline{X})$ it would be preferable to calculate instead the dimension of the group of automorphisms of the Dieudonné space of $\underline{X}$. For this we define Dieudonné spaces with PEL-structure over arbitrary schemes which allows us to speak of the moduli space of these Dieudonné spaces $\mathtt{D}$. Using the crystalline Dieudonné functor we get a morphism $D\colon \mathtt{B} \longrightarrow \mathtt{D}$ which induces a homeomorphism of the underlying Zariski spaces. It follows that the stratification by isomorphism classes of Dieudonné spaces is equal to the Oort stratification. Further we get

$$\dim(\underline{\mathrm{Aut}}(\underline{X})) = \dim(\underline{\mathrm{Aut}}(D(\underline{X})).$$

We conclude from the smoothness of the morphism $\mathtt{A} \longrightarrow \mathtt{B}$ that if the minimal Oort strata are nonempty, all Oort strata are nonempty. As a final application it follows that if the μ-ordinary stratum in the sense of [We1] corresponds to a maximal Oort stratum, this μ-ordinary stratum is open and dense.

I will now give an overview of the structure of this work: In the first chapter truncated Barsotti-Tate groups with PEL-structure and their moduli space are defined. We show that this moduli space is an algebraic stack by defining an explicit smooth cover. In the second chapter we show that the canonical morphism of the deformation functor of an abelian variety (or, what amounts to the same by the theorem of Serre-Tate, its Barsotti-Tate group) into the deformation functor of its p^n-torsion is formally smooth for $p > 2$ (or more precisely a hull in the sense of Schlessinger) (Theorem 2.8 and Corollary 2.17). The main technical tool is the theory of obstructions for deforming group schemes developed by Illusie. In fact, he proves this result in [Ill] for the case of Barsotti-Tate groups without any additional structure. The main problem of generalizing this result is the fact that deformations of truncated Barsotti-Tate groups are not rigid and that forgetting this additional structure does not define a monomorphism of the corresponding deformation functors.

The third chapter starts with the assertion that every truncated Barsotti-Tate group with additional structure over a complete local noetherian ring with perfect residue class field is the p-torsion of a Barsotti-Tate group with additional structure (Proposition 3.2). We deduce that the moduli space of truncated Barsotti-Tate groups is smooth over $\mathbb{Z}_{(p)}$ and that its fibres are of dimension zero. We note that this does not imply that the special fibre is étale; in fact being only an Artin stack the assertion of being étale over the base does not make any sense.

In the fourth chapter we endow the set of isomorphism classes of truncated Barsotti-Tate groups with a topology using the general language of algebraic stacks and in the fifth chapter we relate truncated Barsotti-Tate groups and Dieudonné spaces and define a morphism between their respective moduli spaces and show that this is a homeomorphism.

The sixth chapter finally gives the definition of the Oort stratification of a scheme S with respect to a truncated Barsotti-Tate group with additional structures over S. In particular we define the Oort stratification of the model of a Shimura variety of PEL-type at a prime of good reduction and give the main theorem (6.9). After that we give some conditions when the Oort strata are nonempty.

In the last chapter we give two examples.

This work would neither have been started nor have been finished if not for J. de Jong's encouragement. It is a pleasure to thank him for giving advice generously. Further I am grateful to S. Orlik and U. Görtz for their remarks on a first version of the manuscript. Part of it has been written by the author while being guest at MIT and the nice atmosphere of MIT and the financial support of the Deutsche Akademischer Austauschdienst (DAAD) are gratefully acknowledged.

1 Truncated Barsotti-Tate groups with endomorphisms and polarizations

Notation: *We fix once and for all a prime p.*

(1.1) Fix an $n \geq 1$. Recall that an *at n truncated Barsotti-Tate group* (or just a BT_n) over a scheme S is a commutative finite locally free S-group scheme X, which is annihilated by p^n such that the following condition holds: If $n > 1$, then X is flat over $\mathbb{Z}/p^n\mathbb{Z}$, and if $n = 1$ then for $S_0 = V(p) \subset S$ and $X_0 = X \times_S S_0$ the sequence

$$X_0 \xrightarrow{F} X_0^{(p)} \xrightarrow{V} X_0$$

is exact where F (resp. V) denotes the relative Frobenius (resp. the Verschiebung).

For all $1 \leq m \leq n$, the p^m-torsion $X[p^m]$ is finite locally free over S, it is a BT_m, and its rank is of the form p^{mh} for some integer h which is called the *height of X*.

(1.2) We recall some definitions and results from [Ill] 2. Let G be an S-group scheme, flat and locally of finite presentation. Denote its neutral section by e and its structure morphism by π. Its cotangent complex is denoted by $L_{G/S}$.

The object

$$\ell_G = \mathbb{L}e^* L_{G/S}$$

in $D(S)$ is called the *co-Lie complex of G*. Further define the *Lie complex*

$$\ell_G^\vee := \mathbb{R}\mathcal{H}om(\ell_G, \mathcal{O}_S).$$

As G is locally a complete intersection over S, ℓ_G is perfect with perfect amplitude contained in $[-1,0]$. Therefore, $\ell_G^\vee$ is perfect with perfect amplitude contained in $[0,1]$.

The functors $G \mapsto \ell_G$ and $G \mapsto \ell_G^\vee$ transform exact sequences in distinguished triangles. In particular if

$$0 \longrightarrow G \longrightarrow G^0 \longrightarrow G^1 \longrightarrow 0$$

is a resolution of G with formally smooth $fppf$-sheaves in groups we have

$$\ell_G \xrightarrow{\sim} (\omega_{G^1} \longrightarrow \omega_{G^0})$$

where $\omega_{G^i} = \ell_{G^i} = e^*\Omega^1_{G^i/S}$ lies in degree $-i$.

The only nonzero cohomology sheaves of ℓ_G and $\ell_G^\vee$ are denoted by

$$\omega_G := H^0(\ell_G), \qquad n_G := H^{-1}(\ell_G),$$
$$t_G := H^0(\ell_G^\vee), \qquad \nu_G := H^1(\ell_G^\vee).$$

We have

$$t_G = \omega_G^*, \qquad n_G = \nu_G^*$$

where $(\ )^*$ denotes the $\mathcal{O}_S$-linear dual.

(1.3) In the sequel $\mathcal{D}$ will be a tuple $(O, {}^*, \varepsilon, h)$ where

- O is a finite free (not necessarily commutative) $\mathbb{Z}_{(p)}$-algebra.
- * is a $\mathbb{Z}_{(p)}$-algebra-involution on O.
- ε is equal to ± 1.
- h is a positive integer.

We denote by $O_p = O \otimes_{\mathbb{Z}_{(p)}} \mathbb{Z}_p$ the p-adic completion of O. The involution induced by * is again called *. We write $\mathcal{D}_p$ for the tuple $(O_p, {}^*, \varepsilon, h)$.

(1.4) Definition: Let S be some scheme. A *Barsotti-Tate group with $\mathcal{D}$-structure over S* (or shorter *a BT with $\mathcal{D}$-structure*) is a tuple (X, λ, ι) where X is a Barsotti-Tate group of height h over S, $\lambda: X \longrightarrow X^\vee$ is an isomorphism ($X^\vee$ denoting the Serre dual of X) such that $\lambda^\vee = -\varepsilon\lambda$ and $\iota: O \longrightarrow \mathrm{End}(X)$ is a $\mathbb{Z}_{(p)}$-algebra homomorphism compatible with the involution * on O and the Rosati-involution given by λ on $\mathrm{End}(X)$.

We get an analogous definition for an *at n truncated Barsotti-Tate group with $\mathcal{D}$-structure* (or shorter *a BT_n with $\mathcal{D}$-structure*). We will also write BT_∞ for a Barsotti-Tate group.

A pair (X, ι) where X is a BT_n (for $1 \leq n \leq \infty$) and $\iota: O \longrightarrow \mathrm{End}(X)$ is a $\mathbb{Z}_{(p)}$-algebra homomorphism is called a BT_n with O-structure. Note that $\mathrm{End}(X)$ is a $\mathbb{Z}_p$-module such that an O-action is the same as an O_p-action on X. Therefore a BT_n with $\mathcal{D}$-structure is the same as a BT_n with $\mathcal{D}_p$-structure where the latter is defined in the obvious way.

If $\underline{X}_n = (X_n, \lambda_n, \iota_n)$ and $\underline{X}_m = (X_m, \lambda_m, \iota_m)$ is a BT_n (resp. a BT_m) with $\mathcal{D}$-structure ($1 \leq n, m < \infty$), a *morphism $\varphi: \underline{X}_n \longrightarrow \underline{X}_m$ of truncated Barsotti-Tate groups with $\mathcal{D}$-structures* is a homomorphism $\varphi: X_n \longrightarrow X_m$ of abelian $fppf$-sheaves which makes the following diagrams commutative:

$$\begin{array}{ccc} X_n & \xrightarrow{\varphi} & X_m \\ {\scriptstyle\lambda_n}\downarrow & & \downarrow{\scriptstyle\lambda_m} \\ X_n^\vee & \xleftarrow{\varphi^\vee} & X_m^\vee, \end{array} \qquad \begin{array}{ccc} X_n & \xrightarrow{\varphi} & X_m \\ {\scriptstyle\iota_n(a)}\downarrow & & \downarrow{\scriptstyle\iota_m(a)} \\ X_n & \xrightarrow{\varphi} & X_m, \end{array} \qquad \text{for } a \in O$$

We also have the analogous definition of a morphism of Barsotti-Tate groups with $\mathcal{D}$-structure.

(1.5) Example: Let A be an abelian scheme of relative dimension g over a base scheme S and let $\lambda: A \longrightarrow A^\vee$ be an isomorphism of A with the dual abelian scheme $A^\vee = \underline{\mathrm{Ext}}^1(A, \mathbb{G}_m)$. Assume that via the canonical identification $A^{\vee\vee} = A$ we have $\lambda^\vee = \varepsilon\lambda$ for some $\varepsilon \in \{\pm 1\}$. Then λ induces isomorphisms $\lambda_n: A[p^n] \xrightarrow{\sim} A^\vee[p^n] = A[p^n]^\vee$ on the p^n-torsion where for a finite locally free commutative group scheme X we denote by $X^\vee = \underline{\mathrm{Hom}}(X, \mathbb{G}_m)$ its Cartier dual. Via the biduality isomorphism $A[p^n] \xrightarrow{\sim} A[p^n]^{\vee\vee}$ we have $(\lambda_n)^\vee = -\varepsilon\lambda_n$. By going to the inductive limit we get an isomorphism $\lambda_\infty: A[p^\infty] \xrightarrow{\sim} A[p^\infty]^\vee$ of the p-divisible group of A to its Serre dual and via biduality we have $(\lambda_\infty)^\vee = -\varepsilon\lambda_\infty$.

If we denote by $\iota_n: \mathbb{Z}_{(p)} \longrightarrow \mathrm{End}(A[p^n])$ the $\mathbb{Z}_{(p)}$-algebra structure of $\mathrm{End}(A[p^n])$, the triple $(A[p^n], \lambda_n, \iota_n)$ is BT_n with $(\mathbb{Z}_{(p)}, \mathrm{id}, \varepsilon, 2g)$-structure and the inductive limit $(A[p^\infty], \lambda[p^\infty], \iota[p^\infty])$ is a Barsotti-Tate group with $(\mathbb{Z}_{(p)}, \mathrm{id}, \varepsilon, 2g)$-structure over S.

(1.6) For $1 \leq n \leq \infty$ let $\underline{X} = (X, \lambda, \iota)$ be a BT_n with $\mathcal{D}$-structure. For $1 \leq m \leq n$ denote by $\underline{X}[p^m]$ the triple consisting of $X[p^m]$ together with the isomorphism $X[p^m] \xrightarrow{\sim} X^\vee[p^m] = X[p^m]^\vee$ induced by λ and the induced O-action. This is a BT_m with $\mathcal{D}$-structure.

If $\underline{X} = (X, \lambda, \iota)$ is a BT_n with $\mathcal{D}$-structure $(1 \leq n \leq \infty)$, the *height* of $\underline{X}$ is the height of X.

(1.7) Let $\underline{X} = (X, \lambda, \iota)$ be a BT_n with $\mathcal{D}$-structure $(1 \leq n \leq \infty)$ over a scheme S. If $f: T \longrightarrow S$ is a morphism of schemes over $\mathbb{Z}_p$ we define

$$f^*\underline{X} = (X_T, \lambda_T, \iota_T)$$

where $X_T = X \times_S T$, $\lambda_T = \lambda \times \mathrm{id}_T: X_T \longrightarrow (X^\vee)_T = (X_T)^\vee$, and $\iota_T(a) = \iota(a) \times \mathrm{id}_T$ for $a \in O$.

This way we get the fibered category of BT_n with $\mathcal{D}$-structure over the category $(\mathtt{Sch}/\mathbb{Z}_{(p)})$ of schemes over $\mathbb{Z}_{(p)}$. It is a stack for any topology on $(\mathtt{Sch}/\mathbb{Z}_{(p)})$ which is coarser than the *fpqc*-topology.

We denote by $\mathtt{BT}_{\mathcal{D},n}$ the stack in groupoids over $(\mathtt{Sch}/\mathbb{Z}_{(p)})_{fppf}$ whose fibre category over some scheme S is the category of BT_n with $\mathcal{D}$-structure of height h where the morphisms are the isomorphisms of BT_n with $\mathcal{D}$-structure.

For $1 \leq n < \infty$ we further define $\widetilde{\mathtt{BT}}_{\mathcal{D},n}$ as the stack in groupoids over $(\mathtt{Sch}/\mathbb{Z}_{(p)})_{fppf}$ whose fibre category over S consists of pairs $(\underline{X}, \alpha)$ where $\underline{X} = (X, \lambda, \iota)$ is a BT_n with $\mathcal{D}$-structure and where α is an isomorphism of $\mathcal{O}_S$-modules $\pi_*(\mathcal{O}_X) \xrightarrow{\sim} \mathcal{O}_S^{p^{nh}}$ if $\pi: X \to S$ denotes the structure morphism. Forgetting α defines a 1-morphism of stacks

$$\omega: \widetilde{\mathtt{BT}}_{\mathcal{D},n} \longrightarrow \mathtt{BT}_{\mathcal{D},n}$$

which makes $\widetilde{\mathtt{BT}}_{\mathcal{D},n}$ into a $GL_{p^{nh}}$-torsor over $\mathtt{BT}_{\mathcal{D},n}$ for the *fppf*-topology (or for any other topology finer than the Zariski topology).

(1.8) Proposition: *Fix an integer* $1 \leq n < \infty$.

(1) *The stack* $\widetilde{\mathtt{BT}}_{\mathcal{D},n}$ *is representable by a quasi-affine scheme of finite type over* $\mathbb{Z}_{(p)}$.

(2) *The stack* $\mathtt{BT}_{\mathcal{D},n}$ *is an Artin stack, i.e. an algebraic stack in the sense of [LM]. It is of finite type over* $\mathbb{Z}_{(p)}$.

Proof: (1) The fibre categories of $\widetilde{\mathtt{BT}}_{\mathcal{D},n}$ are discrete and we have to show that the associated sheaf is representable by a quasi-affine scheme of finite type over $\mathbb{Z}_{(p)}$.

Let S be a $\mathbb{Z}_{(p)}$-scheme. To endow the $\mathcal{O}_S$-module $\mathcal{A} = \mathcal{O}_S^{p^{nh}}$ with the structure of a BT_n with $\mathcal{D}$-structure amounts to define the following: First we endow $\mathcal{A}$ with a Hopf algebra structure (which is given by matrices, defining multiplication, unit, comultiplication, counit and antipode, satisfying certain closed compatibility conditions) and this gives us a finite locally free group scheme $X = \operatorname{Spec}(\mathcal{A})$. Then we choose a $\mathbb{Z}_{(p)}$-basis $(a_1, \dots, a_r)$ of O and the O-action is then given by r endomorphisms of $\mathcal{A}$ satisfying certain closed compatability conditions making them into a $\mathbb{Z}_{(p)}$-algebra homomorphism $\iota: O \longrightarrow \operatorname{End}(X)$ (here we need the fact that all our schemes are defined over $\mathbb{Z}_{(p)}$). Finally, the polarization is given by an $\mathcal{O}_S$-bilinear form on $\mathcal{A}$, using the fact that $X^\vee = \operatorname{Spec}(\mathcal{H}om(\mathcal{A}, \mathcal{O}_S))$, again satisfying certain closed compatibility conditions with the Hopf algebra condition and with ι. The condition that the bilinear form is perfect is open (see e.g. assertion (2) of the lemma below).

This way we get the quasi-affine $\mathbb{Z}_{(p)}$-scheme Y' of finite type classifying finite locally free group schemes $X = \operatorname{Spec}(\mathcal{A})$ with $\mathcal{D}$-structure together with isomorphisms $\mathcal{A} \xrightarrow{\sim} \mathcal{O}_S^{p^{nh}}$. The condition that a group scheme X is killed by p^n defines a closed subscheme Y_n of Y'.

We claim that the canonical morphism $\widetilde{\mathrm{BT}}_{\mathcal{D},n} \longrightarrow Y_n$ of schemes is representable by an immersion. The defining condition to be checked is the following: Let $\underline{X} = (X, \iota, \lambda)$ be an S-valued point of Y_n. For $n > 1$ this is an S-valued point of $\widetilde{\mathrm{BT}}_{\mathcal{D},n}$ if and only if X is flat over $\mathbb{Z}/p^n\mathbb{Z}$ which is equivalent by [Gr] III, 2.2 to the fact that the complex

$$X \xrightarrow{p} X \xrightarrow{p^{n-1}} X$$

is exact. For $n = 1$ this is an S-valued point of $\widetilde{\mathrm{BT}}_{\mathcal{D},n}$ if and only if over $S_0 = V(p) \subset S$ the complex

$$X \times_S S_0 \xrightarrow{F} (X \times_S S_0)^{(p)} \xrightarrow{V} X \times_S S_0$$

is exact. Therefore the claim follows from assertion (3) of the lemma below.

(2) By definition we have $\mathrm{BT}_{\mathcal{D},n} = [GL_{p^{hn}} \backslash \widetilde{\mathrm{BT}}_{\mathcal{D},n}]$ in the sense of stacks. Therefore (2) follows from (1) via [LM] 4.6.1.

(1.9) Lemma: *Let S be a scheme.*

(1) *Let $\mathcal{K}_\bullet = (\ldots \to \mathcal{K}_{m+1} \to \mathcal{K}_m \to \ldots)$ be a complex of finitely presented $\mathcal{O}_S$-modules and let n be an integer. Assume that $\mathcal{K}_{n-1}$ is locally free. Then the set U consisting of $s \in S$ where $H^n((\mathcal{K}_\bullet)_s) = 0$ is open in S.*

(2) *In the situation of (1) define the functor $\mathrm{Ex}_n(\mathcal{K}_\bullet)$ on the category of S-schemes $f: T \longrightarrow S$ by*

$$\mathrm{Ex}_n(\mathcal{K}_\bullet)(T) = \begin{cases} \{f\}, & \text{if } H_n(f^*\mathcal{K}_\bullet) = 0; \\ \emptyset, & \text{otherwise.} \end{cases}$$

If $\mathcal{K}_{n-1}$ and $\mathcal{K}_n$ are locally free, $\mathrm{Ex}_n(\mathcal{K}_\bullet)$ *is representable by the open subscheme U of S.*

(3) *Let $\mathcal{H}_\bullet$ be a finite complex of finite locally free S-group schemes and consider the functor* $\mathrm{Ex}(\mathcal{H}_\bullet)$ *on the category of S-schemes $f\colon T \longrightarrow S$ by*

$$\mathrm{Ex}(\mathcal{H}_\bullet)(T) = \begin{cases} \{f\}, & \text{if } f^*\mathcal{H}_\bullet \text{ is exact;} \\ \emptyset, & \text{otherwise.} \end{cases}$$

Then $\mathrm{Ex}(\mathcal{H}_\bullet)$ *is representable by an open subscheme of S.*

Proof: (1) and (2) follow from [EGA IV] 12.3.3. Let us show (3). We may assume that $\mathcal{H}$ is of the form

$$X' \xrightarrow{f} X \xrightarrow{g} X''.$$

Let $f'\colon X' \longrightarrow \mathrm{Ker}(g)$ be the homomorphism induced by f. The set U of points $s \in S$ where f'_s is an epimorphism equals the set of $s \in S$ such that f'_s is faithfully flat. This set is open: Indeed, we may assume that $S = \mathrm{Spec}(R)$ is affine. We have $X' = \mathrm{Spec}(A')$ and $\mathrm{Ker}(g) = \mathrm{Spec}(B)$ where A' and B are R-algebras which are of finite presentation as R-modules and where A' is locally free. Denote by $\varphi'\colon B \longrightarrow A'$ the homomorphism given by f'. Then $U = \{\, s \in S \mid \varphi'_s \text{ is injective} \,\}$ and U is open by (1). We claim that U represents $\mathrm{Ex}(\mathcal{H})$. As the property of being exact is stable under base change it suffices to show that $\mathcal{H}_U$ is exact, i.e. that f'_U is faithfully flat. But this follows from the fibre criterion for flatness.

(1.10) We call $\mathcal{D} = (O, {}^*, \varepsilon, h)$ *unramified*, if O_p is a maximal order in an unramified $\mathbb{Q}_p$-algebra, i.e. O_p is isomorphic to a product of matrix algebras over the rings of integers of unramified field extensions of $\mathbb{Q}_p$.

If $\mathcal{D}$ is unramified, then via Morita equivalence for hermitian modules the category of BT_n with $\mathcal{D}$-structure over some scheme S is equivalent to the product of categories of BT_n with $\mathcal{D}'_p$-structure where $\mathcal{D}'_p = (O'_p, {}^*, \varepsilon', h')$ is in of one of the following forms (for some unramified extension K of $\mathbb{Q}_p$):

(I) $O = O_K \times O_K$ and $(a,b)^* = (b,a)$ for $a, b \in O_K$.

(II) $O = O_K$ and * is an automorphism of order 2.

(III) $O = O_K$ and * is trivial.

When we are in case (I) the category of BT_n with $\mathcal{D}'_p$-structure over S is further equivalent to the category of BT_n with O_K-structure. We call this also the *linear case*.

We call case (II) and (III) the *hermitian case*.

2 Deformation of truncated Barsotti Tate groups with $\mathcal{D}$-structure

(2.1) We fix the following notations for this chapter: Denote by k a perfect field of characteristic p. Denote by $\mathcal{C}$ the category of pairs (R, φ) where R is a local Artinian ring and $\varphi\colon \kappa(R) \xrightarrow{\sim} k$ is an isomorphism. Let $\underline{X}_0 = (X_0, \lambda_0, \iota_0)$ be a BT_n $(1 \leq n \leq \infty)$ with $\mathcal{D}$-structure over k.

We consider the functor $\mathrm{Def}(\underline{X}_0)\colon \mathcal{C} \longrightarrow \mathbf{Sets}$ whose value at some (R, φ) in $\mathcal{C}$ is given by the set of isomorphism classes of pairs $(\underline{X}, \alpha)$ consisting of a BT_n $(1 \leq n \leq \infty)$ with $\mathcal{D}$-structure $\underline{X} = (X, \lambda, \iota)$ over R and an isomorphism $\alpha\colon \underline{X} \otimes_{R,\varphi} k \xrightarrow{\sim} \underline{X}_0$ of BT_n with $\mathcal{D}$-structure over k. In the same way we define the functors $\mathrm{Def}(X_0, \iota_0)$ and $\mathrm{Def}(X_0)$.

(2.2) Let A be some commutative ring. For an integer $n \geq 1$ we set $A_{/n} = A/p^n A$ and $A_{/\infty} = \varprojlim A/p^n A$.

(2.3) Lemma: *Let O be a finite étale $\mathbb{Z}_p$-algebra. For $1 \leq n \leq \infty$ let X, Y be abelian $fppf$-sheaves of $O_{/n}$-modules on some scheme S. Then the canonical map*

$$(*) \qquad \mathcal{E}xt^i_{O_{/n}}(X, Y) \longrightarrow \mathcal{E}xt^i_{\mathbb{Z}_{/n}}(X, Y)$$

is injective for all $i \geq 0$.

Proof: The map (*) is induced by the composition

$$\begin{aligned} \mathbb{R}\mathcal{H}om_{O_{/n}}(X, Y) &\longrightarrow \mathbb{R}\mathcal{H}om_{O_{/n}}(X, \mathbb{R}\mathcal{H}om_{\mathbb{Z}_{/n}}(O_{/n}, Y)) \\ &= \mathbb{R}\mathcal{H}om_{O_{/n}}(X, \mathcal{H}om_{\mathbb{Z}_{/n}}(O_{/n}, Y)) \\ &\xrightarrow{\sim} \mathbb{R}\mathcal{H}om_{\mathbb{Z}_{/n}}(X, Y). \end{aligned}$$

Therefore it suffices to show that

$$\mathcal{E}xt^i_{O_{/n}}(X, Y) \longrightarrow \mathcal{E}xt^i_{O_{/n}}(X, \mathcal{H}om_{\mathbb{Z}_{/n}}(O_{/n}, Y))$$

is injective. As $O_{/n}$ is finite étale over $\mathbb{Z}_{/n}$ there exists a finite étale $\mathbb{Z}_{/n}$ algebra A_n such that the A_n-algebra $B_n = A_n \otimes_{\mathbb{Z}_{/n}} O_{/n}$ is isomorphic to the product of copies of A_n. As A_n is finite locally free over $\mathbb{Z}_{/n}$ we have functorial isomorphisms

$$\mathcal{E}xt^i_{O_{/n}}(X, Y) \xrightarrow{\sim} \mathcal{E}xt^i_{B_n}(X_{B_n}, Y_{B_n})$$

$$\mathcal{E}xt^i_{O_{/n}}(X, \mathcal{H}om_{\mathbb{Z}_{/n}}(O_{/n}, Y)) \xrightarrow{\sim} \mathcal{E}xt^i_{B_n}(X_{B_n}, \mathcal{H}om_{A_n}(B_n, Y_{B_n})).$$

Now the A_n-linear section $\varphi \mapsto \varphi(1_{B_n})$ of the canonical map

$$Y_{B_n} \hookrightarrow \mathcal{H}om_{A_n}(B_n, Y_{B_n})$$

is automatically B_n-linear because B_n is a product of copies of A_n. This implies the injectivity of $\mathcal{E}xt^i_{B_n}(X_{B_n}, Y_{B_n}) \longrightarrow \mathcal{E}xt^i_{B_n}(X_{B_n}, \mathcal{H}om_{A_n}(B_n, Y_{B_n}))$ and therefore the claim.

(2.4) Lemma: *Let O be a commutative flat $\mathbb{Z}$-algebra, let X be a finite locally free scheme of O-modules over some scheme S and let M be a quasicoherent $\mathcal{O}_S \otimes_{\mathbb{Z}} O$-module (which we consider as an abelian fppf-sheaf). Then there is a canonical functorial isomorphism*

$$\mathbb{R}\mathcal{H}om_{\mathcal{O}_S \otimes_{\mathbb{Z}} O}(\ell_X, M) \xrightarrow{\sim} \tau_{\leq 1} \mathbb{R}\mathrm{Hom}_O(X^\vee, M)$$

in $D(S, \mathcal{O}_S \otimes_{\mathbb{Z}} O)$.

Proof: For $O = \mathbb{Z}$ this is proven in [MM] §14. For general O the assertion follows from this because of functoriality: O-linear homomorphisms on the left hand side correspond to O-linear homomorphisms on the right hand side.

(2.5) Lemma: *Let O be a finite étale $\mathbb{Z}_p$-algebra, let $n \geq 1$ be an integer and let S_0 be an affine scheme such that $p^n \mathcal{O}_{S_0} = 0$ and let J be a quasicoherent $\mathcal{O}_{S_0}$-module. Let X_0 and Y_0 be BT_n with O-structure over S_0. Then we have canonical functorial isomorphisms*

$$\mathrm{Hom}_{O/p^n}(X_0, \ell_{Y_0}^\vee \overset{\mathbb{L}}{\otimes} J) \cong \mathrm{Hom}_{\mathcal{O}_{S_0} \otimes_{\mathbb{Z}} O}(\omega_{X_0^\vee}, t_{Y_0} \otimes J),$$

$$\mathrm{Ext}^1_{O/p^n}(X_0, \ell_{Y_0}^\vee \overset{\mathbb{L}}{\otimes} J) \cong \mathrm{Hom}_{\mathcal{O}_{S_0} \otimes_{\mathbb{Z}} O}(\omega_{X_0^\vee}, \nu_{Y_0} \otimes J).$$

Proof: The proof is the same as the one of lemma 4.1 in [Ill] if one uses (2.4) instead of loc. cit. (2.1.5) and if one combines loc. cit. 2.2.4 with (2.3).

(2.6) Proposition: *Let $i: S \hookrightarrow S'$ be a nilimmersion of affine schemes. Let O be a finite étale $\mathbb{Z}_p$-algebra. Let $1 \leq n \leq \infty$ and let $\underline{X} = (X, \iota)$ be a BT_n with O-structure over S. Then there exists a deformation $\underline{X}'$ of $\underline{X}$ over S'.*

Proof: As in [Ill] 4.5.3 one is reduced to the case that $S \hookrightarrow S'$ is given by a small surjection $R' \longrightarrow R$ of local Artinian rings with algebraically closed residue field k of characteristic p (use the fact that we already know that the existence of a lifting of a BT_n is *fpqc*-local).

The case $n = \infty$ follows easily from Grothendieck-Messing theory and we will prove only the case $n < \infty$.

Denote by $\underline{X}_0 = (X_0, \iota_0)$ the reduction of $\underline{X}$ to $S_0 = \mathrm{Spec}(k)$. The obstruction to lift (X, ι) as a BT_n with O-module structure is an element $o(X, \iota) \in \mathrm{Ext}^2_{O/p^n}(X_0, \ell_{X_0}^\vee \overset{\mathbb{L}}{\otimes} J)$ and the obstruction $o(X)$ to lift X as a BT_n is the image of $o(X, \iota)$ under the canonical map $\mathrm{Ext}^2_{O/p^n}(X_0, \ell_{X_0}^\vee \overset{\mathbb{L}}{\otimes} J) \longrightarrow$ $\mathrm{Ext}^2_{\mathbb{Z}/p^n}(X_0, \ell_{X_0}^\vee \overset{\mathbb{L}}{\otimes} J)$ [Ill] 3.3. But this map is injective (2.3) and $o(X) = 0$ by [Ill] 4.4. Therefore we can lift (X, ι) to S'.

(2.7) Proposition: *Assume that O is a finite étale $\mathbb{Z}_p$-algebra. Let $S_0 \hookrightarrow S \hookrightarrow S'$ be closed immersions defined by ideals $J \subset J_0 \subset \mathcal{O}_{S'}$ such that $JJ_0 = 0$.*

Let $1 \leq n \leq \infty$ be such that $p^n \mathcal{O}_{S_0} = 0$ (for $n = \infty$ we mean by this that p is locally nilpotent on S_0) and let $\underline{X} = (X, \iota)$ be a BT_n with O-structure over S.

1. *The set of isomorphism classes of extensions $\underline{X}'$ of $\underline{X}$ to S' forms a principal homogeneous space under*

$$\mathrm{Hom}_{\mathcal{O}_{S_0} \otimes_{\mathbb{Z}} O}(t^*_{X_0^\vee}, t_{X_0}) \otimes J.$$

2. *The group of automorphisms of an extension $\underline{X}'$ (inducing the identity on $\underline{X}$) is isomorphic to*

$$\mathrm{Hom}_{\mathcal{O}_{S_0} \otimes_{\mathbb{Z}} O}(t^*_{X_0^\vee}, t_{X_0}) \otimes J.$$

Proof: Again we consider only the case $n < \infty$. Denote by $\mathrm{Def}_{S'}(\underline{X})$ the set of isomorphism classes of extensions of $\underline{X}$. By (2.6) we know that $\mathrm{Def}_{S'}(\underline{X})$ is non empty. Choose a fixed lift $\underline{X}'_0 = (X'_0, \iota'_0)$ of $\underline{X}$. Combining (2.5), [Ill] 3.3, and [Ill] 2.2.5, the choice of (X'_0, ι'_0) gives an identification

$$\mathrm{Def}_{S'}(X, \iota) = \mathrm{Hom}_{\mathcal{O}_{S_0} \otimes_{\mathbb{Z}} O}(t^*_{X_0^\vee}, t_{X_0}) \otimes J.$$

The same references give (2).

(2.8) Now assume that $\underline{X}_0$ is a Barsotti Tate group with $\mathcal{D}$-structure over k then for $1 \leq n \leq m \leq \infty$ we have a canonical morphism of functors

$$\Psi\colon \mathrm{Def}(\underline{X}_0[p^m]) \longrightarrow \mathrm{Def}(\underline{X}_0[p^n]).$$

Theorem: *Assume that $p > 2$ and that $\mathcal{D}$ is unramified. Then Ψ is formally smooth.*

Proof: We have the following formal facts: Let

$$\mathcal{F} \xrightarrow{f} \mathcal{G} \xrightarrow{g} \mathcal{H}$$

be morphisms of deformation functors on $\mathcal{C}$. Then if f is surjective and $g \circ f$ is formally smooth, g is formally smooth. Further, every formally smooth morphism is surjective. Using these two facts we can assume that $m = \infty$. Further we can assume that n is finite. The proof of the formal smoothness will be given in several steps in the following sections.

(2.9) We can assume from now on that we are in the linear or in the hermitian case (1.10). We first consider the linear case, i.e. we assume that (X_0, ι_0) is a Barsotti-Tate group with O_K-action where K is an unramified extension of $\mathbb{Q}_p$.

Proposition: *Let $S_0 \hookrightarrow S \hookrightarrow S'$ be defined by ideals $J \subset J_0 \subset \mathcal{O}_{S'}$ with $JJ_0 = 0$ and let $n \geq 1$ be an integer such that $p^n \mathcal{O}_{S_0} = 0$. Let (X, ι) be a Barsotti-Tate group with O_K-structure over S and denote by $\mathrm{Def}_{(X,\iota)}(S')$ the* *set of isomorphism classes of deformations of* (X, ι) *over* S'. *Then the canonical map*

$$\Phi\colon \mathrm{Def}_{(X,\iota)}(S') \longrightarrow \mathrm{Def}_{(X,\iota)[p^n]}(S'), \qquad (X', \iota') \mapsto (X', \iota')[p^n]$$

is bijective.

Proof : This follows verbatim as in [Ill] 4.7 using (2.6) and (2.7). Note that the proof in loc. cit. shows the following more precise assertion: Choosing a lift (X', ι') of (X, ι) (which exists by (2.6)) we get by (2.7) identifications

$$\mathrm{Def}_{(X,\iota)}(S') = \mathrm{Hom}_{\mathcal{O}_{S_0} \otimes_{\mathbb{Z}} O}(t^*_{X_0^\vee}, t_{X_0} \otimes J),$$
$$\mathrm{Def}_{(X,\iota)[p^n]}(S') = \mathrm{Hom}_{\mathcal{O}_{S_0} \otimes_{\mathbb{Z}} O}(t^*_{X_0^\vee[p^n]}, t_{X_0[p^n]} \otimes J)$$

where the second identification is given by $(X', \iota')[p^n]$. But as $p^n \mathcal{O}_{S_0} = 0$ we have $t_{X_0[p^n]} = t_{X_0}$ and Φ is then given by the identity of $\mathrm{Hom}_{\mathcal{O}_{S_0} \otimes_{\mathbb{Z}} O}(t^*_{X_0^\vee}, t_{X_0} \otimes J)$.

(2.10) Corollary: *Let* K *be an unramified extension of* $\mathbb{Q}_p$ *of degree* m *and let* (X_0, ι_0) *be a Barsotti-Tate group with* O_K*-structure over* k.

(1) *The functor* $\mathrm{Def}_{(X_0,\iota_0)}$ *is formally smooth and pro-representable.*

(2) *The functor* $\mathrm{Def}_{(X_0,\iota_0)[p^n]}$ *is formally smooth and the canonical map* $\Phi\colon \mathrm{Def}_{(X_0,\iota_0)} \longrightarrow \mathrm{Def}_{(X_0,\iota_0)[p^n]}$ *is a hull.*

(3) *The tangent spaces of* $\mathrm{Def}_{(X_0,\iota_0)}$ *and* $\mathrm{Def}_{(X_0,\iota_0)[p^n]}$ *are both canonically isomorphic to* $t_{X_0^\vee} \otimes_{(k \otimes_{\mathbb{Z}_p} O_K)} t_{X_0}$ *and via these identifications* Φ *induces the identity.*

(2.11) From now on we are in the hermitian case, i.e. we assume that $\underline{X}_0 = (X_0, \lambda_0, \iota_0)$ is a Barsotti-Tate group with $\mathcal{D}$-structure where $\mathcal{D}$ is of the form $\mathcal{D} = (O_K, \dots)$ for some unramified extension K of $\mathbb{Q}_p$. To shorten notations we set $\underline{Y}_0 := (Y_0, \lambda_0, \iota_0) := \underline{X}_0[p^n]$ which is a BT_n with $\mathcal{D}$-structure for some integer $1 \leq n < \infty$. The goal is to show that

$$\mathrm{Def}(\underline{X}_0) \longrightarrow \mathrm{Def}(\underline{Y}_0)$$

is formally smooth. For this we will use the fact that $\mathrm{Def}(X_0, \iota_0) \longrightarrow \mathrm{Def}(Y_0, \iota_0)$ is formally smooth (2.10).

We first look at the canonical morphism of functors

$$\mathrm{Def}(\underline{Y}_0) \longrightarrow \mathrm{Def}(Y_0, \iota_0).$$

(2.12) We define an involution Λ on $\mathrm{Def}(Y_0, \iota_0)$ as follows: Let R be in $\mathcal{C}$ and $(Y, \iota, \alpha) \in \mathrm{Def}(Y_0, \iota_0)(R)$ and set

$$(Y, \iota, \alpha)^\Lambda = (Y^\vee, \iota^\vee, \alpha^\Lambda = \lambda_0^{-1} \circ (\alpha^\vee)^{-1})$$

where $\iota^\vee$ sends $a \in O$ to $\iota(a^*)^\vee$. Then if $(Y,\iota) \xrightarrow{\text{can}} (Y^{\vee\vee},\iota^{\vee\vee})$ denotes the canonical isomorphism we have a commutative diagram

$$\begin{array}{ccc} (Y,\iota)\otimes_R k & \xrightarrow{\alpha} & Y_0 \\ {\scriptstyle \varepsilon\mathrm{can}\otimes\mathrm{id}}\downarrow & & \| \\ (Y^{\vee\vee},\iota^{\vee\vee})\otimes_R k & \xrightarrow{\alpha^{\Lambda\Lambda}} & Y_0. \end{array}$$

Therefore as elements in $\mathrm{Def}(Y_0)(R)$ we have $(Y,\iota,\alpha)^{\Lambda\Lambda} = (Y,\iota,\alpha)$.

(2.13) Lemma: *We keep the notations of the previous section.*

(1) *The image of* $\mathrm{Def}(\underline{Y}_0)$ *in* $\mathrm{Def}(Y_0,\iota_0)$ *is the functor of* Λ*-invariants* $\mathrm{Def}(Y_0)^{\Lambda=1}$.

(2) *On tangent spaces,* $t_{\mathrm{Def}(\underline{Y}_0)} \longrightarrow t_{\mathrm{Def}(Y_0,\iota_0)}$ *is an injection.*

Proof : Let us prove (1). It is obvious that the image is contained in $\mathrm{Def}(Y_0)^{\Lambda=1}$. Now let (Y,ι,α) be an element in $\mathrm{Def}(Y_0)^{\Lambda=1}(R)$ for some R in $\mathcal{C}$. This means there exists an isomorphism of deformations

$$\lambda' \colon (Y,\iota,\alpha) \longrightarrow (Y^\vee,\iota^\vee,\alpha^\Lambda).$$

By definition of the involution Λ to give λ' is the same as a lifting of λ_0 to some isomorphism $(Y,\iota) \xrightarrow{\sim} (Y^\vee,\iota^\vee)$. Defining $\lambda = \frac{1}{2}(\lambda' + \varepsilon\lambda'^\vee)$ we get $\lambda^\vee = \varepsilon\lambda$ (this makes sense as the category of groups which are annihilated by some power of p is a $\mathbb{Z}_p$-linear category and because we have $p > 2$), and we have $((Y,\lambda,\iota),\alpha) \in \mathrm{Def}(\underline{Y}_0)$.

To show (2) let (Y,ι,λ) and (Y',ι',λ') be two elements in $t_{\mathrm{Def}(\underline{Y}_0)} = \mathrm{Def}(\underline{Y}_0)(k[T]/T^2)$ which have the same image in $t_{\mathrm{Def}(Y_0,\iota_0)}$. This means there exists an isomorphism $(Y,\iota) \xrightarrow{\sim} (Y',\iota')$ inducing the identity over k. Using this isomorphism we identify (Y,ι) and (Y',ι') and we have to show that there exists an automorphism φ of the lifting (Y,ι) such that $(\lambda^{-1}\circ\varphi^\vee\circ\lambda)\circ(\lambda^{-1}\circ\lambda')\circ\varphi = \mathrm{id}$. But the automorphisms of the lifting (Y,ι) form the additive group of some finite dimensional k-vector space V. In particular it is commutative and $\varphi = -\frac{1}{2}(\lambda^{-1}\circ\lambda')$ does the job.

(2.14) Corollary: *We have a canonical isomorphism*

$$t_{\mathrm{Def}(\underline{Y}_0)} \cong \{\, u \in \mathrm{Hom}_{O\otimes_{\mathbb{Z}_p}k}(t^*_{Y_0^\vee}, t_{Y_0}) \mid (t^{-1}_{\lambda_0})\circ(-u^*)\circ(t^*_{\lambda_0}) = u \,\}$$

and $\Psi\colon \mathrm{Def}(\underline{X}_0) \longrightarrow \mathrm{Def}(\underline{Y}_0)$ *induces an isomorphism on tangent spaces.*

(2.15) It follows from Grothendieck-Messing theory that $\mathrm{Def}(X_0,\iota_0)$ is pro-representable by some formally smooth $W(k)$-algebra R' and by (2.10) we know that $\mathrm{Def}(X_0,\iota_0) \longrightarrow \mathrm{Def}(Y_0,\iota_0)$ is a hull. Further by rigidity of Barsotti-Tate groups, $\mathrm{Def}(\underline{X}_0) \longrightarrow \mathrm{Def}(X_0,\iota_0)$ is injective and hence we can identify

$$\mathrm{Def}(\underline{X}_0) = \mathrm{Def}(X_0,\iota_0)^{\Lambda=1}.$$

As 2 is invertible, this functor is pro-representable by the formally smooth $W(k)$-algebra $R = R'_{\Lambda=1}$ of Λ-coinvariants.

By $\dim(t_{\mathrm{Def}(\underline{Y}_0)}) < \infty$ it follows from the Schlessinger criterion that $\mathrm{Def}(\underline{Y}_0)$ has a hull, say $h_S \longrightarrow \mathrm{Def}(\underline{Y}_0)$. By the definition of a hull the morphism $\Psi\colon \mathrm{Def}(\underline{X}_0) = h_R \longrightarrow \mathrm{Def}(\underline{Y}_0)$ lifts (non uniquely) to a morphism $h_R \longrightarrow h_S$ which induces an isomorphism on tangent spaces. By the lemma below, $h_R \longrightarrow h_S$ is an isomorphism, therefore $\Psi\colon \mathrm{Def}(\underline{X}_0) \longrightarrow \mathrm{Def}(\underline{Y}_0)$ is a hull and in particular formally smooth. This finishes the proof of (2.8).

(2.16) Lemma: *Let A be a local noetherian ring with perfect residue field k, let R and S be two complete local noetherian A-algebras with residue field k, and let $\varphi\colon S \longrightarrow R$ be a local homomorphism of A-algebras. Assume that R is formally smooth over A. Then if φ induces an isomorphism $\mathfrak{m}_S/\mathfrak{m}_S^2 \xrightarrow{\sim} \mathfrak{m}_R/\mathfrak{m}_R^2$ of k-vector spaces, φ is an isomorphism.*

Proof: Set $n = \dim_k(\mathfrak{m}_S/\mathfrak{m}_S^2) = \dim_k(\mathfrak{m}_R/\mathfrak{m}_R^2)$. By [EGA IV] **0**, 19.7 we can assume that $A = k$ and that $R = k[[t_1, \dots, t_n]]$. It is enough to show that S is formally smooth over k and for this it suffices to prove that S is regular (as k is perfect). Let $a_1, \dots, a_n \in \mathfrak{m}_S$ be elements which lift a basis of $\mathfrak{m}_S/\mathfrak{m}_S^2$ and define a local homomorphism $R \longrightarrow S$ by $t_i \mapsto a_i$. The composition with φ is a local homomorphism $R \longrightarrow R$ which induces an isomorphism on $\mathfrak{m}_R/\mathfrak{m}_R^2$ and is therefore an isomorphism (as R is regular and complete). It follows that

$$\dim(R) \leq \dim(S) \leq \dim_k(\mathfrak{m}_S/\mathfrak{m}_S^2) = \dim_k(\mathfrak{m}_R/\mathfrak{m}_R^2) = \dim(R),$$

hence S is regular.

(2.17) Assume that $p > 2$ and that $\mathcal{D}$ is unramified and let $\underline{X}_0$ be a BT with $\mathcal{D}$-structure over k. Then we showed during the proof of (2.8):

Corollary: *The canonical functor* $\mathrm{Def}(\underline{X}_0) \longrightarrow \mathrm{Def}(\underline{X}_0[p^n])$ *is a hull for all $n \geq 1$.*

3 Smoothness of the moduli space of truncated Barsotti-Tate groups

(3.1) From now on we assume that p is odd and that $\mathcal{D}$ is unramified. Let O' be the center of O_p. As $\mathcal{D}$ is unramified, $O' \otimes_{\mathbb{Z}_p} \mathbb{F}_p$ is a product of finite fields $k_1, \dots, k_r$. We call

$$f = \text{l.c.m.}\{\, [k_i : \mathbb{F}_p] \mid i = 1, \dots, r \,\}$$

the *inertia index* of $\mathcal{D}$. Further let n be a fixed positive integer.

(3.2) Proposition: *Let $\underline{X}$ be a BT_n with $\mathcal{D}$-structure over a complete local noetherian ring R with perfect residue field k of characteristic p. Assume that*

k contains the field $\mathbb{F}_{p^f}$ where f is the inertia index of $\mathcal{D}$. Then there exists a BT with $\mathcal{D}$-structure $\underline{\tilde{X}}$ over R such that $\underline{\tilde{X}}[p^n] \cong \underline{X}$.

Proof : We can assume by (2.8) that $R = k$ is a perfect field of characteristic p containing $\mathbb{F}_{p^f}$. We prove the theorem by using covariant Dieudonné theory. This gives an equivalence of the category of BT_n of height h with $\mathcal{D}$-structure over k and tuples $(M, F, V, \iota, \langle\ ,\ \rangle)$ of the following form: M is a free $W_n(k)$-module of rank h and F (resp. V) is a σ-linear (resp. σ^{-1}-linear) endomorphism of M such that $FV = VF = p\,\mathrm{id}_M$ (here σ denotes the Frobenius automorphism of $W_n(k)$) and such that $\mathrm{Ker}(V) = \mathrm{Im}(F)$ for $n = 1$; further, $\iota\colon O_K \longrightarrow \mathrm{End}_{W_n(k)[F,V]}(M)$ is a $\mathbb{Z}_p$-algebra homomorphism, and $\langle\ ,\ \rangle\colon M \otimes_{W_n(k)} M \longrightarrow W_n(k)$ is a ε-symmetric bilinear form such that $\langle Fm, m'\rangle = \langle m, Vm'\rangle^\sigma$ and $\langle \iota(a)m, m'\rangle = \langle m, \iota(a^*)m'\rangle$ for all $m, m' \in M$ and $a \in O_K$. It suffices to deal with the three special cases of (1.10) separately:

Case (I) (cf. [Ill] 1.7): Let (Y, ι) be the BT_n with O_K-structure corresponding to $\underline{X}$ and denote by (M, ι) its Dieudonné module. As k contains $O_K/pO_K = \mathbb{F}_{p^f}$ the action ι corresponds to a decomposition

$$M = \bigoplus_{j \in \mathbb{Z}/f\mathbb{Z}} M(j)$$

where $M(j) = \{ m \in M \mid \iota(a)m = \sigma^{-j}(a)m \text{ for } a \in O_K \text{ and } m \in M \}$. Then F (resp. V) is homogeneous of degree -1 (resp. $+1$). The $W_n(k)$-rank of $M(j)$ is $g := \mathrm{ht}(Y)/f$. Let $(\bar{M} = \bigoplus \bar{M}(j), F, V)$ be the reduction modulo p of (M, F, V, ι). Define

$$T(j) := V\bar{M}(j-1) = \mathrm{Ker}(F|_{\bar{M}(j)}).$$

Further choose a complement

$$L(j)$$

of $T(j)$ in $\bar{M}(j)$ and set

$$r(j) = \dim_k(L(j)) = g - \dim_k(T(j)).$$

For all $j \in \mathbb{Z}/f\mathbb{Z}$ choose elements $\bar{d}_{r(j)+1}(j-1), \ldots, \bar{d}_g(j-1) \in \bar{M}(j-1)$ such that $V\bar{d}_{r(j)+1}(j-1), \ldots, V\bar{d}_g(j-1)$ is a basis of $T(j)$. Further choose a basis $\bar{e}_1(j), \ldots, \bar{e}_{r(j)}(j)$ of $L(j)$. Finally denote by $d_i(j-1)$ and $e_i(j)$ liftings of $\bar{d}_i(j-1)$ resp. $\bar{e}_i(j)$ to $M(j-1)$ resp. $M(j)$.

Then

$$e_1(j), \ldots, e_{r(j)}(j), Vd_{r(j)+1}(j-1), \ldots, Vd_g(j-1)$$

is a basis $(e'_i(j))$ of $M(j)$. Further $F\bar{e}_1(j), \ldots, F\bar{e}_{r(j)}(j)$ is a basis of $F\bar{M}(j) = \mathrm{Ker}(V|_{\bar{M}(j)})$, hence

$$Fe_1(j), \ldots, Fe_{r(j)}(j), d_{r(j)+1}(j-1), \ldots, d_g(j-1)$$

is a basis $(e_i''(j-1))$ of $M(j-1)$. Let $A(j)$ be the matrix of the change of basis from $(e_i''(j))$ to $(e_i'(j))$. In the bases $(e_i'(j))$ and $(e_i''(j-1))$, F has the $(g \times g)$-matrix $\begin{pmatrix} 1_{r(j)} & 0 \\ 0 & p \end{pmatrix}$, and in the bases $(e_i''(j-1))$ and $(e_i'(j))$, V has the matrix $\begin{pmatrix} p1_{r(j)} & 0 \\ 0 & 1 \end{pmatrix}$.

Let $\tilde{A}(j) \in GL_g(W(k))$ be a lifting of $A(j)$, let $\tilde{M}(j)$ be a free $W(k)$-module with basis $(\tilde{e}_i''(j))$, and let $(\tilde{e}_i'(j))$ be the basis of $\tilde{M}(j)$ which is the image of $(\tilde{e}_i''(j))$ under $\tilde{A}(j)$. Define $F\colon \tilde{M}(j) \longrightarrow \tilde{M}(j-1)$ as the σ-linear map with matrix $\begin{pmatrix} 1_{r(j)} & 0 \\ 0 & p \end{pmatrix}$ with respect to the bases $(\tilde{e}_i'(j))$ and $(\tilde{e}_i''(j-1))$, and $V\colon \tilde{M}(j-1) \longrightarrow \tilde{M}(j)$ as the σ^{-1}-linear map with matrix $\begin{pmatrix} p1_{r(j)} & 0 \\ 0 & 1 \end{pmatrix}$ with respect to the bases $(\tilde{e}_i''(j))$ and $(\tilde{e}_i'(j-1))$. Then $(\tilde{M} = \bigoplus \tilde{M}(j), F, V)$ gives a BT with O_K-structure $\underline{\tilde{X}}$ over k and the isomorphism $\tilde{M}/p^n\tilde{M} \xrightarrow{\sim} M$ which sends the class of $\tilde{e}_i''(j)$ to $e_i(j)$ defines an isomorphism of $\underline{\tilde{X}}[p^n]$ with $\underline{X}$.

Case (II): Let $\underline{M} = (M, F, V, \iota, \langle\, , \rangle)$ be the Dieudonné module of $\underline{X}$. As in case (I) the O_K-action is given by a decomposition $M = \bigoplus_{j \in \mathbb{Z}/f\mathbb{Z}} M(j)$ and as the involution is trivial, this is an orthogonal sum. We define g, $\bar{M}(j)$, and $T(j)$ as in case (I) and denote the perfect pairing on $\bar{M}(j)$ induced by $\langle , \rangle$ again by $\langle , \rangle$. As the induced pairing on $T(j) \times \bar{M}(j)/T(j)$ induces a perfect duality (which follows from $T(j) = \operatorname{Ker}(F|_{\bar{M}(j)})$), $r := g - \dim(T(j)) = g/2$ is independent of j and $T(j)$ is totally isotropic and hence maximal totally isotropic. Choose as $L(j)$ a maximal totally isotropic complement of $T(j)$. Then we can choose elements $\bar{d}_i(j-1)$ and $\bar{e}_i(j)$ as in case (I) such that the matrix of $\langle\, , \rangle$ with respect to $\bar{e}_1(j), \dots, \bar{e}_r(j), V\bar{d}_{r+1}(j-1), \dots, V\bar{d}_g(j-1)$ is given by

$$J(\varepsilon) = \begin{pmatrix} & & & & 1 \\ & & & \cdots & \\ & & 1 & & \\ & \varepsilon & & & \\ \cdots & & & & \\ \varepsilon & & & & \end{pmatrix} \in GL_g.$$

We can find liftings $d_i(j-1)$ and $e_i(j)$ such that $\langle\, , \rangle$ is still given by $J(\varepsilon)$ with respect to the basis $(e_i'(j))$ (for $\varepsilon = 1$ we use the hypothesis that $p \neq 2$). Define $e_i''(j)$ and $A(j)$ as in case (I). Then the compatibility of $\langle , \rangle$ with F and V implies that $J(\varepsilon) = (A(j)J(\varepsilon)^t A(j))^\sigma$, i.e. $A(j)A(j)^* = 1$ for $A(j)^* = J(\varepsilon)^t A(j) J(\varepsilon)^{-1}$. If we lift $A(j)$ to a matrix $\tilde{A}(j) \in GL_g(W(k))$ with $\tilde{A}(j)\tilde{A}(j)^* = 1$ we can define $(\tilde{e}_i'(j))$ and $(\tilde{M} = \bigoplus \tilde{M}(j), F, V)$ as in case (I). Further, we define a lifting of $\langle\, , \rangle$ as the bilinear form which is given by $J(\varepsilon)$ with respect to $(\tilde{e}_i'(j))$. Then $(\tilde{M} = \bigoplus \tilde{M}(j), F, V, \langle\, , \rangle)$ is a lifting of $\underline{M}$. Indeed, we have to check that $\langle Fm, m' \rangle = \langle m, Vm' \rangle^\sigma$ but this is equivalent to $\langle Fm, Fm' \rangle = p\langle m, m' \rangle^\sigma$ which follows from $\tilde{A}(j)\tilde{A}(j)^* = 1$ by checking this for m and m' in $\{e_i'(j)\}$.

Case (III): Because of the existence of the involution on O_K, f is even, say $f = 2f'$. If $\underline{M} = (M, F, V, \iota, \langle\, , \rangle)$ is the Dieudonné module of $\underline{X}$, the O_K action is given by a decomposition $M = \bigoplus_{j \in \mathbb{Z}/2f'\mathbb{Z}} M(j)$ such that F (resp.

V) is homogeneous of degree -1 (resp. of degree $+1$) and such that $M(j)$ is orthogonal to $M(j')$ for all $j' \neq j + f'$ and such that $M(j)$ and $M(j+f')$ are in perfect duality for all j. Therefore we use the same construction as in case (II) by replacing $M(j)$ by $N(j) = M(j) \oplus M(j+f')$ and choosing the elements $\bar{d}_i(j-1)$ and $\bar{e}_i(j)$ homogeneous with respect to the decomposition of $N(j)$. We omit the details.

(3.3) Corollary: *The algebraic stack* $\mathtt{BT}_{\mathcal{D},n}$ *is smooth over* $\mathbb{Z}_{(p)}$.

Proof : The 1-morphism $\widetilde{\mathtt{BT}}_{\mathcal{D},n} \longrightarrow \mathtt{BT}_{\mathcal{D},n}$ is smooth and surjective and we have to show that $\widetilde{\mathtt{BT}}_{\mathcal{D},n}$ is smooth over $\mathbb{Z}_{(p)}$. It suffices to prove that $\mathrm{Hom}_{\mathbb{Z}_{(p)}}(\mathrm{Spec}(R'), \widetilde{\mathtt{BT}}_{\mathcal{D},n}) \longrightarrow \mathrm{Hom}_{\mathbb{Z}_{(p)}}(\mathrm{Spec}(R), \widetilde{\mathtt{BT}}_{\mathcal{D},n})$ is surjective where $R' \longrightarrow R$ is a surjection of local Artinian rings with an algebraically closed residue field k (as the residue fields of $\mathbb{Z}_{(p)}$ are perfect). If k is of characteristic 0, the smoothness is clear because all BT_n in characteristic zero are étale. If k is of characteristic p, the smoothness follows from (3.2) and (2.8).

(3.4) Lemma: *The generic fibre* $\mathtt{BT}_{\mathcal{D},n} \otimes \mathbb{Q}$ *is a Deligne-Mumford stack, which is étale over* $\mathbb{Q}$.

Proof : In characteristic zero every Barsotti-Tate group is finite étale, hence locally for the étale topology it is a constant $\mathbb{Z}/p^n\mathbb{Z}$-module and, being flat over $\mathbb{Z}/p^n\mathbb{Z}$, it is étale locally isomorphic to the constant sheaf $(\mathbb{Z}/p^n\mathbb{Z})^h$. Hence we have

$$\mathtt{BT}_{\mathcal{D},n} \otimes \mathbb{Q} = [\mathrm{Aut}((\mathbb{Z}/p^n\mathbb{Z})^h) \backslash \mathrm{Spec}(\mathbb{Q})]$$

in the sense of stacks.

(3.5) Corollary: *The smooth scheme* $\widetilde{\mathtt{BT}}_{\mathcal{D},n}$ *has relative dimension* $p^{2nh} = \dim(GL_{p^{nh}}/\mathbb{Z}_{(p)})$ *over* $\mathbb{Z}_{(p)}$.

4 Isomorphism classes of truncated Barsotti-Tate groups

(4.1) Isomorphisms between BT_n over a scheme S are given by certain linear isomorphisms of the underlying $\mathcal{O}_S$-modules. This implies that if we have two $\bar{\mathbb{F}}_p$-valued points $(\underline{X}, \alpha)$ and $(\underline{X}', \alpha')$ of $\widetilde{\mathtt{BT}}_{\mathcal{D},n}$, $\underline{X}$ and $\underline{X}'$ are isomorphic as BT_n with $\mathcal{D}$-structure if and only if they lie in the same $GL_{p^{nh}}(\bar{\mathbb{F}}_p)$-Orbit. This provides a bijection of the set $\Xi_{\mathcal{D},n}$ of isomorphism classes of BT_n with $\mathcal{D}$-structure and the set of $GL_{p^{nh}}$-orbits of $\widetilde{\mathtt{BT}}_{\mathcal{D},n}$.

(4.2) To study the set $\Xi_{\mathcal{D},n}$ we first recall some general facts from the action of algebraic groups on schemes (cf. [DG]):

Let k be an arbitrary field, denote by G a smooth group scheme of finite type over k and denote by X an k-scheme of finite type with left G-action. Denote the set of G-orbits on X by Ξ.

(1) Every G-orbit in X is locally closed.
(2) If Ξ is finite, then there exists a bijection between Ξ and the set of $\mathrm{Gal}(\bar{k}/k)$-equivalence classes of $G(\bar{k})$-orbits of $X(\bar{k})$ for any algebraic closure $\bar{k}$ of k. Further, if K is any algebraically closed extension of $\bar{k}$ the canonical map of the set of $G(K)$-orbits of $X(K)$ to the set of $G(\bar{k})$-orbits of $X(\bar{k})$ is bijective.
(3) The closure of an orbit is a union of orbits.
(4) There exists a closed orbit.

We define a partial order on Ξ as follows: For $O, O' \in \Xi$ we set

$$O \leq O' :\Leftrightarrow O \subset \bar{O}' \Leftrightarrow O \cap \bar{O}' \neq \emptyset.$$

(4.3) If Δ is a partially ordered set, we can define the right directed topology on Δ. It is the topology which has as a basis the subsets $U_\delta = \{\, \epsilon \in \Delta \mid \epsilon \geq \delta \,\}$ for $\delta \in \Delta$. We denote by Δ_{top} this topological space. This defines a functor of the category of partially ordered sets (morphisms being those maps which respect the order) into the category of topological spaces (morphisms being continuous maps). This functor is fully faithful. Its essential image consists of the Kolmogoroff spaces where every intersection of open sets is again open.

(4.4) We keep the notations of (4.2). Consider the quotient stack $[G\backslash X]$. This is an Artin stack and the canonical morphism $X \longrightarrow [G\backslash X]$ is smooth and surjective.

Lemma: *Assume that Ξ is finite. The underlying Zariski space $[G\backslash X]_{\mathrm{top}}$ of $[G\backslash X]$ in the sense of [LM] is homeomorphic to Ξ.*

Proof: It follows from (4.2)(2) and the fact that every G-torsor over an algebraically closed field is trivial that the underlying set of $[G\backslash X]_{\mathrm{top}}$ is equal to Ξ. Note that $[G\backslash X]_{\mathrm{top}}$ carries the quotient topology of the topology of X ([LM] 5.6.1). Let U be a subset of Ξ_{top} such that its inverse image V under the canonical map $\pi\colon X \longrightarrow [G\backslash X]$ is open in X. The subscheme V carries an induced G-action. Denote by S the set of closed G-orbits in V. We have $S \neq \emptyset$ by (4.2)(4) and $U = \pi(V)$ is the union of the sets $\{\, \xi \in \Xi \mid \xi \geq s \,\}$ for $s \in S$. Hence U is open. Now let U be an open subset of Ξ. Then $V = \pi^{-1}(U)$ is the union of finitely many locally closed subsets. Hence it is constructible. Further it is stable under generization, hence V is open.

(4.5) Note that if the number of G-orbits on X is finite it follows that every point of $[G\backslash X]_{\mathrm{top}}$ is locally closed. In general points are still constructible by the theorem of Chevalley for algebraic stacks ([LM] 5.9.4).

(4.6) Using Dieudonné theory it follows from the classification of finite dimensional representations of the (commutative) algebra $k[X,Y]/(XY)$ for an algebraically closed field k by Gelfand and Ponomarev [GP] applied to Dieudonné modules by Kraft [Kr] that there exist only finitely many isomorphism classes of BT_1 over k of a given height. Using this description of BT_1 B. Moonen has given a precise description of the set of isomorphism classes of BT_1 with $\mathcal{D}$-structure over an algebraically closed field k [Mo] if $\mathcal{D}$ is unramified. His results in particular show that $\Xi_{\mathcal{D},1}$ is finite and independent of k in the unramified case.

5 Truncated Barsotti-Tate groups and Dieudonné spaces

(5.1) We fix the following notation: If S is an $\mathbb{F}_p$-scheme and $\mathcal{M}$ is an $\mathcal{O}_S$-module we denote by $\mathcal{M}^{(p)}$ the pullback $\mathrm{Frob}_S{}^*\mathcal{M}$ where Frob_S denotes the absolute Frobenius on S.

(5.2) Let S be a scheme of characteristic p. A *Dieudonné space over* S is a triple $(\mathcal{M}, F, V)$ where $\mathcal{M}$ is a locally free finitely generated $\mathcal{O}_S$-module and where $F: \mathcal{M}^{(p)} \longrightarrow \mathcal{M}$ and $V: \mathcal{M} \longrightarrow \mathcal{M}^{(p)}$ are $\mathcal{O}_S$-linear homomorphisms such that $FV = 0_{\mathcal{M}}$ and $VF = 0_{\mathcal{M}^{(p)}}$. Such a Dieudonné space is called *regular* if $\mathrm{Im}(V)$ and $\mathrm{Im}(F)$ are locally direct summands of $\mathcal{M}^{(p)}$ resp. $\mathcal{M}$ (note that this implies that $\mathrm{Ker}(V)$ and $\mathrm{Ker}(F)$ are locally direct summands of $\mathcal{M}$ resp. $\mathcal{M}^{(p)}$) and that $\mathrm{Ker}(V) = \mathrm{Im}(F)$ (or equivalently $\mathrm{Ker}(F) = \mathrm{Im}(V)$). The locally constant function $S \longrightarrow \mathbb{Z}$ sending $s \in S$ to the rank of $\mathcal{M}$ in s is called the *height of* $\mathcal{M}$. We have the obvious notion of a morphism of Dieudonné spaces over S.

If $(\mathcal{M}, F, V)$ is a Dieudonné space over S, its *dual* $(\mathcal{M}, F, V)^\vee$ is given by $(\mathcal{M}^\vee, F^\vee, V^\vee)$ where $\mathcal{M}^\vee$ is the $\mathcal{O}_S$-linear dual of $\mathcal{M}$ and where $F^\vee$ (resp. $V^\vee$) is the $\mathcal{O}_S$-linear dual of V (resp. of F).

(5.3) For a scheme S of characteristic p a *Dieudonné space with $\mathcal{D}$-structure over* S is a tuple $\underline{M} = (\mathcal{M}, F, V, \iota, \langle\,,\rangle)$ such that:

- $(\mathcal{M}, F, V)$ is a regular Dieudonné space over S of height h.
- $\iota: O/pO \longrightarrow \mathrm{End}_{\mathcal{O}_S}(\mathcal{M})$ is an $\mathbb{F}_p$-algebra homomorphism such that $\iota(a) \circ F = F \circ \iota(a)^{(p)}$ and $\iota(a)^{(p)} \circ V = V \circ \iota(a)$ for all $a \in O/pO$.
- $\langle\,,\rangle$ is a perfect $\mathcal{O}_S$-bilinear form such that the corresponding isomorphism $\lambda: \mathcal{M} \to \mathcal{M}^\vee$ is an isomorphism $(\mathcal{M}, F, V) \xrightarrow{\sim} (\mathcal{M}, F, V)^\vee$ of Dieudonné spaces and such that $\lambda \circ \iota(a) = \iota(a^*)^\vee \circ \lambda$. If ε is equal to 1 (resp. -1) we assume that $\langle\,,\rangle$ is symmetric (resp. alternating).

(5.4) Let $\underline{M}$ be a Dieudonné space with $\mathcal{D}$-structure over an $\mathbb{F}_p$-scheme S. If $f: T \longrightarrow S$ is a morphism of schemes we have the obvious notion of the pullback $f^*\underline{M}$. This way we get a fibered category over $(\mathtt{Sch}/\mathbb{F}_p)$. If we admit in the fibre categories only isomorphisms as morphisms, we get a fibre category fibered in groupoids which is a stack if we endow $(\mathtt{Sch}/\mathbb{F}_p)$ with some topology coarser than the *fpqc*-topology, say the *fppf*-topology to fix notations. We call this stack $\mathtt{DS}_{\mathcal{D}}$.

We also define a smooth covering $\widetilde{\mathtt{DS}}_{\mathcal{D}}$ of $\mathtt{DS}_{\mathcal{D}}$ as the scheme representing the functor on $(\mathtt{Sch}/\mathbb{F}_p)$ which associates to an $\mathbb{F}_p$-scheme S a pair $(\underline{M}, \alpha)$ where $\underline{M} = (\mathcal{M}, F, V, \iota, \langle\,,\rangle)$ is a Dieudonné space with $\overline{\mathcal{E}}$-structure and where α is an isomorphism $\mathcal{M} \xrightarrow{\sim} \mathcal{O}_S^h$ of $\mathcal{O}_S$-modules. The map $(g, (\underline{M}, \alpha)) \mapsto (\underline{M}, g\alpha)$ defines an action of $GL_{h,\mathbb{F}_p}$ on $\widetilde{\mathtt{DS}}_{\mathcal{D}}$ and we have

$$\mathtt{DS}_{\mathcal{D}} = [G \backslash \widetilde{\mathtt{DS}}_{\mathcal{D}}]$$

in the sense of stacks. In particular, $\mathtt{DS}_{\mathcal{D}}$ is an Artin stack.

(5.5) Let S be an $\mathbb{F}_p$-scheme and let $\underline{X} = (X, \iota, \lambda)$ be a BT_1 with $\mathcal{D}$-structure. We denote by

$$\mathbb{D}(X) = \mathcal{E}xt^1_{S/\Sigma}(X^\vee, \mathcal{O}_{S/\Sigma})$$

the associated crystal in $\mathcal{O}_{S/\Sigma}$-modules of the Cartier dual of X. Here $\Sigma = (\mathbb{F}_p, 0, \gamma)$ is the PD-scheme Spec$(\mathbb{F}_p)$ endowed with the trivial PD-structure. By [BBM] 3.3.10, $\mathbb{D}(X)$ is locally free of rank h. Evaluating $\mathbb{D}(X)$ in the trivial PD-thickening $(S, S, 0)$ gives a locally free $\mathcal{O}_S$-module $D(X)$. This is functorial and compatible with base change. Hence relative Frobenius, Verschiebung, the O/pO-action ι, and the isomorphism $\lambda\colon X \longrightarrow X^\vee$ induce maps $F\colon D(X)^{(p)} \to D(X)$, $V\colon D(X) \to D(X)^{(p)}$, $\iota\colon O/pO \longrightarrow \mathrm{End}_{\mathcal{O}_S}(D(X))$, and $\lambda\colon D(X^\vee) = D(X)^\vee \longrightarrow D(X)$. Let $\langle\,,\rangle$ be the perfect bilinear form on $D(X)$ defined by λ and set

$$D(\underline{X}) := D_S(\underline{X}) := (D(X), F, V, \iota, \langle\,,\rangle).$$

If S is the spectrum of a perfect field k, we claim that we have a canonical isomorphism $M(\underline{X}^\vee)^{(p)} \xrightarrow{\sim} D(\underline{X}^\vee)$ where $M(\underline{X})$ is the usual contravariant Dieudonné module with $\mathcal{D}$-structure whose underlying k-vector space is given by $\mathrm{Hom}_k(X, CW)$ where CW is the sheaf of Witt covectors. Indeed, there is a canonical functorial isomorphism

$$M(X)^{(p)} \cong \Gamma(S/\Sigma, \mathbb{D}(X)) = \varprojlim_n \Gamma(\mathrm{Spec}(W_n(k)), \mathbb{D}(X)_{(S,\mathrm{Spec}(W_n(k)))})$$

by [BBM] 4.2.14 and 4.2.9. But as X and hence $\mathbb{D}(X)$ is killed by p we have for all $n \geq 0$ isomorphisms $\Gamma(\mathrm{Spec}(W_n(k)), \mathbb{D}(X)_{(S,\mathrm{Spec}(W_n(k)))}) \xrightarrow{\sim} \Gamma(S, \mathbb{D}(X)_{(S,S)})$ which proves the claim.

In particular, we see that over a perfect field k the functor $D(\ )$ provides an equivalence of the category of BT_1 with $\mathcal{D}$-structure and the category of Dieudonné spaces with $\mathcal{D}$-structure.

For an arbitrary $\mathbb{F}_p$-scheme S, F is given by $D(F_{X^\vee/S})$ and its cokernel is $\mathrm{Lie}(X^\vee)$ which is locally free as S is of characteristic p, hence $\mathrm{Im}(F)$ is locally a direct summand. Further, the cokernel of V can be identified with $\mathrm{Lie}(X^{(p)})$ which implies that $\mathrm{Im}(V)$ is locally a direct summand. As $\mathrm{Im}(V)$ is locally free, $\mathrm{Ker}(V)$ is locally a direct summand. Hence we have $\mathrm{Im}(F) = \mathrm{Ker}(V)$ if this holds after base change to any geometric point $\bar{s} \longrightarrow S$. As the functor D

commutes with base change this follows from the fact that over an algebraically closed field D is an equivalence of categories.

Altogether it follows that the functor D defines a 1-morphism of algebraic stacks

$$D\colon \mathtt{BT}_{\mathcal{D},1} \longrightarrow \mathtt{DS}_{\mathcal{D}}$$

which induces an equivalence of fibre categories over perfect fields.

(5.6) Proposition: *The 1-morphism $D\colon \mathtt{BT}_{\mathcal{D},1} \longrightarrow \mathtt{DS}_{\mathcal{D}}$ is a homeomorphism (i.e. the induced map on the underlying Zariski spaces $(\mathtt{BT}_{\mathcal{D},1})_{\mathrm{top}} \longrightarrow (\mathtt{DS}_{\mathcal{D}})_{\mathrm{top}}$ is a homeomorphism).*

Proof: We know already that D is bijective. Therefore it suffices to show that D is closed. We show the valuative criterion in the form of [LM] 7.10 for discrete valuation rings R of equal characteristic p. We may assume that R equals $k[[t]]$ where k is algebraically closed. We denote by K its field of fractions. Further set $\tilde{R} = W_2(k)[[t]]$ and $\tilde{K} = W_2(k)((t))$. We lift the Frobenius of R (resp. K) to $\tilde{R}$ (resp. $\tilde{K}$) by taking the usual Frobenius on $W_2(k)$ and by sending t to t^p. We endow the kernels of $\tilde{R} \longrightarrow R$ and $\tilde{K} \longrightarrow K$ with their canonical PD-structures γ_R and γ_K. Then de Jong [dJ] has defined a contravariant functor $M_{(R,\tilde{R},\gamma_R)}$ (resp. $M_{(K,\tilde{K},\gamma_K)}$) from the category of BT_1 over R (resp. K) to the category of regular Dieudonné spaces over R (resp. K) which commutes with dual and base change $(R,\tilde{R},\gamma_R) \longrightarrow (K,\tilde{K},\gamma_K)$. Via functoriality we can extend this functor to objects with $\mathcal{D}$-structures. Applying this to the Cartier dual we get covariant functors

$$M'_R\colon \mathtt{BT}_{\mathcal{D},1}(R) \longrightarrow \mathtt{DS}_{\mathcal{D}}(R), \qquad M'_K\colon \mathtt{BT}_{\mathcal{D},1}(K) \longrightarrow \mathtt{DS}_{\mathcal{D}}(K)$$

which commute with base change $R \longrightarrow K$. By [dJ] 7.1 there are functorial isomorphisms $M'_R(\underline{X}^{(p)}) \xrightarrow{\sim} D_R(\underline{X})$ and $M'_K(\underline{Y}^{(p)}) \xrightarrow{\sim} D_K(\underline{Y})$.

Now let $\underline{X}_\eta$ be a BT_1 with $\mathcal{D}$-structure over K and let $\underline{M}$ be a Dieudonné space over R such that $D_K(\underline{X}_\eta) \cong \underline{M}_K$. As M'_R is an equivalence of categories (loc. cit. 10), there exists a BT_1 $\underline{X}'$ with $\mathcal{D}$-structure such that $M'_R(\underline{X}') = \underline{M}$. Further it follows from loc. cit. 9 and from the existence of the principal polarization that M'_K is fully faithful. As we have $M'_K(\underline{X}'_K) \cong \underline{M}_K \cong D_K(\underline{X}_\eta) \cong M'_K(\underline{X}_\eta^{(p)})$ it follows that $\underline{X}'_K \cong \underline{X}_\eta^{(p)}$. Let K' be the perfect closure of K and denote by R' a valuation ring of K' which dominates R. Then R' contains $k[[t^{1/p}]]$, hence there exists a BT_1 with $\mathcal{D}$-structure $\underline{X}$ such that $\underline{X}^{(p)} \cong \underline{X}'_{R'}$. It follows that $\underline{X}_{K'}^{(p)} \cong (X_\eta)_{K'}^{(p)}$ and therefore $\underline{X}_{K'} \cong (X_\eta)_{K'}$.

(5.7) Let S be a scheme of characteristic p and let $\underline{X}$ be a BT_1 with $\mathcal{D}$-structure over S. Let $\underline{\mathrm{Aut}}(\underline{X})$ be the affine S-group scheme of automorphisms of $\underline{X}$. If $f\colon T \longrightarrow S$ is an S-scheme, a T-valued point of $\underline{\mathrm{Aut}}(\underline{X})$ is given by an automorphism α of $f^*\underline{X}$. Then $D(\alpha)$ is an automorphism of the Dieudonné space $D(f^*\underline{X}) = f^*D(\underline{X})$. As D is a functor, this construction is compatible with composition of automorphisms. Hence we get a homomorphism of affine S-group schemes

$$\Delta_{\underline{X}}\colon \underline{\mathrm{Aut}}(\underline{X}) \longrightarrow \underline{\mathrm{Aut}}(D(\underline{X})).$$

This is a bijection on k-valued points if k is a perfect field over S.

Proposition: *Let $\underline{X}$ be a BT_1 with $\mathcal{D}$-structure over a field k of characteristic p and denote by $\underline{M}$ the associated Dieudonné space. Then we have an equality*

$$\dim(\underline{\mathrm{Aut}}(\underline{X})) = \dim(\underline{\mathrm{Aut}}(\underline{M})).$$

Proof : We know already that $\Delta_{\underline{X}}$ is universally bijective. Hence the assertion follows from the following general fact:

Lemma: *Let $f\colon X \longrightarrow Y$ be a universally bijective morphism of schemes which are of finite type over a field k. Then* $\dim(X) = \dim(Y)$.

Proof of the Lemma: In our case (i.e. $X = \underline{\mathrm{Aut}}(\underline{X})$ and $Y = \underline{\mathrm{Aut}}(\underline{M})$) X_{red} and Y_{red} are smooth and the assertion is obvious. We leave the general case to the reader.

6 Dimensions of Oort strata

(6.1) In this chapter we denote by $\mathcal{E} = (B, {}^*, V, \langle\ ,\ \rangle, O_B, \Lambda, X_\infty)$ a Shimura-PEL-datum, integral and unramified at $p > 2$, G its associated reductive group over $\mathbb{Q}$, $[\mu]$ the associated conjugacy class of cocharacters of G, and E its field of definition, the reflex field of $\mathcal{E}$ (see [We2] 3.1 for the precise definitions). Fix an embedding of the algebraic closure $\bar{\mathbb{Q}}$ of $\mathbb{Q}$ in $\mathbb{C}$ into some algebraic closure $\overline{\mathbb{Q}}_p$ of $\mathbb{Q}_p$. Via this embedding we can consider $[\mu]$ as a $G(\overline{\mathbb{Q}}_p)$-conjugacy class of cocharacters. Denote by $v|p$ the place of E given by the chosen embedding $\bar{\mathbb{Q}} \hookrightarrow \overline{\mathbb{Q}}_p$ and let $\kappa(\nu)$ be its residue class field.

Further fix an open compact subgroup $K^p \subset G(\mathbf{A}_f^p)$ and denote by $\mathcal{A}_{\mathcal{E},K^p}$ the associated moduli space, defined by Kottwitz [Ko]. We assume that K^p is sufficiently small such that $\mathcal{A}_{\mathcal{E},K^p}$ is representable. It is then a smooth equidimensional quasi-projective scheme over the localization of O_E in p. It classifies tuples $(A, \bar{\lambda}, \iota, \bar{\eta})$ where A is an abelian scheme up to prime-to-p-isogeny, $\bar{\lambda}$ is a $\mathbb{Q}$-homogeneous polarization of A containing a polarization $\lambda \in \bar{\lambda}$ of degree prime to p, $\iota\colon O_B \to \mathrm{End}(A) \otimes_{\mathbb{Z}} \mathbb{Z}_{(p)}$ is an involution-preserving $\mathbb{Z}_{(p)}$-algebra homomorphism where the involution is * on O_B and the Rosati-involution given by $\bar{\lambda}$ on $\mathrm{End}(A) \otimes_{\mathbb{Z}} \mathbb{Z}_{(p)}$, and where $\bar{\eta}$ is a K^p-level structure. Further $(A, \bar{\lambda}, \iota, \bar{\eta})$ satisfies the determinant condition, i.e. we have an identity of polynomial functions on O_B

$$\det(b|\,\mathrm{Lie}(A)) = \det(b|V_0)$$

where $V_0 \subset V_{E_\nu}$ is the weight zero space of some $\mu \in [\mu]$ (see [Ko] §5 or [RZ] 3.23 a) for a precise formulation of the determinant condition). We denote by $\mathcal{A}_0$ the reduction $\mathcal{A}_{\mathcal{E},K^p} \otimes \kappa(\nu)$ at ν.

(6.2) We denote by $\bar{G}$ the reductive $\mathbb{F}_p$-group of O_B/pO_B-linear symplectic similitudes of $\Lambda/p\Lambda$. Via the canonical bijection of $G(\bar{\mathbb{Q}}_p)$-conjugacy classes of cocharacters and $\bar{G}(\bar{\mathbb{F}}_p)$-conjugacy classes of cocharacters we consider $[\mu]$ as a $\bar{G}(\bar{\mathbb{F}}_p)$-conjugacy class of cocharacters. Its field of definition is $\kappa = \kappa(\nu)$. We denote by $(X^*, R^*, X_*, R_*, \Delta)$ the based root datum of $\bar{G}$.

(6.3) To $\mathcal{E}$ we associate the tuple $\mathcal{D} = (O_B, {}^*, -1, \mathrm{rk}_{\mathbb{Z}_p}(\Lambda))$ and we set $(\mathtt{BT}_{\mathcal{D},n})_0 = \mathtt{BT}_{\mathcal{D},n} \otimes_{\mathbb{Z}_{(p)}} \kappa(\nu)$ and $(\widetilde{\mathtt{BT}}_{\mathcal{D},n})_0 = \widetilde{\mathtt{BT}}_{\mathcal{D},n} \otimes_{\mathbb{Z}_{(p)}} \kappa(\nu)$. For every S-valued point $(A, \bar{\lambda}, \iota, \bar{\eta})$ of $\mathtt{A}_0$ we have $\mathrm{Lie}(A[p^n]) = \mathrm{Lie}(A)$ (because $p\mathcal{O}_S = 0$). We have the obvious analogue of the determinant condition for BT_n with $\mathcal{D}$-structure $\underline{X} = (X, \iota, \lambda)$ which we can formulate as requiring an identity of polynomial functions on O_B

$$\det(b|\,\mathrm{Lie}(X^\vee)) = \det(b|\Lambda_1)$$

where Λ_1 is the weight one space in $\Lambda/p\Lambda_\kappa$ with respect to some $\mu \in [\mu]$ which is defined over κ. We call an S-valued point of $\mathtt{BT}_{n,0}$ a *BT_n with $\mathcal{E}$-structure over S*.

As the PEL-datum is unramified, the determinant condition is given, after some finite étale base change, by requiring that certain direct summands of the locally free finite module $\mathrm{Lie}(A)$ (given by ι) have a certain fixed rank. This implies that the determinant condition defines an open and closed substack of $(\mathtt{BT}_{\mathcal{D},n})_0$. We denote this substack by $\mathtt{BT}_{n,0}$ and set

$$\mathtt{BT}_0 = \mathtt{BT}_{1,0}.$$

If we define $\widetilde{\mathtt{BT}}_{n,0}$ in an analogous way, $\widetilde{\mathtt{BT}}_{n,0} \longrightarrow \mathtt{BT}_{n,0}$ is still a $GL_{p^{nh}}$-torsor.

Taking p^n-torsion and forgetting the level-structure defines a canonical morphism

$$\Psi\colon \mathtt{A}_0 \longrightarrow \mathtt{BT}_{n,0}.$$

(6.4) Theorem: The morphism Ψ is smooth.

Proof: We use Grothendieck's smoothness criterion via infinitesimal liftings. Let $R' \twoheadrightarrow R$ be an infinitesimal thickening of local Artinian rings with perfect residue class field, let $(A, \bar{\lambda}, \iota, \bar{\eta})$ be an R-valued point of $\mathtt{A}_0$. By the rigidity of étale covers, any lifting of $(A, \bar{\lambda}, \iota)$ to R' will give a unique R'-valued point of $\mathtt{A}_0$. Hence the smoothness of Ψ follows from the theorem of Serre-Tate and from (2.8).

(6.5) We write $\mathtt{DS}_0$ for the algebraic stack of Dieudonné spaces $(\mathcal{M}, F, V, \iota, \langle\,,\rangle)$ with $\mathcal{D}$-structure over schemes S such that

$$\det(b|\mathcal{M}/\,\mathrm{Im}(F)) = \det(b|\Lambda_1)$$

as polynomial functions on O_B. This amounts to the fact that the parabolic subgroup of G_S which stabilizes $\mathrm{Im}(F)$ has in every geometric point the same type $I \subset \Delta$ as the opposite of the type defined by $[\mu]$.

By the same arguments as above, $\mathtt{DS}_0$ is an open and closed substack of $\mathtt{DS}_{\mathcal{D}}$. We call an S-valued point of $\mathtt{DS}_0$ a *Dieudonné space with $\mathcal{E}$-structure over S*.

(6.6) Denote by $\Xi = \Xi(\mathcal{E})$ the set of isomorphism classes of BT_1 with $\mathcal{E}$-structure $\underline{X}$ over $\bar{\mathbb{F}}_p$ (or any other algebraically closed field of characteristic p (4.6)). By (4.1) and (5.6) we have an identification

$$\Xi = (\mathtt{BT}_0)_{\mathrm{top}} = (\mathtt{DS}_0)_{\mathrm{top}}$$

which endows Ξ with a topology which is derived from a (unique) partial order.

(6.7) Let S be a scheme of characteristic p and $\underline{X}$ be a BT_1 with $\mathcal{E}$-structure over S. Let $f_{\underline{X}}: S \longrightarrow \mathtt{BT}_0$ be the corresponding 1-morphism. For every $\xi \in \Xi$ the inverse image $S_\xi = f_{\underline{X}}^{-1}(\{\xi\})$ is a locally closed subset of S (4.5). Hence $S = \bigcup_\xi S_\xi$ is a stratification where we have

$$\bar{S}_\xi \subset \bigcup_{\xi' \leq \xi} S_{\xi'}.$$

Definition: *This stratification is called the* Oort stratification of S with respect to $\underline{X}$.

(6.8) As a special case of (6.7) we can consider the case $S = \mathtt{A}_0$ and $\underline{X} = (A^{\mathrm{univ}}, \lambda^{\mathrm{univ}}, \iota^{\mathrm{univ}})[p]$ where $(A^{\mathrm{univ}}, \bar{\lambda}^{\mathrm{univ}}, \iota^{\mathrm{univ}}, \bar{\eta}^{\mathrm{univ}})$ is the universal abelian scheme over $\mathtt{A}_0$ and where $\lambda^{\mathrm{univ}} \in \bar{\lambda}^{\mathrm{univ}}$ is a polarization of order prime to p (1.5). This defines the Oort stratification

$$\mathtt{A}_0 = \bigcup_{\xi \in \Xi} \mathtt{A}_\xi.$$

It follows from the smoothness of $\Psi: \mathtt{A}_0 \longrightarrow \mathtt{BT}_{1,0}$ that we have in fact an equality

$$\overline{\mathtt{A}_\xi} = \bigcup_{\xi' \leq \xi} \mathtt{A}_{\xi'}.$$

(6.9) Let $\tilde{\mathtt{A}}_0$ be the fibre product of $\mathtt{A}_0$ and $\widetilde{\mathtt{BT}}_0$ over $\mathtt{BT}_0$. This is a smooth quasi-projective scheme over $\kappa(\nu)$. It is a $GL_{p^{nh}}$-torsor over $\mathtt{A}_0$. We have a diagram of smooth morphisms

$$\mathtt{A}_0 \xleftarrow{\Pi} \tilde{\mathtt{A}}_0 \xrightarrow{\Phi} \widetilde{\mathtt{BT}}_0.$$

(6.10) For a point x in $\widetilde{\mathtt{BT}}_\xi$ the stabilizer of x in GL_{p^h} is the group of automorphisms of the BT_1 with $\mathcal{E}$-structure corresponding to x. Its dimension $d(\xi)$ depends only on the isomorphism class ξ. As the morphisms Π and Φ are smooth, we have

$$\operatorname{codim}(\mathtt{A}_\xi, \mathtt{A}_0) = \operatorname{codim}(\widetilde{\mathtt{BT}}_\xi, \widetilde{\mathtt{BT}}_0) = \widetilde{\mathtt{BT}}_0 - \dim(GL_{p^h}) + d(\xi)$$

if $\mathtt{A}_\xi \neq \emptyset$. Hence we get by (3.5):

Theorem: *Let ξ be an isomorphism class of a BT_1 with $\mathcal{D}$-structure over $\bar{\mathbb{F}}_p$, let $d(\xi)$ be the dimension of the automorphism group of ξ, and let $\mathtt{A}_\xi$ be the corresponding Oort stratum. Then $\mathtt{A}_\xi$ is equi-dimensional, and if it is nonempty we have*

$$\operatorname{codim}(\mathtt{A}_\xi, \mathtt{A}_0) = d(\xi).$$

(6.11) Consider now the following two assumptions:

(BT) In each connected component of $(\mathtt{DS}_0)_{\mathrm{top}}$ there exists a unique closed point.

(AV) Every closed point of $(\mathtt{DS}_0)_{\mathrm{top}}$ lies in the image of the morphism $\Psi\colon \mathtt{A}_0 \longrightarrow \mathtt{BT}_0$.

We conjecture that these conditions are always fulfilled. For examples where (BT) and (AV) are known to hold, see chapter 7.

(6.12) Theorem: *Assume that (AV) holds. Then $\mathtt{A}_\xi \neq \emptyset$ for all $\xi \in \Xi$.*

Proof : By (6.3) the image of Ψ in Ξ is open. Further by hypothesis it contains every closed point. As Ξ is sobre and of finite Krull dimension, this implies the theorem.

(6.13) Proposition: *In each connected component of Ξ there exists a unique open point.*

Proof : As Ξ is finite, we know that there exist open points in each connected component of Ξ. Further it follows from the smoothness of $\mathtt{BT}_0$ (3.3) that each connected component of Ξ is irreducible. This proves the claim.

(6.14) Denote by $\mathtt{A}_0^{\mu-\mathrm{ord}}$ the μ-ordinary locus of $\mathtt{A}_0$ (see [Wel] for the definition). Because of the smoothness of Ψ the union of maximal Oort strata is open and dense in $\mathtt{A}_0$. Hence if the μ-ordinary locus equals the union of maximal Oort strata, the main result of [Wel] is implied.

For examples where this holds see chapter 7.

7 Examples

(7.1) We keep the notations of the previous chapter. Further fix integers $r \geq 2$ and $n \geq 0$ and denote by $\mathtt{DS}_0^{r,\leq n}$ the closed substack of $\mathtt{DS}_0$ which classifies Dieudonné spaces with $\mathcal{E}$-structure $\underline{M} = (\mathcal{M}, F, V, \iota, \langle\ ,\ \rangle)$ such that each geometric fibre of

$$V^{(p^{r-1})} \circ \ldots \circ V\colon \mathcal{M} \longrightarrow \mathcal{M}^{(p^r)}$$

has rank $\leq n$.

Then $\mathtt{DS}_{\mathcal{D}}^{r,=n} := \mathtt{DS}_{\mathcal{D}}^{r,\leq n} \backslash \mathtt{DS}_{\mathcal{D}}^{r,\leq n-1}$ is open in $\mathtt{DS}_{\mathcal{D}}^{r,\leq n}$. For fixed r this defines a stratification

$$\mathtt{DS}_0 = \bigcup_{n=0}^{h} \mathtt{DS}_{\mathcal{D}}^{r,=n}$$

such that the closure of $(\mathtt{DS}_{\mathcal{D}}^{r,=n})_{\mathrm{top}}$ is the union of the $(\mathtt{DS}_{\mathcal{D}}^{r,=m})_{\mathrm{top}}$ for $m \leq n$.

For each scheme S of characteristic p and each BT_1 with $\mathcal{E}$-structure $\underline{X}$ this defines a stratification

$$S = \bigcup_{n=0}^{h} S^{r,=n}.$$

Note that for $r = h$ this stratification is the stratification by the height of the étale part of X which we call the *p-rank stratification of S with respect to $\underline{X}$*.

(7.2) Generalized Hilbert-Blumenthal case: Let F be a totally real extension of $\mathbb{Q}$ of degree n, let V be an F-vector space of dimension $2d$ which is endowed with a symplectic $\mathbb{Q}$-valued form $\langle\ ,\ \rangle$ such that $\langle fv, w\rangle = \langle v, fw\rangle$ for all $f \in F$ and $v, w \in V$. Let G be the associated group of F-linear symplectic similitudes. We fix an isomorphism

$$G_{\mathbb{R}} \cong \{\, (g_1, \ldots, g_n) \in (\mathrm{GSp}_{2d,\mathbb{R}})^n \mid c(g_1) = \cdots = c(g_n) \,\}$$

where c denotes the multiplicator homomorphism. Denote by X_∞ the $G(\mathbb{R})$-conjugacy class of the homomorphism

$$\mathbb{C}^\times \longrightarrow G(\mathbb{R}),$$

$$x + \sqrt{-1}y \mapsto \left(\begin{pmatrix} x \cdot I_d & -y \cdot I_d \\ y \cdot I_d & x \cdot I_d \end{pmatrix}, \ldots, \begin{pmatrix} x \cdot I_d & -y \cdot I_d \\ y \cdot I_d & x \cdot I_d \end{pmatrix} \right).$$

Then (G, X_∞) is a Shimura datum. Its reflex field E is $\mathbb{Q}$.

Note that for $F = \mathbb{Q}$ this is just the usual Siegel case.

Let $p > 2$ be a prime such that $F_p = F \otimes_{\mathbb{Q}} \mathbb{Q}_p$ is a product of unramified extensions of $\mathbb{Q}$. To simplify notations let us assume that this is indeed a field, i.e. that p is inert (and unramified) in F. The general case can be easily reduced to this case. We choose a selfdual O_{F_p}-lattice Λ in $V \otimes_F F_p$ and set $\mathcal{E} = (F, \mathrm{id}, V, \langle\,,\rangle, O_F \otimes \mathbb{Z}_{(p)}, \Lambda, X_\infty)$. By the choice of a sufficiently small open compact subgroup K^p of $G(\mathbf{A}_f^p)$ we get a model $\mathtt{A}$ over $\mathbb{Z}_{(p)}$ of the Shimura variety associated to (G, X_∞, K) where $K = K^p K_p$, with K_p the stabilizer of Λ in $G(\mathbb{Q}_p)$.

We want to show that conditions (BT) and (AV) of (6.11) hold. We first prove (BT), more precisely we show:

Lemma: *There exists a unique closed point in Ξ and the subset consisting of this point is equal to $(\mathtt{DS}_{\mathcal{D}}^{2,=0})_{\mathrm{top}}$.*

Proof : We have to show that over an algebraically closed field k there exists up to isomorphism a unique Dieudonné space $(M, F, V, \iota, \langle\ ,\ \rangle)$ with $\mathcal{E}$-structure such that $V^2 M = (0)$. Here we consider V (resp. F) as σ^{-1}-linear (resp. σ-linear) map $M \longrightarrow M$ where σ denotes the Frobenius of k. We can describe such a Dieudonné space as

$$(M = \bigoplus_{j \in \mathbb{Z}/n\mathbb{Z}} M(j), F, V, \langle\ ,\ \rangle)$$

as in case (II) of the proof of (3.2). By hypothesis we have $FM(j+1) = VM(j-1) =: T(j)$ for all j. Further $T(j)$ is maximal isotropic in $M(j)$. We choose a maximal isotropic complement $L(j)$ of $T(j)$ in $M(j)$. Then $V|_{T(j)}$ is zero and $V|_{L(j)}$ induces an isomorphism $L(j) \xrightarrow{\sim} T(j+1)$. Via this isomorphism we identify $L(j) = T(j+1)$. Then F is determined by the σ^2-linear bijections $L(j) \xrightarrow{\sim} L(j-2)$ given by this identification. Hence the lemma follows from the theorem of Dieudonné which says that up to isomorphism there exists for every nonzero integer a and every positive integer d' a unique pair (L, F) where L is a d'-dimensional vector space and F is a bijective σ^a-linear map.

Now the hypothesis (AV) follows from the fact that over $\bar{\mathbb{F}}_p$ there exists an n-fold product A of supersingular elliptic curves such that we can embed F into $\mathrm{End}(A) \otimes \mathbb{Q}$. As a corollary we get:

Corollary: *Let k be an algebraically closed field of characteristic p and let $\underline{X}$ be a BT_1 with $\mathcal{E}$-structure over k. Then there exists a point $(A, \bar{\lambda}, \iota, \bar{\eta}) \in \mathtt{A}(k)$ such that $(A, \lambda, \iota)[p] \cong \underline{X}$ where $\lambda \in \bar{\lambda}$ is a polarization of order prime to p.*

We remark that it is not difficult to show that an analogous assertion holds if $\underline{X}$ is a BT_n with $\mathcal{E}$-structure for some finite $n \geq 1$.

Further it is easy to see that there exist BT_1 with $\mathcal{E}$-structure $\underline{X}$ whose p-rank is equal to one half their height: Let N be a k-vector space of dimension d and endow $N \oplus N^*$ with the canonical symplectic form; for $j \in \mathbb{Z}/n\mathbb{Z}$ define $M(j) = N \oplus N^*$; let $F(j)$ be a σ-linear map $M(j) \longrightarrow M(j-1)$ which respects the decomposition in N and N^* and which is a bijection on N and zero on N^* and set $F = \bigoplus F(j)$; let V be the adjoint map to F (i.e. the unique σ^{-1}-linear map such that $\langle Fm, m' \rangle = \langle m, Vm' \rangle^\sigma$. Hence its isomorphism class defines the unique open point of Ξ which implies:

Corollary: *The ordinary locus of $\mathtt{A}$ is open and dense in the special fibre.*

(7.3) Unitary case: Let F be a quadratic imaginary extension of $\mathbb{Q}$ with complex conjugation *, let V be an F-vector space of dimension d, denote by $\beta: V \times V \longrightarrow F$ an $F/\mathbb{Q}$-hermitian non-degenerate sesquilinear form. Let $G = GU(V, \beta)$ be the general unitary group of F-linear automorphisms of V which respect β up to a scalar in $\mathbb{Q}$. We consider G as a reductive group over $\mathbb{Q}$. We fix an embedding $\epsilon: F \to \mathbb{C}$ and an isomorphism $G_{\mathbb{R}} \cong GU(r, d-r)$ for an integer $0 \leq r \leq d$ using the identification $F \otimes_{\mathbb{Q}} \mathbb{R} = \mathbb{C}$ given by ϵ. Denote by X_∞ the $G(\mathbb{R})$-conjugacy class of the homomorphism

$$\mathbb{C}^\times \longrightarrow G(\mathbb{R}), \qquad z \mapsto -\operatorname{diag}(z^r, \bar{z}^{d-r}).$$

Then (G, X_∞) is a Shimura datum. Its reflex field E is $\mathbb{Q}$ if $r = d-r$ and it is F otherwise.

Choose a unit δ in the ring O_F of integers of F such that $\delta = -\delta^*$ and denote by $\langle\, , \rangle$ the skew-hermitian bilinear form $\langle v, v' \rangle = \mathrm{Tr}_{F/\mathbb{Q}}\, \delta\beta(v, v')$ on V. Further let $p > 2$ be a prime such that we are in one of the following cases:

(a) The prime p is split in F. In this case we have $F \otimes \mathbb{Q}_p = \mathbb{Q}_p \times \mathbb{Q}_p$ and this induces a decomposition $V_{\mathbb{Q}_p} = V' \times V''$ where V' and V'' are totally isotropic and in perfect duality with respect to $\langle\, , \rangle$. The choice of a $\mathbb{Z}_p$-lattice Λ' in V' defines via the duality of V' and V'' a lattice Λ which is O_F-invariant and selfdual. Hence the tuple $\mathcal{E} = (F, {}^*, V, \langle\, , \rangle, O_F, \Lambda, X_\infty)$ is a Shimura-PEL-datum, integral and unramified at p.

(b) The prime p is inert and unramified in F and the Witt index of the hermitian form induced by β on $V_{\mathbb{Q}_p}$ is equal to $[h/2]$. In fact, the condition on the Witt index is equivalent to the assumption that we can choose an O_F-invariant self-dual $\mathbb{Z}_p$-lattice Λ in $V_{\mathbb{Q}_p}$, and $\mathcal{E} = (F, {}^*, V, \langle\, , \rangle, O_F, \Lambda, X_\infty)$ is a Shimura-PEL-datum, integral and unramified at p.

Fixing a sufficiently small open compact subgroup K^p in $G(\mathbf{A}_f^p)$ we get the associated moduli scheme $\mathsf{A}_{\mathcal{E},K^p}$ which is quasi-projective and smooth of relative dimension $r(h-r)$ over $O_{E,(p)}$. Further let ν be a place of E over p and denote by A_0 the reduction of $\mathsf{A}_{\mathcal{E},K^p}$ modulo ν. From now on, assume that $r = 1$ and consider the two cases (a) and (b) separately:

Case (a): We set $\mathcal{D} = (O_F, {}^*, -1, 2d)$. As we have $O_F \otimes \mathbb{Z}_p = \mathbb{Z}_p \times \mathbb{Z}_p$, to give a BT_1 with $\mathcal{D}$-structure is the same as to give simply a BT_1 X of height d. The determinant condition is equivalent to the condition that $\mathrm{rk}(\mathrm{Lie}(X)) = r = 1$. If k is an algebraically closed field of characteristic p, every BT_1 of height d and dimension 1 is isomorphic to

$$\xi(s) = (\mathbb{Z}/p\mathbb{Z})^s \times G_s$$

for $s = 0, \dots, d-1$ and where G_s is the BT_1 whose covariant Dieudonné module (M, F, V) is given by the free k-vector space M with basis $e_1, \dots, e_s$ and $Fe_i = e_{i+1}$ for $i = 1, \dots, s-1$ and $Ve_1 = e_s$ (because of $FV = VF = 0$ this determines

F and V). As G_s is infinitesimal (F acts nilpotent), we have $\mathrm{Aut}((\mathbb{Z}/p\mathbb{Z})^s \times G_s) = \mathrm{Aut}((\mathbb{Z}/p\mathbb{Z})^s) \times \mathrm{Aut}(G_s)$. The first factor is clearly zero-dimensional over k and an easy calculation using (5.7) shows that $\dim(\mathrm{Aut}(G_s)) = s-1$. Hence we get a stratification

$$\mathtt{A}_0 = \mathtt{A}_{\xi(0)} \cup \ldots \cup \mathtt{A}_{\xi(d-1)},$$

in particular the hypothesis (BT) of (6.11) holds. Further, we have $\dim \mathtt{A}_{\xi(s)} = s$ under the assumption that all Oort strata are nonempty.

To show that this is indeed the case, we have to prove that the hypothesis (AV) holds ((6.12)), i.e. that $\mathtt{A}_{\xi(0)}$ is nonempty. This follows easily from Honda-Tate theory (cf. [HT] 5.2 and 8).

Note that in this case the Oort stratification coincides with the Newton polygon stratification (and with the p-rank stratification). It follows that all Newton polygon strata are non-empty.

Case (b): We set $\mathcal{D} = (O_F, {}^*, -1, 2d)$. As we have $O_F \otimes \mathbb{Z}_p = O_K$ where K is "the" unramified quadratic extension of $\mathbb{Q}_p$, to give a BT_1 with $\mathcal{D}$-structure is the same as to give a BT_1 X with $\mathcal{D}_p$-structure where $\mathcal{D}_p = (O_K, {}^*, -1, 2d)$. If X is defined over an algebraically closed field k of characteristic p the determinant condition is equivalent to the condition that $\mathrm{rk}(L_0) = r = 1$ where $L_0 = \{\, x \in \mathrm{Lie}(X) \mid \iota(a)x = ax \,\}$.

In this case it is not as easy as in case (a) to give a complete description of the possible isomorphism classes of BT_1 with $\mathcal{D}$-structure. We state only the results and refer to [BW] for the proofs. Every such $\underline{X}$ is isomorphic to

$$\xi(s) = \underline{B}(s) \oplus \underline{S}^{d-s},$$

for some $s = 1, \ldots, d$ where $\underline{S}$ and $\underline{B}(s)$ are defined as follows:

As a k-vector space $\underline{S} = (M = M_0 \oplus M_1, F, V, \langle\,,\rangle)$ has a basis e and f. We set $M_0 = ke$ and $M_1 = kf$. The alternating form is given by $\langle e, f\rangle = 1$, and we set $Fe = -f$ and $Ve = f$.

The underlying k-vector space of $\underline{B}(s) = (M = M_0 \oplus M_1, F, V, \langle\,,\rangle)$ is the vector space generated by the basis $(e_1, \ldots, e_s, f_1, \ldots, f_s)$ and the grading is given by $M_0 = ke_1 \oplus \ldots \oplus ke_s$ and $M_1 = kf_1 \oplus \ldots \oplus kf_s$. The alternating form is defined by $\langle e_i, f_j\rangle = (-1)^{i-1}\delta_{ij}$, and F and V are given by

$$\begin{aligned} Ve_i &= f_{i+1}, && \text{for } i = 1, \ldots, s-1,\\ Vf_s &= e_1, &&\\ Fe_i &= f_{i-1}, && \text{for } i = 2, \ldots, s,\\ Ff_1 &= (-1)^s e_s. && \end{aligned}$$

For $1 \le s \le d$ we have

$$\dim \underline{\mathrm{Aut}}(\bar{\underline{B}}(s) \oplus \bar{\underline{S}}^{d-s}) = \begin{cases} \frac{s}{2} - 1, & \text{if } s \text{ is even},\\ d - \frac{s+1}{2}, & \text{if } s \text{ is odd}. \end{cases}$$

Therefore we get a stratification

$$\mathtt{A}_0 = \mathtt{A}_{\xi(2)} \cup \mathtt{A}_{\xi(4)} \cup \ldots \cup \mathtt{A}_{\xi(2[d/2])} \cup \mathtt{A}_{\xi(2[(d+1)/2]-1)} \cup \ldots \cup \mathtt{A}_{\xi(3)} \cup \mathtt{A}_{\xi(1)}.$$

It follows that the assumption (BT) holds. Further if $\mathtt{A}_{\xi(1)} \neq \emptyset$ it follows that all strata are nonempty (6.12). But it is clear that we can endow the d-fold product of supersingular elliptic curves over $\bar{\mathbb{F}}_p$ with an O_F-action such that we get a point in our moduli space and this will be a point in $\mathtt{A}_{\xi(1)}$.

The dimension from one stratum to the next one drops by 1. Further, this is a refinement of the Newton polygon stratification which is given by

$$\mathtt{A}_0 = \mathtt{A}_{\xi(2)} \cup \ldots \cup \mathtt{A}_{\xi(2[d/2])} \cup \bigcup_{i=1}^{[(d+1)/2]} \mathtt{A}_{\xi(2i-1)}.$$

Hence it follows from [dJO] in this case that all Newton polygon strata are again non-empty. Moreover the μ-ordinary stratum in the sense of [We1] is equal to $\mathtt{A}_{\xi(2)}$, hence it is open and dense.

Bibliography

[BBM] P. Berthelot, L. Breen, W. Messing: "Théorie de Dieudonné cristalline", LNM **930**, Springer (1982).

[BW] O. Bültel, T. Wedhorn: "On the reduction of Shimura varieties for some unitary groups", in preparation.

[DG] M. Demazure, P. Gabriel: "Groupes algébriques", North-Holland Publishing Company - Amsterdam (1970).

[EGA] A. Grothendieck, J. Dieudonné: "Eléments de géométrie algébrique", I Grundlehren der Mathematik **166** (1971) Springer, II-IV Publ. Math. IHES **8** (1961) **11** (1961) **17** (1963) **20** (1964) **24** (1965) **28** (1966) **32** (1967).

[dJ] A.J. de Jong: "Finite locally free group schemes in characteristic p and Dieudonné modules", Inventiones Math. **114** (1993), 89–137.

[dJO] A.J. de Jong, F. Oort: "Purity of the stratification by Newton Polygons", J. AMS. **13** (2000), 209–241.

[GO] E. Z. Goren, F. Oort: "Stratifications of Hilbert modular varieties", J. Alg. Geom. **9** (2000), 111–154.

[GP] I.M. Gelfand, V.A. Ponomarev: "Indecomposable representations of the Lorentz group", Uspehi Mat. Nauk **23** (1968) no. 2 (140), 3–60.

[Gr] A. Grothendieck: "Groupes de Barsotti-Tate et cristaux de Dieudonné", Presses de l'Université de Montréal (1974).

[HT] M. Harris, R. Taylor: "On the geometry and cohomology of some simple Shimura varieties", preprint Harvard (1999).

[Ill] L. Illusie: "Déformations de Groupes de Barsotti-Tate" in "Séminaire sur les Pinceaux Arithmétiques: la conjecture de Mordell," Astérisque **127**, 151–198 (1985).

[Ko] R. Kottwitz: "Points on some Shimura varieties over finite fields", J. AMS **5** (1992), 373–444.

[Kr] H. Kraft: "Kommutative algebraische p-Gruppen (mit Anwendungen auf p-divisible Gruppen und abelsche Varietäten)", lecture manuscript, Universität Bonn, 1975.

[LM] G. Laumon, L. Moret-Bailly: "Champs algébriques", Ergebnisse der Mathematik und ihrer Grenzgebiete (3. Folge) **39**, Springer (2000).

[MM] B. Mazur, W. Messing: "Universal Extensions and One Dimensional Crystalline Cohomology", LNM **370**, Springer (1974).

[Mo] B. Moonen: "Group schemes with additional structure and Weyl group cosets", this volume, 255–298.

[Oo1] F. Oort: "Newton polygons and formal groups: conjectures by Manin and Grothendieck," to appear: Ann. Math. **152** (2000).

[Oo2] F. Oort: "Newton polygon strata in the moduli space of abelian varieties", this volume, 417–440.

[Oo3] F. Oort: "A stratification of a moduli space of abelian varieties", this volume, 345–416.

[RR] M. Rapoport, M. Richartz: "On the classification and specialization of F-isocrystals with additional structure," Comp. Math. **103** (1996), 153–181.

[RZ] M. Rapoport, T. Zink: "Period Spaces for p-divisible Groups", Annals of Mathematics Studies **141**, Princeton University Press, Princeton (1996).

[SGA3] A. Grothendieck et al.: "Schémas en groupes", LNM **151**, **152**, **153**, Springer (1970).

[vGe] G. van der Geer: "Cycles on the moduli space of abelian varieties", in: Moduli of curves and abelian varieties, Aspects Math. **33**, Vieweg, Braunschweig (1999), 65–89.

[We1] T. Wedhorn: "Ordinariness in good reductions of Shimura varieties of PEL-type", Ann. Sci. de l'ENS **32** (1999), 575–618.

[We2] T. Wedhorn: "Congruence relations for some Shimura varieties", J. reine angew. Math. (Crelle) **524** (2000), 43–71.

Progress in Mathematics, Vol. 195, © 2001 Birkhäuser Verlag Basel/Switzerland

HYPERELLIPTIC JACOBIANS AND MODULAR REPRESENTATIONS

YURI G. ZARHIN

1. Introduction

In [17] the author proved that in characteristic 0 the jacobian $J(C) = J(C_f)$ of a hyperelliptic curve

$$C = C_f : y^2 = f(x)$$

has only trivial endomorphisms over an algebraic closure K_a of the ground field K if the Galois group $\mathrm{Gal}(f)$ of the irreducible polynomial $f \in K[x]$ is "very big". Namely, if $n = \deg(f) \geq 5$ and $\mathrm{Gal}(f)$ is either the symmetric group $\mathbf{S}_n$ or the alternating group $\mathbf{A}_n$ then the ring $\mathrm{End}(J(C_f))$ of K_a-endomorphisms of $J(C_f)$ coincides with $\mathbf{Z}$. The proof was based on an explicit description of the Galois module $J(C_f)_2$ of points of order 2 on $J(C_f)$. Namely, the action of the Galois group $\mathrm{Gal}(K)$ factors through $\mathrm{Gal}(f)$ and the $\mathrm{Gal}(f)$-module $J(C_f)_2$ could be easily described in terms of the (transitive) action of $\mathrm{Gal}(f)$ on the set $\mathfrak{R}_f$ of roots of f.

It turns out that if $\mathrm{Gal}(f)$ contains $\mathbf{A}_n$ then the Galois module $J(C_f)_2$ enjoys the following property ([17]):

(*): each subalgebra in $\mathrm{End}_{\mathbf{F}_2}(J(C_f)_2)$ which contains the identity operator and is stable under the conjugation by Galois automorphisms either consists of scalars or coincides with $\mathrm{End}_{\mathbf{F}_2}(J(C_f)_2)$.

Applying (*) to the subalgebra $\mathrm{End}(J(C_f)) \otimes \mathbf{Z}/2\mathbf{Z}$, one concludes that it consists of scalars, i.e., $\mathrm{End}(J(C_f)$ is a free abelian group of rank 1 and therefore coincides with $\mathbf{Z}$. (The case of $\mathrm{End}(J(C)) \otimes \mathbf{Z}/2\mathbf{Z} = \mathrm{End}_{\mathbf{F}_2}(J(C_f)_2)$ could not occur in characteristic zero.)

The proof of (*) was based on the well-known explicit description of $J(C_f)_2$ [14], [12] and elementary properties of $\mathbf{A}_n$ and its simplest nontrivial representation in characteristic 2 of dimension $n-1$ or $n-2$ (depending on whether n is odd or even).

In this paper we study property (*) itself from the point of view of representation theory over $\mathbf{F}_2$. Our results allow, in principle, to check the validity of (*) even if $\mathrm{Gal}(f)$ does not contain $\mathbf{A}_n$. We prove that $\mathrm{End}(J(C_f)) = \mathbf{Z}$ for an infinite series of $\mathrm{Gal}(f) = \mathbf{L}_2(2^r) := \mathrm{PSL}_2(\mathbf{F}_{2^r})$ and $n = 2^r + 1$ (with $r \geq 3$

Partially supported by EPSRC grant GR/M 98135 and NSF grant DMS 0070664.

and $\dim(J(C_f)) = 2^{r-1}$) or when $\mathrm{Gal}(f)$ is the Suzuki group $\mathbf{Sz}(2^{2r+1})$ and $n = 2^{2(2r+1)} + 1$ (with $\dim(J(C_f)) = 2^{4r+1}$).

We refer the reader to [10], [11], [6], [7], [9], [17] for a discussion of known results about, and examples of, hyperelliptic jacobians without complex multiplication.

The paper is organized as follows. In §2 we state the main results and begin the discussion of linear representations for which an analogue of the property (*) holds true; we call such representations *very simple*. In §3 we prove that the very simplicity of the Galois module X_ℓ of points of prime order ℓ on an abelian variety X implies in characteristic zero that X does not have nontrivial endomorphisms. In §4 we recall basic facts about permutation groups and corresponding ordinary representations and modular representations over $\mathbf{F}_2$. We use them in §5 in order to restate the main results as assertions about the very simplicity of certain permutation modules using an explicit description of points of order $\ell = 2$ on hyperelliptic jacobians. It turns out that all these permutation modules are Steinberg representations. In §6 we prove that the Steinberg representations are the only absolutely irreducible nontrivial representations (up to an isomorphism) over $\mathbf{F}_2$ for the groups $\mathbf{L}_2(2^r)$ and $\mathbf{Sz}(2^{2r+1})$. In §7 we study very simple linear representations; in particular, we prove that all the (modular) Steinberg representations discussed in Section 6 are very simple. This ends the proof of the main results.

2. Main results

Throughout this paper we assume that K is a field. We fix its algebraic closure K_a and write $\mathrm{Gal}(K)$ for the absolute Galois group $\mathrm{Aut}(K_a/K)$. If X is an abelian variety of dimension g defined over K then for each prime $\ell \ne \mathrm{char}(K)$ we write X_ℓ for the kernel of multiplication by ℓ in $X(K_a)$. It is well-known that X_ℓ is a $2g$-dimensional $\mathbf{F}_\ell$-vector space provided with a natural structure of $\mathrm{Gal}(K)$-module. We write $\mathrm{End}(X)$ for the ring of K_a-endomorphisms of X and $\mathrm{End}^0(X)$ for the corresponding finite-dimensional $\mathbf{Q}$-algebra $\mathrm{End}(X) \otimes \mathbf{Q}$.

The following notion plays a crucial role in this paper and will be discussed in detail in §7.

Definition 2.1. Let V be a vector space over a field $\mathbf{F}$, let G be a group and $\rho : G \to \mathrm{Aut}_{\mathbf{F}}(V)$ a linear representation of G in V. We say that the G-module V is *very simple* if it enjoys the following property:

If $R \subset \mathrm{End}_{\mathbf{F}}(V)$ be an $\mathbf{F}$-subalgebra containing the identity operator Id such that

$$\rho(\sigma) R \rho(\sigma)^{-1} \subset R \quad \forall \sigma \in G$$

then either $R = \mathbf{F} \cdot \mathrm{Id}$ or $R = \mathrm{End}_{\mathbf{F}}(V)$.

Remarks 2.2. (i) Clearly, the G-module V is very simple if and only if the corresponding $\rho(G)$-module V is very simple.

(ii) Clearly, if V is very simple then the corresponding algebra homomorphism

$$\mathbf{F}[G] \to \mathrm{End}_{\mathbf{F}}(V)$$

is surjective. Here $\mathbf{F}[G]$ stands for the group algebra of G. In particular, a very simple module is absolutely simple.

(iii) If G' is a subgroup of G and the G'-module V is very simple then the G-module V is also very simple.

(iv) Let G' be a normal subgroup of G. If V is a faithful very simple G-module then either $G' \subset \mathrm{Aut}_{\mathbf{F}}(V)$ consists of scalars (i.e., lies in $\mathbf{F} \cdot \mathrm{Id}$) or the G'-module V is also very simple.

Lemma 2.3. *Let X be an abelian variety of positive dimension g over K. Let ℓ be a prime different from* $\mathrm{char}(K)$. *Assume that the* $\mathrm{Gal}(K)$*-module X_ℓ is very simple. Then either* $\mathrm{End}(X) = \mathbf{Z}$ *or* $\mathrm{char}(K) > 0$ *and X is a supersingular abelian variety.*

We prove Lemma 2.3 in §3.

Theorem 2.4. *Let K be a field with* $\mathrm{char}(K) \neq 2$, *$f(x) \in K[x]$ an irreducible separable polynomial of degree $n \geq 5$. Let $\mathfrak{R} = \mathfrak{R}_f \subset K_a$ be the set of roots of f, let $K(\mathfrak{R}_f) = K(\mathfrak{R})$ be the splitting field of f and* $\mathrm{Gal}(f) := \mathrm{Gal}(K(\mathfrak{R})/K)$ *the Galois group of f, viewed as a subgroup of* $\mathrm{Perm}(\mathfrak{R})$. *Let C_f be the hyperelliptic curve $y^2 = f(x)$. Let $J(C_f)$ be its jacobian,* $\mathrm{End}(J(C_f))$ *the ring of K_a-endomorphisms of $J(C_f)$. Assume that n and* $\mathrm{Gal}(f)$ *enjoy one of the following properties:*

(i) *$n = 2^m + 1 \geq 9$ and the Galois group* $\mathrm{Gal}(f)$ *of f contains a subgroup isomorphic to* $\mathbf{L}_2(2^m)$;

(ii) *For some positive integer k we have $n = 2^{2(2k+1)} + 1$ and the Galois group* $\mathrm{Gal}(f)$ *of f is isomorphic to* $\mathbf{Sz}(2^{2k+1})$;

Then:

(a) *The* $\mathrm{Gal}(K)$*-module $J(C)_2$ is very simple;*

(b) *Either* $\mathrm{End}(J(C_f)) = \mathbf{Z}$ *or* $\mathrm{char}(K) > 0$ *and $J(C_f)$ is a supersingular abelian variety.*

Remark 2.5. It follows from Lemma 2.3 that in order to prove Theorem 2.4, it suffices to check only assertion a).

3. Proof of Lemma 2.3

Recall that $\dim_{\mathbf{F}_\ell}(X_\ell) = 2g$. Since X is defined over K, one may associate with every $u \in \mathrm{End}(X)$ and $\sigma \in \mathrm{Gal}(K)$ an endomorphism ${}^\sigma u \in \mathrm{End}(X)$ such that

$${}^\sigma u(x) = \sigma u(\sigma^{-1}x) \quad \forall x \in X(F_a).$$

Let us put

$$R := \mathrm{End}(X) \otimes \mathbf{Z}/\ell\mathbf{Z} \subset \mathrm{End}_{\mathbf{F}_\ell}(X_\ell).$$

Clearly, R satisfies all the conditions in Definition 2.1. This implies that either $R = \mathbf{F}_\ell \cdot \mathrm{Id}$ or $R = \mathrm{End}_{\mathbf{F}_\ell}(X_\ell)$. If $\mathrm{End}(X) \otimes \mathbf{Z}/\ell\mathbf{Z} = R = \mathbf{F}_\ell \cdot \mathrm{Id}$ then the free abelian group $\mathrm{End}(X)$ has rank 1 and therefore coincides with $\mathbf{Z}$. If $\mathrm{End}(X) \otimes \mathbf{Z}/\ell\mathbf{Z} = R = \mathrm{End}_{\mathbf{F}_\ell}(X_\ell)$ then the free abelian group $\mathrm{End}(X)$ has rank $(2\dim(X))^2 = (2g)^2$ and therefore the $\mathbf{Q}$-algebra $\mathrm{End}^0(X)$ has dimension $(2g)^2$.

Now Lemma 2.3 becomes an immediate corollary of the following assertion proven in [17] (see Lemma 3.1).

Lemma 3.1. *Let Y be an abelian variety of dimension g over an algebraically closed field K_a. Assume that the semisimple $\mathbf{Q}$-algebra* $\mathrm{End}^0(Y) = \mathrm{End}(Y) \otimes \mathbf{Q}$ *has dimension $(2g)^2$. Then* $\mathrm{char}(K_a) > 0$ *and Y is supersingular.*

4. Permutation groups and permutation modules

Let B be a finite set consisting of $n \geq 5$ elements. We write $\mathrm{Perm}(B)$ for the group of permutations of B. A choice of ordering on B gives rise to an isomorphism

$$\mathrm{Perm}(B) \cong \mathbf{S}_n.$$

Let G be a subgroup of $\mathrm{Perm}(B)$. For each $b \in B$ we write G_b for the stabilizer of b in G; it is a subgroup of G.

Remark 4.1. Assume that the action of G on B is transitive. It is well-known that each G_b is a subgroup of index n in G and all the G_b's are conjugate to each other in G. Each conjugate of G_b in G is the stabilizer of a point in B. In addition, one may identify the G-set B with the set of cosets G/G_b with the standard action by G.

Let $\mathbf{F}$ be a field. We write $\mathbf{F}^B$ for the n-dimensional $\mathbf{F}$-vector space of maps $h : B \to \mathbf{F}$. The space $\mathbf{F}^B$ is provided with a natural action of $\mathrm{Perm}(B)$ defined as follows. Each $s \in \mathrm{Perm}(B)$ sends a map $h : B \to \mathbf{F}$ into $sh : b \mapsto h(s^{-1}(b))$. The permutation module $\mathbf{F}^B$ contains the $\mathrm{Perm}(B)$-stable hyperplane

$$(\mathbf{F}^B)^0 = \{h : B \to \mathbf{F} \mid \sum_{b \in B} h(b) = 0\}$$

and the $\mathrm{Perm}(B)$-invariant line $\mathbf{F} \cdot 1_B$ where 1_B is the constant function 1. The quotient $\mathbf{F}^B/(\mathbf{F}^B)^0$ is a trivial 1-dimensional $\mathrm{Perm}(B)$-module.

Clearly, $(\mathbf{F}^B)^0$ contains $\mathbf{F} \cdot 1_B$ if and only if $\mathrm{char}(\mathbf{F})$ divides n. If this is *not* the case then there is a $\mathrm{Perm}(B)$-invariant splitting

$$\mathbf{F}^B = (\mathbf{F}^B)^0 \oplus \mathbf{F} \cdot 1_B.$$

Clearly, $\mathbf{F}^B$ and $(\mathbf{F}^B)^0$ carry natural structures of G-modules. Their characters depend only on the characteristic of $\mathbf{F}$.

Let us consider the case of $\mathbf{F} = \mathbf{Q}$. Then the character of $\mathbf{Q}^B$ sends each $g \in G$ into the number of fixed points of g ([15], ex. 2.2, p. 12); it is called the *permutation character*. Let us denote by $\chi = \chi_B : G \to \mathbf{Q}$ the character of $(\mathbf{Q}^B)^0$. It is known that the $\mathbf{Q}[G]$-module $(\mathbf{Q}^B)^0$ is absolutely simple if and only if G acts doubly transitively on B ([15], ex. 2.6, p. 17). Clearly, $1 + \chi$ is the permutation character.

Now, let us consider the case of $\mathbf{F} = \mathbf{F}_2$. It is well-known that one may view $\mathbf{F}_2^B$ as the $\mathbf{F}_2$-vector space of *all* subsets of B with symmetric difference as a sum. Namely, a subset T corresponds to its characteristic function $\chi_T : B \to \{0, 1\} = \mathbf{F}_2$ and a function $h : B \to \mathbf{F}_2$ corresponds to its support $\mathrm{supp}(h) = \{x \in B \mid h(x) = 1\}$. Under this identification each $s \in G \subset \mathrm{Perm}(B)$ sends T into $s(T) = \{s(b) \mid b \in T\}$.

Under this identification the hyperplane $(\mathbf{F}_2^B)^0$ corresponds to the $\mathbf{F}_2$-vector space of *all* subsets of B of *even* cardinality with symmetric difference as a sum.

If n is even then let us define the $\mathrm{Perm}(B)$-module

$$Q_B := (\mathbf{F}_2^B)^0 / (\mathbf{F}_2 \cdot 1_B).$$

If n is odd then let us put

$$Q_B := (\mathbf{F}_2^B)^0.$$

When n is even, the quotient Q_B corresponds to the $n-2$-dimensional $\mathbf{F}_2$-vector space of *all* subsets of B of *even* cardinality with symmetric difference as a sum where each subset $T \subset B$ of even cardinality is identified with its complement $B \setminus T$.

Remark 4.2. Clearly, $\dim_{\mathbf{F}_2}(Q_B) = n - 1$ if n is odd and $\dim_{\mathbf{F}_2}(Q_B) = n - 2$ if n is even. In both cases Q_B is a faithful G-module.

Let $G^{(2)}$ be the set of 2-regular elements of G. Clearly, the Brauer character of the G-module $\mathbf{F}_2^B$ coincides with the restriction of $1 + \chi_B$ to $G^{(2)}$. This implies easily that the Brauer character of the G-module $(\mathbf{F}_2^B)^0$ coincides with the restriction of χ_B to $G^{(2)}$.

Remark 4.3. Let us denote by $\phi_B = \phi$ the Brauer character of the G-module Q_B. One may easily check that ϕ_B coincides with the restriction of χ_B to $G^{(2)}$ if n is odd and with the restriction of $\chi_B - 1$ to $G^{(2)}$ if n is even.

Remark 4.4. Assume that $n = \#(B)$ is *even*. Let us choose $b \in B$ and let $G' := G_b$ and $B' = B \setminus \{b\}$. Then $n' = \#(B') = n - 1$ is odd and there is a canonical isomorphism of G'-modules $Q_{B'} \cong Q_B$ defined as follows. First, there is a natural G'-equivariant embedding $\mathbf{F}_2^{B'} \subset \mathbf{F}_2^B$ which can be obtained by extending each $h : B' \to \mathbf{F}_2$ to B by letting $h(b) = 0$. Second, this embedding identifies $(\mathbf{F}_2^{B'})^0$ with a hyperplane of $(\mathbf{F}_2^B)^0$ which does not contain 1_B. Now the desired isomorphism is given by the composition

$$Q_{B'} = (\mathbf{F}_2^{B'})^0 \subset (\mathbf{F}_2^B)^0 \to (\mathbf{F}_2^B)^0 / (\mathbf{F}_2 \cdot 1_B) = Q_B.$$

This implies that if the G'-module $Q_{B'}$ is very simple then the G-module Q_B is also very simple.

Remark 4.5. Assume that G acts on B doubly transitively, $\#(B)$ is odd and $\#(B)-1=\dim_{\mathbf{Q}}((\mathbf{Q}_B)^0)$ coincides with the largest power of 2 dividing $\#(G)$. Then it follows from a theorem of Brauer-Nesbitt ([15], Sect. 16.4, pp. 136–137; [3], p. 249) that Q_B is an absolutely simple $\mathbf{F}_2[G]$-module. In particular, Q_B is (the reduction of) the Steinberg representation [3].

5. Points of order 2 on hyperelliptic jacobians

We keep all the notations of Section 2. In addition, we assume that K is a field of characteristic different from 2. Let C be a hyperelliptic curve over K defined by an equation $y^2=f(x)$ where $f(x)\in K[x]$ is a polynomial of degree $n\geq 5$ without multiple roots. The rational function $x\in K(C)$ defines a canonical double cover $\pi: C\to \mathbf{P}^1$. Let $B'\subset C(K_a)$ be the set of ramification points of π (Weierstraß points). Clearly, the restriction of π to B' is an injective map $\pi: B'\hookrightarrow \mathbf{P}^1(K_a)$, whose image is either the set $\mathfrak{R}=\mathfrak{R}_f$ of roots of f if n is even or the disjoint union of ∞ and $\mathfrak{R}$ if n is odd. By abuse of notation, we also denote by ∞ the ramification point lying above ∞ if n is odd and by ∞_1 and ∞_2 two unramified points lying above ∞ if n is even. Clearly, if n is odd then $\infty\in C(K)$. If n is even then the 2-element set $\{\infty_1,\infty_2\}$ is stable under the action of $\mathrm{Gal}(K)$.

Let us put

$$B=\{(\alpha,0)\mid f(\alpha)=0\}\subset C(K_a).$$

Then π defines a bijection between B and $\mathfrak{R}$ which commutes with the action of $\mathrm{Gal}(K)$. If n is even then B coincides with B'. In the case of odd n the set B' is the disjoint union of B and ∞.

Theorem 5.1. *Suppose n is an integer which is greater than or equal to 5. Suppose $f(x)\in K[x]$ is a separable polynomial of degree n, $\mathfrak{R}\subset K_a$ the set of roots of f, let $K(\mathfrak{R})$ be the splitting field of f and $\mathrm{Gal}(f):=\mathrm{Gal}(K(\mathfrak{R})/K)$ the Galois group of f.*

Suppose C is the hyperelliptic curve $y^2=f(x)$ of genus $g=[\frac{n-1}{2}]$ over K. Suppose $J(C)$ is the jacobian of C and $J(C)_2$ is the group of its points of order 2, viewed as a $2g$-dimensional $\mathbf{F}_2$-vector space provided with the natural action of $\mathrm{Gal}(K)$. Then the homomorphism $\mathrm{Gal}(K)\to \mathrm{Aut}_{\mathbf{F}_2}(J(C)_2)$ factors through the canonical surjection $\mathrm{Gal}(K)\twoheadrightarrow \mathrm{Gal}(K(\mathfrak{R})/K)=\mathrm{Gal}(f)$ and the $\mathrm{Gal}(f)$-modules $J(C)_2$ and $Q_{\mathfrak{R}}$ are isomorphic. In particular, the $G(K)$-module $J(C)_2$ is very simple if and only if the $\mathrm{Gal}(f)$-module Q_B is very simple.

Remark 5.2. Clearly, $\mathrm{Gal}(K)$ acts on B through the canonical surjective homomorphism $\mathrm{Gal}(K)\twoheadrightarrow \mathrm{Gal}(f)$, because all points of B are defined over $K(\mathfrak{R})$ and the natural homomorphism $\mathrm{Gal}(f)\to \mathrm{Perm}(B)$ is injective. Clearly,

$\pi : B \to \mathfrak{R}$ is a bijection of $\mathrm{Gal}(f)$-sets. This implies easily that the $\mathrm{Gal}(f)$-modules Q_B and $Q_{\mathfrak{R}}$ are isomorphic. So, in order to prove Theorem 5.1 it suffices to check that the $\mathrm{Gal}(K)$-modules Q_B and $J(C)_2$ are isomorphic.

Proof of Theorem 5.1. Here is a well-known explicit description of the group $J(C)_2$ of points of order 2 on $J(C)$. Let us denote by L the K-divisor $2(\infty)$ on C if n is odd and the K-divisor $(\infty_1) + (\infty_2)$ if n is even. In both cases L is an effective divisor of degree 2. Namely, let $T \subset B'$ be a subset of even cardinality. Then ([14], Ch. IIIa, Sect. 2, Lemma 2.4; [12], pp. 190–191; see also [11]) the divisor $e_T = \sum_{P \in T}(P) - \frac{\#(T)}{2}L$ on C has degree 0 and $2e_T$ is principal. If T_1, T_2 are two subsets of even cardinality in B' then the divisors e_{T_1} and e_{T_2} are linearly equivalent if and only if either $T_1 = T_2$ or $T_2 = B' \setminus T_1$. Also, if $T = T_1 \Delta T_2$ then the divisor e_T is linearly equivalent to $e_{T_1} + e_{T_2}$. Hereafter we use the symbol Δ for the symmetric difference of two sets. Counting arguments imply easily that each point of $J(C)_2$ is the class of e_T for some T. We know that such a choice is not unique. However, in the case of odd n if we demand that T does not contain ∞ then such a choice always exists and unique. This observation leads to a canonical group isomorphism

$$Q_B = (\mathbf{F}_2^B)^0 \cong J(C)_2, \quad T \mapsto \mathrm{cl}(e_T)$$

in the case of odd n. Here cl stands for the linear equivalence class of a divisor. In the case of even n we are still able to define a canonical surjective group homomorphism

$$(\mathbf{F}_2^B)^0 \to J(C)_2, \quad T \mapsto \mathrm{cl}(e_T)$$

and one may easily check that the kernel of this map is the line generated by the set B, i.e., the line generated by the constant function 1_B. This gives rise to the injective homomorphism

$$Q_B = (\mathbf{F}_2^B)^0/(\mathbf{F}_2 \cdot 1_B) \to J(C)_2,$$

which is an isomorphism, by counting arguments. So, in both (odd and even) cases we get a canonical isomorphism $Q_B \cong J(C)_2$, which obviously commutes with the actions of $\mathrm{Gal}(K)$. In other words, we constructed an isomorphism of $\mathrm{Gal}(K)$-modules Q_B and $J(C)_2$. In light of Remark 5.2, this ends the proof of Theorem 5.1. □

Combining Theorem 5.1 and Lemma 2.3 (for $\ell = 2$), we obtain the following corollary.

Corollary 5.3. *Let K be a field with $\mathrm{char}(K) \neq 2$, K_a its algebraic closure, $f(x) \in K[x]$ an irreducible separable polynomial of degree $n \geq 5$. Let $\mathfrak{R} = \mathfrak{R}_f \subset K_a$ be the set of roots of f, let $K(\mathfrak{R}_f) = K(\mathfrak{R})$ be the splitting field of f and $\mathrm{Gal}(f) := \mathrm{Gal}(K(\mathfrak{R})/K)$ the Galois group of f, viewed as a subgroup of $\mathrm{Perm}(\mathfrak{R})$. Let C_f be the hyperelliptic curve $y^2 = f(x)$. Let $J(C_f)$ be its jacobian, $\mathrm{End}(J(C_f))$ the ring of K_a-endomorphisms of $J(C_f)$. Assume that the $\mathrm{Gal}(f)$-module $Q_{\mathfrak{R}}$ is very simple. Then either $\mathrm{End}(J(C_f)) = \mathbf{Z}$ or $\mathrm{char}(K) > 0$ and $J(C_f)$ is a supersingular abelian variety.*

Notice that in order to prove Theorem 2.4, it suffices to check the following statement.

Theorem 5.4. *Let n be a positive integer, B a n-element set, $H \subset \mathrm{Perm}(B)$ a permutation group. Assume that (n, H) enjoy one of the following properties:*

(i) *$n = 2^m + 1 \geq 9$ and H contains a subgroup isomorphic to $\mathbf{L}_2(2^m)$;*
(ii) *For some positive integer k we have $n = 2^{2(2k+1)} + 1$ and H contains a subgroup isomorphic to $\mathbf{Sz}(2^{2k+1})$;*

Then the H-module Q_B is very simple.

Proof of Theorem 2.4 modulo Theorem 5.4. Let us put

$$n = \deg(f), B = \mathfrak{R}, H = \mathrm{Gal}(f).$$

It follows from Theorem 5.4 that the $\mathrm{Gal}(f)$-module $\mathbf{Q}_{\mathfrak{R}}$ is very simple. Now the result follows readily from Corollary 5.3. □

We prove Theorem 5.4 at the end of §7.

6. Steinberg representation

In this section we prove that the Steinberg representation is the only nontrivial absolutely irreducible representation over $\mathbf{F}_2$ (up to an isomorphism) of the groups $\mathbf{L}_2(2^m)$ and $\mathbf{Sz}(2^{2k+1})$. We refer to [3] for basic properties of Steinberg representations.

Let us fix an algebraic closure of $\mathbf{F}_2$ and denote it by $\mathcal{F}$. We write $\phi : \mathcal{F} \to \mathcal{F}$ for the Frobenius automorphism $x \mapsto x^2$. Let $q = 2^m$ be a positive integral power of two. Then the subfield of invariants of $\phi^m : \mathcal{F} \to \mathcal{F}$ is a finite field $\mathbf{F}_q$ consisting of q elements. Let q' be an integral positive power of q. If d is a positive integer and i is a non-negative integer then for each matrix $u \in \mathrm{GL}_d(\mathcal{F})$ we write $u^{(i)}$ for the matrix obtained by raising each entry of u to the 2^ith power.

Recall that an element $\alpha \in \mathbf{F}_q$ is called *primitive* if $\alpha \neq 0$ and has multiplicative order $q - 1$ in the cyclic multiplicative group $\mathbf{F}_q^*$.

Lemma 6.1. *Let $q > 2$, let d be a positive integer and let G be a subgroup of $\mathrm{GL}_d(\mathbf{F}_{q'})$. Assume that there exists an element $u \in G \subset \mathrm{GL}_d(\mathbf{F}_{q'})$, whose trace α lies in $\mathbf{F}_q^*$ and has multiplicative order $q - 1$. Let $V_0 = \mathcal{F}^d$ and let $\rho_0 : G \subset \mathrm{GL}_d(\mathbf{F}_{q'}) \subset \mathrm{GL}_d(\mathcal{F}) = \mathrm{Aut}_{\mathcal{F}}(V_0)$ be the natural d-dimensional representation of G over $\mathcal{F}$. For each positive integer $i < m$ we define a d-dimensional $\mathcal{F}$-representation*

$$\rho_i : G \to \mathrm{Aut}(V_i)$$

as the composition of

$$G \hookrightarrow \mathrm{GL}_d(\mathbf{F}_{q'}), \quad x \mapsto x^{(i)}$$

and the inclusion map

$$\mathrm{GL}_d(\mathbf{F}_{q'}) \subset \mathrm{GL}_d(\mathcal{F}) \cong \mathrm{Aut}_{\mathcal{F}}(V_i).$$

Let S be a subset of $\{0,1,\dots,m-1\}$. Let us define a $d^{\#(S)}$-dimensional $\mathcal{F}$-representation ρ_S of G as the tensor product of the representations ρ_i for all $i \in S$. If S is a proper subset of $\{0,1,\dots,m-1\}$ then there exists an element $u \in G$ such that the trace of $\rho_S(u)$ does not belong to $\mathbf{F}_2$. In particular, ρ_S could not be obtained by extension of scalars to $\mathcal{F}$ from a representation of G over $\mathbf{F}_2$.

Proof. Clearly,

$$\operatorname{tr}(\rho_i(u)) = (\operatorname{tr}(\rho_0(u))^{2^i} \quad \forall u \in G.$$

This implies easily that

$$\operatorname{tr}(\rho_S(u)) = \prod_{i\in S} \operatorname{tr}(\rho_i(u)) = (\operatorname{tr}(\rho_0(u))^M$$

where $M = \sum_{i\in S} 2^i$. Since S is a *proper* subset of $\{0,1,\dots,m-1\}$, we have

$$0 < M < \sum_{i=0}^{m-1} 2^i = 2^m - 1 = \#(\mathbf{F}_q^*).$$

Recall that there exists $u \in G$ such that $\alpha = \operatorname{tr}(\rho_0(u))$ lies in $\mathbf{F}_q^*$ and the exact multiplicative order of α is $q-1 = 2^m - 1$.

This implies that $0 \ne \alpha^M \ne 1$. Since $\mathbf{F}_2 = \{0,1\}$, we conclude that $\alpha^M \notin \mathbf{F}_2$. Therefore $\operatorname{tr}(\rho_S(u)) = (\operatorname{tr}(\rho_0(u))^M = \alpha^M \notin \mathbf{F}_2$. □

Theorem 6.2. *Let $q \ge 8$ be a power of 2 and $G = \mathbf{L}_2(q) = \mathrm{PSL}_2(\mathbf{F}_q) = \mathrm{SL}_2(\mathbf{F}_q)$. Let $\rho : G \to \mathrm{Aut}(V)$ be an absolutely irreducible faithful representation of G over $\mathcal{F}$. If the trace map $\operatorname{tr}_\rho : G \to \mathcal{F}$ takes on values in $\mathbf{F}_2$ then $\dim_{\mathcal{F}}(V) = q$. In particular, ρ is the Steinberg representation of G.*

Proof. Let us put $q' = q$. We have

$$G = \mathrm{SL}_2(\mathbf{F}_q) \subset \mathrm{GL}_2(\mathbf{F}_q).$$

Clearly, for each $\alpha \in \mathbf{F}_q$ one may find a 2×2 matrix with determinant 1 and trace α. This implies that G satisfies the conditions of Lemma 6.1.

The construction described in Lemma 6.1 allows us to construct a $d^{\#(S)}$-dimensional $\mathcal{F}$-representation ρ_S of G for each subset S of of $\{0,1,\dots,m-1\}$. It is well-known ([1], pp. 588–589) that the ρ_S's exhaust the list of all absolutely irreducible $\mathcal{F}$-representations of $G = \mathrm{SL}_2(\mathbf{F}_q)$ and therefore ρ is isomorphic to ρ_S for some S. It follows from Lemma 6.1 that either S is empty or $S = \{0,1,\dots,m-1\}$. The case of empty S corresponds to the trivial 1-dimensional representation. Therefore $S = \{0,1,\dots,m-1\}$ and ρ is $2^m = q$-dimensional. □

Suppose $m = 2k+1 \ge 3$ is an odd integer. Let $q = 2^m = 2^{2k+1}$ and $d = 4$. Recall ([4], pp. 182–194) that the *Suzuki group* $\mathbf{Sz}(q)$ is the subgroup of $\mathrm{GL}_4(\mathbf{F}_q)$ generated by the matrices $S(a,b), M(\lambda), T$ defined as follows. For each $a, b \in \mathbf{F}_q$ the matrix $S(a,b)$ is defined by

$$S(a,b) = \begin{pmatrix} 1 & 0 & 0 & 0 \\ a & 1 & 0 & 0 \\ b & a^{2^{k+1}} & 1 & 0 \\ a^{2^{k+1}+2} + ab + b^{2^{k+1}} & a^{2^{k+1}+1} + b & a & 1 \end{pmatrix}$$

and for each $\lambda \in \mathbf{F}_q^*$ the matrix $M(\lambda)$ is defined by

$$M(\lambda) = \begin{pmatrix} \lambda^{1+2^k} & 0 & 0 & 0 \\ 0 & \lambda^{2^k} & 0 & 0 \\ 0 & 0 & \lambda^{-2^k} & 0 \\ 0 & 0 & 0 & \lambda^{1+2^k} \end{pmatrix}.$$

The matrix T is defined by

$$T = \begin{pmatrix} 0 & 0 & 0 & 1 \\ 0 & 0 & 1 & 0 \\ 0 & 1 & 0 & 0 \\ 1 & 0 & 0 & 0 \end{pmatrix}.$$

Notice that the trace of $S(0,b)T$ is $b^{2^{k+1}}$. This implies easily that for each $\alpha \in \mathbf{F}_q$ one may find an element of $\mathbf{Sz}(q) \subset \mathrm{GL}_4(\mathbf{F}_q)$ with trace α. This implies that

$$G = \mathbf{Sz}(q) \subset \mathrm{GL}_4(\mathbf{F}_q)$$

satisfies the conditions of Lemma 6.1. Notice also that $\#(\mathbf{Sz}(q)) = (q^2+1)q^2(q-1)$ ([4], p. 187).

Theorem 6.3. *Let $\rho : \mathbf{Sz}(q) \to \mathrm{Aut}(V)$ be an absolutely irreducible faithful representation of $\mathbf{Sz}(q)$ over $\mathcal{F}$. If the trace map $\mathrm{tr}_\rho : \mathbf{Sz}(q) \to \mathcal{F}$ takes on values in $\mathbf{F}_2$ then $\dim_{\mathcal{F}}(V) = q^2$. In particular, ρ is the Steinberg representation of G.*

Proof. Let us put $q' = q$. We know that $G = \mathbf{Sz}(q) \subset \mathrm{GL}_4(\mathbf{F}_q)$ satisfies the conditions of Lemma 6.1.

The construction described in Lemma 6.1 allows us to construct a $4^{\#(S)}$-dimensional $\mathcal{F}$-representation ρ_S of G for each subset S of $\{0,1,\dots,m-1\}$. It is known ([8], pp. 56–57) that the ρ_S's exhaust the list of all absolutely irreducible $\mathcal{F}$-representations of $G = \mathrm{SL}_2(\mathbf{F}_q)$ and therefore ρ is isomorphic to ρ_S for some S. It follows from Lemma 6.1 that either S is empty or $S = \{0,1,\dots,m-1\}$. The case of empty S corresponds to the trivial 1-dimensional representation. Therefore $S = \{0,1,\dots,m-1\}$ and ρ is $4^m = q^2$-dimensional. □

Remark 6.4. Assume in the case 5.4(i) (resp. 5.4(ii)) that $H = \mathbf{L}_2(2^m)$ (resp. $\mathbf{Sz}(2^{2k+1})$). It follows from Remark 4.5 that Q_B is the Steinberg representation of H.

7. Very simple representations

Examples 7.1. (i) If $\dim(V)=1$ then V is always very simple.

(ii) Assume that there exist G-modules V_1 and V_2 such that $\dim(V_1)>1, \dim(V_2)>1$ and the G-module V is isomorphic to $V_1\otimes_{\mathbf{F}} V_2$. Then V is *not* very simple. Indeed, the subalgebra

$$R=\mathrm{End}_{\mathbf{F}}(V_1)\otimes \mathbf{F}\cdot \mathrm{Id}_{V_2}\subset \mathrm{End}_{\mathbf{F}}(V_1)\otimes_{\mathbf{F}}\mathrm{End}_{\mathbf{F}}(V_2)=\mathrm{End}_{\mathbf{F}}(V)$$

is stable under the conjugation by elements of G but coincides neither with $\mathbf{F}\cdot\mathrm{Id}$ nor with $\mathrm{End}_{\mathbf{F}}(V)$. (Here Id_{V_2} stands for the identity operator in V_2.)

(ii)bis Let $X\to G$ be a central extension of G. Assume that there exist X-modules V_1 and V_2 such that $\dim(V_1)>1,\dim(V_2)>1$ and V, viewed as X-module, is isomorphic to $V_1\otimes_{\mathbf{F}}V_2$. Then V is *not* very simple as an X-module. Since X and G have the same images in $\mathrm{Aut}_{\mathbf{F}}(V)$, the G-module V is also not very simple.

(iii) Assume that there exists a subgroup G' in G of finite index $m>1$ and a G'-module V' such that the $\mathbf{F}[G]$-module V is *induced* by the $\mathbf{F}[G']$-module V'. (In particular, m must divide $\dim(V)$.) Then V is *not* very simple. Indeed, one may view W as a G'-submodule of V such that V coincides with the direct sum $\oplus_{\sigma\in G/G'}\sigma W$. Let $R=\oplus_{\sigma\in G/G'}\mathrm{End}_{\mathbf{F}}(\sigma W)$ be the algebra of all operators sending each σW into itself. Then R is stable under the conjugation by elements of G but coincides neither with $\mathbf{F}\cdot\mathrm{Id}$ nor with $\mathrm{End}_{\mathbf{F}}(V)$.

Example 7.2. Let $n\geq 5$ be an integer, B a n-element set. Suppose G is either $\mathrm{Perm}(B)\cong \mathbf{S}_n$ or the only subgroup in $\mathrm{Perm}(B)$ of index 2 (isomorphic to $\mathbf{A}_n$). Then the G-module Q_B is very simple. If n is odd then this assertion is proven in [17], Th. 4.1. If n is even then $n\geq 6$, $n'=n-1\geq 5$ is odd and the result follows from the odd case combined with Remarks 2.2 and 4.4.

Remarks 7.3. Assume that there exist two G-modules V_1 and V_2 such that $\dim(V_1)>1,\dim(V_2)>1$ and the G-module V is isomorphic to $V_1\otimes_{\mathbf{F}}V_2$.

(i) If V is simple then both V_1 and V_2 are also simple. Indeed, if say, V' is a proper G-stable subspace in V_1 then $V'\otimes_{\mathbf{F}}V_2$ is a proper G-stable subspace in $V_1\otimes_{\mathbf{F}}V_2=V$.

(ii) If V is absolutely simple then both V_1 and V_2 are also absolutely simple. Indeed, assume that say, $R_1:=\mathrm{End}_G(V_1)$ has $\mathbf{F}$-dimension greater than 1. Then $\mathrm{End}_G(V)=\mathrm{End}_G(V_1\otimes_{\mathbf{F}}V_2)$ contains $R_1\otimes\mathrm{Id}_{V_2}\cong R_1$ and therefore also has dimension greater than 1.

Lemma 7.4. *Let H be a group, $\mathbf{F}$ a field and V a simple $\mathbf{F}[H]$-module of finite $\mathbf{F}$-dimension N. Let $R\subset\mathrm{End}_{\mathbf{F}}(V)$ be an $\mathbf{F}$-subalgebra containing the identity operator* Id *and such that*

$$uRu^{-1}\subset R\quad \forall u\in H.$$

Then:

(i) *The faithful R-module V is semisimple.*
(ii) *Either the R-module V is isotypic or there exists a subgroup $H' \subset H$ of index r dividing N and a H'-module V' of finite $\mathbf{F}$-dimension N/r such that $r > 1$ and the H-module V is induced by V'. In addition, if $\mathbf{F} = \mathbf{F}_2$ then $r < N$.*

Proof. We may assume that $N > 1$. Clearly, V is a faithful R-module and

$$uRu^{-1} = R \quad \forall u \in H.$$

Step 1. V is a *semisimple* R-module. Indeed, let $U \subset V$ be a simple R-submodule. Then $U' = \sum_{s \in H} sU$ is a non-zero H-stable subspace in V and therefore must coincide with V. On the other hand, each sU is also a R-submodule in V, because $s^{-1}Rs = R$. In addition, if $W \subset sU$ is an R-submodule then $s^{-1}W$ is an R-submodule in U, because

$$Rs^{-1}W = s^{-1}sRs^{-1}W = s^{-1}RW = s^{-1}W.$$

Since U is simple, $s^{-1}W = \{0\}$ or U. This implies that sU is also simple. Hence $V = U'$ is a sum of simple R-modules and therefore is a semisimple R-module.

Step 2. The R-module V is either isotypic or induced. Indeed, let us split the semisimple R-module V into the direct sum

$$V = V_1 \oplus \cdots \oplus V_r$$

of its isotypic components. Dimension arguments imply that $r \leq \dim(V) = N$. It follows easily from the arguments of the previous step that for each isotypic component V_i its image sV_i is an isotypic R-submodule for each $s \in H$ and therefore is contained in some V_j. Similarly, $s^{-1}V_j$ is an isotypic submodule obviously containing V_i. Since V_i is an isotypic component, $s^{-1}V_j = V_i$ and therefore $sV_i = V_j$. This means that s permutes the V_i; since V is H-simple, H permutes them transitively.

This implies that all V_i have the same dimension N/r and therefore r divides $\dim(V) = N$. Let $H' = H_i$ be the stabilizer of V_i in H, i.e.

$$H_i = \{s \in H \mid sV_i = V_i\}.$$

The transitivity of the action of H on the V_j's implies that $[H : H_i] = r$.

If $r = 1$ then $H = H' = H_i$. This means that $sV_i = V_i$ for all $s \in H$ and $V = V_i$ is isotypic.

Assume that $r > 1$ and consider the H'-module $W = V_i$. Clearly, $[H : H'] = [H : H_i] = r$ divides N and the H-module V is induced by W.

Step 3. Assume that $r = N$ and $\mathbf{F} = \mathbf{F}_2$. Then each V_i is one-dimensional and contains exactly one non-zero vector say, v_i. Then the sum $\sum_{i=1}^{N} v_i$ is a non-zero H-invariant vector which contradicts the simplicity of the H-module V. □

Theorem 7.5. *Suppose H is a group and*

$$\rho : H \to \mathrm{Aut}_{\mathbf{F}_2}(V)$$

is an absolutely simple $\mathbf{F}_2[H]$-module of finite dimension N. Suppose there exists an $\mathbf{F}_2$-subalgebra $R \subset \mathrm{End}_{\mathbf{F}_2}(V)$ containing the identity operator Id *and such that*

$$uRu^{-1} \subset R \quad \forall u \in H.$$

Assume, in addition, that H does not have nontrivial cyclic quotients of order dividing N. If the R-module V is isotypic then there exist $\mathbf{F}_2[H]$-modules V_1 and V_2 such that V, viewed as H-module, is isomorphic to $V_1 \otimes_{\mathbf{F}_2} V_2$ and the image of $R \subset \mathrm{End}_{\mathbf{F}_2}(V)$ under the induced isomorphism

$$\mathrm{End}_{\mathbf{F}_2}(V) = \mathrm{End}_{\mathbf{F}_2}(V_1 \otimes_{\mathbf{F}_2} V_2) = \mathrm{End}_{\mathbf{F}_2}(V_1) \otimes_{\mathbf{F}_2} \mathrm{End}_{\mathbf{F}_2}(V_2)$$

coincides with $\mathrm{End}_{\mathbf{F}_2}(V_1) \otimes \mathrm{Id}_{V_2}$. *In particular, if both V_1 and V_2 have dimension greater than* 1 *then the H-module V is not very simple.*

Proof. Since V is isotypic, there exist a simple R-module W, a positive integer d and an isomorphism

$$\psi : V \cong W^d$$

of R-modules. Let us put

$$V_1 = W, \quad V_2 = \mathbf{F}_2^d.$$

The isomorphism ψ gives rise to the isomorphism of $\mathbf{F}_2$-vector spaces

$$V = W^d = W \otimes_{\mathbf{F}_2} \mathbf{F}_2^d = V_1 \otimes_{\mathbf{F}_2} V_2.$$

We have

$$d \cdot \dim(W) = \dim(V) = N.$$

Clearly, $\mathrm{End}_R(V)$ is isomorphic to the matrix algebra $\mathrm{Mat}_d(\mathrm{End}_R(W))$ of size d over $\mathrm{End}_R(W)$.

Let us put

$$k = \mathrm{End}_R(W).$$

Since W is simple, k is a finite-dimensional division algebra over $\mathbf{F}_2$. Therefore k must be a finite field.

We have

$$\mathrm{End}_R(V) \cong \mathrm{Mat}_d(k).$$

Clearly, $[k : \mathbf{F}_2]$ divides $\dim_{\mathbf{F}_2}(W)$ and therefore divides $\dim_{\mathbf{F}_2}(V) = N$. Clearly, $\mathrm{Aut}(k/\mathbf{F}_2)$ is always a cyclic group of order $[k : \mathbf{F}_2]$ and therefore has order dividing N.

Clearly, $\mathrm{End}_R(V) \subset \mathrm{End}_{\mathbf{F}_2}(V)$ is stable under the adjoint action of H. This induces a homomorphism

$$\alpha : H \to \mathrm{Aut}_{\mathbf{F}_2}(\mathrm{End}_R(V)) = \mathrm{Aut}_{\mathbf{F}_2}(\mathrm{Mat}_d(k)).$$

Since k is the center of $\mathrm{Mat}_d(k)$, it is stable under the action of H, i.e., we get a homomorphism $H \to \mathrm{Aut}(k/\mathbf{F}_2)$, which must be trivial, since H is perfect and $\mathrm{Aut}(k/\mathbf{F}_2)$ is a cyclic group of order dividing N and therefore the kernel of the homomorphism must coincide with H. This implies that the center k of $\mathrm{End}_R(V)$ commutes with H. Since $\mathrm{End}_H(V) = \mathbf{F}_2$, we have $k = \mathbf{F}_2$. This implies that $\mathrm{End}_R(V) \cong \mathrm{Mat}_d(\mathbf{F}_2)$ and one may rewrite α as

$$\alpha : H \to \mathrm{Aut}_{\mathbf{F}_2}(\mathrm{Mat}_d(\mathbf{F}_2)) = \mathrm{Aut}(\mathrm{End}_{\mathbf{F}_2}(V_2)) = \mathrm{Aut}_{\mathbf{F}_2}(V_2)/\mathbf{F}_2^* = \mathrm{Aut}_{\mathbf{F}_2}(V_2).$$

It follows from the Jacobson density theorem that $R = \mathrm{End}_{\mathbf{F}_2}(W) \cong \mathrm{Mat}_m(\mathbf{F}_2)$ with $dm = N$.

The adjoint action of H on R gives rise to a homomorphism

$$\beta : H \to \mathrm{Aut}_{\mathbf{F}_2}(\mathrm{End}_{\mathbf{F}_2}(W)) = \mathrm{Aut}_{\mathbf{F}_2}(W)/\mathbf{F}_2^* = \mathrm{Aut}_{\mathbf{F}_2}(W).$$

Clearly, α and β provide V_2 and V_1 respectively with the structure of H-modules. Notice that

$$R = \mathrm{End}_{\mathbf{F}_2}(V_1) = \mathrm{End}_{\mathbf{F}_2}(V_1) \otimes \mathrm{Id}_{V_2} \subset \mathrm{End}_{\mathbf{F}_2}(V_1) \otimes_{\mathbf{F}_2} \mathrm{End}_{\mathbf{F}_2}(V_2) = \mathrm{End}_{\mathbf{F}_2}(V).$$

Now our task boils down to the comparison of the structures of H-module on $V = V_1 \otimes_{\mathbf{F}_2} V_2$ defined by ρ and $\beta \otimes \alpha$ respectively. I claim that

$$\rho(g) = \beta(g) \otimes \alpha(g) \quad \forall g \in H.$$

Indeed, notice that the conjugation by $\rho(g)$ in $\mathrm{End}_{\mathbf{F}_2}(V) = \mathrm{End}_{\mathbf{F}_2}(V_1 \otimes_{\mathbf{F}_2} V_2)$ leaves stable $R = \mathrm{End}_{\mathbf{F}_2}(V_1) \otimes_{\mathbf{F}_2} \mathrm{Id}_{V_2}$ and coincides on R with the conjugation by $\alpha(g) \otimes \mathrm{Id}_{V_2}$. Since the centralizer of $\mathrm{End}_{\mathbf{F}_2}(V_1) \otimes \mathrm{Id}_{V_2}$ in

$$\mathrm{End}_{\mathbf{F}_2}(V) = \mathrm{End}_{\mathbf{F}_2}(V_1) \otimes_{\mathbf{F}_2} \mathrm{End}_{\mathbf{F}_2}(V_2)$$

coincides with $\mathrm{Id}_{V_1} \otimes \mathrm{End}_{\mathbf{F}_2}(V_2)$, there exists $u \in \mathrm{Aut}_{\mathbf{F}_2}(V_2)$ such that

$$\rho(g) = \beta(g) \otimes u.$$

Since the conjugation by $\rho(g)$ leaves stable the centralizer of R, i.e. $\mathrm{Id}_{V_1} \otimes \mathrm{End}_{\mathbf{F}_2}(V_2)$ and coincides on it with the conjugation by $\mathrm{Id}_{V_1} \otimes \alpha(g)$, there exists a non-zero constant $\gamma \in \mathbf{F}_2^*$ such that $u = \gamma\beta(g)$. This implies that

$$\rho(g) = \beta(g) \otimes u = \gamma \cdot \beta(g) \otimes \alpha(g).$$

Now one has only to recall that $\mathbf{F}_2^* = \{1\}$ and therefore $\gamma = 1$. $\square$

Remark 7.6. In the notations of Th. 7.5 the H-modules V_1 and V_2 must be absolutely simple. This follows easily from Remarks 7.3.

Lemma 7.4 and Theorem 7.5 together with Remark 7.6 imply easily the following criterion of very simplicity over $\mathbf{F}_2$.

Theorem 7.7. *Let H be a group and V be a $\mathbf{F}_2[H]$-module of finite dimension N over $\mathbf{F}_2$. Assume, in addition, that H does not have nontrivial cyclic quotients of order dividing N (e.g., H is perfect).*

Then V is very simple if and only if the following conditions hold:

(i) *The H-module V is absolutely simple;*
(ii) *There do not exist a subgroup $H' \neq H$ of H and a $\mathbf{F}_2[H']$-module V' such that V is induced by V';*
(iii) *There do not exist absolutely simple $\mathbf{F}_2[H]$-modules V_1 and V_2, both of dimension greater than 1 and such that the H-module V is isomorphic to $V_1 \otimes_{\mathbf{F}_2} V_2$.*

Combining Theorem 7.7 with Lemma 7.4 and Theorem 7.5 we get easily the following corollary.

Corollary 7.8. *Let H be a group and V be a $\mathbf{F}_2[H]$-module of finite dimension N over $\mathbf{F}_2$. Then V is very simple if the following conditions hold:*
(i) *The H-module V is absolutely simple;*
(ii) *H does not contain a subgroup of finite index r with $r \mid N$ and $1 < r < N$. In addition, H does not have cyclic quotients of order N, i.e., H does not have a normal subgroup H' of index N with cyclic quotient H/H';*
(iii) *There do not exist absolutely simple $\mathbf{F}_2[H]$-modules V_1 and V_2, both of dimension greater than 1 and such that the H-module V is isomorphic to $V_1 \otimes_{\mathbf{F}_2} V_2$.*

The following assertion follows easily from Lemma 7.4 and Theorem 7.5.

Corollary 7.9. *Suppose a positive integer $N > 1$ and a group H enjoy the following properties:*
- *H does not contain a subgroup of index dividing N except H itself.*
- *Let $N = ab$ be a factorization of N into a product of two positive integers $a > 1$ and $b > 1$. Then either there does not exist an absolutely simple $\mathbf{F}_2[H]$-module of $\mathbf{F}_2$-dimension a or there does not exist an absolutely simple $\mathbf{F}_2[H]$-module of $\mathbf{F}_2$-dimension b.*

Then each absolutely simple $\mathbf{F}_2[H]$-module of $\mathbf{F}_2$-dimension N is very simple. In other words, in dimension N the properties of absolute simplicity and very simplicity over $\mathbf{F}_2$ are equivalent.

The next two theorems provide examples of very simple Steinberg representations.

Theorem 7.10. *Let $q = 2^m \geq 8$ be an integral power of 2, let B be a $(q+1)$-element set. Let G' be a group acting faithfully on B. Assume that G' contains a subgroup G isomorphic to $\mathbf{L}_2(q)$. Then the G'-module Q_B is very simple.*

Proof. We have $\mathbf{L}_2(q) = G \subset G' \subset \mathrm{Perm}(B)$. Clearly, it suffices to check that the $\mathbf{L}_2(q)$-module Q_B is very simple.

First, notice that $\mathbf{L}_2(q)$ acts doubly transitively on B. Indeed, each subgroup of $\mathbf{L}_2(q)$ (except $\mathbf{L}_2(q)$ itself) has index $\geq q+1 = \#(B)$ ([16], (6.27), p. 415). This implies that $\mathbf{L}_2(q)$ acts transitively on B. If the stabilizer G_b of a point $b \in B$ has index $q+1$ then it follows easily from Th. 6.25 on p. 412 of

[16]) that G_b is conjugate to the (Borel) subgroup of upper-triangular matrices and therefore the $\mathbf{L}_2(q)$-set B is isomorphic to the projective line $\mathbf{P}^1(\mathbf{F}_q)$ with the standard action of $\mathbf{L}_2(q)$ which is well-known to be doubly (and even triply) transitive. By Remark 4.5, this implies that the $\mathbf{F}_2[\mathbf{L}_2(q)]$-module Q_B is absolutely simple. Recall that

$$\dim_{\mathbf{F}_2}(Q_B) = \#(B) - 1 = q = 2^m.$$

By Theorem 6.2, there are no absolutely simple nontrivial $\mathbf{F}_2[\mathbf{L}_2(q)]$-modules of dimension $< 2^m$. This implies that Q_B is *not* isomorphic to a tensor product of absolutely simple $\mathbf{F}_2[\mathbf{L}_2(q)]$-modules of dimension > 1. Recall that all subgroups in $\mathbf{L}_2(q)$ different from $\mathbf{L}_2(q)$ have index $\geq q+1 > q = \dim_{\mathbf{F}_2}(Q_B)$. It follows from Corollary 7.8 that the G-module Q_B is very simple. Since $G \subset G'$, the G'-module Q_B is also very simple. □

Theorem 7.11. *Let k be a positive integer and $q = 2^{2k+1}$, let B be a (q^2+1)-element set. Let G' be a group acting faithfully on B. Assume that G' contains a subgroup G isomorphic to $\mathbf{Sz}(q)$. Then the G'-module Q_B is very simple.*

Proof. We have $\mathbf{Sz}(q) = G \subset G' \subset \mathrm{Perm}(B)$. First, notice that $\mathbf{Sz}(q)$ acts doubly transitively on B. Indeed, the classification of subgroups of Suzuki groups ([4], Remark 3.12(e), p. 194) implies that each subgroup of $\mathbf{Sz}(q)$ (except $\mathbf{Sz}(q)$ itself) has index $\geq q^2+1 = \#(B)$. This implies that $\mathbf{Sz}(q)$ acts transitively on B. If the stabilizer G_b of a point $b \in B$ has index q^2+1 then it follows easily from the same classification that G_b is conjugate to the subgroup $\mathfrak{F}\mathfrak{H}$ generated by all $S(a,b)$ and $M(\lambda)$ and therefore the $\mathbf{Sz}(q)$-set B is isomorphic to an *ovoid* $\mathcal{O} = \mathbf{Sz}(q)/\mathfrak{F}\mathfrak{H}$ where the action of $\mathbf{Sz}(q)$ is known to be doubly transitive ([4], Th. 3.3 on pp. 184–185 and steps g) and i) of its proof on p. 187). By Remark 4.5, this implies that the $\mathbf{F}_2[\mathbf{Sz}(q)]$-module Q_B is absolutely simple. Recall that $\dim_{\mathbf{F}_2}(Q_B) = \#(B)-1 = q^2 = 2^{2(2k+1)}$. By Theorem 6.3, there are no absolutely simple nontrivial $\mathbf{F}_2[\mathbf{Sz}(q)]$-modules of dimension $< 2^{2(2k+1)}$. This implies that Q_B is *not* isomorphic to a tensor product of absolutely simple $\mathbf{F}_2[\mathbf{Sz}(q)]$-modules of dimension > 1. Recall that all subgroups in $G = \mathbf{Sz}(q)$ (except $\mathbf{Sz}(q)$ itself) have index $\geq q^2+1 > q^2 = \dim_{\mathbf{F}_2}(Q_B)$. It follows from Corollary 7.8 that the G-module Q_B is very simple. Since $G \subset G'$, the G'-module Q_B is also very simple. □

Proof of Theorem 5.4. The cases (i) and (ii) of Theorem 5.4 follow from Theorems 7.10 and 7.11 respectively applied to $G' = H$. □

In light of Corollary 5.3 it would be interesting to classify all permutation subgroups $G \subset \mathrm{Perm}(B)$ with very simple G-modules Q_B. We finish the paper by examples of very simple Q_B attached to the Mathieu groups M_{11} and M_{12} and to the related group $\mathbf{L}_2(11) = \mathrm{PSL}_2(11)$.

Theorem 7.12. *Let n be a positive integer, B a n-element set, $G \subset \mathrm{Perm}(B)$ a permutation group. Assume that (n, G) enjoy one of the following properties:*

(i) $n = 11$ *and* G *is isomorphic either to* $\mathbf{L}_2(11)$ *or* M_{11}*;*
(ii) $n = 12$ *and either* $G \cong \mathrm{M}_{12}$ *or* $G \cong \mathrm{M}_{11}$ *and* G *acts transitively on* B.

Then the G*-module* Q_B *is very simple.*

Proof. Assume that $n = 11$. Since M_{11} contains a subgroup isomorphic to $\mathbf{L}_2(11)$ ([2], p. 18), it suffices to check the case of $G = \mathbf{L}_2(11)$, in light of Remark 2.2(iii).

The group $G = \mathbf{L}_2(11)$ has two conjugacy classes of maximal subgroups of index 11 and all other subgroups in G have index greater than 11 ([2], p. 7). Therefore all subgroups in G (except G itself) have index greater than 10 and the action of G on the 11-element set B is transitive. The permutation character (in both cases) is (in the notations of [2], p. 7) $1+\chi_5$, i.e., $\chi = \chi_5$. The restriction of χ_5 to the set of 2-regular elements coincides with the absolutely irreducible Brauer character φ_4 (in the notations of [5], p. 7). In particular, the corresponding G-module Q_B is absolutely simple and has dimension 10. Since $10 = 2 \cdot 5$ and 5 is a prime, the very simplicity of the G-module Q_B follows from Th. 5.4 of [17]. This proves case (i).

Assume that $n = 12$.

Suppose $G = \mathrm{M}_{11}$ and the action of G on the 12-element set B is transitive. All subgroups G_b in G of index 12 are isomorphic to $\mathbf{L}_2(11)$ ([2], p. 18). It follows from the already proven case (i) for $\mathbf{L}_2(11)$ and Remark 4.4 that the G-module Q_B is very simple.

Suppose $G = \mathrm{M}_{12}$. The action of G on the 12-element B is transitive, since all subgroups in G (except G itself) have index ≥ 12. All subgroups G_b in G of index 12 are isomorphic to M_{11} ([2], p. 33). It follows from the already proven case (i) for M_{11} and Remark 4.4 that the G-module Q_B is very simple. □

Combining Corollary 5.3 and Theorem 7.12 (with $B = \mathfrak{R}, G = \mathrm{Gal}(f)$ and taking into account that the irreducibility of f means that $\mathrm{Gal}(f)$ acts transitively on $\mathfrak{R}$), we obtain the following statement.

Theorem 7.13. *Let* K *be a field with* $\mathrm{char}(K) \neq 2$, K_a *its algebraic closure,* $f(x) \in K[x]$ *an irreducible separable polynomial of degree* $n \geq 5$. *Let* $\mathfrak{R} = \mathfrak{R}_f \subset K_a$ *be the set of roots of* f, *let* $K(\mathfrak{R}_f) = K(\mathfrak{R})$ *be the splitting field of* f *and* $\mathrm{Gal}(f) := \mathrm{Gal}(K(\mathfrak{R})/K)$ *the Galois group of* f, *viewed as a subgroup of* $\mathrm{Perm}(\mathfrak{R})$. *Let* C_f *be the hyperelliptic curve* $y^2 = f(x)$. *Let* $J(C_f)$ *be its jacobian,* $\mathrm{End}(J(C_f))$ *the ring of* K_a*-endomorphisms of* $J(C_f)$.

Assume that n *and* $\mathrm{Gal}(f)$ *enjoy one of the following properties:*

(i) $n = 11$ *and* $\mathrm{Gal}(f)$ *is isomorphic either to* $\mathbf{L}_2(11)$ *or to* M_{11}*;*
(ii) $n = 12$ *and* $\mathrm{Gal}(f)$ *is isomorphic either to* M_{11} *or to* M_{12}.

Then either $\mathrm{End}(J(C_f)) = \mathbf{Z}$ *or* $\mathrm{char}(K) > 0$ *and* $J(C_f)$ *is a supersingular abelian variety.*

References

[1] R. Brauer, C. Nesbitt, *On the modular characters of groups.* Ann. of Math. **42** (1941), 556–590.

[2] J. H. Conway, R. T. Curtis, S. P. Norton, R. A. Parker, R. A. Wilson, Atlas of finite groups. Clarendon Press, Oxford, 1985.

[3] J. E. Humphreys, *The Steinberg representation.* Bull. AMS (N.S.) **16** (1987), 247–263.

[4] B. Huppert, N. Blackburn, Finite groups III. Springer-Verlag, Berlin Heidelberg New York, 1982.

[5] Ch. Jansen, K. Lux, R. Parker, R. Wilson, An Atlas of Brauer characters. Clarendon Press, Oxford, 1995.

[6] N. Katz, *Monodromy of families of curves: applications of some results of Davenport-Lewis.* In: Séminaire de Théorie des Nombres, Paris 1979-80 (ed. M.-J. Bertin); Progress in Math. **12**, pp. 171–195, Birkhäuser, Boston-Basel-Stuttgart, 1981.

[7] N. Katz, *Affine cohomological transforms, perversity, and monodromy.* J. Amer. Math. Soc. **6** (1993), 149–222.

[8] R. P. Martineau, *On 2-modular representations of Suzuki groups.* Amer. J. Math. **94** (1972), 55–72.

[9] D. Masser, *Specialization of some hyperelliptic jacobians.* In: Number Theory in Progress (eds. K. Györy, H. Iwaniec, J. Urbanowicz), vol. I, pp. 293–307; de Gruyter, Berlin-New York, 1999.

[10] Sh. Mori, *The endomorphism rings of some abelian varieties.* Japanese J. Math, **2**(1976), 109–130.

[11] Sh. Mori, *The endomorphism rings of some abelian varieties.* II, Japanese J. Math, **3**(1977), 105–109.

[12] D. Mumford, *Theta characteristics of an algebraic curve.* Ann. scient. Éc. Norm. Sup. (4) **4** (1971), 181–192.

[13] D. Mumford, Abelian varieties, Second edition, Oxford University Press, London, 1974.

[14] D. Mumford, Tata Lectures on Theta. II. Progress in Math. **43**, Birkhäuser, Boston-Basel-Stuttgart, 1984.

[15] J.-P. Serre, Linear representations of finite groups, Springer-Verlag, 1977.

[16] M. Suzuki, Group theory I, Springer Verlag, Berlin Heidelberg New York, 1982.

[17] Yu. G. Zarhin, *Hyperelliptic jacobians without complex multiplication.* Math. Res. Letters **7**(2000), 123–132.

Progress in Mathematics, Vol. 195, © 2001 Birkhäuser Verlag Basel/Switzerland

WINDOWS FOR DISPLAYS OF p-DIVISIBLE GROUPS

THOMAS ZINK

INTRODUCTION

The starting point of this work was the classification of p-divisible groups over a discrete valuation ring of characteristic 0 with perfect residue field of characteristic $p \geq 3$ obtained by C. Breuil in his note [B]. We will show that such a classification holds under quite general circumstances. We prove this by showing that the category used by Breuil to classify p-divisible groups is equivalent to the category of Dieudonné displays, which we defined in [Z-DD]. Breuil obtains his result by a very useful classification of finite flat group schemes over a discrete valuation ring as above. We have no generalization of such a classification.

Our results are much more general and easier to prove, if we exclude p-divisible groups which have an étale part. Therefore this case is of independent interest. We will describe it first.

We fix a prime number p. Let R be a commutative ring with 1, such that p is nilpotent in R.

Definition 1. *A frame for R consists of the following data:*

1) *A p-adic ring A which is torsion-free as an abelian group.*
2) *An endomorphism σ of A.*
3) *A surjective ring homomorphism $A \to R$, such that the kernel J is an ideal with divided powers.*

We require that σ induces the Frobenius endomorphism on A/pA.

The next objects we define give the possibility to operate a display.

Definition 2. *Let (A, J, σ) be a frame over R. A Dieudonné A-window over R consists of the following data:*

1) *A finitely generated projective A-module M.*
2) *A submodule $M_1 \subset M$ which contains JM.*
3) *A σ-linear map $\Phi : M \to M$.*

The following conditions are satisfied:

(i) *M/M_1 is a projective R-module.*

(ii)

$$\Phi M_1 \subset pM,$$

and M is generated by the union of ΦM and $\frac{1}{p}\Phi M_1$ as an A-module.

Let (M, M_1, Φ) be a Dieudonné A-window. One can show that there is a unique A-linear map $\Psi : M \to A \otimes_{\sigma,A} M$ such that $\Psi(\Phi(m)) = p \otimes m$ for all $m \in M$.

Definition 3. *We will say that (M, M_1, Φ) satisfies the nilpotence condition, if there is a number N, such that $\Psi^N(M) \subset J \otimes_{\sigma^N, A} M$. A Dieudonné A-window which satisfies the nilpotence condition is called an A-window.*

Theorem 4. *Let R be an excellent ring. Then the category of A-windows over R is equivalent to the category of formal p-divisible groups over R.*

The proof of the theorem uses the classification of formal p-divisible groups by displays given in [Z-DFG]. We will assume that the reader is acquainted with the introduction of this paper. Notations and results in this introduction will be freely used.

Let us again consider the case where R is an arbitrary ring with p nilpotent. There is a Cartier morphism (compare [BAC] IX §1, 2 prop.2) $\delta : A \to W(A)$ which is characterized by the property:

$$\mathbf{w}_n(\delta(a)) = \sigma^n(a).$$

Here $\mathbf{w}_n$ denotes the nth Witt polynomial. Let us denote the composite of δ with the canonical map $W(A) \to W(R)$ by $\varkappa$. Then we may associate to a window (M, M_1, Φ) a display $\mathcal{P} = (P, Q, F, V^{-1})$ over R as follows. We set:

$$P = W(R) \otimes_{\varkappa, A} M.$$

We define Q as the kernel of the canonical map $P \to M/M_1$. Moreover F is the ^{F}-linear extension of Φ to P. Finally V^{-1} is uniquely determined by the following equations:

$$\begin{array}{rcll} V^{-1}(\xi \otimes m_1) & = & {}^F\xi \otimes \Phi_1 m_1, & \text{for } \xi \in W(R),\ m_1 \in M_1; \\ V^{-1}({}^V\xi \otimes m) & = & \xi \otimes \Phi m, & \text{for } m \in M. \end{array}$$

In this way we obtain a functor Dsp from the category of A-windows over R to the category of displays over R. The quasi inverse of this functor is defined by the Dieudonné crystal $\mathcal{D}_\mathcal{P}$ which is associated to a display $\mathcal{P}$. Indeed we find the window associated to $\mathcal{P}$ if we set:

$$M = \mathcal{D}_\mathcal{P}(A).$$

This shows that the functor Dsp is an equivalence of categories. Therefore the theorem above follows from [Z-DFG].

As a typical example of a frame we can take for R the polynomial ring $k[T_1, \ldots, T_d]$ over a field of characteristic p and for A the ring of restricted power series $C_k\{T_1, \ldots, T_d\}$ over a Cohen ring for k. For σ we take an endomorphism which induces on C_k any lifting of the Frobenius on k, and which acts on the indeterminates by $\sigma(T_i) = T_i^p$.

We consider now the case of an arbitrary p-divisible group. In this case we restrict ourself to the category $\mathcal{Z}$ of rings R with the following property: R is a local ring with maximal ideal $\mathfrak{m}$ and perfect residue field k of characteristic $p \geq 3$. Moreover we assume that there is a number N such that $x^N = 0$ for each $x \in \mathfrak{m}$. The exact sequence:

$$0 \to W(\mathfrak{m}) \to W(R) \to W(k) \to 0,$$

has a unique ^{F}-equivariant section $W(k) \to W(R)$. We set $\hat{W}(R) = \hat{W}(\mathfrak{m}) \oplus W(k)$, where $\hat{W}(\mathfrak{m}) \subset W(\mathfrak{m})$ is the subring of Witt vectors whose components are almost all zero. In the sense of the splitting of the sequence above $\hat{W}(R)$ is a subring of $W(R)$, which is stable by the Frobenius ^{F} and the Verschiebung ^{V}. We note that the stability by ^{V} doesn't hold in the case $p = 2$ (compare [Z-DD] Lemma 2).

Consider a pd-thickening $S \to R$ in the category $\mathcal{Z}$. This is a surjection of rings, such that the kernel $\mathfrak{a}$ is equipped with divided powers. The Witt polynomials $\mathbf{w}_n : W(\mathfrak{a}) \to \mathfrak{a}$ may be divided by p^n in the sense of these divided powers. These divided Witt polynomials provide an isomorphism of additive groups:

$$\log : W(\mathfrak{a}) \to \mathfrak{a}^{\mathbb{N}}.$$

The inverse image of the infinite direct sum $\mathfrak{a}^{(\mathbb{N})} \subset \mathfrak{a}^{\mathbb{N}}$ by this homomorphism is denoted by $\widetilde{W}(\mathfrak{a})$. We have $\hat{W}(\mathfrak{a}) \subset \widetilde{W}(\mathfrak{a})$. This last inclusion is an equality if the divided powers on $\mathfrak{a}$ are pointwise nilpotent. Let us denote by $\widetilde{W}(S)$ the subring of $W(S)$ generated by $\hat{W}(S)$ and $\widetilde{W}(\mathfrak{a})$. Of course the ring $\widetilde{W}(S)$ depends on the divided powers, but we omit this in the notation.

We call a p-adic ring A a $\hat{\mathcal{Z}}$-ring if for each number n the ring $A/p^n A$ is a $\mathcal{Z}$-ring. We set $\hat{W}(A) = \varprojlim \hat{W}(A/p^n A)$. If we have a pd-thickening $A \to R$ we set $\widetilde{W}(A) = \varprojlim \widetilde{W}(A/p^n A)$, where the last projective limit goes over numbers n, such that p^n is zero in R. The rings $\hat{W}(A)$ and $\widetilde{W}(A)$ may be identified with subrings of $W(A)$.

Definition 5. *A Dieudonné frame (A, J, σ) for a $\mathcal{Z}$-ring R consists of the following data:*

1) *A $\hat{\mathcal{Z}}$-ring A which is torsion-free as an abelian group.*
2) *An endomorphism σ of A.*
3) *A surjective ring homomorphism $A \to R$, such that the kernel J is an ideal with divided powers.*

We require that σ induces the Frobenius endomorphism on A/pA, and that the Cartier morphism $A \to W(A)$ factors through $\widetilde{W}(A)$.

Let us denote by $\varkappa$ the composite of the morphisms $A \xrightarrow{\delta} \widetilde{W}(A) \to \hat{W}(R)$. If we tensorize a Dieudonné A-window by $\varkappa$ we obtain a Dieudonné display. Again this functor is an equivalence from the category of Dieudonné

A-windows over the $\mathcal{Z}$-ring R to the category of Dieudonné displays over R. As a consequence of [Z-DD] we obtain:

Theorem 6. *Let R be a $\mathcal{Z}$-ring such that its maximal ideal is nilpotent. Let (A, J, σ) be a Dieudonné frame for R. Then the category of Dieudonné A-windows over R is equivalent to the category of p-divisible groups over R.*

We will now give some examples of Dieudonné frames.

Example 1: Let k be a perfect field of characteristic $p \geq 3$. Let d and s be arbitrary numbers. Then we consider the $\mathcal{Z}$-ring:

$$R = k[T_1, \dots, T_d]/(T_1^s, \dots, T_d^s)$$

We set $A = W(k)[T_1, \dots, T_d]/(T_1^s, \dots, T_d^s)$. The obvious surjection $A \to R$ has as kernel the pd-ideal pA. We define σ as the ^{F}-linear endomorphism of A such that $\sigma(T_i) = T_i^p$. Then we obtain a Dieudonné frame (A, J, σ).

We let s vary, and denote the objects just defined by R_s respectively A_s. We set $R = k[[T_1, \dots, T_d]]$ and $A = W(k)[[T_1, \dots, T_d]]$. Let us define the endomorphism $\sigma : A \to A$ in the same way as before. Then we obtain a frame $A \to R$, which is not a Dieudonné frame. But it is a limit of Dieudonné frames. Indeed, to give a Dieudonné A-window over R is the same thing as to give a Dieudonné A_s-window $\mathcal{M}_s$ over the ring R_s for each number s together with isomorphisms:

$$A_s \otimes_{A_{s+1}} \mathcal{M}_{s+1} \cong \mathcal{M}_s \, .$$

As a result we obtain that the category of p-divisible groups over the ring $R = k[[T_1, \dots, T_d]]$ is equivalent to the category of Dieudonné A-windows over R.

Example 2: Let k be as in the previous example. Let S be a flat and finite local $W(k)$-algebra with residue field k. We set $R = S/p^u S$, where u is some fixed number. We choose a presentation:

$$0 \to \mathfrak{c} \to W(k)[T_1, \dots, T_d] \to S \to 0,$$

such that each T_i for $i = 1, \dots, d$ is mapped to the maximal ideal of S. Let $A_0 \subset W(k)[T_1, \dots, T_d] \otimes \mathbb{Q}$ be the subring generated over $W(k)[T_1, \dots, T_d]$ by all elements of the form $c^n/n!$ for $c \in \mathfrak{c}$, $n \in \mathbb{N}$. We obtain an obvious surjection $A_0 \to S$, whose kernel is a pd-ideal. Let us denote by A the p-adic completion of A_0. Then $A \to S$ is a pd-thickening. Consider the ^{F}-linear endomorphism σ of $W(k)[T_1, \dots, T_d] \otimes \mathbb{Q}$ such that $\sigma(T_i) = T_i^p$ for $i = 1, \dots, d$. It induces an endomorphism of A_0 and A. The pd-thickening $A \to R$ together with the endomorphism σ of A is a Dieudonné frame. As in example 1 this may be used to classify p-divisible groups by Dieudonné A-windows over S.

Example 3: Let $k[[T_1, \dots, T_d]]$ be as in the last example. Let $f_1, \dots, f_r \in k[[T_1, \dots, T_d]]$ be elements such that $f_1, \dots, f_r, T_{r+1}, \dots, T_d$ is a system of parameters of this local ring. We set

$$R = k[[T_1, \dots, T_d]]/(f_1, \dots, f_r, T_{r+1}^s, \dots, T_d^s),$$

where s is some fixed number.

We choose arbitrary liftings $\tilde{f}_1, \dots, \tilde{f}_r \in W(k)[[T_1, \dots, T_d]]$. Consider the following morphism of flat $W(k)[[T_{r+1}, \dots, T_d]]$-algebras:

$$W(k)[[T_1, \dots, T_d]] \to W(k)[[T_1, \dots, T_d]]/(\tilde{f}_1, \dots, \tilde{f}_r).$$

We denote by B the p-adic completion of the pd-hull of this morphism (compare [BBM] lemme 2.3.3). We set $A = B/(T_{r+1}^s, \dots, T_d^s)$. This is a p-adic ring which is a pd-thickening of R. The endomorphism σ of $W(k)[[T_1, \dots, T_d]]$ used in the last example extends to A. This gives a Dieudonné frame for R.

We may use Dieudonné windows over $S = k[[T_1, \dots, T_d]]/(f_1, \dots, f_r)$ with frame $\hat{B} = \varprojlim_s B/(T_{r+1}^s, \dots, T_d^s) \to S$ to classify p-divisible groups over S as in the end of example 1.

The main part of this work was done during a visit at the University of Minnesota. I would like to thank W. Messing who helped me with the definition of a Dieudonné frame and also with technical points in the proofs.

1. The Case of a Formal p-Divisible Group

Definition 1.1. *Let R be a ring such that p is nilpotent in R. A quasiframe for R consists of the following data:*

1) *A p-adic ring A, and a surjective ring homomorphism $\alpha : A \to R$ whose kernel will be denoted by J.*
2) *A ring endomorphism $\sigma : A \to A$ and a σ-linear map $\sigma_1 : J \to A$, such that $\sigma\,|_J = p \cdot \sigma_1$.*
3) *A ring homomorphism $\varkappa : A \to W(R)$.*

We require that the following conditions are satisfied:

(i) *The image of the ideal J in A/pA consists of nilpotent elements.*
(ii) *The following diagram is commutative:*

$$\begin{array}{ccc} A & \xrightarrow{\ \varkappa\ } & W(R) \\ & {\scriptstyle\alpha}\searrow \quad \swarrow{\scriptstyle \mathbf{w}_0} & \\ & R & \end{array}$$

(iii) *The following relations are satisfied:*

$$\varkappa(\sigma(a)) = {}^F\varkappa(a) \quad \textit{for } a \in A;$$
$${}^V\varkappa(\sigma_1(u)) = \varkappa(u) \quad \textit{for } u \in J.$$

We are going to make some comments on this definition.

The ideal J is contained in the radical of A. This is clear.

The first relation of (iii) implies:

$$\alpha(\sigma(a)) = \alpha(a)^p \mod pR.$$

Any finitely generated projective R-module L_0 may be lifted to a finitely generated projective A-module L.

Indeed, for big m the morphism α factors $A \to A/p^m \to R$. Then (i) implies that any element in the kernel of the second arrow is nilpotent. Therefore finitely generated projective R-modules lift with respect to that surjection (compare e.g. H. Bass: Algebraic K-Theory, Benjamin 1968, Chapt. III, 2.10). For the first arrow we obtain liftings by a limit argument.

Example: Let R be a complete discrete valuation ring of characteristic 0 with residue field k. Let C_k be a Cohen ring for k. We choose an endomorphism σ of C_k which lifts the Frobenius endomorphism on k. By the structure theorem for complete local rings R is an Eisenstein extension of C_k, i.e. we have an exact sequence:

$$0 \to E(u)C_k[u] \to C_k[u] \to R \to 0,$$

where $E(u) \in C_k[u]$ is an Eisenstein polynomial of degree e. We denote the image of u in R by π.

We consider the ring $A = C_k[u, \frac{u^{ep}}{p}]$. We extend σ to an endomorphism of A by the rule $\sigma(u) = u^p$. Because $E(u) = u^e + pF(u)$ for some polynomial $F(u)$ the morphism $C_k[u] \to R$ extends to a homomorphism $A \to R$. One checks that σ maps the kernel of this homomorphism to pA. This allows to define σ_1 as in 1.1. Finally we define a morphism

$$\varkappa : A \to W(R),$$

as follows. On the subring C_k this is the composite $C_k \to W(C_k) \to W(R)$, where the first arrow is the Cartier morphism associated to σ. To complete the definition we set:

$$\varkappa(u) = [\pi], \quad \varkappa(\frac{u^{ep}}{p}) = \frac{[\pi^{ep}]}{p}.$$

Here $[\pi]$ denotes the Teichmüller representative in $W(R)$. It is easy to check that $(A \to R, \sigma, \varkappa)$ is a quasiframe.

If we use the fact that the Witt ring $W(R)$ is p-adic ([Z-DFG] Prop.1.3), we obtain that $\varkappa$ extends to the completion $\hat{A} = C_k[[u, \frac{u^{ep}}{p}]]$. Hence we have also a quasiframe if we replace A by $\hat{A}$. Breuil used this quasiframe to prove Fontaines conjecture for p-divisible groups in the case where k is a finite field of characteristic $p > 2$.

Let $\tilde{P}$ be a finitely generated projective A-module, and set $P = R \otimes_A \tilde{P}$. Then any direct summand L of P may be lifted to a direct summand $\tilde{L}$ of $\tilde{P}$, and hence any direct decomposition $P = L \oplus T$ may be lifted to a direct decomposition $\tilde{P} = \tilde{L} \oplus \tilde{T}$.

Definition 1.2. *Let A be a quasiframe over R. An A-window over R consists of the following data:*

1) *A finitely generated projective A-module M.*
2) *A submodule $M_1 \subset M$ which contains JM.*
3) *Two σ-linear maps $\Phi : M \to M$ and $\Phi_1 : M_1 \to M$.*

The following conditions are satisfied:

(i) *M/M_1 is a projective R-module.*

(ii)

$$\Phi(m_1) = p\Phi_1(m_1), \ \text{for } m_1 \in M_1;$$
$$\Phi_1(um) = \sigma_1(u)\Phi(m), \ \text{for } m \in M, u \in J.$$

(iii) *The union of $\Phi_1(M_1)$ and $\Phi(M)$ generates M as an A-module.*

(iv) *Let $\Psi : M \to A \otimes_{\sigma,A} M$ be the unique A-linear map with the property that $\Psi(\Phi_1(m_1)) = 1 \otimes m_1$, for $m_1 \in M_1$, and that $\Psi(\Phi m) = p \otimes m$ for $m \in M$. Then there is an integer N, such that $\Psi^N(M) \subset J \otimes_{\sigma^N,A} M$.*

We will refer to (iv) as the nilpotence condition. We have to show the existence of Ψ in order to justify this definition. This is done exactly as in [Z-DFG] Lemma 1.10:

We may lift the direct summand $M_1/JM \subset M/JM$ to a direct summand L of M, by the remark after Definition 1.1. Let T be a complement to L. Then we obtain decompositions of A-modules:

$$M = L \oplus T, \qquad M_1 = L \oplus JT.$$

A decomposition of M which satisfies the last equation is called normal. We consider the following σ-linear map:

$$\Phi_1 \oplus \Phi : L \oplus T \to M.$$

It follows from (iii) and (ii) of Definition 1.2 that this is a σ-linear isomorphism, i.e. the linearization $(\Phi_1 \oplus \Phi)^\sharp$ is an isomorphism. Then we set:

$$\Psi = (id_L \oplus p\, id_T)((\Phi_1 \oplus \Phi)^{\#})^{-1}.$$

One verifies that with this definition Ψ has the property required by (iv) above. Therefore this Ψ is independent of the chosen normal decomposition.

The obvious example of a quasiframe is $A = W(R), \alpha = \mathbf{w}_0, \varkappa = id$. The pd-ideal J is $I_R = {}^V W(R)$ with the natural divided powers ([Z-DFG] §2.3). The operators Φ, Φ_1, resp. Ψ are in this case denoted by F, V^{-1}, resp $V^\sharp$.

Next we define a functor from the category of A-windows to the category of displays over R. Let us start with an A-window (M, M_1, Φ, Φ_1). Then we obtain a display (P, Q, F, V^{-1}) as follows:

We set $P = W(R) \otimes_{\varkappa,A} M$. The kernel of the map

$$W(R) \otimes_{\varkappa,A} M \xrightarrow{\mathbf{w}_0 \otimes id} R \otimes_{\alpha,A} M \to M/M_1$$

is by definition Q. The map $F : W(R) \otimes_{\varkappa,A} M \to W(R) \otimes_{\varkappa,A} M$ is defined by

$$F(\xi \otimes m) = {}^F\xi \otimes \Phi m.$$

Finally we claim that there is a unique ^{F}-linear map $V^{-1} : Q \to P$, which satisfies the following relations:

$$(1) \qquad \begin{aligned} V^{-1}(\xi \otimes m_1) &= {}^F\xi \otimes \Phi_1 m_1, && \text{for } \xi \in W(R), m_1 \in M_1; \\ V^{-1}({}^V\xi \otimes m) &= \xi \otimes \Phi m, && \text{for } m \in M. \end{aligned}$$

If we show that V^{-1} exists it is obvious that (P, Q, F, V^{-1}) is a display. For the existence of V^{-1} we choose a normal decomposition $M = L \oplus T$ and $M_1 = L \oplus JT$. Then we obtain $P = W(R) \otimes_{\varkappa,A} L \oplus W(R) \otimes_{\varkappa,A} T$ and $Q = W(R) \otimes_{\varkappa,A} L \oplus I_R \otimes_{\varkappa,A} T$. We define V^{-1} on the first direct summand of Q by the first equation of (1) and on the second direct summand by the second equation of (1).

The first equation of (ii) shows that this definition is possible. We have to verify that with this definition the properties (1) are satisfied. For this it is enough to show that the first equation of (1) holds for $m_1 \in JT$ and the second equation of (1) holds for $m \in L$. Indeed, let $u \in J$ and $t \in T$. Then we obtain using (ii):

$$\begin{aligned} V^{-1}(\xi \otimes ut) &= V^{-1}(\xi\varkappa(u) \otimes t) = V^{-1}(\xi^V \varkappa(\sigma_1(u)) \otimes t) \\ &= V^{-1}({}^V({}^F\xi\varkappa(\sigma_1(u))) \otimes t) = {}^F\xi\varkappa(\sigma_1(u)) \otimes \Phi t \\ &= {}^F\xi \otimes \sigma_1(u)\Phi t = {}^F\xi \otimes \Phi_1(ut). \end{aligned}$$

Finally we check the second equation (1) for $m = l \in L$:

$$V^{-1}({}^V\xi \otimes l) = {}^{FV}\xi \otimes \Phi_1(l) = p\xi \otimes \Phi_1(l) = \xi \otimes \Phi(l).$$

Proposition 1.3. *The construction above provides a functor from the category of A-windows over R to the category of displays over R.*

Hence we have also a functor from the category of A-windows to the category of formal p-divisible groups.

We will now give conditions for the quasiframe A, which assure that the functor from A-windows to displays is an equivalence of categories.

Definition 1.4. *A frame over R consists of the following data:*

1) *A torsion-free (as an abelian group) p-adic ring A.*
2) *A surjective homomorphism $A \xrightarrow{\alpha} R$, whose kernel will be denoted by J.*
3) *An endomorphism $\sigma : A \to A$.*

We require the following properties

(i) *σ lifts the Frobenius on A/pA.*
(ii) *The ideal J has divided powers.*

We associate to any A-frame an A-quasiframe as follows. According to Cartier there is a morphism:

$$\delta : A \to W(A),$$

which is uniquely determined by the property that

$$\mathbf{w}_n(\delta(a)) = \sigma^n(a).$$

The divided powers on J define an inclusion (see (17) below) $J \to W(A)$, whose image will be denoted by $\breve{J} \subset W(A)$. Clearly $\breve{J}$ is a pd-ideal and so is $\breve{J} \oplus I_A \subset W(A)$. The ideal $\mathfrak{c} = \breve{J} \oplus I_A$ is the kernel of $W(A) \to R$. It follows that δ is a morphism of pd-extensions (since we have no p-torsion):

$$\begin{array}{ccc} A & \longrightarrow & W(A) \\ & \searrow \quad \swarrow & \\ & R & \end{array}$$

We extend the ^{F}-linear homomorphism ${}^{V^{-1}} : I_A \to W(A)$ to an ^{F}-linear homomorphism ${}^{V^{-1}} : \mathfrak{c} \to W(A)$, by setting ${}^{V^{-1}}\breve{J} = 0$.

We also introduce the σ-linear map:

$$\sigma_1 = \frac{1}{p}\sigma : J \to A.$$

Then we have the relation:

$$\delta(\sigma_1(a)) = {}^{V^{-1}}\delta(a), \qquad a \in J. \tag{2}$$

Since $W(A)$ has no p-torsion we may multiply equation (2) by p to verify it. Since $p\ {}^{V^{-1}} = {}^F$ holds on the ideal $\mathfrak{c}$, we obtain:

$$\delta(\sigma(a)) = {}^F\delta(a). \tag{3}$$

But this follows directly from the definition of the Cartier morphism.

We define $\varkappa$ as the composition

$$\varkappa : A \xrightarrow{\delta} W(A) \xrightarrow{W(\alpha)} W(R).$$

Since $W(\alpha)(\breve{J}) = 0$ it follows that $W(\alpha)$ is a morphism of pd-thickenings of R.

$$\begin{array}{ccc} W(\alpha) : W(A) & \longrightarrow & W(R) \\ & \searrow \quad \swarrow & \\ & R & \end{array}$$

Proposition 1.5. *The data $\alpha : A \to R$, σ, σ_1, $\varkappa$ define a quasiframe over R.*

Proof: Since J has divided powers we have $u^p = p\alpha_p(u) \equiv 0 \mod pA$ for $u \in J$ and therefore (i) is satisfied. The next condition (ii) is trivially satisfied. The first relation of (iii) is satisfied because of (3). From (2) we conclude:

$$\varkappa(\sigma_1(u)) = {}^{V^{-1}}\varkappa(u) \text{ for } u \in J.$$

We note here that $VV^{-1} \neq id$ on the ideal $\mathfrak{c}$, so that the desired relation doesn't follow immediately. But since $\varkappa(u) \in W(\alpha)(\mathfrak{c}) \subset I_R$ we have indeed ${}^{VV^{-1}}\varkappa(u) = \varkappa(u)$. This proves the proposition. *Q.E.D.*

We note that the definition of a window greatly simplifies if A is a frame. Indeed, since A has no p-torsion the operator Φ_1 is uniquely determined by $p\Phi_1 = \Phi$. The second relation (ii) of Definition 1.2 follows automatically from the fact that Φ is σ-linear. Hence Definition 3 of an A-window over R is equivalent to Definition 1.2.

In practice we have frames which satisfy the additional condition that $\sigma(J)$ generates pA as an A-module. With this assumption the condition (ii) of Definition 2 is equivalent to the requirement that $\Phi(M_1)$ generates pM as an A-module.

We can also define a frame for a p-adic ring R. If (A, J, σ) is such a frame then $(A, J + p^m A, \sigma)$ is a frame for $R/p^m R$ for each number m. Then we define a window exactly as above, but we require the nilpotence condition (iii) only modulo pR or equivalently modulo $p^m R$ for any m. With this definition a window over R is the same thing as a compatible system of windows $\{\mathcal{M}_m\}$, where $\mathcal{M}_m$ is a window over $R/p^m R$, such that $\mathcal{M}_m$ is obtained from $\mathcal{M}_{m+1}$ by base change.

Let us consider an A-window $\mathcal{M} = (M, M_1, \Phi)$. On M_1 we will set $\Phi_1 = \frac{1}{p}\Phi$. We have associated a display to $\mathcal{M}$ by setting:

$$(4) \qquad \begin{aligned} P &= W(R) \otimes_{\varkappa, A} M, \\ Q &= \operatorname{Ker} W(R) \otimes_{\varkappa, A} M \to M/M_1\,, \\ F(\xi \otimes x) &= {}^F\xi \otimes \Phi(x), \quad \text{for } \xi \in W(R), x \in M. \end{aligned}$$

Finally the operator $V^{-1} : Q \to P$ is uniquely determined by the relations:

$$(5) \qquad \begin{aligned} V^{-1}(\xi \otimes m_1) &= {}^F\xi \otimes \Phi_1 m_1, && \text{for } m_1 \in M_1\,, \\ V^{-1}({}^V\xi \otimes m) &= \xi \otimes \Phi m, && \text{for } m \in M. \end{aligned}$$

Hence we have functors:

$$(A\text{-windows}) \xrightarrow{\mathrm{Dsp}} (\text{displays}/R) \xrightarrow{\mathrm{BT}} (\text{formal } p\text{-divisible groups } /R).$$

Here BT is the functor from [Z-DFG] chapt. 3. The composition with Dsp will also be denoted by BT.

Theorem 1.6. *If A is a frame the functor* Dsp *is an equivalence of categories.*

Proof: We construct a quasiinverse functor. Let $\mathcal{P} = (P, Q, F, V^{-1})$ be a display over R. We set $\overline{R} = R/pR$ and we denote by $\overline{\mathcal{P}} = \mathcal{P}_{\overline{R}}$ the display obtained by base change. Then we will associate to it an A-window as follows:

In [Z-DFG] Definition 2.6 we have associated to the display $\mathcal{P}$ a crystal $\mathcal{D}_{\mathcal{P}}$ which was called the Dieudonné crystal. It is defined for pd-thickenings $S \to R$, such that p is nilpotent in S. For the pd-thickening $A \to R$ we set:

$$\mathcal{D}_{\mathcal{P}}(A) = \varprojlim \mathcal{D}_{\mathcal{P}}(A/p^n A).$$

This makes sense because for big n we have a pd-thickening $A/p^n A \to R$.

We define the finitely generated projective A-module M:

$$M = \mathcal{D}_{\mathcal{P}}(A) = \mathcal{D}_{\overline{\mathcal{P}}}(A).$$

Then the maps $F^{\#} : \overline{\mathcal{P}}^{(p)} \to \overline{\mathcal{P}}$ respectively $V^{\#} : \overline{\mathcal{P}} \to \overline{\mathcal{P}}^{(p)}$ induce A-linear maps by evaluating the Dieudonné crystal:

$$\begin{aligned} \Phi^{\#} &= \mathcal{D}(F^{\#}) : \quad A \otimes_{\sigma,A} M \to M, \\ \Psi^{\#} &= \mathcal{D}(V^{\#}) : \quad M \to A \otimes_{\sigma,A} M, \end{aligned} \tag{6}$$

such that $\Phi^{\#} \circ \Psi^{\#} = p$ and $\Psi^{\#} \circ \Phi^{\#} = p$.

Since $\varkappa$ is a morphism of pd-extensions we obtain a canonical isomorphism:

$$W(R) \otimes_{\varkappa,A} M \cong P \cong \mathcal{D}_{\mathcal{P}}(W(R)). \tag{7}$$

The last isomorphism follows from [Z-DFG] Proposition 2.12. Here again we set:

$$\mathcal{D}_{\mathcal{P}}(W(R)) = \varprojlim \mathcal{D}_{\mathcal{P}}(W(R)/p^n W(R)).$$

The isomorphism (7) takes the maps $\Phi^{\#}$ respectively $\Psi^{\#}$ to the maps $F^{\#} : W(R) \otimes_{F,W(R)} P \to P$ respectively $V^{\#} : P \to W(R) \otimes_{F,W(R)} P$, which are associated to the display $\mathcal{P}$ by [Z-DFG] Corollary 2.17.

We define Φ to be the σ-linear map associated to $\Phi^{\#}$. We define M_1 as the kernel of the natural map:

$$M \to P/Q,$$

which is induced from the map $\mathcal{D}_{\mathcal{P}}(A) \to \mathcal{D}_{\mathcal{P}}(R) \cong P/I_R P$.

Our next aim is to show that $\Phi(M_1) \subset pM$. We note that for any display $\mathcal{P}$ there is an exact sequence:

$$P \xrightarrow{V^{\#}} W(R) \otimes_{F,W(R)} P \to W(R) \otimes_{F,W(R)} P/Q \to 0. \tag{8}$$

Indeed, P is generated by elements of the form $\xi \cdot V^{-1} y$, where $y \in Q$. Hence the image of $V^{\#}$ is generated by $V^{\#}(\xi V^{-1} y) = \xi \otimes y$, for $y \in Q$. This shows the exactness of (8).

We consider the $\overline{R}$-module:

$$H = \overline{R} \otimes_A M = \overline{R} \otimes_{W(R)} P.$$

The image of Q or equivalently M_1 is a direct summand $H_1 \subset H$. If we tensor the exact sequence (8) with $\overline{R} \otimes_{W(R)}$ we obtain an exact sequence:

$$H \xrightarrow{\overline{V}^{\#}} \overline{R} \otimes_{Frob,\overline{R}} H \to \overline{R} \otimes_{Frob,\overline{R}} H/H_1 \to 0.$$

The map $\overline{V}^{\#}$ is also obtained if we tensorize $\Psi^{\#} : M \to A \otimes_{\sigma,A} M$ with $\overline{R} \otimes_A$. Since the kernel of the map $A \to \overline{R}$ is $J + pA$ it follows that

$$\text{Image } (A \otimes_{\sigma,A} M_1) \subset \Psi^{\#} M + J \otimes_{\sigma,A} M + pA \otimes_{\sigma,A} M, \tag{9}$$

where the inclusion takes place in $A \otimes_{\sigma,A} M$. Since $\Phi^{\#}\Psi^{\#} = p\, id_M$, and since $\sigma(J) \subset pA$, we obtain

$$\Phi(M_1) \subset pM,$$

if we apply $\Phi^{\#}$ to (9). Hence we have defined a map $\Phi_1 : M_1 \to M$. We note that

$$\Phi_1(um) = \sigma_1(u)\Phi m, \qquad \text{for } u \in J.$$

This equation is verified by multiplying it with p.

Next we describe the $\mathcal{P}$-triple over the pd-extension $A \to R$ in terms of $\mathcal{M} = (M, M_1, \Phi, \Phi_1)$. Let us denote this triple by $(\widetilde{P}, F, V^{-1})$. Because $\widetilde{P}$ is the value of the Dieudonné crystal $\mathcal{D}_{\mathcal{P}}$ evaluated at the pd-thickening $W(A) \to R$, and since moreover δ is a morphism of pd-thickenings of R, we find a canonical isomorphism:

$$W(A) \otimes_{\delta,A} M \simeq \widetilde{P}. \tag{10}$$

The maps Φ on M and F on $\widetilde{P}$ are induced by applying the Dieudonné crystal to $F^{\#} : \overline{\mathcal{P}}^{(p)} \to \overline{\mathcal{P}}$. Since (10) is canonical this shows that

$${}^{F}\xi \otimes \Phi m = F(\xi \otimes m), \quad \text{for } m \in M,\ \xi \in W(A). \tag{11}$$

The map $F : \widetilde{P} \to \widetilde{P}$ already determines V^{-1} uniquely because $W(A)$ has no p-torsion. The domain of definition $\hat{Q}$ of V^{-1} is the kernel of the map:

$$W(A) \otimes_{\delta,A} M \to M/M_1,$$

induced by $W(A) \xrightarrow{\mathbf{w}_0} A \xrightarrow{\alpha} R$ on the first factor. The operator V^{-1} is determined on this kernel by the following equations:

$$\begin{aligned} V^{-1}(u \otimes m) &= {}^{V^{-1}}u \otimes \Phi m, \quad \text{for } u \in \mathfrak{c},\ m \in M; \\ V^{-1}(\xi \otimes m_1) &= {}^{F}\xi \otimes \Phi_1 m_1, \quad \text{for } \xi \in W(A),\ m_1 \in M_1. \end{aligned} \tag{12}$$

These equations are verified by multiplying them with p. Hence we have obtained a full description of the triple $(\widetilde{P}, F, V^{-1})$ in terms of (M, M_1, Φ, Φ_1).

We are now able to show the commutativity of the diagram:

$$\begin{array}{ccc} M_1 & \xrightarrow{\Phi_1} & M \\ \downarrow & & \downarrow \\ Q & \xrightarrow{V^{-1}} & P, \end{array} \tag{13}$$

where $M \to P$ is induced by (7) and $M_1 \to Q$ is its restriction to M_1. Indeed, this diagram is the composition of two commutative diagrams

$$\begin{array}{ccc} M_1 & \xrightarrow{\Phi_1} & M \\ \downarrow & & \downarrow \\ \hat{Q} & \xrightarrow{V^{-1}} & \tilde{P} \\ \downarrow & & \downarrow \\ Q & \xrightarrow{V^{-1}} & P. \end{array}$$

The commutativity of the upper square was just proved. The point was that there is no p-torsion in this diagram, and therefore we could multiply the upper and lower horizontal arrow by p to verify it. The lower square is commutative by the definition of a triple.

Now we will show that (M, M_1, Φ, Φ_1) is a window. The only condition we have to verify is that the union of $\Phi_1 M_1$ and ΦM generates the A-module M. For this we choose a normal decomposition of M:

$$M = L_0 \oplus T_0, \qquad M_1 = L_0 \oplus JT_0.$$

We consider the A-linear map:

$$U_0^\# = \Phi_1^\# \oplus \Phi^\# : A \otimes_{\sigma,A} L_0 \oplus A \otimes_{\sigma,A} T_0 \to M. \tag{14}$$

It suffices to show that this is an isomorphism. This is in fact equivalent to the condition (ii) of Definition 2.

We set $L = W(R) \otimes_{\varkappa,A} L_0$ and $T = W(R) \otimes_{\varkappa,A} T$. By the commutative diagram (13) we see that $U^\# = W(R) \otimes_{\varkappa,A} U_0^\#$ is the map:

$$(V^{-1})^\# \oplus F^\# : W(R) \otimes_{F,W(R)} L \oplus W(R) \otimes_{F,W(R)} T \to P. \tag{15}$$

Since $P = L \oplus T$ is a normal decomposition the map (15) is an isomorphism by the definition of a display. The equality of determinants of endomorphisms of $R \otimes_{\mathbf{w}_0, W(R)} P = R \otimes_A M$:

$$\det(R \otimes_{\mathbf{w}_0, W(R)} U^\#) = \det(R \otimes_{\alpha,A} U_0^\#),$$

shows that the right hand side is a unit in R. From the assumption that J is in the radical of A it follows that $\det U_0^\#$ is a unit too. Hence $\mathcal{M} = (M, M_1, \Phi, \Phi_1)$ is an A-window.

Therefore we have defined a functor:

$$\mathrm{Win} : (\text{displays}/R) \to (A\text{-windows}).$$

It is obvious from our considerations that this functor is quasiinverse to Dsp. Indeed assume that either $\mathcal{P} = \mathrm{Dsp}\,\mathcal{M}$ or that $\mathcal{M} = \mathrm{Win}\,\mathcal{P}$. Then we have defined a canonical isomorphism $P \simeq W(R) \otimes_{\varkappa, A} M$, which is compatible with F and Φ resp. V^{-1} and Φ_1. *Q.E.D.*

Example: Let R be a ring, such that $pR = 0$, and assume that R admits a p-basis in the sense of [BM]. This means that R regarded as an R-module via the Frobenius $Frob : R \to R$ is a free R-module. It is shown in [BM] that there is a torsion-free p-adic ring A, such that A/pA is isomorphic to R. Moreover the Frobenius endomorphism of R lifts to an endomorphism σ of A. If we set $J = pA$ we obtain a frame (A, J, σ) over R. Clearly $(A, p^m A, \sigma)$ is also a frame over $A/p^m A$ for any number m. By [Z-DFG] we obtain:

Corollary 1.7. *Let R be an excellent ring. Then the category of formal p-divisible groups over R is equivalent to the category of A-windows over R.*

Example: Let S be a torsion-free p-adic ring. We set $R = S/p^m S$ for some number m. Then we construct a frame for R as follows:

Let I be an index set. We denote by $C = \mathbb{Z}_p\{X_i\}_{i \in I}$ the ring of restricted power series, i.e. the p-adic completion of the polynomial ring $\mathbb{Z}_p[X_i]_{i\in I}$. We consider the endomorphism σ of C given by $\sigma(X_i) = X_i^p$. Assume we are given an epimorphism of rings:

$$C \to S,$$

whose kernel will be denoted by $\mathfrak{c}$. We define B_0 as the subring of $C \otimes \mathbb{Q}$ which is generated by C and by the elements $\frac{c^n}{n!}$ for $c \in \mathfrak{c}$. We obtain a surjection $B_0 \to S$ whose kernel J_0 is generated by the elements $\frac{c^n}{n!}$ for $c \in \mathfrak{c}$. Then J_0 is a pd-ideal and consequently $J_0 + pB_0$ is a pd-ideal too. Since $\sigma(\mathfrak{c}) \subset J_0 + pB_0$, we obtain $\sigma(J_0) \subset J_0 + pB_0$. Therefore σ induces an endomorphism of B_0. We claim that $\sigma(b) = b^p \mod pB_0$ for $b \in B_0$. It is enough to verify this for generators of B_0. For $b \in C$ the congruence is trivial. For $c \in \mathfrak{c}$ we have to verify that:

$$\sigma(\frac{c^n}{n!}) = \left(\frac{c^n}{n!}\right)^p = 0 \mod pB_0,$$

where the last congruence follows since we have divided powers on J_0. We set $\sigma(c) = c^p + pu$ for $u \in C$. Then we obtain:

$$\sigma(\frac{c^n}{n!}) = \sum_{i+j=n} \frac{c^{pi}}{i!}\frac{p^j}{j!}u^j = 0 \mod pB_0.$$

We denote by B the p-adic completion of B_0. By continuity σ extends to an endomorphism of B, and the morphism $B_0 \to S$ gives a pd-thickening of $B \to S$, whose kernel will be denoted by J. Then $(B, p^m + J, \sigma)$ is a frame for

$S/p^m S$, and (B, J, σ) is a frame for the p-adic ring S (compare the remarks after proposition 1.5). Then we obtain:

Proposition 1.8. *Let S/pS be an excellent ring. The category of p-divisible groups over S whose reduction modulo p has no étale part is equivalent to the category of B-windows over S.*

2. Variants of the Witt Ring

As a preparation for the next section we will make some comments on the Witt ring and in particular on the Cartier morphism.

In [Z-DFG] §1.4 we applied the Witt ring functor W to a pd-thickening $S \to R$ of $\mathbb{Z}_{(p)}$-algebras, whose kernel will be denoted by $\mathfrak{a}$. The divided powers are uniquely determined by a function $\alpha_p : \mathfrak{a} \to \mathfrak{a}$, which defines a^p divided by p. The divided Witt polynomials $\mathbf{w}_n/p^n$ make sense on $W(\mathfrak{a})$ and they define an isomorphism of additive groups:

$$\log : W(\mathfrak{a}) \to \prod_{\mathbb{N}} \mathfrak{a}. \tag{16}$$

An element of the right hand side is denoted by $[a_0, a_1, \ldots, a_n, \ldots]$. We call this the logarithmic coordinates on $W(\mathfrak{a})$. For $a \in \mathfrak{a}$ we set:

$$\exp a = \log^{-1}[a, 0, \ldots, 0, \ldots]. \tag{17}$$

Then $\exp : \mathfrak{a} \to W(a)$ is an injective ring homomorphism. If we regard $\mathfrak{a}$ as a $W(S)$-module via restriction of scalars $\mathbf{w}_0 : W(S) \to S$ the morphism exp becomes a $W(S)$-module homomorphism. We often say that the divided powers define $\mathfrak{a}$ as an ideal of $W(S)$. We write $\mathfrak{a} \subset W(S)$, without mentioning that this inclusion is defined by exp. A basic property is that ${}^F\mathfrak{a} = 0$.

We will consider commutative rings without unit. These are often denoted by $\mathcal{N}, \mathcal{M}$ etc.

Definition 2.1. *We call $\mathcal{N}$ pointwise nilpotent, if each element $n \in \mathcal{N}$ is nilpotent. We call $\mathcal{N}$ bounded nilpotent, if there is a number $k \in \mathbb{N}$, such that $n^k = 0$ for all $n \in \mathcal{N}$. If there is a number k such that $\mathcal{N}^k = 0$, we call $\mathcal{N}$ nilpotent.*

Let $\hat{W}(\mathcal{N}) \subset W(\mathcal{N})$ be the subset of Witt vectors with only finitely many non-zero components. If $\mathcal{N}$ is pointwise nilpotent $\hat{W}(\mathcal{N})$ is an ideal in $W(\mathcal{N})$. Moreover $\hat{W}(\mathcal{N})$ is again pointwise nilpotent.

Lemma 2.2. *Assume that $\mathcal{N}$ is bounded nilpotent and annihilated by some power of p. Then $W(\mathcal{N})$ is bounded nilpotent and annihilated by some power of p.*

Proof: By considering the ideals $p^k\mathcal{N}$ one reduces to the case where $p\mathcal{N} = 0$. Then one finds a number M, such that ${}^{F^M}W(\mathcal{N}) = 0$.

Let us consider the set S of all Witt vectors in $W(\mathcal{N})$, which have the form:

$$[n_0] + {}^{V}[n_1] + \ldots + {}^{V^{M-1}}[n_{M-1}],$$

where $n_i \in \mathcal{N}$ for $i = 1, \ldots, M-1$ are arbitrary elements. It is clear that there is a number N, such that $\xi^N = 0$ for each $\xi \in S$.

Next we note that each $\xi \in W(\mathcal{N})$ has a unique expression:

$$\xi = \xi_0 + {}^{V^M}\xi_1 + \ldots + {}^{V^{kM}}\xi_k + \ldots,$$

where $\xi_k \in S$. It follows from the definition of M that the summands in the last sum are pairwise orthogonal. Hence we obtain $\xi^N = 0$. *Q.E.D.*

We call $\mathcal{N}$ modulo p bounded nilpotent, if $\mathcal{N}/p\mathcal{N}$ is bounded nilpotent. In this case $\mathcal{N}/p^k\mathcal{N}$ is bounded nilpotent too.

Proposition 2.3. *Assume that $\mathcal{N}$ is modulo p bounded nilpotent. Then the p-adic topology on $W(\mathcal{N})$ coincides with the linear topology, which has as a fundamental system of neighbourhoods of zero the subgroups $W(p^k\mathcal{N})$, where k runs through the natural numbers.*

Proof: Fix a number n. Then by the last lemma there is a number r, such that p^r annihilates $W(\mathcal{N}/p^n\mathcal{N})$. This shows:

$$p^r W(\mathcal{N}) \subset W(p^n\mathcal{N}).$$

Hence the p-adic topology is finer. On the other hand the natural divided powers on $p^r\mathcal{N}$, for some number $r \geq 1$ induce an isomorphism of additive groups:

$$W(p^r\mathcal{N}) \xrightarrow[\sim]{\log} \prod p^r\mathcal{N}.$$

This implies $p^m W(p\mathcal{N}) = W(p^{m+1}\mathcal{N})$. Therefore we obtain:

$$W(p^n\mathcal{N}) = p^{n-1}W(p\mathcal{N}) \subset p^{n-1}W(\mathcal{N}).$$

Q.E.D.

If R is a p-adic ring then the ring $W(R)$ is p-adic too (see: [Z-DFG] Prop.1.3). Hence $W(\mathcal{N})$ is a p-adic ring in the situation of the last proposition.

We extend the functor $\hat{W}$ to rings $\mathcal{N}$ which are p-adic and modulo p bounded nilpotent:

$$\hat{W}(\mathcal{N}) = \varprojlim_r \hat{W}(\mathcal{N}/p^r\mathcal{N}) \subset W(\mathcal{N}).$$

Since this may contradict our old definition of $\hat{W}$, we say that we take $\hat{W}$ in the topological sense. Obviously $\hat{W}(\mathcal{N})$ is a closed subgroup for the topology on $W(\mathcal{N})$ given by the subgroups $W(p^r\mathcal{N})$ and hence also for the p-adic topology. By definition we have:

$$\hat{W}(\mathcal{N}) \cap W(p^r\mathcal{N}) = \hat{W}(p^r\mathcal{N}).$$

Let $1 \leq u < r$ be number. Then the divided powers on $p^u\mathcal{N}/p^r\mathcal{N}$ provide a homomorphism:

$$\log : \hat{W}(p^u\mathcal{N}/p^r\mathcal{N}) \to \oplus p^u\mathcal{N}/p^r\mathcal{N}.$$

This is an isomorphism, if the divided powers are nilpotent in the sense of [Z-DFG] (3.4) (i.e. pointwise), which is the case if $p \geq 3$ or if $u \geq 2$. Hence with these assumptions we obtain an isomorphism:

$$\log : \hat{W}(p^u\mathcal{N}) \to \hat{\oplus} p^u\mathcal{N},$$

where the right hand side denotes the set of those elements of $\prod p^u\mathcal{N}$, whose components converge p-adically to zero. Using this we can modify the proof of the last proposition to obtain:

Proposition 2.4. *Assume that $\mathcal{N}$ is p-adic and modulo p bounded nilpotent. Let us consider $\hat{W}(\mathcal{N})$ in the topological sense. Then the p-adic topology on $\hat{W}(\mathcal{N})$ coincides with the linear topology which has as a fundamental system of neighbourhoods of zero the subgroups $\hat{W}(p^r\mathcal{N})$.*

Moreover $\hat{W}(\mathcal{N}) \subset W(\mathcal{N})$ is a closed subgroup for the p-adic topology on $W(\mathcal{N})$, and the p-adic topology on $\hat{W}(\mathcal{N})$ is induced by the p-adic topology on $W(\mathcal{N})$. In particular the group $\hat{W}(\mathcal{N})$ is p-adic.

Next we consider the Cartier morphism:

$$\Delta : W(\mathcal{N}) \to W(W(\mathcal{N})). \tag{18}$$

It is the unique functorial ring homomorphism, such that:

$$\mathbf{w}_n(\Delta\xi) = {}^{F^n}\xi, \quad \text{for } \xi \in W(\mathcal{N}).$$

Assume that $\mathcal{N}$ is pointwise nilpotent. Then $\hat{W}(\mathcal{N})$ is pointwise nilpotent too. Therefore $\hat{W}(\hat{W}(\mathcal{N}))$ makes sense. We have inclusions

$$\hat{W}(\hat{W}(\mathcal{N})) \subset W(\hat{W}(\mathcal{N})) \subset W(W(\mathcal{N})).$$

Lemma 2.5. *Assume that $\mathcal{N}$ is pointwise nilpotent. Then the Cartier morphism (18) induces a morphism*

$$\Delta : \hat{W}(\mathcal{N}) \to \hat{W}(\hat{W}(\mathcal{N})).$$

Proof: We begin to verify, that $\Delta(\hat{W}(\mathcal{N})) \subset W(\hat{W}(\mathcal{N}))$. By functoriality it suffices to prove this if $\mathcal{N}$ and hence $W(\mathcal{N})$ has no p-torsion. Indeed consider any element $\underline{n} = (n_0, n_1, \ldots) \in \hat{W}(\mathcal{N})$. Then we have for suitable numbers r and s a homomorphism of pointwise nilpotent algebras:

$$\mathfrak{a} = (\mathbb{Z}[X_0, \ldots, X_s]/(X_0^r, \ldots, X_s^r))^+ \to \mathcal{N},$$

such that X_i is mapped to n_i and $n_i = 0$ for $i > s$. Here the index $+$ denotes the polynomials without constant term. This reduces our assertion to the case, where $\mathcal{N}$ is the torsion-free algebra $\mathfrak{a}$.

The Frobenius endomorphism on $\hat{W}(\mathcal{N})$ satisfies the congruence:

$$^F\xi = \xi^p \bmod p\hat{W}(\mathcal{N}).$$

Hence there is a Cartier morphism:

$$\hat{W}(\mathcal{N}) \to W(\hat{W}(\mathcal{N})).$$

Since this coincides with Δ our claim above is shown.

Let $\hat{I}_{\mathcal{N}} = {}^V\hat{W}(\mathcal{N})$. This ideal is equipped with divided powers:

$$\alpha_p({}^V\xi) = p^{p-2}\,{}^V(\xi^p).$$

These divided powers are pointwise nilpotent in the sense of [Z-DFG] (3.4) because $\hat{W}(\mathcal{N})$ is pointwise nilpotent. Hence they define an isomorphism:

$$\log : \hat{W}(\hat{I}_{\mathcal{N}}) \to \bigoplus_{\mathbb{N}} \hat{I}_{\mathcal{N}}. \tag{19}$$

Let us consider an element $\xi \in \hat{W}(\mathcal{N})$, such that $\Delta(\xi) \in \hat{W}(\hat{W}(\mathcal{N}))$. We claim that this implies $\Delta({}^V\xi) \in \hat{W}(\hat{W}(\mathcal{N}))$.

Indeed by [DFG] Lemma 2.11 we have the formula in $\hat{W}(\hat{W}(\mathcal{N}))$:

$$\Delta({}^V\xi) = {}^V\Delta(\xi) + \exp\ {}^V\xi$$

in the notation of (16). This shows our claim. Finally we know that Δ acts on a Teichmüller representative $[n] \in W(\mathcal{N})$ as follows:

$$\Delta([n]) = [[n]] \in \hat{W}(\mathcal{N}).$$

Hence we have $\Delta({}^{V^r}[n]) \in \hat{W}(\mathcal{N})$ by our claim above. This finishes the proof of the lemma. *Q.E.D.*

Let us consider a ring $\mathcal{N}$ which is p-adic, and modulo p bounded nilpotent. Then the ring $\hat{W}(\mathcal{N})$ taken in the topological sense has exactly the same properties. By proposition 2.4 we obtain:

$$\begin{aligned}\hat{W}(\hat{W}(\mathcal{N})) &= \varprojlim_m \hat{W}(\hat{W}(\mathcal{N})/p^m\hat{W}(\mathcal{N}))\\ &= \varprojlim_m \hat{W}(\hat{W}(\mathcal{N}/p^m\mathcal{N})).\end{aligned}$$

Again we find that $\hat{W}(\hat{W}(\mathcal{N})) \subset W(W(\mathcal{N}))$ and that the Cartier morphism induces a ring homomorphism:

$$\Delta : \hat{W}(\mathcal{N}) \to \hat{W}(\hat{W}(\mathcal{N})). \tag{20}$$

Let R be a local ring with maximal ideal $\mathfrak{m}$ and with perfect residue field of characteristic $p \geq 3$. We assume that there is a number M, such that $x^M = 0$ for any $x \in \mathfrak{m}$. Let us denote the category of these rings R by $\mathcal{Z}$.

Let k be the residue field of R. The natural map $W(R) \to W(k)$ has a Teichmüller section [Z-DD] §2. Therefore we may write canonically:

$$W(R) = W(\mathfrak{m}) \oplus W(k).$$

We note that $\hat{W}(\mathfrak{m})$ is a $W(R)$-submodule of $W(\mathfrak{m})$.

We define a subring $\hat{W}(R) \subset W(R)$:

$$\hat{W}(R) = W(k) \oplus \hat{W}(\mathfrak{m}). \tag{21}$$

We will also consider p-adic local rings A, such that $A/p^n A$ is in $\mathcal{Z}$ for any number n. We call them $\hat{\mathcal{Z}}$-rings. The ring $\hat{W}(R)$ is an example of a $\hat{\mathcal{Z}}$-ring.

We define:

$$\hat{W}(A) = \varprojlim_n \hat{W}(A/p^n A).$$

The maximal ideal $\mathfrak{n}$ of A is modulo p bounded nilpotent. We have a decomposition which is similar to (21):

$$\hat{W}(A) = W(A/\mathfrak{n}) \oplus \hat{W}(\mathfrak{a}). \tag{22}$$

The ring $\hat{W}(A)$ is a closed subring of $W(A)$ for the p-adic topology, and moreover the p-adic topology on $\hat{W}(A)$ coincides with the topology induced by the p-adic topology on $W(A)$. The ring $\hat{W}(A)$ is again a $\hat{\mathcal{Z}}$-ring. From (20) we deduce that the Cartier morphism induces a map:

$$\Delta : \hat{W}(A) \to \hat{W}(\hat{W}(A)). \tag{23}$$

Consider a surjection $S \to R$ in $\mathcal{Z}$, and assume that its kernel $\mathfrak{a}$ is equipped with divided powers, which are compatible with the natural divided powers on pS. We will denote the inverse image of $\bigoplus_{\mathbb{N}} \mathfrak{a}$ by the homomorphism (16) by $\widetilde{W}(\mathfrak{a})$. By [Z-DFG] (3.4) we have $\hat{W}(\mathfrak{a}) \subset \widetilde{W}(\mathfrak{a})$, and this inclusion is an equality, if the divided powers are pointwise nilpotent. We will denote by $\widetilde{W}(S)$ the subring of $W(S)$ generated by $\widetilde{W}(\mathfrak{a})$ and $\hat{W}(S)$. We note that $\hat{W}(S) = \widetilde{W}(S)$ if the divided powers are pointwise nilpotent. In general we have an exact sequence:

$$0 \to \widetilde{W}(\mathfrak{a}) \to \widetilde{W}(S) \to \hat{W}(R) \to 0.$$

The composite $\widetilde{W}(S) \to \hat{W}(R) \xrightarrow{\mathbf{w}_0} R$ is again a pd-thickening, if we define the pd-structure on the kernel $\mathfrak{c}$ of $\widetilde{W}(S) \to R$ as follows:

If we embed $\mathfrak{a} \subset \widetilde{W}(S)$ by exp (16), we find a direct decomposition:

$$\mathfrak{c} = \mathfrak{a} \oplus {}^V\widetilde{W}(S). \tag{24}$$

Since there are divided powers defined on each direct summand of (24), we obtain divided powers on $\mathfrak{c}$. If the natural number n is big enough we obtain a pd-thickening in $\mathcal{Z}$:

$$\widetilde{W}(S)/p^n\widetilde{W}(S) \to R. \tag{25}$$

Remark: We verify that the divided powers on $\mathfrak{c}$ and on $p\widetilde{W}(S)$ are compatible. We will even show that the same is true if we replace $\widetilde{W}(S)$ by $W(S)$. Let us denote by $\exp a$ the image of an element $a \in \mathfrak{a}$ by the embedding $\mathfrak{a} \subset W(S)$. Assume we are given an equation:

$$p\xi = \exp a + {}^V\eta, \qquad a \in \mathfrak{a},\ \xi, \eta \in W(S). \tag{26}$$

Let $\alpha_p : \mathfrak{a} \to \mathfrak{a}$ be the function given by the divided powers (see above (16)). Since $\exp a\ {}^V\eta = 0$ our assertion says exactly that:

$$p^{p-1}\xi^p = \exp \alpha_p(a) + p^{p-2}\ {}^V(\eta^p). \tag{27}$$

We set $\xi_0 = \mathbf{w}_0(\xi)$. From (26) we obtain $p\xi_0 = a$ and therefore $\alpha_p(a) = \xi_0 a^{p-1}$, since the divided powers on $\mathfrak{a}$ are compatible with those on pS. The left hand side of (27) is:

$$\begin{aligned} \xi(\exp a + {}^V\eta)^{p-1} &= \xi \exp a^{p-1} + \xi p^{p-2}\ {}^V(\eta^{p-1}) \\ &= \exp \xi_0 a^{p-1} + p^{p-2}\ {}^V({}^F\xi\eta^{p-1}). \end{aligned}$$

Hence it is sufficient to show that ${}^F\xi = \eta$. This is a consequence of the following general assertion:

Let S be a ring, and let $\exp : pS \to W(S)$ be the map defined by the natural divided powers of pS. Then we have for $\xi \in W(S)$ the identity:

$$^{FV}\xi - {}^{VF}\xi = \exp p\xi_0. \tag{28}$$

We obtain this by multiplying the equation $p - {}^V1 = \exp p$ with ξ. The last equation may be verified in $W(\mathbb{Z})$ by applying the Witt polynomials. Comparing the equation (28) to (26) we obtain ${}^{VF}\xi = {}^V\eta$ as desired.

Consider a $\hat{\mathcal{Z}}$-ring A. Then a pd-thickening $A \to R$ is supposed to be compatible with the natural divided powers on pA. It induces pd-thickenings in $\mathcal{Z}$:

$$A/p^mA \to R,$$

if m is large. In this case we define:

$$\widetilde{W}(A) = \varprojlim_m \widetilde{W}(A/p^mA). \tag{29}$$

This is a subring of $W(A)$ because A is p-adic.

Consider again the pd-thickening $S \to R$. Then $\widetilde{W}(S)$ is a $\hat{\mathcal{Z}}$-ring and $\widetilde{W}(S) \to R$ a pd-thickening. Therefore we may form $\widetilde{W}(\widetilde{W}(S))$. This is a subring of $W(W(S))$.

Lemma 2.6. *The Cartier morphism induces a map:*

$$\widetilde{W}(S) \to \widetilde{W}(\widetilde{W}(S)).$$

Proof: By (23) it is enough to show that the Cartier morphism induces a map $\widetilde{W}(\mathfrak{a}) \to \widetilde{W}(\widetilde{W}(\mathfrak{a}))$. In [Z-DFG] Lemma 2.14 we expressed the map $W(\mathfrak{a}) \to W(W(\mathfrak{a}))$ in terms of logarithmic coordinates. From this the assertion about $\widetilde{W}$ is obvious. *Q.E.D.*

This lemma has also a version if we replace S by a $\hat{\mathcal{Z}}$-ring A. Then the ring $\widetilde{W}(A)$ defined by (29) is again a $\hat{\mathcal{Z}}$-ring and the kernel of the composition:

$$\widetilde{W}(A) \xrightarrow{\mathbf{w}_0} A \to R,$$

has a natural pd-structure (compare (24)).

Let us verify that $\widetilde{W}(A)$ is indeed a $\hat{\mathcal{Z}}$-ring. We denote by $\mathfrak{n} \subset A$ the maximal ideal and by $k = A/\mathfrak{n}$ the residue field. With the obvious notation we have a direct decomposition:

$$\widetilde{W}(A) = W(k) \oplus \widetilde{W}(\mathfrak{n}). \tag{30}$$

Let $u \geq 2$ be a number such that $p^u\mathfrak{n} \subset \mathfrak{a}$. Then we have $p^n\widetilde{W}(p^u\mathfrak{n}) = \widetilde{W}(p^{n+u}\mathfrak{n})$ for each number n. As in the proof of proposition 2.4 we conclude that the p-adic topology on $\widetilde{W}(\mathfrak{n})$ coincides with the topology of projective limit used to define $\widetilde{W}(\mathfrak{n})$:

$$\widetilde{W}(\mathfrak{n}) = \varprojlim_m \widetilde{W}(\mathfrak{n}/p^m\mathfrak{n}). \tag{31}$$

In particular $\widetilde{W}(A)$ is a p-adic ring. Then we may apply $\widetilde{W}$ to the pd-thickening $\widetilde{W}(A) \to R$ in $\hat{\mathcal{Z}}$.

From the fact that the p-adic topology on $\widetilde{W}(\mathfrak{n})$ is given by (31) and from the decomposition (30) we deduce:

$$\widetilde{W}(\widetilde{W}(A)) := \varprojlim \widetilde{W}(\widetilde{W}(A)/p^m\widetilde{W}(A)) = \varprojlim \widetilde{W}(\widetilde{W}(A/p^mA)).$$

For each m the lemma 2.6 gives a Cartier morphism:

$$\widetilde{W}(A/p^mA) \to \widetilde{W}(\widetilde{W}(A/p^mA)).$$

If we pass to the projective limit we obtain a Cartier morphism for a pd-thickening $A \to R$ where A is a $\hat{\mathcal{Z}}$-ring:

$$\widetilde{W}(A) \to \widetilde{W}(\widetilde{W}(A)). \tag{32}$$

3. The Case of a General p-Divisible Group

Definition 3.1. *Let R be a $\mathcal{Z}$-ring. A Dieudonné frame over R is a frame $(A \xrightarrow{\alpha} R, J, \sigma)$, such that A is a $\hat{\mathcal{Z}}$-ring and such that the Cartier morphism given by σ:*

$$\delta : A \to W(A),$$

factors through the subring $\widetilde{W}(A) \subset W(A)$.

The map $\alpha : A \to R$ induces a map $\widetilde{W}(A) \to \hat{W}(R)$. We define $\varkappa$ to be the composite of the following maps:

$$\varkappa : A \xrightarrow{\delta} \widetilde{W}(A) \to \hat{W}(R). \tag{33}$$

Let us assume that $\mathcal{M} = (M, M_1, \Phi)$ is a Dieudonné A-window over R in the sense of Definition 2. We may replace in the equations (4) the ring $W(R)$ by $\hat{W}(R)$ to obtain a Dieudonné display since for such objects no nilpotence condition is required. Therefore we obtain a functor:

$$\mathbb{DD} : (\text{Dieudonné } A\text{-windows}) \to (\text{Dieudonné displays}). \tag{34}$$

Theorem 3.2. *Let R be a $\mathcal{Z}$-ring and let $A \to R$ be a Dieudonné frame. Then the functor* $\mathbb{DD}$ *is an equivalence of categories.*

From this theorem we may prove our main theorem 6 of the introduction: For an artinian ring R we may simply refer to theorem 20 of [Z-DD], which says that in this case the category of Dieudonné displays is equivalent to the category of p-divisible groups. Now let R be any $\mathcal{Z}$-ring with nilpotent maximal ideal and residue class field k. Then for any p-divisible group X over R, we find an artinian local subring $R' \subset R$, with the same residue class field k, such that X is defined over R'. Indeed, this follows from the existence of the universal deformation of X_k over a power series ring over $W(k)$ proved by Grothendieck and Messing. Moreover let X' and Y' be p-divisible groups over R'. We denote by X and Y the objects obtained by base change over R. Then the base change map is bijective:

$$\operatorname{Hom}(X', Y') \to \operatorname{Hom}(X, Y). \tag{35}$$

Indeed, let $\mathfrak{m}$ be the maximal ideal of R and consider the commutative diagram:

$$\begin{array}{ccc} R'/R' \cap \mathfrak{m}^n & \longrightarrow & R/\mathfrak{m}^n \\ \downarrow & & \downarrow \\ R'/R' \cap \mathfrak{m}^{n-1} & \longrightarrow & R/\mathfrak{m}^{n-1}. \end{array}$$

The vertical arrows are pd-extensions. Hence to lift a homomorphism of p-divisible groups from downstairs is the same as to lift the Hodge filtration of the crystal ([M] Chapt. V Thm. 1.6). Using this, one shows easily that (35) is a bijection for the upper horizontal monomorphism, if it is bijective for the lower one.

We note that Theorem 3 and Theorem 4 of [Z-DD] hold for any $\mathcal{Z}$-ring with nilpotent maximal ideal (see theorem 3.4 below). Then we may apply the arguments for p-divisible groups above to Dieudonné displays. This reduces the general case of theorem 6 to the case where R is artinian. Hence the proof of theorem 6 is finished.

The proof of theorem 3.2 is based on the construction of a crystal associated to a Dieudonné display. In [Z-DD] we have constructed such a crystal on the nilpotent crystalline site. We will now extend this crystal to the whole crystalline site, without assuming that the divided powers are nilpotent.

We consider a pd-thickening $S \to R$ in the category $\mathcal{Z}$. Let us denote by $\mathfrak{a} \subset S$ the kernel. Using exp we consider $\mathfrak{a} \subset W(S)$ also as an ideal of $W(S)$.

Definition 3.3. *Let $\mathcal{P}$ be a Dieudonné display over R. A $\mathcal{P}$-triple over S is a triple $(\widetilde{P}, F, V^{-1})$ such that:*

1) *$\widetilde{P}$ is a finitely generated free $\widetilde{W}(S)$-module, equipped with an isomorphism*

$$\hat{W}(R) \otimes_{\widetilde{W}(S)} \widetilde{P} \xrightarrow{\sim} P.$$

2) *An ^{F}-linear operator $F : \widetilde{P} \to \widetilde{P}$ which lifts $F : P \to P$.*
3) *An ^{F}-linear operator $V^{-1} : \hat{Q} \to \widetilde{P}$, which is defined on the inverse image $\hat{Q}$ of Q by the map $\widetilde{P} \to P$, and which lifts $V^{-1} : Q \to P$.*

The following properties hold:

(i) *$V^{-1}({}^V\tilde{\xi}\tilde{x}) = \tilde{\xi}F\tilde{x}$, for $\tilde{x} \in \widetilde{P}, \tilde{\xi} \in \widetilde{W}(S)$;*
(ii) *$V^{-1}(\mathfrak{a}\widetilde{P}) = 0$.*

We could obviously define a "display" with coefficients in $\widetilde{W}(S)$. Then [Z-DD] theorem 3 and its proof holds for any pd-thickening $S \to R$ (and not just nilpotent ones), if we replace $\widetilde{W}(S)$ by $\hat{W}(S)$. The essential point is that an extension of V^{-1} to $\hat{Q}$ may be defined over $\widetilde{W}(S)$ but not over $\hat{W}(S)$. From this we obtain the following result:

Theorem 3.4. *Let $\alpha : \mathcal{P}_1 \to \mathcal{P}_2$ be a morphism of Dieudonné displays over R. Let $\mathcal{T}_i$ be a $\mathcal{P}_i$-triple over S for $i = 1, 2$. Then there exists a unique morphism of triples $\tilde{\alpha} : \mathcal{T}_1 \to \mathcal{T}_2$ which lifts α.*

We associate to a Dieudonné display $\mathcal{P}$ over R a crystal as follows: Let $S \to R$ be a pd-thickening in $\mathcal{Z}$. Then we find a $\mathcal{P}$-triple $\mathcal{T} = (\widetilde{P}, F, V^{-1})$ over S. This is done by writing down structural equations for $\mathcal{P}$ (see [Z-DD] (3)) and lifting them to S. By the last theorem $\mathcal{T}$ is unique up to canonical isomorphism. Therefore the following definition makes sense:

$$\mathcal{D}_{\mathcal{P}}(S) = S \otimes_{\mathbf{w}_0, \widetilde{W}(S)} \widetilde{P}.$$

We call $\mathcal{D}_{\mathcal{P}}$ the Dieudonné crystal of $\mathcal{P}$. If $S \to R$ is a nilpotent pd-thickening, we have $\widetilde{W}(S) = \hat{W}(S)$, and therefore $\mathcal{D}_{\mathcal{P}}(S)$ coincides with the Dieudonné crystal defined in [Z-DD] in this case.

We note that $\mathcal{D}_{\mathcal{P}}$ commutes with base change in the following sense: Assume we are given a morphism of pd-thickenings in the category $\mathcal{Z}$:

$$\begin{array}{ccc} S & \longrightarrow & S' \\ \downarrow & & \downarrow \\ R & \longrightarrow & R'. \end{array}$$

The vertical arrows are pd-thickenings with kernels $\mathfrak{a} \subset S$ resp. $\mathfrak{a}' \subset S'$. Assume we are given a Dieudonné display $\mathcal{P}$ over R, and let $\mathcal{P}' = \mathcal{P}_{R'}$ be the Dieudonné display obtained by base change. Let $\mathcal{T} = (\widetilde{P}, F, V^{-1})$ be a $\mathcal{P}$-triple over S.

Then we define a $\mathcal{P}'$-triple $\mathcal{T}' = (\widetilde{P}', F, V^{-1})$ over S' with coefficients in $\widetilde{W}(S')$ as follows:

$$\widetilde{P}' = \widetilde{W}(S') \otimes_{\widetilde{W}(S)} \widetilde{P}.$$

The operator $F : \widetilde{P}' \rightarrow \widetilde{P}'$ is the obvious ^{F}-linear extension of $F : P \rightarrow P$. Finally the ^{F}-linear operator $V^{-1} : \hat{Q}' \rightarrow \widetilde{P}'$ is uniquely determined by the formulas:

$$\begin{aligned} &V^{-1}(\xi \otimes y) = {}^F\xi \otimes V^{-1}y, \text{ for } \xi \in \widetilde{W}(S'),\ y \in \hat{Q}; \\ &V^{-1}({}^V\xi \otimes x) = \xi \otimes Fx, \text{ for } x \in \widetilde{P}; \\ &V^{-1}(a' \otimes x) = 0 \text{ for } a' \in \mathfrak{a}' \subset \widetilde{W}(S'). \end{aligned}$$

This construction of the triple $\mathcal{T}'$ provides a canonical isomorphism:

$$\mathcal{D}_{\mathcal{P}'}(S') \cong S' \otimes_S \mathcal{D}_{\mathcal{P}}(S). \tag{36}$$

If A is a $\hat{\mathcal{Z}}$-ring and $A \rightarrow R$ is a pd-thickening we set

$$\mathcal{D}_{\mathcal{P}}(A) = \varprojlim_n \mathcal{D}_{\mathcal{P}}(A/p^n A),$$

where n is large, such that $A/p^n A \rightarrow R$ is a pd-thickening.

Proof of theorem 3.2: We are going to repeat the arguments of the proof of theorem 1.6 in the new context.

Let us begin with the construction of a quasiinverse functor:

$$\text{Win} : (\text{Dieudonné displays}) \rightarrow (\text{Dieudonné } A\text{-windows}). \tag{37}$$

We consider a Dieudonné display $\mathcal{P}$ over R. Then we have to associate to this an A-window (M, M_1, Φ). We set:

$$M = \mathcal{D}_{\mathcal{P}}(A).$$

We define M_1 as the kernel of the canonical map:

$$M = A \otimes_{\widetilde{W}(A)} \widetilde{P} \rightarrow R \otimes_{\hat{W}(R)} P \rightarrow P/Q,$$

where $(\widetilde{P}, F, V^{-1})$ is the $\mathcal{P}$-triple over A.

Finally we define Φ. To do this we consider the ring $\overline{R} = R/pR$. Then $A \rightarrow \overline{R}$ is again a pd-thickening with kernel $\mathfrak{a} + pA \subset$A.

If we apply base change to the morphisms of pd-thickenings

$$\begin{array}{ccc} A & \xrightarrow{id} & A \\ \downarrow & & \downarrow \\ R & \longrightarrow & R', \end{array}$$

we obtain a canonical isomorphism:

$$\mathcal{D}_{\mathcal{P}}(A) \simeq \mathcal{D}_{\widetilde{\mathcal{P}}}(A).$$

We note that the last diagram involves a priori two different rings $\widetilde{W}(A)$, one with respect to $A \to R$ and the second with respect $A \to \overline{R}$. In fact these two rings coincide, but this fact is irrelevant for us.

The same considerations as in the proof of theorem 1.6 lead to homomorphisms:

$$\Phi^{\#} : A \otimes_{\sigma,A} M \to M,$$
$$\Psi^{\#} : M \to A \otimes_{\sigma,A} M.$$

The argument following equation (8) shows:

$$\Phi M_1 \subset pM.$$

We set $\Phi_1 = \frac{1}{p}\Phi : M_1 \to M$. To show that (M, M_1, Φ) is a Dieudonné window, we still have to verify that the union of ΦM and $\Phi_1 M_1$ generates M as an A-module.

Before we prove this we will describe the $\mathcal{P}$-triple $(\widetilde{P}, F, V^{-1})$ over A in terms of (M, M_1, Φ). We will describe a canonical isomorphism:

$$\widetilde{P} \cong \mathcal{D}_{\mathcal{P}}(\widetilde{W}(A)), \tag{38}$$

in the manner of [Z-DFG] Proposition 2.12. Let us postpone the proof of (38).

From this we may finish the proof of the theorem as follows: The morphism of pd-thickenings

$$\begin{array}{ccc} A & \xrightarrow{\quad \delta \quad} & \widetilde{W}(A), \\ & \searrow \quad \swarrow & \\ & R & \end{array}$$

provides a canonical isomorphism:

$$\widetilde{W}(A) \otimes_{\delta,A} M \cong \mathcal{D}_{\mathcal{P}}(\widetilde{W}(A)) \cong \widetilde{P}. \tag{39}$$

Since the isomorphisms are functorial they commute with the Frobenius:

$$F(\xi \otimes m) = {}^{F}\xi \otimes \Phi m, \text{ for } m \in M, \; \xi \in \widetilde{W}(A).$$

Here F is from the triple $(\widetilde{P}, F, V^{-1})$. Since $\widetilde{W}(A)$ is p-torsion free the isomorphism (39) is also compatible with V^{-1}. From (39) we deduce an isomorphism:

$$\hat{W}(R) \otimes_{\varkappa,A} M \simeq P.$$

It follows that this isomorphism is compatible with F and V^{-1} too. From this isomorphism we deduce as in the last part of theorem 1.6 that the union of ΦM and $\Phi_1 M_1$ generates M as an A-module. This shows that we have an isomorphism of functors:

$$\mathbb{D}\mathbb{D} \circ \mathrm{Win} \cong id.$$

On the other hand let us start with a Dieudonné window (M, M_1, Φ). We denote by $\mathcal{P}$ the Dieudonné display associated by the functor $\mathbb{D}\mathbb{D}$. Then it is clear that

$(\widetilde{P}, F, V^{-1})$ described above in terms of (M, M_1, Φ) is a $\mathcal{P}$-triple over A. From this we deduce an isomorphism:

$$\mathcal{D}_{\mathcal{P}}(A) \simeq M,$$

which is compatible with the Frobenius. Hence the functors $\mathbb{D}$ and Win are indeed quasi-inverse. *Q.E.D.*

Finally we give a precise definition of the isomorphism (38). Let us consider a pd-thickening $S \to R$ in $\mathcal{Z}$.

Proposition 3.5. *Let $\mathcal{P}$ be a Dieudonné display over R. Consider a $\mathcal{P}$-triple $\mathcal{T} = (\widetilde{P}, F, V^{-1})$ over S. We have defined a pd-thickening in the category $\hat{\mathcal{Z}}$ (compare (24)) :*

$$\widetilde{W}(S) \to R.$$

We denote the associated Cartier morphism by Δ: $\widetilde{W}(S) \to \widetilde{W}(\widetilde{W}(S))$ which is defined by lemma 2.6. Then a $\mathcal{P}$-triple over $\widetilde{W}(S)$ is obtained as follows:

$$\widetilde{\mathcal{T}} = (\widetilde{W}(\widetilde{W}(S)) \otimes_{\Delta, \widetilde{W}(S)} \widetilde{P}, F, V^{-1}).$$

Here $F : \widetilde{W}(\widetilde{W}(S)) \otimes_{\Delta, \widetilde{W}(S)} \widetilde{P} \to \widetilde{W}(\widetilde{W}(S)) \otimes_{\Delta, \widetilde{W}(S)} \widetilde{P}$ denotes the ^{F}-linear extension of $F : \widetilde{P} \to \widetilde{P}$.

The operator V^{-1}, which is defined on the inverse image of Q by the canonical map $\widetilde{W}(\widetilde{W}(S)) \otimes_{\Delta, \widetilde{W}(S)} \widetilde{P} \to P$ is uniquely determined by the following equations:

$$V^{-1}({}^V\hat{\xi} \otimes x) = \hat{\xi} \otimes Fx, \quad \text{for } x \in (\widetilde{W}(\widetilde{W}(S)),$$
$$V^{-1}(\hat{\xi} \otimes y) = {}^F\hat{\xi} \otimes V^{-1}y, \quad \text{for } y \in \hat{Q} \subset \widetilde{P}.$$

Proof: The proof is the repetition of the proof of proposition 2.12 in [Z-DFG], which we omit. *Q.E.D.*

As a corollary we obtain the canonical isomorphism (38).

We will now discuss the examples of Dieudonné frames in the introduction in more detail.

Example: Let k be a perfect field of characteristic $p \geq 3$. We fix numbers d and s and consider the ring

$$R = k[T_1, \ldots, T_d]/(T_1^s, \ldots, T_d^s). \tag{40}$$

We set $A = W(k)[T_1, \ldots, T_d]/T_1^s, \ldots, T_d^s)$. Then the canonical map:

$$A \to R,$$

is a pd-thickening with nilpotent divided powers on pA. We define $\sigma : A \to A$ by the equations:

$$\sigma(\xi) = {}^F\xi, \text{ for } \xi \in W(k), \tag{41}$$
$$\sigma(T_i) = T_i^p, \text{ for } i = 1, \ldots, d. \tag{42}$$

We claim that A is a Dieudonné frame. For this we have to show that the Cartier morphism associated to σ:

$$\delta : A \to W(A),$$

factors through $\widetilde{W}(A)$. We note that $\widetilde{W}(A) = \hat{W}(A)$ in this example because the divided powers are nilpotent. Clearly $\delta(T_i)$ is the Teichmüller representative of T_i and therefore

$$\delta(T_i) = [T_i] \in \hat{W}(A).$$

It remains to be shown that $\delta(\xi) \in \hat{W}(A)$ for $\xi \in W(k)$. But $\delta : W(k) \to W(A)$ is the Teichmüller section of the homomorphism $W(A) \to W(k)$ induced by the canonical map $A \to k$. Since this Teichmüller section takes by definition values in $\hat{W}(A)$, we have proved that A is a Dieudonné frame.

Let us change the notation: Let $R = k[[T_1, \ldots, T_d]]$ be the power series ring and $A = W(k)[[T_1, \ldots, T_d]]$. We define $\sigma : A \to A$ as before. Then A becomes a frame over R. Then we obtain as explained in example 1 of the introduction:

Proposition 3.6. *The category of p-divisible groups over $R = k[[T_1, \ldots, T_d]]$ is equivalent to the category of Dieudonné A-windows.*

Example: Let k be as above. Let S be a local finite flat $W(k)$-algebra, with residue field k. We set $R = S/p^u S$, where u is some fixed number. We choose a presentation:

$$0 \to \mathfrak{c} \to W(k)[T_1, \ldots, T_d] \to S \to 0,$$

such that each T_i is mapped to the maximal ideal of S. We consider the subring $A_0 \subset W(k)[T_1, \ldots, T_d] \otimes \mathbb{Q}$ which is generated over $W(k)[T_1, \ldots, T_d]$ by all elements of the form $\frac{c^n}{n!}$, for $c \in \mathfrak{c}$, $n \in \mathbb{N}$. The natural map $A_0 \otimes \mathbb{Q} \to S \otimes \mathbb{Q}$ induces a map $A_0 \to S$, whose kernel J_0 is generated by the element $\frac{c^n}{n!}$ as an ideal.

Let σ be the ^{F}-linear $W(k)$-algebra endomorphism of $W(k)[T_1, \ldots, T_d]$, such that $\sigma(T_i) = T_i^p$. Then σ leaves the ideal $\mathfrak{c} + pW(k)[T_1, \ldots, T_d]$ stable. Since $J_0 + pA_0 \subset A_0$ is a pd-ideal, it follows that σ extends to an endomorphism of A_0. As in the example of proposition 1.8 one verifies:

$$\sigma(a) \equiv a^p \mod pA_0\,, \qquad a \in A_0.$$

Let A be the p-adic completion of A_0. Then we obtain a pd-thickening $A \to S$, and hence also a pd-thickening $A \to R$. The endomorphism σ extends to A.

We claim that (A, σ) is a frame for R. Again all we need to show is that the Cartier morphism

$$\delta : A \to W(A)$$

factors through $\widetilde{W}(A)$. Since $\delta(T_i) = [T_i] \in \hat{W}(A)$ it follows that

$$\delta(h) \in \hat{W}(A), \tag{43}$$

for $h \in W(k)[T_1, \ldots, T_d]$.

Taking into account that $W(A)$ is p-adic, it suffices to show that $\delta(J_0) \subset \widetilde{W}(A)$. By (43) we know that $\delta(\mathfrak{c}) \subset \widetilde{W}(A)$. Clearly $\delta(\mathfrak{c})$ is in the kernel of $\widetilde{W}(A) \to A \to S$. The divided power structure on $J \subset A$ defines an embedding $J \subset \widetilde{W}(A)$. Hence the kernel of the map $\widetilde{W}(A) \to S$ is $J \oplus {}^V\widetilde{W}(A)$. This kernel is a pd-ideal because each direct summand is. Hence $\delta(c) \in J \oplus {}^V\widetilde{W}(A)$ for $c \in \mathfrak{c}$ implies $\delta\left(\frac{c^n}{n!}\right) \in J \oplus {}^V\widetilde{W}(A)$. Since J_0 is generated over $W(k)[T_1, \ldots, T_d]$ by products of elements of the form $\frac{c^n}{n!}$ for $c \in \mathfrak{c}$, we conclude that $\delta(J_0) \subset \widetilde{W}(A)$. This concludes the proof that (A, σ) is a Dieudonné frame for $R = S/p^uS$. Hence A-windows for the ring R classify p-divisible groups over R.

As in the last example we obtain the same result for the frame $A \to S$ which is not a Dieudonné frame.

Proposition 3.7. *The category of Dieudonné A-windows is equivalent to the category of p-divisible groups over S.*

This result was proved by Breuil in the case where S is a discrete valuation ring by using the crystalline Dieudonné theory of Berthelot, Breen, and Messing [BBM].

Example: Consider Example 3 of the introduction. We are given elements $f_1, \ldots, f_r \in k[[T_1, \ldots, T_d]]$, such that $f_1, \ldots, f_r, T_{r+1}, \ldots, T_d$ is a system of parameters in $k[[T_1, \ldots, T_d]]$. We set $R = k[[T_1, \ldots, T_d]]/(f_1, \ldots, f_r)$. Then we have defined in the introduction a pd-thickening $\hat{B} \to R$. Moreover we have defined an endomorphism σ of $\hat{B}$. With this structure $\hat{B}$ becomes a frame over R. This is a frame but not a Dieudonné frame. Nevertheless we obtain by the limit argument given in the introduction:

Proposition 3.8. *The category of Dieudonné $\hat{B}$-windows over R is equivalent to the category of formal p-divisible groups over R.*

REFERENCES

[B] Breuil, Ch.: Schémas en groupe et module filtré, C.R. Acad. Sci. Paris Sér.I Math. **328** (1999) no.2, 93-97.

[BAC] Bourbaki, N.: Algèbre Commutative, Chapt. 8 et 9, Masson 1983.

[BBM] Berthelot, P., Breen, L., Messing, W.,: Théorie de Dieudonné Cristalline II, Springer Lecture Notes No. 930, 1982.

[BM] Berthelot, P., Messing, W.,: Théorie de Dieudonné Cristalline III: Théorèmes d'Equivalence et de Pleine Fidelité, in: The Grothendieck Festschrift Volume I, Birkhäuser 1990.

[M] Messing, W.: The crystals associated to Barsotti-Tate groups, LNM **264**, Springer 1972.

[Z-DFG] Zink,Th.: The display of a formal p-divisible group, to appear in: Semestre p-Adique, Astérisque SFM.

[Z-DD] Zink,Th.: A Dieudonné Theory for p-Divisible Groups, http://www.mathematik.uni-bielefeld.de/~zink

MIX
Papier aus verantwortungsvollen Quellen
Paper from responsible sources
FSC® C105338

If you have any concerns about our products,
you can contact us on
ProductSafety@springernature.com

In case Publisher is established outside the EU,
the EU authorized representative is:
Springer Nature Customer Service Center GmbH
Europaplatz 3, 69115 Heidelberg, Germany

Printed by Libri Plureos GmbH
in Hamburg, Germany